Contributions to the History of Herpetology

Volume 2

Das.XIIII Capitel seit von den wunden der vergifftigen thier biß/es sey von einem wüttende hund/schlangen oder scorpion

THE EARLIEST ILLUSTRATIONS AND DISCUSSIONS OF herpetological subjects in printed books were about vipers because of the practical importance of these venomous reptiles to physicians and the general public. (See biographies beginning on pages 11 and 12.) This illustration is reproduced from "Buch der C[h]irurgia" (Book of Surgery), published in Strassburg (now Strasbourg, France) in 1497. This incunable, the first illustrated surgical book, was intended as a manual for apprentices to barbers and surgeons and includes surgical techniques and drug treatments. The author—Hieronymus Brunschwig or, in English, Jerome of Brunswick (ca. 1450–ca. 1512)—was one of the first German surgeons. The book combines treatments from ancient and medieval authorities with the author's own extensive experience.

Brunschwig's book also contains an extensive section on snake envenomation. In this image three doctors are seen attending to a patient—the traditional "wound man" of early medical texts—who has been bitten by a viper that still encircles his leg. The heading in German Fraktur typeface reads: "The 14th Chapter shall deal with the wounds caused by the bite of the venomous animal / may it be a mad dog / snakes or scorpion." The illustration was printed from a woodcut that has been colored by hand. The hats of the doctors and one of their gowns have been highlighted with gold. (Original in the Adler collection.)

Contributions to the History of Herpetology

Volume 2

Edited by Kraig Adler

Issued to Commemorate the
Society's 50th Anniversary Meeting
Saint Louis 2007

Society for the Study of Amphibians and Reptiles

Editor's Note

This volume had its genesis in 1982 when the Society for the Study of Amphibians and Reptiles commissioned David M. Dennis and Kraig Adler to produce an audiovisual program to celebrate the Society's 25th Anniversary Meeting, which was held that year. This brief program—entitled "Herpetologists Then and Now"—chronicled the development of herpetology in North America. This show utilized six computer-coordinated projectors and a 10-meter-wide screen and was set to music. The production was revised and expanded to a 40-minute show during which hundreds of images of people, institutions, habitats, and herpetological illustrations were projected. This program was shown at meetings in the United States through 1988.

A frequent comment at those showings was that no permanent record existed of the hundreds of portraits of herpetologists that flashed past the audience so quickly. Volume 1 of "Contributions to the History Herpetology" was created to satisfy that need but in the meantime the audiovisial program was revised again. In 1989 Dennis and Adler re-wrote the entire program for the occasion of the First World Congress of Herpetology, which was held that year in Canterbury, United Kingdom. It was titled "Herpetology Past and Present" and enlarged to a 120-minute-long production. The content was internationalized and copies of Volume 1 were presented to the congress delegates. This new audiovisual program was later shown at many herpetological meetings in the United States and Canada over the next decade. Unfortunately, only a small proportion of the persons featured in the audiovisual program could be included in Volume 1, so plans were made for this second volume.

Volume 1 also included compilations by John S. Applegarth and Ronald Altig on taxonomic authors and doctoral lineages in herpetology, respectively, and these sections have been greatly augmented in Volume 2. Their cooperation in revising these sections to produce a book of enhanced reference value is acknowledged and greatly appreciated. This volume commemorates the 50th Anniversary Meeting of the Society for the Study of Amphibians and Reptiles, which was founded as The Ohio Herpetological Society in 1958. The 50-year history of the Society has been published in *Herpetological Circulars*, number 36, 2007.

Because Volume 1 has been out-of-print since 2000, the section "Herpetologists of the Past [Part 1]" has been put on the web for public access at www.ssarherps.org (click on "Publications"). A comprehensive index to both volumes has been included in the present volume (see pages 259–273).

Publication costs for this book have been partially funded from the Joseph R. Bailey Endowment and the Robert E. Gordon Endowment.

CONTRIBUTIONS TO HERPETOLOGY, VOLUME 21

Kraig Adler, *Editor* • Timothy D. Perry, *Associate Editor*

Volumes in the *Contributions to Herpetology* series can be purchased from the Publications Secretary: Breck Bartholomew, P. O. Box 58517, Salt Lake City, Utah 84158–0517, USA (*telephone and fax*: area code 801, 562–2660; *e-mail*: ssar@herplit.com). A list of all Society publications, including those of The Ohio Herpetological Society and the *Catalogue of American Amphibians and Reptiles*, is printed at the end of this book. Volumes in the *Contributions* series are published occasionally and ordered by separate subscription, although Society members receive a pre-publication discount. Authors who wish to have manuscripts considered for publication in the *Contributions* series should contact the Editor: Kraig Adler, Cornell University, Department of Neurobiology and Behavior, Seeley G. Mudd Hall, Ithaca, New York 14853–2702, USA.

All members of the Society receive a quarterly technical journal (*Journal of Herpetology*), a quarterly news-journal (*Herpetological Review*), and, optionally, the *Catalogue of American Amphibians and Reptiles*. Society members receive special pre-publication discounts on *Herpetological Circulars* and on books in the *Facsimile Reprints in Herpetology*, *Contributions to Herpetology*, and *Herpetological Conservation* series. The *Catalogue* is available by separate subscription. Apply to the Society's Treasurer: Dr. Kirsten Nicholson, Central Michigan University, Department of Biology, 217 Brooks Hall, Mount Pleasant, Michigan 48859, USA (*telephone*: area code 989, 774–3578; *e-mail*: kirsten.nicholson@cmich.edu; *SSAR web address*: www.ssarherps.org). Overseas customers can make payments in USA funds or by International Money Order. All persons may charge to MasterCard, VISA, American Express, or Discover Card (include account number and expiration date). Details concerning SSAR membership may also be obtained from the web site.

ISBN: 978–0–916984–71–7. Library of Congress Control Number: 2007921985.

Production specifications, including place and date of publication, are given on the last page of this book.

Contributions to the History of Herpetology

edited by Kraig Adler

Table of Contents

Herpetologists of the Past, Part 2, by Kraig Adler

*Biographies are arranged according to the time period in which each person's herpetological work began, typically the appearance of the first publications. An alphabetical list covering volumes 1 and 2 is on the next page.

Index of Authors in Taxonomic Herpetology, Second Edition, by John S. Applegarth

Academic Lineages of Doctoral Degrees in Herpetology, Second Edition, by Ronald Altig

BENEFACTOR

RONALD A. JAVITCH

Ronald A. Javitch Natural History
Rare Book Foundation
Montreal

The Society is pleased to acknowledge the continuing generosity of Mr. Javitch, who has supported several of its publication programs.

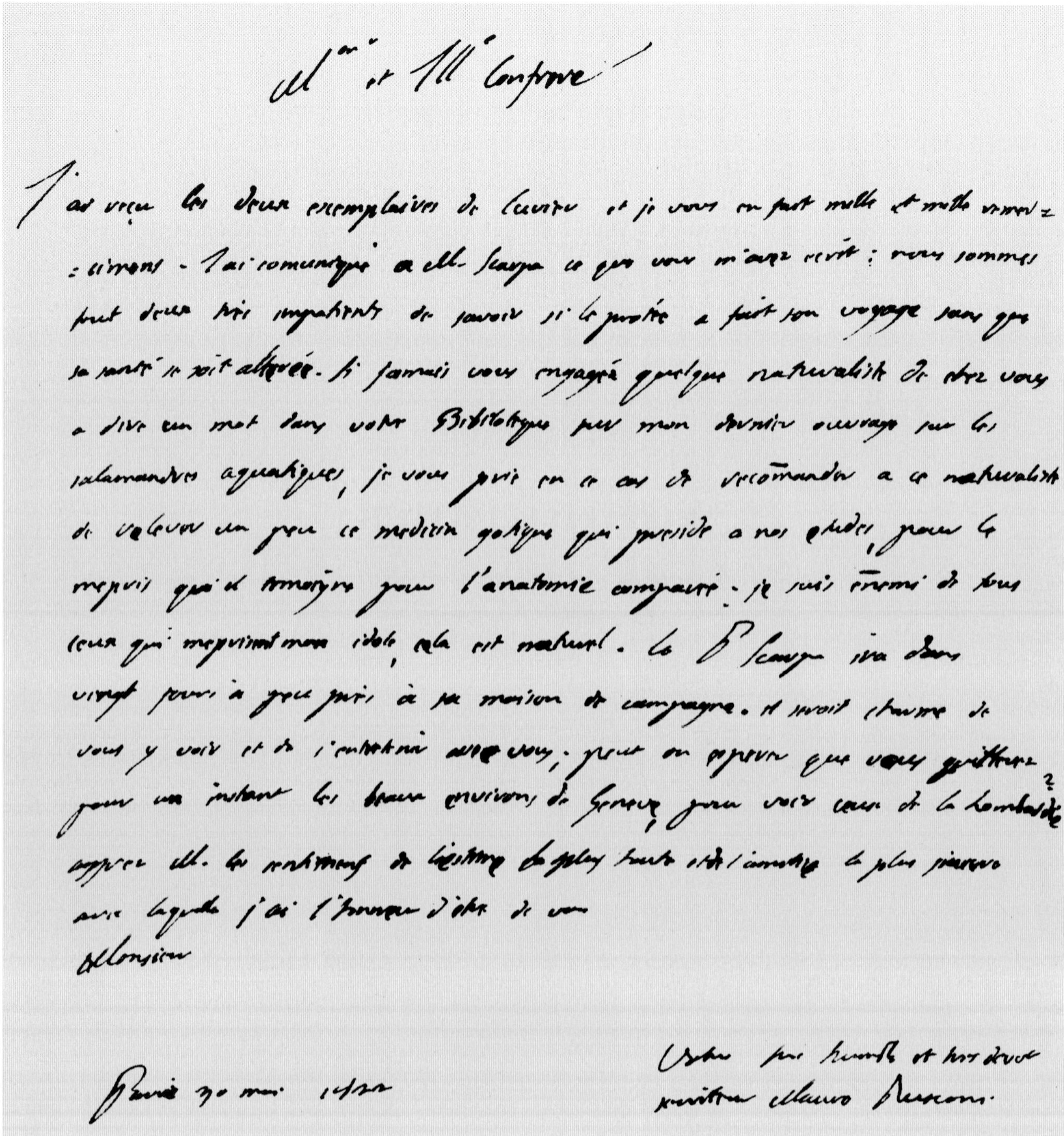

MAURO RUSCONI (1776–1849), AN ITALIAN ANATOMIST and one of the founders of modern embryology, may be considered as the first great amphibian biologist. (For his biography, see page 44.) He was a pupil of the French comparative anatomist and paleontologist, Georges Cuvier. Rusconi authored seven books on the anatomy, development, and behavior of frogs and salamanders and the lymphatic system of reptiles.

Rusconi had a special research interest in the Olm (*Proteus anguinus*). He observed and collected them in subterranean grottos in Slovenia beginning in 1816, and he published all or part of two books and several papers on the species. In this letter in French (reproduced above) to Jean-Pierre Maunoir, an anatomist and surgeon at the Académie de Genève in Switzerland, Rusconi asks Maunoir whether a live olm ("le protée," line 4) he had sent had arrived in good health. He suggests that Maunoir's institutional library might wish to purchase a copy of his latest book ("dernier ouvrage sur les salamandres aquatiques," lines 6–7). This reference is to his book on the reproductive behavior of *Triturus carnifex* (1821) in which he published the first illustrations of courtship and egg-laying behavior in any salamander. Further on in the letter, Rusconi refers to someone who refuses to study anatomy comparatively as old-fashioned ("got[h]ique," line 8). At the bottom, the letter is dated Pavia 30 May 1822 and signed by Rusconi.

Rusconi drew the illustrations for his books and personally hand colored each plate. The small edition sizes of 25 to 150 copies make his books among the rarest of all herpetological treatises. (Original letter in the Adler collection.)

Herpetologists of the Past
Part 2

by Kraig Adler

THE 50TH ANNIVERSARY OF THE FOUNDING OF THE SOCIETY for the Study of Amphibians and Reptiles is an occasion to take stock of the dynamic discipline of herpetological science and of the people—past and present and from throughout the world—who have contributed to its development. Like its predecessor volume, which marked the First World Congress of Herpetology in 1989, this one references several thousand persons. This biographical section, "Herpetologists of the Past, Part 2," focuses on 285 individuals—from the late 15th Century to the present day. All of them are now deceased, and all have played roles in building the science of herpetology. Many other individuals are also noted in these accounts. The present volume complements Volume 1, which covered 152 biographees and numerous associated persons. Together these volumes provide materials for a comprehensive history of our discipline. Such a book is still needed for our field.

The individuals included here represent a diverse range of careers including priests and clergymen, physicians and dentists, members of the nobility, a diplomat, poets, military officers, administrators, museum and zoo curators, and academics with primary specialties in fields other than herpetology. Some have made scientific contributions exclusively to herpetology whereas the work of others was of broader importance to biology. One, an expert on concealing coloration in amphibians and reptiles, designed the camouflage for British tanks that were used to defeat German forces in North Africa during World War II. Another helped to smuggle people across a national border at great personal risk. Some were gifted amateurs with no formal education and others were highly trained professionals. Several died in their twenties or thirties. One perished at the age of barely 18 in pursuit of a new species, and another was murdered in the field by his guide. Others continued to make contributions despite long and severe illnesses. Several faced political persecution and some were imprisoned; one was executed. Yet all made significant, even sometimes seminal, contributions to our knowledge of the biology of amphibians and reptiles. The examples of their lives can still enlighten and inspire.

Content and Organization.—Choosing the men and women to feature in these biographies was a difficult task. They represent only a small proportion of the many individuals who are worthy of this recognition. Although I consulted with colleagues around the world, probably none would agree with all of my final choices.

The biographies are arranged in an approximately chronological order so that the development of the discipline of herpetology can be seen in perspective. This also allows each person to be viewed in a broader scientific context. Connections between and among herpetologists are noted where possible. An attempt has also been made to place events in a broader historical context including political, religious, and military events that had an impact on the work of the biographees. Since some biographees had short careers of just a few years but others were active for many decades, each person is positioned in sequence according to the period in which he or she first became prominent in herpetology. Usually this coincides with the appearance of their first publications. This organization is reflected in the Table of Contents and in the running heads at the top of each page. Admittedly, this procedure is somewhat arbitrary. Because of this arrangement, some persons at first may seem to be oddly placed, especially when their herpetological activity did not begin until late in their lives. Because of this sequencing and because numerous other persons are mentioned in these biographies, I have included an exhaustive index (pages 259–273). This index also covers the biographies in Volume 1 so that it is comprehensive for the series to date. The text of Volume 1 is available on the web. Visit www.ssarherps.org and click on "Publications." Names abbreviated in the text are given in full in the index where known.

Format of Biographies.—Each person's name is given in the title of their biography in the form most often used in their own books and papers. All names are spelled in their Romanized form. If alternative spellings of a name are known these are noted but preference is given to the form used by the person as indicated by the style of his or her signature.

In reviewing a person's contributions, emphasis has been given to his or her herpetological work. Thus, non-herpetological aspects of a person's life, which in several instances were more prominent than their work with amphibians and reptiles, are given minor discussion here. Since herpetology as a biological discipline is a 20th Century phenomenon, the earlier contributors generally were naturalists or zoologists for whom herpetology was only one of several interests.

References to obituaries or other biographical information are provided for each person, although for some no information has been published. In other instances the references cited provide few details and I have had to depend upon primary sources. In many instances such information has come from family members or former colleagues of the biographee. Thus, much information in this chapter is published for the first time. References abbreviated to author and date are given in full in the Literature Cited section (pages 256–258) or in that person's biography. Only references to printed documentation are cited even though electronic resources were often used to locate materials.

Dates.—Virtually all dates given in these biographies are reckoned according to the modern calendar now in general use throughout the world. This calendar—the Gregorian calendar or sometimes termed the reformed Julian calendar—first came into use in 1582 but was not then adopted universally. By that time the accumulated discrepancy between the old Julian and

Gregorian calendars was ten days; thus, ten days of 1582 were omitted to bring the calendars into correspondence again. However, Great Britain and her colonies did not adopt the new calendar until 1752 and the former Soviet Union and some neighboring countries did not do so until 1918 by which times the discrepancies were 11 and 12 days, respectively. Sweden on the other hand switched over gradually during the period 1696 to 1753. For events in these countries before these dates, therefore, it is common in the literature to see double dates (for example, the date "1/13 April 1839" in Russia, with the second being that according to the modern calendar).

In this book I have given dates in their modern Gregorian format. However, in the biographical information available to me it was not always possible to determine whether the old or modern form was used. Thus, readers must use such dates with caution if they occur before the calendar was changed to the Gregorian format in a particular country.

Literature Cited.—I provide a brief list of literature beginning on page 256, which is intended as an introduction to the historical literature of herpetology. It is supplemental to the list given in Volume 1 except that references are repeated if they are cited in Volume 2.

Acknowledgments.—The compilation of this sort of book requires the active cooperation of many professional colleagues and library personnel who have generously acceded to my numerous requests for reviews, information, and materials. In particular I am indebted to three colleagues who critiqued the entire manuscript—Ronald Altig, John S. Applegarth, and Aaron M. Bauer—and who also sent references and other materials.

The following persons reviewed one or more biographies and in some instances also provided materials such as obituaries, portraits, or signatures: Luis Albuja, Michele Aldrich, Ana Almendáriz, Natalia Ananjeva, María Cristina Ardila Robayo, J. W. Arntzen, Ray Ashton, Stuart Baldwin, Emilio Balletto, Bernard Bechtel, Daniel Beck, Wolfgang Bischoff, Wolfgang Böhme, Leo J. Borkin, Roger Bour, William R. Branch, Alvin Braswell, Bayard H. Brattstrom, Alvin Breisch, Henrik Bringsøe, William S. Brown, Silvio Bruno, Édouard-Raoul Brygoo, Brooks Burr, James R. Buskirk, Mario Cabrera, Jonathan Campbell, Charles C. Carpenter, Gustavo Casas Andreu, the late José M. Cei, Ellen Censky, Charles Chantell, Juan Carlos Chebez, John Cloudsley-Thompson, Emma Cochrane, Harold G. Cogger, Joseph T. Collins, Luis A. Coloma, Francis R. Cook, John E. Cooper, James Corgan, Brian Crother, David Cundall, David M. Damkaer, J. C. Daniel, Ilya S. Darevsky, Indraneil Das, Patrick David, Anslem de Silva, James R. Dixon, Herndon G. Dowling, Robert Drewes, William E. Duellman, Tatjana Dujsebayeva, Evgeniy Dunayev, Georg Džukić, Richard Etheridge, Howard E. Evans, Neal Evenhuis, Julián Faivovich, John W. Ferner, Oscar Flores Villela, Xosé A. Fraga-V., Dieter Fuchs, H. Carl Gerhardt, Richard Goris, George Gorman, Harry W. Greene, Kurt Grossenbacher, Rainer Günther, Jakob Hallermann, Monique Halloy, James Hanken, David L. Hardy, Sr., Barbara Hawgood, Robert Henderson, Hans-Joachim Herrmann, W. Ronald Heyer, Richard Highton, Tsutomu Hikida, Lipke B. Holthuis, René Honegger, Victor Hutchison, John Iverson, Ronald I. Johnstone, Le Vu Khoi, Arnold G. Kluge, Zoltán Korsós, Michael Kroniger, Sergius Kuzmin, William Lamar, José A. Langone, Benedetto Lanza, Esteban O. Lavilla, Robin Lawson, John Legler, Rafael Lemaitre, Jean Lescure, Alan E. Leviton, Ernest A. Liner, Jonathan Losos, Wilson R. Lourenço, Jeffrey E. Lovich, John D. Lynch, Dennis Magee, Masafumi Matsui, Colin McCarthy, Roy W. McDiarmid, Fernando Mendoza Quijano, Andrés Merino-Vitteri, Amy R. M. Meyers, Joseph Mitchell, Edward Moll, Alvaro Mones, Thomas Moore, P. le Fras N. Mouton, James B. Murphy, Robert W. Murphy, Charles W. Myers, Quang Truong Nguyen, Max A. Nickerson, Göran Nilson, Midori Nishioka, Ettore Olmo, Fernando Ortiz-Crespo, Hidetoshi Ota, Robert Peck, Hans-Albert Pederzani, Edmundo Perez Ramos, John B. Phillips, Harald Pieper, José P. Pombal Júnior, Robert Powell, Andrew Price, Peter Pritchard, Martha Ramírez-Pinilla, Christopher Raxworthy, Anders Rhodin, Werner Rieck, Zbyněk Roček, José Rosado, Findlay Russell, Michael Ryan, Alfredo Salvador, Jay M. Savage, Chuck Schaffer, Andreas Schlüter, Josef F. Schmidtler, Thomas Schöttler, Magno Segalla, Takeshi Seto, Glenn Shea, Richard Shine, John Simmons, Jack Sites, Gerald R. Smith, Hobart M. Smith, the late Margaret M. Stewart, Zbigniew Szyndlar, Wilmer W. Tanner, Roger Thorpe, Franz Tiedemann, Grace Tilger, Stephen G. Tilley, Michihisa Toriba, Daniel Townsend, Linda Trueb, Thomas Uzzell, P. E. Vanzolini, Judit Vörös, Richard Wahlgren, David B. Wake, Marvalee H. Wake, Harlan D. Walley, Pei-chao Wang, Robert G. Webb, John Weigel, Kent Wells, Yehudah L. Werner, Romulus Whitaker, Jorge Williams, Beryl Wilson, Hsiang-k'uei Yeh, Er-mi Zhao, Ken-tang Zhao, George R. Zug, and Richard G. Zweifel.

For assistance in locating biographical materials, I thank Michele L. Aldrich, Natalia B. Ananjeva, Vladimir Bobrov, the late Charles M. Bogert, Wolfgang Böhme, William R. Branch, Donald G. Broadley, the late Leo D. Brongersma, Gustavo Casas Andreu, Harold G. Cogger, Ronald I. Crombie, David M. Damkaer, Ilya S. Darevsky, Patrick David, Anslem de Silva, Laurence Dorr, Evgeniy Dunayev, the late Josef Eiselt, Oscar Flores V., Carl Gans, Armin Geus, Bernd Görmer, Arnold B. Grobman, Rainer Günther, William Holmström, Lipke B. Holthuis, Marinus S. Hoogmoed, Adel Ibrahim, Koji Iizuka, Benedetto Lanza, the late Raymond F. Laurent, Masafumi Matsui, the late C. J. McCoy, Jr., Shab Mohammadi, Charles W. Myers, Göran Nilson, Harald Pieper, Zbyněk Roček, José Rosado, Wolfgang Schneider, Margaret S. Shaw, the late Zdeněk V. Špinar, Christine Stocker, Franz Tiedemann, P. E. Vanzolini, Richard Wahlgren, David B. Wake, Pei-chao Wang, Yehudah L. Werner, and Er-mi Zhao.

Individual photographs and signatures are credited where appropriate, but I do wish to acknowledge the following persons for sending important collections of these materials: Natalia B. Ananjeva, Wolfgang Bischoff, the late Charles M. Bogert, Harold G. Cogger, the late Olivér Gy. Dely, Tatjana Dujsebayeva, the late Hajime Fukada, the late Howard K. Gloyd (via Kathryn J. Gloyd, the late Roger Conant, and Jonathan Campbell), Arnold B. Grobman, Lipke B. Holthuis, Marinus S. Hoogmoed, the late Toshijiro Kawamura, Zoltán Korsós, Alan E. Leviton, the late Walter L. Necker (via Kaethe Necker), the late M. Graham Netting, Andrew F. Stimson, Richard Wahlgren, and the late Heinz Wermuth.

Library and archives personnel at numerous institutions cooperated by sending relevant materials. I thank the following for their help: Academy of Natural Sciences, Philadelphia (Ned S. Gilmore, Marsha Gross), Australian Museum, Sydney (Rose Decker, Ian Loch, Alison C. Miller), Biblioteca Lancisiana, Rome (Marco Fiorilla), Cornell University Libraries, Ithaca (David W. Corson, Medha H. Devare, Mary Ochs, Carolyn T. Spicer), Deutsches Entomologisches Institut, Müncheberg (Editha Schubert), Hunt Institute for Botanical Documentation, Carnegie Mellon University, Pittsburgh (Anita L. Karg, Angela Todd), Institut National Agronomique, Paris (Nadine Pontal), Linnean Society of London (Gina Douglas), Museum of Comparative Zoology, Harvard University, Cambridge (Roxane Combs), Muséum National d'Histoire Naturelle, Paris (Virginie Bouetel, Véronique van de Ponseele), Museum für Naturkunde, Humboldt Universität, Berlin (Sabine Hackethal, Hannelore Landsberg), Natural History Museum, London (Katie Anderson, Ann Datta, James Hatton, Polly Tucker), Percy FitzPatrick Institute of African Ornithology, University of Cape Town (Margaret Sandwith), Smithsonian Institution Archives, Washington (William Cox), Staatsbibliothek zu Berlin (Wolfgang Hamm), and Western Australian Museum, Perth (Margaret Triffitt).

Several natural history booksellers assisted by sending useful references or answering questions. I thank Breck Bartholomew, Andreas Brahm, Russell Ciochon, Andrew Isles, Betty and the late John Johnson, Rudolph Wm. Sabbot, Allard Schierenberg, Dieter Schierenberg, Jeroen Schierenberg, Ernst Spek, Julius Steiner, the late Howard K. Swann, and Clarence Wolf.

For translations into English I thank Satoshi Amagai, Natalia B. Ananjeva, K. E. Bättig von Wittelsbach, Vladimir Bobrov, Marta del Campo, Frank Dodd, Evgeniy Dunayev, Davina Eisenberg, Maria Eisner, Thomas Eisner, the late Hajime Fukada, Richard Goris, T. Ulmar Grafe, Lipke B. Holthuis, Minoru Koyama, John Simmons, Zbigniew Szyndlar, Richard Wahlgren, Er-mi Zhao, and Guisheng Zhong.

Susan Prince of the Department of Classics at the University of Colorado advised on the Latinization of personal names.

Much of the biographical material that I have assembled was accumulated during my term as Editor of the *Facsimile Reprints in Herpetology* series of the Society for the Study of Amphibians and Reptiles from 1961 to 2001. Over that period, many persons acting incidental to their duties as research technician or student assistant have located valuable materials. There have been too many to acknowledge them individually but I do want to offer them my grateful thanks. I also acknowledge my debt to Dawn Potter, my administrative aide, for handling the voluminous correspondence associated with this project as well as for locating important library materials. I thank my longterm editorial associate, Timothy D. Perry, for his numerous contributions to this project, most especially his expert skills in turning my manuscript into a presentable volume.

I am grateful to my wife, Dolores Adler, for her support, her comments on the text, and for her understanding for my long absences to work on this project.

Corrections and Missing Materials.—I am aware of the possibility that some information presented here might be incorrect, although every effort has been made to insure accuracy. Since further volumes of this series might be undertaken I earnestly solicit comments and corrections with supporting documentation wherever possible. I was successful in finding signatures for each biographee but not each person's portrait. Some—like those of Mark Catesby, Jan Swammerdam, and Edward Topsell—have been sought by historians for decades. These and the likenesses of P. J. Bonnaterre, Giovanni Caldesi, J. G. Fischer, Alphonse Guichenot, Edward Hallowell, Catherine Hopley, Fritz Nieden, Michael Oppel, and Charles Owen thus remain prime desiderata.

Correspondence may be addressed to: Kraig Adler, Cornell University, Department of Neurobiology and Behavior, Seeley G. Mudd Hall, Ithaca, New York 14853–2702, USA or to kka4@cornell.edu. I earnestly solicit contributions of portraits, handwritten or signed correspondence, and other materials relating to herpetologists and to the discipline. My archives will eventually be deposited in a public institution for open access, but in the meantime, I am able to serve requests from colleagues for their own publications and other projects.

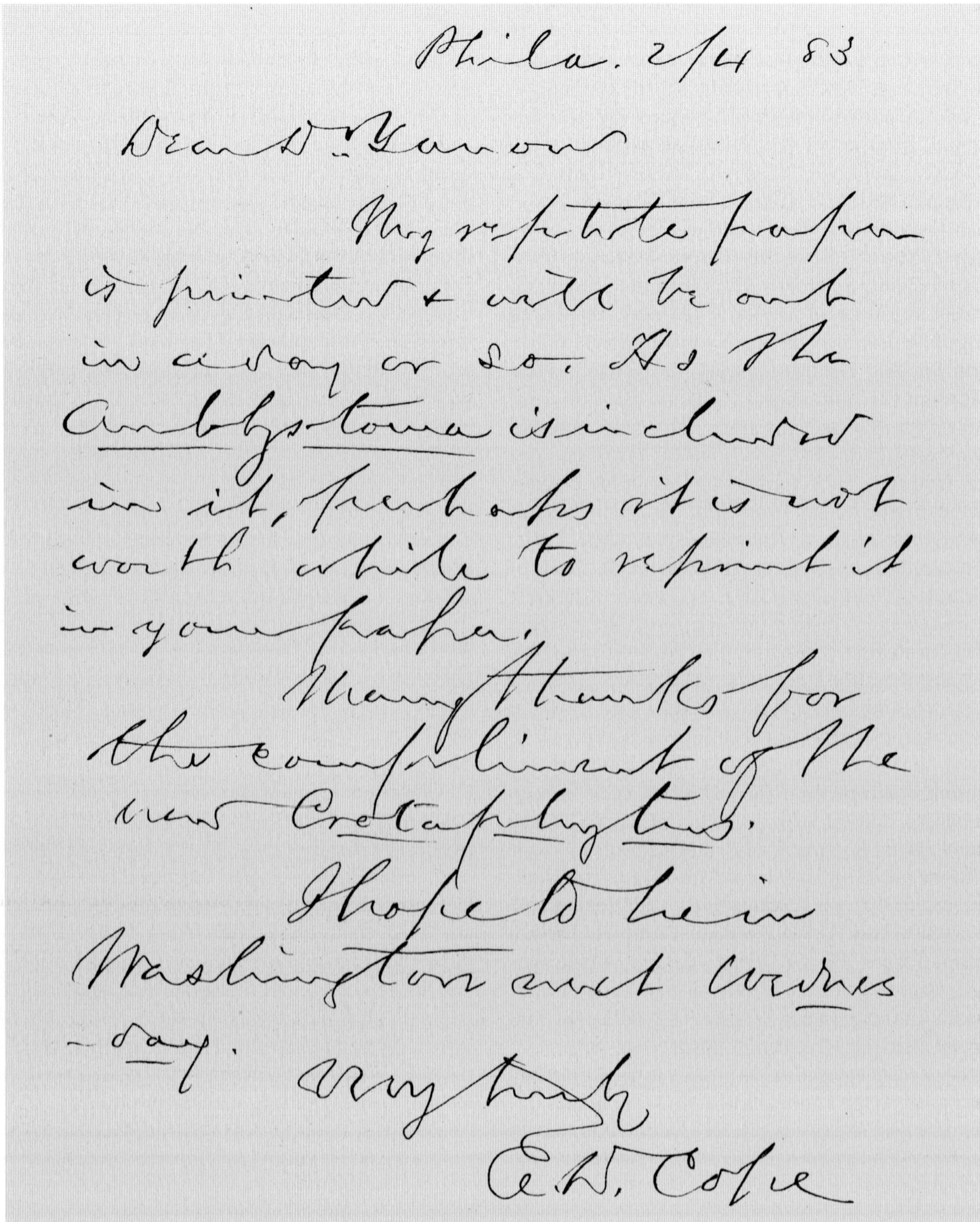

Phila. 2/4 83

Dear Dr Yarrow

My reptile paper is printed + will be out in a day or so. As the Amblystoma is included in it, perhaps it is not worth while to reprint it in your paper.

Many thanks for the compliment of the new Crotaphytus.

I hope to be in Washington next Wednesday.

Very truly
E. D. Cope

CORDIAL RELATIONSHIPS AMONG HERPETOLOGISTS HAVE always been important to the development of the discipline. In this letter (transcribed at the right), Edward Drinker Cope, then the preeminent herpetologist in the United States, writes to Henry C. Yarrow, M.D., honorary curator of reptiles and amphibians at the United States National Museum in Washington. Cope gives Yarrow advance notice of publication of his new species of salamander from Idaho, *Ambystoma epixanthum* (now a synonym of *A. macrodactylum*).

In the next paragraph, Cope thanks Yarrow for naming a new lizard in his honor, *Crotaphytus* (now *Gambelia*) *copeii*, from Baja California. Yarrow was returning the gesture that Cope had made in 1875 when he named a new lizard from Arizona for Yarrow, *Sceloporus jarrovii*. For Yarrow's biography, see page 100 in this volume; for that of Cope see page 46 in volume 1. (Original letter in the Adler collection.)

Phila[delphia] *2/4* [=February 4] [18]*83*

Dear Dr. Yarrow

My reptile paper is printed + will be out in a day or so. As the Amblystoma [now spelled *Ambystoma*] *is included in it, perhaps it is not worth while to reprint it in your paper.*

Many thanks for the compliment of the new Crotaphytus.

I hope to be in Washington next Wednesday.

Very truly
E. D. Cope

BIOGRAPHIES

LEONICENO, Nicolò (1428–1524).

Nicolò Leoniceno (see *Note*), a celebrated Italian physician and Greek scholar, authored the first printed work on reptiles. He was born in Vicenza in 1428. His father was a physician and a member of the Leoniceni, an ancient noble family; his mother was the daughter of the secretary to Pope Alexander V. Nicolò was a boy of keen intelligence who delighted in his studies and mastered Latin and Greek by the age of 12. He was subject to attacks of epilepsy, which continued through his early manhood and were said to have motivated his interest in a medical career. His knowledge of the classical languages was so complete that he knew by rote all of the orations of Demosthenes and of Cicero, among others, but such memorization of whole tracts was not rare in those days before printed books. In 1446 Leoniceno entered the university in Padua to study philosophy and medicine (doctorates in philosophy and medicine, both about 1453). After some years of travel that included England, he became a teacher in Padua. In 1464 he was called to Ferrara as a professor, first of mathematics and then of moral philosophy. He taught in Greek and his mastery of this language was a key to his major scholarly achievements. Soon he was named professor of theoretical medicine at Ferrara and there he remained, except for a year in Bologna (1508–1509), for the rest of his very long life.

Teaching then was based on Greek texts, especially those of Galen, but as interpreted and systematized by Arabic authors who introduced distortions because of their poor translations into Latin. Leoniceno pioneered the effort to revive Greek medicine in its pristine form by recovering the original texts and he prepared accurate translations. Università degli Studi di Ferrara, founded 1391, thus became the main center for studying a true Galenic medicine and was the greatest medical school in Europe. He discovered how errors in transcribing from manuscript to manuscript or in translating even single words could distort the meaning of entire passages and lead to a medical system that was a threat to human life.

Leoniceno noted similar errors in the transcriptions of the great early encyclopedia of botany, zoology, and other subjects, the "Naturalis Historiæ" of Pliny the Elder, which remained in manuscript form and was copied for 14 centuries until the first printed edition was issued in 1469. Pliny, who died from asphyxiation while observing the eruption of Mount Vesuvius in 23 A.D., was not a naturalist and had no practical knowledge of any branch of the subject. He led Roman cavalry units in Germany and then practiced law in Rome before finally becoming commander of the Roman fleet based at Misenum (now Miseno). He was also a writer, however, and his natural history books were based on his extracts from Roman and mostly Greek sources. Because of Pliny's lack of scientific knowledge and sometimes also simple carelessness, his translations could be inaccurate and occasionally unintelligible. The copyists further compounded the problem. Nevertheless, teaching was based on Pliny. His statements, like those of Aristotle and others that he translated, came to be regarded during the Middle Ages as infallible.

Since he could read the original Greek, Leoniceno knew that Pliny made mistakes. In 1492 he published a tract in which he pointed them out, although deftly assigning much of the blame to others. He furthermore charged Pliny with many errors based on verbal confusion. Given the slavish reverence in which this native son of Italy was held, these criticisms represented an act of the greatest intellectual courage. Among others, Pandolfo Collenuccio, an Italian lawyer and amateur naturalist, immediately attacked him. Leoniceno defended himself in three additional tracts, finally even calling Aristotle's authority into question, but attitudes towards the classical authors were thus irrevocably changed. Leoniceno made numerous other contributions, notably the first scholarly treatise on syphilis (1497). He continued to teach and do scholarly work in Ferrara until his death there on 9 June 1524 at the age of 96. He never married.

Leoniceno's work on snakes—the first printed tract solely on a herpetological subject—is a booklet of 14 unnumbered pages entitled "De Tiro, seu Vipera" and was published anonymously in Venice in 1498. Tirus is a viper previously described by Galen and Avicenna. In this work Leoniceno discusses whether vipers give birth to living young or hatch them from eggs. This pamphlet was reprinted on the final pages of his larger work, "De Serpentibus Opus Singulare ac Exactissimum" (1518), a book of 108 unnumbered pages that is dedicated to the most intriguing woman of the Italian Renaissance, Lucrezia Borgia, Duchess of Ferrara, illegitimate daughter of a future pope and infamous for having been complicit in the murder of one of her husbands. The first part of the book is devoted to the dipsade (a serpent whose bite causes violent thirst), the viper, the adder, the basilisk, a spotted serpent, the asp (an African cobra), and the crocodile. He later

discusses the toxicology of snakebites, a matter of the greatest interest to the medical profession, and then compares rabies in analogy to snake venoms. As with his earlier works on the errors of Pliny, the main text is replete with marginal references to errors by classical and medieval scholars including Aristotle, Galen, and Albertus Magnus.

• *References*: "Della Vita e Delle Opere di Nicolò Leoniceno Vicentino," by D. Vitaliani, Tipolitografia Sordomuti, Verona, (6), 312 pages, 1892; "Nicolò Leoniceno," by J. J. Bylebyl, Dict. Sci. Biogr., 8: 248–250, 1973; "Landmarks of Botanical History," by E. L. Greene, Stanford Univ. Press, Stanford, part 2, 1983 (Leoniceno on pp. 528–543, 979–982). • *Portrait*: Università degli Studi di Ferrara, courtesy Hunt Institute, Pittsburgh. • *Signature* (Latin form): From Vitaliani, 1892. • *Note*: His Latin name was Nicolaus Leonicenus, as it appears on his publications.

COLLENUCCIO, Pandolfo (1444–1504).

Pandolfo Collenuccio (see *Note 1*), an Italian lawyer, historian, and early writer on vipers, was born on 7 January 1444 in Coldinoce, a small town near Rocca Contrada at the edge of the Duchy of Urbino. His family was of fairly high rank and soon moved to Pesaro, which he always regarded as his native country. He studied law at the university in Padua (J.D. 1465) and came to hold various public offices in a number of places in Italy. Collenuccio was a voluminous writer and is best known for his posthumously published history of the Kingdom of Naples (1539), which went through 18 editions by 1618. He was also a rash and bold person who had upset the governor of Pesaro, Duke Giovanni Sforza, which accounts for his spending so much time away from his native city. Ironically, it was Collenuccio who had earlier petitioned the pope to cede Pesaro, a fief of the Holy See, to Sforza. The Duke of Valentino (Cesare Borgia) succeeded Sforza and Collenuccio wrote to him, incautiously, to complain about Sforza, but these letters fell into the hands of his antagonist who then became governor again. Sforza had been Borgia's brother-in-law, by his former marriage to Borgia's notorious sister, Lucrezia Borgia. Pretending to have forgiven Collenuccio, Sforza induced him to come home to Pesaro on the promise of help. Collenuccio returned, hoping to spend his life there in study, but Sforza had him arrested in front of his family. He was informed that he must die in five days. After being imprisoned, he was strangled to death by an executioner on 11 July 1504 (see *Note 2*), the victim of a petty, vindictive tyrant.

Following Nicolò Leoniceno's published challenge in 1492 to the infallibility of Pliny the Elder's "Naturalis Historiæ," Collenuccio, then living in Ferrara, was the first to come to the Roman's defense in a book entitled "Pliniana Defensio . . . adversus Nicolai Leoniceni Accusationem" (1493). It was a rude, if cleverly disguised attack that called Leoniceno a Pliniomastex (Latin for Plinywhipper). Many of Leoniceno's examples were botanical and, although Collenuccio was the better botanist of the two, his specious arguments ultimately failed. To defend Pliny's statements ascribing meaning to the similarities of plant leaves to the shapes of certain animals was, in reality, indefensible but an idea then popular among botanists and the general public who revered Pliny.

Collenuccio wrote a small book on vipers entitled "De Vipera Libellus" that was issued in Venice in 1506. It was published posthumously under the editorship of his son, Hannibal Collenuccio, and arose out of his controversy with Leoniceno. It consists of 36 unnumbered pages. In the book he reviews classical knowledge on vipers based on the works of Aristotle, Herodotus, Nicander, Galen, Pliny, and others. He discusses the differences between the common snake and the viper but depends on information from previous authors rather than his own observations. Although he mentions Leoniceno in his introduction, Collenuccio pointedly makes no reference to Leoniceno's earlier work (1498) on the same subject.

• *References*: "Memoirs Connected with the Life and Writings of Pandolfo Collenuccio da Pesaro," by W. MacD. Tartt, privately printed, Cheltenham, U.K., vii, 329, (2) pages, 1868; "Pandolfo Collenuccio Umanista Pesarese del Secolo XV," by A. Saviotti, Ann. Scuola Norm. Super. Pisa. Filos. Filol., 5: 33–328, 1888; "Pandolfo Collenuccio," by E. Melfi, Diz. Biograf. Ital., 27: 1–5, 1982; "Landmarks of Botanical History," by E. L. Greene, Stanford Univ., Press, Stanford, part 2, 1983 (Collenuccio on pp. 544–552, 982–984). • *Portrait*: Courtesy Hunt Institute, Pittsburgh. The rope around his neck is in apparent allusion to the method of his execution. • *Signature* (Latin form): Biblioteca e Musei Oliveriani, Pesaro. • *Note 1*: His Latin name was Pandulphus Collenucius, sometimes spelled Pandolphus Collenutius. • *Note 2*: Melfi (1982) gives the month of death as June.

ALDROVANDI, Ulisse (1522–1605).

Ulisse Aldrovandi (see *Note 1*), the Italian encyclopedist of nature, founded the earliest modern natural history museum and, as the first professor of natural history, established the subject as a discipline for study. He was born in Bologna on 11 September 1522. Both parents were from noble families. His father served as secretary to the Senate of Bologna, then a papal state, and his mother was first cousin to a future pope, Gregory XIII. Both connections were to be of profound importance to his career. He was a restless youth especially after his father's early death in 1529. He ran away from home several times, traveling mostly by foot and as far away as Spain. By 1548 he had studied mathematics, Latin, law, and philosophy for seven years at the university of Padua and then he turned to medicine. Returning home in 1549, he was one of a group accused of heresy, arrested, and briefly imprisoned, and he had to go to Rome to clear his name. There he chanced to meet the French naturalist, Guillaume Rondelet, who at that time was the personal physician to a cardinal at the Vatican. Aldrovandi was mentored by Rondelet and accompanied him to fish markets to get specimens for study, which led to Rondelet's book, "De Piscibus Marinis" (1554–1555).

Aldrovandi decided to make natural history his own special area for study and began to gather specimens for his personal collection. Eventually he went to the university in Pisa to complete his medical studies and then returned to Bologna where he was granted his degree (M.D. 1553). He became a professor at Bologna, first of logic, but by 1561 he was named professor of natural history, the first of its kind in any university. He had been teaching the subject since at least 1556. Technically he was professor of the history of the "simples," by which Aldrovandi meant not only plants but also animals and minerals, and whether or not they were of medicinal value. Thus he was not an herbalist or a pharmacologist. This distinction is important because previously lectures had been devoted to single species and only those of medicinal value. He founded the botanical garden at the university in 1567, but when he was asked to prepare an official pharmacopoeia, one that would establish the exact characteristics of medicinal substances that druggists were required to use in filling prescriptions, he received much opposition from the medical faculty and the apothecaries. The opposition was finally quelled by the intervention and patronage of Pope Gregory, his mother's cousin, and his successor, Sixtus V, as well as by the Senate of Bologna.

Aldrovandi's rapidly growing museum was housed in several rooms in his family's palace and, like modern museums, it had an extensive library and research activities associated with it. The museum was also open to visitors: scholars, nobles, and even the general public. One visitor, the Italian botanist Pietro Andrea Mattioli, wrote to him in 1553 (in translation): "Thus I remain always with heart aflutter and with baited breath until I see all the 'simples' you have collected, and really I would like to come to Bologna only for this end." On Aldrovandi's death, the museum was bequeathed to the city of Bologna where it remains today as the nucleus of their civic museum.

It had been Aldrovandi's plan to extend Aristotle's work, especially in zoology, because botany was covered by the herbalists. He was the first since Aristotle to open the eggs of chickens during the incubation period and to describe the development of the embryo (published 1599) for which work he is regarded as one of the founders of embryology. He spent his lifetime and fortune acquiring enormous numbers of specimens from all parts of the world, studying them, and having them illustrated in huge volumes of watercolor drawings. All of this was in

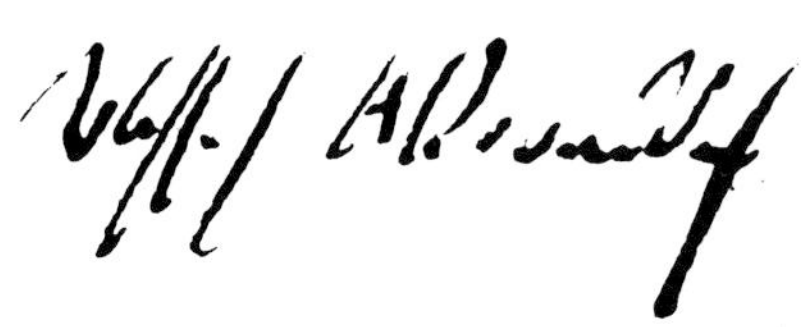

preparation for a series of books that became the most elaborate illustrated iconography of Renaissance zoology. His books are profusely illustrated with woodcut drawings, some of which occupy entire folio pages. Allowing for the method of printing, the quality of the printed illustrations was high but still far from the accuracy and beauty of the original watercolors, which still exist today in Bologna, bound in 20 thick volumes (many were reprinted in 2007). Conrad Gessner, the Swiss naturalist with whom Aldrovandi had corresponded in his youth, had completed an earlier compendium of zoology in five volumes (1551–1587), but Aldrovandi's effort was larger and more grandiose—a total of 13 folio volumes published from 1599 to 1668—but just four of them were issued during his lifetime and the first when he was already 77. By 1602 he had become blind and it was left to his students and associates to finish and edit the volumes for publication. Aldrovandi died in Bologna on 4 May 1605 (see *Note 2*).

Two of the volumes are herpetological in nature, at least in part, and both were revised and edited by Bartolomeo Ambrosino, one of Aldrovandi's pupils and successors at the university. The first, "De Quadrupedibus Digitatis Viviparis Libri Tres, et de Quadrupedibus Digitatis Oviparis Libri Duo," covers mammals, reptiles (lizards, crocodiles, turtles), and amphibians. It was issued in 1637 (reissued five times from 1642 to 1665). The second, "Serpentum, et Draconum Historiæ Libri Duo," covers snakes and some worms plus a wide variety of so-called fabulous animals (winged dragons, crowned serpents, eight-legged basilisks) that were believed to exist. This was published in 1639 (reissued 1640; see *Note 3*). Altogether there were 11 volumes on zoology (1599–1642) and one each on minerals (1648) and plants (1668). Some of the information in these volumes was compiled from dubious sources, both ancient and medieval, or represented developmental monstrosities of little value in comprehending the natural world. Unlike Gessner, however, who organized his kinds of animals alphabetically, Aldrovandi grouped them by similarities and thus began the partitioning of organisms into useful, if not natural, categories. He was not very discriminating, however, and all kinds of animals known to him were included, whether real or imagined, yet much of value could be gleaned from his works. Carl Linnaeus, in fact, cited him in his "Systema Naturæ" and made numerous references to his texts and illustrations in describing species. Aldrovandi's volumes on animals remained the most comprehensive illustrated work on zoology for more than a century until the advent of Georges Buffon's "Histoire Naturelle," which started to appear in Paris in 1749.

• *References*: "Memorie delle Vita di Ulisse Aldrovandi," by G. Fantuzzi, L. Volpe, Bologna, vi, 263 pages, 1774; "Ulisse Aldrovandi," by G. Montalenti, Diz. Biograf. Ital., 2: 118–124, 1960; "Ulisse Aldrovandi," by C. Castellani, Dict. Sci. Biogr., 1: 108–110, 1970; "Possessing Nature. Museums, Collecting, and Scientific Culture in Early Modern Italy," by P. Findlen, Univ. Calif. Press, Berkeley, xviii, 449 pages, 1994; Alessandrini and Ceregato, 2007. • *Portrait*: Adler collection. • *Signature* (1594, Latin form): Biblioteca Universitaria, Bologna, by permission of the Ministero per i Beni Culturali e Ambientali, and courtesy Benedetto Lanza. • *Note 1*: Aldrovandi's forename is sometimes spelled Ulissi. His Latin name was Ulyssis Aldrovandus. • *Note 2*: The death date is sometimes given as 10 May or 10 November and the year as 1607. • *Note 3*: The volume on serpents is generally dated 1640. However, it appeared in 1639 as dated on the colophon page. In the version of 1640, the text has been reset and the full page of errata and colophon omitted.

TOPSELL, Edward (1572–1625).

Edward Topsell, a Church of England clergyman and author, wrote the first book in English on amphibians and reptiles. He was born in 1572 (baptized 3 February) in Sevenoaks (Kent), just southeast of London where he received his early schooling. In 1587 he entered Christ's College at Cambridge (B.A. 1591 or 1592, M.A. probably 1595), was ordained, and after preaching at Hartfield and Framfield (both Sussex) became rector at East Hoathly (Sussex) in 1596. Topsell had a form of paralysis in one arm; as he wrote later in life, ". . . sometimes enfeebled to write through a paraliticke righte arme, even from my birth . . . . " Because of his father's death when he was one year old he was also handicapped by extreme poverty, which may explain why he began to write books as early as 1596 to provide a supplemental source of income. He later held religious posts in several places in eastern and southern England until April 1604 when he became perpetual curate at St. Botolph's in Aldersgate, London, although he continued to preach at other places. He died in London where he was buried at St. Botolph's in 1625 (see *Note 1*).

Topsell's many books began with "The Reward of Religion" (1596), consisting of lectures on the Book of Ruth. It was popular enough to justify a third edition by 1613. He published two others in 1599, "Commentary upon the Lamentations of Jeremy" and "Time's Lamentation"; the latter is an exposition on the prophet Joel in which Topsell denounces fashionable vices and frivolities. Additional titles of expository comments upon scripture followed. He was a careful interpreter, although lacking in imagination and deep insight. His lessons were often superficial and included many anecdotes and quotations. Such a writer would seem well suited to produce the natural history books on which his fame now rests, the works from which Shakespeare probably (see *Note 2*) and Isaak Walton certainly derived much of their knowledge of the natural world.

Topsell was not a naturalist. It seems that his interest in writing natural history was to glorify the works of God and at the same time to earn a living. The public was eager for such picture books and Topsell's "The Historie of Foure-footed Beastes" was issued in 1607. For the most part it is a rework of Conrad Gessner's volume on mammals (1551) in his series "Historia Animalium." As one writer has noted, " . . . the actual text of Gesner is treated with freedom, paraphrased, rearranged, condensed, expanded in a way only possible in an age when plagiarism was universal and copyright unknown." Gessner's illustrations were also copied but generally badly. Topsell used some other sources and added a few observations of his own. He regarded his book as suitable reading matter in the godly home on Sundays.

Topsell's book "The Historie of Serpents" (1608; reprinted 1973) repeated the pattern of his book on mammals. As was generally held in those days, Topsell's "Serpents" included a variety of animals besides amphibians and reptiles: spiders, scorpions, honeybees, and wasps, all of which possessed "poysons," but also worms, sea serpents, certain fishes, the Hydra killed by Hercules, winged dragons, and caterpillars. He was also prone to extreme exaggeration. On page 111, he described the cattle-eating boas, "which Serpent was an hundred and twenty foote in length." Most of the book is based on Gessner's two herpetological volumes (1554, 1587) but also on some other sources. There are very few statements in the book that are based on Topsell's own observations: comments on toads (pages 191, 314–315) and newts (page 213). A second edition of "Serpents" was issued posthumously (1658; reprinted 1967). This then was the great picture book on reptiles that the English-speaking public read as its major sourcebook for the next century and more.

Later, probably in 1613–1614, Topsell began to write a companion volume on birds, "The Fowles of Heaven," which was a translation of Ulisse Aldrovandi's three volumes on birds (1599–1603), in turn based partly on Gessner. The manuscript was finally published in 1972.

• *References*: "Edward Topsell," by S. Lee, Dict. Natl. Biogr., 19: 986–987, 1921; "Edward Topsell," p. 217–226 *in* C. E. Raven, English Naturalists from Neckam to Ray, Cambridge Univ. Press, Cambridge, 1947; "Edward Topsell," by G. Lewis, Oxford Dict. Natl. Biogr., 55: 43–44, 2004. • *Portrait*: No portrait is known to exist. • *Signature* (probably 1613 or 1614): This item is reproduced by permission of The Huntington Library, San Marino, California. • *Note 1*: Lee (1921) and some other authorities give the date of Topsell's death as 1638 when his successor as curate at St. Botolph's was appointed. However, it seems probable that the post was simply vacant for several years. Recent authorities accept the 1625 date. • *Note 2*: Topsell's printer, William Jaggard, was also the printer of Shakespeare's First Folio.

JONSTON, John (1603–1675).

John (or Jan) Jonston (see *Note*), a Polish physician and natural history compiler and writer, was born on 3 September 1603 in Samter (now Szamotuły), near Poznań, Poland, of a Scottish immigrant father and a German mother. He received his early education in German schools and then spent the next decade traveling in Europe or on estates in Poland, often as tutor to the children of aristocrats, and as a student at several universities in Germany, Britain, and Holland. He took degrees at some of them—at St. Andrews in Scotland (M.A. 1623), where he studied theology and philosophy, and at Franeker in Holland (M.A. 1629)—and then studied botany and medicine at Cambridge before taking his medical degree in Leiden with a dissertation on fever (M.D. 1634; and *ad eundem*, a courtesy M.D. at Cambridge, also in 1634).

After receiving his graduate degrees Jonston practiced medicine in Leiden, but from 1636 he lived in Lissa (now Leszno), near Samter, to work under the long-term patronage of a noble family, the Leszczynskis. He was a Calvinist as were so many early scientists and naturalists including Gessner, Ray, Scheuchzer, Sloane, and Swammerdam. Jonston was also active in the Czech Brethren community in Leszno. During this period Jonston declined several positions offered to him by Dutch and German universities including the professorship of medicine at Leiden in 1640 and that at Frankfurt an der Oder in 1642. He preferred private study to lecturing. In 1652 he purchased an estate in Ziebenburg but in 1665 events following the Polish-Swedish War forced him to move to another estate, Zybendorf, near Liegnitz (now Legnica), Poland, where he lived until his death. He maintained a medical practice throughout, yet spent much of his life writing extensively on medicine, natural history, and other subjects. He died in Składowice, near Liegnitz, on 8 June 1675.

Jonston's natural history books were compilations. They were derived almost entirely from the works of Gessner, Aldrovandi, and Nieremberg as well as Aristotle, Pliny, and other classical authors. Jonston's works were carefully written and he became a popular writer. His first such title, "Thaumatographia Naturalis" (1632, with several later editions; English translation in 1657), covered the ten wonders of nature and included many kinds of animals.

His most famous work on animals, however, was the "Historiæ Naturalis," first issued in Frankfurt am Main in six parts (1650–1653). This was the first of the zoological encyclopedias to use copper-engraved plates. This method of printing captured finer detail than was possible with the woodcuts of previous authors, and the book became the most widely read illustrated zoology of the 17th Century. Snakes, caecilians, amphisbaenians, and legless lizards are included in the part entitled "Serpentibus et Draconibus" (1653), with the remaining amphibians and reptiles in "Quadrupedibus" (1650). The entire work is extensively illustrated on 249 uncolored plates. These include excellent likenesses of many species and also figure the so-called fabulous animals, fanciful creatures based on misshapen and misinterpreted specimens or merely on hearsay and myth. Several additional Latin printings were issued plus a Dutch translation in 1660. The most elaborate edition, however, was that published posthumously in Amsterdam in 1718, "Theatrum Universale Omnium Animalium," under the editorship of Hendrik Ruysch. This edition contains 270 plates and sometimes occurs in a hand-colored version; it includes Ruysch's treatise on the fishes of the East Indies, which accounts for the extra plates.

• *References*: "Jan Jonston," by T. Bilikiewicz, Mianowskiego, Warsaw, 233 pages, 1931; "John Jonston," by J. K. Crellin, Dict. Sci. Biogr., 7: 164–165, 1973; "John Johnstone," by T. Seccombe and Y. Foote, Oxford Dict. Natl. Biogr., 30: 405, 2004. • *Portrait*: Courtesy L. B. Holthuis. • *Signature* (1623, Latin form): Reproduced by kind permission of the Keeper of Muniments, University of St. Andrews Library (UY 305/3/191). • *Note*: In his books Jonston's name generally appears in its Latin form, Johannes Jonstonus. In biographies it is occasionally given as Jean, Johann, or Joannes and as Johnston or Johnstone. His father's name was spelled Johnstone.

SEVERINO, Marco Aurelio (1580–1656).

Marco Aurelio Severino (see *Note 1*) was an Italian physician, surgeon, and author of the first comprehensive treatise on comparative anatomy. He was an early student of the anatomy of reptiles. Born in Tarsia (Calabria) on 2 November 1580, he received his education in the classics and law at schools in Calabria. On attending the university in Naples he switched to medicine. Although most of his study was in Naples, he completed his medical training in Salerno (M.D. 1606). Then he opened a practice in Tarsia before returning to Naples in 1609 to study surgery with Giulio Iasolino. There, beginning in 1610, he taught surgery and anatomy privately before accepting a chair in these subjects at the university in 1615. Severino's surgical teachings are said to have been "... imbued with all the terrors of torture and mutilation." Nevertheless, his

fame as a surgeon attracted many students. He had unorthodox views for his time, both religious and philosophical. At one point he was called before the Inquisition, although he was eventually acquitted. In Naples on 12 July 1656 he died of the plague that he had attempted to help stamp out.

Most of Severino's writings are on anatomy and surgery. He held strongly anti-Aristotelian views especially as expressed in his most famous work, "Zootomia Democritæa" (1645). This is regarded as the first comprehensive treatment of comparative anatomy (he called it zootomy) in which human anatomy is related to that of other animals. In his writings he puts emphasis on observation and experience, especially on his dissections of animals, but he also accepted teleology in nature.

Reptiles and occasionally amphibians were included in Severino's books on vertebrate anatomy. In his "Zootomia" he notes that he first dissected the viper in 1616. He discovered that its heart had two auricles and one ventricle, which may be the first recognition of the existence of two auricles in any lower vertebrate. He noted how the shapes of the visceral organs conform to the serpentine shape of the body. He described and figured the basic internal anatomy of a colubrid snake including the larynx, lungs, heart, stomach, gall bladder, spleen, and genitalia. He dissected a variety of amphibians and reptiles: a toad, a lizard, marine and terrestrial turtles, some colubrids, and the viper.

Severino's only strictly herpetological book is "Vipera Pythia" (1650, reissued with slight modifications in 1651; see *Note 2*) that was published when he was elderly. It is a strange combination of facts based on actual observations by Severino and others and also includes prevailing myths. He discusses whether venomous snakes were allowed on Noah's ark and, if so, how these dangerous animals were controlled. He also accurately describes the anatomy and reproduction of vipers.

• *References*: Cole, 1944 (Severino on pp. 132–149); "Marco Aurelio Severino," by C. B. Schmitt, Dict. Sci. Biogr., 12: 332–334, 1975. • *Portrait* (about 1630): Adler collection. • *Signatures* (above, 1638): Bodleian Library, Oxford University (ms. d'Orville 51, fol. 153r); (below, 1647): Biblioteca Lancisiana, Rome (ms. 75, c. 437 recto), both in Latin form. • *Note 1*: Severino's Latin name was Marcus Aurelius Severinus, as it appears on his books. • *Note 2*: "Pythia" is from the Greek name, Latinized to *Python*, for the mythical serpent slain by Apollo near Delphi.

SWAMMERDAM, Jan (1637–1680).

Jan Jacobz Swammerdam, a Dutch microscopist and perhaps the greatest comparative anatomist of the 17th Century, made fundamental discoveries in amphibian anatomy and development. He was born on 12 February 1637 in Amsterdam, the son of a prosperous apothecary and collector of curios (including animals) from around the globe. The family had originated in the Dutch village of Swammerdam (now Zwammerdam), which they adopted as the family name. Jan Swammerdam went to medical school in Leiden in 1661 and during 1663–1665 he studied privately with biologists in France. On returning to Amsterdam he joined a group of physicians including the anatomist Gerhard Blaes (Blasius in Latin), which met irregularly until 1672 to conduct dissections. Swammerdam finished his medical education (M.D. 1667, Leiden) but apparently never practiced. He lived in several towns in Holland, occasionally visiting other scientists such as Anton van Leeuwenhoek in Delft.

Swammerdam devoted his short and troubled life to private scientific studies, which was interrupted by illness, periods of depression, religious anxiety, and mental instability. He burned many of his works before his early death from malaria on 17 February 1680 in Amsterdam. In 1674 he had appealed to a mystic for permission to publish his research on the mayfly. One of his friends from college days reported in 1675 that Swammerdam had destroyed his manuscript on silkworm anatomy and that his friends should pray for him. Swammerdam finally joined a religious sect and abandoned scientific investigation altogether, but left instructions in his will for publication of his remaining manuscripts. This was not accomplished for nearly half a century under the title "Bybel der Natuur," edited by Herman Boerhaave (1737–1738; German edition, 1752; English edition, 1758, entitled "Bible of Nature").

Swammerdam made discoveries of great originality and importance, as much for his invention of new techniques as for his biological insights. He was the first to observe red blood cells (1658, in frogs) and he discovered the valves in the lymphatic system (1664). He made the first description (1667) of the buccal force-pump mechanism of respiration in salamanders. He developed a technique for injecting blood vessels with wax and dyes so as to observe them in fine detail. He showed experimentally that muscular contraction involves

a change in the shape, but not in the volume, of muscle. Swammerdam is also remembered for his pioneering work on insects, in particular the detailed anatomy of honeybees and his studies on metamorphosis published in his "Historia Insectorum Generalis" (1669).

Frogs were an essential element in Swammerdam's comparative studies as a point of comparison for insects and humans. Thus he made detailed studies of their anatomy and development and as a result made many key discoveries. One of these, noted earlier, was the existence of red blood cells. He discovered that frog's eggs pass from the ovary to the oviduct via the coelom. He gave the first account of cleavage in a frog's zygote, the first for any animal, but did not understand its significance. It was much later that two Italian biologists furthered our understanding: Lazzaro Spallanzani (1785) pointed out the significance of cleavage and saw the first two furrows crossing at right angles and Mauro Rusconi (1826) noted that cleavages cut up the yolk and were not simply surface phenomena.

Swammerdam's dissection of tadpoles was equally remarkable. He discovered that the gills, which were originally external, became overgrown with skin and thereafter functioned internally. He carefully described metamorphosis—a particular point of comparison with his work on insects—including the dramatic changes in the mouthparts, tail, and gut. He conducted experiments on muscle contraction in frogs as early as 1665 using muscle-nerve preparations and discovered that when the cut nerve was mechanically stimulated the muscle contracted. He illustrated these anatomical observations and experiments in finely detailed plates. It must be remembered that all of Swammerdam's observations on frogs were published posthumously in 1737–1738, but that the research was done during 1658–1673.

• *References*: Cole, 1944 (Swammerdam on pp. 270–305); "Jan Swammerdam, Zijn Leven en zijn Werken," by A. Schierbeek, N.V. Uitgeversmaatschappij "De Tijdstroom," Leiden, 280 pages, 1947 (English ed., Swets & Zeitlinger, Amsterdam, 1967); "Jan Swammerdam," by M. P. Winsor, Dict. Sci. Biogr., 13: 169–175, 1976. • *Portrait*: No portrait is known to exist. That commonly reproduced as his and said to be by Rembrandt was painted five years before Swammerdam was born (see Cole, 1944, p. 305, for further details). • *Signature*: From Schierbeek, 1947.

CHARAS, Moyse (1619–1698).

Moyse Charas (see *Note*) was a French apothecary, chemist, physician, and an early student of vipers. He was born in Uzès (Gard) on 2 April 1619 and educated in pharmacy and Latin poetry in Marseille, Montpellier, and Lyon. He opened an apothecary shop in Orange but soon moved to Paris (1646). After taking a chemistry course from an apothecary, Nicaise Le Febvre, he opened his own boutique under the sign "Aux Vipères d'Or" (At the Golden Vipers). During the 17th Century, the source of the viper's poison and the remedies for its bite were popular topics and especially appropriate for study by apothecaries. In 1667 the Compagnie des Apothicaires entrusted him with publication of a book on precise methods for the preparation of drugs, "La Thériaque d'Andromachus" (1668).

At this time Charas joined Christophe Glaser in teaching chemistry at the Jardin du Roy (the full name was Jardin Royale des Herbes Médicinales, founded in 1635 to promote pharmaceutical botany and which, after the French Revolution, became the Muséum d'Histoire Naturelle), but Glaser quit in 1671 leaving Charas to carry on. Charas became apothecary to Louis XIV and demonstrator in chemistry at the king's garden. He often used living vipers when discussing poisons in his classes. His book, "La Pharmacopée Royale, Galénique, et Chymique" (1676), was published by order of the king. It was widely translated and remained popular for the next century, ranking Charas among the great apothecaries.

Unfortunately, because of his religion Charas had to leave the country in 1680 when the king began to turn against the Protestants. He first went to England where he became apothecary to Charles II. While in England he received his M.D. from the faculty in Orange (1681). He then moved to Holland in 1682 and finally to Spain where he intended to practice medicine. However, on account of his faith he fell afoul of the Inquisition and was imprisoned in 1688. At one point he was threatened with death for contradicting the archbishop of Toledo on the subject of the viper's delivery of venom. He only escaped by converting to Catholicism. In 1689 he returned to Holland and finally to Paris in 1691 where he returned to royal favor and became a member of the Académie des Sciences (1692). He died in Paris on 21 January 1698.

Charas spent many years studying vipers, their venom, and remedies for their bites, and was actually bitten twice during his studies. His views were challenged in 1664 in a book by an Italian physician, Francesco Redi, who claimed that his experiments showed that the yellow venom had to be injected

beneath the skin to cause death. Charas published his own book in response, "Nouvelles Experiences sur la Vipere" (1669), in which he stated (in translation) "The venom is not found in the viper's body when the [animal] is calm, but appears when the viper is angry: its enraged spirit then escapes through the fangs when it bites; the effect is, therefore, spirituous . . . The yellow juice is merely harmless saliva." That snakes must be enraged to be able to kill was a popular view at the time. Shakespeare, in the final scene of "Antony and Cleopatra" (1608), has Cleopatra saying to the asp: "Poor venomous fool, Be angry, and dispatch." Charas relied on what he saw as incontestable facts: the harmlessness of many bites and of the venom when it is placed on the skin or taken into the mouth and ingested.

At the time there was much myth and superstition surrounding these topics and it was difficult to separate fact from fiction, but Redi had experimentation on his side. Redi and Charas then proceeded to debate through a series of books (Redi's in 1670, 1685, and 1686, with an English edition in 1673; Charas's in 1672, 1677, and 1694, with English versions in 1670, 1673, and 1677). The matter was not resolved until Redi's claims were confirmed by the work of an Italian biologist, Felice Fontana (1767, 1781).

• *References*: "Moyse Charas et les Vipères au Jardin du Roy," by M. Phisalix, Arch. Mus. Natl. Hist. Nat., sér. 6, 12: 469–472, 1935; "Les Vipères au Jardin du Roy et a l'Académie des Sciences," by M. Phisalix, Bull. Soc. Zool. France, 62: 5–20, 1937; "Moïse Charas," by R. d'Amat, Dict. Biogr. Franç., 8: 464, 1959; Jaussaud and Brygoo, 2004 (Charas on pp. 130–131). • *Portrait*: From Phisalix, 1937. • *Signature* (1669): Bibliothèque Municipale de Lyon (ms. 1723, pièce 201). • *Note*: Charas's forename is sometimes spelled Moïse.

WURFFBAIN, J. P. (1655–1711).

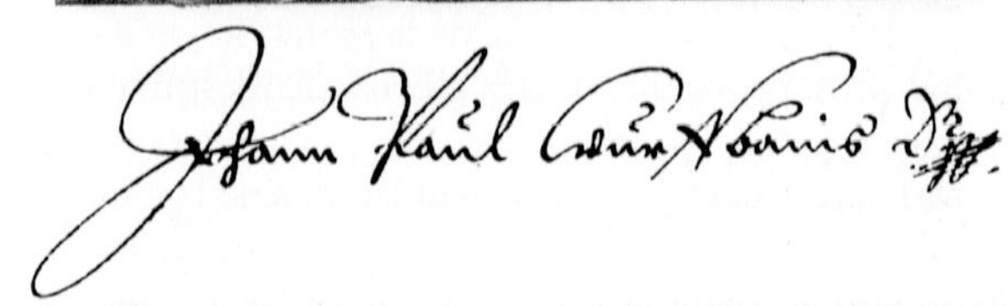

The German physician, Johann Paul Wurffbain (see *Note*), who wrote the first book on salamanders, was born on 14 December 1655 in Nuremberg. His father was a merchant with business in the East Indies. Johann was orphaned at the age of six and raised by an aunt who was the widow of a physician. After attending the Egidienschule in Nuremberg he went to the university in Altdorf in 1673 where he studied philosophy as a basis for his studies in medicine. He finished his dissertation by 1678 and took a study trip to the Netherlands and England before returning to Altdorf where he received his degree (M.D. 1679). Wurffbain set up a medical practice in Nuremberg and conducted his research privately. He is the namesake for a human medical condition involving a cerebellar tumor, Wurffbain's syndrome. He authored some 35 titles and his reputation earned him membership in Leopoldina, the German academy of sciences (1682). He turned down several invitations to take the chair in medicine at Altdorf, preferring instead to stay in Nuremberg. He died there from a stroke in January 1711, probably on the 14th.

Wurffbain had a special interest in salamanders, which formed the basis for his medical dissertation. In his "Naturalis Historiæ," the Roman scholar Pliny the Elder of the first century A.D. states that salamanders have no sex and are generated from putridity; that they are invulnerable to fire; and that their saliva destroys human hair. These and other myths prevailed during the Middle Ages and greatly retarded progress in understanding the biology of these animals. Wurffbain decided to study salamanders (*Salamandra*) and newts (*Triturus*), their anatomy, food habits, behavior, reproduction, medicinal uses, and mythology. He collected all of the available citations from the existing literature. As to the salamander's supposed life in fire, Wurffbain cited 50 authors in favor and ten against. He dissected salamanders and described the food contents in the stomach. He confirmed that *Salamandra* is live bearing and found 34 larvae in one female. He also observed the shedding of the animal's epidermis. And he proceeded to debunk most of the prevailing myths.

All of this information is contained in Wurffbain's published dissertation, "Salamandra," a 98-page booklet with three plates (1677), and an expanded version issued as a book, "Salamandrologia," with five plates (1683). The plates illustrate the external appearance of several species, the larvae, eggs, skeleton, internal organs including the urogenital system, and mythical figures from emblems and hieroglyphics. These two titles represent the first comprehensive basis for our modern understanding of salamander biology.

• *References*: "Johann Paul Wurffbain," by [J. Pagel], p. 1005 *in* W. Haberling et al. (eds.), Biograph. Lexikon hervor. Ärzte, 2nd ed., vol. 5, Urban & Schwarzenberg, Berlin, 1934; "Johann Paul Wurffbain (1655–1711) Leben und Werk," by W. Seitz, Inaugural-Dissert., Philipps-Univ., Marburg, (5), 81, (17) pages, 1982. • *Portrait*: National Library of Medicine, Bethesda. • *Signature* (1691): From Seitz, 1982. • *Note*: In Latin his surname was customarily spelled Wurfbainius with a single *f*, in conformance with Classical Latin rules.

TYSON, Edward (1651–1708).

Edward Tyson was a physician and comparative anatomist who produced the first anatomical study of a pitviper. He was born to a well-to-do family in Clevedon, near Bristol, England, on 20 January 1651. He went to college in Oxford (B.A. 1670, M.A. 1673, Magdalen Hall) followed by his medical studies (M.B. 1677). He then he set up a practice in London. (Much later, in 1680, he took his M.D. at Cambridge.) He quickly developed a relationship with the Royal Society, in particular with Robert Hooke, and was made a fellow in 1679. One of his duties at society meetings was to arrange demonstrations, which often consisted of his own dissections of animals.

Tyson's research focus was to describe the anatomy of selected types of animals, recognizing that a general natural history of animals could not be completed until such detailed studies were accumulated. While this approach may seem self-evident, most authors in Tyson's day simply copied the prior work of others. In this approach, Tyson's efforts were in parallel to certain anatomists in Europe, notably Claude Perrault in Paris.

Putting aside his medical writings, Tyson produced a long series of anatomical studies on a wide range of animals. He contributed the description of a shark embryo to Francis Willughby's "Icthyographia [*sic*]" that was written mostly by John Ray (1686). His most elaborate and detailed anatomies were those of the porpoise (1680), rattlesnake (1683), tapeworm (1683), opossum (1698, 1704), and chimpanzee (1699). The last-named is Tyson's classic for he was the first to notice the close resemblance in their detailed anatomy between a great ape and a human and its implication for relationships.

He carried on his medical practice to the very end of his life, which ended suddenly on 1 August 1708 in London. He was skilled at dissection and a very keen observer but, unfortunately, he left many studies unpublished.

The anatomy of the European vipers already had been described several times—by Baldo Abati (1589), M. A. Severino (1645, 1650), Johann Vesling (1651), and Moyse Charas (1669)—but Tyson was the first to write an anatomy of a pitviper. His subject was a large male *Crotalus horridus* that was received alive by the Royal Society in December 1682. It soon died and Tyson quickly made the most of it by producing what was then the most complete anatomy of any reptile. He had studied its locomotion when it was alive and later, on studying the muscles and ribs, could visualize how the body glided. His descriptions of the viscera, the urogenital system, and the anal glands were detailed and accurate, as were those of the tongue, glottis, and the respiratory system.

His most intriguing discovery, however, was the pair of facial pits. At first he thought them to be ears, since the absence of external ears in serpents had long puzzled anatomists. Even as late as the 1930s, a prominent author wrote that the pit organ might be involved in hearing. Tyson noted that the opening led to a blind cavity unconnected with any internal organ. Only in the 1950s did the Americans Theodore Bullock and Raymond Cowles demonstrate that it is an infrared receptor.

• *References*: "Edward Tyson, M.D., F.R.S., 1650–1708 and the Rise of Human and Comparative Anatomy in England," by M. F. Ashley Montague, Amer. Philos. Soc., Philadelphia, xxix, 488 pages, 1943; Cole, 1944 (Tyson on pp. 198–221); "Edward Tyson: Comparative Anatomy of the Timber Rattlesnake," by R. H. Ahrenfeldt, Herpetologica, 11: 49-69, 1955; "Edward Tyson," by W. C. Williams, Dict. Sci. Biogr., 13: 526–528, 1976; "Edward Tyson," by A. Guerrini, Oxford Dict. Natl. Biogr., 55: 818–819, 2004. • *Portrait* (about 1695): Royal College of Physicians, London. • *Signature* (1705): Wellcome Library for the History & Understanding of Medicine, London.

RAY, John (1627–1705).

John Ray (originally spelled Wray; see *Note*) was England's greatest naturalist before Darwin and produced the first modern classification of animals. He was born in Black Notley (Essex), a small village in rural East Anglia, on 29 November 1627. His interests in botany and religion were fostered by his mother, the village herbalist, who took him on her plant-collecting forays. He entered college at Cambridge in 1644 (B.A. 1648, M.A. 1651) and eventually held a series of academic and administrative positions in Trinity College where he taught classics and math and, like many of the faculty, gave sermons. At Cambridge Ray embraced the teachings of the Platonists—to study nature as a way to reveal the works of God—which deeply influenced his philosophy of nature. While recuperating from a serious illness in 1650, Ray started to explore the surrounding countryside, which rekindled his boyhood interest in botany. In 1658 he began the first of many field excursions into all parts of Britain, often on horseback, to observe nature. This exploration led to his first book, a catalogue of the plants of Cambridgeshire, the first flora of a British county, which was published anonymously in 1660.

In the meantime Ray was ordained deacon and priest in December 1660. However, he soon lost all his posts at Cambridge when, after the restoration of Charles II, all clergy were required to take an oath of allegiance. Ray refused on principle, resigned, and left Cambridge in 1662. He took temporary employment as a private tutor but soon resumed his field excursions. Many of these trips were in the company of one of his former pupils, Francis Willughby, the only son of an English nobleman with whom he struck a lifelong friendship that was to lay the foundation for the modern classifications of plants and animals.

In partnership with Willughby, Ray decided in 1662 to write a natural system for all plants and animals, a *systema natura* based on everything then known and on their future researches. Ray was responsible for the plants and Willughby the animals. During 1663–1666 they traveled to Europe and the Mediterranean where they visited the leading naturalists and museums, purchased books and illustrations, examined specimens, and made detailed notes that Ray later utilized to the very end of his life. As one biographer put it, this trip was to Ray what the voyage of H.M.S. "Beagle" was to Darwin.

On returning from Europe, Ray lived at Willughby's estate and they began writing their respective volumes. Ray's major innovation was to realize that, in classifying organisms, it was necessary to use the whole structure and not just one or two characters (as was then in vogue), and this included knowledge of internal anatomy, physiology, and of habits. Ray's detailed studies became widely respected and he was elected to the Royal Society in 1667. He had just published his second major work, a catalogue of English plants (1670), when disaster struck. Willughby died in July 1672 at the age of 37. Although Ray was temporarily permitted to continue his work at the estate, he was finally turned out in 1676. Fortunately Willughby had provided an annuity for Ray in his will, and this became his main source of income and allowed him the relative luxury to pursue the study of natural history for the rest of his life.

Ray soon returned to his ancestral home to live and to write his books. He died at his home in Black Nottley on 17 January 1705. He is buried in the local churchyard in full view of the rural landscape that first fired his interest in botany.

Ray and Willughby's scientific achievements are contained in a series of books published during Ray's lifetime and beyond. Ray's botanical work was capped by his magisterial "Historia Plantarum Generalis" (1686–1704), a 2860-page, folio-sized trio of volumes covering the 18,600 species then known for which he provided the first modern classification. It was Ray, for example, who devised the criteria by which species are defined, a concept later adopted by Carl Linnaeus. Ray was also the first to divide flowering plants fundamentally into monocotyledons and dicotyledons. His botanical contributions were prodigious and these alone would have guaranteed him an honored place in the pantheon of biology, but he had made a commitment to carry on Willughby's zoological work as well.

In many respects Ray's contributions to zoology were even greater than those he made to botany. Georges Cuvier, speaking of Ray's zoological work as a whole, once called it "the basis for all modern zoology" and of Ray, "the first true systematist of the animal kingdom." At the time of his death, Willughby had written only small portions of two books, both of which were completed by Ray: "Ornithologiæ" (1676) and "Icthyographia [*sic*]" (1686). Only Willughby's name appears on the title pages even though Ray wrote most of both books. These were extensively illustrated with plates. In preparing these books, as with the others, Ray collated all published names and descriptions, rigorously excluded all that he believed to be irrelevant or superstitious, to which he added his own direct observations made in England and Europe. In his preface to the volume on fishes he states that he has visited "all the chief fishing ports of England, and the markets of Belgium, Germany, Italy, and France; have bought all the species new to us." His classifications were practical and designed to help beginners to identify specimens but these classifications, like those that followed, also provided the foundation for Linnaeus's catalogues.

The next volume to be published was on mammals and reptiles (including amphibians) entitled "Synopsis Methodica Animalium Quadrupedum et Serpentini Generis" (1693; reprinted 1978). His introduction covers fundamental points such as spontaneous generation (which he rejected) and whether the eggs of all future generations were created "in the beginning" or anew at each generation (he decided for the latter). Most of the book consists of his classification, a dichotomous system in which most everything falls into one of only two categories. He first divided all animals into *Sanguinea* (vertebrates) and *Exanguia* (invertebrates), based on the presence or absence of red blood. The vertebrates are divided on the basis of their respiration into *Pulmone respirantia* (air breathers) and *Branchiis respirantia* (gills), and the air breathers further subdivided on the basis of the number of ventricles in the heart: *Duobus* (quadrupeds, cetaceans, and birds, with two) and *Unico* (amphibians and reptiles, with one). The last-named category is subdivided into three groups: the amphibians and turtles (a strange but then-common association, today preserved in the German word for turtle, Schildkröte, or shield-toad), lizards (including crocodiles), and snakes (including caecilians, amphisbaenians, and *Anguis*).

Ray's correct placement of cetaceans was partly based on his detailed dissection of a porpoise (1671), just as his study of swim bladders (1675) guided his earlier classification of fishes. Despite his botanical preferences, Ray was also a keen zoological anatomist. His skills were developed through practice during his student days at Cambridge when, for example, he dissected birds in his rooms at Trinity College. In his reptile book he mentions his dissections of frogs, watersnakes, and vipers. The list of amphibians and reptiles that he examined is a long one and includes a large number of exotics he saw during his European travels in the 1660s. He also observed living animals in Europe and in the fields near his home. Since the volume on mammals and reptiles was written at the very end, it benefited most from his long experience and mature judgment.

The "Synopsis" was the last zoological book by Ray to be issued during his lifetime, although he lived to see four other books published. For the last 15 years of his life Ray was in poor health and pain, mainly from uncurable ulcerated sores on his legs made worse by his sedentary work. Often he was able to work only two hours a day. Even the constant attention of his friend, the physician and naturalist Hans Sloane, provided no relief. His "Historia Insectorum," the last of his zoological works, was issued posthumously (1710) and incorporated Willughby's notes. Yet despite working mostly alone and without the resources of an academic institution, Ray was able to produce an enormous volume of work of the highest order. As his chief biographer wrote, "The output of those years is probably the largest in sheer mass of material that any one man had produced since Aristotle."

• *References*: "John Ray," by G. S. Boulger, Dict. Natl. Biogr., 16: 782–787, 1921; "John Ray Naturalist His Life and Works," 2nd ed. by C. E. Raven, Cambridge Univ. Press, Cambridge, xx, 506 pages, 1950; "John Ray," by C. Webster, Dict. Sci. Biogr., 11: 313–318, 1980; "John Ray (1627–1705) Essex Naturalist," by S. A. Baldwin, Baldwin's Books, Witham, Essex, (4), 80 pages, 1986; "John Ray," by S. Mandelbrote, Oxford Dict. Natl. Biogr., 46: 178–183, 2004. • *Portrait*: Adler collection. • *Signatures* (Wray form, 1667): From "Royal Society Charter Book," 1980; (Ray form): From Baldwin, 1986. • *Note*: Ray changed the spelling of his family name from Wray in 1669, on the grounds that it was not possible to Latinize the initial *W*. His name appears in several Latin forms in his books including Joannis Raii, Joannis Raji, and as in his reptile book of 1693, Joanne Raio.

CALDESI, Giovanni (1650–ca. 1732).

Giovanni Battista Caldesi, who wrote the first monograph devoted to turtles, was a physician in the court of the Medici family of Florence, which ruled Tuscany for two and a half centuries until 1737. What little is known of his life is intimately tied to that family, specifically to the last Grand Duke of Tuscany, Gian Gastone de' Medici. The Medicis were a family of great wealth and political influence whose cultural achievements in literature, science, and the arts were rivaled only by the Golden Age of Classical Athens. During this enlightened period, Florence counted among its citizens Dante, Galileo, Leonardo da Vinci, and Michelangelo.

Caldesi was born in Arezzo, 60 km southeast of Florence, on 7 January 1650. He was a student of Francesco Redi, also from Arezzo and personal physician to the grand dukes during 1666-1697. Redi was renown for his studies disproving spontaneous generation and for the first experiments on viper bites. Caldesi was a surgeon and, like Redi, interested in the natural sciences. He was also a confidant of Marcello Malpighi who consulted him on anatomical issues. In 1682, on Redi's recommendation, Caldesi became personal physician to Prince Gian Gastone who was then only 11 years old. Unfortunately, Florence was already in decline under the rule of the prince's grandfather and father. The prince, who became the last Medici grand duke of Tuscany in 1723, led a life of indolence, drunkenness, and depravity. Although he was homosexual, a marriage was arranged to the young widowed daughter of the duke of Saxe-Lauenberg in order to produce a male heir. Thus the prince, together with a party including Caldesi, went to Reichstadt (Bohemia) in 1696 and was duly married. It was intended that the prince live in Bohemia for a while but he ran away from his new wife, first to Paris, then Prague, and finally returned to Florence in 1698. She could never be induced to join him in Florence and, with no heir to the throne, the Grand Duchy of Tuscany passed to the Austrian Empire upon the death of Gian Gastone in 1737. Caldesi never lost the favor of the prince, however, who appointed him to the nobility of Arezzo and raised him in rank on several occasions.

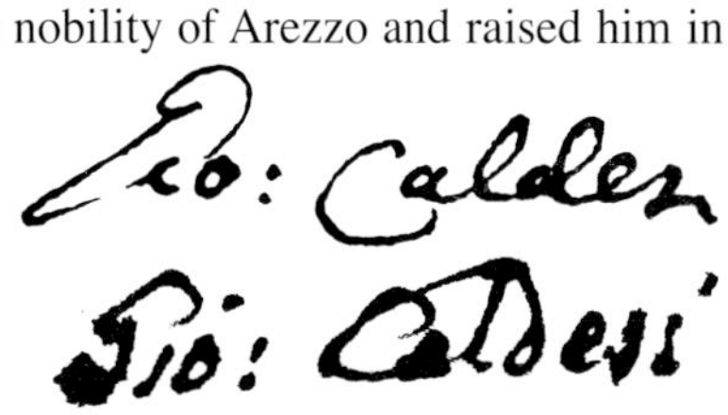

The last known record of Caldesi is a petition addressed by him to the city government of Arezzo, dated 1732, by which time he would have been about 82 years old.

Caldesi's sole published work was a book, "Osservazioni Anatomiche . . . Intorno alle Tartarughe Marittime, d'Acqua Dolce, e Terrestri" (1687), which reviews the internal anatomy of a sea turtle (*Caretta caretta*), a freshwater turtle (*Emys orbicularis*), and a land tortoise (*Testudo hermanni*). It is written in the form of a long letter to Redi and includes nine highly detailed, engraved plates depicting the internal structure of the shell, the appendicular skeleton, and the anatomy of the internal organs including the reproductive and circulatory systems. Caldesi distinguishes the two auricles of the heart but was unclear about the number of ventricular cavities; he noted that the heart continued to beat long after its removal from the body. He also conducted experiments on the source of bile and the pattern of blood flow. To view Caldesi as an experimentalist is consistent with his training by Redi.

• *References*: "Di Giovanni Caldesi Naturalista Aretino," by F. Mercanti, Atti Regia Accad. Petrarca Sci. Lett. Arti, Arezzo, 7: 93–104, 1885; "Giovanni Caldesi," by P. Cristofolini, Diz. Biograf. Ital., 16: 622–623, 1973. • *Signatures* (1688, above): Biblioteca Universitaria di Bologna (ms. 2085, XI); (1730, below): Biblioteca Universitaria di Pisa (ms. 86, c. 216).

GOTTWALD, Christoph (1636–1700).

Christoph Gottwald (see *Note*), a physician and collector of natural history curiosities, produced an early atlas of the anatomy of the sea turtle. He was born in July 1636 in Danzig (now Gdańsk, Poland). His father, a merchant, was originally from Silesia. Christoph attended public schools and had private tutorials including instruction in drawing for which he had considerable aptitude. He went to Strassburg (now Strasbourg, France) to study medicine and natural sciences and in 1660 matriculated in Leiden to finish his medical training (M.D. 1662). After travel in Italy and France, he returned to Danzig in 1663 to open his medical practice.

Gottwald was an assiduous collector and amassed an enormous museum of art and natural history objects, one of the largest private collections of its time. His intention was to produce an illustrated catalogue, especially of shells, for which he drew and etched the plates himself. Unfortunately, he died—on 12 January 1700 in Danzig—before he could issue any books. The collections passed to his son, Johann Christoph Gottwald, who added to them. After his death in 1713 his heirs had the collection catalogued for auction in 1716, but Peter the Great bought the entire lot on his second trip abroad in 1716–1717 during which he also acquired the cabinets of two Dutch collectors, Albertus Seba and Frederick Ruysch.

Gottwald's single herpetological contribution was his book, "Physikalisch-anatomische Bemerkungen über die Schildkröten" (1781), published long after the author's death. It consists of a short text of 32 pages, which in actuality is a set of extensive captions to the ten copper-etched plates, all of which are on large foldout sheets. These depict external views and the internal anatomy and osteology of a sea turtle (*Caretta caretta*) except for the final plate, which shows external views of a freshwater turtle (*Emys orbicularis*) and several exotic tortoises. He wrote the original text in Latin, but an anonymous reviewer (1783) took Gottwald's translator to task for poor choice of terminology and German orthography and also criticized the illustrations (probably taken from originals drawn by Gottwald no later than the 1690s). In fact, the illustrations are superior in content and draftsmanship to contemporaneous illustrations of the same species in Giovanni Caldesi's book (1687) to which the reviewer compared Gottwald's book. The same publisher also issued his books on the anatomy of the beaver and on his extensive shell collection, the latter edited by Johann S. Schröter, both published in 1782 and illustrated with plates drawn by Gottwald nearly a century earlier.

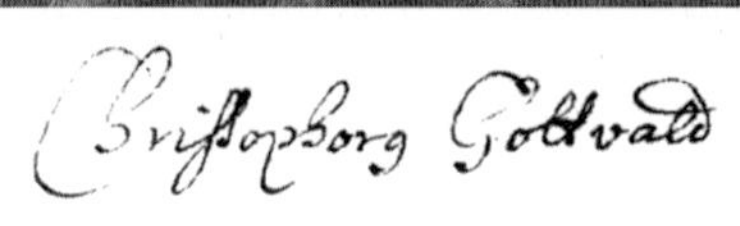

• *References*: "D. Christoph Gottwaldts physikalisch-anatomische Bemerkungen über die Schildkröten [book review]," anonymous, Allg. Deutsch. Biblioth., Anh. 1771–1791, 1: 504–505, 1783; "Christoph Gottwald," by Schwarz, p. 226 *in* C. Krollmann, Altpreussische Biographie, vol. I. N. G. Elwert Verlag, Marburg, 1974. • *Portrait*: From J. S. Schröter, Musei Gottwaldiani Testaceorum, Nuremberg, 1782. • *Signature* (1698, Latin form): Deutsche Akademie der Naturforscher, Halle (Leopoldina-Archiv, M1 Matrikel, 231). • *Note*: His Latin name was Christophorus Gottvald. Several other free variations of his surname were used at that time including Gottwaldt and Gottvaldt.

VALLISNIERI, Antonio (1661–1730).

Antonio Vallisnieri (see *Note 1*), an Italian physician, naturalist, and author of the first monograph on chameleons, was born in Tresilico (now Trassilico, in Lucca), about 35 km north of Pisa, on 3 May 1661. His father had been governor of Camporgiano (Lucca) before moving the family south to Tresilico and was responsible for his son's early education. Antonio attended Jesuit schools in neighboring Modena and then a Roman Catholic college in Reggio (Reggio nell'Emilia) where he graduated with a bachelor's degree in Aristotelian philosophy (1682). Advised to follow his interests in biology at the university in Bologna, Vallisnieri spent 1682–1685 with the comparative anatomist and experimentalist, Marcello Malpighi, who was on the medical faculty, but he took his doctorate in medicine and philosophy at Reggio (1684). His medical training was completed in Venice and Parma (1687–1688) before he settled in Reggio in 1689 and established a medical practice there. During his personal time Vallisnieri began collecting, studying, and experimenting with animals. His first studies were on the reproduction of insects, in which he confirmed Francesco Redi's experiments (published 1668) that disproved spontaneous generation; beginning in 1700 he authored several titles on the topic. In that same year, he was invited to be professor of medicine at the university in Padua where, ten years later, he accepted the chair of theoretical medicine. He held this position until his death, in Padua, on 18 January 1730.

In 1720 Vallisnieri had declined an offer to become first physician to the pope in order to continue his teaching and research at the university. At that time, most medical instruction was according to the Scholastic philosophy: the authority of the ancient scholars combined with a method of logic based on Aristotelian principles. Vallisnieri, however, had a modern view and came to reject Scholastic knowledge in favor of direct ob-

servation and experimentation, a method he learned from Malpighi that is traceable to Redi and ultimately to Galileo. He made contributions to diverse topics including teratology, parasitology, embryology, and geology. He developed the so-called "chain of beings," the connection of all living creatures including man and a forerunner of evolutionary concepts, although he interpreted this as an expression of God's design.

Vallisnieri's sole contribution to herpetology was his book, "Istoria del Camaleonte Affricano e di Varj Animali d'Italia" (1715; see *Note 2*). In this book he describes the anatomy and biology of *Chamaeleo chamaeleon* and then compares it to various animals, including frogs, salamanders, and other lizards. In salamanders, for example, he described the white skin secretions, the elongate lungs, and fish and frog eggs in their stomachs, but he also claimed that they had paired penes.

• *References*: "Notizie della Vita e Degli Studi del Kavalier Antonio Vallisnieri," by G. di Porcia, p. xli–lxxx *in* A. Vallisnieri, Opere Fisico-Mediche, vol. 1, S. Coleti, Venice, 1733; "Antonio Vallisnieri (or Vallisneri)," by G. Montalenti, Dict. Sci. Biogr., 13: 562–565, 1976. • *Portrait and signature* (1726): Adler collection. • *Note 1*: Also spelled Vallisneri. His name is sometimes followed by "seniore" to distinguish him from his son who succeeded him as professor at Padua. • *Note 2*: The dedication is dated 20 August 1714. The book was reissued in Vallisnieri's collected works in 1715 and 1733. Editions of 1713 and 1714 have been cited but are unconfirmed.

SCHEUCHZER, Johann Jacob (1672–1733).

Johann Jacob Scheuchzer was a Swiss physician, historian, and one of the founders of paleontology who discovered the first fossil salamander. He was born on 2 August 1672 in Zürich. An earnest pupil by the age of three, he later became a brilliant student at the Carolinum, a theological academy in Zürich. He was already interested in natural history and in 1691 entered the university in Altdorf, about 25 km east of Nuremberg where he studied science and medicine for two years before finishing his medical work at Utrecht (M.D. 1694). By 1690 Scheuchzer had begun to collect fossils, which he regarded as inorganic products, but soon realized that they were the remnants of organisms that he believed died as the result of the biblical Great Flood. On returning to Zürich in 1694, he began his explorations of the Swiss Alps. He became the assistant municipal physician and medical supervisor at an orphanage in Zürich and in 1710 was appointed a professor of mathematics at the Carolinum. He also became the director of the city's natural history museum, the Kunsthammer. Despite this busy schedule he wrote many books, among them the 29-volume "Histoire Suisse." In 1733 he was promoted to chief municipal physician, but died a few months later, on 23 June.

Scheuchzer amassed an enormous collection of fossil plants and animals and wrote many works about them. He was also renown for his medical studies on the effects of high elevations, for descriptions of glaciers and their movements, and for making detailed topographic maps of Switzerland. His reputation as a herpetologist rests on two of his publications. In 1726 he published the description of a 1-m-long fossilized skeleton from Germany (see *Note*) that he believed to be a human who had died in the Flood. He called it "Homo Diluvii Testis" (Human Witness to the Flood), the title of his book. Reflecting his religious zeal, the book was not dated 1726 on the title page as would have been customary, but MMMM XXXII or 4032 years after the Flood! In 1811, having seen Scheuchzer's illustration, Georges Cuvier, the French comparative anatomist, declared it to be a salamander despite its large size. (The first Asiatic giant salamander was not

described until 1837.) He later proved his point by uncovering more of the skeleton from the matrix and J. J. von Tschudi placed the animal in his new genus *Andrias* in 1837. Most paleontologists today believe that this Miocene salamander is so similar in its skeletal anatomy to the living species in China and Japan that they put them all in the same species, *A. scheuchzeri*. This is the first record of a fossil salamander.

During 1731–1735 Scheuchzer authored a massive four-volume folio entitled "Kupfer-Bibel . . . Physica Sacra," comprising 750 full-page copperplate engravings showing the natural history behind all biblical events with accompanying text. Amphibians and reptiles figure prominently on many of the plates, especially snakes. In particular, in volume four (1735) there are 66 individual snakes illustrated on 24 plates from the Musei Linckiani, collected by Heinrich Linck of Leipzig and his son and grandson, both named Johann Heinrich Linck. The drawings are as good as those in any contemporary zoological publication and, in fact, better than most. What is of scientific importance is that Carl Linnaeus, in the editions of his "Systema Naturæ," cited several of these illustrations—as he did for those of some other earlier authors such as Ulisse Aldrovandi, Mark Catesby, and Albertus Seba—when he described his new species.

• *References*: "Johann Jakob Scheuchzer (1672–1733)," by R. Steiger, Beibl. Vierteljahr. Naturf. Gesell. Zürich, 21: 1–75, 1933; "Johann Jakob Scheuchzer," by P. E. Pilet, Dict. Sci. Biogr., 12: 159, 1975; Müsch, 2000; "Johann Jakob Scheuchzer and the *Physica Sacra*," by R. Wahlgren, Bull. Internatl. Soc. Hist. Biblio. Herpetol. (now Biblioth. Herpetol.), 4(2): 5–6, 2003. • *Portrait*: Adler collection. • *Signature* (1715, Latin form): From Candollea, 33: 167, 1978. • *Note*: The fossil beds are said to be at Oeningen, a former Augustinian monastery. In fact, the locality is about 3 km away, just above the village of Wangen (Baden-Württemberg) and overlooking Lake Constance on the German-Swiss border. The locality is sometimes incorrectly given as Oensingen, a town just south of Zürich.

CATESBY, Mark (1683–1749).

Mark Catesby was the first acclaimed illustrator of the fauna and flora of British North America, Spanish Florida, and the Bahamas (see *Note 1*). He was born at the home of his mother in the village of Castle Hedingham (Essex) on 3 April 1683 (see *Note 2*). His father was a lawyer, landowner, and longtime mayor of nearby Sudbury (Suffolk). His introduction to botany came through an acquaintance, arranged by his grandfather who was a horticulturist, with the leading English naturalist of the day, John Ray, who lived but 15 km away. His life as a naturalist, however, began in 1712 at the age of 29, when he joined his sister's family in Williamsburg, capital of colonial Virginia. He explored Virginia as far west as the Blue Ridge Mountains and also Bermuda and Jamaica. He collected specimens, took notes, and made sketches. On returning to England in 1719, a group of naturalists including Sir Hans Sloane, whose collections later formed the nucleus of the British Museum, decided to sponsor Catesby for a return trip in order to secure animal and plant specimens for their collections and gardens. Sloane particularly wanted more snakes, which Catesby earlier had shipped to him in jars of rum; sometimes the rum was consumed by the British sailors while en route to England and the specimens were ruined.

Catesby arrived in South Carolina in 1722 and over the next three years he explored the piedmont and coastal regions of the Carolinas, Georgia, and Florida. Late in 1725 he also visited the Bahamas. He made colored drawings of plants and animals and recorded observations which, on his return to England in 1726, he transformed into one of the classics of natural history, "The Natural History of Carolina, Florida, and the Bahama Islands," issued in two folio volumes together consisting of ten 20-plate parts, plus an appendix with 20 additional plates, over the period 1730–1747 (reprinted 1974). The most lasting element of this book are the plates; all 220 of them were etched personally by Catesby since he could not afford an engraver and had to learn the craft largely by himself. There are 36 herpetological plates (numbers 38 to 72 in volume 2 plus plate 10 in the appendix, showing a great blue heron with a spotted salamander in its beak). 156 sets were issued, so a total of some 34,320 plates had to be colored by hand. In the beginning Catesby did much of the coloring, but toward the end merely did the supervising and touch up. Production of this book occupied Catesby for the rest of his life, which ended in London on 23 December 1749 at the age of 67. In 1768 King George III purchased Catesby's working drawings from which the copper printing plates were etched. Today these drawings are housed at Windsor Castle.

The folio size of the book allowed Catesby to illustrate animals and plants in nearly natural size, a method that was to be used by John James Audubon in his enormous folio atlas of birds nearly a century later. Catesby combined plant and animal subjects on the same plate. These juxtapositions were not random but based on actual information. For example, one snake was entwined in a potato vine because farmers had unearthed this species when harvesting that crop in Virginia. His illustrations appear rather stiff and flat, but the artistic layout and overall effect are stunning. Generally the drawings are detailed enough that the species can be identified. In his text Catesby often debunked prevailing myths, many of them about snakes. He was also a keen observer and, for example, accurately distinguished the two species of green snakes native to the Carolinas, both in the text and in the plates, before either species had been described and named.

The scientific nomenclature in Catesby's work is not binomial and has no nomenclatural standing because it is pre-Linnaean. However, because Carl Linnaeus frequently cited Catesby's plates and based a number of his species on them, his work is of considerable taxonomic importance. After Catesby's death the book was reissued in London in 1754 and 1771, under the supervision of the ornithologist George Edwards, who had worked with Catesby during the last ten years of his life and learned from him how to etch plates. There were several European editions, among them a German edition in 1750 that is limited to the fishes, amphibians, and reptiles.

Catesby was not formally trained as either an artist or a naturalist. He did, nevertheless, receive much recognition for his accomplishments, most notably by election to the Royal

Society in 1733. This was a time when serious amateurs like Ray and Catesby could still be members of this august society. Catesby regularly attended its meetings where he was an active participant in discussions. He is immortalized in herpetological nomenclature by the specific name of the American bullfrog (*Rana catesbeiana*, named by George Shaw in 1802), a species that is native to eastern North America and now introduced in many other parts of the world.

• *References*: "Mark Catesby: The Colonial Audubon," by G. F. Frick and R. P. Stearns, Univ. Illinois Press, Urbana, x, 137 pages, 1961; "Mark Catesby," by G. F. Frick, Dict. Sci. Biogr., 3: 129–130, 1971; "Empire's Nature: Mark Catesby's New World Vision," ed. by A. R. M. Meyers and M. B. Pritchard, Univ. North Carolina Press, Chapel Hill, xviii, 272 pages, 1998; "Mark Catesby," by K. B. Sterling, Amer. Natl. Biogr., 4: 568–570, 1999; "Mark Catesby," by F. N. Hepper, Oxford Dict. Natl. Biogr., 10: 531–532, 2004. • *Portrait*: No portrait is known. • *Signature*: From "Royal Society Charter Book," 1980. • *Note 1*: The earliest surviving drawings of reptiles from this region, or for any part of the Western Hemisphere for that matter, are five drafted by John White in North Carolina (the ill-fated "Lost Colony" of Roanoke Island) and the West Indies during the 16th Century (see "John White and the Earliest [1585–1587] Illustrations of North American Reptiles," by H. M. Smith, M. J. Preston, R. B. Smith, and E. F. Irey, Brimleyana, 16: 119–131, 1990). These drawings were widely copied; they are now in the British Museum. Catesby copied several of them in his "Natural History," including one of a West Indian iguana. • *Note 2*: Sterling (1999) gives Catesby's birth date as 31 December 1682, without explanation.

OWEN, Charles (ca. 1660–1746).

Charles Owen, a non-conformist Presbyterian minister and writer, authored a well-known 18th Century text on amphibians and reptiles. He was born sometime after 1654, probably in Carmarthenshire, England (see *Note*). He became minister of the Cairo Street Chapel located in Warrington just east of Liverpool in 1696 where he maintained a large congregation for nearly 50 years. His two brothers and son all became non-conformist ministers. Owen educated students for the ministry but was more of a political dissenter than a theologian. He published numerous sermons and political treatises, one of which, the notorious "Plain Dealing" (1715), addressed the Schism Act of 1714, which forbade all public education to dissenters from the Church of England. His writings were regarded as reflecting scandalously on the church and its clergy and he was indicted by a grand jury. There was no conviction but he was heavily fined. He received a Doctor of Divinity diploma from Edinburgh University in 1728, probably as a tribute to some of his recent writings. He died in Warrington on 17 February 1746.

Owen wrote (and also published) a single book on natural history, "An Essay Towards a Natural History of Serpents" (1742), which is a notable exception to the theme of all of his other writings. His reason for writing it may be explained in the book's preface in which he states: "I have attempted to give a short Display of the Divine Perfections, which, as they appear eminent in the System of the Creation in general, so in the Serpent they may be seen in particular; and if it produces in the Reader a more exquisite Perception of God in all his Works, I have my End . . . ." In studying nature as a revelation of God's work Owen was not unique for his day.

Serpents, in Owen's sense and that of the day, covered a broad category of animals in addition to snakes: lizards, salamanders, some turtles, and crocodiles but even a number of invertebrates "in whose Veins I find something of the Serpent's Blood" (spiders, scorpions, honeybees, and wasps, among others). In fact, Owen does a credible job of summarizing the literature—especially the works of Mark Catesby, Moyse Charas, Conrad Gessner, John Jonston, Richard Mead, Pliny the Elder, John Ray, and Francesco Redi—and, where it was correct, his summary is useful. Curiously, Owen makes no reference to the only book on reptiles then published in English, that by Edward Topsell (1608, 1658). On the other hand he seems to have accepted at face value everything that he read. Included are such creatures as sea serpents, winged dragons, and basilisks (eight-legged beasts with crowns), which are even figured on the plates along with perfectly good representations of rattlesnakes, salamanders, and chameleons. The final chapter is devoted to the primeval serpent and its relationship to the devil, animal gods, and the reasons for worshiping serpents.

• *References*: "Charles Owen, D.D.," by A. Gordon, Dict. Natl. Biogr., 14: 1294–1295, 1921; "Some Quaint Conceptions of North African Natural History," by A. Loveridge, Sci. Month., 57: 398–400, 1943; "Charles Owen," by D. L. Wykes, Oxford Dict. Natl. Biogr., 42: 186–187, 2004. • *Portrait*: No portrait is known. • *Signature* (1709): Harris Manchester Coll., Oxford Univ., courtesy Susan Killoran (William Shepherd manuscripts, vol. 2). • *Note*: Owen's older brother, James, was born in November 1654 at the farmhouse in Bryn, Abernant Parish (Carmenthenshire), and took his early schooling in the county until 1670. He is said to have "educated his younger brother Charles" ("James Owen," by J. H. Westaway, Oxford Dict. Natl. Biogr., 42: 215–216, 2004). It seems reasonable to infer that Charles was also born there a few years after 1654.

FONTANA, Felice (1730–1805).

Gasparo Ferdinando Felice Fontana (see *Note 1*), Italian microscopist and experimentalist and founder of modern toxinology, conducted the first definitive work on the mode of action of snake venom. He was born 15 April 1730 (see *Note 2*) in Pomarolo, near Rovereto in the foothills of the Alps. He was educated in several schools including the University of Padua (1750–1752) and then spent several years at universities in Bologna and Rome. In 1760 he began his work at the University of Pisa. Grand Duke Francis Stephen, who was greatly interested in natural science, ruled Tuscany from Austria. In 1766, by a change in the laws of royal succession, his son, Peter Leopold, became Grand Duke of Tuscany and took up residence in Florence. He awarded Fontana the chair in natural science at Pisa and summoned him to his court in Florence to create a great museum. This was opened in 1755 and housed the finest anatomical collections then known, including the famous colored wax models of human anatomy constructed under Fontana's direction. These can be seen in the museum of "La Specola" today.

At the end of 1775 Fontana began a three-year tour of France and England, mainly to make new acquisitions for the

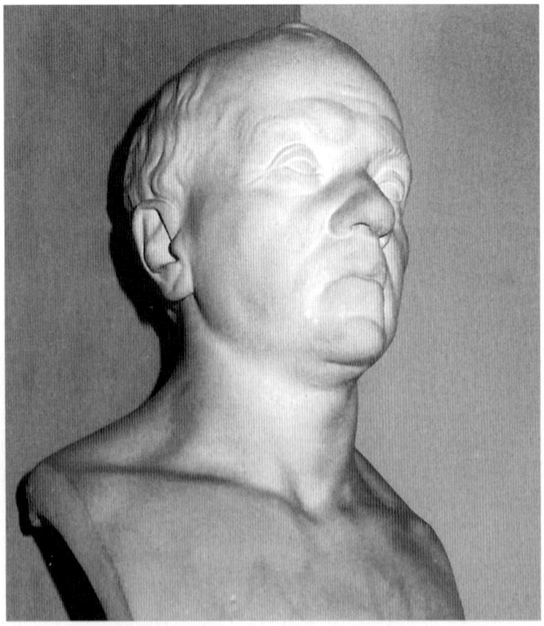

museum. Napoleon Bonaparte was in Florence in 1796 during the Franco-Austrian War and, upon seeing the wax models, commissioned Fontana to make a set for France. These were turbulent times politically. After the French were forced to leave the city in 1799, Fontana, who sympathized with French Republican causes, was briefly imprisoned by Tuscan partisans. He died in Florence on 10 March 1805.

Fontana's general interest was in the irritability and sensitivity of living tissues, following the lead of Albrecht von Haller of Göttingen. He conducted microscopic investigations and made numerous discoveries in nerve anatomy and cell structure. He was one of the first to recognize cell nuclei, kidney tubules, and nerve sheaths. He discovered the nucleolus and the canal or space in the ciliary body of the eye that bears his name. He studied the blight, which devastated grain crops in Tuscany, and determined its fungal origin. His interest in the venom of vipers was tied to his interest in irritability, since the action of the venom to coagulate the blood altered the irritability of all parts of the organism including the nerves.

He began his studies on venom in about 1764, motivated partly by research by an English physician, Richard Mead, who conducted the first *in vitro* experiments on viper venom and who claimed (1702) a "mechanical" basis for the action of venom in which the poison consisted of needle-like particles that pierced the red blood corpuscles and caused them to release fluid. This was at variance with the conclusions of Francesco Redi who a century earlier had performed the first scientific experiments on viper bites. Fontana decided to resolve this controversy and extend the work using microscopic methods (in translation) "not withstanding the risk that attends the handling [of] such dangerous animals."

Fontana's experiments and observations yielded numerous results. He was the first to describe the histology of the venom gland. He concluded that the venom was a hemotoxin and acted directly on the blood. He rejected erroneous ideas about the mode of action: mechanical effects, chemical corrosion, and nerve-mediated action. He also established a dose-effect relationship—by allowing test animals to be bitten once, twice, or several times and, later, by using modified glass capillaries to deliver a quantity of venom into a varying number of incisions—and thereby provided a possible explanation for the theory proposed a century earlier by Moyse Charas (1669) who claimed that only an "enraged" viper could cause death. Fontana suggested that the more enraged a viper, the larger the dose of venom it delivered. He determined minimum doses of venom necessary to kill prey and he concluded that it would require two bites in most instances to cause the death of a full-grown person. Thus he concluded that recovery from a bite was not due to the antidotes given, but to the fact that the amount of venom received was not lethal.

To compensate for the possibility of error in the reactions of living animals, Fontana believed in using large sample sizes. He estimated that in all of his studies on vipers, he conducted 6,000 experiments using 3,000 vipers and 4,000 animal subjects. In order to further reduce variation resulting from the length of time the fangs were inserted into the prey, he sometimes decapitated snakes and manually inserted the fangs. This way he could also choose the exact position of the bite and control the volume of venom injected. His results were published in a book in 1767, entitled "Richerche Fisiche sopra il Veleno della Vipera" (second edition 1781 in French, based on much new research, with German and English editions in 1787 and 1795, respectively).

• *References*: "The Experimental Investigation of Viper Venom by Felice Fontana (1730–1805)," by M. P. Earles, Ann. Sci., 16: 255–268, 1960; "Felice Fontana," by L. Belloni, Dict. Sci. Biogr., 5: 55–57, 1972; "Felice Fontana, Life and Works," by P. K. Knoefel, Soc. Stud. Trentini Sci. Stor., Trento, xix, 422 pages, 1984; "Abbé Felice Fontana (1730–1805): Founder of Modern Toxinology," by B. J. Hawgood, Toxicon, 33: 591–601, 1995; "Gasparo Ferdinando Felice Fontana," by R. G. Mazzolini, Diz. Biograf. Ital., 48: 663–669, 1997. • *Portrait*: Museo Zoologica "La Specola," Univ. Florence, photograph by Marco Borri, courtesy Benedetto Lanza. • *Signature*: From Riv. Stor. Sci. Med. Nat., Florence, 20: 128, 1977. • *Note 1*: Fontana was an abbate (abbé), a lay abbot who discharges no spiritual functions. • *Note 2*: Some authors give the birth year as 1720, but recent authorities agree on 1730.

PALLAS, Peter S. (1741–1811).

The German explorer and naturalist, Peter (in Russian, Pyotr) Simon Pallas, who spent over 40 years studying the fauna, flora, and geology of the Russian Empire, described many Eurasian species of amphibians and reptiles. Born in Berlin on 3 October 1741, he was interested in natural history as a boy and by the age of 15 he was making collections of insects. He was educated at home and then, in 1754, at the Collegium Medico-Chirurgicum in Berlin where his father was a professor of surgery. Later he attended the universities in Halle and Göttingen and finished his dissertation on the worm-caused illnesses of humans, in Leiden (M.D. 1759) at the age of 19. Despite his father's repeated calls to settle in Berlin and practice medicine, Pallas spent 1761–1766 studying museum collections in England and Holland. He published his first major works at this time ("Miscellanea Zoologica," 1766; "Spicilegia Zoologica," [1767–] 1774–1780). By the mid-1760s he was regarded as the most able young naturalist in Europe, but was still unable to find a suitable academic position in England, Germany, or Holland.

In 1767 he was invited by Empress Catherine II (Catherine the Great) to come to St. Petersburg to work at the Imperial Academy of Sciences. Soon he was appointed to lead the so-called "Physical Expeditions," a group of naturalists and astronomers who would travel—sometimes together and sometimes separately in order to cover more territory—throughout the rapidly expanding empire to study its natural history. This group included the naturalists Johann G. Georgi, Samuel G. Gmelin, Johann von Güldenstädt, Ivan Lepechin, and a former pupil of Carl Linnaeus, J. P. Falck. The expeditions retraced some of the routes of the earlier explorers of Siberia, Johann G. Gmelin and Georg Steller. Pallas departed St. Petersburg in June 1768 bound for Moscow, then went southeast to Orenburg and, after a side trip to the Caspian Sea, went eastward through the Ural Mountains to Tobolsk, the Altai Mountains, and finally to Krasnoyarsk by 1771. Pallas, who generally traveled with a military escort and servants, then moved further east to Lake Baikal and beyond to the mountains of Transbaikalia as far east as the Onon and Ingoda rivers near the easternmost border of Mongolia, a journey of some 9000 km from St. Petersburg including several side trips. He returned via Krasnoyarsk to the Urals, thence to the lower reaches of the Volga River and finally north to Moscow and St. Petersburg by July 1774.

It was easiest to travel during the harsh winters when the rivers and swamplands were frozen. After the spring thaw, travel was generally impossible and in summer it was slow unless they were floating downstream. Large collections were made which, together with Pallas's notes, were dispatched to St. Petersburg from time to time. It was an arduous journey. The temperatures in Transbaikalia sometimes froze the mercury in the thermometers and several members of the original party died including Samuel Gmelin. The report on Pallas's expedition started to appear before his return, based on his notes and collections sent ahead. The book eventually comprised three volumes, entitled "Reisen durch Verschiedene Provinzen des Russischen Reichs" (1771–1776). The descriptions of amphibians and reptiles occur in all three volumes and include many new and well-known species, among them *Rana ridibunda, Eremias arguta, Phrynocephalus helioscopus, Agkistrodon* (now *Gloydius*) *halys, Elaphe dione,* and *Eryx miliaris*. Pallas's most remarkable lizard, *Ophisaurus apodus,* was also named from specimens collected on this journey in what was his only strictly herpetological paper, "Lacerta Apoda Descripta" (1775).

Later, at the age of 52 and tired of city life, Pallas mounted a new expedition, to the southern provinces of the empire, in particular to the Crimea, which had been annexed by Russia in 1783. Pallas departed in February 1793 and went south to Zarizyn (now Volgograd) and to Astrakhan on the northern coast of the Caspian Sea, then west via the River Kuma to Georgiyesk in the northern Caucasus and later to the Crimea, returning to St. Petersburg by September 1794. Catherine rewarded him with estates near Simferopol in the Crimea, which he felt obliged to accept. He spent nearly 15 miserable years there, in swampy, unhealthy lands remote from the libraries and many of the collections he needed. After Catherine died in 1796, Pallas, now in failing health, was granted permission by Emperor Alexander I, Catherine's grandson and his former pupil from years before, to leave Russia. He returned to Berlin in 1810 to continue writing his masterwork, "Zoographia Rosso-Asiatica" (1811–1831 [–1842], with herpetology in volume 3 [1827]; see *Note*), but he did not live to see the book published. Pallas died in Berlin on 20 September 1811. Many additional herpetological taxa were described and named, among them *Testudo graeca ibera, Agama* (now *Trapelus*) *sanguinolenta, Alsophylax pipiens,* and *Lacerta* (now *Podarcis*) *taurica*. Pallas's few surviving herpetological collections are at the zoological museum in Berlin.

Pallas was one of the great naturalists of his age, a peer of Charles Linnaeus, Count de Buffon, and Georges Cuvier. He pioneered the use of internal anatomical characteristics to classify animals. His views on the distribution of plants and animals and their relationship to the environment led to the development of biogeography as a scientific discipline. In 1766 he was the first to show the sequential development of animals in the form of a family tree. His contributions to geography and geology were no less profound. He was the first to propose a general hypothesis for the formation of mountains, and his ideas helped lead to the development of theoretical geology. Moreover, his recognition that the earth had a prolonged geological history was influential in the development of ideas about the evolution of nature, as noted by Charles Darwin.

• *References*: "Peter Simon Pallas," by F. Ratzel, Allg. Deutsch. Biogr., 25: 81–98, 1887; "Pyotr Simon Pallas," by V. A. Esakov, Dict. Sci. Biogr., 10: 283-285, 1974; ["The Voyages of Peter-Simon Pallas,"] by V. Muravyev, Mysl Publ., Moscow, 94 pages, 1977; "Peter Simon Pallas (1741–1811). Materialien einer Biographie," by F. Wendland, W. de Gruyter, Berlin and New York, xviii, 1176 pages, 1992; ["Petrus Simon Pallas, Botanicus,"] by A. Sytin, KMK Sci. Press, Moscow, 338 pages, 1997; Rieck et al., 2001 (Pallas by W. Bischoff on pp. 544–545); ["The Imperial Academy's 'Physical' Expeditions (1768–1775) and the Formation of Herpetology in Russia,"] by L. J. Borkin, p. 21–45 *in* E. I. Kolchinsky (ed.), Russian–German Links in Biology and Medicine, Borei Art Press, St. Petersburg, 2001. • *Portrait*: From I. Dörfler, Botaniker-Porträts, F. Jasper, Vienna, 1907. • *Signature* (1776): Museum für Naturkunde, Berlin. • *Note*: The exact dates of publication of the four text volumes and the atlas of the "Zoographia" are not known definitively. Some authors have claimed that Pallas's book was not available for purposes of zoological nomenclature before 1827 when a few copies were distributed, since prior references to it were made only from two sets of proof copies (E. Stresemann, Ibis, 93: 316–318, 1951; for an opposing view see A. N. Svetovidov, Zool. Jour., Moscow, 1976: 596–599, 1976). Apparently volume 3, which contains the section with amphibians and reptiles, was printed in 1811, possibly 1814, but not released to the general public until 1831 owing to delays caused by engraving the plates, hence the citation of all of these dates in the literature. The natural history addenda by Eduard Eversmann were issued in three parts (1835–1842; reprinted 1876); these contain no herpetology.

WALBAUM, Johann Julius (1724–1799).

Johann Julius Walbaum, a German physician, was an amateur zoologist who specialized on birds, fishes, turtles, and frogs. He was born on 30 June 1724 in Wolfenbüttel (Lower Saxony), about 10 km south of Braunschweig. His father was a brewer and died when Johann was 13. Johann had to interrupt his schooling to help his mother run the family brewery. He taught himself botany in his free time and returned to school in 1741. Four years later he went to the university in Helmstedt to study medicine and later attended Göttingen (M.D. 1748) where one of his teachers was the renown medical scholar, Albrecht von Haller. The next year he began a medical practice in Lübeck, just northeast of Hamburg and near the sea. He became one of the city's most distinguished physicians, specializing in obstetrics. He also built a private museum, which after his death in Lübeck on 21 August 1799 (see *Note*), was donated to a local organization from which the natural history museum of Lübeck developed. The museum was destroyed in 1942 during World War II.

All of Walbaum's research was done privately and he authored nearly 30 titles. Eight are herpetological (1782–1795). Except for one book, these publications are short papers published in the *Schriften* of the Berlin Society of Natural History Friends, of which he was a member. These discuss local and exotic species of frogs and turtles. He named *Rana gigas* (1784) but it is a synonym of *Bufo marinus*. He illustrated and described in detail a new Asiatic turtle (1785) but did not name it. J. F. Gmelin (1789) gave it a name, based entirely on Walbaum's description. The species is native to China and Indochina and is known today as *Geoemyda spengleri*.

Walbaum's major contribution to herpetology is his book, "Chelonographia oder Beschreibung einiger Schildkröten nach natürlichen Urbildern verfertiget" (undated but published in 1782). It covers six species of turtles from different parts of the world, described in great detail in both the German and Latin sections. All but one were well-known species. Although Walbaum gave them Latin names, he was not consistently binomial and his book therefore has been rejected for nomenclatural purposes. The one new species described in his book is now known as *Homopus signatus*, from southern Africa. It was properly named by J. D. Schoepff (1801) who copied his illustrations and his text from Walbaum, word for word, and to whom he gives full recognition. Although Walbaum is not credited with the names of any amphibians and reptiles, he did name some fishes, most notably the rainbow trout (*Oncorhynchus mykiss*), one of the most important commercial and sport fishes in the world, which has been introduced globally.

• *References*: "Der Ichthyologe und Schildkrötenforscher Johann Julius Walbaum," by H.-P. Müller, Die Heimat, Neumünster, 80: 195–197, 1973; "Johann Julius Walbaum," p. 93 *in* L. Gebhardt, Die Ornithologen Mitteleuropas, vol. 3 (Jour. Ornithol., 115), 1974; "Johann Julius Walbaum," anonymous, Dict. German Biogr., 10: 311, 2006. • *Portrait*: From Müller (1973), courtesy Hans-Peter Müller. • *Signature* (1783): Museum für Naturkunde, Berlin. • *Note*: Müller (1973) gave Walbaum's death date as 26 August 1799, without explanation.

DAUBENTON, Louis (1716–1800).

The French naturalist and anatomist, Louis Jean-Marie Daubenton (sometimes D'Aubenton), a colleague of Georges Buffon and first director of the Muséum d'Histoire Naturelle, authored the first scientific review of the amphibians and reptiles of the world. He was born on 29 May 1716 in Montbard

(Côte-d'Or), Buffon's birthplace. His father intended him for the priesthood, but when he was sent to study theology at the Sorbonne in Paris he also attended the anatomy lectures at the Jardin du Roi, the king's garden, apparently unknown to his father. After his father died in 1736, Daubenton completed his medical training (M.D. Rheims 1741) and returned to Montbard to open his practice. The next year, however, he was summoned to Paris by Buffon to be his assistant in managing the garden. Three years later Daubenton was put in charge of the natural history collections and became Buffon's chief collaborator in the production of the monumental encyclopedia of nature, the "Histoire Naturelle." Daubenton, in fact, was co-author of the first 15 volumes (1749–1767).

Daubenton was primarily an anatomist. Georges Cuvier once remarked that he was (in translation) ". . . the first who united anatomy and natural history in a continuous manner." He displayed a mastery of comparative anatomy, recognizing the importance of using the same term to refer to corresponding parts of animals, and was the first to apply the comparative method to fossil forms. He was also interested in museum methods of preservation and display. During the French Revolution he played a key role in transforming the natural history collections of the king's garden into the Muséum d'Histoire Naturelle, which was created in June 1793. The next month his colleagues elected him the first director, a post he held until 1797 except for one year. He was appointed senator in 1799, but he died on 1 January 1800 in Paris from a cerebral stroke suffered during his first session in the senate chamber.

In addition to his contributions to Buffon's "Histoire Naturelle," Daubenton authored numerous works on the anatomy of mammals, museum methods, minerals, and botany. His sole contribution to herpetology was the volume entitled "Les Animaux Quadrupèdes Ovipares, et les Serpens" (1784), part of a series entitled "Tableau Encyclopédique et Méthodique des Trois Règnes de la Nature." This was the first comprehensive and scientific review of the world's amphibians and reptiles, but it was unillustrated. In contrast, his colleague Count de Lacepède, who also worked with Buffon at the king's garden, published his well-known review of amphibians and reptiles in 1788–1789. It was extensively illustrated. Daubenton's publisher soon followed suit, having commissioned Pierre-Joseph Bonnaterre, a French abbot and naturalist, to superintend the production of two sets of plates and write extensive captions for them. Thus Bonnaterre's "Erpétologie" with 26 plates (1789) and "Ophiologie" with 43 (1790) were issued as a complement to Daubenton's text and are often bound with it.

• *References*: "Louis-Jean-Marie Daubenton," by Y. Chatelain, Dict. Biogr. Franç., 10: 241–244, 1965; "Louis-Jean-Marie Daubenton," by C. Limoges, Dict. Sci. Biogr., 15: 111–114, 1978; Jaussaud and Brygoo, 2004 (Daubenton on pp. 165–167). • *Portrait*: Adler collection. • *Signature*: Muséum National d'Histoire Naturelle, Paris.

BONNATERRE, P.-J. (1751–1804).

The abbot and naturalist, Pierre-Joseph Bonnaterre, produced one of the earliest illustrated encyclopedias of the amphibians and reptiles of the world and was the first to use the French word "Erpétologie." He was born on 24 November 1751 in Saint-Geniez (Aveyron) in southern France. His father was a surgeon, but the son chose to enter the seminary in nearby Rodez in 1772 and in February 1779 was ordained a priest. He became a vicar briefly at Cassuéjouls, near Rodez, before moving to Paris. At some point thereafter he became an abbé (a lay abbot who discharges no spiritual duties). A publisher in Paris employed him in the production of Archbishop François Fénelon's works, which began to appear in 1787.

Bonnaterre was also employed by the publisher Charles-Joseph Panckoucke to write the captions and superintend the production of the plates for several volumes in the series "Encyclopédie Méthodique," an enormous encyclopedia composed of 196 volumes (1782–1832). Some of France's most distinguished naturalists contributed to the natural history section, including J.-B. Bory de Saint-Vincent, Louis Daubenton, P.-A. Latreille, and G.-A. Olivier. It is not clear why Bonnaterre was invited to write several parts because there is no evidence that he had an earlier interest in natural history. Nevertheless, he became quite knowledgeable about the subject and wrote a significant portion of the section of the encyclopedia entitled "Tableau Encyclopédique et Méthodique des Trois Règnes de la Nature," namely the parts on ichthyology (1788), cetology, or whales and their relatives (1789), the first part of ornithology (1790–1791), and the two parts of herpetology ("Erpétologie," covering all amphibians and reptiles except snakes [1789] and "Ophiologie," covering snakes [1790]). The herpetological parts include 26 and 43 plates, respectively; all figures were copied from existing books, but especially from the works of Mark Catesby, Count de Lacepède, A. J. Roesel von Rosenhof, and Albertus Seba. The "text" actually consists of extensive explanations for the plates. It was mostly copied and reworked from Lacepède's two-volume summary of amphibians and reptiles, which had just been published (1788–1789).

It was commonplace in the days before copyright to have works plagiarized for both text and illustrations.

Bonnaterre did introduce some new names in these herpetological captions, but most are not recognized today. A Mexican amphisbaenid (*Bipes canaliculatus*), a North American toad (*Bufo terrestris*), and a West Indian iguana (*Cyclura cornuta*) are the sole exceptions. Recently, Bonnaterre has been credited with other names that were first introduced by Lacepède (see *Note*). The actual herpetological text for this encyclopedia appeared in 1784—"Les Animaux Quadrupèdes Ovipares, et les Serpens," by Daubenton. Since this text had no illustrations, it is often found bound together with the plates and captions produced by Bonnaterre.

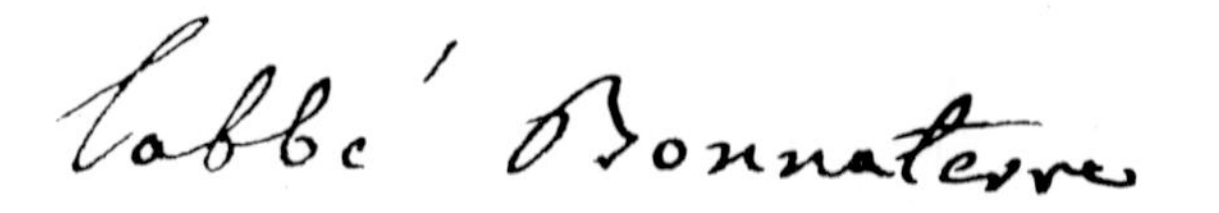

Bonnaterre's work for Panckoucke quickly came to an end. The French Revolution had begun in 1789 and the clergy became special targets. Many were massacred in 1792 and because Bonnaterre had refused to swear to the constitutional oath that year, he returned to his Aveyron homeland and remained hidden there during the Reign of Terror, emerging only in December 1794. In 1796 he founded a botanical garden in Rodez and in May of the next year he became a professor at the Ecole Centrale de l'Aveyron. He died in Saint-Geniez on 21 September 1804.

• *References*: "(Pierre-Joseph) Bonnaterre," by S. Le Tourneur, Dict. Biogr. Franç., 6: 986, 1954; Additional information from Édouard-Raoul Brygoo, 1997. • *Signature*: Muséum National d'Histoire Naturelle, Paris. • *Note*: By action of the International Commission on Zoological Nomenclature (1987, 2002), most names used in Lacepède's "Histoire Naturelle" have been declared unavailable for nomenclatural purposes. Many of Lacepède's names, however, were later used in Bonnaterre, which is a binomial work, and are now credited to the latter.

THUNBERG, C. P. (1743–1828).

The physician and botanist, Carl Peter (or Pehr) Thunberg, a student of and successor to Carl Linnaeus, wrote on amphibians and reptiles. He was born on 11 November 1743 at Jönköping (Småland) in southern Sweden. He began his studies in natural science and medicine at Uppsala University under Linnaeus in 1761 and took his Ph.D. there in 1767; his M.D. was conferred *in absentia* in 1772. Linnaeus, who by this time had completed all twelve editions of his catalogues of the plants and animals of the world, had always exhorted his students to explore unknown regions and to collect specimens for study and description. Thus inspired, Thunberg, while visiting Amsterdam in 1771, got an opportunity to visit Japan, which was then closed to all outsiders except for Dutch traders. He joined the Dutch East Indies Company as a surgeon and during 1772–1775 spent time in Dutch colonies in South Africa to collect with his fellow Swede, Anders Sparrman, another Linnaean student, and to learn the language so he could pass for Dutch in Japan. South Africa at this time was a perilous place to explore. As Thunberg later wrote (in translation): "I met the dangers of life; I prudently eluded ferocious tribes and beasts, and for the sake of discovering the beautiful plants . . . I joyfully ran, sweated, and chilled."

Thunberg then sailed to Japan via Batavia (now Djakarta) in Dutch Java and arrived in Nagasaki Bay in August 1775. The Dutch were required to live and work on a small (75 by 120 m) artificial island in the bay but Thunberg, who served as the station's surgeon, was sometimes permitted to botanize ashore. He also collected along the route of the annual Dutch pilgrimage to Edo (now Tokyo), a long trip on foot that took 50 days, to pay tribute to the shogun. He departed Japan in December 1776 and returned via Batavia, Ceylon (now Sri Lanka, which was then under Dutch control), and the Cape of Good Hope and arrived in Amsterdam in October 1778. He went home to Sweden via London in March 1779 and never left the country again. Thunberg became an assistant to Linnaeus's son, Carl Linnaeus *filius*, who had taken his father's professorial chair in medicine and botany on his death in 1778. Thunberg himself occupied this chair from 1784 upon the early death of Linnaeus *filius* and held it until his own death on 8 August 1828 at his summer residence near Uppsala. He was the last of the pupils of Linnaeus *pater* in Sweden.

Thunberg's reputation rests primarily on his botanical studies. His most important books on regional floras were those on Japan and South Africa, but he also wrote a summary of his nine years of travel, in four volumes. His zoological work was minor by comparison, but because of the large collections then available to him in Sweden and the early period

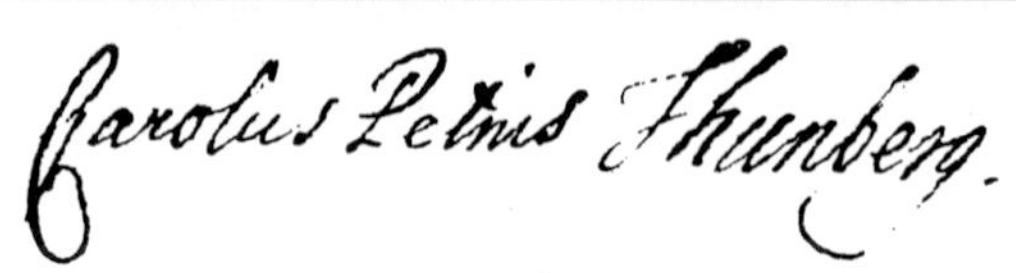

in which he worked, these studies are of historical importance. As Linnaeus's successor he inherited the collections made by the master's students and also had access to private collections held by the Swedish nobility. He followed Linnaeus's tradition to have students work up these collections but as was true for Linnaeus, Thunberg also wrote each student's thesis, although the student or "respondent" had to defend it and pay for its printing.

Thunberg published his first strictly herpetological papers in 1787. One contained descriptions and an excellent plate of new species of turtles from his voyage. Two of the taxa are recognized today as *Chelonia mydas japonica* from Japan and *Homopus areolata* from the Cape. In 1807 he described a new boa from the East Indies but it is a synonym of *Candoia carinata*. Thunberg's other herpetological contributions are mostly contained in the theses of his students, among the most notable of which are the following: reviews of the collections from Count Carl Gyllenborg, Crown Prince Adolph Friderick, and Claes Grill (1787, student respondent F. W. Radloff); collections from Johan Alströmer and C. P. Thunberg including description of the South American snake now known as *Thamnodynastes strigilis* (1787, L. M. Holmer); "Fauna Cayanensis" (1823, A. Kjeller); and "Fauna Guineensis" (1823, M. Kahn). In addition, Thunberg wrote as small dissertations a series on animals and plants described in the Bible (1825–1829). The part on Amphibia (at that time covering all amphibians and reptiles) was published in 1827. It contains remarks on collections made in the Middle East by Fredrik Hasselquist who, like so many other Linnaeus students, died on his expedition.

• *References*: "Linnaeus and the Linnaeans," by F. A. Stafleu, A. Ossthoek, Utrecht, xvi, 386 pages, 1971 (Thunberg on pp. 152–155); "Carl Pehr (Peter) Thunberg," p. 347–350 *in* M. Gunn and L. E. Codd, Botanical Exploration of Southern Africa, A. A. Balkema, Cape Town, 1981; "Carl Peter Thunberg, 1743–1828," by L. Wallin, Almqvist & Wiksell Internatl., Stockholm, 64 pages, 1993. • *Portrait* (1808): From I. Dörfler, Botaniker-Porträts, F. Jasper, Vienna, 1907. • *Signature* (1792, Latin form): From Candollea, 33: 436, 1978.

BLUMENBACH, J. F. (1752–1840).

Johann Friedrich Blumenbach was a German naturalist, anthropologist, and comparative anatomist, and one of the first great teachers of herpetologists. He was born to a wealthy, cultured family on 11 May 1752 in Gotha (Thuringia), about 50 km southeast of Göttingen. From his youth he had been exposed to natural science and after completing his secondary education in 1769, he studied at the university in Jena and then in Göttingen where he took his medical degree (M.D. 1775). His epochal doctoral dissertation (1776 [1775]), the first survey of the characteristics and distribution of the human races, is the starting point for physical anthropology and began with his attempt to bring order to the university's museum collection of human skulls. In 1776 he was appointed a professor and first curator of the zoological museum at Göttingen, positions he held for the rest of his life.

Blumenbach was also a comparative anatomist, the first in Germany to lecture on the subject. Both his lectures and textbooks were highly influential. He was widely regarded as the finest teacher of natural science of his era and drew students

from afar. Many acknowledged his role in the formation of their own ideas. Among those of herpetological interest were Heinrich Boie, J. L. C. Gravenhorst, Alexander von Humboldt, J. J. Kaup, Blasius Merrem, Lorenz Oken, Heinrich Rathke, J. G. Schneider, and Prince Maximilian of Wied-Neuwied. Benjamin Smith Barton came from America and A. A. Berthold became his successor as curator of the museum in 1840 (see *Note*). Another Blumenbach student was Baron Georg von Langsdorff, later the Russian envoy and explorer of Brazil (1813–1820, 1824–1829) who facilitated the work of many herpetological collectors in Brazil (Edouard Ménétriés, J. C. Mikan, Johann Natterer, Giuseppe Raddi, J. B. von Spix). The influence of Blumenbach through these and other students who made early contributions to the biology of amphibians and reptiles is comparable only to that of Georges Cuvier.

Blumenbach's authority was also expressed through his extensive writings. His "Handbuch der Naturgeschichte" (1779–1780, with 13 editions by 1832 and several foreign language editions) was based on the Linnaean system, but it combined morphological, physiological, and ecological elements into a more modern biological concept of organisms. Other works on natural history followed: his ten-part serial "Abbildungen naturhistorischer Gegenstände" (1796–1810, later editions in 1810, 1827, and 1830), of which eight of the 100 accounts, each with a plate, are herpetological, and his "Beyträge zur Naturgeschichte" (1790, second edition 1806–1811), in which he more fully developed his ideas that animals and plants must have had a very long history on earth. His great text on comparative anatomy (1805, second edition 1815, third 1824) was also widely read.

His writings on amphibians and reptiles were mostly contained within these and other more comprehensive works. For example, in "Specimen Physiologiæ Comparatæ inter Animantia Calidi et Frigidi Sanguinis" (1787), he was the first to describe the features that characterize larval salamanders (compressed tail, four limbs, external gills). This is obvious today but even in 1800 P.-A. Latreille, in his monograph of the salamanders of France, denied the existence of gills. Blumenbach also noted that female *Salamandra salamandra* can give birth to young even when isolated from males for many months, leading him to conclude that this salamander is truly live-bearing and that a single fertilization is sufficient for a long time. He also studied regeneration in salamanders and conducted experiments in which he noted that tails would re-grow completely, albeit slowly, after amputation. His only strictly herpetological book was "Afbildninger med Vedføiede Beskrivelser over Amphibierne" (1832, in Danish), which covers 68 species of amphibians and reptiles from throughout the world in 42 hand-colored, engraved plates and accompanying text.

Blumenbach's sole successful contribution to herpetological taxonomy was naming the North American scarletsnake (*Cemophora coccinea*), but he did name numerous other animals including the groundhog (*Marmota*) and the African elephant (*Loxodonta africana*). He became the most distinguished member of the medical faculty at Göttingen—he was named Professor Primarius in 1816—and was honored at home and abroad for his many contributions to biology and anthropology. He died in Göttingen on 22 January 1840.

• *References*: "Éloge Historique de Jean-Frédéric Blumenbach," p. 197–228 *in* P. Flourens, Recueil des Éloges Historiques, G. Frères, Paris, 1856; "Joh. Friedrich Blumenbach," by O. Schmidt, Allg. Deutsch. Biogr., 2: 748–751, 1875; "Johann Friedrich Blumenbach," by W. Baron, Dict. Sci. Biogr., 2: 203–205, 1970. • *Portrait* (1823) *and signature* (1780): Adler collection. • *Note*: In 1977 the entire Göttingen zoological collection was transferred to the Museum Alexander Koenig in Bonn.

GMELIN, Johann Friedrich (1748–1804).

"Gmelin" is familiar to herpetologists as the name of the compiler and editor of the 13th edition of Carl Linnaeus's "Systema Naturæ." Twelve editions of this catalogue of the animals of the world—which give only the briefest information for each species—had been produced during Linnaeus's lifetime. With his death in 1778 someone else would have to produce any new edition. That person was Johann Friedrich Gmelin, born into one of the greatest family dynasties of leading German lawyers, theologians, physicians, and naturalists (see *Note 1*). He was born on 8 August 1748 in Tübingen (Baden-Württemberg), about 30 km south of Stuttgart, where his father was a professor of botany and chemistry at the university. He studied natural science there with his father and also medicine (M.D. 1768) and then traveled abroad for additional study in Holland, England, and Austria before returning to Tübingen in 1772 as a professor of medicine. He moved to the university in Göttingen (Lower Saxony) in 1775 as professor of philosophy and medicine, and by 1778 he became professor of medicine and chemistry and also lecturer on botany and mineralogy. From his titles alone it would seem that he had unusually broad knowledge. He also published important original works on poisons and chemistry.

Gmelin was also a translator, compiler, and editor. From 1777 to 1785 he produced a German edition of Linnaeus's system of minerals from the Latin original. Perhaps as a result of this experience with Linnaeus's works, he was led to produce a new and enlarged edition, the 13th (1788–1793), of Linnaeus's "Systema Naturæ." It was enlarged because after the 12th edition many new species had been discovered and, besides, Linnaeus had missed many others in the prior literature and all of these had to be incorporated. Moreover, many of these new species, although they might have been described and even illustrated, had not yet been named according to the Linnaean system of binomial nomenclature. Gmelin was thus in a position to name them and he did. He was not a zoologist and never had a research program on that topic, but solely in his role as compiler he named a large number of animals. The amphibians and reptiles were included in part 3 of volume 1 ([1789]; see *Note 2*). He named, for example, a European tortoise (*Testudo hermanni*), an Asian lizard (*Phrynocephalus guttatus*) and snake (*Dendrelaphis pictus*), and the African python (*Python sebae*). The Indian gharial (*Gavialis gangeticus*) is another example, a species that had been described in the literature and even illustrated as early as 1734 by Albertus Seba, but it never had a binomial Latin name until Gmelin.

One other herpetological work has appeared under Gmelin's authorship, but only after his death in Göttingen on 1 November 1804. This is volume 12 ("Amphibia," undated but probably published 1809 or 1810) in a series of 14 volumes issued in London under the title "A Genuine and Universal System of Natural History . . . By the late Sir Charles Linnaeus . . . Improved, Corrected, and Enlarged, By J. Frid. Gmelin, M.D." (1794–1810; see *Note 3*). Volume 12 bears no resemblance whatsoever to Gmelin's 13th edition of Linnae-

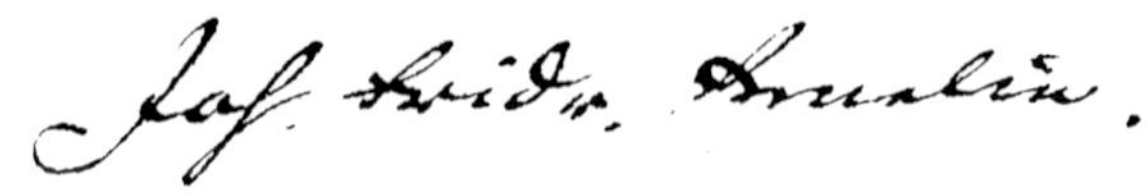

us, but is a new work reviewing the amphibians and reptiles of the world and is accompanied by 21 hand colored plates. Judging from additional information given on the title page, it is clear that the editors added liberally to what Gmelin provided years before, if anything. Since he had been dead for at least five years, the editors may simply have used his name (and that of Linnaeus) to help sell their books. Some plates are dated as early as 1794 and appeared first in *Encyclopædia Londinensis* (see *Note 3*), which was produced by the same editors. The editors of this serialized encyclopedia, produced in weekly parts over the period 1796 to 1829, were known to take great liberties in plagiarizing texts and altering illustrations from published sources.

• *References*: "Johann Friedrich Gmelin," by M. Gmelin, Allg. Deutsch. Biogr., 9: 270, 1879 (for Gmelin family, see same volume, pp. 264–266); Rieck et al., 2001 (Gmelin by W. Bischoff and G. Hallmann on p. 463); "Johann Friedrich Gmelin," anonymous, Dict. German Biogr., 4: 42, 2003. • *Portrait*: Bibliothèque Nationale, Paris. • *Signature* (1797): Museum für Naturkunde, Berlin. • *Note 1*: Three different Gmelins named amphibians and reptiles during the late 18th Century and they are commonly confused or their roles ascribed to a single person. The progenitor of the line was Johann Georg Gmelin the Elder (1674–1728), an apothecary and chemist in Tübingen. Two of his sons became naturalists: Johann Georg Gmelin the Younger (1709–1755), professor of chemistry and natural history in St. Petersburg and explorer of Siberia, and Philipp Friedrich Gmelin (1721–1768), professor of botany and chemistry in Tübingen. Johann Friedrich Gmelin, the main subject of this biography, was the son of Philipp Friedrich Gmelin and thus the nephew of Johann Georg Gmelin the Younger. Samuel Gottlieb Gmelin (1744–1774), son of Johann Conrad Gmelin (1707–1759), apothecary and physician in Tübingen who was another son of Johann Georg Gmelin the Elder, was an older cousin of Johann Friedrich Gmelin. Samuel Gmelin explored Siberia, as his uncle had done earlier, and the Caspian Sea region. From the latter expedition he named the turtle now known as *Mauremys caspica*. Yet another, more distantly related Gmelin was Carl Christian Gmelin (1762–1837), professor of natural history and director of the natural history museum in Carlsruhe. His main contribution to zoology was a five-volume series on vertebrates, worldwide in coverage, entitled "Gemeinnützige systematische Naturgeschichte." The herpetological part (Volume 3, "Amphibien," 1815; second edition 1839) includes 36 beautifully hand-colored plates. He described two new snakes in *Der Naturforscher* (1799) that are today synonyms of *Coluber viridiflavus* and *Elaphe scalaris*. • *Note 2*: The original printing of the 13th edition was in Leipzig. A second, line-by-line reprint, probably pirated, was issued in Leiden, also in 1789. An English edition, translated by William Turton, appeared in 1802. • *Note 3*: *Encyclopædia Londinensis* and "A Genuine and Universal System of Natural History" have recently been discussed in detail (see R. Wahlgren, Newsl.-Bull. Internatl. Soc. Hist. Biblio. Herpetol. [now Biblio. Herpetol.], 2(1): 20–26, 2000).

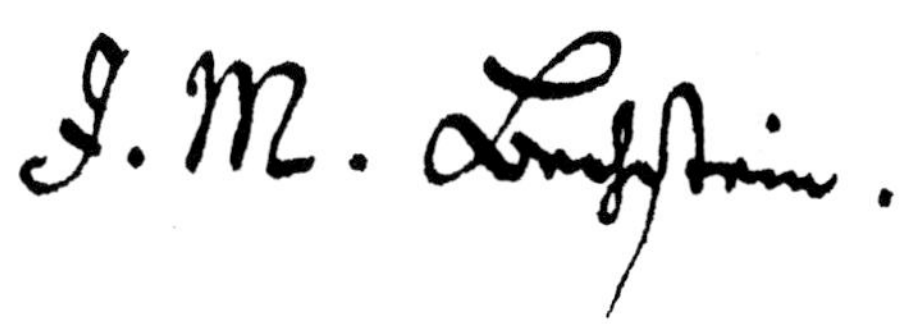

BECHSTEIN, J. M. (1757–1822).

The German naturalist, Johann Matthäus (or Matthaeus) Bechstein, was a forester and ornithologist, and the father of herpetoculture. He was born on 11 July 1757 in Waltershausen (Thuringia) about 15 km southwest of Gotha. He enjoyed nature as a child and collected animals to bring into the house. He attended secondary school in Gotha and then the university in Jena (1778–1781) to read theology. He also studied natural science and forestry. Beginning in 1785 he became a teacher of mathematics and natural history at a school in Schnepfenthal but by 1793 taught mostly forestry. He took a public examination in 1795 at the forestry school in Berlin and then founded his own forestry research institute in Waltershausen, but it closed in 1799 for lack of funds. Fortunately, that same year the reigning duke of Saxe-Meiningen, Georg I, founded a forestry school in Dreissigacker (near Meiningen in Thuringia) and appointed Bechstein director, a post he held for life. It opened in 1801 and became a forestry academy in 1803. For his many public contributions Bechstein was awarded a doctorate by the University of Erlangen (Ph.D. 1806). He died in Dreissigacker on 23 February 1822.

Bechstein's official duties were in forestry, but his personal interests were much broader. He was one of the first to be concerned with wildlife conservation and he called for the protection of animals that were even considered to be pests or harmful such as bats and venomous snakes. He was also a prolific writer and published a large number of books. Perhaps his best known was a book on the natural history of birds for the home (1792), but he also wrote several multi-volume texts on German or general natural history. His reputation as a herpetologist rests primarily on two works. The purpose of the first, "Naturgeschichte der Stubenthiere" (1797), is better understood from a translation of its longer title: "Natural History or Introduction to the Knowledge and Maintenance of Mammals, Amphibians [including reptiles], Fishes, Insects, and Worms Which One Can Hold in the Household." In it he described in great detail (under six subheadings: description, residence, food, reproduction, capture, best features) how five species—*Emys orbicularis, Hyla arborea, Lacerta agilis, Natrix natrix,* and *Triturus cristatus*—could be maintained in captivity. He notes, for example, that frogs will not touch dead insects; and they announce weather changes by loud croaking but that the calling is very unpleasant. This is the first manual for keeping amphibians and reptiles in captivity. The first such guide in the English language appeared a century later, Gregory Bateman's "The Vivarium" (1897).

Bechstein's other contribution to herpetology was much more technical. During 1800–1802 he issued a five-volume work, "Herrn De la Cepede's Naturgeschichte der Amphibien," being a translation of Count B.-G.-É. Lacepède's "Histoire Naturelle des Quadrupèdes Ovipares et des Serpens" (1788–1789). Lacepède's was one of the most important herpetological books of its time. Volumes one and two of Bechstein correspond to Lacepède's first volume, covering all amphibians and reptiles except snakes, and his volumes three, four, and five equate to Lacepède's second volume on snakes including caecilians and amphisbaenians.

Bechstein's version is far more than a mere translation because he revised it with the addition of new information that appeared subsequent to Lacepède, particularly from the works of German authors such as Blasius Merrem, J. G. Schneider, and Johann David Schoepff. He added extensive footnotes, some of them occupying almost the entire page, and additional accounts of new species; all of these are signed with the letter "B." Bechstein's 168 plates are newly engraved and, in fact, he copied only some of them from Lacepède (63 of 360 figures). Several were original. Instead, he turned to the best available sources including illustrations from Mark Catesby, Carl Linnaeus, Peter S. Pallas, A. J. Roesel von Rosenhof, Patrick Russell, J. J. Scheuchzer, and Albertus Seba among others. Given these and the minor authors who are cited by Bechstein, he must have had access to a large amount of literature and his text accounts indicate that he had a sophisticated knowledge of contemporary herpetology. A large part of this book is original and perhaps Bechstein was simply using Lacepède's name merely to publish what is almost his own book and in a field where he had no standing.

• *References*: "Joh. Matthäus Bechstein," by [J.] Carus, Allg. Deutsch. Biogr., 2: 205–206, 1875; Rieck et al., 2001 (Bechstein by W. Rieck on pp. 415–418); "Johann Matthäus Bechstein: The Father of Herpetoculture," by L. Heichler and J. B. Murphy, Herpetol. Rev., 35: 8–13, 2004. • *Portrait and signature* (1794): Museum für Naturkunde, Berlin.

BARTRAM, William (1739–1823).

William Bartram, an American botanist and ornithologist, made the first herpetological survey of the southeastern United States. Billy Bartram, as he was called as a boy, was born on 9 April 1739 at the home of his parents in Kingsessing on the outskirts of Philadelphia. He was the third son of John Bartram—a horticulturist who in 1765 was appointed by King George III as "Royal Botanist" for his American dominions. He was the only one of the sons who shared his father's interests in nature and exploration. During 1752–1755 Billy was enrolled to study history and languages at the Academy of Philadelphia (see *Note*), but he much preferred to travel with his father. In 1753, when Billy was 14, he accompanied his father to the Catskill Mountains of New York State and contributed by making drawings of plants, animals, and scenery. Two years later his father took him to explore Connecticut. His first really extensive trip was from July 1765 to April 1766 to the Carolinas, Georgia, and East Florida. His father was then 66 and contracted malaria on the trip, which was his last.

During his teenage years Bartram was put in touch with some of his father's English correspondents, notably the ornithologist George Edwards and the merchant Peter Collinson, a fellow Quaker who wished to introduce American plants into English gardens. John Bartram sent some of Billy's drawings to Edwards and Collinson who provided advice on improving his work. They even published some of them. His first drawing (a bird) was made in 1753. As early as 1756 he was drawing turtles and these animals became a primary object of his artistry. Collinson had a particular interest in them and asked John Bartram to have Billy draw them and even re-draw them when he thought something was wrong. On one occasion he wrote to the father ". . . I wish he would paint the Pond Turtle [*Chrysemys picta*] over again. It is the most indifferently performed; the shell is made almost white,–whereas it is black." Collinson knew the animal alive from specimens sent to him by the Bartrams. Collinson went on to say he wanted drawings of the plastron also, ". . . for there is always something remarkable there." This advice helped Bartram to be more observant and accurate in his work.

During 1767–1772 Bartram tried farming and various businesses but they all failed. At one point he had to flee to escape his debtors, yet throughout this difficult period he continued to send drawings to Collinson and to another English patron, the physician John Fothergill. In October 1772 Fothergill offered Bartram financial support to collect plants and seeds and to draw plants and animals in Canada, but Bartram insisted on going instead to the Southeast again. Accordingly, from April 1773 to October 1777 (not 1778 as stated wrongly by Bartram who was poor with dates), he journeyed through the Carolinas, Georgia, and Louisiana and portions of East and West Florida that are now part of the states of Alabama, Florida, and Mississippi. This trip, mostly on foot and often alone, covered more than 3,800 km. The expedition began as an English-initiated journey, but with the Revolutionary War beginning in April 1775 and independence of the new American nation declared in July 1776, it ended up as an American one. On his return Bartram began writing a book from the extensive notes and drawings made on his trip, the epic "Travels Through North & South Carolina, Georgia, East & West Florida . . . ." (1791), which includes long passages about amphibians and reptiles. He recorded some 48 species (10 frogs, 14 turtles, 5 lizards, 18 snakes, and the alligator) but no salamanders.

Considering the state of knowledge of American amphibians and reptiles at the time and the fact that Bartram was a home-trained naturalist specializing in botany, his herpetological observations were surprisingly good. (John R. Forster had issued the first Linnaean catalogue of North American animals only in 1771 and the section on reptiles was a scant three pages long.) Bartram's most celebrated observations were those on the alligator. Although he unquestionably got some things wrong and occasionally exaggerated, time has shown that he was mostly right. He witnessed the young alligators following their mother ". . . just like a hen does her brood of chickens, and she is equally assiduous and courageous in defending the young . . . you may hear the young ones continually whining and barking, like young puppies." His observations on maternal behavior, for which he was maligned by John Holbrook and other later writers, have been largely vindicated, but only after more than a century had passed. He noted the ophiophagus eating habits of *Coluber constrictor* and the sounds produced by the tail of *Pituophis melanoleucus* when vibrated in dry leaves. He also remarked on the puffing up of *Heterodon* and body flattening behavior of *Sistrurus* when these snakes are disturbed. Remarkably, he seems to be the first person to record hissing behavior in snakes, even though it is now recorded in at least five families. He noted ". . . a terrible hiss, resembling distant thunder," which emanated from a Pinesnake at a distance of 45 m.

Bartram made an attempt to give Latin names to some animals. His method was not uniformly binomial, however, and thus his book was rejected for nomenclatural purposes by the International Commission on Zoological Nomenclature (1957). Nevertheless, other authors based new species of reptiles on Bartram's descriptions. Two are still recognized, both named by the French naturalist F.-M. Daudin: the Gopher Tortoise (*Gopherus polyphemus*) and the Pinesnake (*P. melanoleucus*).

Bartram's book was reissued in London and Dublin and translated into German, Dutch, and French. Extracts, including herpetological ones, were published in various journals. This notoriety made him America's most famous naturalist of the day, especially in Europe. Natural history then was practiced mostly by amateurs like Bartram, but soon it was to become an academic subject taught in schools and colleges. Since about 1789 Bartram had been tutoring a young Philadelphia naturalist, Benjamin Smith Barton, who had just received his medical training in Europe and who, at the College of Philadelphia, became America's first professor of natural history (see *Note*) in 1790. Bartram, recognizing that the task of covering all of American natural history was too much for one person, turned over many of his notes and drawings to Barton in the hope that he would publish them. Although Barton did make use of some of them, he also made his own (and more scientific) contributions to American natural history, but he predeceased Bartram by eight years. Some other Bartram protégés were the ornithologist Alexander Wilson and the entomologist and herpetologist Thomas Say, who was Bartram's great nephew.

Bartram, who never married, died on 22 July 1823 at his home in Kingsessing, once his father's home where the elder Bartram himself had died in 1777.

• *References*: "William Bartram," by R. McC. Peck, Amer. Natl. Biogr., 2: 297–299, 1999; "William Bartram," by M. B. Simpson, Jr., Oxford Dict. Natl. Biogr., 4: 217–219, 2004; "America's First Herpetological Expedition: William Bartram's Travels in Southeastern United States," by K. Adler, Bonner Zool. Beitr., 52: 275–295, 2004. • *Portrait and signature*: From Goode (1901). • *Note*: Benjamin Franklin proposed a new educational system for Philadelphia in 1749 that included a College of Philadelphia and a preparatory Academy of Philadelphia for younger students, all designed to educate men for the business and governing classes rather than for the clergy. The classes were to be taught in English instead of Latin and the curriculum was to include natural history, geography, and modern languages. The academy was opened in 1751 and William Bartram was thus one of its earliest students. When the trustees opened a second college in 1765, America's first medical school, the college with its associated academy, in effect became the country's first university. It did not adopt its current name, the University of Pennsylvania, until 1791.

BARTON, Benjamin Smith (1766–1815).

Benjamin Smith Barton, physician, ethnographer, and the first professor of natural history in the United States, wrote the earliest books devoted exclusively to American herpetology. He was born in Lancaster (Pennsylvania) on 10 February 1766. His father, an Episcopal clergyman, was an amateur botanist and a correspondent of Carl Linnaeus, and his mother was the sister of David Rittenhouse, a noted astronomer. As a boy young Benjamin spent time in the woods and along Conestoga Creek observing birds, insects, and plants. Both of his parents were dead by the time he was 14 and in 1782 he moved to Philadelphia to live with an older brother. He studied science and medicine at the College of Philadelphia and, like many young medical students of the day, went on to Europe to complete his education. Natural history was a requisite part of the medical curriculum at that time and Barton had the good fortune to study with some of the best physician-naturalists including John Hunter at Edinburgh, Joseph Banks in London, and finally with J. F. Blumenbach in Göttingen. Much was expected of someone with such advantages of family and of education.

Barton's most important teacher, as far as the natural history of the United States is concerned, was William Bartram whose celebrated "Travels" was published in 1791. Bartram was Barton's senior by 27 years and had extensive field experience from New York to Florida. He regularly tutored the solicitous Barton beginning about 1789 and graciously turned over many of his drawings and notes to him in the hope that they would be published. Unfortunately, that wish went mostly unrealized and, in fact, Bartram outlived Barton by eight years. Barton himself taught natural history to others, among them the Ohio naturalist, Jared P. Kirtland, and in 1803 at the request of President Thomas Jefferson, Meriwether Lewis, who was about to make his expedition to the Pacific Ocean with William Clark.

On returning from Europe in 1789 Barton began a medical practice (see *Note*) and in 1790 accepted the newly created professorship of natural history and botany at the College of Philadelphia, which in 1791 became the University of Pennsylvania (see *Note* for William Bartram, p. 35). He remained on this faculty for the rest of his life. His only significant field trip was with his uncle Rittenhouse to survey the western border of Pennsylvania in 1789. As Barton wrote, "in this wilderness [my uncle] first fostered my love and zeal for natural history." Because he was frail in constitution and later suffered from frequent attacks of gout, Barton was unable to take long field excursions thereafter.

Nevertheless, Barton was a man of great industry, often working by midnight oil, and wrote several important books including the first American textbook of botany (1803, new edition in 1814), one on ethnography (1797), and a materia medica of medicinal plants (1798–1804). He published an important work on goiter (1800) that was dedicated to his former teacher, Blumenbach. He founded the *Philadelphia Medical and Physical Journal* in 1805 to which he contributed numerous articles on botany, zoology, anthropology, earthquakes, and many other topics. His critics often accused him of having too many interests, yet he accomplished a great deal, especially given that he died at the age of 49 of pulmonary tuberculosis from which he had suffered for many years. He died at his home in Philadelphia on 19 December 1815.

Although he proposed names for several salamanders, none of the currently valid names of American amphibians and reptiles is credited to Barton. He did publish extensively on these animals and wrote several papers for the *Transactions of the American Philosophical Society*. The first one was on remedies for rattlesnake bites (1793). In another, on the natural history of *Ambystoma maculatum* (1803), he described its noxious skin secretions based on experiments in which he put the secretions on his tongue and those of his pupils, anticipating the chemical bioassays of the 20th Century. He authored two other herpetological papers, one on snakebite (1803) and the other on torpidity in alligators (1805).

Barton issued privately a number of short books on herpetological topics. His most famous was a 70-page book, without illustration, on the natural history of American snakes, especially of *Crotalus horridus,* entitled "A Memoir Concerning the Fascinating Faculty Which Has Been Ascribed to the Rattlesnake, and Other American Serpents" (1796; German edition issued by E. A. W. von Zimmermann in 1798). Barton had observed rattlesnakes on his trip to western Pennsylvania in 1789 and kept live specimens for observation. This book represents the first separate publication ever issued exclusively on American herpetology. A journal version was published (1799) and later serialized in at least two other journals (1803–1804). Barton also issued a supplement to the rattlesnake book (1800) and a new and enlarged edition (1814, with a plate).

He privately issued two other herpetological booklets, one on *Siren* (1808; new edition in 1812) and the other on *Cryptobranchus* (1812 [1814]), each with a plate. Barton's choice of topics was significant. Europeans had been perplexed about these unusual American salamanders and thought that they must be the larval forms of some as-yet-undiscovered species (thus implying that American naturalists were not fully capable of studying their own fauna). Barton intended to show that these animals were indeed adults. Amphibians and reptiles were mentioned in passing in some other booklets, most notably in one entitled "A Discourse on Some of the Principal Desiderata in Natural History" (1807), which contributed importantly to arousing an intelligent public interest in natural history.

• *References*: "An Account of the Life and Writings of [Benjamin Smith Barton]," rev. ed. by W. P. C. Barton, *in* B. S. Barton, Elements of Botany, R. Desilver, Philadelphia, 325 pages, 1836; "Benjamin Smith Barton," by G. Blumer, Dict. Amer. Biogr., 1: 17–18, 1927; "Benjamin Smith Barton (1766–1815)," p. 88–95 *in* H. A. Kelly, Some American Medical Botanists, D. Appleton, New York, 1929; "Benjamin Smith Barton's Influence on Trans-Allegheny Natural History," by J. Ewan, Bartonia, 54: 28–38, 1988; "Benjamin Smith Barton," by P. D. Thomas, Amer. Natl. Biogr., 2: 287–288, 1999. •*Portrait* (1800–1805): College of Physicians of Philadelphia. • *Signature*: Academy of Natural Sciences, Philadelphia. • *Note*: His medical degree, an honorary one, was only received years later (Universität Kiel, 1796).

PALISOT DE BEAUVOIS, A.-M.-F.-J. (1752–1820).

Ambroise-Marie-François-Joseph Palisot, Baron de Beauvois (see *Note*), French traveler and botanist, made important contributions to North American herpetology. Palisot de Beauvois was born to an aristocratic family in Arras, north of Paris, on 27 July 1752. He was educated at Harcourt College in Paris and admitted to the bar, but instead was eager to travel and conduct botanical exploration. In 1786 he began a period of 12 years overseas—in West Africa, Haiti, and the United States—during which time he built, and lost, three great natural history collections. The African materials were burned when the English destroyed the trading station at which he worked (1791). Those in Haiti perished during a slave insurrection (1793) and his largest American collections were lost in a shipwreck enroute to France (1798). Despite these devastating setbacks, after his return to France he spent the remaining 22 years of his life publishing papers and books based on smaller collections that did reach Paris and on his surviving notes. His most important titles are a book on the insects of West Africa, Haiti, and the United States (1786–1805) and one on the flora of West Africa (1805–1821). He died in Paris on 21 January 1820.

Palisot de Beauvois resided in Philadelphia at four different times between 1791 and 1798. He was an elected member of the American Philosophical Society beginning in January 1792, and he also had an association with the privately owned Peale Museum in Philadelphia. Although he was sometimes cut off from his funds at home because of the French Revolution, he was able to travel widely in the new nation from New York west to Ohio, and south through the Appalachian Mountains to Georgia. He published only two titles on herpetology (both 1799), based on observations and specimens from the United States. In one, he described as new, under the name *Siren operculata*, a larval salamander from New Jersey that seems referable to *Ambystoma tigrinum*, a species named by Jacob Green in 1825. In the other, which is entirely on snakes, he described two of the most distinctive snakes of eastern North America—*Crotalus adamanteus* and the northern race of *Agkistrodon contortrix*—in addition to erecting the genus *Agkistrodon* itself. This paper also covers the natural history of venomous snakes and includes some of the prevailing myths.

One other North American snake is now credited to Palisot de Beauvois, *Farancia erytrogramma*, which was published in volume 4 of C.-S. Sonnini and P.-A. Latreille's "Histoire Naturelle des Reptiles" (1801). Francis Harper (1940) has explained why this species should be thus attributed and not to the authors of the book.

• *References*: "Palisot de Beauvois as an Overlooked American Botanist," by E. D. Merrill, Proc. Amer. Philos. Soc., 76: 899–920, 1936; "Palisot de Beauvois (1752–1820)," p. 209–221 *in* A. Lacroix, Figures de Savants, vol. 4, Gauthier-Villars, Paris, 1938. • *Portrait*: From Lacroix, 1938. • *Signature* (1813): From H. M. Burdet, Candollea, 32: 201, 1976. • *Note*: The family name is Palisot (sometimes spelled Pallisat), although the baron's works are often incorrectly cited under Beauvois.

BRONGNIART, Alexandre (1770–1847).

Alexandre Brongniart was a geologist, mineralogist, and zoologist who founded stratigraphic paleontology and made important contributions to the classification of amphibians and reptiles. He was born on 5 February 1770 in Paris where his father was a prominent architect. Alexandre had obvious talent for science as a boy and went to study mineralogy at the École des Mines in Paris and then took a master's degree at the Sorbonne in 1789. In England he visited Derbyshire, famous for its lead mines, and then returned to Paris to study medicine. He also assisted his uncle who was a chemist at the Jardin des Plantes, originally the king's garden, which was transformed into the Muséum d'Histoire Naturelle during the French Revolution. Brongniart joined the National Guard during the Revolution, was posted to the Pyrénées in 1793, and was even briefly imprisoned.

Later, Brongniart sought a position at the museum but in 1796 he settled for a teaching post at the École des Mines. Then he taught natural history at Collège des Quatre-Nations, one of the constituent units of the Sorbonne. In 1802, at the age of only 32, Brongniart was named director of the porcelain factory at Sèvres on the western edge of Paris. This plant, then as now, manufactured dishes and vases renown for their deep blue colors. By 1802 he was also teaching a course on the natural history of reptiles at Collège du Plessis (now Faculté des Sciences). Finally, in 1822, he became a professor at the museum and took charge of mineralogy. He held this post for life and died in Sèvres on 7 October 1847.

Brongniart was a man of great intellect and made important scientific contributions especially to mineralogy and geology.

He authored a fundamental text on mineralogy in 1807 and the next year published a classic study on the stratigraphic exploration of the Paris Basin, co-authored with Georges Cuvier, the leading scientist at the museum. Through a careful study of aquatic fossils Brongniart was able to propose a correct chronology of the earth's upper layers (1829), thereby providing the foundation for the field of stratigraphic paleontology. It was Brongniart who introduced the geological term Jurassic, named for the Jura Mountains on the French-Swiss border.

Amphibians and reptiles were a minor part of Brongniart's work, but he made major contributions to their classification using the comparative methods of Cuvier. He regularly dissected amphibians and reptiles with Cuvier and Constant Duméril, an associate of Cuvier's who later became professor in charge of reptiles and fishes at the museum. Previously, amphibians and reptiles had been mixed in a single class—called Amphibia by Carl Linnaeus and Reptilium by J. N. Laurenti. However, in two journal papers (both 1800) Brongniart was the first to recognize that frogs and salamanders belonged together and were different from other reptiles. He noted that these animals lay eggs without shells and the larvae have gills, among other features in common. He erected the name Batraciens (from the Greek word for frog, βάτραχος) for this order of reptiles, the others being Cheloniens, Sauriens, and Ophidiens. He added that caecilians could not be properly placed because they were still too poorly known. He also named a genus of sea turtles (*Chelonia*), the Fijian iguana (*Brachylophus fasciatus*), and a Madagascan chameleon (*Furcifer bifidus*) in one of these papers.

Later, Brongniart issued a much-expanded version of his classification in the form of a 53-page book entitled "Essai d'une Classification Naturelle des Reptiles" (1805). This was a preprint of his journal article of the same title (1806). His arrangement was quickly adopted by Duméril who used it in his general classification of animals, "Zoologie Analytique" (1806), and later in his monumental, nine-volume "Erpétologie Générale" (1834–1854). It was only later that a student of Duméril's, Henri de Blainville, first gave amphibians the same rank in classification as reptiles (1816).

• *References*: "Alexandre Brongniart," by J. S. de Sacy, Dict. Biogr. Franç., 7: 419, 1956; Jaussaud and Brygoo, 2004 (Brongniart on pp. 105–109). • *Portrait*: Hunt Institute, Pittsburgh. • *Signature* (1829): Adler collection.

Geoffroy Saint-Hilaire, Étienne (1772–1844).

The French anatomist and zoologist, Étienne-François Geoffroy Saint-Hilaire, published extensively on reptiles especially crocodilians and turtles. He also made important contributions to evolutionary theory and to the concept of homology. He was born on 15 April 1772 in Étampes (Essonne), located about 50 km south of Paris. His father was a lawyer to the king and, when he purchased a house in the village of Saint-Hilaire, he named his son after it. The son studied for an ecclesiastical career and became a canon at age 15, but then went to Paris to study law and later took classes in natural history from Louis Daubenton at the Collège de France.

In 1793 Geoffroy accepted a position at the Jardin des Plantes, originally the king's garden, which was transformed into the Muséum d'Histoire Naturelle during the French Revolution. Later that year, at the age of 21, Geoffroy became professor of zoology at the museum with responsibility for vertebrates. The next year he took charge of the museum's menagerie. In 1795 his chair was split when Count de Lacepède returned to Paris from his self-imposed exile to escape the Reign of Terror. Thereafter, Geoffroy's official responsibilities were mammals and birds, but he continued his interest in reptiles. In 1795 it was Geoffroy who made the fateful decision to bring to the museum from Normandy a zoologist who was three years his senior and with whom he would collaborate early on, but who became his chief rival and the greatest naturalist of France, Georges Cuvier.

In 1798, at the age of 26, Geoffroy accompanied Napoleon Bonaparte's three-year, ill-fated expedition to conquer Egypt. He was one of 100 members of the Commission Scientifique et Artistique who discovered the glories of ancient Egypt, studied the colossal ruins of Luxor and Karnak, found the Rosetta Stone, and surveyed the Isthmus of Suez with an idea to build a canal. The savants often worked under desperate conditions—Bedouin raids, bubonic plague, and being cut off from supplies. Like many others Geoffroy contracted dysentery and briefly lost his sight from ophthalmia, yet he amassed enormous natural history collections including crocodiles, turtles, and other reptiles that were to become fodder for his debates with Cuvier on the pace of evolutionary change.

Geoffroy also played a role in the confiscation of natural history collections in lands occupied by the French Army. These included important herpetological specimens and manu-

script material from the Netherlands, Italy and the Pontificial States, German principalities, and Austria. Geoffroy was personally sent by Napoleon to Portugal and Spain in 1808 for this purpose. Together with France's overseas voyages in later decades, it is no surprise that by mid-century the museum in Paris had by far the world's largest natural history (and herpetological) collections. Geoffroy became blind in 1840. The next year he retired his professorship and was succeeded by his son, Isidore Geoffroy Saint-Hilaire, who continued some of his father's work including that on reptiles. The elder Geoffroy became paralyzed in 1842 and died in Paris on 19 June 1844.

Geoffroy became one of the museum's greatest scientists. He authored some 300 titles on zoology, anatomy, physiology, embryology, and teratology. His theoretical work was particularly important. He believed in the transmutation of species from a single organizational plan, which brought him into conflict with Cuvier who held that species were immutable and that each one had its own unique and permanent form. The eight public debates early in 1830 between these two erstwhile colleagues became one of the most furious scientific exchanges of the century and reptiles played a key role.

When Geoffroy was in Egypt he obtained living and fossil specimens as well as 3,000-year-old mummies of crocodiles and turtles. He and Cuvier made detailed studies of these and other living species in order to find evidence for their positions on transformation of species. Coincidentally they named new taxa. Geoffroy described the turtle genus *Trionyx* and the West African race of the Nile crocodile (*suchus*). Of some 35 herpetological titles (1803–1838) by Geoffroy, 21 are on the anatomy, osteology, and biology of crocodilians. Whereas Cuvier's views on evolution of species became obsolete after Charles Darwin, many of Geoffroy's are regarded as prescient. In 1833, for example, Geoffroy wrote (in translation): "The external world is all-powerful in alteration of the form of organized bodies . . . these [modifications] are inherited, and they influence all the rest of the organization of the animal, because if these modifications lead to injurious effects, the animals which exhibit them perish and are replaced by others of a somewhat different form, a form changed so as to be adapted to the new environment." Darwin cited Geoffroy in his works and noted his insights on the importance of determining homologies.

The most elaborate herpetological work authored by Geoffroy, in this instance by father and son, was the extensive amphibian and reptile section (1827; second edition 1829) in the great series "Description de l'Egypte," published by order of Napoleon, which consisted of nine folio volumes of text and 14 elephant folio volumes of plates—all produced in a grandiose format. The plates are among the largest ever produced for a natural history work and measure 55 by 70 cm. There are 13 herpetological plates including one with a magnificent lifesize cobra, *Naja haje*. Besides the Geoffroys, other parts of this section were written by two assistants at the museum, J.-V. Audouin who was the son-in-law of Alexandre Brongniart, and J. C. Savigny who participated in the expedition to Egypt. Many new reptiles were named in this work including the gecko *Tarentola annularis* and the snakes *Coluber florulentus, Echis pyramidum,* and *Macroprotodon cucullatus*. This contribution represents the first summary of the herpetology of Egypt.

• *References*: "Vie, Travaux et Doctrine Scientifique d'Étienne Geoffroy Saint-Hilaire," by I. Geoffroy Saint-Hilaire, P. Bertrand, Paris, 479 pages, 1847; "Éloge Historique d'Étienne Geoffroy Saint-Hilaire," p. 229–284 *in* P. Flourens, Recueil des Éloges Historiques, vol. 1, G. Frères, Paris, 1866; "The Confiscation and Replacement of Herpetological Collections by France . . . ," by P. Daszkiewicz and A. M. Bauer, News-Bull. Internatl. Soc. Hist. Biblio. Herpetol. (now Biblio. Herpetol.), 4(1): 15–27, 2003; Jaussaud and Brygoo, 2004 (Geoffroy on pp. 243–246). • *Portrait*: Adler collection. • *Signature* (1809): Courtesy Roger Bour.

DUVERNOY, G. L. (1777–1855).

The French physician and anatomist, Georges Louis Duvernoy (see *Note*), made significant contributions to the knowledge of amphibian and reptilian anatomy. He was born on 6 August 1777 in Montbéliard (now in Doubs), the birthplace of Georges Cuvier. (Montbéliard was ruled by the dukes of Württemberg for four centuries, but in 1793 it was annexed to France.) He was the son of a Lutheran pastor and a distant cousin of Cuvier who was the major influence in his career. After studying classics in Montbéliard, he attended the Caroline Academy in Stuttgart (1792–1794), as did Cuvier, but after the annexation of his homeland he studied medicine in Strasbourg and Paris. Following military duty, he defended his thesis (M.D. 1801) and then set up a medical practice in Montbéliard, but in 1802–1805 worked with Cuvier in Paris before resuming his practice. He returned to Paris in 1809 to teach, arranged by Cuvier, but his wife refused to move and he was forced to return home. Upon the death of his first wife in 1827 he joined the faculty at the university in Strasbourg, a post also arranged by his cousin, first in zoology, then in medicine, and by 1832 he was dean of the science faculty. After

his mentor's death in 1832, Duvernoy was passed over as his successor to be professor of comparative anatomy at the Museum d'Histoire Naturelle in favor of Henri de Blainville. In 1837 Duvernoy returned to Paris as professor of natural history at the Collège de France and finally, in 1850, he succeeded to Cuvier's chair on the death of Blainville. He died in Paris on 1 March 1855.

Duvernoy made contributions to the anatomy, physiology, and paleontology of animals. He collaborated with Constant Duméril in collecting and editing Cuvier's "Leçons d'Anatomie Comparée" ([1800–] 1805). He conducted a diverse array of studies, for example, on the teeth of shrews, the nervous system of mollusks, osteology of cows, fossil rhinoceros, and reproduction in animals generally. His herpetological contributions were equally diverse (about 20 titles, 1804–1849) and included the functional morphology of tongues (especially in chameleons), respiration in turtles, the systems involved in swallowing, and heart structure in crocodilians. During 1831–1833 he published several papers on the anatomical differences between venomous and non-venomous serpents. As part of this work, he named the genus *Dispholidus*, the boomslangs (1832). His major technical work, "Fragments sur les Organes Génito-Urinaires des Reptiles et Leurs Produits" (1848, reissued 1851), despite its title, deals almost exclusively with salamanders (*Salamandra, Triturus*). The two exquisitely detailed, hand colored plates show the anatomy of the urogenital systems and the structure of sperm. There is also a short section on concretions in the urinary tracts of softshell turtles.

Duvernoy's most sumptuous herpetological book was the section entitled "Les Reptiles" in the so-called disciples's edition of Cuvier's "Le Règne Animal," a series of 20 volumes (1836–1849) issued after Cuvier's death by 11 of his former students and associates. The reptile section, which includes amphibians and is normally bound in two volumes, was issued in 13 parts (1837–1842). It includes 46 mostly hand colored plates that rank among the most outstanding illustrations ever produced on the subject, as would be expected in tribute to France's finest naturalist.

• *References*: "Georges Louis Duvernoy 1777–1855," p. 1309 *in* P. Tort (ed.), Dictionnaire du Darwinisme et de l'Èvolution, vol. 1, Presses Univ. France, Paris, 1996; Jaussaud and Brygoo, 2004 (Duvernoy on pp. 203–204). • *Portrait* (1838): Muséum National d'Histoire Naturelle, Paris. • *Signature* (1847): Natural History Museum, London. • *Note*: Sometimes his surname is spelled Duvernois or DuVernoy and his given names reversed, although "G. L. Duvernoy" is how he routinely signed his name on his publications.

OKEN, Lorenz (1779-1851).

The German philosopher of nature, Lorenz Oken (born Okenfuss), a major contributor to the development of fundamental biological concepts during the first half of the 19th Century, also published on amphibians and reptiles. He was born to poor farmers in the Black Forest on 1 August 1779 in Bohlsbach bei Offenburg (Baden-Württemberg) near the French border and Strasbourg. He went to the university in Freiburg (M.D. 1804). After a year in Würzburg, he became a lecturer at the university in Göttingen (1805–1807) where he came under the influence of Johann Blumenbach. He had already published an important book, which made him a leader of the "Naturphilosophie" group in Germany, in which he extended the philosophical principles of Immanuel Kant to the natural sciences. In "Grundriss der Naturphilosophie" (1802) he organized all animals into five classes, stating that (in translation): "the animal classes are virtually nothing else than a representation of the [five human] sense-organs, and that they must be arranged in accordance with them." Reptiles and amphibians composed his class Rhinozoa in which the nose opens for the first time into the mouth and inhales air. This division of organisms into philosophically based categories was carried to lower levels of classification, which led to the establishment of categories and names to complete the predetermined pattern. Among herpetologists, Leopold Fitzinger, J. J. Kaup, and J. B. von Spix were the most notable adherents of this absurd idea.

In 1807 Oken became a professor at the university in Jena but his strong political views caused his academic career to be disrupted. In 1817 Oken founded a new journal, *Isis*, or as it is often cited, *Isis von Oken*. Although Oken wrote most of the articles the journal was open to all and was an outlet for papers by herpetologists from the German-speaking parts of Europe: the Boie brothers (Friedrich and Heinrich), Fitzinger, Kaup, Blasius Merrem, Hermann Schlegel, A. F. A. Wiegmann, and Oken himself. It became the primary forum for several decades where authors could publish and defend their work and critique that of others. But the *Isis* also carried Oken's political commentary on the revolutionary youth movements sweeping across Germany, which were severely repressed by the state governments. One state, Weimar, called upon Oken to censor his work or resign from his professorship. Rather than submit to censorship, he left Jena in 1819 and eventually accepted a lower-ranking professorship in

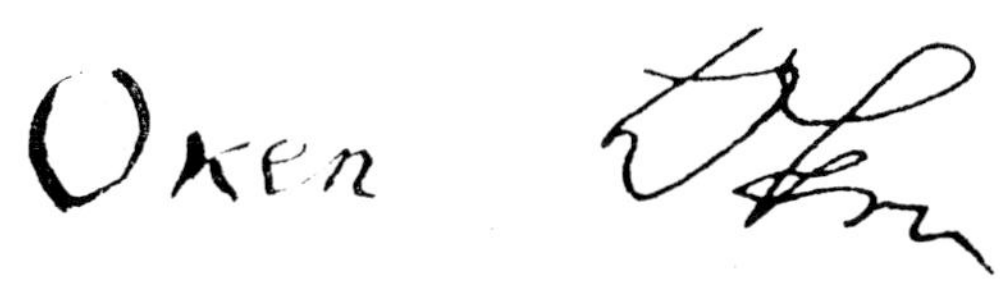

Munich. He re-established his teaching and research there, had doctoral students such as Louis Agassiz, and counted J. G. Wagler among his colleagues, but he continued his political pronouncements. When the Bavarian government requested that he move to a more isolated provincial university in the kingdom, he resigned in 1832 and found his final refuge in Switzerland at the newly formed university in Zürich where the zoologist, H. R. Schinz, was one of his colleagues. He resumed his academic work, had doctoral students including J. J. von Tschudi, and edited *Isis* until it was terminated in 1848. He died in Zürich on 11 August 1851.

Oken focused on fundamental concepts in biology at a time when the field was in ferment, before Theodor Schwann proposed the cell theory (1839) and Charles Darwin and Alfred Wallace enunciated the mechanisms of evolution through natural selection (1858). Oken wrote extensively on the origin of life, how organisms developed from simple to more complex forms, and on the structural units of living organisms. His particular views have not prevailed but through his writings he brought attention to these and other important questions. One of his theories, first announced in his inaugural lecture at Jena in 1807, was that the vertebrate skull was formed through fusion of several vertebrae, an idea to which Johann Wolfgang von Goethe, the great German polymath and politician of Weimar who had heard Oken's lecture, later claimed priority. This idea, as well as its reincarnation and reformulation by Richard Owen, were later disproved by Thomas H. Huxley (1858), with essential support from W. K. Parker. Again, Oken's notion was false, but it served the cause of science by focusing attention on evolutionary-developmental questions.

In 1822 Oken was the first to propose that general meetings and congresses of biologists and other scientists be held annually. The first such meeting was held in Leipzig in 1822 and the British Association for the Advancement of Science was declared to be organized according to Oken's idea. He was acclaimed the father of scientific meetings at the 1842 science congress in Strassbourg. Oken also contributed to natural history through his three-volume textbook, "Lehrbuch der Naturgeschichte" (1812–1826), and an illustrated encyclopedia, "Allgemeine Naturgeschichte für Alle Stände," in seven volumes (bound in 14) plus an atlas (1833–1842). Part three of volume six comprises fishes, amphibians, and reptiles (1836), with 16 herpetological plates in the atlas. Oken published a few papers strictly on amphibians or reptiles: the crocodile skull, shoulder musculature of turtles, and the nostrils of *Proteus*, among others. He named the herpetological genera *Anilius, Bombina,* and *Hemidactylus*.

• *References*: "Lorenz Oken (eigentlich Okenfuss)," by A. Lang, Allg. Deutsch. Biogr., 24: 216–226, 1887; "Lorenz Oken," by R. O[wen], Encyclop. Brit., 11th ed., 20: 55–57, 1911; "Lorenz Oken (or Okenfuss)," by M. Klein, Dict. Sci. Biogr., 10: 194–196, 1974. • *Portrait*: Adler collection. • *Signatures* (Printed form, about 1840): Adler collection; (Script form, 1847): Natural History Museum, London.

OPPEL, Michael (1782–1820).

The zoologist and artist, Nikolaus Michael Oppel, came from the Upper Palatinate, a region near the current Czech border. He was born on the farm of his parents at Schönficht (now part of Plößberg), 12 km northeast of Neustadt/Waldnaab, on 7 December 1782. His early education was in Latin, and by the time he went to Amberg for his secondary schooling in 1796 he was already demonstrating a self-taught talent for drawing. There he studied philosophy, theology, and natural sciences and briefly considered a theological career but was inspired by the writings of Georges Buffon, the French naturalist, to specialize in zoology. During walks in the countryside, he became particularly drawn to butterflies and reptiles.

In 1806 he moved to Munich and worked at the Bavarian Academy of Sciences. Recognizing his multiple talents the academy arranged for a scholarship, which allowed Oppel to study in Paris (1807–1810) under the sponsorship of the German naturalist Alexander von Humboldt. At that time Napoleonic France was the dominant European power and Paris the leading center for natural history study, led by Georges Cuvier, Constant Duméril, Étienne Geoffroy Saint-Hilaire, Count de Lacepède, and P.-A. Latreille. Oppel attended the lectures of Cuvier and Duméril and soon became the latter's student. Duméril, the greatest taxonomic herpetologist of his time, had recently published his general classification of animals, "Zoologie Analytique" (1806), and was beginning to plan his masterwork, the multi-volume "Erpétologie Générale" (1834–1854). He assigned Oppel the task of arranging the genera. What Oppel produced, however, was a more elaborate classification of amphibians and reptiles. Other Duméril students were covering turtles (A. F. Schweigger) and higher classification and anatomy (Henri de Blainville), so Oppel's first publications concentrated on snakes (published in two long papers in 1810 [1811]; re-paginated author's edition with added caption title [1811]).

The expanded and re-worked German edition (1811)—a book entitled "Die Ordnungen, Familien und Gattungen der Reptilien" and covering all amphibians and reptiles—was published at least one month later.

In these works Oppel established the orders Caudata, Squamata, and Testudinata; the families Anguidae, Cheloniidae, Colubridae, Gekkonidae, Lacertidae, Scincidae, and Viperidae; the snake genera *Tortrix, Trigonocephalus,* and *Typhlops*; and the crocodilian genus *Gavialis*. He was the first to classify *Anguis* and *Ophisaurus* as lizards because they have a sternum. His most important contribution, however, was to recognize that caecilians were amphibians and not snakes as they had long been classified. Duméril later contended that Oppel's ideas were derived entirely from his lectures in 1809, but there is a great difference between Duméril's arrangement as published in his 1806 book and that of Oppel. It seems clear that while Oppel profited from Duméril's tutelage he worked mostly independently, as Cuvier himself later acknowledged.

Early in 1811 after a study tour of France, Spain, and Italy, Oppel returned to Munich as an assistant to the academy's newly appointed curator of zoology, J. B. von Spix. Oppel resigned his post in 1818, however, to become professor of natural history at the royal lyceum in Munich and was succeeded by J. G. Wagler. Tragically, all three of these young, talented, and highly ambitious zoologists would soon be dead: Spix in 1826 from a disease contracted during his explorations in the interior of Brazil and Wagler from a self-inflicted gunshot wound suffered during a hunting accident in 1832. While still on the academy's staff Oppel had begun work on an illustrated atlas of herpetology, "Naturgeschichte der Amphibien," co-authored with Friedrich Tiedemann and Joseph Liboschitz who contributed the anatomical sections of the work. With his artistic talents Oppel was in charge of the illustrations. The first part, published in 1817, was entirely on crocodilians. This part alone, however, consists of nearly 100 folio pages of text and 15 engraved plates (12 of which are hand-colored) so the projected length of the full book can be imagined. Up to that time it would have been the most elaborate illustrated work on amphibians and reptiles, but it was not to be.

Because of the high cost of producing illustrations as well as the desire to produce the most detailed and accurate representations, Oppel learned the art of etching. This involves making a drawing in a chemical film on the surface of a copper printing plate and then incising the image with nitric acid. Apparently it was Oppel's exposure to these chemicals that eventually killed him (see *Note 1*). He died on 16 February 1820 in Munich at the age of 37. He had just married the year before. By the time of his death he had already drawn many of his subjects and these drawings are still preserved in a Munich library, but no further parts of the book were ever issued. Except for the few works mentioned his only other publications were some poems that were published posthumously (1833).

The quality of Oppel's artistry is not appreciated today because of the rarity of his illustrated works. The part on crocodilians was produced in an edition of only 150 copies and, therefore, it is one of the rarest of all herpetological titles. The drawings in it are extremely detailed, finely hand-colored, and life-like even if somewhat stylized. In 1798 a new process had been invented for printing illustrations—lithography—a method that allows fine gradations of shading and thus a better sense of depth and contour than is possible with engravings. It quickly became the predominant method for illustrating natural history books in the 19th Century. The first herpetological book entirely illustrated by lithography was issued in Munich in 1819 (see *Note 2*). Ironically, Oppel was thus a victim of a printing technique that was then being replaced by another, safer one.

With the death of Oppel in 1820 and the deaths of Spix and Wagler shortly thereafter, Munich, which briefly had become Berlin's rival as a center for herpetological research in Germany, became dormant. The academy's rich collections of amphibians and reptiles remained largely unstudied until the arrival almost a century later of Lorenz Müller, like Oppel originally an artist, who became curator in 1912.

• *References*: "[Nicol. Michael Oppel]," anonymous, Münchener Polit. Zeit., 1820(80): unpaginated and 1820(81): unpaginated, 1820; "Sonette von Bayerischen Dichtern, vol. 3" by F. A. Greger, privately publ., Regensburg, 1833 (Oppel on pp. 172–176); "Oppel," by [K.] Halm, Allg. Deut. Biogr., 24: 392, 1887; "Reptiles. History of Herpetology," by A. C. G[ünther] and A. S. Wo[odward], Encyclop. Brit., ed. 11, 23: 136–141, 1911. • *Signature* (1811): Bavarian Academy of Sciences, courtesy Bernd Görmer. • *Note 1*: The chemical products used for traditional intaglio print making produce toxic fumes and may also be absorbed through the skin. These can cause life-threatening health problems such as chemical pneumonia, and many are suspected carcinogens. In Oppel's case, it was apparently the mixture of nitric acid with copper oxide that led to his death (Greger, 1833). • *Note 2*: The first herpetological book illustrated entirely with lithographs was Karl Schmid's "Naturhistorische Beschreibung der Amphibien" (1819), which contains 54 uncolored plates. According to Nissen (1969, p. 369; repeated in Haines [2000, p. 74]), Oppel drew these plates. To the contrary, Schmid ("Naturhistorische Beschreibung der Insekten," 1822, in the introduction to the entire series, pp. ii and v) states that his brothers drew most of the plates and that he followed Oppel for the systematic arrangement (see Schmidtler, 2007, p. 116).

SCHWEIGGER, August Friedrich (1783–1821).

The German physician and naturalist, August Friedrich Schweigger, was a noted botanist and an early specialist on turtles. He was born on 8 September 1783 in Erlangen (Bavaria) about 15 km north of Nuremberg. As a boy he was an avid reader and interested in butterflies. He was educated at the local gymnasium and in 1800 entered the university in Erlangen to study medicine, botany, and zoology. He was a student of J. C. D. Schreber, a former pupil of Carl Linnaeus. Schweigger received his doctorate (M.D. 1804) and then proceeded to Berlin to take practical courses in medicine. With sponsorship from the Prussian government, he went to Paris in 1806. Despite Napoleon Bonaparte's defeat of Prussia at the battle of Jena in October 1806, he remained there to study and publish (1809) on the state of French hospitals. While in

Paris he became associated with the naturalists at the Museum d'Histoire Naturelle including Georges Cuvier, Ètienne Geoffroy Saint-Hilaire, and Constant Duméril. The latter was arranging the genera of reptiles and his students—Henri de Blainville, Michael Oppel, and J. B. von Spix—were working with him. Schweigger decided to review the turtles. He made a particular study of the skeleton and of the plates covering the shell. At that time the museum in Paris had the largest herpetological collection in the world. It had recently been augmented as a result of Napoleon's conquests when the museum authorities in Paris summarily requisitioned the collections of the conquered nations.

Schweigger's choice of group to study may have been because his former teacher, Schreber, had also been the teacher of J. D. Schoepff who earlier had monographed the world's turtles (1792–1801). Schoepff, however, had died prematurely in 1800 and Schreber edited the final part of his book. Thus, as Schweigger arrived at the university in 1800, his teacher would have been actively working on Schoepff's manuscript and had some of his specimens.

In 1809 Schweigger completed his manuscript on turtles but was called to the university in Königsberg (Prussia, now Kaliningrad, Russia) as professor of botany and director of the botanical gardens. Most of his publications from this point were botanical except for two titles on turtles and one on unsegmented invertebrates. He made numerous tours abroad—to Denmark, England, France, Italy, and Sweden—to examine collections, but his real interest was fieldwork and collecting new materials for study. He especially wanted to visit the Orient. In 1820 he planned a trip to Sicily and Greece and on 28 June 1821, while on a field excursion at Cammarata near Girgenti (now Agrigento) near the southwest coast of Sicily, he was murdered and robbed by his guide who was soon caught. He confessed, was condemned to death, and executed. Schweigger was not quite 38 years old.

Schweigger published just two titles in herpetology, both derived from his manuscript of 1809. These are nothing less than comprehensive reviews of the turtles of the world. The first, published in 1812, is a 255-page review entitled "Prodromus Monographiae Cheloniorum," published in two parts in the journal *Königsberger Archiv* and in Latin. In the span of only 37 pages it contains the original descriptions of 24 new taxa, of which half are recognized today: the tortoises *Chersine angulata* and *Kinixys erosa* and two species of sidenecks (*Pelusios*) from Africa; five species of South American sidenecks (*Phrynops, Peltocephalus,* and *Podocnemis*); the pond turtles *Melanochelys trijuga* of South Asia and *Mauremys leprosa* of the Mediterranean region; and the genus *Chelydra*, the snapping turtles of the Americas. For over a century, Schweigger's specific name, *gigantea*, was used for the giant tortoise of the Seychelles Islands. Almost all of his types are at the museum in Paris. The 1814 title, published as a small book of 64 pages, is merely a synopsis of the earlier papers but without the detailed descriptive information and with a few corrections to the earlier work. It was also published in Königsberg, but since the book is in Latin the city is called by its Roman name, Regiomonti. He had wanted to include illustrations in the format used earlier by Schoepff but his finances were too meager to permit this.

• *References*: "Schweigger," p. 415–419 *in* A. M. C. Duméril and G. Bibron, Erpétologie Générale, vol. 1. Encyclop. Roret, Paris, 1834; "August Friedrich Schweigger," by E. Wunschmann, Allg. Deutsch. Biogr., 33: 332–333, 1891; Rieck et al., 2001 (Schweigger by R. Bour and M.-N. Uhl on pp. 595–601); Damkaer, 2002 (Schweigger on pp. 198–200); "August Friedrich Schweigger," anonymous, Dict. German Biogr., 9: 241, 2005; Additional information from Roger Bour, 2006. • *Portrait and signature* (1817): Museum für Naturkunde, Berlin.

BLAINVILLE, Henri-Marie Ducrotay de (1777–1850).

Henri de Blainville, anatomist, zoologist, and successor to Georges Cuvier, wrote important works on amphibians and reptiles. He was born to lesser Norman nobility in Arques, France, on 12 September 1777. His early life was difficult, with the death of his father when he was five years old and the imprisonment of his mother until the fall of Robespierre. At the age of 28 he attended lectures at the Collège de France in Paris and discovered his love for learning. This event changed his life and for two years he devoted himself to medicine (M.D. 1808). He worked in Cuvier's laboratory and also with Constant Duméril. Blainville became an excellent comparative anatomist, but he came to oppose Cuvier and his views on animal classification. Whereas Cuvier believed in separate and multiple creations of groups without transitions, Blainville, who was Roman Catholic, held that the series of groups (*série*), were linked according to God's plan. Although Blainville's ideas about transitions may seem close to our understanding of relationships today, they flowed from his concept

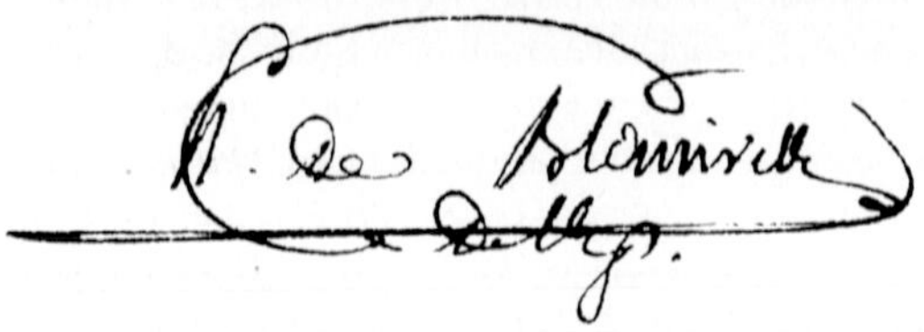

of God's plan of creation rather than from objective study. He believed that fossil forms could fill the apparent gaps in the series of living organisms.

Nevertheless, Blainville had a bold intellect and became a professor of comparative anatomy and physiology in 1812. On their respective deaths, he replaced B.-G.-È. Lacepède as an elected a member of the Académie des Sciences (1825) and then succeeded to J.-B. Lamarck's professorship (1829). Finally, despite much opposition, he gained Cuvier's chair in comparative anatomy (1832). He was a brilliant lecturer, and it was after presenting such a lecture at the Sorbonne on 1 May 1850 that he died, while traveling by train from Paris to Dieppe to visit a sick relative. His body was found in a train car just after he had boarded.

Blainville published few works exclusively devoted to herpetology but amphibians and reptiles figured importantly in much of his writing. His most important contribution to herpetology was the separation of amphibians, which he called Icthyoïdes or Nudipellifères (the scale-less reptiles), from the scaled reptiles or Squammifères (1816, 2 papers) with which they had been grouped since Aristotle. Michael Oppel, who had also been a student of Cuvier and Duméril, had only recently (1811) moved the caecilians from being a group of snakes to a kind of amphibian, and Blainville continued this arrangement, although he called them Pseudophidiens. Blainville published a number of short papers on herpetology. In 1816 he named the genus of leatherback sea-turtles (*Dermochelys*) and became the first to separate the crocodilians—which he named Emydo-Sauriens—from the lizards. He authored a paper on the caecilians and their classification in 1839. His masterwork, "Ostéographie ou Description Iconographique Comparée" issued in 24 parts (1839–1864), was never completed but contains a major section on crocodiles (1855).

His largest contribution purely on herpetology was a report (1835; reprinted 1978) on the reptiles collected in California by P.-E. Botta, the ship's surgeon, during the circumnavigation of the French ship "Héros" in 1825–1829. The ship stopped at various ports from Bodega Bay north of San Francisco to San José del Cabo at the tip of the Baja California peninsula from October 1826 to July 1828. This major paper, which begins with a detailed classification of the world's amphibians and reptiles, also contains descriptions of many new and distinctive species of western North America: the lizards *Callisaurus draconoides, Elgaria multicarinata,* and *Phrynosoma coronatum* and the snakes *Charina bottae* (named for the collector), *Lampropeltis zonata, Pituophis catenifer,* and *Tantilla planiceps*. Blainville also named the California race of *L. getula*, based on its striped phase. The new forms were well illustrated on four quarto plates drawn by the great French natural history artist, J. G. Prêtre.

• *References*: "Étude sur la Vie et les Travaux de M. Ducrotay de Blainville," by H. C. M. Nicard, J.-B. Bailliere, Paris, 253 pages, 1890; "Henri Marie Ducrotay de Blainville," by W. Coleman, Dict. Sci. Biogr., 2: 186–188, 1973; Damkaer, 2002 (Blainville on pp. 241–249); Jaussaud and Brygoo, 2004 (Blainville on pp. 79–81). • *Portrait and signature*: Library, Muséum National d'Histoire Naturelle, Paris.

RUSCONI, Mauro (1776–1849).

The comparative anatomist and amphibian biologist, Mauro Rusconi, who was one of the founders of modern embryology, became arguably the best-known Italian scientist of his generation. He was born on 18 November 1776 in Pavia to a merchant family of modest means. After local schooling, in 1796 he joined an army that came to serve the Cisalpine Republic (a satellite republic of France established in 1797 by Napoleon Bonaparte in northern and central Italy with its capital at Milan). He quickly attained the rank of artillery captain. After the surrender of the French garrison at Mantua in 1799 to Austro-Russian forces, he abandoned his military career to study medicine at the Università di Pavia.

At that time Pavia was one of Europe's greatest universities where the biologist Lazzaro Spallanzani had recently introduced the experimental method into the curriculum. Rusconi learned to make anatomical preparations that involved not only careful dissections but also extremely fine injections of vessels. He also practiced microscopical techniques and how to draw in fine detail. He received his medical degree at Pavia in 1806 and a license to practice in 1809, but soon he abandoned clinical work to become preceptor (or demonstrator) in physiology and comparative anatomy at the university in 1811. He spent the next year in Paris studying with Georges Cuvier, then the leading comparative anatomist in the world. On returning to Pavia his training and position should have assured him of an illustrious academic career, but this was not

to be. In 1819 the university's curriculum was changed and Rusconi lost his post. He was not offered a new one.

Without the benefits of an academic affiliation, Rusconi retired to private life and personal study at his home. There he continued his field studies and laboratory work on amphibians. Since he was extremely sensitive to cold, he would spend entire winters closed up in his house. This permitted long periods for concentrated study and the production of several books and many papers on the natural history, anatomy, and development of amphibians and reptiles. Collectively these works earned him an international reputation as Italy's finest contemporary biologist, but he was appreciated more in other countries—where he earned prizes, honorary degrees, and other accolades—than in his own homeland.

Rusconi's observations and discoveries concerning the development of frogs and, by extrapolation, of all vertebrates were of the highest order. Jan Swammerdam, the noted Dutch anatomist, observed early cleavage in a frog embryo, the first for any vertebrate animal, but he did not understand its significance. Rusconi was the first to note that cleavage was not simply a surface phenomenon but that it cut up the yolk. He wrote (in translation): "These divisions and subdivisions represent an activity by which nature distributes the elementary molecules of the main systems." This led to the discovery by others of the relationship between cleavage planes and the embryonic axes. He discovered that the neural folds, which meet, close, and thus form a canal, were the primordia of the central nervous system, brain, and spinal cord. Rusconi also discovered that the blastopore, which results from gastrulation, becomes the future anus. Indeed, for many years the blastopore was called "Rusconi's anus" in medical dictionaries. As he first described it (in translation): "At 17 hours after laying, an arc-like furrow [the blastopore] . . . marks the boundary between the brown and whitish parts of the ovum. This furrow is the first trace of the anus."

Unfortunately Rusconi was exposed to dramatic political and military events toward the end of his life. Most of northern Italy was controlled by Austria and, following the February 1848 insurrection in Paris and proclamation of the French Republic, revolt spread across Germany and reached Vienna by March. Italian partisans soon realized this could be the moment to force the preoccupied Austrians to leave. Austrian troops withdrew eastward, but by August the so-called "First War of Italian Independence" was lost. Rusconi left Pavia in August 1848 to move northwest and away from Austrian-controlled areas and to the shores of Lake Como, located north of Milan near the Swiss border, where he spent the remaining months of his life. He had suffered great pain from a bladder disease during his last years. On the morning of 27 March 1849, having been delirious with fever the night before, he died alone in Cadenabbia.

Rusconi published more than 40 titles, 27 of them specifically on herpetological topics (1817–1854) and these included seven books. He drew and personally hand colored the highly detailed plates in each copy. His herpetological work covered a broad range of topics, among them the development of frog and salamander embryos and the metamorphosis of their larvae; artificial fertilization in frogs; courtship and fertilization in salamanders; reproduction in the Olm (*Proteus anguinus*); respiratory organs and respiration in frog tadpoles; the lymphatic systems of salamanders and frogs as well as sea turtles and some other reptiles; and the prey-capturing behavior of chameleons. His first book (1817) covered the descriptive anatomy of the circulatory system of newts (*Triturus carnifex*). His second book (1819; English summaries in two papers in 1821), co-authored with his former university colleague, Pietro Configliachi, dealt with the olm, which was a subject of interest to Rusconi throughout his life (see page 6). Rusconi and Configliachi initiated their fieldwork in August 1816 when, furnished with torches and a small net, they descended into the cavern of Maddalena (now Črna Jama), located in Slovenia between Trieste and Ljubljana. There, in a pool nearly 140 m below the surface and at air and water temperatures of about 12°C, they saw their first *Proteus* in the wild. In their book, they described the natural history and anatomy of this cave-adapted salamander, which retains its gills throughout life, and confirmed Cuvier's claim that it reproduced in this gilled state. Rusconi's third book (1821), "Amours des Salamandres Aquatiques," describes the courtship and egg-laying behavior of newts (*T. carnifex*) and includes the first illustrations of these behaviors ever published for any salamander.

Rusconi's fourth and most important book (1826) describes the development of the Edible Frog (*Rana esculenta*). It was in this short book (61 pages and 4 plates) that he reported his critical discoveries about frog embryos noted earlier. It won for him the gold medal of the Institut de France in 1831. In 1837 his fifth book was issued, on the anatomy of the eel-like *Siren intermedia*, in which he challenged Cuvier's classification of this animal as the larval stage of an as-yet undiscovered species. His sixth book (1845), on the lymphatic system of reptiles, resulted from a public controversy with an Italian anatomist, Bartolomeo Panizza, about whether blood vessels are inside the lymphatics or not. Rusconi was wrong (they are not, for they are only invaginated into the lymphatics), but the controversy spurred him to conduct detailed studies and resulted in this book. His seventh and last book (1854) was issued posthumously under the editorship of Giuseppe Morganti. This treatise reported Rusconi's research, completed by 1839 and published only in a short journal article that year, on the natural history, development, and metamorphosis of *Salamandra salamandra*. Rusconi was the first to discover that mating occurred on land, but he did not understand how fertilized eggs could be present in the oviducts before mating took place. (The discovery of the function of spermathecae, which permit internal fertilization, was not published until 1858 by Karl T. E. von Siebold.)

Rusconi's books were issued in very small editions of from 25 to 100 copies (except for the first, of which there were apparently 150 copies), probably because he personally hand colored all of the illustrations. They are not widely consulted today, in part because of their rarity (see *Note*), but his reputation as the first great amphibian biologist is secure.

• *References*: "Sulla Vita Scientifica e Sulle Opere di Anatomia e Fisiologia Comparata del Dottor Mauro Rusconi," by S. Biffi, Soc. Edit. Ann. Univ. Sci. Indust., Milan, 167 pages, 1853; "La Vita e le Opere di Mauro Rusconi (1776–1849)," by G. Cattaneo, Monit. Zool. Ital., 36: 1–5, 1925. • *Portrait*: From Biffi (1853), courtesy Division of Rare and Manuscript Collections, Cornell University Library. • *Signature* (1822): Adler collection. • *Note*: W. Junk, "Rara Historico-Naturalia," vol. 2, W. Junk, Berlin, 1926 (p. 156), announced a full-color facsimile of all of Rusconi's books on amphibians and reptiles. It was never published.

LESUEUR, Charles-Alexandre (1778–1846).

Charles-Alexandre Lesueur (see *Note*), French zoologist and artist, was a student of North American turtles and fishes. He was born on 1 January 1778 in the port city of Le Havre on the north coast of France. His father was a naval officer and Charles himself began a military career, but being interested in art and natural history, he volunteered for the French voyage to explore Australia and Tasmania (1800–1804) led by the infamous Nicolas Baudin. Few expeditions had such an auspicious start. The naturalists on board included the botanist André Michaux and the zoologists J.-B. Bory de Saint-Vincent, François Péron, and Lesueur, who was officially a gunner but in effect the ship's artist. Bory and Michaux jumped ship in Mauritius because of the cruelty of Baudin. Half the crew of the two ships eventually died during the voyage as well as Baudin himself before the ships returned. Lesueur was severely tested, yet his artistic talents were honed and he and Péron gathered enormous collections that they took to Paris. Lesueur collaborated with Louis Bosc, A. G. Desmarest, and Constant Duméril at the museum in Paris during this period.

In 1815 Lesueur accepted an invitation from an American who was visiting Paris, William Maclure, to join him first in the West Indies and then in Philadelphia where Maclure lived. Thus, this seasoned zoologist-artist and world traveler came to join the Philadelphia scientific community in 1816. Lesueur, in fact, became one of the founders of the Academy of Natural Sciences, which was started in 1812 and incorporated in 1817. This founding group included several persons of herpetological interest: Thomas Say, Gerard Troost, and Titian R. Peale, as well as Maclure and Lesueur. Other naturalists, such as John J. Audubon, C. L. Bonaparte, and C. S. Rafinesque, also attended the early meetings of the academy.

It was Maclure who encouraged Lesueur, Say, and Troost to join the utopian community at New Harmony in southern Indiana where Lesueur resided during 1826–1837. Lesueur traveled widely during this period, especially along the Mississippi River from Saint Louis to New Orleans. He returned to Paris in 1837 and in 1845 became director of the new museum in Le Havre, but he died in Le Havre the next year (12 December).

Lesueur's contributions to herpetology were secondary to his interests in aquatic animals, particularly fishes, so it is probably not coincidental that his main work with reptiles concerned freshwater turtles. He named three widely distributed turtles of eastern United States: the map turtle, *Graptemys geographica*, and two softshells, *Apalone mutica* and *A. spinifera*. The map turtle was described in the very first issue of *Journal of the Academy of Natural Sciences* (1818) and included a fine engraving made from Lesueur's own drawing. He published a half dozen other herpetological papers (1818–1839), mostly in Parisian journals, which included the descriptions of the two softshells (1827).

• *References*: "Charles Alexander Lesueur," anonymous, Natl. Cyclop. Amer. Biogr., 8: 475–476, 1898; "Charles-Alexandre Lesueur Voyageur et Peintre-Naturaliste Havrais (1778–1846)," by A. Maury, French Amer. Rev., Washington, pp. 161–171, 1948 (July-Sept.); "Charles Alexandre Lesueur," by P. T. Stroud, Amer. Natl. Biogr., 13: 523–525, 1999; Damkaer, 2002 (pp. 216–222); "Charles Alexandre Lésueur en Amérique du Nord (1816–1837)," by J. Bonnemains, Ann. Mus. Havre, 74: 1–81, 2005. • *Portrait* (1818): From "Sketch of Charles A. Le Sueur," by D. S. Jordan, Pop. Sci. Month., 46(February): 547–550, 1895, courtesy David Damkaer. • *Signature* (1827): Academy of Natural Sciences, Philadelphia. • *Note:* The family name is often spelled Le Sueur or LeSueur, but on his own publications he spelled it "Lesueur."

GREEN, Jacob (1790–1841).

Although primarily a chemist by profession, Jacob Green became the first specialist on North American salamanders. He was born in Philadelphia on 26 July 1790, the son of a Presbyterian clergyman who in 1812 became president of the College of New Jersey (now Princeton University) and grandson of a Revolutionary patriot who earlier had been president of the same institution. Green received his education at the University of Pennsylvania (A.B. 1807, A.M. 1810), then studied medicine (see *Note 1*) and law. He practiced the latter briefly before becoming professor of chemistry and natural history in Princeton (1818–1822), finally returning to Philadelphia where he had the same title at Jefferson Medical College (1825–1841). Although he was interested in botany as a boy, as a professional he published on a wide variety of subjects including chemistry, paleontology (especially trilobites), and ethnology. He was informally associated with the Academy of Natural Sciences and died in Philadelphia on 1 February 1841.

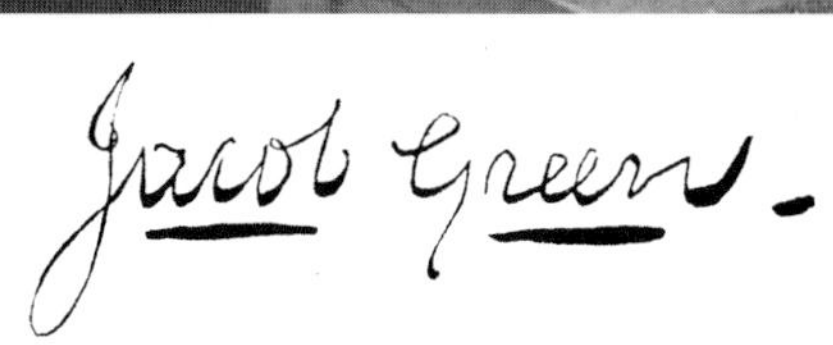

Green published eight papers on herpetology (1818–1831). Except for one on the bones of rattlesnakes and another on Mexican lizards co-authored with the artist-naturalist Titian R. Peale, all of the rest focused on salamanders. Nine well-known species found in eastern North America were named by Green: *Ambystoma jeffersonianum* (see *Note 2*), *A. tigrinum*, *Desmognathus fuscus* (see *Note 3*), *Eurycea bislineata*, *E. cirrigera*, *E. longicauda*, *Gyrinophilus porphyriticus*, *Plethodon cinereus* (see *Note 4*), and *P. glutinosus*. Two local naturalists, Thomas Say and Richard Harlan, called Green's taxonomic judgment into question in print, but he knew salamanders better than they. He had the advantage of seeing them in the field in their natural habitats and with their living colors and shapes. Green planned to do a comprehensive monograph on the subject, as he boldly announced in an open letter to Georges Cuvier published in 1830, but it was never issued.

• *References*: "Jacob Green, 1790–1841, Chemist," by E. F. Smith, published by author, Philadelphia, 34 pages, 1923 (abridged in Jour. Chem. Educ., 20: 418–427, 1943); "Jacob Green," by H. B. Baker, Dict. Amer. Biogr., 4: 548–549, 1931; "Jacob Green," by W. D. Miles, Dict. Sci. Biogr., 5: 517–518, 1972. • *Portrait and signature*: From Goode, 1901. • *Note 1*: Baker (1931) noted that Green was "turned aside from medicine by the crude surgery of the day" and that his M.D. from Yale University in 1827 was honorary. However, as early as 1818 Green listed himself as an M.D. in his publications. • *Note 2*: Although Green (1827) did not say so explicitly, because he collected the animals near the campus it seems likely that this species was named for Jefferson College of Canonsburg, located in westernmost Pennsylvania (merged with Washington College in 1869 to become Washington and Jefferson College of nearby Washington, Pennsylvania). Thus, only indirectly is the salamander named for Thomas Jefferson, third president of the United States. Coincidentally, Jefferson Medical College in Philadelphia, where Green taught for most of his career, was founded in 1825 under the charter of Jefferson College. Green was one of its founders and sometimes taught chemistry during the summer at Jefferson College, which likely explains his field work along Chartiers Creek near Canonsburg (see "Old Jakey Green at Canonsburg," by G. W. Bennett, Proc. Penn. Acad. Sci., 23: 218–221, 1949). • *Note 3*: Some authors (e.g., B. I. Crother et al., Herpetol. Rev., 34: 198, 2003) regard Green's *Salamandra fusca* as unidentifiable and credit the name instead to C. S. Rafinesque ("Annals of Nature, no. 1," publ. by author, Lexington, Kentucky, 1820; reprinted 1962). Rafinesque's specimens were from northern New York State, thus his description could also apply to the sympatric *Desmognathus ochrophaeus*. Crother et al. were perhaps influenced by Rafinesque's words "tail acute" as applying to the tail's edge, but taken literally, they mean ending in a sharp point, not keeled. • *Note 4*: In his description of *Salamandra erythronota*, which is a synonym of *Plethodon cinereus*, Green (1818) credits the name to Rafinesque. Only in 1960 was Rafinesque's paper of 1818 re-discovered ("Unrecorded Papers of Rafinesque and Jacob Green," by G. H. Goodwin, Jr., Syst. Zool., 9: 35–36, 1960). Rafinesque's *erythronota* clearly has priority over Green's *cinereus*. It was suppressed by the International Commission on Zoological Nomenclature (Opinion 665) in the interest of stability.

RATHKE, M. Heinrich (1793–1860).

Together with Karl von Baer, Christian Pander, and Mauro Rusconi, Martin Heinrich Rathke is recognized as one of the founders of modern embryology. He was born into a wealthy family in Danzig, Prussia (now Gdańsk, Poland) on 25 August 1793. After studying with J. F. Blumenbach at Göttingen, he went to Berlin where he completed his medical degree in 1818. His dissertation, published the same year, comprised the first comparative study of the urogenital system of urodeles including detailed anatomy of the ovary and testis as well as a "gland" in females that is now known as the spermatheca. Rathke failed, however, to notice sperm inside or to understand its function in storing sperm for later reproduction, discoveries made later by Franz von Leydig (1853) and Karl T. E. von Siebold (1858), respectively. His dissertation also contained the first description of the development of fat bodies (and their relationship to general metabolism) and the fragmented adrenals of salamanders. Although his research was primarily embryological, amphibians and reptiles continued to figure prominently in his research.

After Berlin he returned to Danzig where he practiced medicine, but he continued his research even when he became district physician in 1826. It was during this period that he made his most seminal discovery, the existence of gill slits and gill arches in the embryos of birds and mammals. In 1829 he moved to Dorpat (today Tartu, Estonia) in the Russian Empire as professor and then, in 1835, to the university in Königsberg (now Kaliningrad, Russia) where he succeeded von Baer and remained for the rest of his life. He died there on 15 September 1860, much honored as a scientist and also for his influence on the development of his university. He studied many subjects including the embryology of air sacs in birds, the regressive development of tadpole tails, the relationship of kidneys to Wolffian bodies (which he discovered), and the true identity of lancets or amphioxus, which previously had been thought to be a larval mollusk. Rathke's pouch, a pocket marking the point of invagination of the hypophysis in vertebrates, is named for him.

Rathke made numerous contributions to the development and anatomy of amphibians and reptiles. Besides his medical thesis on salamanders noted previously, he published books on the development of snakes (1839), turtles (1848), and crocodiles (1866), the sternum of lizards (1853), and wrote long journal articles on the circulatory systems of snakes and lizards (1856, 1857, 1861). He discovered the paired exocrine organs—Rathke's glands—embedded in the bridge of most aquatic turtles (1848). His book on the Crimean fauna, "Zur Fauna der Krym" (1836), has significant herpetological content. Rathke's other notable contribution of interest was his editorship (and primary authorship) of part 5 of "Zoologischer Atlas" (1833), a series begun by his then-deceased university colleague, J. Friedrich Eschscholtz. This part deals almost entirely with Californian salamanders and includes Rathke's description of the newt, *Taricha torosa*.

• *References*: "Heinrich Rathke,"by G. Zaddach, Neue Preuss. Provinc. Blätt., ser. 3, 3: 271–312, 1860; "Rathke," by L. Stieda, Allgem. Deutsch. Biogr., 27: 352–355, 1888; "Martin Heinrich Rathke," by V. L. Bullough, Dict. Sci. Biogr., 11: 307–308, 1975; Menz, 2000. • *Portrait*: Tartu University Library, courtesy Mare Rand (photo 4005:1A). • *Signature* (1844): Adler collection.

BORY DE SAINT-VINCENT, J.-B. G. M. (1778-1846).

Jean-Baptiste Geneviève Marcellin Bory de Saint-Vincent (see *Note 1*), naturalist, explorer, and soldier who made significant contributions to our knowledge of islands and their biota, published on amphibians and reptiles. He was born a baron into a respectable family of lawyers on 6 July 1778 in Agen (Lot-et-Garonne), southwestern France. As a young boy he developed an interest in plants and insects, and by the age of 16 he was corresponding with Count de Lacepède and other leading naturalists in Paris. Because of these connections he was able to play a key role in the release of the priest-naturalist P.-A. Latreille from prison in 1795 by conveying information from Latreille to influential naturalists. Soon, however, his own family got caught up in the Reign of Terror; his father and uncle were imprisoned but later freed. Jean-Baptiste found refuge in southwestern France. He joined the army in 1799 and remained an officer until he retired as colonel in 1815, but was often seconded for other duties. In 1800, for example, as a zoologist he joined the French expedition to Australia led by Nicolas Baudin, but because of Baudin's despicable behavior he jumped ship in Mauritius and spent the next year exploring the Mascarene Islands in the western Indian Ocean. After returning to France in 1802, Bory served in campaigns in Austria, Germany, and Spain (1803–1813), where he served in the cavalry—he even made botanical collections on the battlefields—and was involved in drafting topographic maps for

Napoleon Bonaparte's general staff. In 1815 he was elected to the Chamber of Deputies from Lot-et-Garonne and actively supported Napoleon. After the emperor's final fall in June 1815, Bory was banished from France. He traveled to Berlin but lived mainly in Brussels and, deprived of his army pension, supported himself by writing and editing books and journals.

After the general amnesty in 1819, Bory returned to Paris where he continued his natural history work, but he did not pay his bills and spent 1825–1828 in debtor's prison. In 1829 he joined the French expedition to Peloponnese, a peninsula in southern Greece known as Morea during the Middle Ages. Bory studied natural history from his teenage years and throughout his life, even when he was engaged in military campaigns or in prison. On Lacepède's recommendation in 1808 Bory was elected a corresponding member of the Académie des Sciences; he became a full member in 1834, in large part because of the success of the Morea expedition. The government asked him to lead an expedition during 1840–1842 to explore the new French possessions in Algeria (the herpetological section of the report, by Alphonse Guichenot, was published in 1850). Suffering from a heart ailment, Bory spent his last days at home still working on his natural history projects. He died in Paris on 22 December 1846 (see *Note 2*).

Bory's written output in natural history is substantial. While in exile during 1819–1821 he co-founded and co-edited the journal *Annales Générales des Sciences Physiques* and from 1822 to 1831 edited the 17-volume "Dictionnaire Classique de l'Histoire Naturelle" for which he also wrote numerous accounts. His most important contributions dealt with the origin of islands and their biotas (1802, 1803). It was Bory, for example, who was the first to note that amphibians were absent from oceanic islands. His book on the zoology of man (1827), following Johann Blumenbach, classified the species of humans and their relationships to other animals. His herpetological titles are few but substantial. Besides a short paper on salamanders (1798) and another describing a new snake from South America (1823), Bory's writings on amphibians and reptiles are contained in larger works. As one of a series of 51 volumes in the pocket-sized "Encyclopédie Portative," Bory contributed the herpetology, entitled "Résumé d'Erpétologie, ou d'Histoire Naturelle des Reptiles" (1828); another version was issued the same year with the title "Précis d'Erpétologie ou d'Histoire Naturelle des Reptiles." This volume includes 52 lithographic plates, but these were also issued separately under the title "Iconographie des Reptiles" (1828) in a hand-colored edition. The book was reissued in 1842 in identical form by a different publisher but with the title "Traité Élémentaire d'Erpétologie ou d'Histoire Naturelle des Reptiles." An Italian edition appeared in 1835. Bory and the French herpetologist, Gabriel Bibron, co-authored the section on reptiles (and amphibians) and fishes in the zoology report of the "Expédition Scientifique de Morée" (1833). In it they reviewed Bory's collections from Greece and named several new reptiles, of which the following are recognized today: *Ablepharus kitaibelii, Algyroides moreoticus, Ophiomorus punctatissimus,* and *Podarcis peloponnesiaca.*

• *References*: "Notice Historique sur Bory de Saint-Vincent," by A. Lacroix, Inst. France, Acad. Sci., Paris, 75 pages, 1916; "Jean Baptiste Georges Marie Bory de Saint-Vincent," by W. George, Dict. Sci. Biogr., 2: 320–321, 1970; Damkaer, 2002 (Bory on pp. 222–227). • *Portrait and signature*: From Lacroix, 1916. • *Note 1*: Bory's forenames are sometimes given in a different order or as Jean Baptiste Georges Marie. • *Note 2*: Some sources give Bory's death date as 26 December.

KUHL, Heinrich (1797–1821). HASSELT, Johan Coenraad van (1797–1823).

Indonesia, formerly the Dutch East Indies with its megadiversity of amphibians and reptiles, has long been a rich prize for herpetologists. The East Indies was originally managed by the Vereenigde Oostindische Compagnie (United Netherlands East Indian Company), but after the final defeat of Napoleon Bonaparte in 1815 and the restoration of The Netherlands as an independent nation, the East Indies became a true colony. In order to improve the nation's economy, which had deteriorated under French rule, the king sent a naturalist, C. G. C. Reinwardt (see *Note*), as director general of agriculture to the Indies to evaluate the situation. It was soon realized that more than a single person was required to properly study the natural products of the Indies and thus a commission was established for the purpose by royal decree, the Natuurkundige Commis-

sie voor Nederlandsch Indië (Commission for the Study of the Natural Sciences of the Netherlands East Indies). The oversight committee of four Dutch naturalists included Reinwardt, C. J. Temminck, and Jan van der Hoeven. It was decided that the first team to be sent would consist of two naturalists, a preparator, and an artist. The naturalists chosen were Heinrich Kuhl, a German, and Johan Coenraad van Hasselt, a Dutchman, who were then fellow medical students at the University of Groningen in The Netherlands. These two young and very promising zoologists, whose names will be forever linked, were duly appointed. Tragically, after reaching Java both were dead within three years.

Kuhl was born 17 September 1797 in Hanau am Main (Hesse). His father was a judge and Heinrich began to study law but switched to medicine because of an interest in natural history. Theodor van Swinderen, professor of natural science at Groningen, had met him and being impressed with his knowledge offered him an assistantship at the Groningen Museum. Here Kuhl met Van Hasselt who was born on 24 June 1797 in Doesburg (Gelderland). The two became fast friends and both entered the medical school at Groningen (both earning M.D.s in 1820). It was Van Swinderen who mentored the two in natural history and Temminck arranged for their appointments to the commission in 1819. During their medical studies Kuhl visited London with Temminck for four months in 1819 to study natural history collections. Later in that year Kuhl and Van Hasselt traveled to Paris to meet Georges Cuvier, J.-B. Lamarck, and other French naturalists at the Muséum d'Histoire Naturelle and to study the collections there. By mid-1820 they were well prepared to depart for Java.

The Dutch ship "Nordloh" left the island of Texel (North Holland) on 11 July 1820 with the commission of four men on board. En route to Java they went ashore at Madeira Island off Morocco, the Cape of Good Hope (then a Dutch colony), and the Cocos Islands in the eastern Indian Ocean where they collected. The ship arrived in Batavia (now Jakarta) at the end of December 1820. A house in Buitenzorg (now Bogor), West Java, was assigned as the commission's headquarters and they immediately set about collecting, taking notes, and making drawings. Less than nine months after arrival Kuhl developed an inflammation of the liver after returning from a strenuous collecting trip. He died in Buitenzorg on 14 September 1821, three days short of turning 24 years old. The commission's artist, Gerardus Keultjes, died four days later. Van Hasselt and the preparator, Gerrit van Raalten, continued until Van Hasselt succumbed to the climate and dysentery on 8 September 1823 in Buitenzorg, age 26. Van Raalten, who was nearly killed by a rhinoceros in Java, packed the collections for shipment and made copies of all the notes for safekeeping. The last of the four members of the original commission, he died of fever in Timor in 1829.

In December 1823 a new commission was established by royal decree that consisted of two zoologists, Heinrich Boie and H. C. Macklot, who arrived in Java in June 1826. Boie, who did not have a strong constitution, succumbed to typhoid fever by 1827 and Macklot was killed during a rebellion in 1832. A French naturalist and pupil of Cuvier, P.-M. Diard, who had previously collected in Indochina, joined the commission in 1829 as Boie's successor. He died in Java apparently of poisoning from the arsenic he used in preparing animal skins. Despite these human costs and the large sums of money consumed to fund the commission's work, the large herpetological collections remained largely unstudied at the Rijksmuseum in Leiden for nearly a century until Nelly de Rooij and P. N. van

Kampen completed their respective summaries of the reptiles and amphibians of the Indo-Australian Archipelago.

Despite their short careers, Kuhl and Van Hasselt made important contributions to herpetology. Each of them published their doctoral dissertations in 1820 but only Van Hasselt's was herpetological (on metamorphosis in frogs). Just prior to departing for Java they issued a book, largely the work of Kuhl, under the general title "Beiträge zur Zoologie und Vergleichenden Anatomie" (1820). This contains several sections on "Amphibien" and includes the anatomy of selected species. There are also descriptions of several new forms, all in a section authored by Kuhl, including reptiles now known as *Psammobates oculifera* from the Cape of Good Hope, *Diploglossus monotropis* from Central America, *Draco fimbriatus* and *Mabuya multifasciata* from Southeast Asia, and *Python molurus bivittatus*, the well-known Burmese python of the pet trade. All of their other herpetological titles resulted from extracts made from their letters that were published in European journals: *Isis von Oken* (Jena), *Algemeene Konst- en Letter-Bode* (Haarlem), and *Bulletin des Sciences Naturelles et de Géologie, sect. 2, Bulletin Universel des Sciences et de l'Industrie* [De Férussac] (Paris) during 1822. Most of these papers are co-authored by Kuhl and Van Hasselt. In one that was not, Kuhl named the gecko genus *Ptychozoon* (1822). In these papers numerous new taxa are described, among them the frog genera *Megophrys*, *Occidozyga*, and *Rhacophorus,* and several new species. The collections, all from West Java, sketches, drawings, and notes are at the museum in Leiden.

• *References on Kuhl*: "Heinrich Kuhl," p. 202 *in* L. Gebhardt, Die Ornithologen Mitteleuropas. Brühlscher Verlag, Giessen, 1967; Bauchot et al., 1990 (Kuhl on p. 96); Fransen et al., 1997 (Kuhl on pp. 256–258). • *References on Van Hasselt*: Bauchot et al., 1990 (Van Hasselt on p. 90); Fransen et al., 1997 (Van Hasselt on pp. 239–240). • *References on both*: "On Two *Rhacophorus* Species Mentioned by Kuhl & Van Hasselt," by L. D. Brongersma, Arch. Néerland. Zool., 6: 341–346, 1942; Fransen et al., 1997 (Natuurkundige Commissie voor Nederlandsch Indië on pp. 270–275). • *Portraits*: Courtesy M. S. Hoogmoed. • *Signatures* (Kuhl, 1816; Van Hasselt, 1822): Courtesy L. B. Holthuis. • *Note*: Caspar Georg Carl Reinwardt (1773–1854), who spent 1816–1822 in the East Indies and later became professor of natural history at the University of Leiden, named several new snakes from Java, among them the new genus and species, *Xenopeltis unicolor* (*in* F. Boie, 1827).

SCHINZ, H. R. (1777–1861).

The Swiss physician and zoologist, Heinrich Rudolf Schinz, who authored one of the great illustrated atlases of amphibians and reptiles, was born in Zürich on 30 March 1777. His father, a minister who worked in Uetikon near Zürich, encouraged his boyhood interest in nature. After finishing secondary school, Heinrich attended the local Zürcherische Medicinische Institut before going to the university of Würzburg and finally to Jena to finish his medical studies (M.D. 1798). After a study trip to Paris, he returned to Zürich to practice medicine. This work did not satisfy him and he dedicated more and more time to zoology. From 1804 to 1833 he was a teacher at the medical institute and when the University of Zürich was established in 1833 he became a professor of natural history there. He held this position until his death. He

also was appointed curator of the zoological museum in Zürich. He died in Zürich on 8 March 1861 (see *Note 1*).

Schinz wrote a large number of illustrated books on animals for use in schools and universities as well as for the general public. His first was on bird nests and eggs, followed by large illustrated atlases on each group: birds, mammals, reptiles, and fishes. These had worldwide scope. The volume on reptiles includes amphibians and is entitled "Naturgeschichte und Abbildungen der Reptilien." The book is nominally dated 1833. Actually it was published in 17 parts from 1833 to 1835 and includes 102 excellent lithographed plates drawn by K. J. Brodtmann. The book was issued in both uncolored and hand-colored versions. Schinz's other major contribution to herpetology is his four-volume German translation of the first edition of Georges Cuvier's "Le Règne Animal" (1817 [1816]), entitled "Das Thierreich . . . dem Herrn Ritter von Cuvier" (1821–1825). There are many additions by Schinz in the form of footnotes; these are all signed with the letters "A. d. U." (see *Note 2*). Cuvier's was the most widely used classification and description of the animals of the world. In the German edition, amphibians and reptiles are in volume 2 (1822).

Schinz introduced some new names in both the "Naturgeschichte" and the "Thierreich," but most are synonyms. A few names, however, are recognized today, such as *Acanthodactylus erythrurus* and *Vipera aspis hugyi* in the former title and *Elaphe scalaris* and *Phyllurus* in the latter.

• *References*: "[Heinrich Rudolf Schinz]," anonymous, Zürcher. Jugend, Zürich, 65: 1–18, 1863; "Heinrich Rudolf Schinz," by Hunziker, Allg. Deutsch. Biogr., 31: 303–305, 1890; "Heinrich Rudolf Schinz," p. 315 *in* L. Gebhardt, Die Ornithologen Mitteleuropas, Brühlscher Verlag, Giessen, 1964; • *Portrait*: From Anonymous, 1863. • *Signature* (1826): Museum für Naturkunde, Berlin. • *Note 1*: Some sources give 1862 as the year of death. • *Note 2*: "Anhang des Uebersetzers" (Addition of the Translator).

LE CONTE, John E. (1784–1860).

The Le Contes (see *Note 1*)—descendents of the Huguenots of Normandy who fled France to escape persecution by Louis XIV—were one of the great families of 19th Century American naturalists (see *Note 2*). Into this distinguished group John Eatton Le Conte, Jr., naturalist, army officer, artist, and specialist on North American amphibians and turtles, was born on 22 February 1784 near Shrewsbury (New Jersey). He attended Columbia University (A.B. 1803) and entered the Corps of Topographical Engineers of the United States Army in 1818, rising to the rank of major before retiring in 1831 because of ill health. He lived most of his life in New York City. On the death of his father in 1822, he and his brother (see *Note 2*) inherited land in the New York region and in Georgia, respectively. He moved to Philadelphia in 1852 and died there on 21 November 1860. His natural history interests were very broad, especially in entomology. He was an active member of the Lyceum of Natural History of New York (now the New York Academy of Sciences), which he co-founded and, later, the Academy of Natural Sciences in Philadelphia.

Le Conte's contributions to herpetology consisted of 12 papers (three reprinted in 1964) that were packed with descriptions of new species and included comprehensive reviews of the frogs and turtles of the United States. Some 14 species of currently recognized and well-known North American amphibians and reptiles were named by him: *Acris gryllus, Hyla gratiosa, H. versicolor, Pseudacris nigrita, Rana capito, R. palustris, R. sylvatica, Pseudobranchus striatus, Siren intermedia* (see *Note 3*), *Glyptemys* (formerly *Clemmys*) *insculpta, Kinosternon integrum, K. sonoriense, Pseudemys concinna*, and *P. rubriventris*.

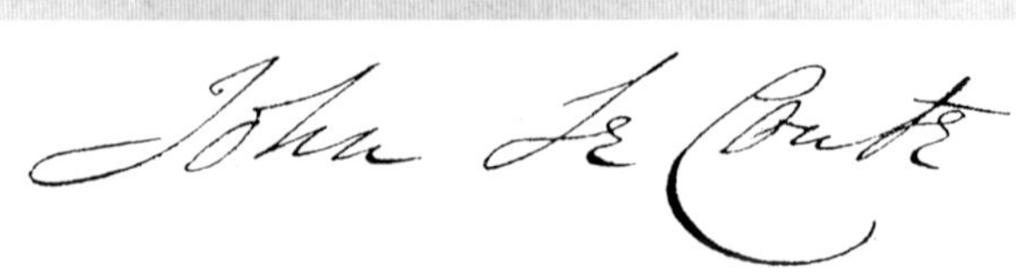

Le Conte was an experienced field naturalist and collected in the middle-Atlantic states as well as in the South where his brother owned a plantation in Liberty County, Georgia (see *Note 2*). Thus, he was able to sample two faunistically different areas. He was also attuned to the distinctive vocalizations of frogs and, in one late paper (1856 [1857]), noted that at least three undescribed species existed in Georgia based on calls he had heard.

The epithet *lecontei* was used in naming two species of North American snakes, but both honor his son, John L. Le Conte (see *Note 4*), who traveled and collected extensively in the West. John E. Le Conte apparently never ventured beyond the states of the Eastern Seaboard.

• *References*: "Memoir of John Lawrence LeConte 1825–1883," by S. H. Scudder, Biogr. Mem. U.S. Natl. Acad. Sci., 2: 261–293, 1886 (for John E. Le Conte, see pp. 269–272, 292); "John Eatton Le Conte, Jr.," p. 153 *in* C. A. Elliott, Biographical Dictionary of American Science, Greenwood Press, Westport, Connecticut, 1979. • *Portrait* (about 1850): Hunt Institute, Pittsburgh. • *Signature* (1817): Academy of Natural Sciences, Philadelphia. • *Note 1*: The family sometimes spelled their name LeConte, Leconte, or LeComte, but John E. Le Conte styled it "Le Conte" on his publications. • *Note 2*: John E. Le Conte's brother, Lewis Le Conte (1782–1838), was a Georgia plantation owner and amateur naturalist who established a botanical garden and natural history library on his property. Lewis's sons, John Le Conte (1818–1891) and Joseph Le Conte (1823–1901), were both Columbia University–trained M.D.s, professors at the University of California (John also became president), and members of the U.S. National Academy of Sciences. John was a physicist (he published a paper on alligator behavior in 1845) whereas Joseph was a geologist. John Eatton Le Conte's only surviving son, John Lawrence Le Conte (1825–1883), was also a Columbia-trained M.D. but never practiced and, instead, became America's greatest entomologist of his era and, like his cousins, a member of the National Academy. • *Note 3*: In recent years *Siren intermedia* has been credited to D. H. Barnes ("A Monograph of the Doubtful Reptils [*sic*]," Amer. Jour. Sci., 11: 268–297, 1826 [*Siren* in a footnote on p. 269; author's edition, p. 5]). He was the recording secretary of the Lyceum of Natural History of New York, which published Le Conte's formal description of the species in 1828. Barnes clearly credits the new name to Le Conte: "discovery of another species of siren, by Capt. Le Conte . . . is called by its discoverer *Siren Intermedia*." In this footnote, dated 15 August 1826, he cites Le Conte's yet-to-be-published figure of *S. intermedia*: "Fig. Annals of the Lyceum, Vol. 2, fig. 1," indicating he had unpublished parts of Le Conte's article available to him. Although the new name originated with Le Conte, there is no indication that the description in Barnes was conveyed by him, as is required by today's International Code of Zoological Nomenclature to credit the name to Le Conte, thus authorship rests with Barnes (for discussion, see "The Authorship and Date of Publication of *Siren intermedia*," by H. M. Smith, R. B. Smith, and H. L. Sawin, Great Basin Nat., 35: 100–102, 1975). • *Note 4*: *Rhinocheilus lecontei* was named for John L. Le Conte and based on a specimen he collected in San Diego ("Catalogue of North American Reptiles. Part I.–Serpents," by S. F. Baird and C. Girard, Smithsonian Inst., Washington, p. 121, 1853). *Crotalus lecontei* (=*C. viridis*) was based on a specimen sent from Oklahoma by John L. Le Conte ("Descriptions of New Species of Reptiles Inhabiting North America," by E. Hallowell, Proc. Acad. Nat. Sci. Philadelphia, 6: 180, 1852).

LICHTENSTEIN, Hinrich (1780–1857).

The German naturalist and explorer, Martin Hinrich (or Heinrich) Carl Lichtenstein, who was born in Hamburg on 10 January 1780, was responsible for building the Zoological Museum in Berlin into a collection of world renown. His father was a professor, theologian, and Orientalist, but young Hinrich decided on a medical career and studied in Jena and Helmstädt (now Helmstedt; Dr.med. 1802). He was then hired as the personal physician to the new Dutch governor of the Cape Colony in 1802 and, until the British wrested control of the colony from the Dutch in 1806, Lichtenstein was able to travel throughout the colony and beyond in southern Africa, primarily collecting insects and plants. He returned to Germany to write up his African experiences for publication ("Reisen im südlichen Afrika," 1811–1812).

In 1810 the University of Berlin was founded and Lichtenstein was chosen for the chair in zoology, but he had been trained as a physician. He quickly completed a doctorate in the natural sciences at Berlin (Dr.phil. 1811) and was appointed professor of zoology the same year. The university had started a museum, directed by the entomologist Carl Illiger, which already possessed the famous fish, amphibian, and reptile collection of Marcus Bloch and specimens obtained by Peter Pallas during his Russian expeditions. Illiger died prematurely in 1813 and Lichtenstein took over temporarily as director. He was made permanent director two years later and held the post until his death in 1857.

Lichtenstein began to build the museum by sending collectors to various parts of the world and by purchasing private collections. In 1813 the museum possessed only about 3,800 specimens of vertebrates (405 were amphibians and reptiles), but by the time of his death the collections numbered in excess of 28,000 specimens (more than 5,000 amphibians and reptiles). During the 19th Century the main object was to have representative collections of species, not large series of the same species for study of variability. Thus, Lichtenstein would sell off duplicates as a way of financing the museum and future collectors. Among his most important collectors herpetologically were Ferdinand Deppe and Count von Sack (Mexico), C. G. Ehrenberg and Wilhelm Hemprich (Middle East), Eduard Eversmann (Central Asia), Juan Gundlach (Cuba), Gerard Krefft (Australia), F. J. F. Meyen (South America, Philippines, China), Wilhelm Peters (Mozambique), R. A. Philippi (Chile), and Richard and Robert Schomburgk (British Guiana, Australia). Lichtenstein issued sale catalogues for disposing of the duplicates but specimens were sometimes sold before his assistants had sufficient time to thoroughly study them. He seemed to have little respect for anatomical criteria or bibliographic priority and was more concerned with getting names on specimens for purposes of cataloging them for the museum or for sale.

The sale catalogues included descriptions of many new species of vertebrates including amphibians and reptiles. The first, "Verzeichniss der Doubletten des zoologischen Museums" (1823; reprinted 2003), included several new taxa such as *Hylodes nasus* from Brazil and *Uromastyx ocellatus* from Egypt, but several others he named in this book had earlier been published in his natural history appendix to Eversmann's "Reise von Orenburg nach Buchara" (1823) including *Eremias grammica* and *Eryx tataricus*. The other sale catalogue of interest here was his "Nomenclator Reptilium et Amphibiorum Musei Zoologici Berolinensis" (1856; reprinted 2003). Many new frogs and lizards are named in this book, among them *Atelopus cruciger* from Venezuela, *Phrynocephalus interscapularis* of Central Asia, and *Ameiva festiva*, which is found throughout Central America. Although only Lichtenstein's name appears on the title pages of these catalogues, his assistants apparently wrote most of them. In his introduction to the "Verzeichniss," Lichtenstein explains that Ferdinand Schultze authored the remarks on lizards and most of the diagnoses. In the "Nomenclator," he notes that David Weinland and Weinland's successor, Eduard von Martens, were primarily responsible for preparing the book. Therefore, the exact authorship of the new taxa in these books remains unclear even today. Perhaps Lichtenstein's only paper strictly on herpetology was his description of *Gastrotheca ovifera* from Venezuela (1854, jointly with Weinland) in which they describe the female's brooding pouch full of eggs.

Given Lichtenstein's own talents and his position at the university, he became well connected socially and politically. He counted Johann Wolfgang von Goethe and Alexander von Humboldt as well as many high officials in the Prussian government among his friends. In 1842 he persuaded the king, Friedrich Wilhelm IV, to donate the grounds of his pheasantry as the location for a public zoo, which was opened in 1844 under the directorship of Lichtenstein. He attracted many students including Eduard Eichwald, J. G. Fischer, F. P. Jarocki, Wilhelm Keferstein, and A. F. A. Wiegmann, and among those mentioned earlier, Eversmann, Hemprich, von Martens, Philippi, Schultze, and Weinland. Lichtenstein died aboard a steamship in the Baltic Sea on 3 September 1857 when he was returning to Kiel from a trip to Sweden.

• *References*: "Martin Heinrich Karl Lichtenstein," by W. Hess, Allg. Deutsch. Biogr., 18: 556–557, 1883; "Hinrich Lichtenstein. Lebensbild des ersten Zoologen der berliner Universität," by E. Stresemann, p. 73–93 *in* Forschen und Wirken, vol. 1, Beiträge zur wissenschaftlichen und politischen Entwicklung der Universität, Deutsch. Verlag Wissensch., Berlin, 1960; Rieck et al., 2001 (Lichtenstein on pp. 359–362; by W. Bishoff on pp. 521–523); A. Bauer *in* A. Bauer (ed.), 2004. • *Portrait*: Courtesy Heinz Wermuth. • *Signature* (1854): Adler collection.

KAUP, Johann Jakob (1803–1873).

The German zoologist and paleontologist, Johann Jakob Michael Kaup (rarely Caup), who named many genera of lizards, was born on 20 April 1803 in Darmstadt (Hesse), 30 km south of Frankfurt am Main. After attending a local gymnasium, he enrolled at the university in Göttingen in October 1822, but his studies with J. F. Blumenbach did not satisfy him, nor did those with H. G. Bronn in Heidelberg the next summer. In the fall of 1823 he went to Leiden at the invitation of C. J. Temminck, director of the Rijksmuseum, and spent most of the next two years working on fishes, amphibians, and reptiles at the museum. It does not appear that he completed his university studies. After travels in Denmark (to meet his father; Kaup was illegitimate) and northern Germany, he returned to Darmstadt in 1825 where he became a private scholar. In 1828 he was hired as a temporary assistant at the Naturalien-Cabinet, a natural history museum located in the palace of Grand Duke Ludewig I, which became his workplace for the rest of his life. Kaup was awarded a doctorate by the university in Giessen (Dr.phil. 1831). While at the museum he made many trips and in 1834 went to Paris to study fossils and reptiles with Henri de Blainville. He was promoted to Assistant "Inspektor" (or director) of the cabinet in 1837 and to inspector in 1840. He died in Darmstadt on 4 July 1873.

Kaup had very broad interests and published extensively on all classes of vertebrates, on insects, and especially on fossils. Although in 1829 he had authored a respectable sketch of the evolutionary development of European vertebrates, he later repudiated this work. After "Origin of Species" appeared in 1859, he became an opponent of Charles Darwin's ideas. Earlier, Kaup had become an adherent of the ideas of "Naturphilosophie," a group led by Lorenz Oken who Kaup referred to as his "Lehrmeister," or master teacher, and who classified animals according to philosophical principles. (For fuller discussion, see p. 40.) The herpetologists, Leopold Fitzinger and J. B. von Spix, were also members of this group. Kaup, for example, believed that there should be no more than five species in each genus, corresponding to the number of human senses!

Kaup authored about 20 titles on amphibians and reptiles (1825–1834, with one much later paper on the morphology of *Heloderma* in 1865). Half were short papers published in *Isis*, which was edited by Oken. Several of these were critiques of recent works by Prince Maximilian of Wied-Neuwied, Blasius Merrem, and Spix, but even in these Kaup sometimes named new genera. His other original work includes reclassifications of lizards in which he also described several new genera, among them the agamids *Gonocephalus, Hydrosaurus*, and *Phrynocephalus* and the iguanids *Uracentron and Uranoscodon*. His most important herpetological novelty was a North African amphisbaenid—he named it *Trogonophis wiegmanni*—in which he made the surprising discovery that it had acrodont teeth, a character on which John E. Gray (1844) established the family Trogonophidae. Kaup also published two books of herpetological interest. His "Gallerie der Amphibien" ([1826]), an atlas of beautiful hand-colored plates of lizards by two Darmstadt artists, was intended to run to 24 parts, but the project failed after the first part for lack of subscriptions. Together with Bronn, Kaup authored a well-illustrated folio on a Jurassic marine crocodile similar to the modern gharial (1841, supplement 1844). He also cooperated with Louis Agassiz in producing the first universal index to herpetological genera, covering both living and extinct species, Agassiz's "Nomina Systematica Generum Reptilium" ([1844]).

• *References*: "Johann Jakob Kaup," by [W. von] Gümbel, Allg. Deutsch. Biogr., 15: 505–506, 1882; "Johann Jakob Kaup," anonymous, Encyclop. Brit., ed. 11, 15: 700, 1911; "Johann Jakob Kaup," by G. Heldmann, published by author, Darmstadt, x, 28 pages, 1955; "Johann Jacob Kaup," p. 178 *in* L. Gebhardt, Die Ornithologen Mitteleuropas, Brühlscher Verlag, Giessen, 1964. • *Portrait*: Adler collection. • *Signature*: From Heldmann, 1955.

RÜPPELL, Eduard (1794–1884).

The German naturalist and explorer, Eduard Wilhelm Peter Simon Rüppell (see *Note 1*), who was born in Frankfurt am Main on 20 November 1794, made early contributions to the zoology of Northeast Africa and Arabia. Although he was interested in natural history from boyhood, he was forced by family tradition to become a banker like his father. In 1812, however, both of his parents died. Later, he contracted tuberculosis and went to Italy in 1816 to recover. While there he was able to make a short visit to Egypt the next year, which inspired him to undertake more ambitious explorations of that region. Abandoning banking, he studied natural sciences at Italian universities, first in Pavia where his teachers included the zoologists, Pietro Configliachi, Bartolomeo Panizza, and Mauro Rusconi. He then briefly studied in Genoa before returning to Frankfurt in 1821. There he met P. J. Cretzschmar who had been one of the founders of the Senckenbergischen Naturforschenden Gesellschaft in 1817. Cretzschmar told him of the society's plans to begin a museum and he offered its support for his explorations.

Rüppell's first expedition lasted from 1822 to 1827 and explored the Sinai Peninsula, Egypt as far south as the El Obeid oasis (in the center of present-day Sudan), and along the shores of the Red Sea. Rüppell was even able to explore and map the coastline of the Gulf of Aqaba, an area that had been closed to Europeans since the Crusades. All previous attempts to reach Aqaba had been defeated by the fanatical Wahhabite Bedouin, but Rüppell succeeded in 1822, and in 1839 was the first foreigner to be awarded the Patron's medal of the Royal Geographical Society in London. He made large zoological collections, which were shipped at intervals to the new museum in Frankfurt. Some of these consignments must have been enormous because they included skins and skulls, and sometimes entire stuffed specimens of giraffes, hippopotamuses, ostriches, and crocodiles. In Rüppell's absence, Cretzschmar, now director of the museum, organized a book under Rüppell's name to report on the collections, entitled "Atlas zu der Reise im nördlichen Afrika" (published in parts, 1826–1828 [–1831], with the herpetological parts in 1827 and 1830). Carl H. G. von Heyden, a military officer who was then the society's honorary curator of herpetology, studied the amphibians and reptiles and identified several new species including *Bufo arabicus, Cyrtopodion scaber,* and *Pseudotrapelus sinaiticus*. Rüppell returned to Frankfurt in 1827, still suffering from a fever contracted at the Red Sea. Although the book was produced in an elaborate format and with excellent hand colored plates, he was not pleased with its overall quality, which began his difficulties with Cretzschmar.

Rüppell, now aged 36, mounted a second expedition from 1830 to 1834 with the primary goal to explore the highlands of Abyssinia (present-day Ethiopia). In fact, he was able to traverse the entire region, mostly by camel and mule, and map Lake Tana in the highlands as well as to re-visit the Sinai and some other localities from his first expedition. On returning to Europe, Rüppell settled in Frankfurt in order to begin working on his new book, "Neue Wirbelthiere zu der Fauna von Abyssinien gehörig" (published in parts, 1835–1840, with herpetology in 1835). Several of the beautiful, hand-colored plates were of animals drawn in the field by Rüppell. During his absence in Africa, Adolph Reuss, a medical doctor who had succeeded to the honorary curatorship of herpetology in 1829 when Heyden moved on to insects, began to work on Rüppell's collections including those from his second expedition. Reuss published two papers (1833, 1834) in which he named many new species, most of them collected by Rüppell, including *Bufo regularis, Latastia longicauda, Coluber nummifer*, and *Malpolon moilensis*. For the new book, Rüppell wrote the herpetological section himself (see *Note 2*) and in so doing described many additional new species, among them *Agama cyanogaster, Pristurus flavipunctatus*, and the geographic race of *Eretmochelys imbricata* found in the Indian and Pacific oceans (*bissa*). Since Reuss had quit the curatorship when he immigrated to the United States in 1834, the post had been ineffectively occupied for more than a decade when Rüppell himself accepted it in 1847. He held the post until 1862. During this period he published several herpetological papers including a catalogue of the entire Senckenberg collection of amphibians and reptiles (1845) in which he named new species. These include *Chamaeleo affinis* from his Abyssinian journey and a well-known Mexican species, the pitviper *Atropoides nummifer*.

In 1849–1850 Rüppell made one last trip to Egypt but collected very little. His productivity began to diminish about 1850 and stopped completely in 1858. He was reputed to be rude and insulting and became increasingly isolated from his colleagues. His rugged individualism had been a vital factor in the success of his expeditions through often-hostile territory, but it was now a liability. Despite his personal problems at the museum and with its former director, he made the Senckenberg society his sole heir. The importance of his collections to the museum cannot be overestimated. They represent the very foundation of the Senckenberg museum including its first types, and Rüppell's books gave the fledgling institution immediate recognition internationally. Rüppell died in Frankfurt on 10 December 1884 at the age of 90.

• *References*: "Eduard Rüppell," by W. Stricker, Allg. Deutsch. Biogr., 29: 707–714, 1889; "Eduard Rüppell. Leben und Werk eines Forschungsreisenden," by R. Mertens, W. Kramer, Frankfurt a/M, 388 pages, 1949; Rieck et al., 2001 (Rüppell by W. Bischoff and G. Hallmann on pp. 561–563). • *Portrait*: From Mertens, 1949. • *Signature*: Adler collection. • *Note 1*: Rüppell's forenames are sometimes given in other orders. • *Note 2*: Mertens (Senckenb. Biol., 48(A): 6, 1967) has stated (in translation): "It was remarkable that in this [second] collection of Rüppell there were no amphibians, crocodiles, or snakes." He then went on to suggest that this part of Rüppell's collection might have been lost in the sinking of the Russian cargo ship "Archip Dimidoff" in France, which was carrying some of his materials. To the contrary, Reuss (1833) described a new toad from Egypt and in his 1834 paper named six new snakes from Rüppell's collections including one from Abyssinia. Thus, Rüppell's snakes had already been published by the time he issued the herpetological section of his Abysinnian report in 1835. Moreover, Rüppell (1845) listed five specimens of the Nile crocodile in the Senckenberg collection, all noted as being his specimens.

EVERSMANN, Eduard (1794–1860).

Eduard Friedrich Alexsandrovich Eversmann was a German naturalist and explorer of the Russian Steppes. He was born on 23 January 1794 in Wehringshausen, a small village that is now part of Hagen (North Rhine–Westphalia) and 40 km southeast of Essen. With an early interest in natural history, at the age of 19 he had a dream of traveling and exploring Tibet and Kashgar (now Kashi, in westernmost Xinjiang, China).

He studied natural science, first at Marburg in 1812, then Halle, and then during 1813–1814 with Hinrich Lichtenstein in Berlin. He received his doctorate (Dr.phil. 1814) at Halle. In 1816 his father moved to Russia to become director of the Imperial Ordnance Factory at Slatovsk (now Zlatoust) in the southern Urals, which supplied the entire Russian army with cold steel. Young Eversmann went to the university at Dorpat (now Tartu, in Estonia), in the Russian Empire, for medical training (M.D. 1816) and then practiced in Slatovsk for a few years. He still pursued his plan to visit China and felt that being a doctor would help him on such a journey. In further preparation he also learned the necessary languages (Tartar, Persian, and Afghan) and studied Muslim religious customs.

In the fall of 1820, disguised as a merchant and accompanied by several traders, Eversmann began his journey by travelling from Orenburg, on the northern border of present-day Kazakhstan, to Bukhara (now in Uzbekistan), famous then as it is now as a trading point for Oriental carpets. He had planned to join a caravan early in 1821 to venture 1000 km east to Kashgar along one of the ancient Silk Road trading routes to China. However, he was denounced as a Russian spy to the local khan who decided to have him killed. Fortunately Eversmann learned of this plot and quickly headed back to Orenburg. There he settled and practiced medicine until 1828. He soon wrote up his recent expedition in book form—"Reise von Orenburg nach Buchara" (1823)—and sent his specimens to his former teacher, Lichtenstein, at the zoological museum in Berlin. Lichtenstein contributed the natural history appendix in which several new species of lizards and a snake were described. Despite Eversmann's years of preparation to visit China, he never tried again.

All of Eversmann's later expeditions were more limited. He went to the Caspian and Aral seas in 1825. In 1825–1826 he joined a military expedition to Khiva, located along the border of today's Turkmenistan about 400 km northwest of Bukhara. Jointly with the traveler, Grigory S. Karelin, in 1827 he made an extensive survey of the territory just north of the Caspian Sea between the Volga and Ural rivers. The Caucasus was visited in 1830, and in 1834 he went to Saratov along the Volga River. Eversmann moved to the university in Kazan as professor of zoology and botany in 1828, 700 km east of Moscow, after which the region between the Volga and Ural rivers became his area of special interest. In Kazan, where the naturalist, Eduard Eichwald, had also been a professor, Eversmann founded the zoological museum and spent the rest of his life studying his natural history collections especially the insects and birds. He died in Kazan on 14 April 1860.

Eversmann published many works including a three-volume series on the natural history of the Orenburg district (1840–1850). After the death of Peter Pallas, Eversmann wrote the addendum (in three parts, 1835–1842; reprinted 1876) to Pallas's "Zoographia Rosso-Asiatica." Eversmann's only strictly herpetological work was a monograph, with four beautifully hand-colored plates, on the lizards of the Russian Empire, "Lacertæ Imperii Rossici" (1834). He listed 14 species, four of them new and two (*Lacerta praticola* and *L. saxicola*) are recognized today. Both species are bisexual, but the latter was made famous when several of its subspecies were reported to be parthenogenetic clones by the Soviet herpetologist, Ilya Darevsky (1958).

• *References*: ["Eduard Alexsandrovich Eversmann"] *in* A. Bogdanov, Materials for the History of Scientific Activity in Russia on Zoology. Newsbull. Imper. Moscow Univ., 70 (Trans. Dept. Zool., 6), 1891; ["Eduard Alexsandrovich Eversmann,"] p. 25–40 *in* F. N. Milkov, The Naturalists of the Orenburg Territory. Chkalov Publ., Ogiz, 1948; "Eduard Eversmann," p. 87–88 *in* L. Gebhardt, Die Ornithologen Mitteleuropas, Brühlscher Verlag, Giessen, 1964; Rieck et al., 2001 (Eversmann by W. Bischoff, on pp. 442–443). • *Portrait*: From A. Bogdanov, Newsbull. Imper. Moscow Univ., 55 (Trans. Dept. Zool., 3), 1888, courtesy Natalia Ananjeva. • *Signature* (1817): Museum für Naturkunde, Berlin.

LESSON, René-Primevère (1794–1849).

René-Primevère Lesson, the zoologist and voyager who described many new species of amphibians and reptiles collected on French naval voyages, was born on 20 March 1794 in Cabane-Carée, near Rochefort, a coastal town in western France where his father was a clerk for the navy. He entered the naval medical school in Rochefort when he was not quite 16 years old and taught himself natural history. In 1811 he was conscripted into the navy and saw action in several sea battles as an auxiliary surgeon. All persons in such posts were dismissed after the fall of Napoleon Bonaparte in 1815, so he became a gardener in Rochefort and learned pharmacy. He was appointed a naval pharmacist in 1816 and came to Paris to get his degree in 1822. That year, Lesson and Prosper Garnot were appointed as medical officers and naturalists aboard the French corvette "Coquille," which visited South America, Australia, the South Pacific, Dutch East Indies, and other places on its circumnavigation (August 1822 to March 1825). They made large collections of plant and animal specimens including amphibians and reptiles, which are today at the Muséum d'Histoire Naturelle in Paris. Lesson, in fact, was the first naturalist to see living birds of paradise. On his return he was granted leave during 1825–1829 to work on his collections at the museum. He returned to Rochefort to teach botany at the naval medical school and in 1831 became a professor of pharmacy there. He was promoted to chief naval pharmacist of his district in 1835. He died in Rochefort on 28 April 1849.

Lesson's natural history interests were broad. Birds were his specialty (he described some 40 species of hummingbirds alone), but he also worked on mammals, mollusks, and other animals. He authored about a dozen papers on amphibians and reptiles (1828–1842). His major herpetological contributions were the "reptiles" sections of three French naval explorations. The first, and largest, was the report from the voyage of the "Coquille," led by L.-I. Duperrey and on which Lesson participated. The reptiles section was issued in 1829–1831 and contains seven beautifully hand-colored plates in folio. The other two reports were based on collections turned over to Lesson for study: the voyage to the "Indes-Orientales" (1825–1829), published under the editorship of Charles Bélanger (reptiles issued 1832–1834, with seven hand-colored plates); and the circumnavigation of the French ships "Thétis" and "Espérance" (1824–1826), led by Baron de Bougainville (reptiles issued 1837, with two folio plates in color).

Mostly in these reports Lesson described about 50 new reptiles and amphibians. Among those recognized today are South American lizards (*Enyalius brasiliensis, Liolaemus chiliensis, Tropidurus peruvianus*); frogs from Australia (*Litoria aurea*) and New Guinea (*Rana papua*); lizards that are widespread on Pacific Islands (*Emoia cyanura, Gehyra oceanica, Lipinia noctua*); and several taxa from South Asia (*Crocodylus palustris, Geoemyda trijuga thermalis, Naja kaouthia, Rana hexadactyla*). He also named a new genus and species of sea snake (*Microcephalophis gracilis*).

• *References*: "René-Primevère Lesson," by A. Berman, Dict. Sci. Biogr., 8: 265–266, 1973; "René Primevère Lesson (1794–1849)," p. 217–219 *in* B. Mearns and R. Mearns, The Bird Collectors. Acad. Press, San Diego, 1998. • *Portrait* (1827): Adler collection. • *Signature* (1831): Academy of Natural Sciences, Philadelphia.

NILSSON, Sven (1787–1883).

Sven Nilsson, the founder of Swedish zoology who wrote the first herpetology of Scandinavia, was born to a family of farmers on 8 March 1787 near Alfastorp (Malmöhus, now Skåne County) in southernmost Sweden. He was interested in nature as a boy, but after local schooling he went to Lund University in 1806 intending to study for the priesthood. One of his teachers, the zoologist A. J. Retzius, encouraged him to study natural history. Nilsson graduated in 1811 and earned a medical degree there seven years later after which he joined the faculty. He became curator of the Natural History Museum in 1819 and professor in 1821. In 1828 he moved to Stockholm as curator of the zoological collections at the

Academy of Natural Sciences (today in the Riksmuseum), but returned to Lund in 1831 as a professor of natural history where he remained until his retirement in 1856. In 1838 he was ordained a priest and later became vicar and dean. He died in Lund on 30 November 1883.

Typical of the period, Nilsson's interests were broad. Beginning with birds they expanded to include all vertebrates, mollusks, geology, and paleontology. He authored pioneering works on each of these subjects, complementing the earlier works on Swedish botany by his compatriot, Carl Linnaeus. As the authority on Swedish mammals, Charles Darwin consulted with Nilsson on the inheritance of reindeer antlers. In addition to the inclusion of amphibians and reptiles in his reviews of the zoological collections in Stockholm (1829–1831), Nilsson wrote treatises on fossil amphibians, turtles, and lizards (1836, 1841, and 1856, respectively). His major contributions to Scandinavian herpetology, however, were in the two editions of "Skandinavisk Fauna" (volume 3, "Amfibierna," with the alternative title, "Skandinavisk Herpetologi," 1842; second edition 1860). These covered Norway and Sweden, because at that time the two were united. These volumes include extensive synonymies, species descriptions, natural history, and distribution, although they are not illustrated. One species, the Moor Frog (*Rana arvalis*), was described as a new species in the 1842 volume.

• *References*: "Sven Nilsson," by S. N[ettelbladt] et al., p. 452–453 *in* Svenska män och kvinnor–Biografisk uppslagsbok, vol. 5. A. Bonniers förlag, Stockholm, 1949; "Sven Nilsson – En lärd i 1800-talets Lund," by G. Regnéll (ed.). Kungl. Fysiograf. Sällskapet, Lund, 219 pages, 1983. • *Portrait and signature* (1834): Lunds Universitet, courtesy Lars Bjørnshauge.

ESCHSCHOLTZ, J. Friedrich (1793–1831).

Johann Friedrich Gustav von Eschscholtz (see *Note*), Russian naturalist and physician, was an early explorer and describer of the fauna of the Pacific Rim. He was born in Dorpat (now Tartu, Estonia) in the Russian Empire on 1 November 1793. In 1815, within days of finishing his training in medicine and natural history at Dorpat University, he went to Kronstadt (near Saint Petersburg) to join the crew of the Russian brig "Rurik," commanded by Otto von Kotzebue on its circumnavigation of 1815–1818. Eschscholtz served officially as the ship's surgeon, but he also assisted the ship's German naturalist, Adelbert von Chamisso, in making collections. The ship visited several localities of herpetological interest including San Francisco Bay and southern Alaska (in what was then Russian America), the Sandwich Islands (now Hawaii), the Philippines, Chile, and Brazil. On his return, Eschscholtz was elected professor at Dorpat University where he began to study his collections.

During 1823–1826 Eschscholtz joined a second voyage under Kotzebue's command, aboard the Russian warship "Predpriaetiё," which retraced much of the same route, but this time he served as chief naturalist. His Californian collections were more extensive and he covered a larger area than before—from Ross (the main Russian settlement, today's Fort Ross) and Bodega Bay north of San Francisco to Santa Clara in the south and eastward as far as about Rio Vista along the lower reaches of the Sacramento River—during a two-month period late in 1824. He discovered some of the most distinctive salamanders of western North America: *Aneides lugubris, Batrachoseps attenuatus, Taricha torosa,* and the first member of what would later be recognized as a new family of salamanders, *Dicamptodon ensatus*.

On returning to Dorpat in 1826 Eschscholtz worked up his collections. During 1829–1833 the results were published in a folio entitled "Zoologischer Atlas," which was issued in Berlin in five parts with 25 beautifully hand-colored plates. The sections containing the new species of amphibians and reptiles, which appear in parts 1, 2, and 5, were reprinted in 1966. Besides the salamanders, the new species included a sea turtle (*Lepidochelys olivacea*) and a large sail-finned agamid (*Hydrosaurus pustulatus*), both from the Philippines. Eschscholtz applied the name *Triton tereticauda* to what was later named *Salamandra lugubris* by Edward Hallowell in 1849. Even though Eschscholtz's name has priority over Hallowell's, *tereticauda* was suppressed by action of the International Commission on Zoological Nomenclature in 1956 in the interests of stability because *lugubris* has long been in use. It is perhaps not coincidental that this was the only new species of salamander not illustrated in the "Atlas" and, since no specimens reached Dorpat, the description had to be based solely on Eschscholtz's brief field notes.

Eschscholtz died of typhus in Dorpat on 7 May 1831 at the age of 37, before completion of the series. Fortunately, his university colleague, the embryologist M. Heinrich Rathke, who had previously published extensively on salamander anatomy, completed part 5 (1833) based on Eschscholtz's notes and his own study. In fact, it was Rathke who described *Taricha torosa* from Eschscholtz's collections. Except for one short paper on sea turtles in 1829, the "Atlas" was Eschscholtz's sole publication in herpetology, but he made more substantial contributions to other zoological and botanical topics, especially to entomology. John E. Gray named a Californian plethodontid salamander, *Ensatina eschscholtzii*, in his honor in 1850.

• *References*: [Untitled obituary of Eschscholtz], by M. H. Rathke, p. iii–viii, *in* J. F. Eschscholtz, Zoologischer Atlas, part 5, G. Reimer, Berlin, 1833; "Johann Friedrich Eschscholtz," p. 617–622 *in* E. O. Essig, A History of Entomology, Macmillan Co., New York, 1931; "Johann Friedrich Eschscholz," by T. A. Lukina, Nauka, Leningrad, 175 pages, 1975. • *Portrait*: From Eschscholtz, Zoologischer Atlas, part 5, 1833. • *Signature* (1828): From Candollea, 29: 498, 1974. • *Note*: The family name is sometimes spelled Eschscholz, without the *t*, which is its Russian form.

EICHWALD, Eduard (1795–1876).

The zoologist and paleontologist, Karl Eduard Ivanovich Eichwald, who described and named many of the reptiles of southern Russia, was born on 4 July 1795 at Mitau (now Jelgava, Latvia) in the Russian Empire. His father was a private tutor to a noble family, and later he taught natural history. Eduard was schooled at home and then educated at a gymnasium. He matriculated at the university in Dorpat (now Tartu, Estonia) in 1814. Later that year he went to study medicine and natural sciences at the university in Berlin where one of his teachers was Hinrich Lichtenstein. For additional study he spent the following year traveling throughout Europe. He then finished his medical degree (M.D. 1819) at the university in Vilna (now Vilnius, Lithuania) where his teacher, the turtle anatomist L. H. Bojanus, was a professor and, later, rector.

After practicing medicine in Mitau for two years, Eichwald became a professor in Dorpat, and in 1823 he moved to a similar post at the university in Kazan, about 700 km east of Moscow. Here he founded a laboratory of comparative anatomy and directed the botanical garden. During his time in Kazan he engaged in his only major expedition (1825–1826), traversing the vast area surrounding the Caspian Sea (western Turkmenistan, northern Persia, and the Caucasus). In 1827 Eichwald returned to Vilna as a professor of zoology. When the university was closed because of an unsuccessful Polish insurrection against Russia, he took a professorship at the city's new Medical-Surgical Academy. Finally, in 1838, he became a professor of zoology at the Medical-Surgical Academy in St. Petersburg where he remained for the rest of his career. He died in St. Petersburg on 16 November 1876.

Eichwald had broad interests in natural history including zoology, botany, geology, and paleontology. For decades he was regarded as Russia's leading paleontologist. He was the first to work out the geological strata between Lithuania and the Black Sea. Contrary to most of his contemporaries, who believed that orogeny was a sudden and widespread process, in the 1830s he wrote that mountains arise from slow and local uplifts that occur at different times. His great work on the paleontology of Russia, "Lethæa Rossica" (1852–1868), extends to 3,500 pages of text plus three atlases. He also wrote texts on mineralogy (1844) and geology (1846) and he became a major proponent of Darwin after reading his "Origin of Species" (1859).

Eichwald made substantial contributions to herpetology. In his three-volume classification of the animals of the world, "Zoologia Specialis" (1829–1831), he grouped animals not just by their similarities but also by genetic relationships. He covered vertebrates in volume 3, and in it he named 26 new species of amphibians and reptiles, all from Russia and half of which are recognized today. Examples include *Lacerta strigata, Phrynocephalus reticulatus, Coluber najadum*, and *Naja oxiana*. He published three papers exclusively on herpetological topics, and in one of them (1839) he described *Eumeces schneiderii princeps* from the Talych Mountains, which are near the Persian border in what today is Azerbaijan.

Eichwald's most elaborate book for herpetology is his "Fauna Caspio-Caucasia" (1841) that resulted from his expedition to the Caspian Sea. The Caucasus was later explored by Edouard Ménétriés in 1829–1830 and again by Alexander von Nordmann in 1835; the latter's results were published in the larger work on the region by Anatoli Demidoff. Thus the fauna of the Caucasus rapidly became one of the best known in Eurasia. The bulk of Eichwald's "Fauna" is herpetological; 29 of the 40 mostly hand-colored plates are of amphibians and reptiles and serve to illustrate many of the species he had briefly named in "Zoologia Specialis" in 1831 as well as the skink he named in 1839. The value of well-executed and finely colored plates in defining his taxa cannot be overestimated and partly explains why such a high proportion of his species are recognized today. The herpetological text in the "Fauna" is also extensive, with re-descriptions of most of his new taxa and a long review of the morphology of *Varanus griseus* with three plates of its osteology. Eichwald's collections are kept at the museums in Kazan and St. Petersburg.

• *References*: ["Eduard Ivanovich Eichwald,"] p. 321–389 *in* B. E. Raikov, Russki Biologi-Evolutionisty do Darvina. Izd. Acad. Nauk SSSR, Moscow-Leningrad, 1951; "Karl Eduard Ivanovich Eichwald," by V. V. Tikhomirov, Dict. Sci. Biogr., 4: 307–309, 1971; Rieck et al., 2001 (Eichwald by W. Bischoff on pp. 438–439). • *Portrait* (1825): From Raikov, 1951. • *Signature* (1832): Museum für Naturkunde, Berlin.

MÉNÉTRIÉS, Edouard (1802–1861).

The French naturalist, Edouard Ménétriés, who wrote the first zoology of the Caucasus of Russia and described its reptiles, was born in Paris on 14 October 1802. He had planned to be a physician but decided on natural history instead. He worked at the Muséum d'Histoire Naturelle with Georges Cuvier and P.-A. Latreille where he came to specialize on insects. When Czar Alexander I sent his Brazilian envoy, Baron Georg von Langsdorff, an amateur naturalist, back to Brazil, he was charged to explore the interior of the country. The museum authorities in Paris nominated Ménétriés to accompany Langsdorff and the two arrived in Rio de Janeiro in March 1822. After some local explorations by Ménétriés, they traveled through the interior, some 500 km due north of Rio to Tejuco (now Diamantina), in Minas Gerais, before they returned to Rio in February 1825. For this service to the Russian government, the czar appointed Ménétriés curator of the Zoological Museum of the Imperial Academy of Sciences in St. Petersburg in 1825, a post he held for life.

In 1829, the new czar, Nicholas I, sent Ménétriés on an 18-month natural history exploration of the Caucasus, which lies between the Black and Caspian seas in southern Russia. Since this was a region still being subjugated, a military escort of Cossacks accompanied Ménétriés and his three scientific companions. By the end of June, the group reached Pyatigorsk in the northern foothills and spent a few weeks around Mt. El'brus on the border of today's Georgia. They then proceeded towards the Caspian Sea along the River Terek and via Mozdok to Baku, which lies on the southwestern coast of the Caspian Sea in present-day Azerbaijan. From there, they rode south via Sal'yany and the River Kura to Lenkoran on the Caspian shore 30 km from the Persian border where they explored the Talych Mountains. By the time they returned to Lenkoran cholera had broken out, so they shipped out to Baku and, finding the disease there too, they quickly rode overland to Kuba and arrived in St. Petersburg in December 1830.

Ménétriés wrote up the zoological collections in what became the first zoology of the Caucasus, his 300-page "Catalogue Raisonné des Objets de Zoologie Recueillis dans un Voyage au Caucase" (1832). Although Eduard Eichwald had explored the area earlier (1825–1826), his results were published much later (1841). The herpetological section in Ménétriés's book lists 40 species including 13 new ones. The species recognized today are the lacertid *Ophisops elegans*, the scincid *Ablepharus bivittatus*, and two colubrids, *Coluber ravergieri* and *Eirenis collaris*. He also established the genus *Ophisops*.

Ménétriés spent the rest of his life in St. Petersburg, mostly working up insect collections made by Russian naturalists who explored Siberia and Russian America (Alaska, northern

California): Friedrich Eschscholtz, Alexander von Middendorff, and Leopold von Schrenck, among others. He died on 22 April 1861 of emphysema.

• *References*: ["Eduard Menetries,"] *in* A. Bogdanov, Materials for the History of Scientific Activity in Russia on Zoology. Newsbull. Imper. Moscow Univ., 55 (Trans. Dept. Zool., 3), 1888; "Edouard Ménétriés," p. 52–54 *in* N. Papavero, Essays on the History of Neotropical Dipterology, vol. I. Mus. Zool., Univ. São Paulo, São Paulo, 1971; ["Entomologist and Zoologist of the Institute,"] by O. L. Kryzanovski, p. 5–11 *in* Zoologists of Our Country, Proc. Zool. Inst. Russ. Acad. Sci., St. Petersburg, vol. 292, 2002. • *Portrait*: From Bogdanov, 1888. • *Signature* (1858): Archives, Ernst Mayr Library, Museum of Comparative Zoology, Harvard University, Cambridge.

SOWERBY, James De Carle (1787–1871). LEAR, Edward (1812–1888).

The names "Sowerby and Lear" are forever linked herpetologically as artist and lithographer, respectively, for the greatest atlas of turtle illustrations and one of the finest productions among all natural history books. Their illustrations were originally made for Thomas Bell's folio work entitled "Monograph of the Testudinata" (1832–1835 [–1836?]; compact disk 1999). Some 40 years later the plates were reissued with additional plates and a short text by John E. Gray, under the names Sowerby and Lear and with the title "Tortoises, Terrapins, and Turtles" (1872; reprinted 1970; compact disk 1999). Production of these books was complicated, but not unlike a number of the great works on natural history issued during the 19th Century that were published in parts over a period of years and that included hand-colored lithographed plates.

Bell had a special interest in turtles. Sowerby, then the finest English natural history artist, was the obvious choice for illustrator. Bell had known Sowerby at least since 1826 when he had him illustrate two of his turtle papers in *Zoological Journal* (1826–1827). Lear was apparently recommended to Bell by Gray. Bell wanted all animals to be drawn from life and started collecting living specimens at his home and at the zoo in London. He imported the first softshell turtles to be seen alive in England and kept them in his backyard pond. Sowerby, however, was involved in too many projects and could not do his work at the desired pace. In a letter to Sowerby dated 2 March 1832, before the appearance of part 1, Bell says: "I am sorry to trouble you again, but I really must urge the finishing of as many tortoises as possible very soon, as Lear has almost overtaken you and I am getting anxious to begin the work — and cannot venture to make even my intention public until a fair prospect exists of bringing it out with regularity." Another serious problem was that animals were dying before they could be drawn and as Bell wrote later to Sowerby, " . . . the colours are lost after death." Totally exasperated, Bell wrote to Sowerby on 18 September 1833: "I have really postponed writing to you day after day, . . . but what am I to do? My work is standing still, . . . and persons are hesitating about taking [the book] lest it should continue to be thus irregular, or not to be completed at all."

Prophetically, after eight parts were issued the project was terminated when Bell's publisher went out of business. Although most bibliographic sources have given the inclusive dates of publication of the eight parts as 1836–1842, recent evidence suggests that the dates are 1832 to 1835 or 1836 (see *Note*). Bell's specimens ended up at Oxford University although a few, mostly duplicate, shells went to Cambridge. At the time of the publisher's bankruptcy only 24 pages of introduction, 82 of text, and 40 plates had been issued although 21 additional plates had been prepared and even printed. No title page was ever issued. The stock was purchased by another London firm, which reissued the full set of plates (60 in the hand-colored edition, 61 in the black-and-white) with a short text by Gray in 1872. All of Sowerby's original drawings, including some never-published ones as well as unpublished pages of text, are in the Natural History Museum in London.

The relative contributions of Sowerby and Lear to the final product can be partly inferred by comparison of the plates prepared by Sowerby alone for Bell's journal papers in 1826–1827 to their counterparts in the book which were drawn on the printing stones by Lear who was an outstanding artist in his own right. The earlier drawings by Sowerby are rather simple and stiff compared to those in the book. Part of the difference, however, may be due to the medium. The plates in the journal are engravings whereas those in the atlas are lithographs, which allow for more delicate shading and details. The latter have a decidedly more lifelike quality, which we may safely attribute to Lear's special talents.

Sowerby and Lear had unequal reputations at the time they worked on Bell's book. James De Carle Sowerby was a sec-

ond-generation member of the greatest dynasty of English naturalist-artists. He was born on 5 June 1787 in Stoke Newington, then about 5 km north of London on the old Roman road to York but now a part of London. His father, James Sowerby, was a famous botanist and artist who also owned a museum. As described by Arthur de Carle Sowerby, a fifth-generation Sowerby and naturalist of China, "The 'Museum' was a family institution, in which every member had his or her place . . . each had his or her table or desk, colours, brushes, and cupboard with drawers, at which to work and in which to store paraphernalia, and each had an appointed task . . . . As the children grew up, they graduated from the 'dirt' as the little strip of ground at the bottom of the specimen in some plates was called, to more responsible work. Patterns were provided by James himself, and of course any work that did not come up to standard was rejected." James De Carle Sowerby's special interest was fossil shells but he illustrated a wide variety of subjects. He also wrote scientific papers on paleontology and zoology, was a fellow of the Linnean Society, and one of the founders, in 1826, of the Zoological Society of London. He died in London on 26 August 1871.

In contrast to Sowerby, Edward Lear was a relative unknown at the time he worked on Bell's plates. He was born on 12 or 13 May 1812 in Upper Holloway, North London, as the next-to-last of 21 children. His father went into debt and was even briefly imprisoned. The family was split up and Edward, at the age of four, was given to his sister Ann, 21 years his senior, to be raised. He was an epileptic with poor eyesight, chronic bronchitis, and asthma, who went to work at the age of 15 and yet rose to become, with Lewis Carroll, the foremost English writer and illustrator of nonsense verse. Lear did not invent limericks (that verse form can be traced to County Limerick in Ireland at least as far back as the mid-1700s), but he was the first to popularize the form when he wrote and illustrated "A Book of Nonsense" (1846) under his pen name, "Derry Down Derry." For present purposes the following verse may be an apropos example:

There was an Old Person of Ickley,
Who could not abide to ride quickly.
He rode to Karnack on a tortoise's back,
That moony Old Person of Ickley.

Lear invented some of the most wonderful nonsense words in the English language including "flumpetty," "borascible," "slobaciously," and "scroobius." The book went through 24 editions in Lear's lifetime and has never been out of print.

Before achieving fame as a writer and landscape painter Lear began as a natural history artist. He issued a series of drawings of parrots in 1830–1832 and, unlike other zoological artists of the day, specialized in drawing living animals, which appealed to Bell. Late in 1832 Lear went to Liverpool after accepting a commission from Lord Stanley, the future 13th Earl of Derby, to illustrate his private menagerie for publication and to tutor his grandchildren in drawing. Lear's eyesight, which was never good, began to seriously fail about 1836 when he switched to painting landscapes. When his health worsened the next year, Stanley, now earl, arranged for him to go to Italy to recuperate but Lear ended up spending much of his life there to paint landscapes, teach drawing, and write. It was here that he wrote his books of nonsense verse and also wrote and illustrated his well-known children's book containing "The Owl and the Pussy-Cat" (1871). His fame as an art teacher was such that, on a return trip to England in the summer of 1846, he tutored young Queen Victoria in drawing. He also illustrated a volume of Tennyson's poems, posthumously (1889).

Lear maintained a lifelong friendship with Bell to whom he sent seeds to plant in his garden as late as 1871. Lear died at San Remo on the Italian Riviera on 30 January 1888.

• *References on Sowerby*: "The Sowerbys and Their Publications in the Light of the Manuscript Material in the British Museum (Natural History)," by R. J. Cleevely, Jour. Soc. Biblio. Nat. Hist., 7: 343–368, 1976; "James De Carle Sowerby," by R. J. Cleevley, Oxford Dict. Natl. Biogr., 51: 731–733, 2004. • *Sowerby's portrait*: Hunt Institute, Pittsburgh. • *Sowerby's signature* (1841): Royal Botanic Gardens, Kew. • *References on Lear*: "Edward Lear a Biography," by P. Levi, Scribner, New York, 362 pages, 1995; "Edward Lear," by V. Noakes, Oxford Dict. Natl. Biogr., 32: 994–1000, 2004. • *Lear's portrait* (1840) *and signature* (1854): Adler collection. • *Note*: K. Adler (Octavo Editions, San Francisco, 1999) has published a detailed reanalysis of the dating for the eight parts of Bell's monograph in the commentary accompanying the release of full-color reproductions, on compact disk, of both versions of the book.

COCTEAU, Théodore (1798–1838).

The French physician, Jean-Théodore Cocteau, a specialist on lizards, was born on 15 March 1798 in Paris. His father, assistant chief of the Ministry of the Interior, sent him to a Catholic boarding school and then to study medicine. He interned at the

Hôpital Saint-Louis, the oldest hospital in Paris, but he later attended lectures by Georges Cuvier at the Muséum d'Histoire Naturelle, which sparked his interest in zoology. When the Allies invaded France early in 1814, Cocteau dressed the wounds of French soldiers at Val-de-Grâce military hospital where he worked. Later he took up arms to serve the national artillery when the conflict reached Paris, which ended in April 1814 with Napoleon Bonaparte's abdication. Cocteau resumed his practice and his medical studies in Paris (M.D. 1824).

Cocteau's scientific work began at the museum under the mentorship of Constant Duméril who had recently become professor in charge of reptiles and fishes. His first subjects were skinks brought back from Egypt in 1829. During 1832–1838 he authored nearly 20 papers, mostly on lizards and especially skinks, but also on frogs and snakes. He named several new species including the Australian skink *Cryptoblepharis plagiocephalus* and the Sri Lankan snake *Uropeltis ceylanica*. His best-known species, *Anolis sagrei*, widespread in the West Indies, has very successfully invaded Florida. Duméril encouraged several of his younger colleagues to develop natural arrangements of the genera of major groups of reptiles. Probably as a result, Cocteau produced "Tabulæ Synopticæ Scincideorum" (1836, 1837).

Cocteau authored two major works. His "Études sur les Scincöides," issued in 1836, was intended to be the first part of a complete monograph of the Scincidae. When no publisher would agree to issue the book, he did so privately on his own meager funds. Thus this book, which contains four intricately detailed and hand-colored engraved plates, was published in a very small edition and is one of herpetology's rarest titles. His other book was the herpetological volume in Ramon de la Sagra's "Histoire . . . de l'Ile de Cuba" (French edition, 1839–1843; Spanish 1843–1853). The book covers 34 species of amphibians and reptiles including several new ones such as the anguid *Diploglossus delasagra* and the teiid *Ameiva auberi*. The folio atlas of 31 plates contains some of the most lifelike, hand-colored images of reptiles and amphibians ever produced. Cocteau authored the sections on turtles, crocodiles, and lizards, but he did not live to see any of it published. Gabriel Bibron, Duméril's chief assistant, wrote the final parts on snakes and frogs.

Cocteau had supported himself from a medical practice that was too small to allow him the luxury to do everything he wanted to do. He quietly cared for the poor without pay even when cholera ravaged Paris. He did science in time stolen from his practice and was very attentive to his friends. When he contracted ataxic fever at the end of March 1838, he was ordered to bed; but in harsh weather and against all advice he went to attend to his sick father. This only worsened his own condition. He died in Paris on 13 May. He had never married.

• *References*: ["Jean-Théodore Cocteau,"] by A[lexandre] L[efebvre], Rev. Mag. Zool., [ser. 1], 1: 91–96, 1838; "Jean-Théodore Cocteau," by S. Le Tourneur, Dict. Biogr. Franç., 9: 92, 1961. • *Portrait*: Muséum National d'Histoire Naturelle, courtesy Roger Bour. • *Signature*: Staatsbibliothek zu Berlin (Preussischer Kulturbesitz, Handschriftenabteilung. Shelfmark: Slg. Darmstaedter 3 d 1832), courtesy Wolfgang Hamm.

HOEVEN, Jan van der (1801–1868).

Jan van der Hoeven, naturalist, physician, comparative anatomist, and popularizer of science, published on amphibians and reptiles. He was born in Rotterdam (South Holland) on 9 February 1801. He was the son of a merchant who died when Jan was two; his mother remarried in 1810 to a physician. By 1819, Van der Hoeven had entered the University of Leiden where he wrote dissertations on the comparative anatomy of fish skeletons (Ph.D. 1822) and diseases of the ear (M.D. 1824). He was an honorary curator at the Rijksmuseum in Leiden for nine months in 1823. He traveled to Paris the next year where he met Georges Cuvier and P.-A. Latreille, and on his return worked as a physician in Rotterdam. From 1826 he became a professor of zoology at the University of Leiden, a position he held for the rest of his life. Here he lectured on a wide range of subjects including comparative anatomy, zoology, anthropology, and geology, reflecting his research interests.

Van der Hoeven was a disciple of Cuvier and adhered to his theory of catastrophism. He believed that multiple creations occurred over time and in different places including islands where the biotas were created independently. Like many earlier naturalists, he felt that by studying animals one could understand the perfection of the Creator. He was a prolific writer on a wide range of subjects, especially zoology and anthropology, and wrote popular texts on zoology for the general public including one for young people. He died in Leiden on 10 March 1868.

Van der Hoeven published several papers on amphibians and reptiles (about a dozen titles, 1832–1867, many of which were republished in other journals). These cover a miscellany

of topics including the carpus and tarsus of the Japanese giant salamander (*Andrias*); heart anatomy in crocodilians; systematics of European toads; and the blood of the mudpuppy of North America (*Necturus*). He authored two books on herpetological topics. The first, "Icones ad Illustrandas Coloris Mutationes in Chamaeleonte" (1831), discusses color change in chameleons and is illustrated with five engraved plates that are beautifully hand-colored; this is one of herpetology's rarest titles, issued in an edition of only 50 copies. The other book (1867) describes and illustrates the anatomy and osteology of an aquatic salamander, *N. maculosus*, with three lithographed plates drawn by Van der Hoeven.

It should be noted that a book on the anatomy of *Andrias japonicus* (1862), written by F. J. I. Schmidt and Q. J. Goddard, both medical doctors in Rotterdam, was co-authored by Van der Hoeven's only son, a physician and surgeon in Rotterdam, who was also named Jan van der Hoeven (see *Note*).

• *References*: "Levensberigt van Jan van der Hoeven," by P. Harting, Jaarb. Koninkl. Akad. Wetensch., 1868: 1–34, 1868; "Jan van der Hoeven," by P. Smit, Dict. Sci. Biogr., 6: 453–455, 1972. • *Portrait:* Nationaal Natuurhistorisch Museum, Leiden, courtesy L. B. Holthuis. • *Signature* (1840): Natural History Museum, London. • *Note*: Jan van der Hoeven, the son, was born in Leiden on 15 February 1834 and died in Rotterdam on 9 February 1900. To distinguish him from his father, his name is usually followed by "Jzn." or "Jan'szn.," abbreviations for "Janszoon," or "son of Jan").

GENÉ, Giuseppe (1800–1847).

Carlo Giuseppe Gené (see *Note*), Italian zoologist and author of the first herpetology of Sardinia, was born in Turbigo (Milan) on 9 December 1800. He attended local schools including the Collegio di Busto Arsizio, and in 1816 went to study philosophy and mathematics at the university in Pavia. There, while convalescing from a long and nearly fatal disease, he read some books on natural history, which changed his career interests. He was also influenced by one of his teachers, the embryologist and amphibian biologist, Mauro Rusconi. After graduation in 1821 Gené returned home and then, for about six years, traveled extensively in Italy (in Liguria, Lombardy, and the Piedmont) and in 1828 went to Hungary. His interests then were primarily in insects and he published extensively on them. From 1827 he held the post of assistant to the natural history chair at Pavia, and he also developed strong ties to naturalists in Turin. In 1831 he became a professor and director of the Museo di Zoologia at the Università degli Studi di Torino, positions he held for the rest of his relatively short life.

Gené's interest in Sardinia—an island kingdom lying in the Mediterranean Sea about 200 km off the west coast of the Italian Peninsula—began as the result of political developments. Carlo Alberto, who became king of Sardinia in 1831, promoted scientific exploration of his kingdom. (Sardinia was not incorporated into Italy until 1861, when Carlo Alberto's son, Vittorio Emanuele II, began the unification of Italy and became its first king.) As a direct result, from 1833 to 1838 Gené made four excursions to the island in order to make zoological collections, of inverte-

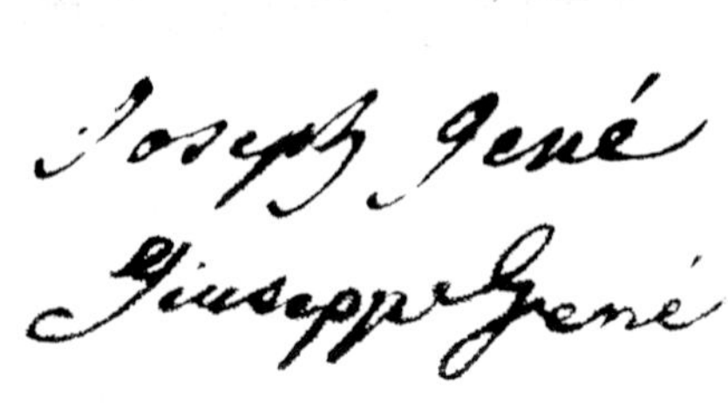

brates and vertebrates, and to publish on the zoology of Sardinia. He died in Turin on 14 July 1847.

Most of Gené's published work is on entomology and other zoological topics. As for herpetology, he published about a half dozen papers (1833–1846). Some of these dealt with his identifications of species in Charles L. Bonaparte's atlas of Italian herpetology (1832–1841). Gené's most important herpetological contribution was a long paper, entitled "Synopsis Reptilium Sardiniae Indigenorum" (1839), illustrated with five highly detailed hand-colored plates. This first review of the amphibians and reptiles of Sardinia lists and describes 20 species, among them several new species including the salamander *Euproctus rusconii* (named for his former teacher, but today known as *E. platycephalus*) and a gecko, *Phyllodactylus europaeus* (now *Euleptes europaea*). Gené also established the genus *Euproctus* in this paper. His herpetological collections are in the museum at the university in Turin.

• *References*: Bruno, 1989 (Gené on pp. 25–26); "Giuseppe Genè [*sic*]," by M. Alippi Cappelletti, Diz. Biograf. Ital., 53: 70–73, 1999. • *Portrait*: National Library of Medicine, Bethesda. • *Signatures* (above, 1842): Natural History Museum, London; (below): National Library of Medicine, Bethesda. • *Note*: Gené's name is given in its Latin form on his paper on Sardinia, Josepho Gené.

TROOST, Gerard (1776–1850).

Gerard Troost, Dutch geologist and mineralogist, was an early naturalist of the American Western Frontier. He was born in 's Hertogenbosch (=Bois-le-Duc), Holland, on 15 March 1776 and studied medicine at the university in Leiden (M.D. 1801). Louis Bonaparte, installed as king of Holland by his older brother, Napoleon Bonaparte, appointed Troost in 1807 to establish a royal collection of minerals. Troost thus came to study in Paris, but with the imminent incorporation of the Netherlands into France he decided to join his brother in the Dutch East Indies. He departed in 1810 and traveled via the United States where instead he decided to stay, settling in Philadelphia. In 1812 he was one of the founders and first president of the Academy of Natural Sciences. During 1825–1827 he joined William Maclure, Thomas Say, and C.-A. Lesueur at the utopian settlement in New Harmony (Indiana) but, disillusioned, Troost moved to Nashville (Tennessee) where he became professor of geology and mineralogy at the University of Nashville and, during 1831–1850, state geologist. He died in Nashville on 14 August 1850.

Troost's work on reptiles was incidental to his main studies in geology and paleontology, which took him and his field apprentices to many previously unsurveyed parts of the so-called Western Frontier, a group of a dozen states stretching from Minnesota to Alabama. He was an indefatigable field naturalist and in 1848, at the age of 72, he completed a grueling tour of about 14,500 km throughout the eastern part of the United States and Canada, of which nearly 1,700 km was on foot.

Troost's most outstanding herpetological discovery was *Macrochelys temminckii*, the largest freshwater turtle in the Western Hemisphere, which he found near the Mississippi River. He named this turtle in 1835 for C. J. Temminck, director of the natural history museum in Leiden, to whom Troost sent specimens throughout his life. His only other contribution on reptiles was a paper on Tennessee snakes (1836) in which he named the western race of *Agkistrodon piscivorus*. In 1836 John E. Holbrook named the Cumberland race of *Trachemys scripta* for Troost, based on a specimen sent by him from eastern Tennessee.

• *References*: "Troost (Gerard)," by J. van Baren, Nieuw Nederl. Biograf. Woordenb., 5: 974, 1921; "Gerard Troost," by L. C. Glenn, Dict. Amer. Biogr., 9: 647–648, 1935; "The Geological Sciences in the Antebellum South," ed. by J. X. Corgan, Univ. Alabama Press, University, Alabama, 195 pages, 1982 (Troost on pp. 39–72); "Gerard Troost," by J. X. Corgan, Amer. Natl. Biogr., 21: 838–839, 1999. • *Portrait* (about 1848): From Corgan, 1982. • *Signature* (1840): Courtesy Marinus S. Hoogmoed.

CANTOR, Theodore (1809–1860).

Theodore Edward Cantor was a Danish physician and naturalist in the employ of the English East India Company who discovered some of the most distinctive species of Asian amphibians and reptiles. Born Theodor Edvard Cantor in Holmens, the naval base at Copenhagen, on 6 February 1809, he attended the Borgerdyd gymnasium (1821–1826), took his medical examinations at the Academy of Surgery in 1832, and defended his thesis at the university in Halle (M.D. 1833). A timely visit by his uncle, a botanist employed at the botanical gardens in Calcutta, aroused his interest in natural history and in tropical Asia.

In 1835 Cantor arrived in Calcutta and took a post with the East India Company as surgeon in the Bengal Medical Service and during 1837–1839 also in the Bengal Marine Survey. During his leisure time he conducted local field surveys. A new opportunity came as the result of the Opium War (1839–1842), which resulted from the company's illegal importation of opium into China. Cantor volunteered to accompany the troops as an assistant surgeon and to collect plants and animals for the company's museum in Calcutta. His regiment bivouacked at Lantau Island (near Hong Kong) during June 1840 where he made collections, and in July they reached Chusan (now Zhoushan) Island, strategically located at the mouth of the Yangtze River just southeast of Shanghai. The Royal Navy bombarded the island and the marines successfully occupied it until March 1841, but increasingly they fell victim to malaria. While treating his patients Cantor developed "a violent cerebral fever," probably malaria, and he was ordered back to the ship in November after he had spent four months making collections and sketches of plants and animals.

From May 1842 to September 1845 Cantor was posted as the civil surgeon in charge of six hospitals on Prince of Wales (or Penang, now Pulau Penang) Island off the southwest coast of the Malay Peninsula and a seventh hospital on the adjacent mainland in Province Wellesley (now Seberang Perai). At that time, Penang, Province Wellesley, Malacca (now Melaka), and Singapore—all located along the Strait of Malacca opposite Sumatra and together known as the Straits Settlements—were administered by the East India Company from Calcutta. Cantor was able to conduct fieldwork throughout the settlements and on the plains in the western part of the peninsula. He spent most of the rest of his life at Fort William, site of the infamous prison cell later known as the "Black Hole of Calcutta." He died on 26 March 1860 but his biographers disagree on the circumstances. According to one he spent his last years working at an insane asylum in India where he himself later became deranged and was institutionalized. The most recent one reported that he died at sea en route from Malacca to Calcutta.

Cantor published about 16 herpetological papers (1836–1848) plus five other papers and a book having major herpetological content. The very first paper announced his most spectacular new species, *Hamadryas* (now *Ophiophagus*) *hannah*, which he found at two localities near Calcutta. This was a surprisingly late discovery given this cobra's huge size and widespread distribution. He published descriptions of numerous other new snakes including *Bungarus bungaroides, B. lividus, Trimeresurus mucrosquamatus*, and several colubrids.

The results of his Chinese collections from Chusan Island were reported in three papers under the series title "General Features of Chusan" (1842), which were re-issued the same year as a single booklet and in an exceedingly rare book, "Zoology of Chusan" (1842), with 12 hand-colored plates from drawings by Cantor. These must be consulted together for the technical descriptions of his new species occur only in the papers while the plates are only in the book. Cantor was forced to sketch his new species for posterity because he had left Calcutta on one day's notice and had no proper preservatives with him when he first arrived in Chusan. Several distinctive new species were named including *Bufo gargarizans, Mauremys mutica*, and *Naja atra*.

Cantor described several new species of reptiles from Malaya in a series of three papers (all 1847; reprinted in 1886, 1966, and again in 1981, the last with the author's name misspelled as "Canton"). These three papers have the general title "Catalogue of Reptiles Inhabiting the Malayan Peninsula and Islands" and were reissued in 1847 as a single booklet. He also wrote catalogues of the fishes of India (1838) and of Malayan mammals (1846) and fishes (1849).

Cantor's papers are filled with natural history notes often based on his own observations as well as information on elevational and latitudinal distributions. Some of his zoological specimens are in the Natural History Museum in London. His colored sketches and some manuscripts are also in the museum in London, the India Office Library in London, and in the Bodleian Library at Oxford University. Unfortunately, some of his specimens are mislabeled as to locality, especially those said to be from Penang Island.

• *References*: "Theodor Edvard Cantor," by J. Collin, Dansk Biograf. Lexikon, 3: 352, 1889; "Theodor Cantor," by R. H. Stamm, Dansk Biograf. Lexikon, 4: 493–494, 1934; "Natural History Drawings in the India Office Library," by M. Archer, H.M. Stationery Office, London, 1962 (Cantor on pp. 51–53, 76–77); "Francis Day (1829–1889) and His Collections of Indian Fishes," by P. J. P. Whitehead and P. K. Talwar, Bull. Brit. Mus. (Nat. Hist.), Hist. ser., 5, 1976 (Cantor on pp. 121–124). • *Portrait* (1841): Copenhagen University Medical-Historical Museum, courtesy Anna-Elisabeth Brade. • *Signature* (1847): Adler collection.

HYRTL, Joseph (1810–1894).

Joseph (or Josef) Hyrtl, who was one of the greatest medical scholars and the most brilliant teacher of anatomy of his era, published on the morphology of amphibians, reptiles, and other vertebrates. He was born on 7 December 1810 in Eisenstadt, Hungary (now in Austria, 12 km from the Hungarian border) where his father was oboist in the orchestra of Nikolaus II, Prince of Esterházy and patron of the classical composer, Franz Joseph Haydn. After the Austrian financial crisis of 1811, the prince's fortunes waned and the Hyrtl family moved to Vienna two years later. Joseph became a choirboy in the Habsburg palace chapel and attended the state boarding school. In 1831 he entered the University of Vienna to study medicine (M.D. 1835) and worked with the anatomist, Joseph Berres. Hyrtl became professor of anatomy at the University of Prague in 1837, but was recalled to his *alma mater* in 1845 to take the chair of Berres who had died. At Vienna he developed an outstanding career as a teacher and a technical anatomist. He is said to have lectured with a histrionic style like that of the Roman orator, Cicero. His anatomical preparations were world renown and were purchased by many academic institutions. One of them, his corrosive technique, involved injection of vessels or bone cavities with a stiffening material and then removal of the surrounding soft tissue or bone by maceration. This technique, for example, allowed his student, Alfonso Corti (see *Note*), an Italian anatomist, to discover the internal structure of the human auditory apparatus in the cochlea.

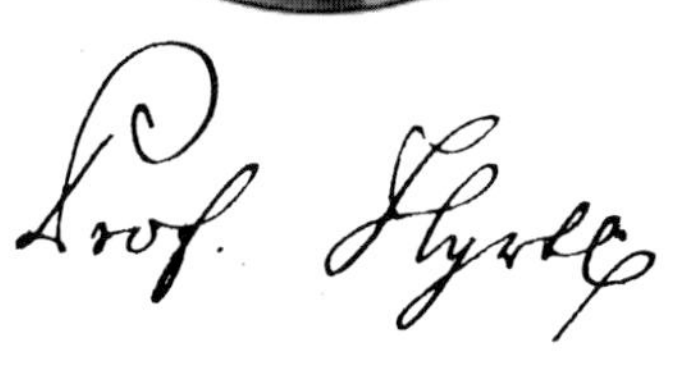

Because of his ambition, irascibility, and contempt for the physiological experimentation of some of his colleagues, Hyrtl became increasingly isolated at the medical school and was, for example, never elected dean. On the university's 500th anniversary in 1864–1865, however, he was named rector, as the most distinguished member of the faculty. In 1874 he retired a bitter but wealthy man to his estate at Perchtoldsdorf on the southern edge of Vienna where he continued his scholarship privately. He died there on 14 July 1894. Austria released postage stamps in his honor in 1937 and again in 1984.

Hyrtl gained international fame for his many important texts including "Lehrbuch der Anatomie des Menschen" (1846), "Handbuch der topographischen Anatomie" (1847), and his book intended to reform anatomical nomenclature, "Onomatologia Anatomica" (1880). He published about 20 papers specifically on amphibians and reptiles (1837–1864) including studies on the circulatory system (especially that of the lungs and heart), the tongues of amphibians, and the vertebrae of lizards. His most extensive herpetological work was his book—"*Cryptobranchus Japonicus*. Schediasma Anatomicum" (1865). This is an anatomical atlas of the Japanese Giant Salamander (*Andrias japonicus*) that was based on a 75-cm female specimen. It includes 14 fine lithographic plates drawn by Hyrtl and a scholarly summary of the animal's natural history.

• *References*: "Joseph Hyrtl. Sein Werk nach 100 Jahren," by V. Patzelt, Anat. Anz., 103: 160–175, 1956; "Professor Josef Hyrtl," by F. Wolf and G. Roth, Wissenschaftliche Reihe 4, Vienna, 151 pages, 1962; "Joseph Hyrtl," by J. Steudel, Dict. Sci. Biogr., 6: 618–619, 1972. • *Portrait*: Museum für Naturkunde, Berlin. • *Signature* (1846): Adler collection. • *Note*: Alfonso Corti (1822–1876) was the author of a monograph on the arterial system of *Varanus griseus* (1847), dedicated to his teacher, Hyrtl.

HALLOWELL, Edward (1808–1860).

Edward Hallowell, Jr., was a Philadelphia physician who, at a time when most naturalists had broader interests, became America's first specialist on herpetology. He was born in Philadelphia on 14 September 1808, the son of an owner of coal lands in Schuylkill County (Pennsylvania). He was a distinguished student at the University of Pennsylvania (A.B. 1827, A.M. and M.D. 1830) and continued his medical studies in Paris before opening a practice in Philadelphia. Pathology was his specialty and he counted among his patients the naturalist, C. S. Rafinesque. Hallowell was highly regarded as a medical practitioner and for his studies on cholera in infants. During his final years, he was invalided from tuberculosis and died in Philadelphia on 20 February 1860.

Hallowell's herpetological career was spent as a volunteer in charge of the collections of amphibians and reptiles at the Academy of Natural Sciences, which was a propitious post to have at that time. This was North America's first public natural history museum, founded in 1812, and during Hallowell's tenure the country's largest. The academy was the recipient of numerous specimens from correspondents and friends throughout the country and around the globe. Many represented species new to science. Between the museum's early period—when Thomas Say, Jacob Green, Richard Harlan, John E. Le Conte, and others were active there—and the arrival of Edward D. Cope in 1859, all of this material came into the hands of one person, Edward Hallowell.

As a result, Hallowell named many new species and genera. In North America, 15 species of amphibians and reptiles currently bear his authorship including some of the most remarkable ones: *Gastrophryne carolinensis, Aneides lugubris, Sceloporus magister,* and *Crotalus cerastes*, plus new genera such as *Dipsosaurus* and *Coniophanes*. He also named the family Ambystomatidae. The museum received specimens from military officers and topographic engineers working in the American West, from government voyages, and from its correspondents, missionaries, and traders in Africa, Central America, the West Indies, Siam, the Philippines, and many other places. He often named new species for his friends, thus explaining the otherwise odd use of the names of American naturalists for some Old World species (for example, West African snakes named for William Blanding and J. P. Kirtland).

In all, he authored about 55 herpetological papers (1839–1861), mostly in Philadelphia journals, and two titles in the government railroad survey volumes (1859; reprinted 1978), which contained large sections on natural history. In addition he reported on the collections made by physician-naturalist Samuel W. Woodhouse on three separate government surveying expeditions during 1849–1851, through a vast region stretching from present-day Oklahoma to southern Arizona. These were reported in his chapter in Lorenzo Sitgreaves's book on the third of these surveys, down the Zuñi and Colorado rivers (1853; reprinted 1962). His final publication, issued posthumously under the editorship of Cope (1860 [1861]), was his most diverse geographically, covering the materials collected by the naturalist, William Stimpson, during a circumnavigation by the United States North Pacific Exploring Expedition of 1853–1856 that visited Java, China, Japan, California, and many Pacific islands.

Although most of his work was taxonomic, Hallowell did make occasional natural history observations. Perhaps his most important was the first recognition that the land and aquatic stages of the eastern newt, which theretofore had been classified by others as separate species and even different genera, were conspecific (1856 [1858]). It was another medical doctor and amateur herpetologist, Howard A. Kelly, who in 1878 provided the final proof by inducing efts to transform into adults.

• *References*: "Edward Hallowell," p. 515–516 *in* H. A. Kelly and W. L. Burrage, Dictionary of American Medical Biography, D. Appleton, New York, 1928; "Edward Hallowell," p. 115 *in* C. A. Elliott, Biographical Dictionary of American Science, Greenwood Press, Westport, Connecticut, 1979; Alumni Records, University Archives, University of Pennsylvania, courtesy Michele L. Aldrich. • *Signature* (1846): Academy of Natural Sciences, Philadelphia.

STORER, D. Humphreys (1804–1891).

David Humphreys Storer, physician, ichthyologist, and medical school dean, was author of the first herpetological survey of a state in the United States. He was born in Portland (Maine, then part of Massachusetts) on 26 March 1804, and attended Bowdoin College (A.B. 1822) and Harvard Medical School (M.D. 1825) before opening a medical practice in Boston, specializing in obstetrics. He later was a professor at Harvard Medical School (1854–1868) and served as the school's dean for most of this period. He died in Boston on 10 September 1891.

Storer was an early member of the Boston Society of Natural History and originally had an interest in mollusks, but it was his appointment to the Massachusetts Zoological and Botanical Survey in 1837 that turned his attention to fishes, amphibians, and reptiles. In 1839 he authored a 68-page report, "Reptiles of Massachusetts," part of a larger work, "Reports on the Fishes, Reptiles and Birds of Massachusetts." This represented the first comprehensive herpetology of any state in the United States. The report on reptiles was republished in 1840, in journal form, by the Boston Society of Natural History, but the supplement—in which he commented on a report (1817) of a marine "serpent" seen in the harbor at Gloucester and which he named *Scoliophus atlanticus*—was wisely omitted. Storer's report, however, was soon overshadowed by publication of the much larger and more elaborate work by James E. De Kay on the amphibians and reptiles of New York State (1842). Other than his state report, Storer's only herpetological titles were a few short papers in the *Proceedings of the Boston Society of Natural History*. He did go on to publish several major works on fishes, however.

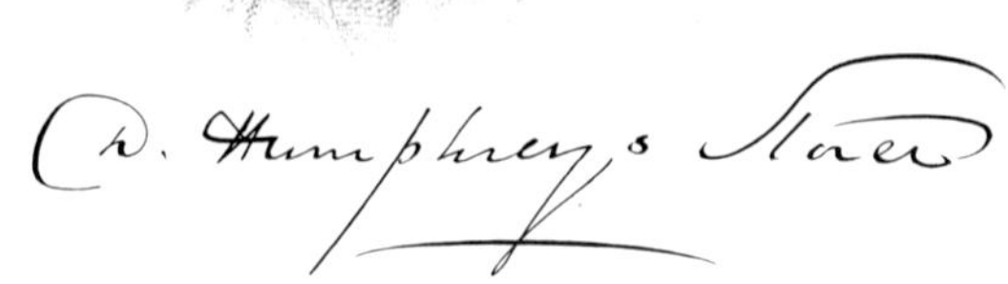

In his state report, Storer described a new snake, *Coluber* (now *Storeria*) *occipitomaculata*, a widespread and common species that somehow had escaped the notice of previous naturalists. In 1853 Spencer F. Baird and Charles Girard honored Storer by erecting the genus *Storeria* for the North American Brownsnakes.

• *References*: "Dr. D. Humphreys Storer," p. 77–80 *in* Anniversary Memoirs of the Boston Society of Natural History, Boston, 1880; "David Humphreys Storer," by H. R. Viets, Dict. Amer. Biogr., 18: 93–94, 1936; "David Humphreys Storer," by C. A. Elliott, Amer. Natl. Biogr., 20: 883–884, 1999. • *Portrait and signature*: Anniv. Mem. Boston Soc. Nat., Hist., opposite p. 80, 1880.

BERTHOLD, A. A. (1803–1861).

The German physiologist and pioneer endocrinologist, Arnold Adolph (sometimes Adolf or Adolphe) Berthold, made significant contributions to systematic herpetology. He was born on 26 February 1803 in Soest (North Rhine–Westphalia) about 50 km east of Dortmund. His father was a cabinetmaker. Berthold began his university studies in Göttingen in 1819 where he was a student of Johann Blumenbach. After finishing (M.D. 1823), he visited Berlin and Paris for further study. He returned to Göttingen as a lecturer in 1825 and was promoted to professor ten years later. Because his university pay was meager, Berthold was also a practicing physician throughout his life. He succeeded Blumenbach in 1840 as curator of the university's zoological museum. Berthold died in Göttingen of typhus on 3 February 1861 at not quite 58 years of age.

Berthold had a wide-ranging and distinguished career and only began his systematic work after becoming curator of the museum. Together with a university colleague, Robert Bunsen, inventor of the burner still used in biology and chemistry laboratories, he discovered that hydrated iron oxide could be used as an antidote for arsenic poisoning (1834). Berthold conducted other research on length of pregnancy, myopia, male hermaphroditism, and the formation of hair and fingernails.

Berthold's most important discovery involved one of the simplest yet most profound biological experiments ever performed. By castrating six young male chickens, he was able to retain all of the male secondary sexual characteristics, which are now known to be under hormonal control, whenever one testis was allowed to lie free in the bird's abdominal cavity. Male characteristics remained even when a testis was transplanted to the body cavity of one of the other birds. This result, which was published in 1849 in a five-page paper, went unnoticed until 1905 when a German physiologist, Moritz Nussbaum, rediscovered it when conducting similar experiments with salamanders. The parallels here to the rediscovery of Gregor Mendel's papers on pea genetics are remarkable. Berthold is regarded today as the founder of experimental endocrinology.

Berthold's first book (1827), ironically, was in systematics, an annotated German translation of P.-A. Latreille's "Familles Naturelles du Règne Animal" (1825). In 1829 he authored a widely used textbook on the physiology of humans and other animals that was reissued several times. Berthold's systematic work with amphibians and reptiles consists of about 15 papers (1840–1859) in which he described 26 new species, among them several that are currently recognized: *Dendrobates histrionicus, Polychrus gutturosus,* and *Bothriechis schlegelii.* All of the materials reported in his taxonomic papers were in the Göttingen collection (see *Note*).

Berthold's physiological interests are reflected in some other papers. He disproved a long-held fallacy, prevalent in Europe and especially in Germany, which claimed that frogs, lizards, and snakes could live for years in the human stomach. For more than seven years he collected frogs that had allegedly been vomited from subjects. He often found freshly ingested insects in the frogs's stomachs, which would be impossible in frogs that had been long resident in a person's stomach. He also conducted experiments in which live frogs were immersed in warm water and discovered they could not survive for more than a few minutes. After his paper was published (1849), no serious scientist ever supported the claims, although newspaper reports continued to appear in Europe and America as late as the 1920s.

• *References*: "Arnold Adolf Berthold," by A. Hirsch, Allg. Deutsch. Biogr., 2: 512, 1875; "Arnold Adolphe Berthold," by M. Klein, Dict. Sci. Biogr., 2: 72–73, 1970; Bondeson, 1990 (see pp. 162–163); "On the Type Specimens of Two Colombian Poison Frogs Described by A. A. Berthold," by C. W. Myers and W. Böhme, Amer. Mus. Novit., 3185: 1–20, 1996; Rieck et al., 2001 (Berthold by W. Böhme on pp. 421–423). • *Portrait*: Courtesy Wolfgang Böhme. • *Signature* (1828): Deutsches entomologisches Institut, Müncheberg. • *Note*: In 1977 the entire Göttingen zoological collection was transferred to the Museum Alexander Koenig in Bonn.

De Kay, James E. (1792–1851).

James Ellsworth De Kay (see *Note*) was a physician, writer, and promoter of the Lyceum of Natural History of New York (now New York Academy of Sciences), and wrote the celebrated series, "Zoology of New York." He was born in Lisbon, Portugal, on 12 October 1792, the son of an American sea captain who moved the family to New York City in 1794. He attended Yale University (A.B. 1812) and the University of Edinburgh (M.D. 1819). On returning to New York City he practiced medicine occasionally, but mostly devoted himself to natural history studies and to the development of the Lyceum of Natural History where he served in many official capacities including editor of its *Annals*. He was also a writer about American art and a member of New York literary circles where he counted among his close friends the poet, William Cullen Bryant, and the novelist, James Fenimore Cooper. He retired to Oyster Bay on the north shore of Long Island (New York) in about 1844 and died there on 21 November 1851.

In 1836 De Kay was asked to direct the zoological portion of the natural history survey of New York, excluding paleontology, which culminated in the most elaborate series of volumes on the topic ever produced about any state in the United States: five large volumes, in quarto, with numerous lithographed plates (published 1842–1844). Amphibians and reptiles were combined with fishes in part 3, issued in two volumes in 1842 (reprinted 1978), with 102 excellent plates drawn by J. W. Hill; 300 sets were issued in which the plates were hand-colored. Zoology was but one part of the series, entitled "Natural History of New York," which ran to 30 volumes (1842–1894) and is one of the great monuments of 19th Century American natural history. De Kay wrote the entire zoology section. The scope of his volumes covers New York as well as surrounding regions and thus lists extralimital species, so the usefulness of the work extends well beyond the state's borders. The species accounts are detailed and, given that De Kay was in contact with the leading American naturalists of the day, were up-to-date.

De Kay produced no other works on living amphibians or reptiles but did write several papers on fossil animals including two on extinct reptiles found in New Jersey. *Storeria dekayi*—a small snake that is widespread in eastern North America—was named by De Kay's friend, John E. Holbrook, in his masterwork, "North American Herpetology" (1836 [1839]). The description was based on a specimen found by De Kay on Long Island.

• *References*: "James Ellsworth De Kay," by W. E. Shaw, Dict. Amer. Biogr., 5: 203–204, 1930; "James Ellsworth De Kay," by K. B. Sterling, Amer. Natl. Biogr., 6: 356–357, 1999; "New York State Natural History Survey 1836–1842," by M. L. Aldrich, Paleontol. Res. Inst. (Spec. Publ. 22), Ithaca, New York, 2000 (De Kay on pp. 163–168); Damkaer, 2002 (De Kay on pp. 206–213). • *Portrait and signature* (above, 1826): American Museum of Natural History, New York; *signature* (below): From H. L. Fairchild, "A History of the New York Academy of Sciences," publ. by author, New York, 1887. • *Note*: He spelled his name "De Kay" on his publications, but it is sometimes written DeKay or even Dekay in the literature.

Fischer, J. G. (1819-1889).

The German vertebrate zoologist, Johann Gustav Fischer (see *Note*), a specialist on the morphology and taxonomy of amphibians and reptiles, was born in Hamburg on 1 March 1819, the son of a bank accountant. After finishing at the Akademisches Gymnasium Johanneum, a school in Hamburg, he studied philosophy and science at the universities in Halle, Leipzig, and Berlin. He earned his doctorate in Berlin (Ph.D. 1843) under the direction of Hinrich Lichtenstein, director of the university's museum. Fischer's dissertation, "Amphibiorum Nudorum Neurologiae" (published in two parts, both 1843), was a study of the cranial nerves in nine species of frogs, salamanders, and caecilians. It is a remarkably detailed and accurate study for its time, and for salamanders and caecilians represented the first study ever undertaken of their nervous systems.

At that time in Germany many zoologists with doctorates found employment at secondary or high schools (gymnasiums) and so it was for Fischer. During 1847–1856 he taught mathematics and science alternately at the two divisions of the Johanneum—the Realschule, a secondary school for science and modern languages, and his former school, the Akademisches Gymnasium—except for 1851–1853, which he spent at home because of illness. In 1856 he co-founded and taught at a private school (Höhere Bürgerschule) in Hamburg. Fischer had a formal association with the Naturhistorischen Museum, which at that time was housed in the same building as the Johanneum. During 1847–1857 and 1877–

1879, he served as a volunteer in charge of the museum's collections of fishes, amphibians, and reptiles. He died in Hamburg on 27 January 1889.

Following up on his thesis research, Fischer's next publication (1852) was a major work on the cranial nerves in 14 species of lizards and crocodilians. His last major anatomical study (1864) was on the mandibular and hyoid arches and their musculature and the cranial nerves of a caecilian and seven species of permanently aquatic salamanders. All of

Fischer's other titles on amphibians and reptiles (about 20 papers, 1855–1888), with the exception of one on the anatomy of *Heloderma horridum* (1882), were taxonomic descriptions of new or poorly known species from throughout the world: Australia, Borneo, California, Cameroon, Colombia, Formosa (now Taiwan), Haiti, Mexico, New Zealand, and Siam (now Thailand), to name a few places. In addition to the Hamburg collection, which today holds most of his collections, Fischer had access to the museums in Berlin, Bremen, Dresden, and Stuttgart, as well as to some private collections. He described about 110 new species of which about one-third are currently recognized.

In three instances, Fischer's papers were published twice, which has caused bibliographic confusion. For example, his monograph of the sea snakes, "Die Familie der Seeschlangen"—in which he included 20 species (two species and three varieties described as new) and his new genus *Astrotia*—was first issued in 1855 in the school program of the Johanneum. It was a standard practice at gymnasiums in Central Europe in the 19th Century to allow staff to publish in school publications, because teachers normally did not have access to the usual outlets available to researchers at scientific institutions. The articles in the school programs were sometimes reprinted elsewhere and often altered, but the school versions have priority of publication. In the case of Fischer's sea snake monograph, it was reprinted in 1856 by a Hamburg scientific society where it was clearly labeled as the second edition. It is ten pages longer than the original and there are three hand-colored plates added, but the new taxa were first named in 1855.

• *References*: "(Johann Gustav) Fischer," p. 314–315 *in* H. Schröder, Lexikon Hamburg. Schriftst. Zeit., 1854; "Johann Gustav Fischer," by C. Gottsche, p. xliv–xlv *in* Jahrb. Hamburg. Wissenschaftl. Anstalt., vol. 6, 1889; "Johann Gustav Fischer," anonymous, Mitt. Hamburg. Zool. Mus. Inst., 90(suppl.): 107–110, 287–288, 313–314, 1993; Additional information from Jakob Hallermann, 2006. • *Signature* (1856): Deutsches Entomologisches Institut, Müncheberg. • *Note*: This Fischer is not to be confused with two others who published on reptiles during the 19th Century: Johann Georg Fischer (1816–1897), a teacher at a school and seminary in Neuzelle near Berlin and author of the series, "Beschreibung aller naturhistorischen Gegenstände" (volume 3, "Die Amphibien und Fische," 1832), and Baron Johann von Fischer (1850–1901), author of "Das Terrarium" (1884; reprinted 1989) and numerous popular articles on amphibians and reptiles, and one-time director of animal parks in Düsseldorf and Montpellier.

TROSCHEL, F. H. (1810–1882).

Franz Hermann Troschel, a German zoologist and editor, was born on 10 October 1810 in Spandau on the western outskirts of Berlin. When he was two, his family moved to Rheinsberg, about 45 km northwest of Berlin, and it was here in the countryside that his interest in nature began. He attended school in nearby Neuruppin and at a boarding school in Berlin. He entered the university in Berlin in 1831 to study natural sciences and mathematics and became a student of A. F. A. Wiegmann and Johannes Müller. After finishing his doctorate in 1834, Troschel taught high school in Berlin during 1835–1849 and from 1838 also assisted at the university's Zoological Museum. In 1842 it was decided to separate responsibility for each vertebrate group, so in 1843 Troschel became the first curator of the herpetological collections. In 1849, however, he left Berlin to become a professor of zoology at the university in Bonn (North Rhine–Westphalia) and also assistant director of the natural history museum in Poppelsdorf at the western edge of Bonn. He became director in 1866.

Troschel's research interests were broad. His main focus was snails, but he also published on insects, spiders, fishes, and fossil animals. Amphibians and reptiles were only a small part of his scope. He taught regularly at the university, became full professor in 1851, and later served as rector. During 1854, 24-year-old Albert Günther, who was soon to begin his storied career at the British Museum, taught in Bonn and was mentored by Troschel. Troschel died in Bonn on 4 November 1882.

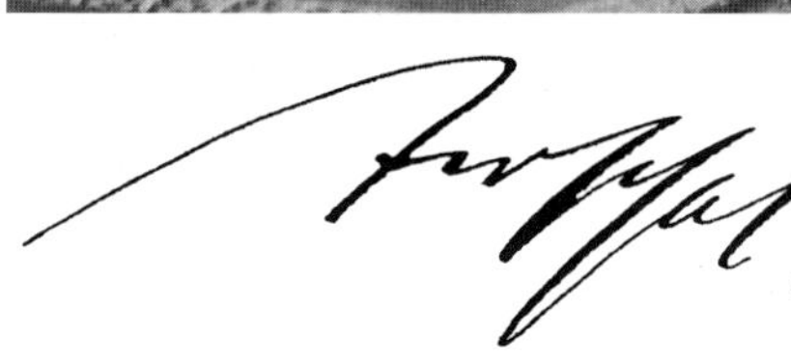

Besides his research program, Troschel had a long career as an editor. His former teacher, Wiegmann, was the co-founder and editor of *Archiv für Naturgeschichte*, a leading German natural history journal that began in 1835. An important innovation begun by Wiegmann—and the forerunner to *Zoological Record,* which was started by Günther in 1864—was to list all of the previous year's literature arranged by taxonomic group. When Wiegmann died prematurely in 1841, Troschel took over editorship of the section on amphibians and reptiles, entitled "Bericht über die Leistungen in der Herpetologie." This section was not a simple list of titles but often included long summaries and commentary by Troschel. In 1848 he became editor-in-chief of the *Archiv*. He held both posts for life.

Excluding his 41 annual literature summaries, Troschel authored about a dozen titles in herpetology (1848–1877). His first was a review of the amphibians and reptiles collected during the expedition of Richard Schomburgk to British Guiana (now Guyana) in 1840–1844. Several new herpetological taxa were named including a sideneck turtle (*Podocnemis unifilis*) and a tree frog (*Hyla calcarata*). Troschel's other herpetological papers are on diverse topics such as the morphology of *Heloderma horridum*; a new genus and species of North American lizard (*Cophosaurus texanus*); the mandibular teeth of snakes; and the natural history of a hynobiid salamander (*Onychodactylus japonicus*). In 1865 he summarized the herpetofauna of Mexico in Baron J. W. von Müller's volume on the history and zoology of that country. Among the 269 species Troschel listed were several new taxa including a rattlesnake (*Crotalus intermedius*). His specimens are at the zoological museums in Berlin and Bonn.

• *References*: "Zur Erinnerung an Dr. Franz Hermann Troschel," by H. von Dechen, Verhandl. Naturhist. Vereins Preuss. Rheinl. Westf. 40(Correspondenzbl. 1): 35–54, 1883; Rieck et al., 2001 (Troschel on p. 362; by W. Bischoff on pp. 627–629). • *Portrait and signature* (1855): Museum für Naturkunde, Berlin (BI/285).

GUICHENOT, Alphonse (1809–1876).

The French herpetologist and ichthyologist, Antoine-Alphonse Théodore Guichenot, reported on some of the great French explorations of the 19th Century. He was born on 31 July 1809 in Paris where his father was a gardener for the menagerie of the Muséum d'Histoire Naturelle. In February 1828 Alphonse became a zoology student at the museum and in 1833 was appointed preparator to the professor in charge of reptiles and fishes, Constant Duméril, then the leading systematic herpetologist in the world. His job was to help Gabriel Bibron, Duméril's principal assistant, to process the large collections of amphibians and reptiles then flowing into the museum from overseas that needed to be incorporated into Duméril and Bibron's nine-volume "Erpétologie Générale" (1834–1854). Specimens had to be measured, scales counted, and literature checked in order that species identifications could be made. Guichenot learned these tasks well and they became the main work of his career. In 1839 he joined a scientific commission that explored in Algeria for three years. On his return, he resumed his post in Duméril's laboratory and in 1856 was duly promoted to assistant naturalist, his final rank. He retired in 1872 and moved to Cluny (Saône-et-Loire) in eastern France where he died on 17 February 1876.

Guichenot exemplified the role that many junior naturalists played as part of Duméril's grand design to monograph the amphibians and reptiles of the world. Once Guichenot worked up particular collections, Duméril would assimilate all of the species—including most of the new ones, which he and Bibron would then name—into the "Erpétologie Générale." This plan left the assistant to publish the expedition reports, which contained the more detailed species descriptions and accompanying illustrations. These routinely appeared much later since expedition reports were usually published in many parts that were spread out over a decade or more. Although most of the new species had been described, these reports are invaluable in providing much more detailed species descriptions, as well as illustrations that rank among the finest ever produced. These reports, therefore, represent essential supplements to the "Erpétologie Générale."

Guichenot wrote the herpetology, and sometimes the ichthyology, sections for five such expeditions. The first was Claude Gay's survey of Chile during 1828–1842, which resulted in Gay's 28-volume "Historia Fisica y Politica de Chile" ([1843–] 1844–1871). This included Guichenot's section on herpetology (1848 [1849?]). It listed 44 species including four new ones of which one, *Bufo rubropunctata,* is recognized today. There are seven exquisite hand-colored plates of amphibians and reptiles in the folio atlas. Guichenot next reported on the herpetology of the Algerian expedition of 1840–1842 in which he participated. His section was published in 1850 and contains four excellent hand-colored plates. Only two new taxa remained to be named, a gecko in the genus *Stenodactylus* and the colubrid genus *Macroprotodon.* Third in order of publication was his report (1852) on the herpetology of the 1839–1843 French expedition to Abyssinia (now Ethiopia) led by Théophile Lefebvre. The folio zoology atlas contains four excellent hand-colored plates but no new species were named.

Guichenot's next report was co-authored with Honoré Jacquinot and summarized the collections obtained in the Pacific Basin during the circumnavigation of the French corvettes "Astrolabe" and "Zélée," the so-called "Voyage au Pôle Sud" commanded by Jules Dumont d'Urville during 1837–1840. Jacquinot was the expedition's surgeon and later edited the scientific report. The herpetology text was published in 1853 but the dozen folio plates, all beautifully hand-colored, were issued during 1842–1852. Duméril and Bibron had already published all of the new species. Finally, the most elaborate herpetological work authored by Guichenot was issued in 1855 as part of the series "Animaux Nouveaux ou Rares . . . de l'Amérique du Sud." This resulted from an expedition in 1843–1847 led by Francis de Castelnau that crossed and recrossed South America, from Rio de Janeiro to Lima and then down the Amazon to Belém. Firmin Bocourt, another Duméril assistant, drew the 18 hand-colored plates. Among the 101 species covered were several new lizards: *Enyalioides laticeps, Gonatodes humeralis*, and *Uracentron flaviceps*.

In addition to these reports, which are the main legacy of this junior naturalist's career, Guichenot also authored a half dozen journal articles on reptiles (1851–1866). In his last paper he named his most remarkable new species, from New Caledonia, a French possession in the South Pacific. This gecko has a distinctive crest of soft spines on the head and nape that immediately distinguishes it from all other geckos—the giant *Rhacodactylus ciliatus*, which has now become popular with herpetoculturists.

• *References*: Bauchot et al., 1990 (Guichenot on p. 89); Jaussaud and Brygoo, 2004 (Guichenot on pp. 267–268); Additional information from Édouard-Raoul Brygoo, 1995. • *Signature* (1864): Deutsches Entomologisches Institut, Müncheberg.

BLEEKER, Pieter (1819–1878).

The Dutch ichthyologist, Pieter Bleeker, who authored the great atlas of the fishes of the Dutch East Indies (now Indonesia), also published on the region's amphibians and reptiles. He was born on 10 July 1819 to a middle-class family in Zaandam (North Holland). He left primary school at the age of 12 to apprentice with an apothecary in Amsterdam for three years during which his interest in anatomy and zoology began. He then attended a clinical school in Haarlem and in 1840 qualified as a surgeon and country doctor. He attempted to get a position at the Rijksmuseum in Leiden. There were no openings, so he went to Paris for six months to attend lectures by Henri-Marie de Blainville, Georges Cuvier's successor as professor of comparative anatomy at the Muséum d'Histoire Naturelle, and to work in local hospitals. In 1841 he received a commission as a military surgeon in the Netherlands East Indies Army and arrived in Batavia (now Jakarta), Java, in March 1842 after six months at sea. During 1842–1847 he began his studies of fishes in Batavia, but after irritating some high government officials, he was transferred to remote hospitals in Java where he conducted no research.

Bleeker returned to Batavia in 1849 to train native students at the medical school there and to continue his research. In 1850 he was the principal founder of the Natuurkundige Vereeniging in Nederlandsch Indië, a society that issued an important journal—*Natuurkundig Tijdschrift voor Nederlandsch Indië*—in which he published most of his papers. He only made one trip outside Java and adjacent Madura: to the Moluccas (now Maluku) and Celebes (now Sulawesi) in September–October 1855. He was dependent on Dutch civil servants and army surgeons to send him specimens from other localities. In September 1860 he returned to the Netherlands, collecting en route home. He first went to Leiden and then in 1863 to 's Gravenhage (called Den Haag, or The Hague, in the spoken language), where he continued his research. He brought with him an enormous collection of fishes and other animals that was auctioned after his death. The herpetological specimens today are at the natural history museums in Leiden and London. He died on 23 January 1878 (see *Note*) in 's Gravenhage.

Bleeker's main research contributions were to ichthyology; he authored more than 500 titles containing 511 new genera and about 1,925 new species. He had the good fortune to be posted at an early date to what is the richest marine and freshwater fish fauna in the world. His major work on this fauna, "Atlas Ichthyologique des Indes Orientales Néerlandaises," in nine volumes (1862–1878), is one of the most magnificent books on fishes ever produced; it was reprinted in 1977–1988 with previously unpublished plates added. By comparison, Bleeker's herpetological output was modest, about 15 formal papers (1850–1863), plus numerous lists of collections received and identified by him. Among these, however, are many new taxa, with several currently recognized species: a caecilian (*Ichthyophis monochrous*), several snakes (including four new *Calamaria*), agamids (*Gonocephalus megalepis* and *Lophocalotes ludekingi*), and a new genus and species of gecko (*Hemiphyllodactylus typus*).

• *References*: "The Autobiography of Pieter Bleeker," by M. Boeseman, p. 1–71, *in* P. Bleeker, Collected Fish Papers of Pieter Bleeker, vol. 1, Smithsonian Inst., Washington, 1973; Fransen et al., 1997 (Bleeker on pp. 212–213). • *Portrait* (1872): Natural History Museum, London, courtesy A. F. Stimson. • *Signature* (1850): Nationaal Natuurhistorisch Museum, Leiden, courtesy L. B. Holthuis. • *Note*: The commonly published death date, 24 January, is incorrect (*fide* 's Gravenhage town archives; L. B. Holthuis, pers. comm.).

CORNALIA, Emilio (1824–1882).

The Italian naturalist, Emilio Baldasasse Maria Cornalia, who was born to a noble family in Milan on 25 August 1824, wrote on amphibians and reptiles. He attended Milanese schools and then, on his father's wish, went to study law at the Università di Pavia in 1842. He soon switched to medicine in order to study natural sciences, in particular geology, and he received his bachelor's degree in medicine in 1848. During his university studies he became an assistant at the university's museum, and

in 1849 took a similar post at the Museo Civico di Storia Naturale di Milano. The museum in Milan, founded in 1838 as Italy's oldest civic natural history museum, began with the donation of two large collections—one amassed by a local nobleman, Giuseppe De Cristoforis, and the other by Giorgio Jan, a botanist originally from Vienna. Jan was the first director and in 1851 Cornalia became the assistant director. By 1853, Jan turned to herpetology and at age 62, jointly with Ferdinando Sordelli, began work on his classic iconography of snakes. Cornalia thus assumed more administrative duties and in 1866 became director at Jan's death. Cornalia was a distinguished if eclectic scientist, publishing in zoology, geology, paleontology, comparative anatomy, and especially in entomology. He traveled widely in Europe and during 1873–1874 in the Middle East. He died in Milan on 8 June 1882.

Herpetology was a small but significant part of Cornalia's scientific output. During 1861–1873 he published five papers on diverse topics including dentition in the snake *Herpeton*, behavior of axolotls, and the frogs of northern Italy. His most significant contribution to herpetology is contained in his book, "Vertebratorum Synopsis in Museo Mediolanense Extantium" (1849). He named eight new species of vertebrates from the museum's collection including a toad (*Atelopus ignescens*) from Ecuador and an Amazonian sideneck turtle (*Podocnemis sextuberculata*).

• *References*: "Commemorazione del Prof. Emilio Cornalia," by L. Maggi, Rend. Reale Istit. Lombardo Sci. Lett., ser. 2, 17: 42–55, 106–155, 1884; "Emilio Cornalia," by F. Di Trocchio, Diz. Biograf. Ital., 29: 114–115, 1983. • *Portrait*: Hunt Institute, Pittsburgh. • *Signature* (1852): Natural History Museum, London.

De Betta, Edoardo (1822–1896).

Francesco Edoardo De Betta was an Italian nobleman, politician, and patriot who produced the first scientific monographs of his country's herpetofauna. He was descended from an eminent Spanish family of Galicia—the Italian branch dates back to 1100—and he was born in the ancestral castle of Malgolo in the town of Rovereto (Trento) on 6 June 1822. As a child he was intensely interested in animals but yielded to his father's demand that he pursue a legal career. He attended secondary school in Milan, finishing in 1839, then went to the university in Pavia to study law and graduated in 1844. He returned to Milan to practice law, first in civil court and then also in criminal court. In February 1848 his cousin, the widow of a nobleman, died in Verona, and De Betta was named sole heir to her large fortune. Since he much preferred natural history to law, he immediately abandoned his legal career and in 1849 moved to Verona to administer the estate and devote his life to natural history. During 1848–1853 he spent all of his free time making collections in the southern Tyrols, Trentino, and in the province of Verona, covering a broad area of northern Italy from approximately the Po River northward to the Austrian border. These collections and his study of them would form the basis of his life's scientific work, but soon politics would intervene.

Since Napoleon's defeat at Waterloo, Verona had been part of the Austro-Hungarian Empire, but nationalism was stirring in the 1850s. In 1854 De Betta was elected a councilor of the Verona city government, a post he held for most of the rest of his life. In 1857 he helped to found the Associazione Agraria Provinciale, ostensibly to promote agricultural interests, but in reality it was a front to unite the citizens of the province against Austria and to prepare them for union with the Kingdom of Italy. Austria objected to the organization and had its

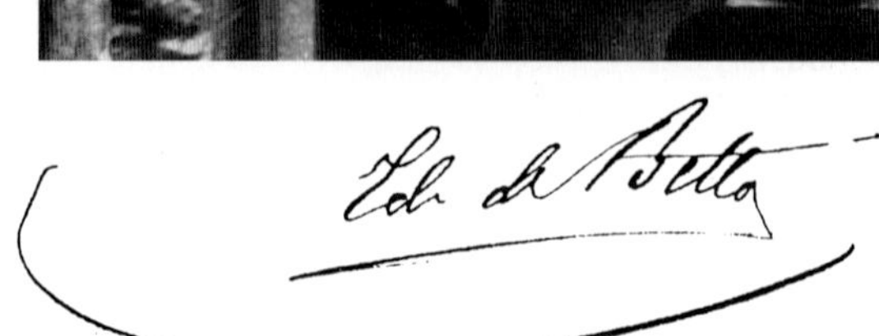

police put the leaders under surveillance. The group had to go underground and everyone, including De Betta, went by pseudonyms (his was "Oen"). He had been elected mayor of Verona in 1865, after which he was bluntly told by the city's Austrian commander (in translation): "Remember, Mr. Mayor, that in case of an uprising, the first ball [bullet] will be for you!" Nevertheless his political work continued, and he eventually had the honor to be one of the three commissioners involved in the transfer of the province to Italy in October 1866. De Betta remained active in city council affairs until 1891. He died on 4 November 1896 in Sogara di Marcellise, a small town of San Martino Buon Albergo (Verona).

De Betta produced about 50 zoological titles covering mollusks, insects, and all vertebrate classes. Nearly 20 of these are on herpetology (1852–1885) and collectively they represent the first comprehensive scientific studies of Italian amphibians and reptiles. Charles L. Bonaparte, a nephew of Napoleon, wrote an earlier monograph on the Italian herpetofauna (1832–1841) that dealt primarily with classification and illustration. De Betta's works, on the other hand, provided in-depth, critical reviews of taxonomic groups and specific regions. Collectively, his catalogue of the reptiles of the Valle di Non and the Italian Tyrol in the eastern Alps (1852), the poisonous snakes of the province of Verona (1856), his masterful 380-page herpetology of the province of Venice and the southern Tyrols (1857), a monograph of Italian salamanders (1864), and another on Italian *Tropidonotus* (1865) all led up to his grand summary, "Rettili ed Anfibi" (1874), part 4 of his series,"Fauna d'Italia." These works included species descriptions, synonymies, keys, distribution, and a critical evaluation of which species were native. De Betta wrote other substantial contributions, notably a review of the herpetofauna of Greece (1868) and another on the distribution of venomous snakes in Europe (1880). He named several new taxa of frogs, lizards, and snakes. His collections, which originally formed his private museum, are now in the Museo di Storia Naturale di Verona.

• *References*: "Edoardo De Betta," by I. Sermonti Spada, Diz. Biograf. Ital., 33: 381–382, 1987; Bruno, 1989 (De Betta on pp. 28–29). • *Portrait*: Hunt Institute, Pittsburgh. • *Signature* (1888): Natural History Museum, London, courtesy A. F. Stimson.

KELAART, Edward F. (1819–1860).

Beginning in 1638, Ceylon (now Sri Lanka) was mostly under Dutch control and among the colonists who arrived from Holland in the 1720s was an ancestor of the island's first native-born naturalist, Edward Frederick (sometimes Fredrick) Kelaart. He was born on 21 November 1819 inside the British fort in Colombo where his father worked as an apothecary for the British Forces. (Great Britain had taken control of Ceylon from the Dutch United East India Company in 1796, after France's invasion of the Netherlands.) His mother died when he was nine; his father, who was an amateur naturalist, raised him. Edward attended Kessen's School in Colombo until he was 14 and then volunteered as a medical assistant in the Colonial Medical Service, which a year later put him on salary as an assistant in the military hospitals. In 1837 he accompanied an Irish regiment returning home in order to reach Scotland where he entered the University of Edinburgh to study medicine

(M.D. 1841). He was the first Ceylonese to gain this degree. During this time he met Andrew Smith, himself an Edinburgh M.D. who had earlier written the first zoology of South Africa. Smith was then with the British Army Medical Service, which Kelaart was about to join as an assistant surgeon, and Smith encouraged him to study and report on the zoology of Ceylon. Kelaart briefly returned to Ceylon but was recalled to England and then posted to Gibraltar (1843–1845).

After spending 1845–1848 working again in England, Kelaart returned to Ceylon in 1849 where he proceeded to study the fauna intensively during his leisure time. In 1852 he was promoted to staff surgeon for the British Forces and in this capacity he was stationed in many parts of the island. This gave him ample opportunities to collect and make observations. The major localities were: Colombo (west coast), Galle (south coast), Trincomalee (east coast), and two sites in the central highlands, Kandy and Newera-Ellia (now Nuwara Eliya). He was never stationed on the north coast, but an ornithologist who lived in Jaffna, Edgar L. Layard, sent him specimens from there. During this period, Kelaart published several papers and issued two volumes of his major work, "Prodromus Faunæ Zeylanicæ" (1852 [1853?] and 1853 [1854], see *Note 1*; volume 1 reprinted 1998).

Kelaart was in England again in 1854–1856, after which he returned to Ceylon for his last tour of duty. In July 1860 he embarked for England. Although ships then traveled by way of Suez, the canal was not yet finished so Kelaart and his companions had to transfer to another steamer, thus requiring that they travel overland some distance in "a blazing tropical sun heating the sandy desert." This was too strenuous for some of Kelaart's companions including a Catholic bishop and the general who had commanded the British troops in Ceylon. Seven members of the party eventually perished over the next two weeks including Kelaart who was the last to die. He dropped dead on the quarterdeck of the S.S. "Ripon" late on 31 August 1860, the night before the ship's arrival at Southampton (see *Note 3*). He was 40 years old.

Kelaart made contributions to all branches of natural history including botany. Most of his personal writing on vertebrates was on mammals and, secondarily, amphibians and reptiles. He sent study collections to the naturalist, Edward Blyth, in Calcutta including most of the birds. He was regularly in contact with other regional naturalists who specialized on herpetology including Theodore Cantor and Thomas C. Jerdon. Kelaart published a number of papers in Colombo and London, but most of this information was included in his "Prodromus," noted earlier. This two-volume work, which is partly a book and partly a periodical, includes his own work and that of several others, among them Blyth. Some of it was original to these volumes and some reprinted or perhaps even preprinted from various journals. Altogether, Kelaart described and named six new species of amphibians and ten reptiles from Ceylon. All six of the snakes were uropeltids, a family of peculiar burrowing species endemic to Ceylon and southern India. Kelaart's herpetological specimens are in the museums in London and Calcutta.

• *References*: "Edward F. Kelaart, M.D., F.G.S.," anonymous, Proc. Linnean Soc. London, 6: xli–xlii, 1862; "A Biographical Sketch of the Late Lieut.-Col. E. F. Kelaart," by E. P. H. Kelaart, published by the author, Colombo, (7), 53 pages, 1932; De Silva, 1996 (Kelaart on p. 17); "The Life and Work of Edward Fredrick Kelaart," by R. Pethiyagoda and K. Manamendra-Arachchi, Jour. South Asian Nat. Hist., 2: 217–246, 1997. • *Portrait and signature*: From Kelaart, 1932. • *Note 1*: Dating of the two volumes of Kelaart's "Prodromus" has been in dispute. Volume 1 is dated 1852 on the title page, but E. H. Taylor (Herpetologica, 4: 26, 1947) notes that the catalogue of birds beginning on page 55 of the appendix is stated to be from the *Journal of the Ceylon Branch of the Royal Asiatic Society* (*JCBRAS*) for January 1853. Thus, if we assume that the appendices were originally issued with the main text of volume 1 and that this section is not a preprint from the journal, then the volume had to be issued in 1853. On the other hand, the reptile sections in this volume including Blyth's paper beginning on page 35 of the appendix, are recorded by F. H. Troschel (Arch. Naturgesch., 19(2): 61, 1853) as having been published in 1852. Moreover, notice was made of the receipt of "Specimens of Dr. Kelaart's Prodromus" at the 15 May 1852 meeting of the Royal Asiatic Society in Colombo, as published in *JCBRAS* (2: xix); this may refer only to the main text. Volume 2 (of which only part 1 was published) is dated 1853 on the title page, but the original wrapper is dated 1854. Some authors, such as Pethiyagoda and Manamendra-Arachchi (1997: 221), claim that it was published in 1855, on the basis that it contains two sections that were published in the *JCBRAS* in 1855. There is nothing in my copy in original wrappers to indicate that these sections were taken from the journal of 1855. It seems more likely that the journal versions were reprinted from the book. • *Note 2*: The month of death is sometimes given as October, probably based on Kelaart (1932: 45). In my copy of this book, however, which is signed by the author and has several corrections in his hand, "October" has been corrected to "August."

JERDON, Thomas C. (1811–1872).

Thomas Caverhill Jerdon was a British Army medical officer and zoologist who conducted the first extensive field surveys of amphibians and reptiles in India. He was born at Biddick House (Durham) in the north of England on 12 October 1811 and attended schools in Yorkshire. He entered Edinburgh University in 1828 as a literature student. He attended lectures on natural history and soon he switched to the study of medicine, which he finished in 1834. After some additional medical training in London, he joined the English East India Company in 1835 as assistant surgeon. Jerdon arrived in Madras in February 1836 and was quickly in the field attending to the troops fighting in the Ganjam District between Madras and Calcutta. This was the first of many assignments that eventually took him to nearly all parts of India and which afforded the opportunity to observe and collect plants and animals. During 1837–1852 he visited much of southern and central India: from Tellicherry on the southwest coast; Narbada (now Narmada) just northeast of Bombay, Jalna, the Dekhan (now Deccan) Plateau, and Neelgherri (now Nilgiri) Hills in the interior; and the Eastern Ghats and Nellore along the eastern coast north of Madras.

Besides his own keen sense of observation, honed during boyhood by a father who was an amateur naturalist, Jerdon sometimes depended upon aboriginal tribes, which have detailed knowledge of local animals. He was able to discover many new species in the wild country between Madras and Nellore that was then occupied by the Yánadis, a semi-nomadic tribe, from information provided by them. Later, he visited other, more distant, parts of northern India (Kashmir and the Punjab in the west to Darjeeling and Assam in the east) and southern Burma (Thayetmyo). He was an enthusiastic, if sometimes careless and impetuous, collector. On one occasion, having been bitten by a new species of pitviper, he applied a ligature and sucked the wound, after which he "set off running immediately . . . and felt no further ill effects."

After the Sepoy Revolt of 1857, which led to the British government taking direct control of India from the East India Company, Jerdon proposed that the government publish handbooks on the vertebrates of India. Lord Canning, the first viceroy, gave Jerdon leave during 1862–1864 to visit parts of India as yet unknown to him and to write the books. This resulted in his books on the birds (1862–1864) and the mammals (1867) of India. In 1868 Jerdon resigned from government service. Following retirement, he went to the Khasia Hills (Assam) in northeastern India where he experienced a severe attack of fever that eventually caused him to return to England in 1870. His health continued to decline and he died at his home in Upper Norwood (Surrey) on 12 June 1872. Jerdon had sent the manuscript of his reptile book to the printer before he left India. It went to press but was never published (see *Note*). Sotheby's auctioned his drawings of reptiles in 1874; their present location is unknown.

Jerdon worked with great industry and had the power of rapid and accurate observation. Rather than keep specimens for later study, however, his habit was to have native assistants make drawings. Then, having sent the specimens off to museums (in London and Calcutta) or misplacing them, he would rely on these drawings and his memory for details. This procedure, combined with his habitual carelessness, made his work of less value. Albert Günther, in his "Reptiles of British India" (1864), noted that many of Jerdon's new species were unrecognizable or were referred to the wrong genera. Jerdon promptly defended himself (1865). Nevertheless, because he explored diverse parts of India for the first time, Jerdon discovered and named a large number of new species (24 of amphibians and 39 reptiles, of which a total of 25, or 40 percent, are recognized today), among them several well-known taxa: *Hyla annectans, Rana crassa, Kachuga sylhetensis, Cyrtodactylus khasiensis, Japalura major,* and the distinctive new genus and species of colubrid snake, *Stoliczkaia khasiensis*. All of his new species were named in three papers, two in 1853 and another in 1870.

• *References*: "Memoir of Dr. T. C. Jerdon," by W. Elliot, p. xlvii–lv *in* T. C. Jerdon, The Birds of India, vol. I, reprint ed., P. S. d'Rozario, Calcutta, 1877; "Thomas Caverhill Jerdon," by C. Brandon-Jones, Oxford Dict. Natl. Biogr., 30: 36, 2004. • *Portrait*: From A. O. Hume, Nests and Eggs of Indian Birds, second ed., vol. I, R. H. Porter, London, 1889. • *Signatures* (above, 1835; bottom, after 1870): © British Library Board, London. • *Note*: The marked proofs of Jerdon's short book, entitled "Catalogue of Reptiles [and Batrachians] Inhabiting the Peninsula of India," were sent home after his death. These have recently been re-discovered, bound in two interleaved volumes, in the library of the Carnegie Museum of Natural History in Pittsburgh.

LEYDIG, Franz von (1821–1908).

Franz Leydig (ennobled to von Leydig in 1869) was a German anatomist and microscopist who conducted important studies on the glandular and sensory organs of amphibians and reptiles. He was born on 21 May 1821 in Rothenburg ober der Tauber, west of Nürnberg, Germany. His father was interested in gardening and beekeeping, which Franz also came to enjoy and this, together with acquisition of a simple microscope at the age of 12, led to his future career. After university studies in Würzburg, he took his medical education at Würzburg and Munich (M.D. 1847, Würzburg; habilitation 1849). He began on the faculty in Würzburg but moved to Tübingen (1857) and finally to Bonn (1875) as professor of comparative anatomy and director of the Anatomical Institute. After becoming emeritus in 1887, he retired to Rothenburg where he died on 13 April 1908.

Leydig had an interest in vertebrates and insects since his boyhood days but a field trip during the winter of 1850–1851 to Sicily, with its rich marine life, greatly enlarged his scope. His special research interests were in the glandular and sensory organs of vertebrates and invertebrates. His major work on the subject, "Lehrbuch der Histologie des Menschen und der Thiere" (1857), laid the foundations for comparative microscopical anatomy. It was Leydig, for example, who introduced the widely used term "fine structure." He discovered the secretory cells in the skin of fishes and aquatic amphibians that lubricate the skin (1848) and first described the interstitial cells of the mammalian testis that produce testosterone (1850). He also found the gland, in the mesonephros of vertebrates, which produces secretions that stimulate the movement of spermatozoa.

Amphibians and reptiles figured prominently in Leydig's research and he made numerous discoveries with them. Although Heinrich Rathke first described the spermatheca of female salamanders (1818), it was Leydig who found that the receptacle contained sperm (1853). He was the first to describe the pedicellate teeth (a basal pedicel plus a distal crown) of living amphibians (1867), which is a key character demonstrating their monophyly. Leydig found the lateral-line sensory organs in larval salamanders (1868). Following the earlier discovery of a French physiologist, Antoine Desmoulins (1824), Leydig studied the sensory nature of the facial pit organ of pitvipers (1868), although its function as an infrared receptor was only determined in the 1950s by two Americans, Theodore Bullock and Raymond Cowles. Leydig first described the Stirnorgan (frontal organ) found in many amphibians and reptiles (1872)—the eye-like structure of which was discovered by a Dutch microscopist, Henri W. de Graaf (1886)—and it became a major topic of Leydig's research program.

All of Leydig's publications are profusely illustrated with highly detailed drawings, most of them drawn by him. He also published several major works on the German herpetofauna including one on the salamanders (1867), the lizards (1872), and another on frogs (1877). In total Leydig wrote more than 30 titles on herpetological subjects (1852–1897).

• *References*: "Franz von Leydig," by O. Taschenberg, Leopoldina, 45: 37–44, 47–52, 57–64, 70–76, 82–88, 1909; "Franz von Leydig," by P. Glees, Dict. Sci. Biogr., 8: 301–303, 1973; Rieck et al. (eds.), 2001 (Leydig by W. Bischoff on pp. 519–521). • *Portrait and signature* (1873): Museum für Naturkunde, Berlin (BI/1467).

SEVERTZOV, Nikolai A. (1827–1885).

Nikolai Alexeevich Severtzov (sometimes Severtsov, Severtzow, or Severtsoff), a Russian zoologist and explorer, was born to a noble family on 5 November 1827 in Voronezh, a city located nearly 500 km south of Moscow. This is a rich agricultural region and the landowners were thus relatively wealthy. The area produced many prominent figures of Russian science including the herpetologist, Jacob von Bedriaga. Severtzov was privately schooled at home and one of his teachers introduced him to birds. He entered Moscow State University in 1843 and during his college years met the Russian traveler, Grigory S. Karelin, who interested him in Middle Central Asia and its natural history. Severtzov graduated in 1847 and received his doctorate in zoology in 1855, but failed to win a position at the university partly because he was a poor lecturer. During 1856 he studied at the Zoological Museum in Berlin.

In 1857 Severtzov began a series of expeditions to Middle Central Asia that were initiated by the Imperial Academy of Sciences and approved by Czar Alexander II. Imperial Russia had been pushing south into this region since the time of Peter the Great. The beginning of Russia's conquest of what was to become Russian Turkestan—today's Kazakhstan, Kyrgyzstan, Tajikistan, Turkmenistan, and Uzbekistan, extending from the Caspian Sea east to the Chinese frontier and south to Afghanistan and Iran—began when Tashkent fell to Russian forces in 1865. Severtzov's first journey was to the lowlands surrounding the Aral Sea (1857–1858, during which he was imprisoned for a month by the Kokands). He explored the Ural River Valley in 1860–1862 and in 1864 reached Alma-Ata and the northern Tian-Shan Mountains (today in Kyrgyzstan) near the Chinese border. During 1865–1868 he made a more extended trip to Turkestan (the so-called Second Tian-Shan Expedition) as far south as Khujand in present-day Tajikistan, to the lands east of the Aral Sea (1874–1875), and to lakes Karakul and Yashikul in the Pamir Mountains near the Chinese and Afghan borders, respectively (1877–1878). His last expedition, in 1879, was to Alma-Ata, Semirechie (in Kazakh, Zhetysu), and Omsk.

Severtzov was one of the first Russians to accept Darwinism, but he independently came to his views on the transformation of species. He published few papers specifically on amphibians and reptiles, but these animals figured in many of his works. His major books include one on the ecology of the mammals, birds, amphibians, and reptiles of the Voronezh Government (1855; reprinted 1950), a large district just east of Ukraine; this book was based on personal observations made in 1844–1853 and served as his doctoral dissertation. This was the first ecological study conducted in Russia. His book on the vertical and horizontal distribution of Turkestan animals (1873; reprinted 1953) had a herpetological focus and was the first zoogeographical analysis of the fauna of Russian Turkestan. He also published a summary of the vertebrate fauna of the Pamir Mountains (1879) and several other works on animal distribution and his travels. His herpetological collections are now in the museum of Moscow State University and the Zoological Institute in St. Petersburg, but some others were sold overseas. His son, Alexei N. Severtzov, also a zoologist, became the founder of evolutionary morphology in Russia and the teacher of Ivan Schmalhausen. The son's many publications on the morphology and development of frogs and lizards (1891–1916) are sometimes erroneously credited to his father.

Nikolai Severtzov retired from his post as senior government adviser in the military department in 1884. On 7 February 1885, while crossing the frozen Ikoretsk River by horse-drawn coach, near its junction with the Don River about 80 km southeast of Voronezh, the ice gave way. Severtzov was pulled from the water alive, but refused to go to the nearest village until the bag containing his manuscripts was found. Within two hours he was dead, at the age of 57, from a heart attack brought on by severe hypothermia. His bag was recovered the next day.

• *References*: ["Nikolai Alexeevich Severtzov Zoologist and Traveler (1827–1885), second ed.,"] by G. P. Dementiev, Izdat. Mosc. Obshch. Ispyt. Prirody, Moscow, 68 pages, 1948; ["Memoir about N. A. Severtzov,"] by A. P. Semenov-Tian-Shanskiy, Priroda, 1980(4): 58–67, 1980; Rossolimo and Dunayev (eds.), 2003 (Severtzov on pp. 227–328, 534–535). • *Portrait*: Museum für Naturkunde, Berlin (Orn. Abt. Bl. 16). • *Signature*: Natural History Museum, London.

KENNICOTT, Robert (1835–1866).

Robert Kennicott, naturalist, co-founder of the Chicago Academy of Sciences, and explorer of Alaska, was the describer of numerous North American snakes. He was born in New Orleans on 13 November 1835. In 1836 his family moved to northeastern Illinois just outside Chicago. Sickly as a child, he was home schooled by his father, a physician and horticulturist. Robert developed a serious interest in natural history and a fondness for snakes. His father arranged for him to spend the winter of 1852–1853 with an old friend, Jared P. Kirtland of Cleveland (Ohio), who was once a student of Benjamin Smith Barton and had a reputation for mentoring young naturalists. The following year the boy was sent to Racine (Wisconsin) to work with another naturalist, Philo R. Hoy, who also had studied with Kirtland. The seminal event of Kennicott's early life, however, was his introduction by Kirtland to Spencer F. Baird, director of the national museum in Washington. By 1853 Kennicott was regularly corresponding with Baird, sending him specimens, and describing to him his experiments on the venom of a massasauga rattlesnake. Illinois at that time was largely unknown to naturalists, so Baird encouraged the 18-year-old to survey the region.

Despite his enthusiasm and obvious gift for natural history, Kennicott was self-conscious about his lack of a formal education. Baird reassured him and guided his development from afar. Kennicott's first publications were 14 papers (1854–1857) in the *Prairie Farmer*, a newspaper devoted to agriculture and horticulture of the Midwest and for which his father often wrote a column. These papers covered many plant and animal subjects and included one on the massasauga (1855). Early in 1855 he went on his first survey, to southern Illinois, and later that year produced his first scientific publication, a catalogue of Illinois animals. That same year, Kennicott sent to Baird specimens of what he thought was a new species of snake. This led Kennicott to publish (1856) what is today known as *Clonophis kirtlandii*, named for his former teacher.

During the winters of 1857–1859, Kennicott traveled to Washington at Baird's request to help organize the herpetological collections. He worked mostly in the cold "snake room" of the "Castle," the original Smithsonian Institution building, which could be reached only "through some hundreds of feet of dark cellar passage," but after work hours he was free to use Baird's office, offering a fireplace and gas lights. Kennicott and several other young protégés of Baird including William Stimpson of herpetological note, actually lived and ate in the Castle, despite attempts by the head of the Smithsonian to evict them! However difficult the conditions, Kennicott's zeal overcame any obstacle and he proceeded to describe large collections of snakes from western United States and Mexico including many novelties. He published three journal papers during 1859–1861, although seven of his new species were named in papers by Baird and James G. Cooper (1859; reprinted 1978), but with credit to Kennicott, in government railroad and Mexican boundary surveys. No less than 16 currently recognized species were named by Kennicott, among them *Arizona elegans*, three species of *Crotalus, Leptodeira septentrionalis, Micruroides euryxanthus*, and five species of *Thamnophis*.

Despite these accomplishments, Kennicott was still restless for adventure. His friend Stimpson had just returned from the Pacific. The Western surveys were already beginning to wind down. The last great continental frontier was the Arctic. During 1859–1862, with the cooperation of Hudson's Bay Company, which administered northwestern Canada for the British government, Kennicott was able to travel as far as Fort Yukon in Arctic Alaska, a survey that brought back large natural history collections to the Smithsonian.

During the mid-1850s, Kennicott had been a primary organizer of the Chicago Academy of Sciences (today the Notebaert Nature Museum), which appointed him as its first curator (in effect, director) in 1864. His father had died in 1863, which sent him into a severe depression. He rallied and in 1865 was the logical person to be chief naturalist on the Russian-American telegraph expedition, which planned to link America and Europe by telegraph cable spanning the Bering Strait and, coincidentally, to further explore Alaska. Alaska then belonged to Russia, but America was negotiating for its purchase. Baird played a key role in providing the information to Congress about Alaska's rich natural resources that would justify the purchase. Most of his information came from Kennicott.

Eventually, the purchase of Alaska was approved in 1867, but Kennicott did not live to enjoy the results of his labor. The person of whom Baird once said, "I know of no American naturalist who gives such promise," had died on 13 May 1866. The medical report attributed his death to heart failure. His closest friends suspected otherwise. One remarked that he was murdered by "slow torture of the mind." Before dawn, Kennicott had left the Russian fort at Nulato and walked alone into the Yukon darkness. His body was found by daylight along the river's shore. One of the most promising naturalists of his era was dead at the age of 30.

• *References*: "Robert Kennicott," anonymous, Trans. Chicago Acad. Sci., 1: 133–226, 1869; "Robert Kennicott, Founder of Museums," by K. P. Schmidt, Progr. Act. Chicago Acad. Sci., 7: 3–9, 1936; "Robert Kennicott," p. 145 *in* C. A. Elliott, Biographical Dictionary of American Science, Greenwood Press, Westport, Connecticut, 1979; "The Early Career of Robert Kennicott, Illinois' Pioneering Naturalist," by R. S. Vasile, Illinois Hist. Jour., 87: 150–170, 1994; "Robert Kennicott," by R. S. Vasile, Amer. Natl. Biogr., 12: 583–585, 1999. • *Portrait*: Hunt Institute, Pittsburgh. • *Signature*: Trans. Chicago Acad. Sci., 1: opposite p. 133, 1869.

SWINHOE, Robert (1836–1877).

Robert Swinhoe was a British diplomat and pioneer naturalist of coastal China and Formosa (now Taiwan), and a specialist on Chinese vertebrates. He was born in Calcutta on 1 September 1836 and brought to England as an infant. He graduated from King's College in London in 1853 and then matriculated at the University of London. He soon became an interpreter for the British Consular Service in China, first at Hong Kong (1854) and then Amoy (now Xiamen, Fujian; 1855). While in Amoy he visited Formosa in March 1856 and amassed the first natural history collections ever made in the island. In June 1858 he circumnavigated Formosa on H.M.S. "Inflexible" as an interpreter and stopped at many ports allowing collections to be made along the coast and inland. He joined the British expeditionary forces under General Sir Robert Napier during the Second Anglo-Chinese War in 1860 and made collections along the route of the march to Peking where Swinhoe witnessed the destruction of the emperor's summer palaces. With the Treaty of Peking, Britain was permitted to begin trade with new Chinese cities and Swinhoe, as vice-consul at the age of 24, became the first foreign diplomat stationed in Formosa. He eventually opened a consulate at Tam-sui (now Tanshui) at the northern tip of the island in 1861. Owing to illness, he was furloughed in London during 1862–1864 but returned to Formosa to initiate another consulate, in Kaohsiung, at the southernmost end of the island.

Throughout his time in Formosa, which totaled about four years, Swinhoe continued to make natural history collections even in the mountainous interior, but in 1866 he took charge of the consulate at Amoy. While stationed there he explored Hainan Island and the adjacent mainland for three months in 1868, returned to Peking in 1868, a visit that included a journey to Kalgan (now Zhangjiakow, Hebei) near the Mongolian border. In 1869 he travelled up the Yangtze River (now Chang Jiang) as far as Chungking (now Chongqing) in eastern Sichuan Province. He returned to London via Japan and California for another furlough (1869–1871) before becoming consul at Ning-po (now Ning-bo, Zhejiang). While visiting Shanghai in February 1873 he was stricken with paralysis of his legs and, as he wrote, "In spite of my paralytic infirmities I have been able to get about [in] wheelbarrows . . . " and sedan chairs, which the Chinese used to carry him around, even into the field to collect. In April of that year he became consul at Che-foo (now Yantai, Shandong) where he continued his natural history work but by October his health gave out. He left China and spent the rest of his life, mortally paralyzed but still writing, in London. He died there on 28 October 1877. Together with the journeys of a French priest, Armand David, who was exploring the interior of China, and those of a Russian, Nikolai M. Przhevalsky, who traveled in northern China during the same time period, Swinhoe's survey of the Chinese littoral permitted the first scientific overview of the fauna and flora of China. In recognition of his accomplishments the Royal Society elected Swinhoe a fellow in 1876.

Swinhoe was a general naturalist and collected plants and insects in addition to vertebrates. Amphibians and reptiles were a small part of his work but significant because his collections were made at an early date. Only Theodore Cantor, who accompanied British Troops during the First Anglo-Chinese War in 1839–1842, preceded him, but Cantor's collecting was confined to Chusan Island (now Zhoushan) southeast of Shanghai. Although Swinhoe discovered many new species of amphibians and reptiles, he never named any of them. His five papers covering herpetology (1858–1873) consist only of lists with annotations based on his observations. His herpetological specimens ended up mostly at the museum in London and also in Calcutta and Vienna where they were described by John E. Gray, Albert

Günther, George A. Boulenger, Edward Blyth, and Franz Steindachner. Included among the new forms were a frog (*Rana swinhoana*), lizards (*Gekko swinhonis, Japalura swinhonis*), snakes (*Bungarus multicinctus, Rhabdophis swinhonis*), and turtles (*Cistoclemmys flavomarginata, Rafetus swinhoei*). What would have been Swinhoe's most spectacular herpetological discovery got away. In 1869 he saw a crocodilian in a Shanghai market but its owner refused to sell it. Ten years later a French customs official and amateur naturalist, A.-A. Fauvel, obtained one and named it *Alligator sinensis* (1879).

• *References*: "Robert Swinhoe," p. 661–670 *in* E. Bretschneider, History of European Botanical Discoveries in China, vol. 2, Sampson Low, Marston Co., London, 1898; "Mr. R. Swinhoe," anonymous, Ibis, ser. 9, 2(Jubilee suppl.): 207–208, 1909; "Biography of Robert Swinhoe 1836–1877," by Y. Takahashi, Quart. Jour. Taiwan Mus., 18: 335–339, 1965; "Robert Swinhoe," by C. T. Fisher, Oxford Dict. Natl. Biogr., 53: 505–506, 2004. • *Portrait* (1864) *and signature* (1870): Natural History Museum, London, courtesy A. F. Stimson.

COOPER, James G. (1830–1902).

Born in New York City on 19 June 1830, James Graham Cooper was a military surgeon and naturalist destined to spend most of his life on the West Coast of the United States. He learned natural history from his parents and his first field experience was in Hoboken (New Jersey) where his family moved when he was seven. His father was a well-known ornithologist and counted John J. Audubon, Spencer Baird, and C. L. Bonaparte among his friends; Bonaparte named Cooper's hawk in his father's honor. James obtained his medical degree at Columbia University (M.D. 1851), but his life as a field naturalist began in 1853 when, as a contract physician-naturalist, he joined the Pacific Railroad Survey Expedition to explore the Pacific Northwest for two years. He accepted this appointment on the advice of Baird who mentored him in Washington before Cooper departed. Following this survey Cooper participated in several other Western expeditions before he was called to active duty during the Civil War as surgeon with the Second Cavalry, California Volunteers. Afterward he joined the Geological Survey of California, but retired from government service altogether one year later.

In 1866 Cooper opened a medical practice in California, first in Santa Cruz, then in Ventura County, and finally in 1875 in Hayward (near Oakland). Throughout this time he wrote numerous articles on zoology (especially conchology and ornithology) as well as paleontology and botany. Beginning in 1860 he developed a long-time association with the fledgling California Academy of Sciences in San Francisco. He served the academy in a number of official capacities including curator of zoology (1862–1868) and director of the museum (1888–1891). He died in Hayward on 19 July 1902. Cooper's ornithological work was honored with the establishment of the Cooper Ornithological Club in 1893. In addition to his catalogue of trees and forest-distribution maps of the continent, he promoted forest reserves and is today recognized as a pioneer forest conservationist.

Cooper's herpetological work was peripheral to his main interest in shells and birds. His report on the herpetology of the railroad survey (1859; reprinted 1978) included a new species of salamander (*Plethodon vehiculum*) and a snake

(*Thamnophis atratus*); Robert Kennicott provided the description for the latter. In this report, which includes 11 fine but irregularly numbered lithographed plates, the field notes by Cooper are labeled "C" and those labeled "S" are by his fellow surgeon, George Suckley. Cooper published some small papers and a book chapter on herpetology, but one paper of special significance was that in 1863 in which he described a new tortoise, *Gopherus agassizii*.

• *References*: "Dr. James G. Cooper," by W. O. Emerson, Bull. Cooper Ornithol. Club, 1: 1–5, 1899; "James Graham Cooper (1830–1902)," p. 38–51 *in* E. E. Hume, Ornithologists of the United States Army Medical Corps, Johns Hopkins Press, Baltimore, 1942; "James Graham Cooper: Pioneer Western Naturalist," by E. Coan, Univ. Press of Idaho, Moscow, 255 pages, 1981; "The California Academy of Sciences 1853–1906," by A. E. Leviton and M. L. Aldrich, California Acad. Sci., San Francisco, xv, 623 pages, 1997 (numerous references to Cooper); "James Graham Cooper," by M. J. Broadhead, Amer. Natl. Biogr., 5: 445–446, 1999. • *Portrait* (1863) *and signature* (1850s): California Academy of Sciences, San Francisco.

MITCHELL, S. Weir (1829–1914).

Silas Weir Mitchell was a leading physician and neurologist and a man of letters, being a well-published poet and novelist. He was also an anatomist, physiologist, and toxicologist, which explains his research interest in reptiles. Mitchell was born in Philadelphia on 15 February 1829. At the age of 15 he entered the University of Pennsylvania but withdrew during his senior year because of the illness of his father. Then he turned to medicine and obtained his degree at Jefferson Med-

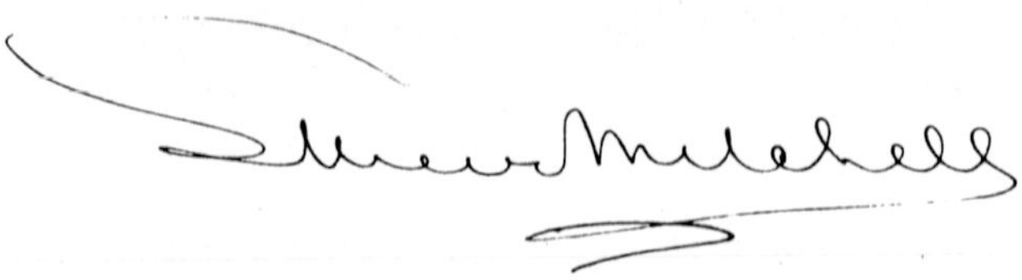

ical College (M.D. 1850) where his father, who informally trained him in chemistry, was a professor. Weir studied in Paris with the physiologist, Claude Bernard, and Charles P. Robin, a microscopist. On returning to Philadelphia late in 1851 he joined his father's medical practice, which he continued for the rest of his life. He also lectured at the Philadelphia Medical Association and at other institutions in the city. When Mitchell died in Philadelphia on 4 January 1914 he was justly revered as the city's foremost citizen.

His medical and physiological research, which was conducted throughout his life, had a broad scope. He was a specialist in neurology and did important work in toxicology, reflex paralysis, and the effects of opium. During the Civil War he was a contract surgeon at an army hospital in the city. While tending to the soldiers he had opportunities to study nerve wounds. His description of erythromelalgia, a neurosis, was one of his masterpieces and his research on the physiology of the cerebellum, carried out during 1863–1865, was also of the highest order. The quality of his work led to his early election to the National Academy of Sciences (1865). Later in life Mitchell wrote books that had wide public impact. "Wear and Tear" (1871) and "Fat and Blood" (1877) called attention to the role of a lack of exercise on nervous disorders and led to what became known popularly as "Weir Mitchell's Rest Cure." In 1886 he became president of the College of Physicians of Philadelphia and the following year co-founded the American Physiological Society.

There was another side to Mitchell's intellect. As a result of overwork during the Civil War he had a nervous breakdown and went to England in 1864. While convalescing he began to write poetry and stories in part based on his medical knowledge. One story was accidentally published, anonymously, in the *Atlantic Monthly* magazine (1866). He had been advised by an old family friend, the senior Oliver Wendell Holmes, author and physician, to give up his literary work until his medical reputation was established, so Mitchell did not begin writing again publicly until 1882. After 1898 most of his time was devoted to literary work. His output was enormous. The novel that made him famous was "Hugh Wynne, Free Quaker" (1897), a Revolutionary War romance set in Philadelphia. His most admired poem, "Ode on a Lycian Tomb" (1899), was written in memory of a daughter who died at the age of 22. He also published psychological novels. By 1896 his works were being published in collected editions and translated into foreign languages.

Mitchell's research with reptiles fits into three areas: anatomy, physiology, and toxicology. During the Civil War he collaborated with George R. Morehouse—a physician specializing in nervous diseases and with whom he worked at an army hospital—in studying the anatomy and physiology of respiration in turtles. Their main experimental subject was *Chelydra serpentina* but they also studied *Apalone mutica*. Turtles were of special interest because of a popular misconception about their method of respiration, which Mitchell knew to be false based on his casual observations with snapping turtles. There was a longstanding claim that they possessed a diaphragm and there was a general curiosity about how turtles respire with the lungs enclosed in a bony case. Mitchell and Morehouse (1863) showed that turtles, like all reptiles, lack a true diaphragm and were able to describe the mechanics of turtle inspiration and expiration.

Mitchell began his work on rattlesnake venom in the late 1850s as an adjunct to his clinical studies. His experimental subject was *Crotalus horridus*. He made the first microscopic description of the degeneration of skeletal muscle and quantitative measurement of the drop in blood pressure that follow a snakebite, work that bears the imprint of his training in Paris with Bernard and Robin. These key results were published in his book, "Researches upon the Venom of the Rattlesnake" (1861). A second major contribution to the topic was made years later (1886), jointly with Edward T. Reichert, a colleague at the University of Pennsylvania. Their main subjects were *Agkistrodon piscivorous* and *C. adamanteus*. By this time, protein chemistry had advanced to allow, through dialysis, separation of two types of toxin having different physiologic effects: one local on muscle, the other on the blood system generally. This was the first demonstration of the heterogeneity of the toxins within snake venoms. Mitchell received acclaim for his work on snake venoms and for which his fellow Philadelphian, Edward D. Cope, had named a new rattlesnake in his honor: *C. mitchellii*.

As late as 1898, Mitchell was still publishing on snake venoms in a paper on the venom of *Crotalus adamanteus* that he co-authored with a physician, Alonzo H. Stewart. Mitchell's line of work on snake venoms was taken up again by Hideyo Noguchi, a Japanese student who arrived in Philadelphia in 1900. It was Mitchell who provided the idea and also the funds to support Noguchi's early work, which used immunological methods.

• *References*: "S. Weir Mitchell Novelist and Physician," by E. Earnest, Univ. Pennsylvania Press, Philadelphia, vii, 279 pages, 1950; "Silas Weir Mitchell," by C. W. Burr and A. H. Quinn, Dict. Amer. Biogr., 7: 62–65, 1957; "S. Weir Mitchell," by D. Wepman, Amer. Natl. Biogr., 15: 629–631, 1999; "Silas Weir Mitchell (1829–1914): Toxicologist, Neurologist and Novelist," by B. J. Hawgood, Jour. Med. Biogr., 8: 63–70, 2000. • *Portrait*: College of Physicians of Philadelphia. • *Signature* (1860): Academy of Natural Sciences, Philadelphia.

GEGENBAUR, Carl (1826–1903).

The German zoologist, Carl Gegenbaur, who was the foremost vertebrate morphologist of the Darwinian era, made significant contributions to knowledge of the anatomy and development of amphibians and reptiles. He was born on 21 August 1826 in Würzburg (Bavaria), but during his childhood moved from place to place in Bavaria because of his father's job as a tax administrator. He finished secondary school in Würzburg, however, and spent much out-of-school time in private study of natural history. In 1845 he entered the university in Würzburg, which had on its faculty some of the finest research biologists then active in Germany including his teachers, Rudolf Kölliker, Franz Leydig, and Rudolf Virchow, from whom he learned microscopical techniques. He earned his medical degree (M.D. 1851) and, on the advice of Johannes Müller, the great anatomist in Berlin, decided to study marine invertebrates. He became a lecturer at Würzburg in 1854 and two years later took a professorial post in zoology at the university in Jena. Here, about 1860, his interests turned to comparative studies of vertebrate myology, osteology, and neural anatomy. Amphibians and reptiles were among his most frequent objects for study.

The Darwinian revolution was stirring at this time and Gegenbaur was able to attract to Jena the zoologist, Ernst Haeckel. Together, they became the strongest continental advocates of Darwinian evolution. Gegenbaur believed in the primacy of comparative anatomy in the process of phylogenetic reconstruction. He had a special interest also in the re-adaptation of structures to new functions. In 1873 Gegenbaur moved to the chair of anatomy at the university in Heidelberg where he spent the rest of his career. He attracted numerous students, among them several of herpetological interest including Jacques von Bedriaga, Max Fürbringer, Hans Gadow, Wilhelm Roux, and Hans Spemann. His fame as a teacher and researcher was disseminated through his writings and books. The most celebrated was his textbook on comparative anatomy, "Grundzüge der vergleichenden Anatomie" (1859; second edition 1870). His powerful influence changed the way biology was organized and taught at universities in Germany and beyond. He retired from his professorship in 1901 and died in Heidelberg on 14 June 1903.

Although Gegenbaur published several journal articles on the morphology of amphibians and reptiles, his major contributions to this subject are contained in four books. The first, on the structure and development of the vertebral column of amphibians and reptiles (1861), focused mainly on the frog. The next year he authored a companion volume on the same topic but emphasized comparative aspects. In 1864 and 1865 Gegenbaur published two parts of a major work on the comparative anatomy of vertebrates in which amphibians and reptiles occupy a prominent place. The first part focused on the anatomy of the carpus and tarsus and in the second, the shoulder girdle. All of these books contain highly detailed plates. (The third part of this series [1872] is devoted exclusively to the head skeleton of sharks.)

• *References*: "Erlebtes und Erstrebtes [autobiography]," by C. Gegenbaur, Engelmann, Leipzig, 114 pages, 1901; "Carl Gegenbaur," by M. Fürbringer, p. 400–466 *in* Heidelberger Professoren, vol. 2, 1903; "Carl Gegenbaur," by W. Coleman, Dict. Sci. Biogr., 15: 165–171, 1978. • *Portrait and signature* (1861): Museum für Naturkunde, Berlin (Orn. Abt. Bl. 176, 2).

BEDDOME, R. H. (1830–1911).

Richard Henry Beddome, a botanist and British army officer, was the first to recognize and describe the great diversity of the herpetofauna of southern India. He was born in Clapham, now part of south London, on 11 May 1830 and educated at Charterhouse (1843–1846). In 1848 he went to India as a junior officer of the 42nd Madras Native Infantry. He rose through the ranks and by 1856 he was the regiment's quartermaster and interpreter. India was governed by the East India Company at that time on behalf of the British government. In 1856 the company organized a Department of Forestry, based

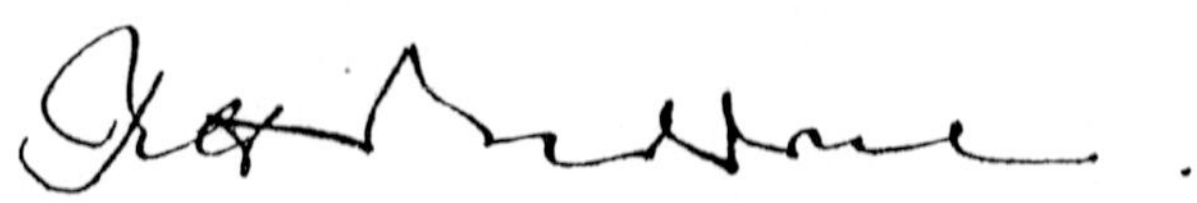

in Madras, to which Beddome was later transferred as chief assistant to the conservator. It was the responsibility of this department to look after the condition of the forests in the vast Madras Presidency. Thus, Beddome conducted tours of the forests, mostly in the hill ranges of southern India, and this gave him opportunity to collect. Ferns and reptiles were his main interests. By 1860 Beddome had become conservator himself, a position he held until he retired from the army in 1882 at the rank of full colonel. He returned to England in 1883 and settled in Putney in the London borough of Wandsworth just east of Kew where he became a frequent visitor at the Royal Botanic Gardens. He died at his home at the age of 81 on 23 February 1911.

Beddome's primary natural history interest was in botany, especially ferns. He published several books on the topic, usually profusely illustrated for purposes of identification. His major work is "Handbook of the Ferns of British India, Ceylon and the Malay Peninsula" (1883, supplement 1892) with 300 plates. He also published on trees and shrubs and, after retirement, turned to horticulture.

Earlier authorities had largely underestimated the high endemism of amphibians and reptiles in southern India. Patrick Russell, for example, who lived in India during 1782–1789 and later published his great atlas on Indian serpents, completely missed the most unusual group of snakes in the region, the shield-tailed snakes (Uropeltidae), since he worked along the east coast. Edward Kelaart had discovered several species in Ceylon, but Beddome on the other hand made these snakes a special interest and he named and described 20 of the 47 currently recognized species. During 1862–1886 he published 19 herpetological papers, which are packed with descriptions of new species, mostly of snakes but also some lizards, frogs, and caecilians. Most of his papers were published in London, but seven of them were in two obscure and short-lived journals issued in Madras (the latter were reprinted in 1940 and three of them were reprinted again in 1962). It should be noted that some species were described twice by Beddome and occasionally with slightly different names and type localities. His collections are mostly in the natural history museum in London but others are in the museums in Calcutta, Paris, and Vienna and at Harvard University in the United States.

• *References*: "Richard Henry Beddome," by J. R. Drummond, Proc. Linn. Soc. London, 1910–1911: 32–34, 1911; "A Facsimile of [Some of] R. H. Beddome's Articles on Indian Reptiles 1862–1870," by M. Smith, Jour. Soc. Biblio. Nat. Hist., 1: 273–334, 1940; "Col. R. H. Beddome's Contribution to the Flora of Peninsular India," by P. Bhargavan, Bull. Bot. Surv. India, 23: 17–24, 1981 (1983); Beddome family records, courtesy of Celia Dibblee and Carol Nuttall, 1998. • *Portrait*: Australian Museum, Sydney, courtesy Ian Loch. • *Signature* (1889): Natural History Museum, London, courtesy A. F. Stimson.

LÜTKEN, Christian (1827–1901).

Christian Frederik Lütken was a Danish marine zoologist who wrote important works on the herpetology of South America and the West Indies. He was born on 4 October 1827 in Sorø (Sjæland) where his father taught philosophy at the Sorø Academy. Lütken attended the academy and was attracted to zoology by one of his teachers, Japetus Steenstrup, during his final year as a student (bachelors degree 1846). He then went to the University of Copenhagen to study zoology with Steenstrup, who had just become professor of zoology there. His university work was soon interrupted, however, when in 1848 he took part in student demonstrations to abolish Denmark's absolute monarchy, to institute a constitution, and to incorporate the German duchies of Schleswig and Holstein. When the duchies rebelled, he was one of seven students who volunteered for the army. As he wrote to his mother, since he was one of the student instigators for war in Germany, it would have been cowardly to let the peasantry alone do the fighting. He served until 1852 but was given leave during 1849–1850 to complete some zoological work.

In 1851 Lütken became Steenstrup's assistant in the university's small museum of zoology while he completed his academic work (magister degree 1853; doctor of zoology 1857, the first such degree awarded in Denmark). In 1864 the university museum was combined with the much larger Royal Natural History Museum to form the present Zoological Museum of the university. Lütken's position did not pay well, so he wrote a popular zoölogy textbook (1855) to supplement his income. It was intended for high schools, but was also used at universities (the last edition, in 1893, was in use until the early 1930s). In 1882 Lütken became inspector (or curator) of vertebrates, succeeding his just-deceased colleague, J. T. Reinhardt, and three years later, when Steenstrup retired, he became professor of zoology and head of the museum's Division of Vertebrates. Lütken suffered a

paralytic stroke in 1898 from which he never fully recovered. He retired at the beginning of 1899 and died on 6 February 1901 in Copenhagen.

Lütken was a descriptive zoologist who specialized on echinoderms and fishes. He authored half a dozen titles on herpetological subjects (1861 [1862]–1894). Reinhardt had explored southeastern Brazil on three trips from 1845 to 1856 and returned with large collections to the Royal Museum. Lütken performed most of the laboratory work on the herpetological specimens in his methodical way and together they co-authored "Bidrag til Kundskab om Brasiliens Padder og Krybdyr" (1861 [1862]), which covers the amphibians, lizards, and crocodilians of the savanna-like cerrados. Many new species were named but no further parts were issued. In 1862, again with Reinhardt, Lütken published a monograph of herpetological collections from the eastern Caribbean, concentrating on the Danish West Indies just east of Puerto Rico (on their purchase by the United States in 1917 they became the U.S. Virgin Islands). Many new species were named. Lütken's other herpetological titles all concern European species except for his beautifully illustrated paper on *Crocodylus intermedius* of Venezuela (1884), which firmly established the validity of this little-known species that had been described in 1819.

• *References*: "Christian Frederik Lütken," by F. A. B[ather], Science, new ser., 13: 540–542, 1901; "Christian Frederik Lütken," by T. Wolff and R. Spärck, Dansk Biograf. Leksikon, 9: 254–255, 1981; "Christian Frederik Lütken, 1827–1901," by F. J. Madsen, p. 349–357 *in* M. Jangoux and J. M. Lawrence (eds.), Echinoderm Studies, vol. 3, A. A. Balkema, Rotterdam, 1989. • *Portrait* (about 1890): From Spärck, Festskr. Københavns Univ., Copenhagen, p. 81, 1962. • *Signature*: Adler collection.

SUMICHRAST, Francisco (1828–1882).
Adrien-Louis-Jean-François (later, Francisco) Sumichrast, a Swiss zoologist and explorer, made early contributions to the natural history of Mexican amphibians and reptiles and conducted the first experiments on the venom of *Heloderma*. He was born on 15 October 1828 in Yvorne (Vaud), a village about 20 km southeast of Lausanne. After studies in Lausanne, Geneva, and Berne, he became dedicated to natural history. His dream had been to explore Mexico and in 1855 he had the opportunity to accompany, as naturalist, the Swiss entomologist, Henri de Saussure, on his trip to eastern Mexico. They arrived in Veracruz in April that year, then traveled to Córdoba where nearby they visited Adolphe Boucard and Auguste Sallé (who were collecting insects) before moving on to Orizaba, Puebla, México, Tampico, and other cities in the interior. Sumichrast stayed with de Saussure for about a year, but the difficulty of travel at that time, when Mexico was in a state of revolution, made de Saussure decide to return to Europe. Sumichrast, however, being exactly where he wanted to be, decided to stay to collect and conduct scientific studies. He made his own explorations in Chiapas, México, Oaxaca, Puebla, Veracruz, and especially in the Isthmus of Tehuantepec.

Reptiles soon became Sumichrast's favorite interest and, perhaps realizing that without a reference collection and associated library, he could not do taxonomic work, decided instead to send his specimens to European and American specialists, and concentrate his own work on other aspects. He authored nearly 20 papers on Mexican amphibians and reptiles (1864–1885), covering their geographic distribution,

coloration in life, behavior, and life history. He made a special study of the Beaded Lizard (*Heloderma horridum*), which had been described less than 40 years earlier and about which nothing was known of its habits. Beginning in 1864, Sumichrast wrote half a dozen papers with information about this species, but his major summary was published in 1875 under the authorship of Firmin Bocourt who had transmitted Sumichrast's notes to the Parisian publisher. Sumichrast reported that Mexicans considered the animal's bite to be extremely dangerous, so he decided to conduct the first experiments on venomous lizards. A chicken bitten on one wing by a young lizard died within 12 hours, but a cat bitten on the leg survived with only a swelling.

Sumichrast's collections were sent to specialists in North America and Europe such as Edward D. Cope in Philadelphia, Albert Günther in London, and Bocourt in Paris. His specimens were utilized for the rival French and British summaries of the Middle American herpetofauna, the "Mission Scientifique au Mexique et dans l'Amérique Centrale" (1870–1909) and the "Biologia Centrali-Americana" (1885–1902), respectively. It is not known how Sumichrast supported himself and it might be reasonable to suppose that his collections were sold for this purpose. However, the secretary of the Smithsonian Institution in Washington, where Sumichrast sent large numbers of specimens, has written (1884) that all of his assistance "was done without any compensation whatever, and solely for the sake of securing a prompt and complete knowledge of the natural history of his adopted country."

Sumichrast married in August 1870 in Oaxaca and the couple had three daughters. Cholera was then rampant in the country. On 25 September 1882, while working at his home in Tonalá (Chiapas) on a collection of reptiles that was to be sent to the Museum of Comparative Zoology at Harvard College, Sumichrast felt ill. He had a premonition of the cause. He informed his family that he was going to die suddenly, drew up his will, and went back to work on the collection. By one o'clock the next afternoon he was dead. He was 55. His youngest daughter, only four, died 12 hours later.

• *References*: "Adrian Luis Jean Francisco Sumichrast," anonymous, Ibis, ser. 5, 1: 584, 1883; "Adrian Luis Jean Francisco Sumichrast," [by E. D. Cope], Amer. Nat., 17: 904–905, 1883; "Notice sur François Sumichrast," by A. Boucard, Bull. Soc. Zool. France, 9: 305–312, 1884 (1885); "Apuntes Biográficos del Señor Francisco Sumichrast," by A. Boucard, La Naturaleza, 7: 312–316, 1885; O. Flores-Villela et al. *in* A. M. Bauer (ed.), 2004. • *Portrait*: From Boucard, 1884. • *Signature* (1868): Academy of Natural Sciences of Philadelphia.

McCoy, Frederick (1823?–1899).

Frederick McCoy (sometimes spelled M'Coy), naturalist and paleontologist and founding director of the National Museum of Victoria (Australia), was born in Dublin, Ireland, probably in 1823 (see *Note*). His father was a physician and a professor at Queen's College, Galway, and early on Frederick also was inclined to a medical career. His education is uncertain but some evidence suggests that he attended a Jesuit school in Dublin and later took medical courses in Dublin and perhaps also at Cambridge. What is certain is that his interests were diverted to natural science and that he published his first pa-

per in 1838 (on gulls). In 1839 McCoy was hired to arrange the museum collections of the Geological Society of Dublin and in 1845 he briefly was the first field staff member of the newly established Geological Survey of Ireland. He went to Cambridge University in 1846 to organize the fossil collections in the Woodwardian Museum before returning to Ireland as professor of mineralogy and geology at Queen's College, Belfast (1850–1854). By this time he had three major books on the fossils of Ireland and England to his credit.

In 1854 the newly established University of Melbourne was seeking to appoint four chairs. Even though McCoy did not have the formal academic qualifications required, he was recognized as one of Great Britain's leading paleontologists and was appointed. He became professor of natural science, which meant that he had to broaden his scope to include chemistry, botany, and zoology. He was a person of enormous industry (he required no more than five hours of sleep each night) and high intellect. He became paleontologist of the Geological Survey of Victoria (1856), founding director of the National Museum of Victoria (1858), a post he held for life, and president of the Royal Society of Victoria (1864). He also began the university's botanical gardens in 1856. As a builder of institutions, McCoy was often at loggerheads with powerful people. One historian referred to him as a "bad tempered redheaded Irishman." However his Australian colleagues felt about him, his work was honored overseas despite his life-long opposition to Darwinian views of evolution. In 1880 he was elected fellow of the Royal Society of London and Cambridge conferred an honorary Sc.D. in 1886. Queen Victoria granted him a knighthood (K.C.M.G. 1891), the first bestowed on an academic in Australia. His health deteriorated in the 1890s and he died at his home in Brighton Beach near Melbourne on 13 May 1899.

Except for a few short articles published in London and Melbourne in the 1860s, McCoy's contributions to herpetology are all contained in a single work in two thick volumes, "Prodromus of the Zoology of Victoria" (1878–1890), in which he illustrated and described living species of vertebrates and invertebrates. The book was issued in 20 parts (called "decades") with 200 plates, 189 of them in color. It covers 447 species of which there are three of amphibians and 38 of reptiles. Many of the taxa included were new to science. Among these are six reptiles named by McCoy, of which three names are used today. One of them is the highly dangerous fierce snake, sometimes called the inland taipan, *Oxyuranus microlepidotus*, which reaches a length of 2.5 m.

• *References*: "Sir Frederick McCoy," by T. G. Bonney, Dict. Natl. Biogr., suppl., 22: 991, 1922; "McCoy Special Issue[s]," by D. McCann et al., Victorian Nat., 118: 145–332, 2001. • *Portrait*: Museum of Victoria, Melbourne, courtesy Jonathan Augier. • *Signature* (1862): Natural History Museum, London. • *Note*: Some sources give McCoy's birth year as 1817, 1821, or 1822.

KEFERSTEIN, Wilhelm (1833–1870).

The German zoologist, Wilhelm Moritz Keferstein, wrote significant papers on the amphibians of Australia and tropical America. He was born on 7 June 1833 in Winsen an der Lühe (Lower Saxony) near Hamburg where his father was a Hannoverian official. After private tutorials, Wilhelm attended grammar school in Lüneburg, then the polytechnic school in Hannover to study hydraulic engineering. Engineering did not interest him, so in 1852 he moved to the university in Göttingen to study natural sciences with the zoologist, A. A. Berthold, among others. His father, however, requested that he choose medicine instead since there were greater career possibilities. Keferstein then studied medicine in Berlin where he worked with Hinrich Lichtenstein and J. P. Müller before returning to Göttingen (M.D. 1854). After his habilitation, he became a professor there. His research interests were very broad including marine invertebrates, fossil shells, and the electric organs of fishes, but he became an early opponent of Darwin's theory. Because of ill health he spent 1858–1859 in Naples, which gave him opportunities to study marine organisms. His masterwork was the volume in H. G. Bronn's series, "Die Klassen und Ordnungen des Thier-Reichs," on the gastropod mollusks (1862–1866). He also made field trips to Norway (1863) and the coast of Normandy (1865) for his work on invertebrates, but developed a chronic pulmonary ailment from which he eventually died at the age of 36 on 25 January 1870 in Göttingen.

On the death of Berthold, who had curated the university's zoological museum (see *Note*), Keferstein took over responsibility in 1861. Fresh herpetological materials came into the museum from Karl von Seebach's journey to Costa Rica in 1864–1865 and from Keferstein's brother-in-law, a physician practicing in Sydney, Australia. Keferstein named 15 new species or varieties from these collections of which several are recognized today. He named the Australian frogs *Pseudophryne coriacea, Uperoleia laevigata*, and two species of *Litoria*. From Costa Rica he described an unusually elongate plethodontid as a new genus and species (*Oedipina uniformis*) and a new genus and species of microhylid (*Hypopachus seebachi*). The American zoologist, Edward D. Cope, had named the frog species one year earlier, but Keferstein's generic name has been used traditionally for two species distributed from Texas to Costa Rica. Keferstein produced five herpetological papers (1866–1868), but even so there is some bibliographic confusion. His two papers published in the *Archiv für Naturgeschichte* in 1868, on Australian and Costa Rican amphibians, were combined in the author's reprint under a new title, "Ueber neue und wenig bekannte Batrachier aus Australien und Amerika" (1868; reprinted 2003).

• *References*: "Wilhelm Moritz Keferstein," by W. Hess, Allg. Deutsch. Biogr., 15: 525–526, 1882; "Die Wirbeltiersammlungen des Museums Alexander Koenig III. Amphibien und Reptilien," by W. Böhme and W. Bischoff, Bonn. Zool. Monogr., 19: 151–213, 1984 (Keferstein on pp. 157–158). • *Portrait*: From Böhme and Bischoff, 1984, courtesy Wolfgang Böhme. • *Signature* (1865): Museum für Naturkunde, Berlin. • *Note*: In 1977 the entire Göttingen zoological collection was transferred to the Museum Alexander Koenig in Bonn.

BOCOURT, Firmin (1819–1904).

The French artist-naturalist and field collector, Marie Firmin Bocourt, who wrote and illustrated one of the greatest expedition reports on reptiles, was born in Paris on 19 April 1819. His father engraved copperplates for the printing of artwork. The Bocourt family lived near the Muséum d'Histoire Naturelle because his father often engraved plates for works by

the museum's naturalists, among them plates for the huge natural history atlas of "Description de l'Égypte" (1827; second edition 1829). Thus, Firmin became familiar with live animals by visiting the museum's menagerie and, by watching his father work, developed dual interests in natural history and drawing. On the recommendation of a family friend, a sculptor who worked at the museum and appreciated his talents, in October 1834 Bocourt, at the age of 15, became preparator for Gabriel Bibron, assistant to the professor in charge of reptiles and fishes, Constant Duméril, then the leading systematic herpetologist in the world.

For someone with no formal education beyond primary school, Bocourt could not have been in a better place to learn herpetology and to draw reptiles. Duméril and Bibron had just begun publication of their monumental world review, the nine-volume "Erpétologie Générale" (1834–1854). Bocourt drew seven plates for the final volumes as well as plates for some of Duméril's other publications. Bocourt's daily interactions were with Bibron, whose knowledge of specimen determination and detailed descriptions of species was unsurpassed, and with Pierre Redouté and Antoine Chazal, leading natural history artists who gave him drawing lessons. He developed his skills for making lifelike drawings by using living reptiles in the museum's menagerie as subjects, yet at the same time included all of the minute details of importance to naturalists. In addition to making drawings for use in publications, Bocourt also decorated the windows and walls of the museum with images of animals. In 1854 he was officially designated "museum painter."

In September 1861 Bocourt was appointed to travel to Siam to take charge of a collection of animals being donated to the menagerie by King Mongkut (Rama IV) who capitalized on Anglo-French rivalry to ensure his country's independence. Thus Bocourt was enabled to conduct his first overseas fieldwork and was even invited on a wild elephant hunt by the king. The voyage lasted 15 months and resulted in Bocourt's first scientific publication (1866), a review of his collections of fishes, amphibians, and reptiles. However, before he could complete his review of these collections, he was assigned to undertake a much greater field expedition, one that was to establish his career as a herpetologist.

In January 1862 French troops arrived in Veracruz on the pretext of collecting France's financial debt from Mexico, which had been in political and military chaos. Napoleon III's true intention, however, was to establish a puppet regime for commercial advantage and to aid the Confederate side in the American Civil War then being fought. Archduke Maximilian of Austria, who would be subservient to French interests, was placed on the newly created throne, but after his arrival in June 1864 he began to lose his conservative Mexican support. By 1866, with the American war over and a Union victory, Napoleon abandoned his Mexican adventure. The next year, during the peasant uprising led by Benito Juárez, the legitimate president of Mexico, Maximilian was captured, tried, and executed. A French scientific commission had accompanied France's original intervention, but by the time that Bocourt was to arrive at the end of 1864 via Jamaica, it was considered unwise for him to proceed to Mexico. Instead, he was diverted to Central America and began his field work in British Honduras. After some inland travel up the Mullins River, Bocourt proceeded to Livingston on the Caribbean coast of Guatemala and then moved upriver to Lake Izabal. From there he spent two years crossing Guatemala in different directions, visiting Cobán in the north, Zacapa and Guatemala (city) in the south, and Sololá and Quezaltenango in the west. He first visited the Pacific coast at Tanesco, via the Rio de Nahualate, then departed Guatemala at San José on the coast and proceeded to El Salvador, Nicaragua, Costa Rica, and the Isthmus of Panama (then part of Colombia), collecting wherever he could. Bocourt departed for France in 1867 via Martinique. He had already sent nine large shipments of specimens to the museum and brought still more collections with him.

In 1848 Bibron had died prematurely and was succeeded by Auguste Duméril who in 1857 succeeded his father, Constant Duméril, as professor in charge of reptiles and fishes. Thus, on his return from Central America, Bocourt worked with Duméril *filius* to work up and publish the herpetological collections from his expedition. It was a complementary collaboration with Duméril primarily responsible for the descriptive work and Bocourt the illustrations. However, since Bocourt had field experience with the animals and had made detailed watercolors of specimens to record their life colors, he was in a unique position to help Duméril enrich the text. The result is one of the great monuments of herpetology, the reptile and amphibian sections of the series, "Mission Scientifique au Mexique et dans l'Amérique Centrale" (issued in 17 parts in 1870–1909; reprinted 1978). The geographic coverage, as the series title states, is Mexico and Central America, but many extralimital species were included from the United States, South America, and the West Indies. The reptile volumes, entitled "Études sur les Reptiles," compose over 1,000 pages of text in large format with 99 highly detailed lithographed plates, all but seven of them drawn by Bocourt. Most of the plates have multiple figures, some as many as 50, so the total number of individual drawings required was enormous.

After publication of the first part of the reptile section, Duméril died in November 1870, leaving Bocourt to carry on alone. The amphibian section was later assigned to Paul Brocchi to write; it was published in 1881–1883 (reprinted 1978). Bocourt wrote and illustrated the next 14 parts of the reptiles by himself. Thus, although this section is generally credited first to Auguste Duméril, who certainly conceived and initiated the book, it was mostly written and illustrated by Bocourt and it should be recognized primarily as his achievement.

In 1874 Bocourt was appointed keeper of the galleries of zoology and mineralogy, a position reserved to honor a worthy museum employee. This provided a salary and relief from most duties, and this allowed him to devote more time to the book. He was pensioned from the keepership in 1892 but still continued his work on the book. By then his eyesight was failing and he developed a cataract on his right eye. A medical operation in November 1903 was not successful, although he continued to work in his laboratory through January. He died in his living quarters at the museum on 3 February 1904. François Mocquard completed the remaining two parts of the reptile section with the help of Léon Vaillant and Fernand Angel.

In addition to his primary work on the "Mission Scientifique" report, Bocourt authored about 35 titles on amphibians and reptiles (1866–1894), mostly based on his own collections in Siam and Central America. He also studied specimens sent to the museum by French and Swiss naturalists including Adolphe Boucard, Alfredo Dugès, Auguste Sallé, and especially Francisco Sumichrast in Mexico; P.-M.-A. Morelet and Juan Rodríguez in Guatemala; and Camille Viguier in Panama, as well as other specimens in the rich collections of the museum. Bocourt named nearly 100 new species of turtles, crocodiles, lizards, and snakes, of which some 60 percent are recognized today, among them *Chelonia agassizi, Chelydra rossignonii*, four species of *Sceloporus*, eight anoles, the xantusiid *Lepidophyma smithii*, and 12 species of snakes including three pitvipers. He also named a few species from beyond the scope of the "Mission Scientifique," from Abyssinia, Brazil, China, Hispaniola, and South Asia, mostly mislabeled specimens that were mixed in with his collections from Mexico and Central America.

• *References*: "Notice Nécrologique sur F. Bocourt, Garde des Galeries Honoraire," by L. Vaillant, Bull. Mus. Hist. Nat., 10: 32–35, 1904 (republished with minor changes but with map of expedition route in Central America, by L. Vaillant, p. ix–xiv *in* A. Duméril et al., Mission Scientifique au Mexique, Rech. Zool., pt. 3, sect. 1, Reptiles, part 17, 1909); Jaussaud and Brygoo, 2004 (Bocourt on p. 83). • *Portrait*: Courtesy of Oscar Flores Villela. • *Signature* (1876): Adler collection.

GRANDIDIER, Alfred (1836–1921). GRANDIDIER, Guillaume (1873–1957).

Madagascar, the world's fourth-largest island, had Arab traders in the 7th Century and was discovered by Portuguese mariners in 1500. It remained a *terra ignota* for its distinctive biota until well into the 19th Century, except for a glimpse provided in a book (1658) written by Étienne de Flacourt, French governor of Madagascar (1648–1654) during an early, failed attempt to colonize the island. The French geographer, explorer, and naturalist, Alfred Grandidier, was the first to make comprehensive scientific collections of Madagascar's vertebrates and he described many of its amphibians and reptiles. He also inspired others including his son, Guillaume Grandidier, to devote their careers to Madagascar. Alfred was born in Paris on 20 December 1836. He was interested in natural history as a boy and was educated privately by tutors. He earned bachelors of letters and of sciences and then, during 1854–1857, attended the Sorbonne and the Collège de France.

In 1858 Grandidier embarked on a 25-month voyage to South America with his brother and a tutor in order to retrace the route of the French naturalist, Alcide d'Orbigny. This trip inspired Alfred to dedicate his life to geographic and scientific research. Tibet, then inaccessible, tempted him. In 1863, alone, he traveled to Ceylon and India in preparation but contracted malaria and to recover had to retreat to Réunion Island, just east of Madagascar. He returned to France and, now attracted to Madagascar, he sailed to the island in the spring

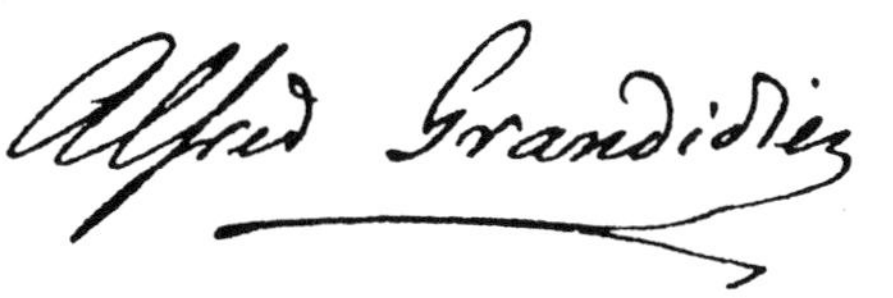

of 1865. He reached Soanianina (now Soanierana-Ivongo) near the Bay of Antongil on the northeastern coast, but was prevented by the Merina kingdom—a people of Indonesian origin who then controlled most of Madagascar—from traveling to the interior. Nevertheless, he had seen the richness of the island's fauna and flora.

From June to November of 1866 Grandidier was again in Madagascar, in the southern and southwestern parts of the island, and this time was able to reach the interior as far as Salobe. After returning to France, he visited Madagascar again (June 1868 to July 1870) and traveled to many parts of the island. He was often exposed to great threats including serious illnesses and pillaging attacks, as well as accusations of witchcraft and spying for the British who then supported Madagascar being independent. At one point he thought it was wise to take a blood oath with the Antanosy king in order to put an end to dangerous rumors and gain favor among the natives. Combining the three visits, he crossed the island three times and traveled some 3,000 km in the interior and 2,500 more along the coast, making huge collections of vertebrates, invertebrates, plants, fossils, minerals, and ethnographic materials.

He returned to France suddenly on hearing of Prussia's declaration of war. At the age of 34 he realized that his place was in France, but by the time he returned Paris was under siege, so he joined a corps of soldiers in Poitiers until the armistice in January 1871. He then proceeded to the Muséum d'Histoire Naturelle where he worked with Alphonse Milne-Edwards, Léon Vaillant, and other professors to work up his collections. He decided to produce and edit an elaborate series, the "Histoire Physique, Naturelle et Politique de Madagascar," which was projected to run to 60 volumes. In addition, he published numerous papers to describe and name the new species he had discovered. He authored only six short herpetological papers (1867–1880), but these are packed with new species. He named a total of 34 species of Madagascar's herpetofauna (8 percent of the 429 currently recognized species). These include six frogs, two turtles (*Pyxis planicauda* and *Erymnochelys madagascariensis*), four chameleons, two of the island's relict oplurine iguanids, as well as 18 gerrhosaurs, geckos, and skinks, but only two snakes.

Alfred married late, in July 1872, and Guillaume—who was destined to explore Madagascar and continue his father's work—was born on 1 July 1873 in Paris. Guillaume studied at the École des Mines in Paris and received his bachelor's degree from the Sorbonne. By 1885 Madagascar was a French protectorate and in 1896, when the French abolished the Merina monarchy by force, it was annexed. Guillaume made two explorations of Madagascar (1898–1899, 1901–1902) and later earned a doctorate in Paris (Dr.Sc. 1905). He joined his father as editor of the "Histoire" series, which eventually extended to 39 volumes (1875–1954).

Originally the herpetological section of this series was to have been composed of two volumes. The first part, by Vaillant and Guillaume Grandidier, appeared in 1910, entitled "Histoire Naturelle des Reptiles. Première Partie: Crocodiles et Tortues." This is a sumptuous work, in quarto, with 28 plates (see *Note*) by several French artists. These are apparently the last plates to be produced using true lithography in a major herpetological work. They are some of the finest illustrations of crocodiles and turtles ever published. The other part covering the amphibians and the rest of the reptiles was to have been by Guillaume Grandidier alone, but it was never issued. His other major zoological work, "Zoologie de Madagascar" (1932, co-authored with Gabriel Petit), which is primarily on vertebrates, contains a 30-page, illustrated section on amphibians and reptiles. Although the "Histoire" series was never completed, it remains one of the great colonial publications, unrivalled for any other French colony.

Alfred died in Paris on 13 September 1921. Guillaume died there on 19 September 1957.

• *References on Alfred Grandidier*: "Alfred Grandidier," anonymous, Comp. Rend. Acad. Sci. France, 173: 485–488, 1921; "A la Memoire de l'Explorateur Alfred Grandidier (1836–1921)," anonymous, Comité du Monument a l'Explorateur Alfred Grandidier, Tananarive, 15 pages, [1938?]; "Alfred Grandidier," by N. Broc, Dict. Biogr. Franç., 16: 960–961, 1985. • *Reference on Guillaume Grandidier*: "Guillaume Grandidier," by N. Broc, Dict. Biogr. Franç., 16: 964, 1985. • *References on both*: L. J. Dorr, Plant Collectors in Madagascar and the Comoro Islands. Royal Bot. Gardens, Kew, 524 pages, 1997 (Alfred on pp. 185–187, Guillaume on pp. 187–189). • *Portraits*: Courtesy G.-G. Aymonin and L. J. Dorr. • *Signature of Alfred Grandidier*: Natural History Museum, London. • *Signature of Guillaume Grandidier*: Courtesy É.-R. Brygoo. • *Note*: There is a total of 28 plates (1–27, 19A). However, these include six double plates, accounting for 12 of the numbered plates (for example, plates 1–2 together is a single set of images). Of these, 16 plates are colored, but 12 of these are doubles.

PARKER, W. K. (1823–1890).

William Kitchen Parker, an English physician and comparative anatomist, specialized on the anatomy of vertebrates including that of amphibians and reptiles. He was born on the family farm at Dogsthorpe, a few kilometers from Peterborough (Northamptonshire) on 23 June 1823. During his early education at the parish school, he developed skill as a draftsman. This was later put to use in his research. He apprenticed to a local druggist and then to a surgeon where he developed an interest in anatomy and in making drawings of animal dissections. He went to London in 1844 as an assistant to a physician in Westminster and then, without any university education, studied medicine at Charing Cross Hospital and attended lectures by Richard Owen at the Royal College of Surgeons. In 1849 he was qualified by the Society of Apothecaries and began a private practice in Pimlico, London, which he maintained until retirement in 1883. He held the Hunterian professorship of comparative anatomy at the Royal College of Surgeons beginning in 1873 but was not a successful lecturer.

Parker's earliest research was on the Foraminifera. That on vertebrates began in 1862 and the major product of his early period was a book, "Monograph on the Structure and Development of the Shoulder-Girdle and Sternum in the Vertebrata" (1868). Parker drafted all of the detailed drawings on the 30 plates in this work and this became the pattern for his later publications. In this book he showed that the forelimb was attached to the trunk by means of the coracoid and scapula, which did not support the views of his teacher. Owen believed that the forelimb was an outlying part of one of the segments of the skull, which he believed consisted of modified vertebrae, an idea traceable to Lorenz Oken in 1807.

Parker's attention then turned to the skull and he wrote and illustrated numerous monographs on the topic that covered about 1,800 pages and 270 quarto plates. He never provided a synthetic work on the topic except for an early book with the writer George T. Bettany entitled "The Morphology of the Skull" (1877). Parker's studies were in support of those by his chief scientific colleague, Thomas H. Huxley, whose work (1858) had dealt the final deathblow to Owen's vertebral theory of the skull. Parker's work was done in parallel with that of several German anatomists, but since his facility with German was almost non-existent, his research was entirely original. It should be remembered that all of the research was done in time spared from his medical practice on which he was totally dependent for his livelihood.

Parker was elected to the Royal Society in 1865 and won its gold medal the next year. Huxley's advice helped him to focus his discursive research program. As a result, Parker's publications remain a mine of useful information. He died suddenly on 3 July 1890 in Cardiff, less than four months after the death of his beloved wife.

Many substantial monographs on the anatomy and development of amphibians and reptiles were produced as part of Parker's larger program on vertebrates. His book on the shoulder girdle and sternum has a large herpetological content. He produced a series of 11 sizeable papers on the skulls of frogs (1871, 1876, 1881), salamanders (1877, 1882 [two papers]), snakes (1878), lizards (1879, 1881), turtles (1880), and crocodiles (1883), together comprising 779 pages of text and 128 plates, all in quarto. In each case a long abstract of the larger work was published separately and in advance.

• *References*: "William Kitchen Parker F.R.S.," by T. J. Parker, Macmillan, London, xv, 145 pages, 1893; "William Kitchen Parker," by G. B. Howes, Dict. Natl. Biogr., 15: 290–292, 1950; "William Kitchen Parker," by R. Hutchins, Oxford Dict. Natl. Biogr., 42: 758–759, 2004. • *Portrait and signature*: From T. J. Parker, 1893.

THEOBALD, William (1829–1908).

William Theobald, Jr., a geologist and naturalist, wrote important works on the herpetology of British India and Burma. Little is known of his early life, except for his year of birth, 1829. It has been assumed that he was English but the Civil Lists refer to him as he does on his earliest publications as "Wm. Theobald, Jr.," which may indicate an American origin. He arrived in Calcutta via Suez in March 1847 aboard the ship "Hindostan" and volunteered for the coal exploration surveys of the upper Damooda and Son (now Soane) valleys of eastern India. In 1848 he became an assistant to the physician-naturalist, John McClelland, who took over the coal survey of India. This survey, started by the East India Company in 1848, was the progenitor of the Geological Survey of India. During 1849–1853, Theobald worked for the geological surveys of Bengal and of Punjab but then transferred to the Geological Survey of India in 1853 where he worked for the rest of his life. He became deputy superintendent for Bengal in

1876 and retired to England in 1881. He died on 31 March 1908 in Ilfracombe (Devon) on the Bristol Channel.

Theobald was a field geologist and naturalist and spent a large portion of his life exploring. During 1855–1856 he toured the Orissa coalfields, assisting William T. Blanford. During the Indian Mutiny in 1857, he was in the field at Mhow where he narrowly missed being murdered. In the Narbada (now Narmada) Valley of western India in 1860 he found stone tools made by palaeolithic humans (see *Note*). In 1862 he took charge of the Pegu survey in Burma, covering as far north as 19° 30' and south through Tenasserim where he once impetuously grabbed a nearly 4-m-long king cobra by the tail thinking it was the colubrid *Ptyas*. On completion of the Pegu survey in 1873, he transferred to Bengal. In 1865 he visited the Sutlej and Spiti valleys (now in Pakistan) in the western Himalayas with his survey colleague, Ferdinand Stoliczka. Other places he explored include the Salt Range and Kangra Valley (Punjab), Kumaun region (in northern India bordering China), Gujarat (Pakistan), and the Pir Panjal Range (Kashmir).

While all of these trips were primarily geological, Theobald also collected plants and animals and these collections led to several publications. He wrote two books on land and freshwater shells (1860, 1876) and collaborated with Sylvanus C. T. Hanley to write "Conchologia Indica" ([1870–] 1876) covering British India. He prepared the third edition of Francis Mason's "Burmah, Its People and Productions" (1882–1883), which contains extensive zoological and botanical sections. In this last-named book Theobald pleads for application of strict priority in scientific names, which was in opposition to the so-called "Kew Rule" of the day. Even though Theobald's position has otherwise prevailed, his opposition to pronouncements made by leading museum scientists, especially those in London, became a pattern in his life.

Theobald published few, but significant, works on the reptiles and occasionally the amphibians of British India and Burma. In 1868 he authored two major catalogues, one of the reptiles (only) of Burma and the other of the herpetological collection of the Asiatic Society of Bengal in Calcutta. He described some new genera and species. He was the first to place the Xenopeltidae with the Boidae and to set up the family Geoemydidae. He also established the family Uromastycidae, which probably represents the only instance in which a religious reference was made in the description of a new family! Noting the peaceful and gentle nature of these animals, he remarked that they are "a very Quaker among lizards."

These two works, particularly the second one, began Theobald's habit of harshly criticizing the relevant authorities in London in print. Albert Günther, who had just published "Reptiles of British India" in 1864, was his first target. Theobald said he had "so superciliously, one might almost say perversely" overlooked earlier work. Theobald's next major work was a catalogue of the reptiles (only) of British India (1876). This time John E. Gray, who had died the previous year, was his target (" . . . owing to the careless use made by Dr. Gray of the materials in his possession . . ." and "Gray . . . falls into extraordinary error . . . ") but Günther receives his share (" . . . his dictatorial assertion."). The truth was that while men like Gray and Günther lacked field experience with fresh specimens, Theobald lacked proper training and comparative material. What was needed was cooperation and mutual respect. Nevertheless, Theobald was an excellent collector and a keen observer and his catalogues are useful compilations. His last herpetological contribution was an extensive section on the topic in his edition of Mason's book on Burma (1882), noted earlier.

• *References*: "William Theobald (1829–1908): Unwitting Reformer of Botanical Nomenclature?" by D. J. Mabberley, Taxon, 34: 152–156, 1985; Leviton and Aldrich, 2000 (Theobald on p. 69). • *Portrait*: Courtesy Indraneil Das. • *Signature* (1890): Natural History Museum, London, courtesy A. F. Stimson. • *Note*: Theobald is often credited, in error, with finding the human skull at Narmada, dated 200,000 B.C.E., that is the oldest human remains discovered on the Indian Subcontinent. The error was due to a mix-up in specimens (for details, see "God-Apes and Fossil Men: Paleoanthropology of South Asia," by K. A. R. Kennedy, Univ. Michigan Press, Ann Arbor, 2000 [Theobald on p. 40]).

LESSONA, Michele (1823–1894).

Michele Lessona was an Italian physician, zoologist, and writer who became an early specialist on the amphibians of Italy. He was born on 20 September 1823 in Venaria Reale (Torino) about 8 km northwest of Turin, the son of a professor of veterinary science. He attended school in Turin and later the Università di Torino where he graduated in medicine and surgery in 1846, and then attended lectures in natural history by the zoologist, Giuseppe Gené. Lessona left Turin suddenly in 1848 to escape the anger of his new wife's

father who had opposed their marriage. The couple first travelled to Greece, then Malta, and finally to Egypt where he took a post at a hospital in Khankah (now El Khânka) just outside Cairo. When his wife died of cholera in 1849, he had to return to Turin with their nine-month-old daughter. He practiced medicine there and taught natural history at a secondary school until 1854 when he became a professor at the university in Genoa. Previously, on his return from Egypt, he met the zoologist, Filippio De Filippi, who had succeeded Gené as director of the university's museum of zoology in Turin, to present him with some lizard specimens he had obtained in Egypt. This friendship led to De Filippi's invitation to accompany him on his expedition to Persia in 1862. In 1864 Lessona moved to the university in Bologna, but then, the following year, he took De Filippi's place at Turin during the latter's extended absence.

De Filippi had given the first lectures in Italy on Darwinian natural selection and Lessona also became an advocate. Together, they translated Darwin's works into Italian. On the death of De Filippi in 1867, Lessona succeeded to his professorship and became director of the museum. In 1869 he published a widely read book on self-help, entitled "Volere è Potere" (Will is Power), which earned him a large audience among the general public. He became rector of the university during 1877–1880 and was elected senator of the Kingdom of Italy in 1892. He died in Turin on 20 July 1894 and was succeeded as director of the museum by his son-in-law and former student, the naturalist and herpetologist, Lorenzo Camerano.

Lessona published broadly on medicine and zoology. He authored about 25 titles specifically on amphibians and some on reptiles (1868–1893) and on diverse topics including the skin and skin glands, reproduction, and distribution. His major technical work on amphibians was a review of the species inhabiting the Piedmont (1877), a region of northwestern Italy. He was also responsible for translating into Italian (1871–1873) the works of Alfred Brehm—Brehm's "Tierleben"—the most widely read global summary of animals of its day and this included a major section on amphibians and reptiles. Lessona published his own atlas of animals in 1893, of which volume 3, on reptiles, amphibians, and fishes, was still being republished as late as 1930.

• *References*: "Michele Lessona," by L. Camerano, Boll. Mus. Zool. Anat. Comp. R. Univ. Torino, 9(188): 1–72, 1894; "La Vita Scientifica di Michele Lessona," by L. Camerano, Atti Accad. Torino, ser. 2, 45: 331–388, 1895. • *Portrait*: From Italian Jour. Zool., 62(4): cover, 1995. • *Signature*: Adler collection.

BLANFORD, W. T. (1832–1905).

William Thomas Blanford, an English geologist, geographer, and zoologist, was an authority on the herpetofauna of South and Southwest Asia. He was born in London on 7 October 1832 and attended schools in Brighton and Paris. In 1852 he went for training at London's Royal School of Mines where Edward Forbes and Thomas Huxley lectured on natural history. Blanford then attended the mining academy in Freiburg (Saxony) before going to India late in 1854 together with his younger brother, Henry Blanford, who was also a trained geologist. In Calcutta they joined the Geological Survey of India, an association that William maintained until 1882 when he retired to England. He was in the field during much of his time with the geological survey, officially exploring for coal fields and lead deposits and compiling geological information about the Indian Subcontinent. These were also opportunities to collect vertebrates and land snails, his special interests.

By 1856 Blanford made his first field excursion, to the Talchir coalfields in Orissa assisted by William Theobald, a naturalist also in the employ of the geological survey. It was here that he recognized signs of ice action and the first evidence for widespread Permian glaciation in Gondwana. Many other field trips followed: Raniganj and Nilgiri Hills (1857–1860), Burma (1860–1862), central provinces of India (1865–1866), Abyssinia in northeastern Africa (1867–1868), Sikkim and the Tibetan frontier (1870), western Baluchistan and Persia (as a member of the India-Persia Boundary Commission, 1871–1872), western Sind (1874–1877), the Indian desert from the Indus River to Jodhpur (1876), and the Punjab and southern Afghanistan (1881–1882). These were arduous trips, usually on foot, horseback, or elephant. The Geological Survey of India, in fact, had its own herd of elephants. Some of them were trained "to pick up pieces of stone and, with a jerk of the trunk, to throw them over [its] head for the rider to catch, thus saving the trouble of dismounting." By 1872 Blanford's health was already seriously affected and furloughs in London became longer. A severe fever late in 1881 led to his retirement the next year. Returning to London, Blanford became an active member of the scientific community. In 1874

he had been elected a fellow of the Royal Society. He continued writing and editing, and died at his home in Campden Hill, London, on 23 June 1905.

Blanford made fundamental contributions in geology and zoogeography, in particular to glacial topography and to defining the faunal subdivisions of the Oriental Region. He was responsible for the creation of the concept of Gondwanaland. He served as editor of the "Fauna of British India" series and wrote the volumes on mammals (1888–1891) and birds (1895–1898). He published books on the expeditions to Abyssinia (1870; reprinted 2003) and Persia (1876), both of which had extensive zoological sections that included amphibians and reptiles. His strictly herpetological publications number about 25 (1870–1881). These include descriptions of numerous new species based on specimens collected on his geological field trips or by his colleagues. He also wrote papers on collections made in Socotra, Mesopotamia, Malaya, and Singapore. He reported on several herpetological collections made by his survey colleague, Ferdinand Stoliczka, who died on the return trip of the Second Yarkand Mission to Kashmir and western China. Blanford's collections are at the natural history museums in London and Calcutta. He published another 150 papers on diverse topics including birds, mammals, mollusks, geology, and zoogeography.

• *References*: "William Thomas Blanford," anonymous, Geol. Mag., new ser., decade 5, 2: 1–15, 1905; "William Thomas Blanford, 1832–1905," by T. G. B[onney], Proc. Royal Soc. London, ser. B, 79: xxvii–xxxi, 1907; "William Thomas Blanford," by T. G. Bonney, Dict. Natl. Biogr., suppl., 1: 178–179, 1927; Leviton and Aldrich, 2000 (Blanford on pp. 65–68); "William Thomas Blanford," by D. T. Moore, Oxford Dict. Natl. Biogr., 6: 176–177, 2004; "The Impact of Travels on Scientific Knowledge: William Thomas Blanford, Henry Francis Blanford, and the Geological Survey of India, 1851–1889," by A. E. Leviton and M. L. Aldrich, Proc. Calif. Acad. Sci., 55(suppl. 2): 117–137, 2004. • *Portrait* (1874) *and signature* (1890): Natural History Museum, London.

Stoliczka, Ferdinand (1838–1874).

Ferdinand Stoliczka (see *Note 1*), a Czech paleontologist and naturalist, was born on 7 July 1838 (see *Note 2*) at Bilany, near Kremsier (now Kroměříž), Moravia. For generations his family had served there as foresters on the estate of the archbishops of Olomouc and from living in this setting Ferdinand developed an early interest in natural history. He attended the German gymnasium in Kremsier (1850–1857). He then went to the university in Vienna to study natural history under Eduard Suess and defended his dissertation at Tübingen (Ph.D. 1861). After working for the Austrian Geological Survey in Vienna during 1861–1862 and publishing papers on the geology and paleontology of Austria-Hungary, he accepted a post as paleontologist at the Geological Survey of India. On arrival in Calcutta in 1863, he immediately began to study the survey's extensive collections of Cretaceous fossils from southern India, fossils that are a useful tool for establishing stratigraphic relationships between regions. This enormous book was published serially in four volumes (1861–1873; volume 1 jointly with his survey colleague, Henry F. Blanford) and runs to nearly 1,500 pages with 176 plates.

Stoliczka's work in British India also involved a series of expeditions to observe geological formations and make collections of fossils. As was also true for some of his other colleagues at the survey, notably William T. Blanford and William Theobald, Stoliczka took the opportunity afforded by these official trips to collect animals, which he studied at home during his personal time. Stoliczka made two early trips to the western Himalayas (1864, 1865) and from the second one wrote a long paper on the fauna (1866). During 1868 he substituted for John Anderson as curator of the new Indian Museum while Anderson was on expedition to Burma and Yunnan. Stoliczka's next trips were to Burma, the Malay Peninsula, and the Andaman and Nicobar Islands (1869); Cutch Peninsula and Darjeeling (1871–1872); and a return trip to the Andamans and Nicobars (1873). Numerous zoological and herpetological papers resulted from these trips.

In 1870 Douglas Forsyth led a diplomatic mission to Yarkand in Chinese Turkestan (now Shache, Xinjiang Uygur Autonomous Region). Anderson described the herpetological collections from this expedition (1872). In May 1873 Forsyth led a second embassy to Yarkand, and Stoliczka, although he had been scheduled to visit Europe at that time, volunteered to serve as naturalist. It was a fateful decision. Forsyth's group, with military escort, reached China from Kashmir using mountain passes as high as 4,477 m. The paths were often very steep and frozen. Temperatures at midnight sometimes dropped to –35°C. Stoliczka, who was a tireless field collector and observer, had a severe attack of spinal meningitis but recovered. By the time the group reached Yarkand in November, his toes had been numb for a month.

After completing their diplomatic work and making collections around Yarkand and nearby at Kashgar (now Kashi Shi), the group was ready to return to India. After an abortive attempt to return via Kabul, the group had to go back to Yarkand before setting out for Kashmir along a more westerly route than the one by which they had originally come. This required crossing the Karakorum Mountains at a pass just east of the world's second highest peak—variously called Dapsang, Godwin-Austen, or K2—at an elevation of 5,575 m. By the time the party had descended to Murghi, Stoliczka had re-contracted meningitis and died there on 19 June 1874 (See *Note 2*). He was 36 years old. His colleagues believed that the effects of the disease had been exacerbated by the high elevations and overwork, not by the cold. His body was taken to Leh where it was interred in a Union Jack–draped coffin with full military honors.

During his brief career Stoliczka published about 150 titles, most of them on paleontology and geology. He wrote two papers on Tertiary frogs and turtles (both in 1869) and 15 herpetological titles (1870–1873) plus a half dozen abstracts of some of them. These papers are generally taxonomic in nature with descriptions of little-known species and of some new genera and dozens of new species, and they also contain notes on natural history, anatomy, and classification. Some of his specimens are in the museums in London and Vienna, but most are at the museum in Calcutta. He kept live animals including snakes at his home for observation and to entertain his friends. His papers often drew attention to zoogeographical patterns and to similarities of living species to fossil forms. Stoliczka's herpetological collections from the Second Yarkand Mission were published by William Blanford (1875, 1878), his colleague at the survey.

• *References*: "Ferdinand Stoliczka, Ph.D.," by W. T. B[lanford], Nature, 10: 185–186, 1874; "Memoir of the Life and Work of Ferdinand Stoliczka, Ph.D.," by V. Ball, p. 1–35 *in* Scientific Results of the Second Yarkand Mission, vol. 2, Govt. India, London, 1886; "Ferdinand Stoliczka (1838–1874): The Life and Work of the Czech Explorer in India and High Asia," by J. Kolmaš, Wiener Stud. Tibetol. Buddhismus., 9: xi, 1–58, 1982. • *Portrait*: Bildarchiv, Österreich Nationalbibliothek, Vienna. • *Signature* (1865): Naturhistorisches Museums, Vienna, courtesy Christa Riedl-Dorn. • *Note 1*: The modern spelling of Stoliczka's name is "Stolička," as it appears in some publications. • *Note 2*: Birth and death dates follow Kolmaš (1982).

DAVID, Armand (1826–1900).

The French Lazarist priest, Jean-Pierre-Armand David, was the first Western naturalist to explore China's interior. He was born on 7 September 1826 (see *Note*) in Esplette near Bayone in the rugged Basque region along the Spanish border. His father was a medical doctor. Armand had an interest in natural history as a boy but decided on an ecclesiastical career and in 1848 entered the Congregation of the Lazarists. He was ordained a priest in 1851 and sent to the College of Savona in northern Italy where he taught science at a Lazarist school. Among his pupils were two future Italian naturalists, Luigi d'Albertis and Giacomo Doria. It was here on the Gulf of Genoa that he began to study natural sciences and dreamed to do missionary work in China. In 1862 he was sent to the Lazarist mission in Peking (now Beijing) to teach science at a school for boys, and he started making field excursions that very first summer. His main interest was zoology and especially ornithology but he also studied botany.

China at this time was ripe for exploration. Prior to the First Opium War of 1840–1842 the country was closed to foreigners, but then a few treaty ports were opened, mostly along the coast. After the Second Opium War of 1860 there were opportunities to visit other parts of the country. During the 1860s three naturalists did so: an English diplomat, Robert Swinhoe, who visited Peking, Ch'ung King (Chongqing), and coastal China; a Russian explorer, Nikolai M. Przhevalsky, who traveled the full east-west extent of northern China; and Armand David who penetrated the center. Father David's first excursions in 1862–1865 were on the plains and in the mountains near Peking. He smuggled out live deer of an unusual new species that was later named for him, Père David's deer. The preserved collections that he sent back to the Muséum d'Histoire Naturelle in Paris were regarded to be of such importance that the museum authorities asked the Superior General of the Lazarists in Paris to allow David to spend all of his time collecting. This permitted him to tour the interior of China where he could often use the existing missionary network for lodging and assistance. At the outset David worried about his prospects. In his diary he lamented (in translation): "Will [Swinhoe] leave anything for me to add?" In fact there was little overlap between the regions they visited.

David's first major expedition (March to October 1866) was from Peking westward—through a region that he called "Mongolia," but is now part of Nei Mongol (Inner Mongolia)—

to the Ordos Plateau, situated in the great bend of the Yellow River (Huang He). His second trip (May 1868 to July 1870) ascended the Yangtze River (Chang Jiang) to Ch'ung King, then by land to Mupin (Baoxing) in western Szechwan (Sichuan), a region that David referred to as "Tibet." In fact, he never visited present-day Tibet. It was on this trip that he discovered many new amphibians, the most spectacular of which was the Chinese Giant Salamander (*Andrias davidianus,* based on a dried skin obtained by David), which Émile Blanchard at the Paris museum named in his honor. David also discovered the giant panda on this journey and succeeded in getting not only a pelt, but also a live animal that was later exhibited at the museum's menagerie. Unfortunately, David's health deteriorated and at one point he was administered the last rites of the Catholic Church. He went home to France during 1870–1872, mainly to recover his health.

David returned to China for one final expedition (October 1872 to March 1874). This route went south from Peking to Honan (Henan) Province, then up the Yellow River to Sian Fu (Xi'an) in Shaanxi Province. He collected extensively in various mountainous regions west of Xi'an near the Gansu border before moving overland southeastward to Kiangsi (Jiangxi) and finally to Fukien (Fujian) on the coast. His health was sometimes so poor that he had to be carried in a sedan chair. David then returned to France in May 1874, but first he spent a period in the hot and dry climate of Algeria to recover his health. He lived the rest of his years in Paris, teaching missionaries, performing priestly duties, and working up his collections at the museum. He published extensively on his travels including both technical papers in the museum's *Archives* and in the journal of the Shanghai branch of the Royal Asiatic Society as well as in books intended for the general public. He described new species of amphibians in each of these places: both salamanders (*Cynops orientalis* and *Batrachuperus pinchonii*) and frogs (*Amolops mantzorum, Polypedates dugritei, Rana chensinensis*, and *R. spinosa*), among those currently recognized.

David died in Paris on 10 November 1900.

• *References*: "Père Armand David," p. 837–856 *in* E. Bretschneider, History of European Botanical Discoveries in China, vol. 2. Sampson, Low, Marston & Co., London, 1898; "Le Lazariste Armand David (1826–1874)," p. 67–91 *in* P. Fournier, Voyages et Découvertes Scientifiques des Missionnaires Naturalistes Français, part 2. P. Lechevalier & Fils, Paris, 1932; "Armand David," by A. Trin, Dict. Biogr. Franç., 10: 340, 1965; "Armand David and His Explorations in China," p. 274–283 *in* B. Mearns and R. Mearns, The Bird Collectors. Acad. Press, San Diego, 1998. • *Portrait* (1882): Congrégation de la Mission, Paris. • *Signature*: Muséum National d'Histoire Naturelle, Paris. • *Note*: Some authorities give his birth date as 27 September 1826.

FAYRER, Joseph (1824–1907).

Joseph Fayrer, an English physician, surgeon, and expert on the venomous snakes of British India, was born in Plymouth on 6 December 1824. He had been a student of natural science in school, but it was an experience while with his father, a naval officer, that directed him to medicine. He assisted during an outbreak of yellow fever on a stationary convict ship commanded by his father at Bermuda in 1843. Later he studied medicine at Charing Cross Hospital (M.R.C.S. 1847) where one of his fellow students was Thomas H. Huxley, the zoologist and future Darwinist. Fayrer continued his medical education at the University of Rome (M.D. 1849). In 1850 he shipped out to Calcutta and a post as assistant surgeon in the Bengal division of the Indian Medical Service, an association with the service that lasted for 45 years. He served in the Pegu war in Burma (1852) and was at Lucknow during the siege (1857), which ignited the Sepoy Revolt and led to dissolution of the East India Company and to direct rule of British India (today India, Pakistan, and Bangladesh) by the British government in 1858.

During 1858–1859 Fayrer was on furlough and gained further medical training in Edinburgh (M.D. 1859) before returning to India. Because of his distinguished medical service and his associated research, he was appointed or elected to several posts including professor of surgery at the Medical College of Calcutta (1859) and president of the Asiatic Society of Bengal (1867). He proposed a scheme for a zoological society and garden in Calcutta, which was finally realized in 1875. In 1871 Fayrer became honorary physician to Queen Victoria. He also served as personal physician to the Prince of Wales (later Edward VII) and accompanied him during the prince's tour of India in 1875. Fayrer retained a close friendship with him for the rest of the king's life.

Because of poor health, Fayrer returned to England in 1872. He became president of the Medical Board of the India Office on retirement from the Indian Army the next year. This board, through the Indian Medical Service, was responsible for all civilian health in British India. Fayrer retired as board president in 1895 at the rank of surgeon-general and received numerous honors including a hereditary knighthood in 1896.

He had been elected a fellow of the Royal Society in 1877. Fayrer died at Falmouth on 21 May 1907.

Fayrer was struck by the appalling number of human deaths caused by venomous snakes each year in India. He recorded a mortality of 11,416 persons for Bengal alone in 1869 and, by extrapolation, estimated more than 20,000 deaths for all of British India. A key point was that in most cases the species involved was unknown. Thus motivated, he had begun studying snake venoms in 1867, work that he continued after he returned to live in England. Before leaving Calcutta, however, he decided to prepare a book for proper identification of the venomous species that also gives the results of his experiments on venoms and recommendations for treatment of bites. His book, "The Thanatophidia [*thanatos*, from the Greek for "death"] of India" (1872), is one of the classics on venomous snakes. It is folio in size with 31 color-printed plates based on Fayrer's own sketches from living specimens that were then rendered by native artists at the Calcutta School of Art. The animals are depicted life size. The double-page illustration of the king cobra is one of the most imposing drawings of a snake ever published. A second edition, revised and enlarged, was issued in 1874 with the same 31 plates.

Fayrer's physiological experiments on venom, which he hoped would lead to an antidote, were published during 1873–1909, mostly in collaboration with Lauder Brunton and sometimes also with Leonard Rogers. A volume of their collected works was issued in 1909 under the title, "On the Poison of Venomous Snakes and the Methods of Preventing Death from Their Bite." Major findings were that *Naja* venom was able to paralyze respiration solely by blocking the neuromuscular junction and that large doses of venom caused cardiac arrest. Alas, although they investigated all of the then-current methods of treatment (vapor of creosote, carbolic acid, potassium permanganate, liquor ammonia), they found all of them lacked effectiveness. They finally recommended (1904) a combination of rapid ligation and suction with the application of potassium permanganate as the best available treatment. (Albert Calmette had announced his anti-*Naja* serum in 1894 but the species-specificity of snake venoms had only been reported in 1901 by Vital Brazil and antivenoms were not available in commercial quantities for some years.)

• *References*: "Recollections of My Life," by J. Fayrer, William Blackwood and Sons, Edinburgh, xii, 508 pages, 1900; "Sir Joseph Fayrer MD FRS (1824–1907) Indian Medical Service: Snakebite and Mortality in British India," by B. J. Hawgood, Toxicon, 34: 171–182, 1996; "Sir Joseph Fayrer," by H. P. Cholmeley and W. F. Bynum, Oxford Dict. Natl. Biogr., 19: 200–202, 2004. • *Portrait* (1881): Wellcome Institute Library, London. • *Signature* (1884): Adler collection.

LATASTE, Fernand (1847–1934).

The zoologist, Vital-Fernand Lataste, who specialized on the natural history and systematics of amphibians and reptiles, was born on 5 March 1847 in Cadillac-sur-Garonne about 30 km southeast of Bordeaux in southwestern France. After attending secondary schools, in 1867 he entered the École Centrale in Paris and then the École Polytechnique before volunteering for military service during the Franco-Prussian war, in 1870–1871. Afterwards he studied law in Bordeaux, obtained his degree, and was admitted to the bar there in 1874. He also earned a bachelor's degree in science and it was at this time that his formal interest in zoology and especially herpetology began. In 1875 he went to Paris for medical study but he never defended his thesis. He supported himself as a tutor at the Collège de France and continued herpetological research during his personal time. He is said to have had an independent, even sometimes aggressive spirit that alienated him from other scholars, yet in 1876 he met a young Belgian student who was 11 years his junior—George A. Boulenger—whom Lataste mentored for the rest of his life.

Most of Lataste's fieldwork was done locally wherever he happened to be living, and he kept live animals and preserved specimens in his apartment for study. In 1880–1881, however, he went to North Africa on a largely self-financed expedition to the Sahara and the Algerian high plateau, venturing unescorted into parts of the desert where few Europeans had explored before. A French civil engineer studying artesian wells in the desert reported how surprised he was to have met a man, all alone and dressed as if he was on a stroll down the streets of Paris, chasing lizards and snakes! In 1884 Lataste went to Tunisia as part of an expedition and in 1885 traveled to Senegal, a mission cut short by his illness. Lataste published a few short papers on his North African collections, but his notes and specimens were later turned over to his protégé, Boulenger, for publication in a large monograph (1891).

By 1889, however, Lataste was at the end of his finances and cut off by his parents who did not approve of his marriage.

Losing all hope to succeed in France, he sailed to Chile to accept positions in Santiago as chair of zoology at the medical school and as assistant to the director of the natural history museum, the zoologist and herpetologist R. A. Phillipi. By 1897 his wife and two of his children were dead. In despair he returned to France and to his place of birth, his scientific career over at the age of 50. After the deaths of his parents, he settled on the family property, surrounded by gardens, vineyards, and woodlands where he continued his keen natural history observations. He became the sage at local scientific meetings and in the 1920s began to publish again on the local fauna. When his legs became paralyzed and he could no longer take long walks, he moved to an apartment in Bordeaux in March 1933 and died there less than a year later on 6 January 1934.

Early in his career Lataste focused on the local herpetofauna and authored numerous life history papers as well as two longer works: "Essai d'une Faune Herpétologique de la Gironde" (1875), a masterful treatise on the biology of the amphibians and reptiles of southwestern France, and "Catalogue des Batraciens et Reptiles des Environs de Paris" (1876), an annotated checklist covering 27 species with a detailed summary of distribution in western France. While in Chile he published a summary of the lizards of that country (1892). He was also interested in systematics and classification and was one of the first in Europe to recognize the importance of osteology and tadpoles in classifying frogs, as best exemplified in his monograph on the discoglossid frogs (1879). In this paper he also noted the value of using the mode of amplexus, vocalizations, and other behavioral characteristics. He named about 25 new taxa including five genera. These include frogs, salamanders, lizards, snakes, and turtles from many parts of the world (Algeria, China, Egypt, Iberian Peninsula, Senegal). His most spectacular novelty was the elapid *Walterinnesia aegyptia*, a widespread desert snake that reaches a length of 1.3 m. Lataste named it for Walter Innes, curator of the medical museum in Cairo who sent him a specimen. Lataste authored about 90 herpetological titles (1874–1894) during his active career and later, in retirement, five more (1925–1933) based on observations made at his country home.

• *References*: "El Profesor Fernand Lataste. Notas Biográficas y Bibliográficas," by C. E. Porter, Rev. Chilena Hist. Nat., 9: 36–45, 1905; "Notice Biographique sur Fernand Lataste," by G. A. Boulenger, Proc.-Verb. Soc. Linn. Bordeaux, 86: 109–113, 1935. • *Portrait*: Courtesy Judith Gibber. • *Signature* (1877): Natural History Museum, London, courtesy Andrew F. Stimson.

MACLEAY, William (1820–1891).

William John Macleay was a pastoralist, politician, naturalist, and benefactor of science in Australia who made significant contributions to the herpetology of his adopted country. He was born in Wick (Caithness), Scotland, on 13 June 1820 and educated at the Edinburgh Academy (1834–1836). He then entered the University of Edinburgh as a medical student where he was attracted to natural science. By 1837, however, he was orphaned and poor, and on the advice of his uncle, Alexander Macleay, who had immigrated to Australia in 1825, he joined him and his uncle's two sons, William Sharpe Macleay and George Macleay, in Sydney in March 1839. Alexander Macleay and William S. Macleay were both gifted entomologists and they and all of William's other Sydney relations encouraged his interest in natural history, but first he had to earn a living. By 1841 he had become a magistrate. He acquired large land holdings, those along the Lower Murrumbidgee River in southern New South Wales, near Wagga Wagga, through squatting. He was a member of Parliament from 1856 through 1874 and served on various national commissions on subjects as diverse as railroads, libraries, and foreign affairs. His personal lands continued to grow and he added vineyards in the 1870s. In 1889 he was knighted for his service to the Crown. Following a carriage accident at the gates of Government House, he was taken to his home at Elizabeth Bay, a Sydney suburb, where he died on 7 December 1891.

In the early 1860s, his cousin, William S. Macleay, reawakened William Macleay's interest in natural history. Recognizing that there were few scientific opportunities in Australia at that time, in 1862 he helped found the Entomological Society of New South Wales and in 1874 he began the Linnean Society of New South Wales of which he was the inaugural president. He became a trustee of the Australian Museum and in 1874 was involved in the disgraceful events that led to the firing of its curator, Gerard Krefft, Australia's founding herpetologist. Macleay built a zoological museum in his home and hired collectors to send specimens from throughout Australia. He even purchased and outfitted a boat, the "Chevert," for an expedition to New Guinea (May–September 1875) that he accompanied. He undertook his own

serious studies of insects, fishes, amphibians, and reptiles and he published some 70 papers, mostly containing descriptions of new species. In 1881 he published a two-volume catalogue of Australian fishes. On Macleay's death his collections including those of his uncle and cousins passed to the University of Sydney (today housed in the Macleay Museum); the insects alone numbered some 600,000 specimens.

Macleay published about 15 papers on amphibians and reptiles (1875–1888). These included numerous new species and genera. *Litoria dorsalis, Hypsilurus boydii, Ctenotus spaldingi, Aspidites ramsayi*, and *Demansia papuensis* are among his taxa recognized today. He was sometimes not a very discerning taxonomist, however, and described the same species more than once, even in the same paper. The most extreme case was his description of a new skink (today *Carlia longipes*) as five different new species and all in the same paper! His major herpetological works include a four-part series on New Guinea based on the "Chevert" expedition (all 1877) and the longest paper, his "Census of Australian Snakes" (1884).

It should be noted that it was Macleay's cousin, William Sharpe Macleay, who described two new geckos from Cuba (1834) when he served as a British diplomat there and sent reptiles to John E. Gray at the museum in London.

• *References*: "The Macleay Memorial Volume," by J. J. Fletcher (ed.), Publ. Linnean Soc. New South Wales, Sydney, li, 308 pages, 1893; "The Society's Heritage from the Macleays," by J. J. Fletcher, Proc. Linnean Soc. New South Wales, 45: 567–635, 1920, and 54: 185–272, 1929; "Sir William John Macleay," by M. Hoare and M. Rutledge, Austr. Dict. Biogr., 5: 185–187, 1974. • *Portrait and signature*: Australian Museum, courtesy Harold Cogger.

WIEDERSHEIM, Robert (1848–1923).

The German comparative anatomist and embryologist, Robert Ernst Eduard Wiedersheim, made important contributions to the knowledge of the anatomy and development of amphibians and reptiles. He was born on 21 April 1848 in Nürtingen (Baden-Württemberg) about 20 km southeast of Stuttgart. His father, a practicing physician, was an amateur naturalist. When Robert was 15, he devoted so much time to his microscopic study of freshwater hydra that his final school report was a near disaster. He wanted to pursue zoology, but at the insistence of his father he entered the university in Tübingen in 1868 to study medicine. Robert's principal teacher there was Franz Leydig. Wiedersheim transferred to Würzburg and then to Freiburg where he completed his medical studies (M.D. 1872). He became a lecturer at the university in Würzburg until 1876 when he returned to Freiburg as a professor and, from 1883, director of the Institute of Anatomy and Comparative Anatomy. One of his colleagues in Freiburg was Alexander Ecker who had authored the classic book on frog anatomy (1864–1882). Wiedersheim soon began the research that led to publication of his widely influential textbook on comparative anatomy of vertebrates, "Lehrbuch der vergleichende Anatomie des Wirbeltiere" (1882, with a seventh edition in 1909). His fame rests on this text, and because of it many students were attracted to him from Europe and beyond. He died on 12 July 1923 in Lindau im Bodensee (Lake Constance) in southernmost Bavaria (see *Note*).

Most of Wiedersheim's anatomical research was with amphibians and reptiles. He authored about 35 titles (1875–1900) on a wide range of topics including the anatomy of *Phyllodactylus europaeus*; development of the urogenital system of crocodiles and turtles; the respiratory system of chameleons; tentacles in caecilians; the parietal eye of lizards; and the development of *Proteus* and *Salamandra*. His books include one published in 1875 on the anatomy of salamanders—*Salamandrina perspicillata* and *Geotriton fuscus* (now *Speleomantes italicus*)—but he completely missed the absence of lungs in the latter, a plethodontid. It was an American, Harris H. Wilder, one of his former students, who in 1894 discovered that plethodontids are lungless. In 1877 Wiedersheim published a monograph on the skulls of salamanders and in 1879 a book on the anatomy of caecilians, based on six species from four of the six currently recognized families. He also completed the book on frog anatomy begun by his colleague, Ecker. Wiedersheim prepared part of the second (1881) and all of the third and last part (1882) of the book.

• *References*: "Lebenserinnerungen," by R. Wiedersheim, J. C. B. Mohr, Tübingen, viii, 207 pages, 1919; "Robert Wiedersheim (Ausserordentlicher Professor zu Freiburg i Br.), die Anatomie der Gymnophiona," by W. Peters, Sitzungs-Ber. Ges. Naturf. Freunde Berlin, 1879: 150–159, 1879; "Robert Wiedersheim," by P. Glees, Dict. Sci. Biogr., 14: 331, 1976; "Robert (Ernst) Wiedersheim," anonymous, Dict. German Biogr., 10: 503, 2006. • *Portrait* (1890) *and signature* (1899): Natural History Museum, London, courtesy A. F. Stimson. • *Note*: Some sources give Wiedersheim's place of death as Bad Schachen, which is near Lindau, and on 12 July.

YARROW, Henry C. (1840–1929).

Henry Crècy Yarrow was an army surgeon and amateur naturalist with a special interest in North American amphibians and reptiles. He was born in Philadelphia on 19 November 1840, received his early education there and in Geneva, Switzerland, and obtained his medical training at the University of Pennsylvania (M.D. 1861). He immediately went to the front lines as assistant surgeon to the Fifth Pennsylvania Cavalry and returned to work at a Philadelphia hospital for the rest of the Civil War. After the war, he served as a surgeon at several posts including Fort McHenry near Baltimore (Maryland) and Fort Macon in North Carolina. Later he was appointed surgeon-naturalist, on the recommendation of Spencer Baird, to the Wheeler Survey for exploring west of the 100th meridian. In 1876 Yarrow moved to Philadelphia to work in an army hospital and then went to Washington (D.C.) and a position in the Surgeon General's Office before retiring from the army in 1891. He had a private medical practice in Washington, primarily in dermatology, and later became a professor at George Washington University.

In 1878 his old friend, Baird, knowing that Yarrow's hobby was herpetology, asked him to serve as honorary curator of reptiles and amphibians at the national museum because the collection had not had regular attention since Charles Girard left the museum in 1860. Yarrow held this post until 1889 when Leonhard Stejneger became curator. Yarrow was recalled to military service in 1908 and reached his final rank of lieutenant colonel before the end of World War I. He died at Old Point Comfort (Virginia) on 2 July 1929 and was buried with full military honors at Arlington National Cemetery.

Yarrow published few titles in herpetology, but they were at a critical time and covered important regions. He formed a friendship with the ornithologist-mammalogist, Elliott Coues, whom he had succeeded as post surgeon at Fort Macon in 1870 and with whom he published several works including five papers (1871–1878) on the natural history of the vicinity of Fort Macon. Both had been naturalists attached to the Wheeler Survey, at different times, but they published their herpetological results as companion papers, in quarto, sharing the same ten outstanding and partly colored lithographed plates (1875; reprinted 1978). Yarrow's section covered the collections made by himself and other naturalists during 1871–1874 in a large region from California to New Mexico and north to Nevada and Colorado, whereas Coues's covered his own collections made in Arizona during 1864–1865. Seven new species were described in Yarrow's report, all authored by Edward D. Cope who named one lizard in Yarrow's honor, *Sceloporus jarrovii* (see *Note*). Yarrow again teamed with Coues to publish a report on the herpetology of Dakota and Montana (1878). Yarrow's final report in the Wheeler Survey, co-authored with Henry W. Henshaw (1878; reprinted 1978), covered collections made in California, Arizona, and Nevada during 1875–1877.

Yarrow later described numerous new species (1880–1883) including one snake, *Elaphe bairdi*, honoring his friend, Baird. He returned the earlier favor by Cope by naming a lizard for him (*Gambelia copeii*; see page 10). Yarrow's other major work in herpetology was his 1882 checklist of North American amphibians and reptiles, which included an annotated list of specimens at the national museum; an abstracted list of the species was issued the next year. He also published a few short herpetological papers in journals (1876–1878) and several medical articles on venomous reptiles (1884–1888).

• *References*: "Dr. Henry C. Yarrow," anonymous, Harper's Weekly, 33: 146–147, 1889 (23 February); "Henry Crècy Yarrow (1840–1929)," p. 533–549 *in* E. E. Hume, Ornithologists of the United States Army Medical Corps, Johns Hopkins Press, Baltimore, 1942. • *Portrait and signature*: From Hume, 1942. • *Note*: In medieval Latin, the conventional equivalents for the *Y* and *w* in Yarrow are *J* and *v*, respectively. The *W* does not exist in Latin, and the letter *Y* was only used as a vowel.

BROCCHI, Paul (1838–1898).

Paul-Louis-Antoine Brocchi, a French naturalist, physician, and agriculturist, was born in Nancy (Meurthe-et-Moselle) on 2 May 1838. Like his father who was an agricultural engineer, Paul dedicated his career to the scientific development of all aspects of agriculture and its transformation into an industry. He was also interested in anatomical research and received his doctor of science degree in 1875 from the Sorbonne (University of Paris) for a dissertation on decapod crustaceans that was completed under the supervision of Henri Milne-Edwards. Brocchi then earned a medical degree after which he became a professor at the Ecole des Hautes-Études for 25 years. He also taught applied zoology at the Institut Agronomique, succeeding Émile Blanchard in that post in 1890, and at other colleges in Paris. He wrote several books including one on the anatomy of decapod crustaceans (1875), but most of them dealt with farming: oyster culture (1883), agricultural zoology including fish farming, beekeeping, and silkworm culture (1886), and cattle (1897). In July 1898 he was appointed to the Legion of Honor for his service to agriculture. He died in Paris the next month.

Since the early 1870s Brocchi had an informal association with the Muséum d'Histoire Naturelle, first to work on decapod crustaceans and later with reptiles and amphibians. He published descriptions of new reptiles from the South Pacific based on the museum's collections, but was soon asked to assume responsibility for writing the amphibian section of one of the great expedition reports, the "Mission Scientifique au Mexique et dans l'Amérique Centrale," edited by Milne-Edwards. This report was a scientific adjunct of the ill-fated attempt by Napoleon III to establish a puppet Mexican empire. (For more details, see p. 88.) When the French effort began to fail in 1864, Firmin Bocourt, the field collector, had to change course and devote his attention to Central America. The herpetological collections were turned over to Auguste Duméril, then professor in charge of reptiles and fishes at the museum, for publication, assisted by Bocourt. Duméril had begun to draft the amphibian text, but died in November 1870. Bocourt was too busy with the reptile section to handle amphibians as well, so Brocchi was eventually invited to do so.

Brocchi published descriptions of many of the new frogs in two papers in 1877, but the bulk of his work on the American collections was reported in his book, "Étude des Batraciens de l'Amérique Centrale" (issued in three parts in 1881–1883; reprinted 1978), as part of the "Mission Scientifique" series, and it was accompanied by 24 excellent lithographic plates. The book covers 153 species, 25 of which were named by Brocchi in addition to a new genus of hylid frogs, *Plectrohyla*. In total he authored a dozen titles on amphibians and reptiles (1876–1883). His most spectacular novelty, however, was not from America but from the French Pacific island of New Caledonia: a half-meter-long giant skink with large, recurved, fang-like teeth that was not seen again by scientists for more than a century—a species now called *Phoboscincus bocourti*.

• *References*: "Brocchi (Paul-Louis-Antoine)," p. 246 *in* T. Lamathière, Panthéon de la Légion d'Honneur, vol. 18, 1895?; "Brocchi (Paul-Louis-Antoine)," by S. Le Tourneur, Dict. Biogr. Franç., 7: 386, 1956. • *Portrait*: Ann. Inst. Agron., ser. 2, 1: 131, 1917. • *Signature* (1877): Adler collection.

SAUVAGE, Henry-Emile (1842–1917).

The French naturalist, Louis César Henry-Emile Sauvage, an anthropologist, paleontologist, and zoologist who specialized on fishes, amphibians, and reptiles, was born on 22 September 1842 in Boulogne-sur-Mer (Pas-de-Calais) on the Strait of Dover. He studied classics at the Institut Haffreingue in Boulogne and then went to the university in Montpellier where he received two baccalaureate degrees in 1862 and began his medical training. In 1868 he moved to Paris and the next year defended his thesis before the Faculté de Médecine.

During the Franco-Prussian War of 1870–1871, he served as an assistant medical officer. In 1872 he joined the staff of the Muséum d'Histoire Naturelle and the laboratory of Léon Vaillant, professor in charge of reptiles and fishes. He was promoted to assistant naturalist in 1875, but he resigned in 1884 and was replaced by François Mocquard. Although he had a productive career at the museum, Sauvage was homesick for Boulogne. He returned there to become curator of the city's museum and to found an aquaculture station in Boulogne that he directed until 1896. He continued to publish extensively on fossil fishes and reptiles, geology, aquaculture, living fishes, reptile anatomy, archeology, and anthropology. He died in Boulogne on 3 January 1917.

Sauvage's research on living amphibians and reptiles (nearly 30 titles, 1876–1913, plus more than 40 herpetological entries in Dechambre's "Dictionnaire des Sciences Médicales") was a small but significant part of his work. Most of his early papers reported on collections made in distant places and sent to the museum in Paris. In these he described several dozen new species of frogs, salamanders, lizards, and snakes including *Elaphe davidi* and *Pachytriton brevipes* from China, *Apostolepis coronata* from Brazil, and *Sphenomorphus maindroni* and *Typhlops bipartitus* from New Guinea. Sauvage also studied the physiological action of skin toxins from French frogs and salamanders, and beginning in 1910 he published extensively on reptilian anatomy. His only herpetological book was in the French edition of A. E. Brehm's "Tierleben," entitled "Merveilles de la Nature" ("Les Reptiles et les Batraciens," [1885]). Sauvage's contributions on fossil reptiles were more extensive (about 45 titles, 1871–1911). These mainly cover Jurassic and Cretaceous turtles, mosasaurs, and plesiosaurs of the Boulogne and Paris regions.

• *References*: "Notice Nécrologique sur le Dr H.-E. Sauvage," by L. Roule, Bull. Mus. Natl. Hist. Nat., 23: 138, 1917; "La Vie et l'Œuvre d'Emile Sauvage," by C. Cépède, Bull. Soc. Acad. Boulogne-sur-Mer, 11: 1–90, 1923; Jaussaud and Brygoo, 2004 (Sauvage by P. Jaussaud and É.-R. Brygoo on pp. 473–475). • *Portrait*: From Cépède, 1923. • *Signature* (1894): Natural History Museum, London, courtesy Andrew F. Stimson.

HOFFMANN, C. K. (1844–1903).

The Dutch anatomist, Christiaan Karel Hoffmann, author of the herpetological sections in H. G. Bronn's classic series, "Klassen und Ordnungen des Thier-Reichs," made numerous original contributions to the anatomy and development of amphibians and reptiles. He was born on 16 July 1844 in Heemstede (North Holland) on the edge of Haarlem. He studied biology and medicine at the University of Amsterdam and took his medical degree at the University of Utrecht (M.D. 1866). Hoffmann received his doctorate in Germany from the University of Göttingen (Ph.D. 1871) for a dissertation on the anatomy of echinoderms. After briefly serving as a physician at the Meerenberg hospital for the mentally retarded located just north of Haarlem and as a prosector in medical anatomy at the universities in Amster-

dam and Leiden, Hoffmann became Curator of Invertebrates at the Rijksmuseum in Leiden in 1872. In December 1874, however, he left the museum to become a professor of zoology at the University of Leiden where his career developed as a specialist in comparative anatomy, morphology, and developmental anatomy. He remained active as a professor until his death in Leiden on 27 July 1903.

Hoffmann authored numerous papers and monographs on the comparative anatomy and development of amphibians and reptiles (about 20 titles, 1876–1902), many of them quarto in size and with numerous detailed illustrations. These cover topics such as the morphology of the shoulder girdle and breast bone, the development of the venous system, the cervical ribs in turtles, the structure of the inner ear, the carpus and tarsus, the development of the urogenital system, and the development of the sympathetic nerve in salamanders. His best-known contributions to herpetology, however, are his sections in Bronn's classic series on the classification and anatomy of the animal kingdom. The volume on amphibians was issued in 23 parts during 1873–1878 and includes 53 fine lithographed plates drawn by Hoffmann, many of them partly colored. There is a detailed description of and key to genera and an extensive description of morphology and development. The corresponding section on reptiles is in three volumes; all are dated 1890 on the title pages but the book appeared in 69 parts over the period 1879–1890. The topical coverage is similar to that for amphibians but is even more extensive with 170 lithographed plates and nearly 2,100 pages of text. This compendium on the morphology of amphibians and reptiles was the most extensive on the topic for nearly a century until the series, "Biology of the Reptilia," edited by Carl Gans and others, began to appear in 1969, and its counterpart, "Amphibian Biology," edited by Harold Heatwole and others that started in 1994.

• *References*: "History of the Carcinological Collections of the Rijksmuseum van Natuurlijke Historie, Leiden, Netherlands (1820–1950)," by L. B. Holthuis, p. 225–242 *in* F. Truesdale (ed.), History of Carcinology, A. A. Balkema, Rotterdam, 1993 (Hoffmann on pp. 231–232); "Christiaan Karel Hoffmann," p. 61–62 *in* L. B. Holthuis, 1820–1958 Rijksmuseum van Natuurlijke Historie, Rijksmus. Natuurl. Hist., Leiden, 1995. • *Portrait*: From Holthuis, 1995. • *Signature* (1893): Courtesy L. B. Holthuis.

CAMERANO, Lorenzo (1856–1917).

Lorenzo Camerano was an Italian naturalist, Darwinist, and senator who made major contributions to the biology of European amphibians and reptiles. He was born on 9 April 1856 in Biella (Vercelli) in the Piedmont of northern Italy, 65 km northeast of Turin. Because of his father's government work his family moved frequently, thus his schooling was in Bresica, Bologna, and finally in Turin where he graduated from the Regio Liceo "Gioberti" in 1874. He had become an accomplished artist in school and his work came to the attention of the naturalist, Michele Lessona, professor at the Università degli Studi di Torino. Camerano's friendship with Lessona and his assistants caused him to abandon painting in favor of natural history and to enter the university to study mathematics and natural sciences.

After graduation in 1878 he became an assistant in the university's museum of zoology—which was directed by Lessona and whose daughter Camerano would soon marry—and he also taught in a local secondary school. In 1889 he competed for and won the chair of comparative anatomy and physiology at the university in Cagliari, but this lasted barely two months because Lessona decided to relinquish his chair at Turin to which Camerano was immediately appointed. In 1894 Camerano took on the added duties of director of the museum of zoology, again following in Lessona's footsteps. He held many distinguished positions reflecting his accomplishments and interests. He was rector of his university (1907–1910) and president of the Italian Alpine Club, and in 1909 he was elected a senator of the Kingdom of Italy. He died in Turin on 22 November 1917.

Camerano's scientific talents were extremely broad. He authored nearly 350 titles of which about 55 are on amphibians and reptiles (1877–1904). His interests were in systematics and agricultural entomology (he wrote three standard texts), ecology (his 1880 paper on community ecology is considered a classic), mammalogy (his special interest was cetaceans), and worm systematics (he was the leading authority on *Gordius*). In herpetology Camerano made contributions to systematics (he described new taxa of frogs, lizards, and snakes), but his main interests were metamorphosis (he conducted numerous experiments on the topic), secondary sexual characteristics (he explained them in terms of Darwinian sexual, and not natural, selection), and ecophysiology (he was especially interested in adaptation to high elevations). Following Harris Wilder's discovery of lunglessness in American plethodontid salamanders (1894), Camerano studied European salamanders and found that some of them also lacked lungs, which led him to studies on respiration in lungless salamanders (1894–1896).

Camerano's most elaborate contribution to European herpetology was a series of five extensive monographs on the amphibians and reptiles of Italy (1883 [1884]–1891). Camerano personally drafted all ten plates in this series, as he did for most of his publications, drawing on his artistic talents first developed as a teenager. His collections are at the museum of zoology in Turin.

• *References*: "Lorenzo Camerano," by E. Giglio-Tos, Boll. Mus. Zool. Anat. Comp. Univ. Torino, 33(725): 1–14, 1918; "Lorenzo Camerano Il Maestro," by L. Cognetti de Martiis, Boll. Mus. Zool. Anat. Comp. Univ. Genova, 7(19): 1–15, 1927 (1928); "Lorenzo Camerano," by B. Baccetti, Diz. Biograf. Ital., 17: 170–172, 1974. • *Portrait*: From Giglio-Tos, 1918. • *Signature* (1890): Adler collection.

BOSCÁ, Eduardo (1843–1924).

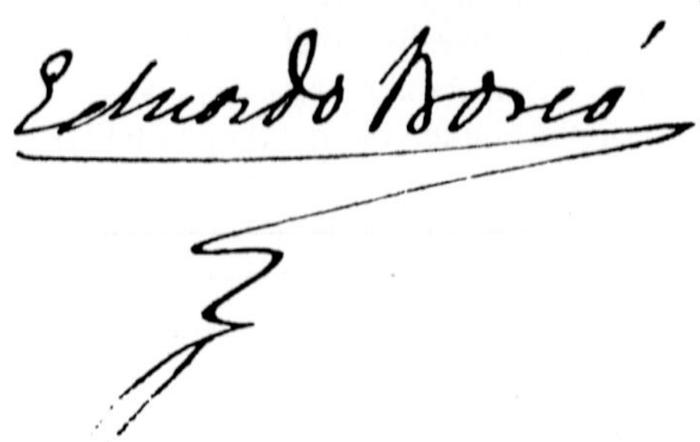

Eduardo Boscá Casanoves (see *Note 1*), the founder of Iberian herpetology, was born on 11 February 1843 in Valencia in eastern Spain. His father was a medical doctor and although this was also Eduardo's destined career, he limited his medical practice in order to devote his life to natural history and to herpetology in particular. After completing secondary school in Valencia at the Instituto Técnico (now Instituto Luis Vives), he attended medical school at the Universidad de Valencia (Licentiate 1869) and then studied natural and veterinary sciences at the Universidad Central de Madrid (now Universidad Complutense de Madrid) where he earned his doctorate (1873). In 1874 he became a professor at the Instituto de Játiva, located in Játiva just south of Valencia. He also had a small medical practice in Játiva. Boscá became a professor at the Universidad de Valencia in both the veterinary school and natural sciences. He directed the university's Museo de Historia Natural and the city's botanical gardens. Boscá also established and directed the Museo Paleontológico and in this capacity visited the major museums of Europe. He then toured South America to study the fauna of the pampas and to observe museums in Argentina and Uruguay. His local fieldwork, collections, and studies were focused on his home province of Valencia, but his interests encompassed all of the Iberian Peninsula. After a lifetime of public service, Boscá died in Valencia at the age of 81 from uremia, on 28 May 1924 (see *Note 2*).

Despite his many contributions to botany, zoology, and paleontology, Boscá's main interest was in herpetology and he authored about a dozen papers on the subject (1877–1919). His most important were three catalogues of the amphibians and reptiles of Spain, Portugal, and the Balearic Islands off the Mediterranean Coast of Spain. The first (1877), in Spanish and full of errors, was followed by a revised version in French (1880) and concluded, in Spanish (1881), with corrections and additions. These were for their day exceptionally detailed works, with lists of specific localities, descriptions, measurements, and comprehensive references to the literature. The 1881 paper includes a very large foldout map on which known localities for all 55 taxa are plotted. This seems to be the first use of this method in herpetology to illustrate species distributions for entire countries. The quality of his work improved markedly as the result of friendly advice from the French herpetologist, Fernand Lataste. Boscá also described and named several of the most distinctive endemic or nearly endemic species of Iberia including *Alytes cisternasii, Chalcides bedriagai, Podarcis pityusensis,* and *Vipera latastei.*

• *References*: "D. Eduardo Boscá," anonymous, Las Provincias, Valencia, 29 May 1924, pages 1–2; "Eduardo Boscá Casanoves, Médico y Naturalista Herpetólogo 1843–1928 [*sic*]," p. 40–41 *in* M. V. Añon (ed.), 101 Hijos Ilustres del Reino Valenciano, vol. 2, Indust. Reprod. Graf., Valencia, 1978; "Eduardo Boscá y el Cultivo de la Historia Natural en la Valencia de la Restauración," by J. I. Catalá Gorgues, Activ. Cient. Real Acad. Med. Comun. Valencia, 6: 51–72, 2005; Additional information from members of the Boscá family, 1991–1992. • *Portrait and signature* (1883): Courtesy Honorato Boscá Berga (senior). • *Note 1*: On some of his publications, Boscá used the form "Boscá y Casanoves." In Catalan the name is written "Eduard Boscà i Casanovas." • *Note 2*: Most biographies give Boscá's year of death, incorrectly, as 1928. His death notice appeared on the front page of Valencia's largest daily newspaper, *Las Provincias,* on 29 May 1924. This date is confirmed by several members of the Boscá family.

SEOANE, Víctor López (1832–1900).

Víctor Ramón López Seoane (see *Note 1*), an amateur naturalist of Galicia in northwestern Spain, was born in Ferrol, a small Atlantic port city 20 km northeast of A Coruña, on 28 September 1832 (see *Note 2*). With a boyhood interest in natural history, he attended schools in Ferrol and the Universidad de Santiago. He received his bachelor of philosophy from the latter in 1851. He then undertook medical studies in Madrid, Andalucía in southern Spain, and finished at Santiago in Galicia (Licentiate 1861). During this period the German ornithologist, Alfred Brehm, encouraged his natural history interests. Seoane taught natural history in A Coruña and returned to Santiago to complete a bachelor of science (1866) before beginning a medical practice at the Hospital de Marina, the military health service in Ferrol.

Seoane's career changed abruptly in 1869 when he married a young widow who belonged to a noble and rich family. He took over the management of her large properties, gave up medicine, and moved to A Coruña. To assist his new work, in 1874 he went to Santiago and then Oviedo to study law. Although Seoane was an amateur, he had extensive academic training and was well connected with leading European herpetologists of the day including his Spanish contemporary, Eduardo Boscá, Fernand Lataste in Paris, and Wilhelm Peters in Berlin. Charles Darwin was another of his correspondents. Studying and working in so many places gave him ample opportunities for fieldwork and he built up a large private museum at his manor house in A Coruña, but in 1985 the collections were moved to a public museum (Casa de las Ciencias–A Coruña). He died in A Coruña on 14 July 1900.

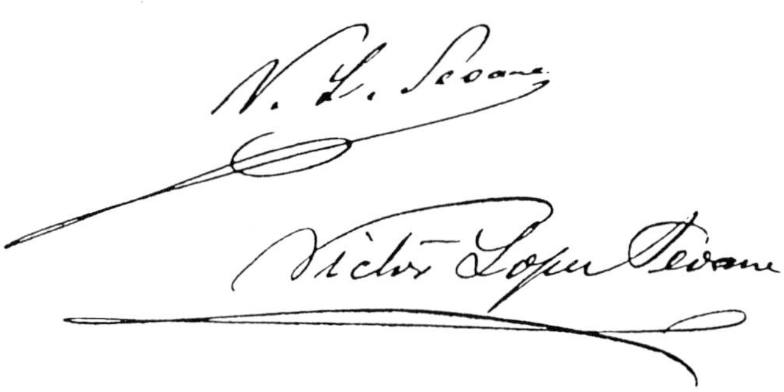

Seoane published broadly on vertebrates and even botany, but his primary interests were in ornithology and herpetology (see *Note 3*). He published five papers on amphibians and reptiles (1877–1890) plus two general works on the Galician fauna (1866, 1878). His main interest was systematics. He described eight new herpetological taxa, mostly from northwestern Spain including a well-known frog, *Rana perezi,* found in southern France, Iberia, and northwestern Africa. Two new species supposedly from the Philippines, a toad and a boa, were actually known species from Puerto Rico. (Before the Spanish-American War of 1898, both places were under Spanish rule.) Seoane also used embryological characteristics to support his taxonomic decisions, especially the development of circulation in the gills of newts, but his notes were not published until 1988.

• *References*: "Biografía del Naturalista Español Don Víctor López Seoane," by M. Ovilo y Otero, Escen. Contemp., Madrid, 1: 5–15, 1883; "A Herpetoloxia a Fins do Século XIX: O Labor de López Seoane," by X. A. Fraga-V., Estud. Hist. Cienc. Téc., Valladolid, 2: 867–879, 1988; "Embrioloxia e Taxonomia a Fins do Século XIX. As Notas Embrioloxicas de López Seoane (1881) no Contexto da Herpetoloxia Europea," by X. A. Fraga, Ingenium, A Coruña, 1: 27–58, 1988; "Víctor López Seoane," by X. A. Fraga-V., Inst. "Jose Cornide" Estud. Coruñ., La Coruña, 31 pages, 1992. • *Portrait and signatures* (1878 above, 1886 below): Courtesy Xosé A. Fraga-Vázquez. • *Note 1*: Although his father's surname was López, Seoane preferred to use his mother's name in his publications. • *Note 2*: Some early sources give Seoane's birth date, incorrectly, as 29 September 1834. • *Note 3*: Seoane apparently backdated several of his ornithological papers in order to gain priority (see C. D. Sherborn, Ann. Mag. Nat. Hist., ser. 6, 14: 154–155, 1894). In one instance he backdated a paper on *Lacerta* to be seen as anticipating Jacques von Bedriaga over a matter of nomenclature (see G. A. Boulenger, Zool. Rec., Rept. Batr., 22: 5, 1886).

BROWN, Arthur Erwin (1850–1910).

Arthur Erwin Brown, an early director of the Philadelphia Zoo, was once the leading authority on the snakes of the United States. He was born on 14 August 1850 in Bucks County just north of Philadelphia. After what has been described as a "liberal education in America and Europe," he worked as deputy administrator of the Delaware and Raritan Canal before his appointment as superintendent (in effect, director) of the Philadelphia Zoological Gardens in April 1876. He was 26 years old.

This was an auspicious year to assume control of America's first zoo, since the nation's Centennial Exposition was being held that year in Philadelphia and was located just across the street from the zoo. Many people thus were attracted to the zoo (nearly 680,000 visited that year), but Brown's organizational and business skills were up to this early test. He served as director for the rest of his life—34 years—until his death at the zoo on 29 October 1910. He became a respected member of the Philadelphia scientific establishment,

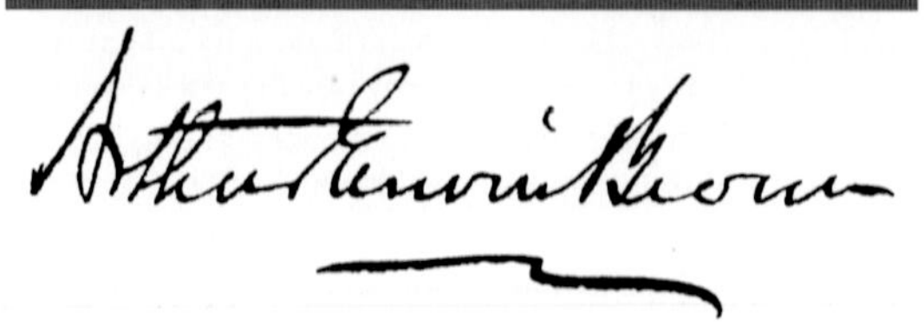

vice president of the Academy of Natural Sciences, and member of the governing board of the Wistar Institute of Anatomy and Biology. The University of Pennsylvania conferred upon him an honorary doctorate (D.Sc. 1908).

Although Brown also published on mammals and on evolutionary theory, his primary interest was snakes. During 1878–1911 he authored 18 papers on amphibians and reptiles, the most important of which was a 100-page review of the genera and species of snakes found north of Mexico (1901). This was a standard reference of its day and regularly consulted as was Brown himself. He described several notable American colubrid snakes: *Bogertophis* (formerly *Elaphe*) *subocularis* and *Lampropeltis alterna* from Texas and from Florida the diminutive *Stilosoma extenuatum*. He also described new taxa of amphibians and reptiles from Mexico, Borneo, and Japan. Several of his herpetological papers dealt with biogeography and evolutionary theory.

• *Reference*: "Arthur Erwin Brown 'Custodian of the Garden' and Naturalist of Note," by R. Conant, Amer. First Zoo, Philad. Zool. Garden, 9(4): 3 pages, 1957. • *Portrait* (1903) *and signature* (1909): Adler collection.

ROCHEBRUNE, Alphonse Trémeau de (1836–1912). Alphonse-Amédée Trémeau de Rochebrune was the author of "Faune de la Sénégambie," a summary of the vertebrates of French West Africa. He was born on 18 September 1836 in Saint-Savin-sur-Gartempe (Vienne) in west-central France about 70 km southwest of Angoulême where his father, a botanist, was curator of the city's museum. Alphonse became a medical student and intern at military hospitals in Angoulême and later served as a surgeon's aide during the Franco-Prussian war of 1870–1871. He defended his medical thesis in Paris in 1874. The next year he was posted to Saint-Louis, Senegal, where he was chief-of-staff at the hospital. During this period Rochebrune traveled throughout the countryside gathering materials for his projected book. "Sénégambie" at that time encompassed much of French West Africa along the coast stretching from present-day Mauritania south to Sierra Leone and inland to western Mali. In 1876, while still in Africa, he offered his services to the museum as a traveling botanist. He was recruited to the museum in 1878 as assistant preparator in anthropology and in 1880 was transferred to zoology, specifically the department covering annelids, mollusks, and zoophytes, as assistant naturalist.

Rochebrune had an original but undisciplined mind, and his studies ranged from botany, his main interest, to zoology, anthropology, and paleontology. Unfortunately, his scope was too broad to make significant contributions to all of them. His work was widely criticized within the museum and the head of his laboratory wrote the museum authorities in March 1910 that he did not wish to see Rochebrune in his department one more day. Rochebrune retired on 1 January 1911 and died in Paris the next year on 23 April.

Rochebrune's herpetological work consists of a few journal papers on fossil snakes, snake vertebrae, and descriptions of new species from Africa (5 papers, 1880–1886). His major work on amphibians and reptiles was contained in "Faune de

la Sénégambie" (1883–1885 [–1887]; the two herpetological parts were issued in 1884). The overall work, which covers 686 species, includes 75 hand-colored lithographed plates. The drawings by Jules Terrier are generally excellent, but the coloring, especially that of amphibians (10 plates) and reptiles (20 plates), can only be characterized as gaudy or even outrageous. The text itself was once described as "The most unreliable work on African herpetology." Rochebrune named 31 new species of living amphibians and reptiles in this book and in three papers, of which only a half dozen frogs and lizards are recognized today. He even included some Australian and Madagascan endemics in the book and gave West African localities for them. Some of his new forms are unrecognizable and have never been allocated to species. Although he claimed in the preface to the "Faune" that his types were placed in the museum in Paris, none of them can be identified in the collections today. Indeed one of the criticisms leveled at Rochebrune by his colleagues was that he did not recognize the importance of designating type specimens.

• *References*: "A. T. de Rochebrune," by [É.] Lamy, Jour. Conchyliol., 60: 361–364, 1913; Jaussaud and Brygoo, 2004 (Rochebrune by P. Jaussaud on pp. 453–454); Additional information from Édouard-Raoul Brygoo, 1995. • *Portrait* (1884) *and signature*: From Rochebrune, Faune de la Sénégambie, 1883–1885.

De Vis, Charles W. (1829–1915).

Charles Walter Devis (from 1882, de Vis) was an English clergyman who became a pioneer Australian zoologist and made early contributions to the herpetology of his adopted country. He was born on 9 May 1829 in Birmingham, England, and attended local schools after which he went to Magdalene College in Cambridge (B.A. 1851, M.A. 1884). In 1851 he became a Church of England deacon and two years later was made rector at St. John's in Brecon, Wales. Sometime after 1862 his enthusiasm for natural history made him give up his pastoral work to become curator of the Queen's Park Museum in Manchester where he wrote his earliest papers.

In June 1870 he immigrated to the Colony of Queensland in northeastern Australia, eventually settling in Rockhampton where he became librarian of the city's School of Arts. During 1880–1882, mostly under the pseudonym "Thickthorn," he published 18 articles on natural history topics (including reptiles) in local newspapers, which brought him to the attention of the trustees of the Queensland Museum in Brisbane. They had been looking to fill the long-vacant post of curator (in effect, director); de Vis was hired in 1882 and assumed the title of director in 1901. The museum was understaffed and poorly funded, but de Vis rapidly built a small library and collections of fossil and extant fauna and flora, minerals, and anthropological materials. In 1891 he founded the museum's journal, *Annals* (now *Memoirs*) *of the Queensland Museum*. He retired from the directorship in 1905 and continued as a consultant until 1912. He died in Enoggera, a northern suburb of Brisbane, on 30 April 1915.

De Vis's 350 titles covered mostly systematic vertebrate zoology and paleontology. He authored about 50 papers on amphibians and reptiles (1881–1911) including two under his early pseudonym and excluding the numerous republications of many of his papers. As was the practice for several Australian scientific journals of the day, abstracts of papers presented at society meetings were published in local newspapers (sometimes without authorship) to advertise the work of the society. Many of de Vis's new taxa were first introduced in this manner and only later formally published in the journals. He described a staggering total of 551 new taxa of Australian, New Guinean, and African animals. Of these, there were new species of extinct turtles (7), lizards (4), and a crocodilian. He named new species of living frogs (7), lizards (48), and snakes (24), most of them Australian, but his success in identifying valid ones varied greatly from group to group: frogs (1), lizards (22), and snakes (3).

George A. Boulenger of the British Museum harshly attacked de Vis's style of work. He wrote (1885): "[He] is no doubt stimulated by the desire of promoting herpetological knowledge in his country, but, through his incompetence and want of care, he will do much harm . . . he has no excuse, and one can only wonder at his daring to write on subjects of which he is so manifestly ignorant." De Vis described a new species of elapid snake (1911), supposedly from Queensland, based on a mislabeled specimen of the African egg-eating snake, *Dasypeltis scabra*. The error was corrected only in 1966, when it was pointed out that, in addition to other features, there were no visible teeth in the mouth. His most flagrant mistake was to describe a South African cordylid as a new genus and species of agamid from the Solomon Islands (1905). The absence of an adequate library and especially of properly identified comparative material had its costs. Nevertheless de Vis did succeed in naming two genera of Australian scincid lizards (*Calyptotis* and *Lygisaurus*) and numerous well-known species such as *Litoria rothii, Nephurus levis, Egernia rugosa,* and *Pseudechis guttatus*.

• *References*: "C. W. DeVis, M.A.," by T. H. Johnston, Proc. Roy. Soc. Queensland, 28: 10–17, 1916; "Charles Walter de Vis," by L. A. Gilbert, Austr. Dict. Biogr., 4: 63–64, 1972; "The Works of Charles Walter de Vis, Alias 'Devis,' Alias 'Thickthorn'," by G. J. Ingram, Mem. Queensland Mus., 28: 1–34, 1990; Additional information from Glenn Shea, 2007. • *Portrait*: Queensland Museum. • *Signature* (1907): Natural History Museum, London.

IHERING, Hermann von (1850–1930).

Hermann Friedrich Albrecht von Ihering, who was born in Kiel (then part of Denmark but later in Schleswig Holstein, Prussia) on 9 October 1850, became one of Brazil's leading naturalists. He attended universities in Giessen, Leipzig, and Berlin. Following service as a physician's assistant during the Franco-Prussian War of 1870–1871, he finished his education at the university in Göttingen (M.D. 1872, Ph.D. 1876). In 1880 he moved to Brazil, which became his home for nearly 45 years. He lived first in Taquara do Mundo Novo, a German colony just northeast of São Leopoldo, then in Porto Alegre, and finally on an island in the delta of the Rio Camaquã in Lagoa dos Patos, all in the state of Rio Grande do Sul in southernmost Brazil. While practicing as a physician he collected animals for European museums and collectors and from 1883 also collected for the Museu Nacional in Rio de Janeiro.

In 1893 Ihering became founding director of the Museu Paulista in São Paulo. He was responsible for building the museum and associated botanical gardens into a world-class institution and he also initiated the museum's journal, *Revista do Museu Paulista*. (In 1939 the zoological collections of the museum became the Departamento de Zoologia of the Secretaria da Agricultura and in 1969 transformed again, into the Museu de Zoologia of the Universidade de São Paulo where they exist today.) In 1916, however, Ihering was removed from the directorship because of Brazil's entry into World War I on the side of the Allies. Four years later, he became director of the state museum in the coastal city of Florianopolis (Santa Catarina) but in 1924 returned to Germany. He resided in Büdingen (Hessen), about 45 km southeast of Giessen, and died there on 24 February 1930.

Ihering was an all-round naturalist and published on a wide range of animals and plants, and also on paleontology and anthropology. His special interest was mollusks, both living and fossil. He published about six titles on amphibians and reptiles (1881–1905)—on snake venoms, a survey of the herpetology of São Paulo, and turtles of the Rio Juruá (Amazonas). Perhaps his most interesting herpetological contribution (1886) was on the egg-laying habits of a hylid frog, *Phyllomedusa iheringii*, named for him by George A. Boulenger in 1885 on the basis of large collections of amphibians and reptiles sent by Ihering to the museum in London. Ihering discovered that these frogs do not lay eggs in water but inside curled-up leaves overhanging the water. This is a behavior that had only been observed once before, in an African rhacophorid. Ihering was especially interested in zoogeography and the origin of the South American fauna, in which studies amphibians and reptiles often figured.

It should be noted that Ihering's son, Rodolpho von Ihering, also published on Brazilian herpetology. (See p. 142.)

• *References*: "Festschrift für Prof. Dr. Hermann von Ihering," Phoenix (Zeitschr. deutsch. Geistearbeit Südamer.), Buenos Aires, [13]: 7–60, 1927; "Prof. Dr. Hermann von Ihering (1850–1930)," anonymous, Revta. Mus. Paulista, 17: 553–566, 1931; "Hermann Friedrich Albrecht von Ihering (1850–1930)," by E. H. Cordero, Physis, 20: 353–363, 1951. • *Portrait and signature*: From Anonymous, 1931.

HAY, Oliver P. (1846–1930).

Oliver Perry Hay was an American paleontologist and herpetologist who published on the living amphibians and reptiles and the fossil turtles of North America. He was born on 22 May 1846 in Saluda (Indiana) near the Ohio River just north of Louisville. The family moved to central Illinois four years later where he was educated in a one-room country school. Wishing to become a minister, Hay attended Eureka College in Illinois (A.B. 1870) but abandoned the plan after giving his first sermon. He then turned to a subject that had interested him in college—natural sciences. From 1870 to 1879, except for a year at Yale University for graduate study (1876–1877), he served as a professor of natural sciences, in succession, at Eureka College, Oskaloosa College (Iowa), and Abingdon College (Illinois). From 1879 to 1892 he was professor of biology and geology at Butler College (now University) in Indianapolis (Indiana) and while there completed his doctorate at Indiana University (Ph.D.

1884). His research interests shifted dramatically when he became Assistant Curator of Zoology at the Field Museum of Natural History in Chicago (1895–1897). He then joined the paleontological staff of the American Museum of Natural History in New York (1900–1907) before conducting private research in paleontology for five years. His final position was as a research associate of the Carnegie Institution of Washington (1912–1926) while he conducted his research at the United States National Museum. He died in Washington on 2 November 1930 at the age of 84.

Hay's research on living amphibians and reptiles consisted of about 30 titles (1881–1928), most of them written when he was in Indiana, but later there were a few papers on softshells and sea turtles and the structure of turtle shells. The Indiana papers dealt with distribution, systematics, behavior, and reproduction; and they culminated in the first checklists of the state's herpetofauna (1887, 2 papers) and the first comprehensive herpetology of Indiana (1893). Hay was the first to describe the breeding habits of *Amphiuma* (1888), which he observed while conducting geological surveys in Arkansas. He discovered that females of this aquatic salamander guard their eggs on land. "Beneath a log . . . I found, to my surprise, a large animal coiled up, which by its smooth glistening skin I immediately saw could not be a snake; but, having never before seen a living amphiuma, it took me some time to realize that I had before me one of these animals." He also described *Storeria victa* (1892), the Florida Brownsnake, perhaps the only species of amphibian or reptile ever named from the stomach contents of another species (a coralsnake in this instance).

After Hay moved to the Field Museum in Chicago in 1895, his interests shifted to studies on the vertebral column and to fossil turtles. In addition to numerous papers describing many new taxa of fossil turtles from North America, he published several major books. The first was "Bibliography and Catalogue of the Fossil Vertebrates of North America" (1902, with a supplement in two volumes, 1929–1930). Hay's *magnum opus* was the book, "The Fossil Turtles of North America" (1908; reprinted 2006), covering 266 species of which 76 were previously not known to science. This is a massive volume with 113 outstanding plates.

Hay's son, William Perry Hay (1871–1947), also published on herpetology. He wrote a checklist of the herpetofauna of the Washington, D.C., region (1902) and a major paper on the systematics of *Malaclemys* (1904) including descriptions of new geographic races as well as a paper on the artificial propagation of these turtles (1913, with later revisions).

• *References*: "Oliver Perry Hay," by R. S. L[ull] and N. E. W[right], Dict. Amer. Biogr., 4: 436–437, 1931; "Oliver Perry Hay," p. 538 *in* Who Was Who in America, vol. 1, A. N. Marquis, Chicago, 1943. • *Portrait*: National Museum of Natural History, Washington, courtesy R. Lemaitre. • *Signatures* (1908 on left and 1921 on right): Adler collection.

GADOW, Hans (1855–1928).

Hans Friedrich Gadow, an anatomist and evolutionary biologist, was a specialist on vertebrates including reptiles and amphibians. He was born on 8 March 1855 in Krakow Schlawe (now Alt-Krakow bei Schlawe), Pomerania, north of Berlin. Gadow was educated first in Frankfurt an der Oder and then in 1875 went on to the universities of Berlin, Jena (Ph.D. 1878), and Heidelberg. At Jena he worked with the evolutionary biologist, Ernst Haeckel, and at Heidelberg with the anatomist, Carl Gegenbaur, both leaders of the Darwinian revolution in Germany. In 1880 Albert Günther, head of zoology at the British Museum, invited Gadow to London to assist in preparation of the ongoing "Catalogue of Birds" (Gadow wrote volumes 8 and 9), but in 1882 Gadow moved to Cambridge University where he spent the rest of his career as Lecturer and Strickland Curator in the university's Museum of Zoology. The quality of his research led to early election to the Royal Society (1892). Gadow died at his home near Cambridge on 16 May 1928.

Gadow's training with Haeckel and especially with Gegenbaur is clearly evident in his earliest publication specifically on reptiles (1882), which was on the abdominal muscles and hind limbs. Other important works followed on the cloaca and copulatory organs (1887), the visceral arches and auditory ossicles (1888), and the vertebral column (1895–1896). In 1898 his influential book, "Classification of the Vertebrata," appeared. In it he outlined a higher classification of vertebrates that remained largely stable until the advent of molecular techniques late in the 20th Century. Gadow wrote a miscellany of titles on amphibians and reptiles, covering topics as diverse as the remains of extinct tortoises, conservation of tuatara, color in amphibians, and embryology of the turtle shell. His major herpetological achievement of this period was his volume in the "Cambridge Natural History" series (volume 8, "Amphibia and Reptiles," 1901; most recent re-

print in 1958). This was the standard summary of the subject and the closest to a textbook of herpetology that existed until the texts of P. V. Terentjev and of Coleman Goin and Olive Goin were published in the early 1960s.

In 1902 and again in 1904 Gadow spent three months in southern Mexico on an expedition that was to redirect his herpetological research. Natural history collectors had been to Mexico before, but none of them had Gadow's training or reputation as a biologist. He traveled mostly by railroad in a car placed at his disposal by the Mexican government and with letters of introduction from General Porfirio Díaz, President of Mexico. He collected, to be sure, but he also observed behavior and ecology, noted elevational as well as geographical distributional patterns, and generally tried to understand the adaptations of the rich Mexican fauna. He published seven papers (1903–1911) on these and other topics, focusing on lizards and snakes. He summarized his trips in a popular book, "Through Southern Mexico" (1908), that is laced with observations on amphibians and reptiles.

One of his most intriguing discoveries in Mexico was a single specimen of what he believed to be the plethodontid *Batrachoseps*, a genus of salamanders then known only from California, which he found on the slopes of the Nevado de Colima at about 2,200 m, but none of the many herpetologists who later trod this very same path have rediscovered it. Gadow's last herpetological papers, still based on his Mexican trips, concern the effect of elevation on distributions (1910) and another on color patterns and mimicry in snakes (1911). His last major work on vertebrates, published posthumously, was "Evolution of the Vertebral Column, A Contribution to the Study of Vertebrate Phylogeny" (1933).

• *References*: "Hans Friedrich Gadow," by T. S. Palmer, Auk, 45: 538–539, 1928; "Hans Friedrich Gadow—1855–1928," by D. M. S. W[atson], Proc. Royal Soc. London, ser. B, 107: i–iii, 1930; "Hans Friedrich Gadow," p. 665 *in* I. Jahn et al. (eds.), Geschichte der Biologie, G. Fischer Verlag, Jena, 1982. • *Portrait*: From Watson, 1930. • *Signature*: Adler collection.

HOPLEY, Catherine C. (1817–1911).

Catherine Cooper Hopley was a writer and naturalist who wrote the first popular book on snakes in English. She was born in Whitstable (Kent), England, on 5 October 1817. Little is known of her family life except that her mother and three siblings (all brothers) also were born in Kent and that she never married. Her oldest brother, Edward Hopley (1816–1869), became a well-known London subject painter and her youngest brother, John Hopley (1821–1904), immigrated to the United States. She is best known as a writer describing her experiences in the United States during the late 1850s and early 1860s where she visited relatives in Indiana and Ohio and, coincidentally, observed the local fauna and flora.

During the spring of 1860, Hopley decided to visit the South and took a position as governess to a family in Richmond (Virginia). She was there when the Civil War broke out in April 1861. The first significant land engagement of the war, the First Battle of Manassas, occurred at Bull Run in July 1861, only 140 km north of Richmond, which became the capital of the Confederate States. Soon after, Hopley decided to travel south to Florida where she was governess to the family of the state's governor. However, she disliked Florida and resigned her post in the summer of 1862. She went back to Richmond, then made her way, with difficulty, through the front lines to Baltimore using an assumed name, "Miss Sarah L. Jones." She used a federal passport to do so, which gave this personal description: 4 feet, 11 inches tall (about 1.5 m); complexion florid; hair carroty; eyes hazel; build robust. In the passport she lied about her name and age (30), but these other particulars are more probably correct. Hopley's sympathies were with the South so she soon departed for England. She returned to the United States at least once, in 1884.

Hopley's first book, "Life in the South" (1863), published anonymously, was a sympathetic view of the South as a slave-owning society. (It was not unusual for a woman to publish anonymously in the 19th Century.) Her next title (1863), also anonymous, was a short biography of Confederate General Thomas J. "Stonewall" Jackson, a victor at the Battle of Manassas. Later she wrote "Rambles and Adventures in the Wilds of the West" (1872), which despite its title is based on her experiences in Ohio and Florida and includes information about the native fauna.

During Hopley's period in London she spent much time at the zoo in Regent's Park, particularly in the reptile house. In the mid-1870s she wrote articles about snakes for a semi-popular journal and a children's magazine. Perhaps even earlier she had decided to write a book on snakes. During her trip to America, she had been impressed with the numerous myths and superstitions she had encountered and vowed to ". . . attempt to combine scientific truths with readable English." The result was her book, "Snakes: Curiosities and Wonders of Serpent Life" (1882), the first popular book on snakes in the English language. She relates her own experiences in the United States (Iowa, Lake Erie, and Virginia) and extensive observations made at the zoo in London. Apparently she spent so much time in the reptile house that the staff worried she was going ". . . to write mischievous letters to the papers," according to a letter she wrote to a journal editor in March 1883. Contemporary reviews of her book were favorable, but one expressed surprise that it was "written by a *lady*" (italics as in the original).

And she was quite observant. For example, although Edward G. Boulenger, curator of the zoo's reptile house and aquarium, is generally credited with the observation that in the colubrid *Xenodon* the maxillary bone moves to enable the snake to erect its teeth in a viper-like manner (1915), Hopley had reported this earlier in her book. She also made significant observations on the way in which the glottis operates when snakes are eating large meals, which she also describes in her book and later in an article in the *American Naturalist* (1884).

In the early 1880s Hopley lived in London, but at some later point moved to Tickenham (Avon), near Bristol. In 1888 she authored a slim volume entitled "British Reptiles and Batrachians," which went through at least three printings by 1911. Hopley died in England on 9 April 1911.

• *References*: "Catherine Hopley," p. 107, 119 *in* M. R. S. Creese, Ladies in the Laboratory? American and British Women in Science, 1800–1900, Scarecrow Press, Lanham, Maryland, 1998; Family records of James Kettel (Ohio, USA) and other relatives of C. C. Hopley, 2006. • *Signature* (1883): Adler collection.

FLETCHER, J. J. (1850–1926).

Joseph James Fletcher, a biologist and editor, made significant early contributions to the distribution and habits of Australian amphibians. He was born in 1850 in Auckland, New Zealand, the eldest son of a Methodist clergyman who had emigrated from England. After the Maori War, the church moved the family to Queensland in 1861 (where Joseph attended Ipswich Grammar School) and then to Sydney. Fletcher enrolled at Newington College and the University of Sydney (B.A. 1870, M.A. 1876). He then taught briefly at Wesley College in Melbourne where he began his interest in natural history. No science degrees were offered in Australia at that time, so he went to London for study at the Royal School of Mines, which sponsored lectures on natural history, and at University College London (B.Sc. 1879). Because there was no recognition for work done at Australian universities, he had to start over. He also studied at Cambridge University and then returned to Australia in 1881 to teach anatomy and physiology at Newington College until 1885. Most of his research during this period was on the embryology of marsupials.

Fletcher became a member of the Linnean Society of New South Wales in 1881 and joined its council two years later. In January 1886 the society's founder and benefactor, the amateur naturalist William Macleay, invited him to become the salaried librarian of the society and in 1893 Fletcher also became secretary (thus director) and for 33 years ran the society's affairs and edited its journal. Fletcher himself was a major contributor to the journal on marsupial reproduction, earthworms, land planarians, peripatus, and on frogs. He retired as secretary in 1919 but continued his local fieldwork that turned to botany. He died suddenly at his home in Sydney on 15 May 1926.

Fletcher contributed ten papers on herpetological subjects (1883–1897 [1898]), all but one of them on frogs. He was particularly interested in determining the natural distributions of species—before these were altered through human development and the introduction of alien species—and produced a five-part series (1891–1897 [1898]) on the subject. These papers also include his observations on behavior and life history, in particular the ability of some frogs to withstand extreme drought. He also published on collections of frogs made by others in more distant places such as Western Australia and Tasmania (1897 [1898]). He described a new genus and four new species of myobatrachid frogs including two species that are recognized today: *Paracrinia haswelli* and *Geocrinia leai*.

• *References*: "Joseph James Fletcher," by W. B. S[pencer], Proc. Linnean Soc. New South Wales, 52: xxxiii–xlii, 1927; "Joseph James Fletcher," by G. P. Walsh, Austr. Dict. Biogr., 8: 525–526, 1981; G. Shea *in* A. Bauer (ed.), 2004. • *Portrait and signature*: From Spencer, 1927.

SARASIN, Paul (1856–1929).
SARASIN, Fritz (1859–1942).

The Sarasins have been wealthy merchants and civic leaders of Basel, Switzerland, since the 18th Century. Two cousins (see *Note 1*), however, Paul Benedikt Sarasin and Karl Friedrich Sarasin, the latter generally known as Fritz Sarasin, became noted zoologists and explorers. They collaborated to produce the most elaborate study of a single species of amphibian or reptile ever produced—a caecilian. Paul was born in Basel on 11 December 1856. His father was a member of the city council. Fritz, whose father was the city's longtime mayor, was born on 3 December 1859, also in Basel. Their academic and research careers were nearly identical beginning with local schools in Basel. They both attended the University of Basel and worked with Ludwig Rütimeyer, an evolutionary biologist who was Professor of Zoology and Comparative Anatomy. Both went to the university in Würzburg for their doctoral work and studied with the zoologist, Carl Semper. Paul's dissertation was on the development of an aquatic snail (Dr.phil. 1882) and Fritz's was on the early development of reptile embryos (Dr.phil. 1883).

Semper had conducted a privately financed zoological expedition to the Philippines during 1858–1865 and, perhaps inspired by this example, the Sarasins decided to conduct overseas explorations using Berlin as their base. They soon departed for Ceylon (now Sri Lanka). One goal was to study the development and anatomy of caecilians, which at that time were poorly known to

biologists. When they found gilled larvae in the brooks, they assumed that oviposition occurred in the water. Months later, when the rainy season began, they were astonished to discover that the female digs a cavity along the bank in which the eggs are laid like beads on a string and then she broods them. They were in Ceylon during the period 1883 to 1886, traveling mostly on foot, and they amassed large collections for study in Berlin. They returned to Ceylon in 1890, 1902, and 1907 and in 1925 Fritz went alone. In 1889 Paul and Fritz traveled to Egypt and Sinai. Another area they studied was the island of Celebes (now Sulawesi) in the East Indies. Because it straddles Wallace's Line, Celebes was of special zoogeographical interest to them. They made trips there in 1893–1896, 1902–1903, and 1905. In 1911–1912, Fritz, together with Jean Roux, herpetologist at the museum in Basel, went to the South Pacific to visit New Caledonia and the more eastward Loyalty Islands. In 1923 Fritz went to Tunisia and in 1931 he traveled to Siam (now Thailand) and Bali. The Sarasins continued to use Berlin as their base until 1896 when they returned to Basel. They both became active in the city's cultural and scientific life. Paul died there on 7 April 1929 and Fritz died in Lugano (Ticino), Switzerland, about 60 km northwest of Milan, on 23 March 1942.

Paul's interests were primarily in developmental biology, anthropology, and zoology generally. Fritz was a zoologist specializing on birds and also an ethnographer. Their herpetological masterwork was a book published in four parts (1887–1890) and entitled "Zur Entwicklungsgeschichte und Anatomie der ceylonesischen Blindwühle *Ichthyophis glutinosus*, L." This is an imposing book, in large quarto, of nearly 300 pages and 24 finely executed lithographed plates, some colored, showing the development and anatomy of the Sri Lankan caecilian. There was at that time and even now nothing to compare with it in detail for any other single species of amphibian or reptile. Unfortunately, the book was produced in a very small edition and has been generally unavailable to zoologists. The text covers all aspects of the animal's natural history including the eggs, fertilization, brooding behavior, and development of the embryo. There is extensive coverage of the lateral line organs of the larvae, the skin and its glands, the skull, the nose and other sense organs including the tentacle, the vascular system, gills, and other aspects of the anatomy. All of their doubly stained histological sections were deposited in the Zoological Institute of the University of Basel. In addition, Paul and Fritz, usually together, authored about 10 herpetological papers (1883–1912), most of them on aspects of the biology and anatomy of the caecilian.

Besides the caecilians needed for their research, the Sarasins made large zoological and anthropological collections in the places they visited. The amphibians and reptiles were reported on by herpetologists at two museums, principally by Fritz Müller in Basel, George A. Boulenger in London, and Roux, Müller's successor in Basel. Müller published on the Ceylon material in four of the supplements (1885–1892) to his catalogue of the amphibians and reptiles in the Basel museum. He also reported on some of the Celebes collection (two papers in 1894). After Muller's death in 1895, all of the collections were turned over to Boulenger for further study (1896, 1897). Roux published on some remaining Celebes materials (1904). Since Roux had participated with Fritz Sarasin in the expedition to New Caledonia and the Loyalty Islands, he published the her-

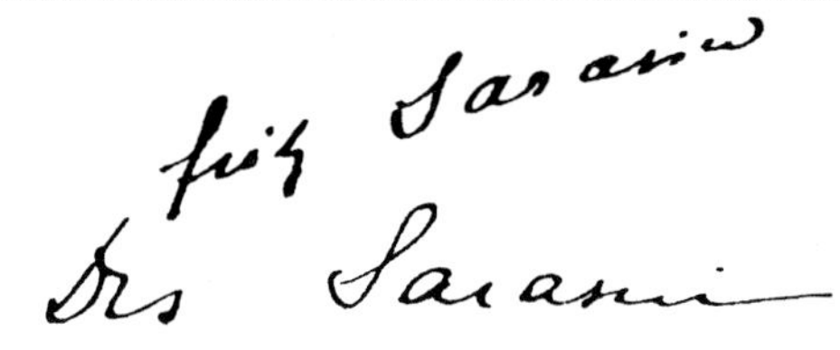

petological results in a large monograph (1913). Numerous new genera and species of amphibians and reptiles were named in these works, often jointly for the two cousins. The specific epithet *sarasinorum*, in the genitive plural, was thus applied by Müller, Boulenger, and Roux to six species. These names honor the contributions of Paul and Fritz Sarasin to the herpetology of Ceylon, Celebes, New Caledonia, and the Loyalty Islands and their great work on the caecilian.

• *References on Paul Sarasin*: "Zur Erinnerung an Paul Benedikt Sarasin 1856–1929," by F. Sarasin, Verh. naturf. Ges. Basel, 40(2): 1–28, 1929; "Paul Sarasin 1856–1929," by H. G. Stehlin, Verh. schweiz. naturf. Ges., 111: 393–407, 1930. • *References on Fritz Sarasin*: "Zur Erinnerung an Herrn Dr. Fritz Sarasin," by [F. Sarasin et al.], privately publ., Basel, 28 pages, [1942]; "Dr. Fritz Sarasin 1859–1942," by F. Speiser, Verh. naturf. ges. Basel, 54: 222–264, 1943; "Fritz Sarasin," p. 186 *in* L. Gebhardt, Die Ornithologen Mitteleuropas (Jour. Ornithol., 111), 1970. • *Portrait of Paul Sarasin*: Universitäts-Bibliothek Basel, courtesy Christine Stocker. • *Signature of Paul Sarasin* (1900): Natural History Museum, London (see *Note 2*). • *Portrait and signature of Fritz Sarasin*: From Speiser, 1943 (see *Note 2*). • *Note 1*: Actually they were second cousins because their fathers were cousins. There was no contact between the families until Paul and Fritz became acquainted as teenage members of the Zofinger-Verein, a national association of Swiss students from well-to-do families. • *Note 2*: Both Sarasins typically signed their letters "DDr. Sarasin," as signed here by Paul in 1888, or as "Drs. Sarasin," signed here by Fritz in 1898 (both courtesy Natural History Museum, London). In such instances it is only by a comparison of the handwriting or from the content of the letter that one can determine which of them wrote a given letter.

WOLTERSTORFF, Willy (1864–1943).

The German zoologist, Wilhelm Georg Wolterstorff, generally known as Willy Wolterstorff, was a specialist on salamanders, and played a leading role in the development of the German herpetocultural community. He was born on 16 June 1864 in Calbe (Saxony-Anhalt) along the river Saale 25 km south of Magdeburg. His family soon moved to Magdeburg where his father became the city's school inspector. At the age of seven Willy suffered an illness that rendered him deaf for life and affected his ability to speak normally. His schooling, therefore, was private and only by special arrangement was he permitted to attend the university in Halle. He studied geology and participated in field trips to collect fossil frogs, which resulted in his first paper (1885). A key moment came in May 1887 when he found live newts during a trip to the Hartz Mountains southwest of Magdeburg. This turned him on to his life's work with salamanders, especially salamandrids. He wrote his dissertation at the university in Erlangen (Dr.phil. 1898) but in the meantime he had returned to Magdeburg in 1891 and to a position at the natural history museum where he worked for the rest of his life. Later he took charge of the herpetological collection and enlarged it greatly; it included what became the largest and most diverse salamander collection in the world.

He retired in 1929 and continued to work at the museum until his death in Magdeburg on 21 January 1943. He did not live to see the museum destroyed when the city was bombed early in 1945 at the end of World War II. The alcoholic collection and his archives had been moved to a salt mine for safekeeping during the war, but tragically these fell victim to arson. Wolterstorff's successor at the museum and his protégé, Günther E. Freytag, attempted to rebuild the collection, but the museum thereafter focused on cultural history.

The German-speaking countries have had a long history of hobbyists keeping amphibians and reptiles in captivity. The naturalist, J. M. Bechstein, first introduced herpetoculture in his book of 1797 that was concerned with keeping animals in the house, but it was Baron Johann von Fischer's book, "Das Terrarium" (1884; reprinted 1989), that spurred real interest among the general public. Several hobbyist magazines soon developed and clubs sprang up in the major cities. One of these magazines, *Blätter für Aquarien- und Terrarienfreunde* (later, *Terrarienkunde*), was begun in 1890, edited by Bruno Dürigen, author of the standard work, "Deutschlands Amphibien und Reptilien" (1892–1897). In 1909 Wolterstorff became editor of the *Blätter*, which merged with another hobbyist magazine, *Wochenschrift für Aquarien- und Terrarienkunde*, in 1938. (It then changed its name in 1951 to become *Die Aquarien- und Terrarien-Zeitschrift* [or DATZ]; see *Note*.) He remained as editor until his death in 1943, a span of 35 years. Although there were efforts to separate aquarists and terrarists, these were resisted, and Wolterstorff maintained close relationships with both the amateur and professional communities in developing the scope of the magazine. He published many of his own research reports in it, made commentaries on other papers, and answered questions from readers. As a byproduct of his voluminous correspondence related to the magazine, many specimens were added to the Magdeburg collections from around the world.

In addition to his work as editor, Wolterstorff was instrumental in developing the hobbyist clubs in Germany. In April 1894 he joined an existing group, "Triton," founded in Berlin in 1888, and by February 1896 he had started a branch in Magdeburg. These clubs sponsored regular meetings in restaurants, university halls, and later on, because of large attendances, even in convention centers where members could share experiences and exhibit their animals. In April 1918 and with the war still raging, Wolterstorff and three colleagues (Günther Hecht in Frankfurt am Main, Fritz Molle in Magdeburg, and Walter Bernhard Sachs in Berlin) founded a national organization called "Salamander," with chapters in the major cities. This is the direct antecedent of today's Deutsche Gesellschaft für Herpetologie und Terrarienkunde, which was founded in September 1964 and began publication of its journal, *Salamandra*, the next year. DGHT is now the largest herpetological-terraristic society in the world. Wolterstorff can rightly be called the foremost person to set the stage for its development, through his half century of devoted labor as editor and organizer.

Because of his editorship and connections with herpetocultural clubs, Wolterstorff corresponded widely, especially with beginners who needed advice and even mentoring. Many of them went on to have notable careers, among them Otto Cyrén in Sweden; Freytag, Wolf Herre, and Lorenz Müller in Germany; Ion Fuhn in Romania; Louis Lantz in Russia and England; and Milutin Radovanović in Yugoslavia. Another Wolterstorff protégé was Paul Kammerer, an Austrian who had an international reputation for his breeding experiments with toads, salamanders, and other organisms, which he claimed proved the inheritance of acquired characteristics until his career ended in scandal and his suicide in 1926.

Wolterstorff's herpetological research interests were in systematics, distribution, behavior, and captive breeding. He described many new taxa of salamandrids from Asia, North America, and especially Europe including the genera *Hypselotriton* and *Mertensiella*. He was particularly interested in hybridization studies. In 1903 he reported that a long-recognized species (*Triturus blasii*) was, in fact, a hybrid between male *T. cristatus* and female *T. marmoratus*. He returned to genetic studies toward the end of his career, work that was carried forward by Freytag and independently by Lantz. Of his approximately 280 publications, about 210 of them (1885–1943) are on amphibians and reptiles.

• *References*: "Ein Meister der Salamanderkunde: Dr. Willy Wolterstorff †," by R. Mertens, Natur und Volk, 73: 122–124, 1943; "Dr. Willy Wolterstorff," by G. Freytag, Copeia, 1947: 215, 1947; "Willy Wolterstorff," by G. E. Freytag, Abhandl. Ber. Mus. Naturk. Vorgesch. Magdeburg, 8: 7–10, 1948; "Willy Wolterstorff und der 'Salamander'," by W. Bischoff, Salamandra, 25: 131–142, 1989; Rieck et al., 2001 (Wolterstorff by W. Bischoff on pp. 378–383, 650–655). • *Portrait* (1934) *and signature* (1935): Natural History Museum, London, courtesy A. F. Stimson. • *Note*: The first three volumes of DATZ were titled *Deutsche Aquarien- und Terrarienzeitschrift*.

SPENCER, Baldwin (1860–1929).

Walter Baldwin Spencer was a biologist, anthropologist, and science administrator who was a specialist on the parietal eye of reptiles and, later, on Australian amphibians. He was born on 23 June 1860 in Stretford (Lancashire), England, and educated at schools in Manchester. He matriculated at Owens College of Victoria University in Manchester where he became a confirmed Darwinist. At Exeter College in Oxford (B.A. 1884) one of his professors, who had visited Australia as a naturalist aboard H.M.S. "Challenger," encouraged him to do research in Australia. At this time Spencer's scientific interests were anatomical and microscopical. He studied the embryology of frogs (1885) and the parietal eye in lizards and tuatara (1886). In 1887 he was appointed to a new chair in biology at the University of Melbourne. He greatly improved the university's research position and his biology department was the first in Australia to appoint female faculty. He held similar leadership positions in various scientific societies. In 1899 he became director of the National Museum, succeeding Frederick McCoy. He was elected to the Royal Society in 1900 and knighted in 1904.

The seminal event of Spencer's career was being named biologist on the 1894 Horn Scientific Expedition to Central Australia, organized by William Horn, a wealthy pastoralist and politician, to explore the MacDonnell Ranges. It became, arguably, the most important Australian expedition of the 19th Century. This journey used camel-drawn wagons to cover some 3,300 km in 100 days. Spencer himself rode a camel throughout the trip. The expedition made large biological collections, but the most significant development for Spencer was meeting Francis J. Gillen, the postmaster at Alice Springs, who had won the trust of the local Aborigines. Thus began

Spencer and Gillen's collaboration, as Spencer's main research interest shifted to anthropology. Their landmark work, entitled "The Native Tribes of Central Australia" (1899, with latest expanded edition in 2005) became the beginning of Australia's understanding of Aboriginality. Spencer retired as professor in 1919 and as director of the museum in 1928. Early in 1929 he traveled to Tierra del Fuego at the southernmost tip of South America to do anthropological research. There, in a snowbound hut on Isla Navarino, Chile, he died from angina pectoris on 14 July 1929.

Besides the works noted earlier, Spencer's herpetological publications are few. He described new frogs from Victoria (1901) including a new genus and species of myobatrachid frog, *Philoria frosti*. His most significant herpetological contribution was the amphibian section of the Horn Expedition report (1896; reprinted 1994). Whereas the companion section on reptiles by A. H. S. Lucas and Charles Frost consists of little more than lists of localities and descriptions of new species, Spencer's is the more comprehensive because of his direct field experience. In this section of the Horn report, he provided extensive data on distribution, habits, and life history, and he also named *Litoria gilleni*.

• *References*: "Sir Baldwin Spencer, K.C.M.G., F.R.S." by J. A. K[ershaw], Victorian Nat., 46: 102–107, 1929; "'So Much That Is New.' Baldwin Spencer, 1860–1929. A Biography," by D. J. Mulvaney and J. H. Calaby, Melbourne Univ. Press, Melbourne, xii, 492 pages, 1985; "Sir Walter Baldwin Spencer," by D. J. Mulvaney, Austr. Dict. Biogr., 12: 33–36, 1990; G. Shea *in* A. Bauer (ed.), 2004. • *Portrait*: From R. T. M. Prescott, Collections of a Century, Natl. Mus. Victoria, Melbourne, 1954, opposite p. 90. • *Signature*: Natural History Museum, London.

MOCQUARD, François (1834–1917).

François Mocquard was a French naturalist who described numerous new species of amphibians and reptiles from all parts of the globe. He was born on 27 October 1834 in Leffond (Haute-Saône) in eastern France. Wanting to become a schoolteacher, he enrolled at the École Normale de Vesoul, graduating in 1855. He then attended the Lycée de Besançon where he took his degree in 1860 followed by further training in physical sciences and mathematics. In the meantime he had taken a post as a laboratory assistant in physics at the Faculté des Sciences in Besançon and studied medicine in time to serve as a junior medical officer in 1870 during the Franco-Prussian War. Following war duty, he returned to complete his doctorate in medicine (1873) and then served as a teacher of physical and natural sciences at the Lycée de Vesoul.

At the age of 43 Mocquard decided to change careers and do pure research in natural history. In 1877 he went to Paris as a student of Alphonse Milne-Edwards who at that time held the chair of mammals and birds at the Muséum d'Histoire Naturelle. Not having the proper academic training to do research work at the museum, Mocquard completed new courses of study (B.Sc. 1879, Ph.D. 1884). In August 1884 he replaced Henry-Émile Sauvage as the aide to the chair of reptiles and fishes, which was then held by Léon Vaillant. Mocquard occupied this post until his retirement in 1908 and was succeeded by his assistant, Fernand Angel. Later Mocquard returned to Leffond where he continued to write scientific papers. He died there on 19 March 1917.

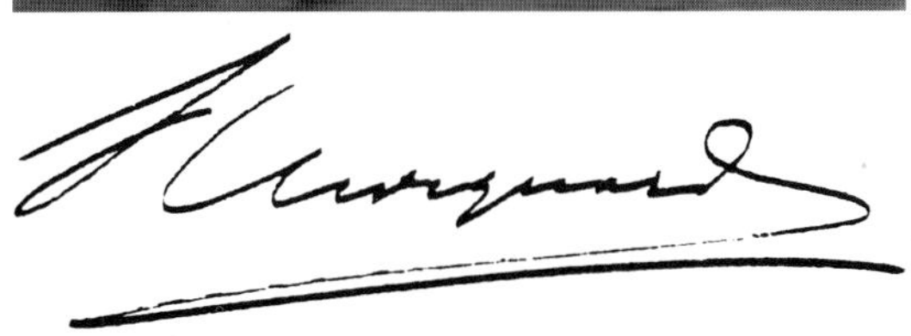

In the beginning, Mocquard's research covered reptiles as well as fish because Vaillant's laboratory included both, but herpetology soon predominated (about 60 titles, 1886–1915). These covered virtually all parts of the world including Borneo, Central America, Mexico, and especially France's overseas colonies in Africa, Indochina, and the highly endemic island of Madagascar. The bulk of his contributions were short papers comprising exact descriptions of new species. At this time Vaillant was heavily involved with administrative work so Mocquard had the opportunity to describe most of the new material that was flowing into the museum.

In addition to his papers, Mocquard published several larger works including comprehensive ones on Borneo and Palawan Island, Philippines (1890), Baja California, Mexico (1899), Indochina (1907), and Madagascar (1909). Of the 268 species of amphibians and reptiles listed for Madagascar in the last-named book, Mocquard had named 46 of them. He also wrote the last two sections of the reptile volumes in the celebrated series, "Mission Scientifique au Mexique et dans l'Amérique Centrale" (1870–1909; reprinted 1978), which was under the senior authorship of Auguste Duméril.

• *References*: "Notice nécrologique sur M. le Dr François Mocquard," by L. Roule, Bull. Mus. Natl. d'Hist. Nat., 23: 315–317, 1917; Jaussaud and Brygoo, 2004 (Mocquard on p. 386). • *Portrait* (1889): Deutsches Entomologisches Institut, Müncheberg. • *Signature* (1893): Natural History Museum, London, courtesy A. F. Stimson.

Ishikawa, Chiyomatsu (1860–1935).

The embryologist and zoologist, Chiyomatsu Ishikawa, who was born in Tokyo on 30 January 1860 (see *Note 1*), was the first to work out the complete life history (see *Note 2*) and development of the Japanese Giant Salamander (*Andrias japonicus*) and thus for any cryptobranchid salamander. He was interested in biology as a boy and studied zoology at Tokyo Imperial University (now University of Tokyo) under the American zoologist, Charles O. Whitman, a former student of Louis Agassiz who in 1888 became first director of the Marine Biological Laboratory at Woods Hole. Ishikawa graduated in 1882 and then became an assistant in the university's Institute of Zoology and Assistant Professor of Zoology. In 1885, however, he went to Germany to study for his doctorate with August Weismann, an experimental biologist at the university in Freiburg (Ph.D. 1889). There he also came into contact with the salamander anatomist, Robert Wiedersheim, Weismann's colleague. Ishikawa returned to his post in Tokyo, was promoted to professor the next year, and awarded a doctorate by the university (D.Sc. 1890). In 1900 he assumed the additional role of chairman of the natural history department at Tokyo Imperial Museum and went on tours of museums and zoology departments in Australia (1906) and in Europe and America (1908–1909). He was elected to the Imperial (now Japan) Academy in 1911 and retired from his professorship in 1923. He died in Taihoku (now Taipei) on 17 January 1935 while on a tour of Formosa (now Taiwan), which was then part of Imperial Japan.

Ishikawa was a zoologist of broad research interests and published more than 200 titles in embryology, cytology, experimental zoology, and systematics. He was also involved in pisciculture and responsible for transplantation of the Ayu fish. He authored about 15 papers on amphibians and reptiles (1886–1940), beginning with one on the formation of the germinal layers in the softshell turtle with Kakichi Mitsukuri, a specialist on reptile embryology at Tokyo Imperial University. His other herpetological papers deal with the giant salamander, next to its Chinese congener the largest living amphibian, which reaches a total length of nearly 1.5 m. In a series of papers based on his own field observations, he determined the natural distribution of the species and recorded its life history including the breeding habits, spermatozoa, the egg, egg deposition, parental care of the eggs (although he mistakenly assumed that it was performed by the female), and the larvae. Ishikawa published a short book on the natural history of the giant salamander in 1903. He also conducted laboratory studies in which he described gastrulation and embryogenesis.

• *References*: "Chiyomatsu Ishikawa (1860–1935)," anonymous, Proc. Imp. Acad., 11: xi, 1935 (reprinted in Biogr. Mem. Members Imp. Acad. 1906–1947, Japan Acad., Tokyo, pp. 191–193, 1979); Hasegawa, 1967 (Ishikawa on pp. 8–9). • *Portrait*: Hunt Institute, Pittsburgh. • *Signature* (Japanese version): Courtesy Japan Academy, Tokyo; (Romanized verison, 1935): Courtesy David Damkaer. • *Note 1*: He was born 8 January 1 Man'en Era, thus 30 January 1860 according to the modern calendar. • *Note 2*: Chujiro Sasaki published a short paper (1887) on field observations of *Andrias japonicus*, but Ishikawa was the first to record its complete life history.

Lidth de Jeude, Th. W. van (1853–1937).

The Dutch zoologist, Theodorus Willem van Lidth de Jeude, who published extensively on the herpetofauna of Holland's overseas colonies, was born on 1 February 1853 in Helmond (North Brabant) about 15 km east of Eindhoven. He attended the University of Utrecht where his grandfather's brother, Th.

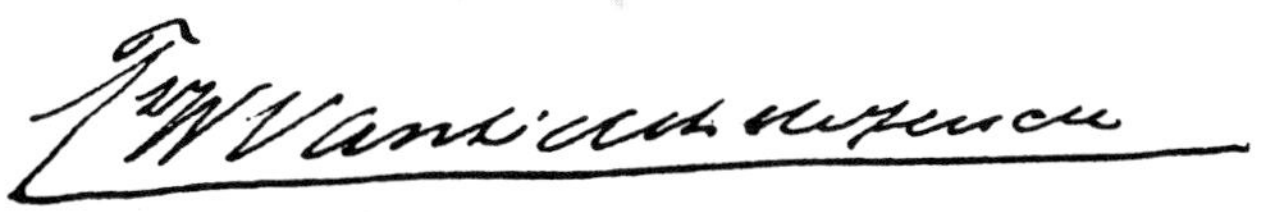

G. van Lidth de Jeude, taught zoology and veterinary science and amassed an enormous private zoological collection (see *Note*). Theodorus Willem studied science and natural history there and received his doctorate (Ph.D. 1882) for a dissertation on coleopteran larvae. While at the university, he served as a zoologist on the second Willem Barents Expedition to the Arctic (1879) and later published on the mollusks collected on this voyage. In 1882 and 1884 he studied fishes at the Zoological Station in Naples and at Kralingen (near Rotterdam), respectively. In 1884 Van Lidth de Jeude became Curator of Lower Vertebrates at the Rijksmuseum in Leiden and when responsibility for fishes was separated six years later, he continued with amphibians and reptiles. He retired from his curatorship in October 1923, worked regularly at the museum until 1931, and died in Leiden on 29 May 1937.

Van Lidth de Jeude's herpetological research focused almost exclusively on the taxonomic study of the large collections made in his country's overseas possessions. Thus the geographic coverage of his work was broad including the West Indies (Aruba, Curaçao), Surinam, South Africa, and especially the East Indies (Borneo, Java, New Guinea, Sumatra, Timor). He wrote the section on snakes from Max Weber's 1888 expedition to the Netherlands East Indies (1890; lizards by Weber [1890] and amphibians by P. N. van Kampen [1907]). Van Lidth de Jeude authored about 25 papers on herpetology (1886–1923), which include descriptions of several dozen new species. Many currently recognized taxa were among them, such as *Eleutherodactylus grandoculis* (Surinam), *Ophisaurus buettikoferi* (Borneo), and *Crotalus unicolor* (Aruba). He also published a book cataloguing the extensive osteological collections of amphibians and reptiles at the Rijksmuseum (1898).

• *References*: "Dr. Th. W. van Lidth de Jeude 1853–1937," by L. D. Brongersma, Jaarb. Gesch. Oud. Leiden, 30: 72–73, 1938; "Theodorus Willem van Lidth de Jeude," p. 80 *in* L. B. Holthuis, 1820–1958 Rijksmuseum van Natuurlijke Historie, Rijksmus. Nat. Hist., Leiden, 1995. • *Portrait* (1900): Nationaal Natuurhistorisch Museum, courtesy L. B. Holthuis. • *Signature* (1898): Natural History Museum, London, courtesy A. F. Stimson. • *Note*: Theodorus Gerardus van Lidth de Jeude's zoological collection, which included many types, was auctioned in 1858 and the remainder in 1867. The British Museum purchased a large part. Among other specimens of herpetological interest, it included duplicates from the collection of Salomon Müller and portions of the collection of Albertus Seba that were not purchased by Peter the Great and removed to St. Petersburg.

OGILBY, J. Douglas (1853–1925).

James Douglas Ogilby, an Irish-born specialist on Australian fishes and reptiles, was born in Belfast on 16 February 1853. His father was a prominent zoologist. Ogilby attended Winchester College, the oldest public school in England (1866–1869), before entering Trinity College, University of Dublin (1871–1872). From 1874 to 1876 he wrote popular articles on Irish vertebrates for *The Zoologist*. Ogilby worked briefly at the British Museum in London and conducted ornithological fieldwork in Texas before taking a post at the Australian Museum in Sydney in February 1885. He was responsible for all vertebrates and specialized on fishes. Unfortunately, he had a great affinity for alcohol. His conduct when drunk caused embarrassment and, despite many warnings, he was fired in 1890; his work was then done on a contract basis outside the

museum. Even during this period he continued his high output of publications including books on Australian mammals (1892) and the edible fishes and crustaceans of New South Wales (1893). He moved to Brisbane (Queensland) in about 1901. Beginning in 1906 he was employed by the Amateur Fisherman's Association of Queensland. In 1912 he took a position as ichthyologist at the Queensland Museum where, happily, the specimens were stored in formalin! There he continued his work on fishes and reptiles and retired in 1920. After a long illness he died in Brisbane on 11 August 1925.

Ogilby's work on amphibians and reptiles, although clearly secondary to that on fishes, was reported in about 25 titles (1887–1906 [1907]). Most of these papers are taxonomic in nature and several of his new taxa are recognized today, such as *Chelodina rugosa, Saltuarius cornutus,* and *Morethia boulengeri*. Continental Australia was his main geographic focus but he also published several papers on New Guinea and the Solomon Islands as well as Lord Howe and Norfolk islands, which are well off Australia's eastern coast. His major paper of this period, "Catalogue of the Emydosaurian and Testudinian Reptiles of New Guinea" (1905), while competently done in other respects, contained his most serious error. In it he announced the discovery of the first cryptodire from the Australasian region, which he named *Devisia mythodes*, the New Guinea Snapping Turtle. Until it was shown in 1947 to be merely a mislabeled North American *Chelydra serpentina*, it perplexed herpetologists and zoogeographers for four decades.

• *References*: "Bibliography of James Douglas Ogilby," by G. P. Whitley, Rec. Austr. Mus., 15: 149–156, 1926; "James Douglas Ogilby," by G. P. Walsh, Austr. Dict. Biogr., 11: 68, 1988. • *Portrait* (1898): George S. Myers/Alan E. Leviton protrait collection, courtesy Alan Leviton. • *Signature* (1890): Natural History Museum, London, courtesy A. F. Stimson.

BEDDARD, Frank E. (1858–1925).

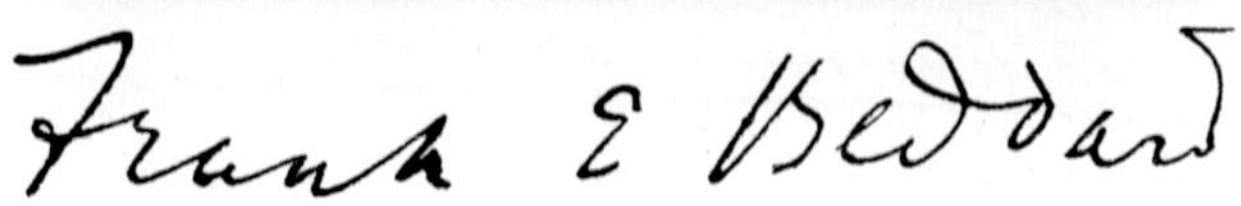

The British naturalist and anatomist, Frank Evers Beddard, published numerous anatomical studies on amphibians and reptiles. He was born the son of a wealthy businessman in Dudley, England, on 19 June 1858 and after schooling at Harrow went to Oxford where he developed an interest in field natural history. On graduation in 1882, he spent two years in Edinburgh working up the collections of isopod crustaceans from the Challenger Expedition. In 1884 he was selected for the post of Prosector at the Zoological Society of London, which operates the zoo in Regent's Park. In this role it was his responsibility, as an anatomist, to dissect animals from the society's gardens, to study and publish on their anatomy, and to present reports and demonstrations at society meetings. Large numbers of mostly vertebrate, animals including amphibians and reptiles, became available for his study and from which he produced a large number of publications, mostly in the society's *Proceedings*. He retired from the post in 1915. In parallel, he conducted a major research program on the systematics of earthworms on which he published his classic monograph (1895).

Beddard was the author of numerous books including "The Structure and Classification of Birds" (1898) and the mammal volume in the "Cambridge Natural History" series (1902). His books, "Animal Coloration" (1892) and "Text-Book of Zoogeography" (1895), include many examples of amphibians and reptiles. His own work at the zoo inspired another volume, "Natural History in Zoological Gardens" (1905). During 1887–1892 he served as editor of *Zoological Record* during a difficult transition in its sponsorship. He was honored for his research by election to the Royal Society (1892), the youngest fellow of his time, and by an honorary D.Sc. from Oxford (1912). He died suddenly of a heart attack at his home in West Hampstead, England, on 14 July 1925.

Beddard wrote about 40 papers on amphibian and reptilian anatomy and osteology (1888–1911), published mostly in the zoo's *Proceedings*. Many of these titles concern internal anatomy, especially of the vascular system. These cover animals as diverse as *Pipa, Rhinoderma, Andrias, Eryx, Ophiophagus, Uromastyx, Chlamydosaurus, Sphenodon,* and *Crocodylus*. Coincidental to his anatomical work he was the first to breed *Xenopus laevis* outside its native habitat (1894)—in a pond at the zoo.

• *References*: "Dr. Frank Evers Beddard," by T. S. Palmer, Auk, 43: 413, 1926; "Frank Evers Beddard—1858–1925," by P. C. M[itchell], Proc. Royal Soc. London, ser. B, 99: xxxvi–xxxvii, 1926. • *Portrait*: From P. C. Mitchell, Centenary History of the Zoological Society of London, Zool. Soc. London, 1929. • *Signature*: Adler collection.

WILDER, Harris H. (1864–1928).
WILDER, Inez W. (1871–1929).

The comparative anatomists, Harris Hawthorne Wilder (see *Note*) and Inez Luanne Whipple Wilder, became leading specialists on North American salamanders. Harris was born on 7 April 1864 in Bangor (Maine) but spent his childhood in several towns in Maine and Massachusetts where his father worked as a chorister and teacher of vocal music. Harris had an early interest in natural history especially in amphibians, and at the age of five perused a book on anatomy that had illustrations of a dissected man. As he later wrote, "This gave me the purpose of my whole life." He attended Amherst (A.B. 1886), a small college in western Massachusetts, and taught high school biology in Chicago before going to Germany in 1889, intending to work with Alexander Ecker, author of the book "Die Anatomie des Frosches" (1864–1882), at the university in Freiburg. However, Ecker had died in 1887 and he ended up studying under Robert Wiedersheim, Ecker's successor, although Wilder was officially the doctoral student of August Weismann. Wilder received his Ph.D. in 1891.

Then after another year of teaching high school in Chicago, Wilder took a position at Smith College, a small all-women's school located near his *alma mater*, Amherst. There he founded and chaired its Department of Zoology. During his career he authored seven books including titles on invertebrate zoology, animal classification, the human body, and anthropology, as well as some 40 scientific papers of which 14 were on amphibians (1890–1921). In July 1906 he married Inez L. Whipple, one of his former master's students and by then a junior faculty member in his department. He led the department, which had a research focus unusual for such a small institution, until his death in Northampton on 27 February 1928.

Inez was born on 19 May 1871 in Diamond Hill (Rhode Island). In 1890 she graduated from Rhode Island State Normal School (now Rhode Island College). She then attended Brown University (Ph.B. 1900) before going to Smith College (M.A. 1904) where she worked with Harris Wilder and began her interest in salamander biology. She was appointed an instructor in the zoology department in 1905 where she later became a professor. Inez succeeded her husband as chairman of the department on his death in February 1928, but she was already terminally ill and died 14 months later in Northampton on 29 April 1929. She had published two books—on mammalian anatomy and on amphibian metamorphosis—as well as 13 papers on salamander biology (1906–1925), the first two under her maiden name, Whipple.

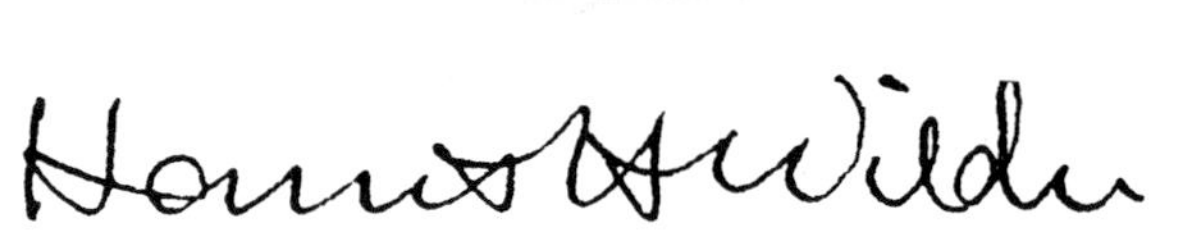

The Wilders, who had no children, worked independently and never published together. At Smith College they developed what was then the most active research program on salamander biology in the world. In 1916 another salamander specialist, Emmett R. Dunn, an expert on systematics and biogeography, joined the faculty. Harris's interests were almost entirely morphological. He published on the anatomies of several salamanders (*Amphiuma, Cryptobranchus, Necturus,* and *Siren*). He also studied the amphibian larynx (1892, 1896) and the early development of *Desmognathus* (1904), but his most extensive contribution was a monograph on the skeletal system of *Necturus maculosus* (1903). Harris discovered that plethodontid salamanders are lungless (1894, 1896). This was a remarkable finding in view of the extensive anatomical work already done on salamanders by European and American specialists. This discovery had important taxonomic, physiological, and ecological implications.

Inez's research program focused on the breeding habits and life history of salamanders with special emphasis on anatomy.

She discovered (and named) the nasolabial groove of plethodontids (1906), another discovery missed by previous workers. She described its structure and function to free the nostrils of water and thus facilitate buccal respiration. She published extensive papers on the life history of *Desmognathus fuscus* (1913) and the metamorphosis of *Eurycea bislineata* (1924). With Dunn, she proposed an explanation for the loss of lungs in plethodontid salamanders (1920). They believed the loss to have originated in swiftly flowing mountain brook habitats where lungs would have been of little value and therefore mostly a liability. They argued furthermore that the origin of the family was in the Appalachian Mountains of eastern North America. Inez's crowning achievement was her book, "The Morphology of Amphibian Metamorphosis" (1925), which featured comparisons of two plethodontids, *D. fuscus* and *E. bislineata*, with the salamandrid, *Notophthalmus viridescens*.

• *References on Harris Wilder*: "Harris Hawthorne Wilder," by H. S. Pratt, Science, 67: 479–481, 1928; "The Early Years of a Zoölogist," by H. H. Wilder, privately printed, Northampton, Massachusetts, xiii, 73 pages, 1930; "Harris Hawthorne Wilder," by H. M. Parshley, Dict. Amer. Biogr., 10: 208–209, 1936. • *References on Inez Wilder*: "Inez Whipple Wilder," p. 883 *in* J. W. Leonard (ed.), Woman's Who's Who of America, 1914–1915, Amer. Commonwealth Co., New York, [1914]; "Inez Whipple Wilder," p. 1346 *in* Who Was Who in America, vol. 1, 1897–1942, A. N. Marquis Co., Chicago, 1942. • *Portraits* (Harris Wilder, 1894; Inez Wilder, about 1910): Smith College Archives. • *Signatures* (Harris Wilder, 1899; Inez Wilder, 1913): Adler collection. • *Note*: According to H. H. Wilder (biographical notes for "Cyclopedia of American Biography" in Smith College Archives), Bert G. Wilder was his "relative" but the exact relationship is not known. Bert Wilder (1841–1925), a former student of Louis Agassiz and longtime professor of zoology at Cornell University, published half a dozen papers (1872–1897) on the anatomy and behavior of amphibians and reptiles.

Lucas, A. H. S. (1853–1936).

Arthur Henry Shakespeare Lucas was a schoolmaster and naturalist who became a specialist on the lizards and marine algae of Australia. He was born, apropos to one of his given names, in Stratford-on-Avon (Warwickshire), England, on 7 May 1853, the son of an itinerant Wesleyan minister who loved natural history. His father introduced him to the seashore where he collected seaweeds and shells as a boy. He attended Balliol College at Oxford (B.A. 1874, M.A. 1877) and the University of London (B.Sc. 1879). He sacrificed his intended medical career to support the children of his widowed brother who, having tuberculosis, was advised to immigrate to Australia. Lucas had taught school in Cambridge and in 1883 took a post as mathematics and science master at Wesley College in Melbourne. He also became associated with the University of Melbourne and was instrumental in its appointment of the biologist and sometimes herpetologist, Baldwin Spencer, to a chair.

During 1884–1892 Lucas edited the *Victorian Naturalist*, the journal of the Field Naturalists' Club of Victoria that had been founded by his brother in 1882. This club had an active role in exploration and one of his frequent field companions was the amphibian specialist, J. J. Fletcher. Lucas moved to Sydney to become headmaster at Newington College (1892–1898) and in 1899 became a senior master and, later, headmaster at Sydney Grammar School for the next 25 years where he became a revered teacher among old Sydneians. After his wife's death in 1919, he lived with his daughter and continued his active fieldwork including trips to Low Island off eastern Queensland and remote Lord Howe Island. On his last expedition at the age of 84, he collected seaweeds at Warrnambool (Victoria); he contracted pneumonia and had to be taken from the train at Albury (New South Wales) where he died three weeks later on 10 June 1936.

Lucas wrote major treatises on Australian natural history including books on algae, botany, and birds. His book, "The Animals of Australia. Mammals, Reptiles and Amphibians" (1909, with W. H. Dudley Le Souëf), contains a 148-page, illustrated section on herpetology. His most important contributions on amphibians and reptiles, however, were a series of 14 papers (1892–1903) in which he reported new distributional data and described new species of frogs, lizards, and a snake. Most of his papers were on lizards and ten of these were co-authored with Charles Frost, an amateur naturalist who had been a fellow member of the Field Naturalists' Club of Victoria for about 30 years before his death in 1915. Together they also published a review of the lizards of New Zealand (1897). Their major work, however, was the reptile section (1896) of the 1894 Horn Expedition to Central Australia, which included five excellent plates, four of them beautifully chromolithographed. Spencer, who had been the biologist on this expedition, which was the first to explore the very center of Australia, contributed the section on amphibians and edited all four volumes of the report (1896; reprinted 1994). All told, Lucas and Frost named 17 currently recognized Australian species including *Pseudophryne semimarmorata, Diplodactylus conspicillatus, Tympanocryptis tetraporophora, Morethia ruficauda,* and three species of *Varanus* (*eremius, gilleni,* and *spenceri*). Their new pygopodid, the monotypic *Ophidiocephalus taeniatus*, was not collected again for nearly 90 years!

• *References*: "Arthur Henry Shakespeare Lucas," by H. C[arter], Proc. Linn. Soc. New South Wales, 62: 243–252, 1937; "Arthur Henry Shakespeare Lucas," by S. Ducker, Austr. Dict. Biogr., 10: 163–164, 1986; G. Shea *in* A. Bauer (ed.), 2004. • *Portrait and signature*: From Carter, 1937.

CALMETTE, Albert (1863–1933).

Léon-Charles-Albert Calmette, a French microbiologist and hygienist, developed the first antiserum for snakebites. He was born in Nice on 12 July 1863 and wished to be a sailor, but a bout of typhoid diverted him to medicine. He attended the Naval Medical School at Brest (1881–1882), graduating as an auxiliary physician, and soon saw duty on a French warship in the South China Sea. He returned to Paris where he completed his medical studies (M.D. 1885). He was posted to Gabon in French West Africa and then to tiny French-occupied islands off the Newfoundland coast where he developed an interest in microbiology. In 1889 Calmette attended microbiology courses at the newly opened Institut Pasteur in Paris.

French Indochina was the richest of all of France's colonies and large investments were made in its infrastructure. The French Colonial Health Service was established in 1890 and Calmette quickly transferred to it. When the service asked Louis Pasteur to organize a branch institute in Saigon (now Ho Chi Minh City), Calmette was the logical choice for director. He had prior experience in Southeast Asia as well as recent training in microbiology at the Institut Pasteur. He departed for Saigon in January 1891. His primary goal was to develop vaccines appropriate for tropical climates especially for smallpox and rabies. In October of that year, southern Vietnam (then called Cochinchine by the French) was inundated by heavy rains. A major flood occurred in the vicinity of Bac Lieu, some 200 km from Saigon, which caused large numbers of cobras to invade the village. Forty people were bitten and four died almost immediately, and the local French administrator sent 19 live cobras to the new institute. Taking advantage of this opportunity, Calmette studied the physiology of envenomation and began work on an antiserum. His first paper on the topic was published in 1892.

Calmette returned briefly to Paris in 1893 and then moved to Lille in 1895 where he founded a new Institut Pasteur. He served as director during 1895–1919. Besides continuing his research on snake venoms, he developed an antituberculosis vaccine, jointly with Camille Guérin, which was called BCG (Bacillus Calmette-Guérin). The two men announced the discovery of a strain of live tubercle bacilli of reduced virulence in 1908. Clinical trials were begun in 1922 and today the vaccine is widely used. Calmette was appointed assistant director of the Institut Pasteur in Paris in 1917, but because northern France was then occupied by the German Army he was not permitted to leave Lille. He finally moved to Paris in 1919. Calmette received many honors including election to the Académie des Sciences in 1904. He died in Paris on 29 October 1933.

In developing an antiserum to neutralize snakebite, Calmette followed the principles successfully used in the preparation of diphtheria antitoxin a few years previously. However, his first attempts in Saigon were unsuccessful. In 1887 the American physiologist, Henry Sewall, had demonstrated that he could render pigeons increasingly more resistant to the venom of rattlesnakes—Massasauga (*Sistrurus catenatus*)—by injecting them with gradually larger doses of venom, so that eventually the pigeons could withstand a dose seven times greater than the minimal lethal dose. Calmette used a modification of this technique in which he reduced the toxicity of cobra venom with a weak solution of hypochlorite. Then, in a series of injections in rabbits using venom from cobras sent alive from Saigon, he progressively reduced the hypochlorite concentration and increased the dose of venom until highly lethal doses could be tolerated. This he announced in 1894 while still in Paris, coincidentally at the same meeting that Césaire Phisalix and Gabriel Bertrand, from the Muséum d'Histoire Naturelle, presented similar evidence from guinea pigs vaccinated with the venom of *Vipera.* In Lille, to procure large quantities of serum, Calmette vaccinated horses and donkeys. On average, it took 16 months to develop an antitoxic serum in what he called his "cobra horses." Eventually, he was able to vaccinate horses so they could survive injection of 2 g of dry cobra venom, 80 times the minimum lethal dose.

From his experiments Calmette believed that his cobra antiserum could neutralize the venom of all species of snake. However, other workers soon showed the antiserum to be ineffective not only against rattlesnake venom but also other elapid venoms. The discovery of the species-specificity of antiserums was first made in Brazil in 1898 by Vital Brazil (published 1901). Calmette authored a major review of animal venoms and serum therapeutics including a summary of his own research in 1907 (English edition 1908).

• *References*: "Léon-Charles-Albert Calmette," by É. Franceschini, Dict. Biogr. Franç., 7: 915–916, 1956; "La Vie et l'Oeuvre de Albert Calmette 1863–1933," by N. Bernard, Albin Michel, Paris, 313 pages, [1961]; "Albert Calmette," by A. Delaunay, Dict. Sci. Biogr., 3: 22–23, 1971; "Doctor Albert Calmette 1863–1933: Founder of Antivenomous Serotherapy and of Antituberculosis BCG Vaccination," by B. J. Hawgood, Toxicon, 37: 1241–1258, 1999. • *Portrait* (1890s): From Arch. Parasitol., 2: 311, 1899. • *Signature* (1897): Adler collection.

NAMIYE, Motoyoshi (1854–1918).

Motoyoshi Namiye (see *Note 1*), who studied the vertebrates of Imperial Japan, was Japan's first herpetologist. He was born on 13 March 1854 (see *Note 2*) in Maruyama-Nishikatamachi, Hongo, which was then on the outskirts of Tokyo (today it is Nishikata, Bunkyo-ku, Tokyo). He was the son of a samurai belonging to the Fukuyama-han and from 1868 he attended school in Fukuyama, 90 km east of Hiroshima. He became a zoology student of Franz Hilgendorf, a German zoologist and ichthyologist who taught at Tokyo Medical College (now part of the University of Tokyo) during 1873–1876 and who named two Japanese snakes. Apparently Namiye never received an academic degree.

In 1879 Namiye was appointed to the staff of the Tokyo Educational Museum (now National Science Museum). In 1889 he became an assistant to Isao Ijima, Professor of Zoology at Tokyo Imperial University (now University of Tokyo) and Director of the university's Zoological Institute. Namiye spent the rest of his career there. His main duties seem to have been curation of the institute's museum collection and the fieldwork associated with it. His first study was birds. A budding interest in amphibians and reptiles was strongly encouraged by an American herpetologist—Leonhard Stejneger, author of the book "Herpetology of Japan and Adjacent Territory" (1907)—who visited Japan in 1896 and again in 1897. They already had common interests in Japanese birds and Stejneger named three new species for Namiye in 1886 and in 1901 a frog (*Rana namiyei*). As the Japanese herpetologist, Yaichirô Okada, once wrote (in translation), " . . . these two men remain in our memories as the founding fathers of our field."

Namiye's interests closely paralleled Japan's enlargement of its national territory through conquest. It annexed Formosa (now Taiwan) in 1895 and Korea in 1910. Sakhalin Island, which is located off the Russian coast north of Japan, had been jointly claimed by the two countries, but after Russia's defeat in the Russo-Japanese War of 1904–1905 Japan took sole control of the southern part of it. Namiye's herpetological publications number about 20 titles (1893–1914). He is known to have conducted fieldwork in many localities on Japan's main island of Hondo (now Honshu), the Ryukyu Islands between Formosa and mainland Japan, Ôshima and other nearby islands to the south off the Izu Peninsula southwest of Tokyo, and Tsushima Island between Korea and Japan. Among his papers are reports on the herpetology of Shikoku, one of the smaller main islands in southern Japan (1903); the snakes of Sakhalin (1906); a three-part series on Formosan venomous snakes (1908–1909); the geckos of the Ryukyus (1912) in which a eublepharid gecko (*Goniurosaurus kuroiwae*) is described and named; and the herpetology of Korea (1913).

Namiye was one of the founders of the Zoological Society of Japan and its journal, *Zoological Magazine*. He died in Tokyo on 24 May 1918.

• *References*: "Motoyoshi Namiye," by I. Ijima, Zool. Mag., 30: unpaginated insert between pp. 268 and 269, 1918; "Motoyoshi Namiye," by N. Kuroda, Auk, 36: 628–629, 1919; ["History of Department of Zoology, Faculty of Science, Tokyo Imperial University, pts. 1–2,"] by N. Yatsu, Science, Tokyo, 8: 340–346, 387–389, 1938 (Namiye on pp. 387–389); Hasegawa, 1967 (Namiye on pp. 59–60). • *Portrait and signature*: From Ijima, 1918. • *Note 1*: Most authorities Romanize the forename as Motoyoshi but some use Motokichi or Motokishi. All are legitimate pronunciations of the characters that Namiye used in his name. The surname has generally been written as "Namiye," which is the form that has been Latinized by both Western and Japanese authors in naming taxa for him. Before the rules for Romanization were laid out, the *eh* sound in Namiye was written with the semivowel *y*; today the name is written "Namie." • *Note 2*: He was born 15 February 1 Ansei Era, thus 13 March 1854 according to the modern calendar.

KOSLOWSKY, Julio (1866–1923).

Julio Germán Koslowsky, the first specialist on Argentine herpetology, was born Julian German Kozłowski on 15 September 1866 in Steinholm, near Riga, in a region of Baltic Russia that

is now in Latvia. He was descended from a noble Polish family and was well educated, first at home and then in Riga and Dorpat (now Tartu, Estonia). He was a polyglot but also studied the natural sciences. Motivated by a desire to know the world, about 1888 he immigrated to Argentina and became a teacher in Buenos Aires. Carlos Berg, a fellow Latvian who was then Director of the Museo Argentino de Ciencias Naturales, hired him in 1893. He was a collector for Berg who wanted to build his museum collection and who sent him on field trips to different parts of Argentina, Paraguay, and the Mato Grosso of Brazil. Besides being a naturalist, Koslowsky was also a rugged explorer who took pleasure in walking and riding horses for long distances. In 1894 he was hired by the Museo de La Plata (now Museo de Ciencias Naturales de La Plata), located nearby in the capital of the province of Buenos Aires, as traveling naturalist, the museum's official collector. He visited extensive regions of Argentina, Bolivia, Paraguay, and southern Brazil, often living with the indigenous tribes, and his collections became the foundation for the museum's herpetology section.

Koslowsky published seven papers (1895–1898) on these materials, all in the *Revista del Museo de La Plata*. Although there were others before him—like Berg and Hermann Burmeister, who preceded Berg as director of the museum in Buenos Aires—who occasionally published on Argentine herpetology, Koslowsky was the first to focus his attention on this subject. He authored 25 new taxa of frogs, lizards, and snakes including the colubrid genus *Pseudotomodon*. His most extensive and synthetic work, also his last in herpetology, was a list of Argentine reptiles with extensive comments on the distribution of species, not only in Argentina but also in Bolivia, Brazil, Chile, Paraguay, and Uruguay. This monograph ended his formal herpetological career at the age of only 33.

In the summer of 1895–1896 Koslowsky joined a commission sent to explore the southwestern region of Chubut Province and to help fix the local boundary between Argentina and Chile. This led to the establishment of colonies in the region in order to stake Argentina's claim; Koslowsky himself settled in one valley, Huemules, at the end of 1896. After two years, however, the winters were so severe that all of the original colonists either died or left, except for Koslowsky who sometimes depended on local tribes for support. The nearest Europeans were several hundred kilometers away. The first telegraph line in the area was installed in his home, Lago Blanco, in 1902. He returned to Buenos Aires for long periods (1905–1907 and 1914–1921). First he worked as a surveyor for the Ministerio de Agricultura and later attempted to start businesses, all of which failed. In 1921 he finally returned to his home in Valle Huemules (Chubut) where he died on 23 September 1923.

• *References*: "Julian German Kozłowski," by S. Lis-Kosłowski, Polski Słownik Biograf., 15: 13–14, 1970; "Historia de la Herpetologia Sudamericana. Julio Koslowsky," by J. D. Williams, Bol. Asoc. Herpetol. Argent., 1(3): 5, 1983; "Julio Germán Koslowsky, Científico, Explorador y Colono," by A. Aguado and J. D. Williams, Museo (Rev. Mus. La Plata), 3(17): 25–30, 2003. • *Portrait* (about 1920): From El Hornero, Buenos Aires, 3: 207, 1923. • *Signature* (1896): Museo de Ciencias Naturales de La Plata.

FLOWER, Stanley S. (1871–1946).

Stanley Smyth Flower was an English zoologist and conservationist who specialized on the herpetology of Southeast Asia and the Middle East. He was born in London on 1 August 1871 and could hardly have had a childhood more predisposed to natural history. His father, Richard Flower, was conservator of the museum of the Royal College of Surgeons, president of the Zoological Society of London and, later, director of the British Museum (Nat. Hist.). Among his father's friends were Darwin, Huxley, and Owen who, along with staff at the zoo, all tutored young Stanley. After schooling at Wellington and King's colleges, he entered the British Army in 1890. He served in India and the Straits Settlements before being seconded as science adviser to the Siamese Government (1896–1898).

In 1898 Flower became director of the zoological gardens at Giza on the western outskirts of Cairo where he served until 1924 except for duty in World War I. During the war, because of his extensive knowledge of mammals, he was appointed commander of the Camel Transport Corps for Palestine, the Sinai, and the Suez Canal region. He once estimated that at one time or another he had responsibility for some 24,000 camels! Even under wartime conditions, however, he continued to make natural history collections that included reptiles.

During his long tenure in Egypt, besides developing the Giza zoo into a respectable institution, he founded the Giza Zoological Museum and the Zoological Survey of Egypt. He was also in charge of the Game Preservation Department of neighboring Sudan. Flower built a reputation as a conservationist (he is credited with saving the egret in Egypt) and because of his high connections in London and in the military, he and his wife sponsored elaborate tea parties at the zoo which attracted major personages of the time including senior

military commanders such as Allenby and Kitchener, and the novelist, Kipling. Throughout this long period in Egypt, Flower continued to collect and publish.

In 1924 Flower retired to Tring, located about 50 km northwest of London, and spent the remainder of his life in association with the Zoological Society of London, arguably the greatest zoo of its day. The centennial publication, "List of the Vertebrated Animals Exhibited in the Gardens of the Zoological Society of London, 1828–1927" (1929), to which Flower contributed the section on lower vertebrates, is a staggering annotated catalogue of the 904 species of amphibians and reptiles that passed through the zoo when the British Empire was at its apogee. During 1925–1928 he also edited the herpetological sections of *Zoological Record*, which was published by the zoo. Flower died at his home in Tring on 3 February 1946.

Flower's herpetological contributions fall into three periods. During his time in Malaya and Siam (1895–1898), he made extensive collections that he reported in three major papers (1896–1899 [1897–1900]). Together, these represent the first comprehensive herpetological review of that region since the work by Theodore Cantor published in 1847. Flower drew the beautiful color plates himself. The second and longest period represents his time in Egypt, during which he produced several technical papers (1923–1931), culminating with his review of the amphibians and reptiles of Egypt (1933), the first summary for Egypt since those of John Anderson (1896, 1898). Flower's work was based on his fieldwork throughout Egypt, and he covered a much larger region than did Anderson.

The third period and Flower's final area of interest began during his directorship of the zoo in Egypt. He catalogued the longevities of vertebrate species and thereby drew attention to the life spans that could be achieved with proper care in captivity. His papers on amphibians and reptiles (1925, supplements in 1936–1937) included extensive literature reviews. These were based largely on records from the London zoo, as well as from those in Paris and other Europeans zoos and his former zoo at Giza. The common theme in these and many of his other publications, as well as his activities generally, was his devotion to the care and conservation of both wild and captive animals.

• *References*: "Major S. S. Flower, O.B.E.," by E. Hindle, Nature, 157: 327–328, 1946; "Major Stanley Smyth Flower," anonymous (probably W. L. Necker), Herpetologica, 3: 85–88, 1946; "Stanley Smyth Flower 1871–1946," by M. A. Smith, Copeia, 1946: 185–187, 1946. • *Portrait and signature* (1927): Natural History Museum, London, courtesy A. F. Stimson.

Brimley, Herbert H. (1861–1946).
Brimley, Clement S. (1863–1946).

The Brimley brothers—Herbert Hutchinson Brimley and Clement Samuel Brimley—became North Carolina's first professional zoologists and their extensive writings included herpetology. They were born to a family of tenant farmers in the midlands of England just north of London, H. H. at the family home in Willington (Bedfordshire) on 7 March 1861 and C. S. at his grandmother's house in Great Linford (Buckinghamshire) on 18 December 1863. Both were educated in local schools and spent much time together in the field collecting bird eggs, fishing, and learning about wildlife. When prices for farm products dropped in the late 1870s, the Brimley family decided to emigrate and a chance meeting with a visiting official of the newly formed North Carolina Department of Agriculture located in Raleigh led them to the United States in 1880. The family tried farming in the Raleigh area with limited success, so the two brothers began collecting natural history specimens for sale to museums, schools, and universities. Later, after learning taxidermy, they started a formal and highly successful business, "Brimley Bros., Collectors and Preparers." H. H. was more the ornithologist and C. S. an entomologist, but both were general naturalists.

In 1884 H. H. was hired by the North Carolina Board of Agriculture to establish a museum and, starting in 1890, to prepare a large display of North Carolina animals for exhibition at the World's Columbian Exposition, Chicago's world fair (1893). By 1895 H. H. became the first full-time curator at the state museum and its sole employee! He later recalled, "I became expert with a feather duster and pushed a wicked carpet sweeper! I had no funds beyond my princely salary of $75 per month." By 1907 or 1908 his duties there became so great that he had to relinquish his business partnership with C. S. who managed the company alone as "C. S. Brimley Zoologist" (see *Note*). In 1919 C. S. joined the Entomology Division of the state's Department of Agriculture and by 1928 H. H.'s curator title was changed to director, a post he held until 1937. The brothers were founders of the North

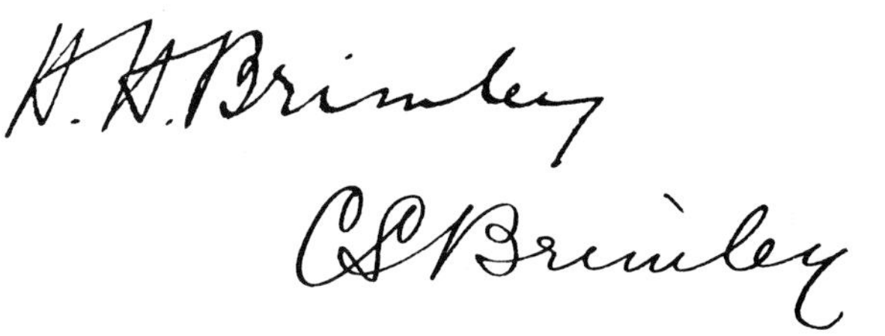

Carolina Academy of Science and C. S. was the only person without a college degree ever to serve as its president. In recognition of his lifetime's contribution to the zoology of the state, C. S. received an honorary Doctor of Laws degree from the University of North Carolina in 1938.

Throughout their years in North Carolina, the Brimley brothers collected and published on the fauna. Their most elaborate contribution, authored jointly with T. Gilbert Pearson, was the book "Birds of North Carolina" (1919; later editions in 1942 and 1959); one of their artists was a very young Roger Tory Peterson who later gained fame for his self-illustrated field guides to birds. The Brimleys's earliest papers, co-authored beginning in 1884, were on birds, but starting in 1895 they wrote some 55 titles on amphibians and reptiles. The vast majority of these were by C. S. alone who was the more academic of the two. These papers covered local lists of species, behavior, reproduction, snakebite, food and feeding habits, and reviews of specific taxonomic groups. C. S. authored the first checklist of North Carolina amphibians and reptiles (1915; revisions in 1918 and 1926) and the first detailed summary with keys of the herpetology of the state (originally published in 32 installments, 1939–1943; reprinted under one cover in 1944). He also described several new herpetological taxa, among them the salamanders *Necturus lewisi, Plethodon metcalfi,* and *Pseudotriton ruber schencki.*

The brothers died in Raleigh in 1946 little more than three months apart, H. H. on 4 April and C. S. on 23 July. *Brimleyana,* the scientific journal of the North Carolina Museum of Natural Sciences, is named in their honor. Leonhard Stejneger named a salamander, *Desmognathus brimleyorum,* for them.

• *References*: "Dr. Clement Samuel Brimley," by H. T. Davis, Herpetologica, 3: 127–130, 1946; "Clement Samuel Brimley," by H. T. Davis, Copeia, 1946: 182, 1946; "Herbert Hutchinson Brimley," by H. T. Davis, Auk, 64: 664–665, 1947; "A North Carolina Naturalist, H. H. Brimley," by E. P. Odum, Univ. North Carolina Press, Chapel Hill, xvi, 205 pages, 1949; "The Brimley Brothers: North Carolina Naturalists," by J. E. Cooper, Brimleyana, 1: 1–14, 1979. • *Portrait* (about 1944, H. H. on left) *and signatures* (H. H. 1898, C. S. 1911): North Carolina State Museum of Natural Sciences, Raleigh. • *Note*: Brimley Bros. was not the forerunner of Carolina Biological Supply Co., which was established in 1927 by T. E. Powell, Jr., then a biology professor at Elon College.

WILLISTON, Samuel W. (1851–1918).

Samuel Wendell Williston was an American vertebrate paleontologist and entomologist who made major contributions to the study of fossil amphibians and reptiles and to reptile osteology. He was born the son of an unschooled blacksmith in Roxbury (Massachusetts) on 10 July 1851 (see *Note*) and had an early passion for books and learning. His family moved to Manhattan (Kansas) in 1857 where he attended school and the local college, now called Kansas State University (B.S. 1872, M.A. 1875). He became a confirmed Darwinist and was denounced at a local church for his lectures on evolution. One of his college professors put him in touch with O. C. Marsh of Yale University who was then collecting vertebrate fossils in western Kansas. Williston worked on Marsh's field crew for two summers; it was Williston who developed the standard technique for protecting fossils prior to transport by wrapping them in strips of strong paper dipped in flour paste.

In 1876 he was hired to be Marsh's laboratory assistant at Yale where he was unavoidably drawn into the bitter competition between Marsh and Edward D. Cope for vertebrate fossils in the American West and the dominance of American vertebrate paleontology. Williston was, in fact, secretly colluding with Cope by writing him with information about Marsh's current research and modes of operation. He was particularly upset by Marsh's insistence that his employees not publish on paleontology, which forced him to consider other career paths, and that led him to medicine and entomology. Accordingly, while at Yale, he earned both medical and research degrees (M.D. 1880, Ph.D. 1885). He had started collecting insects in 1876 and eventually settled on dipteran flies, quite arbitrarily, and became the leading American authority on the group. His "Manual of North American Diptera" (1888) was the standard and by 1908 went through three editions. A South American species of fruit fly, *Drosophila willistoni*, a species that is routinely used by geneticists, is named in his honor. He authored nearly 100 titles in entomology.

Williston left Marsh's employ in 1885, writing in his letter of resignation, "My duties are too irksome, and it is impossible for me to do justice to them." He then taught anatomy at Yale (1886–1890) and concurrently held public health positions for the City of New Haven and the State of Connecticut.

In 1890 he returned to Kansas to become a professor of geology and paleontology at the University of Kansas. He soon began his own expeditions for fossil vertebrates, focusing on Cretaceous deposits in Kansas and, later, Wyoming. He described many new species of early reptiles and mammals and revised their classifications. His research at this time concentrated on mosasaurs, plesiosaurs, and pterodactyls, but also included fossil birds, crocodiles, and turtles. Williston was involved in establishing the university's medical school and served as its first dean (1898–1902).

Williston's research horizon changed when he accepted a professorship at the University of Chicago in 1902, succeeding Georg Baur who had also been a Marsh assistant. Baur had built up a substantial Permian collection of reptiles, but with his premature death in 1898 these remained mostly unstudied. Williston added additional Permian collections from the Red Beds of Texas and New Mexico, which yielded numerous specimens of labyrinthodont amphibians and early reptiles. He died in Chicago on 30 August 1918.

In vertebrate paleontology Williston authored about 140 titles (1878–1918), which include descriptions of many new genera and species of extinct amphibians and reptiles. He published two books of particular interest to herpetologists. His book, "Water Reptiles of the Past and Present" (1914), covers the mosasaurs, plesiosaurs, and other extinct forms as well as aquatic representatives of living groups including squamates, crocodiles, and turtles. His "Osteology of the Reptiles" (1925; reprinted 1971), issued posthumously under the editorship of William K. Gregory, was the standard work on the subject until the appearance of Alfred S. Romer's volume of the same title (1956). Williston's book is extensively illustrated with still-useful drawings by the author.

• *References*: "Samuel Wendell Williston," by H. F. Osborn, Science, 49: 274–278, 1919; "Samuel Wendell Williston. 1852–1918," by R. S. Lull, Biogr. Mem. U.S. Natl. Acad. Sci., 17: 113–141, 1924; "Samuel Wendell Williston," by E. N. Shore, Dict. Sci. Biogr., 14: 409–411, 1976. • *Portrait*: Adler collection. • *Signature*: Bull. Geol. Soc. Amer., 30: opposite p. 66, 1930. • *Note*: Williston's year of birth was long given as 1852, but recent evidence confirms that it is 1851 (see Shore, 1976).

GARCÍA, Evaristo (1845–1921).

Evaristo García Piedrahita, a physician and specialist on the venomous snakes of western Colombia, was born in Cali (Cauca, today in Valle) on 17 November 1845. After completing secondary school at the Colegio de Santa Librada in Cali, he attended the Escuela de Medicina y Cirugía at the Universidad Nacional de Colombia in Bogotá, from which he received the degree of doctor of medicine and surgery (1872). He became a professor at the medical school and headed the hospital clinic but later went to Paris and London for advanced training. After returning to Bogotá he soon moved back to Cali in 1877 to practice medicine. He became an expert on a wide range of diseases including leprosy, elephantiasis, malaria, and various parasites. He was the first to identify beriberi in Colombia, and he described the first cases of viral spastic paraplegia in the region. García was a leader in the Colombian medical commu-

nity and in 1874 was one of the founders of an organization that became the Academia Nacional de Medicina. The Universidad del Valle in Cali later named its hospital in his honor. He died in Cali on 19 June 1921.

Beginning in the late 1860s, Garcia's first research interest was the venomous snakes of his home state of Cauca, which included Cali and extended to the Pacific Ocean. ("Cauca" of García's era is today two adjacent departments bordering the coast, Cauca and Valle.) Snakebite was a logical research topic at the time for a physician devoted to public health and living in a country inhabited by a variety of venomous snakes. There was a high incidence of human snakebites among the rural population. Although there were numerous purported treatments for snakebite then—both medical and traditional—the development of antiserums by Albert Calmette and Vital Brazil had not yet occurred. All that could be done was to write a manual describing and illustrating all of the native venomous snakes and giving a catalogue of the medicinal plants that were available locally for "treatment" of bites. García produced just such a treatise, which became his medical school thesis and was also his first book, "Los Ofidios Venenosas del Cauca." The manuscript was presented to the medical society of Cauca in 1892, printed in Paris to a high standard, and published in Cali (1896; reprinted 1994). The book includes 15 chromolithographic plates, 11 of them illustrating pitvipers, coralsnakes, and coralsnake mimics (the rest are of medicinal plants). One of the snakes—a large, semi-arboreal pitviper found in the Pacific foothills and coastal plain of Colombia—he described as a new species known today as *Bothrops punctata*. This book was García's sole contribution to the herpetological literature.

• *Reference, portrait, and signature* (1894): "Evaristo García. Escritos Escogidos," Fundación E. García, Gobernac. Valle del Cauca, Cali, 190 pages, 1994, courtesy William Lamar.

TORNIER, Gustav (1858–1938).

The German zoologist, Gustav Tornier, had two interests concerning amphibians and reptiles—developmental anatomy and systematics—and he became a specialist on the herpetofauna of former German East Africa. He was born on 9 May 1858 in Damerau, "East Prussia" (see *Note*). Nothing is known about his childhood or early education. He attended the university in Heidelberg where he received his doctorate (Dr.phil. 1892) and by 1891 he had already become an assistant in the Zoological Museum of the University of Berlin. At first he was involved with anatomical preparations and from 1893 worked with the herpetology collection. When its curator, Paul Matschie, was transferred to the mammalogy collection in 1895, Tornier succeeded him as curator of herpetology. Tornier became a professor of zoology at the university in 1902. At the museum he took on the additional duties of administrative manager of the library in 1903, assistant director of the museum in 1921, and acting director during 1922–1923. He retired in October 1923 and died in Berlin on 25 April 1938.

Tornier's interest in African herpetology came at a time when the German Empire, created in 1871, was developing overseas colonies and especially in Africa. German East Africa, which received an imperial charter in 1885, included what are now Burundi, Rwanda, and Tanganyika (the mainland part of present-day Tanzania). Germany's colonies along the western coast of Africa were Cameroon (today Cameroon and easternmost Nigeria) and Togoland (now Togo and easternmost Ghana) in West Africa and German South-West Africa (now Namibia). Tornier's focus was East Africa, but he also published on Cameroon and Togoland as well as on collections from Somaliland, Arabia, and islands in the Indian Ocean. He described numerous new taxa—especially of chameleons (he named six of the recognized species of *Chamaeleo*). His most spectacular novelty was the Tanzanian bufonid, *Nectophrynoides viviparus* (1905), which was believed to be the first viviparous anuran. In fact, since nutrition is provided to the embryos only from their yolk and not by oviducal secretions, their mode of development is technically ovoviviparous. (Two species of *Nectophrynoides* that were discovered later by others exhibit true viviparity.)

Tornier's major work on African herpetology, entitled "Die Reptilien und Amphibien Ost-Afrikas" (1896; reissued by the same publisher as a book in 1897 under the title "Die Kriechthiere Deutsch-Ost-Afrikas"), was first published as part 4 of volume 3 in Karl Möbius's series, "Die Thierwelt Ost-Afrikas." It is illustrated with five excellent plates that demonstrate Tornier's interest in pattern variation and its developmental history. Another of his contributions was to provide the extensive captions for the herpetological plates in the series, "Symbolæ Physicæ," which resulted from the German expedition to Egypt and the Middle East during 1820–1825 involving Wilhelm

Hemprich and C. G. Ehrenberg. Hemprich died in Cairo in 1825 and Ehrenberg returned to Berlin with their huge zoological collections (including 436 amphibian and reptile specimens). The results were published over a very long period (1828–1900). The four beautiful plates in the herpetological section were apparently printed and hand colored in the late 1820s, but these, together with Tornier's captions, were only published in 1899. He authored about 25 titles on herpetological systematics and faunas (1897–1927).

Tornier's other area of interest was in the development and structure of amphibians and reptiles (about 25 titles, 1896–1910). Of special interest to him were variation in skin and scutellation patterns and their developmental histories; regeneration; the development of extra digits and limbs; and the functional morphology of head flaps and neck air sacs in chameleons.

• *References*: "The *Symbolae Physicae* and the Herpetology of Hemprich and Ehrenberg's Expedition to Egypt and the Middle East," by A. Bauer, News-Bull. Internatl. Soc. Hist. Biblio. Herpetol. (now Biblioth. Herpetol.), 2(1): 8–16, 2000; Rieck et al., 2001 (Tornier on p. 365; by W. Bischoff on pp. 624–625), 2001. • *Portrait*: Museum für Naturkunde, Berlin (BI/518). • *Signature* (1895): Natural History Museum, London, courtesy A. F. Stimson. • *Note*: Tornier is said to have been born in East Prussia, a region located along the southeastern coast of the Baltic Sea and extending as far east as Lithuania. Between 1824 and 1878, which includes the year of his birth, East Prussia did not exist separately and was joined with West Prussia to form the Province of Prussia. There were at least eight places named Damerau in East Prussia during the 19th Century. Prussia became part of the German Empire in 1871.

STRECKER, John K., Jr. (1875–1933).

John Kern Strecker, Jr., was an amateur-turned-professional naturalist of Texas who became the first authority on the herpetology of his state. He was born in Waterloo (Illinois) about 30 km south of St. Louis on 10 July 1875. The family soon moved to Fort Scott in eastern Kansas where John Jr. spent his boyhood years. His father's side of the family had been amateur naturalists. There was a family tradition to collect, identify, and preserve specimens from nature, and his father had a particular interest in reptiles. John Jr. attended public schools only through the primary grades and he clearly benefited from his father's hobby interests. The family moved to Waco (Texas) in 1889 where John Jr. apprenticed to his father who was a stonecutter. In Waco John Jr.'s broad natural history interests continued to develop further. By age 16 he published his first paper—on birds. In the 1890s he conducted local field trips for specimens and by the turn of the century began corresponding with professional zoologists across the country.

When Baylor University, the oldest university in Texas and located in Waco, was looking for a replacement curator (in effect, director) for its museum, Strecker was the obvious choice. In October 1903, at the age of 28, he accepted a post that he would hold until his death in Waco on 9 January 1933. During this 30-year period he built the museum to a position of national prominence and became a beloved figure on campus, known for his sense of humor and creative pranks. He was the only one allowed to smoke (a pipe) on this Baptist campus besides the university president. In 1919 he took on additional duties as chief librarian. Baylor saluted him with an honorary M.S. degree in 1925. When he died, the university's flags were flown at half-mast and students were excused from classes to attend his funeral service. In 1940 the museum was re-named the "Strecker Museum."

Above all, Strecker was a field collector and a careful observer. Most of his work was in Texas—especially East and Central Texas, the Gulf Coast, and the mountains of the Big Bend region—and he also collected in the mountains of Arkansas and the swamps of Louisiana. His field attire invariably included leather leggings (for protection from snakebite), a hat, and a tie, the latter so that, as he explained it, he would not be "placed in the despised hobo class." Yet he never learned to drive a car. His herpetological publications (about 65, 1898–1935) reflect his interests: reports on county or regional surveys, life history, behavior, reptile myths and folklore, and systematics. He named *Scaphiopus hurterii* for Julius Hurter, Sr., a mechanical engineer and amateur herpetologist of St. Louis with whom Strecker coauthored "The Amphibians and Reptiles of Arkansas" (1909). Strecker's major contribution to herpetology was "Reptiles and Amphibians of Texas" (1915).

• *References*: "John Kern Strecker," by T. S. Palmer, Auk, 52: 491–492, 1935; "Strecker—Naturalist and Man," by A. N. Bragg, Bios, 32: 177–181, 1961; "Chasing the Bone Pile," by K. Gerhardt, Occas. Pap. Strecker Mus., 2: 1–90, 1995 (Strecker on pp. 9–11, 20–37); "John Kern Strecker, Jr.," by R. W. Neck, p. 120–121 *in* R. Tyler et al. (eds.), New Handbook of Texas, Texas State Hist. Assoc., Austin, 1996. • *Portrait*: Texas Collection, Baylor University. • *Signature* (1912): Baylor University.

PELLEGRIN, Jacques (1873–1944).
Jacques Pellegrin, who specialized on the fishes, amphibians, and reptiles of Africa, was born in Paris on 12 June 1873. As a boy he was fascinated with nature especially fishes. He attended the Lycée Condorcet and then the Sorbonne (University of Paris), and graduated from the latter in 1894 with a degree in natural sciences. He joined the staff of the Muséum d'Histoire Naturelle in 1897 as preparator for Léon Vaillant who was the professor in charge of reptiles and fishes. Before his military service, Pellegrin went on study trips to Greece where his father was French consul. He defended his doctoral thesis in medicine in 1899 and then another in science in 1904, both at the Sorbonne. He was promoted to be Vaillant's chief assistant in 1908 and continued in that role when Louis Roule succeeded Vaillant in 1910. During World War I, Pellegrin was mobilized as a doctor's aide-major. Starting in 1920 he made numerous collecting trips abroad on a nearly annual basis, for government ministries or the museum. In 1937 he succeeded Roule as professor and retired in 1943. Pellegrin died the next year, in Paris, on 12 August.

Pellegrin had a prodigious output of more than 600 publications, mostly on fishes. He was a specialist on African fishes especially those of Lake Chad. He was interested in pisciculture and introduced many fish species to France including trout, carp, and perch. He conducted a dozen expeditions to French Morocco, and also went to Romania and Yugoslavia. Pellegrin reported on these collections as well as many others made by other French personnel. He described nearly 340 new species of fishes. He also named about ten new lizards and snakes and, surprisingly, most of these were not from Africa, but from Australia, Bolivia, Peru, and French Tonkin (the northern part of Vietnam). His herpetological publications number about 40 (1900–1939). Besides his systematic and faunal studies, he published several papers on the longevity of reptiles and amphibians, primarily based on specimens living in the museum's menagerie.

• *References*: "Le Dr. Jacques Pellegrin," anonymous, Pêche à Paris, unpaginated, 1928; Jaussaud and Brygoo, 2004 (Pellegrin on pp. 411–412). • *Portrait* (1936): George S. Myers/Alan E. Leviton Portrait Collection, courtesy Alan Leviton. • *Signature* (1917): Natural History Museum, London, courtesy Andrew F. Stimson.

BRAZIL, Vital (1865–1950).
Vital Brazil Mineiro da Campanha (see *Note 1*) was a Brazilian physician and founder of the Instituto Butantan who developed the first polyvalent antiserum for pitvipers. He was born in Campanha (Minas Gerais) on 28 April 1865 but in 1880, because of finances, his family moved to São Paulo. He wanted to become a physician and, after holding a series of menial jobs to raise the funds, in 1886 began his medical studies at the Faculty of Medicine in Rio de Janeiro (M.D. 1891). He returned to São Paulo and became an inspector for the State Sanitary Department before setting up his own medical practice (1895) in Botucatú, a small rural town surrounded by farms and ranches that afforded numerous opportunities for snakebites. He set up his own laboratory to study bites by pitvipers (*Bothrops* and *Crotalus*) and noticed differences in the responses of patients to these two kinds of bites, for which he had no effective treatments.

Then in 1896 he read about the work of Albert Calmette, the French microbiologist who had developed an antiserum for cobra bites, and recognized that serotherapy could be a treatment for pitviper bites. He closed his practice in 1897, moved to São Paulo, and took a post as assistant to the director of the Instituto Bacteriologico, which gave him the opportunity to develop antisera for pitviper bites. The institute was headed by one of Brazil's most distinguished public health servants and naturalists, Adolpho Lutz, who also happened to be a specialist on frog systematics. With a sympathetic supervisor, Brazil began to inject dogs and goats with small doses of snake venom to develop a protective antiserum.

By 1898 Brazil had obtained evidence for the species-specificity of antivenoms (published 1901). He also found that Calmette's anti-cobra serum provided no protection against rattlesnakes. His work with snakes, however, was interrupted by a national emergency, which, ironically, was to greatly benefit his research on snake antivenoms. An epidemic of bubonic plague broke out in the port city of Santos in 1899 and swept through major Brazilian cities and elsewhere in South America. Brazil was sent to study it and actually contracted the disease himself but was able to survive. To develop anti-plague serum, the government set up a small laboratory under Brazil's directorship. This was in the small town of Butantan (see *Note 2*) on the outskirts of São Paulo, 9 km from the city center. There, in February 1901, the

Instituto Serumtherapico Butantan was begun, first as a center to produce antisera for the plague and secondarily as a place to produce antivenin for snakebites.

At first the institute produced monovalent sera separately for *Bothrops* and *Crotalus* using horses but soon produced a mixed or polyvalent serum. This was crucial because it was often not known in the field which kind of snake was actually involved. An anti-*Micrurus* serum was developed in 1910. Brazil, being a master organizer, used an exchange system by which the institute would send kits with ampules of serum, syringes, and needles to farmers and ranchers in exchange for live snakes. He arranged with the railroad lines—there were 17 different ones in the state of São Paulo alone!—to provide free shipping. Special collecting equipment was also provided, such as sticks with nooses, in order for untrained people to collect snakes with little risk. By the year 1914 he estimated that 2,000 farmers were involved each year and had sent in some 5,000 snakes, meaning also that 5,000 kits had been distributed. Numerous identification manuals were produced and conferences held in order to publicize this program. Brazil also promoted the conservation of the snake-eating Mussurana (*Clelia clelia*) as yet another element in his public campaign. As a result, the number of deaths from snakebite was dramatically reduced; mortality decreased from 25–30% to a mere 2%. Serum was also produced for export. For example, all of the antiserums used in the United States before 1926 came from Butantan.

Brazil remained as director of the institute—renamed Instituto Butantan in 1915—until 1919 when, because of political interference, he resigned. He was reinstated in 1924 but finally ended his directorship in 1927. In the meantime he had started his own animal venoms research center, Instituto Vital Brazil, in Niterói, a suburb of Rio de Janeiro. He left a legacy at both institutions and trained many biologists and herpetologists of the next generation including Afrânio do Amaral, J. F. Gomes, Arthur Neiva, and J. A. Vellard, as well as his own son, Oswaldo Vital Brazil.

Brazil published several major works on protective measures against venomous snakes and their bites (1911, Portuguese and French editions; second editions in 1914). During his lifetime he received numerous honors for his scientific and public contributions, and in 1993 the government issued a new 10,000-cruzeiro bank note with his portrait (see illustration on page 255). He died in Rio de Janeiro on 8 May 1950.

• *References*: "Vital Brazil," memorial issue of An. Paul. Med. Cirurg., 60(5): 337–634, 1950; "Vital Brazil," by G. Rosenfeld, Mem. Inst. Butantan, 34: x–xvi, 1969 (1970); "History of the Primordia of Snake-Bite Accident Serotherapy," by O. V. Brazil, Mem. Inst. Butantan, 49: 7–20, 1987; "Pioneers of Anti-Venomous Serotherapy: Dr Vital Brazil (1865–1950)," by B. J. Hawgood, Toxicon, 30: 573–579, 1992. • *Portrait*: Instituto Butantan, courtesy Denise Maria Mariotti. • *Signature* (1925): Adler collection. • *Note 1*: As a sign of nationalist affirmation, during the 19th Century many Brazilians changed their names to those of rivers, regions, and ancient tribal words. Vital Brazil was the first child of José and Mariana dos Santos Pereira Junior and the name they chose for him was to show his stength of life (Vital) and to honor his country (Brazil), his state (Minas Gerais), and his place of birth (Campanha). In 1865 the name of the country was spelled with a z. • *Note 2*: Butantan means "place of strong wind" in Tupy, language of the Tupian people, now extinct, who were once native to the Brazilian coast. (The word "tegu" means lizard in Tupy.)

NOGUCHI, Hideyo (1876–1928).

Hideyo Noguchi (see *Note 1*), a Japanese bacteriologist and specialist on snake venom, was born in Sanjogata (Fukushima Prefecture), a small village in the mountains of northern Japan, on 9 November 1876 (see *Note 2*). He was born to an impoverished family and had the misfortune to fall into the household fireplace as an infant. He severely burned his left hand, producing a deformity that was a handicap throughout his life. Helped financially by friends, he completed his medical examinations in Tokyo (1897) and later took a post at the government's Institute for Infectious Diseases in Tokyo. Noguchi was extremely industrious, disciplined himself to sleep for short intervals when necessary, and had a natural facility with foreign languages. By chance he mentioned to Simon Flexner, an American visitor to the institute, his desire to study in the United States. This led to Noguchi's arrival in December 1900, unannounced and virtually penniless, at the University of Pennsylvania where Flexner worked.

Although he became Flexner's student, it was his initial meeting with S. Weir Mitchell, Flexner's colleague, that introduced him to research on snake venoms. Mitchell, who had done seminal work on snake venoms, believed that further studies using the newer immunological methods could be fruitful and Noguchi quickly accepted the idea. Mitchell

also provided the early funding for the work. Noguchi began his research quickly and in 1902 started to publish his results, jointly with Flexner, on the composition of snake venom. This was followed by Noguchi's own work on the action of snake venom on lower vertebrates (1904). Later he completed his degree at Pennsylvania (M.S. 1907). His crowning achievement during this period, during which he published some 15 papers on snake venom, was his book, "Snake Venoms, an Investigation of Venomous Snakes with Special Reference to the Phenomena of Their Venoms" (1909). This book includes a review of the venomous snakes of the world, the anatomy of the poison apparatus, a summary of all previous work on venom poisoning and composition, and concludes with the results of Noguchi's own immunological research. He reviewed his work on species-specificity of antivenoms and the role of protease in snake venoms. Kyoto Imperial University awarded him a doctorate in the medical sciences for this work (1911).

By this time Flexner had moved to the Rockefeller Institute of Medical Research (now Rockefeller University) as its first director (1904) and Noguchi accepted his offer of a position at the institute which he held for the rest of his life. His research shifted to studies of syphilis and he was the first to determine the causative agent, a spirochete (*Treponema*), in the brains of paretics. In 1914 he was promoted to full membership in the institute and was awarded a doctoral degree (D.Sc. Tokyo). He proceeded to study a variety of other infectious diseases with success, but finally focused on yellow fever, which led to his undoing and to his own death.

Because of his success with *Treponema* he believed that yellow fever might have a similar cause. He conducted several field studies in Ecuador, Mexico, and elsewhere, but he confused the symptoms of another disease with those of yellow fever. He incorrectly concluded that yellow fever was caused by a spirochete-like organism (*Leptospira*) when, in fact, it is caused by a virus. He went to study yellow fever again, in November 1927, this time in the Gold Coast (now Ghana) of West Africa, but he contracted the disease there and died in Accra on 21 May 1928. At the time there was speculation on the manner of Noguchi's fatal infection and whether he courted infection because of despondency or ill health.

Despite his misjudgments, Noguchi made many important contributions and received awards and honorary degrees from governments and universities around the world. The emperor of Japan decorated him twice and, more recently, his country honored him with his image on the 1,000-yen bank note that was issued in 2004. The Hideyo Noguchi Memorial Association still operates a memorial hall opposite the house in which he was born. In Ghana the Noguchi Memorial Institute for Medical Research is named for him.

• *References*: "(Seisaku) Hideyo Noguchi," by C. E. Dolman, Dict. Sci. Biogr., 10: 141–145, 1974; "Noguchi and His Patrons," by I. R. Plesset, Fairleigh Dickinson Univ. Press, Rutherford, New Jersey, 414 pages, 1980; "Hideyo Noguchi," by K. Nakamura, Doctor Noguchi Mem. Assoc., Tokyo, (12), ii, 79 pages, 1983; ["Hideyo Noguchi and Venomous Snakes,"] by M. Toriba, Japan Snake Inst., Nitta-gun, 16 pages, 1990 (in Japanese); "Hideyo Noguchi," by D. M. Fox and M. L. Meldrum, Amer. Natl. Biogr., 16: 473–474, 1999. • *Portrait* (1903): Carnegie Institution of Washington. • *Signatures* (Japanese form, undated; Romanized form, 1909): Hideyo Noguchi Memorial Association, Tokyo. • *Note 1*: Noguchi's original given name was Seisaku, which, according to custom in Japan at the time, was changed when he reached manhood (1898). • *Note 2*: Noguchi's birth date is often given as 24 November. He was born on 24 September according to the lunar calendar, which in 1876 corresponded to 9 November in the solar calendar.

TAGO, Katsuya (1877–1943).

Katsuya Tago, who wrote the first book on all of Japan's salamanders, was born on 26 May 1877 in Misaka-mura (now Iwaki, in Fukushima Prefecture) in the north of Japan's largest main island, Honshu. He began his salamander studies in 1903 at nearby Nikko, today site of Nikko National Park in the mountains about 120 km north of Tokyo. He studied with the zoologist, Seitaro Goto, at Tokyo Imperial University (now University of Tokyo) and graduated in 1904 with a bachelor's degree. Tago then joined the staff of the fisheries bureau in the government's Ministry of Agriculture and Commerce. Later he was a professor at the Imperial Fisheries Institute (now Tokyo Kaiyo University) until he retired in 1925. In the early 1930s he became a professor at the Imperial (now Tokyo) Women's Medical College in Omori, located in the southern part of Tokyo, and then professor of medicine at Tokyo's Nihon University, now Japan's largest university.

Tago also had a long association with the Interior Ministry where he served on a committee that established "natural monuments." One such designation made during Tago's ten-

ure on this committee was the Japanese Giant Salamander (*Andrias japonicus*). This had the perverse effect of ending ongoing research on this species, but more enlightened recent interpretations of the law have permitted new studies that will aid in rescuing this animal from extinction. Tago conducted fieldwork on salamanders throughout Japan and in 1925 also visited Japanese-held Korea and Manchuria. He kept a large giant salamander alive at his home in Tokyo, which so embarrassed his wife that she told neighbors he was studying whales! Tago died in Tokyo on 19 December 1943.

Tago authored about ten herpetological titles (1903–1930 [1931]), all of them on salamanders. He made numerous studies of salamander life history including egg-laying and feeding habits. He had a special interest in the biology of the giant salamander and was the first to report its presence on Kyushu, the southernmost of Japan's main islands. His last publication was a book, "Newts and Salamanders" (1930 [1931]), in Japanese, in which he reviewed the 18 species then known from Imperial Japan, which included Formosa (now Taiwan) and Korea. The accounts give extensive details of life history, photographs of habitats, and have maps showing distribution. The 28 plates, of which 22 are colored and carefully drawn, depict the external anatomy of adults as well as some egg masses and larvae. All four of Tago's new species of *Hynobius* are included in the book, but only two of them (*H. dunni, H. tokyoensis*) are recognized today. His book was soon superseded by Ikio Sato's more elaborate work on the salamanders of Imperial Japan (1943), which covers 25 species.

• *References*: ["Obituary of Professor Katsuya Tago,"] by H. Takashima, Zool. Mag., 56: 74–75, 1944; Ikoma, 1973 (Tago on p. 231); ["Biography of Tomizo Yoshida (Tago's brother-in-law),"] by N. Yoshida, Bungei Shinjyu Co., Tokyo, 310 pages, 1992 (Tago on pp. 75–84). • *Portrait*: From Ikoma, 1973. • *Signatures* (Japanese version): From Ikoma, 1973; (Romanized version, 1930): Adler collection.

ROSÉN, Nils (1882–1970).

Nils Valfrid Rosén, a Swedish zoologist and fisheries administrator, was also a specialist on reptiles early in his career. He was born on 11 June 1882 (see *Note*) in Hälsingborg (now Helsingborg, Skåne County) in southern Sweden. He attended Lund University (B.A. 1903) where he later wrote a dissertation on parasitic worms (Fil.Dr. 1910). He held various positions at Lund as research assistant, associate professor of zoology, and as teacher in a training college (1903–1914). After holding a scholarship in fisheries during 1910–1914, he became inspector of fisheries, first for the upper northern district of Sweden (in Luleå) and then the west coast district (in Göteborg), after which he retired in 1949. During this period he held several administrative positions. He also published his own research on fishes and wrote several books including a Swedish fisheries dictionary. He died on 1 May 1970 in Ängelholm (Skåne) in southern Sweden.

Rosén published six titles on herpetology (1904–1911). His first notable papers were two (both 1905) listing the specimens of snakes and lizards in the Zoological Museum of Lund University, which had accumulated them over many

years and from all parts of the world. Rosén had little training in systematics and none in herpetology, but he proceeded to describe many new species. George A. Boulenger of the British Museum, who was then the leading authority on the snakes of the world, immediately criticized his paper on snakes. He remarked (1905): "The naming of snakes from [published] descriptions is a difficult task for beginners, who should not be encouraged to publish descriptions of so-called new species, which only go to swell the synonymy." He estimated that probably none of Rosén's species was valid. Rosén quickly replied (1905), but the damage was done. In fact, of the seven new snakes named by Rosén, one small Australian elapid, *Denisonia* (now *Suta*) *fasciata*, is recognized today. The widespread Australian skink, *Egernia inornata*, is the only one of six lizards he named that is recognized. Rosén did publish one other significant paper on reptiles (1911), based on his yearlong expedition to the Bahama Islands in 1908–1909. This is an extensive and competent review of Bahamian reptiles and includes a general account of the fauna and physiography of the islands. It marked his exit from herpetology.

• *References*: "Nils Valfrid Rosén," by T. F[lensburg], p. 342 *in* Svenska män och kvinnor–Biografisk uppslagsbok, vol. 6, A. Bonniers förlag, Stockholm, 1949; "Nils Rosén (1882–1970)," by S. Sahlin, Jour. Conseil Perm. Internatl. Explor. Mer, 34: 3–5, 1971; "Nils Rosén," p. 27 *in* H. Ackefors et al., 100 Years of ICES Personalities from Sweden, Internatl. Council Explor. Sea, Stockholm, 1999. • *Portrait*: From Sahlin, 1971. • *Signature:* Courtesy Richard Wahlgren. • *Note*: Ackefors et al. (1999) give his birthdate as 11 April 1882 without explanation.

KAMPEN, P. N. van (1878–1937).

The Dutch zoologist, Pieter Nicolaas van Kampen, who specialized on the amphibian fauna of the Netherlands East Indies, was born in Amsterdam on 30 June 1878. He attended school in the city and then entered the University of Amsterdam to study biology. He received a doctorate (Ph.D. 1904) for a dissertation on the tympanal region of mammal skulls. That same year he became the chief assistant to Max Weber, professor of zoology, and was appointed a lecturer. From 1905 to 1911 he was in Batavia (now Jakarta), Java, where he primarily conducted fisheries research at the Department of Agriculture. During his last year in the East Indies, he participated as zoologist for the Knut Gjellerup Expedition to northern New Guinea. On returning home in 1911, he became a lecturer once again in the zoology department at the University of Amsterdam, and in 1917 moved to the University of Leiden as a professor of zoology where he taught zoology, comparative anatomy, and physiology. At Leiden he was largely responsible for modernizing the Zoological Laboratory, despite resistance from the university administration. Struck by serious illness in 1929, which forced him to retire in 1931, he endured physical and mental suffering for many years before his death in Leiden on 3 July 1937.

Van Kampen's herpetological titles number about 20 (1905–1931), all but one on the amphibians of the East Indies. These are primarily taxonomic in focus. His interests also included life history (especially tadpoles) and zoogeography. He named numerous new species including many that are currently recognized, such as *Calluella volzi* (Sumatra), *Limnonectes microtympanum* (Sulawesi), *Philautus jacobsoni* (Java), and *Xenobatrachus giganteus* (New Guinea). He summarized his work in the East Indies in his book, "The Amphibia of the Indo-Australian Archipelago" (1923), which was a companion to Nelly de Rooij's two volumes on the reptiles (1915, 1917). His book covers 47 genera and 256 species of frogs and caecilians, but this is a fraction of the rich amphibian fauna now known to inhabit the region. Van Kampen's sole contribution to local Dutch herpetology was his booklet on amphibians and reptiles in the "Fauna van Nederland" series (1927, co-authored with Jacobus Heimans).

• *References*: "Dr. Pieter Nicolaas van Kampen," p. 129 *in* Wie is Dat?, M. Nyhoff, vol. 1, M. Nijhoff, The Hague, 1931; "In Memoriam Prof. Dr. P. N. van Kampen 30 Juni 1878–3 Juli 1937," by C. J. van der Klaauw, Leidsch Univ., 7(1): 1–2, 1937; "In Memoriam P. N. van Kampen," by H. Boschma, Almanak Leidsche Student., 124: 102–103, 1937. • *Portrait*: From Boschma (1937), courtesy L. B. Holthuis. • *Signature* (1917): Natural History Museum, London, courtesy Andrew F. Stimson.

SCHMALHAUSEN, I. I. (1884–1963).

The evolutionary morphologist, Ivan Ivanovich Schmalhausen (or Shmal'gauzen), who became a victim of Lysenkoism in the late 1940s, made significant contributions to our knowledge of the structure and development of amphibian limbs and vertebrae and of the monophyly of amphibians. He was born in Kiev, Ukraine, on 23 April 1884, the son of one of the founders of paleobotany who inspired him to study nature. Young Schmalhausen graduated from the local gymnasium in 1901 and en-

tered Kiev University where, after two years, he became a student of Alexei N. Severtzov, founder of evolutionary morphology in Russia and son of Nikolai A. Severtzov, a noted explorer and zoologist. Both Severtzovs studied amphibians and reptiles. Schmalhausen's first paper, on the embryonic development of lungs in snakes (1905), which included issues such as ontogeny and phylogeny, set the pattern of his future research. His master's thesis at Kiev, defended in 1914, on the fins of fishes and their phylogenetic development, further displayed what would become a lifelong interest: the appendages of vertebrates and the origin of terrestrial limbs. In the meantime, he had moved to Moscow University in 1912 where Severtzov had gone the year before. There, Schmalhausen earned a doctorate in zoology (Ph.D. 1916) for a dissertation on the development of appendages in amphibians and their significance for the origin of limbs of terrestrial vertebrates.

In 1917 he took a professorship at Yuryev University in Estonia, but when German troops occupied that area the faculty were evacuated to Voronezh University located some 500 km south of Moscow. By 1920 he had returned to his *alma mater* in Kiev as a professor. After Severtzov's death, Schmalhausen moved to Moscow to become professor and director of the Institute for Evolutionary Morphology, which Severtzov had founded just before he died in 1936. By 1939, Schmalhausen was chairman of the new Department of Darwinism at the university and, through his research and training of students, had become one of the leading scientists in the Soviet Union. He had been elected to the Academy of Sciences in 1935. He was attempting nothing less than a holistic concept of an organism, a synthesis between development and morphology on the one hand with genetics and evolutionary theory on the other. This period, however, became the zenith of his career, for late in 1948 he was removed from his positions including his professorship. He was later assigned to a new position administered by the Zoological Institute of the Academy of Sciences in Leningrad (now St. Petersburg), but permitted to continue working at the Institute for Evolutionary Morphology in Moscow.

By the early 1930s Mendelian genetics research in the Soviet Union had become quite advanced, but work by Trofim Lysenko, an agricultural biologist who studied the effects of temperature on plants, caught the attention of the Soviet leadership and got the support of Joseph Stalin. Lysenko's research, which even denied the existence of genes and chromosomes, suggested that new "forms" could be created in a few, or even one, generation—the inheritance of acquired characteristics—and, importantly, it conformed to communist ideology, which was interested in molding individuals into communists. Under extreme pressure, the Academy of Sciences was forced to endorse Lysenko's ideas, which dominated Soviet genetics from 1934 to 1964. It was dangerous to criticize him or his views on genetics. Nikolai Vavilov, the founder of modern genetics in the Soviet Union, did so and was sent to prison in 1940 and died there three years later. Schmalhausen's "crime" was that he discussed Western ideas of genetics yet ignored Lysenko in his classic book, "Factors of Evolution" (1946 [1947]; English edition 1949, new printing 1986).

All Mendelists were at risk. As the leading Soviet evolutionary biologist, Schmalhausen was publicly attacked by Lysenko in *Pravda* and, shortly thereafter, at a meeting of the Lenin All-Union Academy of Agricultural Sciences early in August 1948. On 23 August 1948, the Soviet Minister of Higher Education issued Order 1208. This led to mass dismissals of university professors. The next month the minister wrote in *Izvestia* (in translation): "Academician Schmalhausen denies the inheritance of acquired characters. He finds that evolution depends upon mutations which originate directly in the germ cells of the organisms . . . ." Schmalhausen was arrested and put on trial. He escaped imprisonment by claiming that he was not a geneticist and that his views were outdated. Nevertheless, his books and research were destroyed and his assistants dismissed. All of his positions were revoked.

With the downfall of Nikita Khrushchev, who was a strong supporter of Lysenko, in October 1964, Lysenkoism came to an abrupt end, but too late to benefit Schmalhausen. He had died a year earlier, on 7 October 1963 in Moscow. None of the obituaries of Schmalhausen published in the Soviet Union just after his death mentioned his removal as professor and director. Indeed, they do not even use the word "genetics." Within months of Khrushchev's ouster, modern genetics research was again acceptable in the Soviet Union. Lysenko was dismissed from the directorship of the Academy of Science's Institute of Genetics in February 1965.

In 1948 Schmalhausen re-focused his work on comparative and evolutionary embryology, a politically neutral area of research. He continued to do important work using fishes, amphibians, and reptiles as he had done long ago in Kiev. Besides his research on limb and vertebral development (especially the centrum), he studied regeneration of limbs, respiration, homologies of the bones of the skull, the axial skeleton of salamanders, transformation of the visceral arch, sound transmission in salamanders, and the issue of monophyletism in the origin of the first terrestrial vertebrates, the amphibians. The last was a fundamental question that had been challenged during the 1930s and 1940s by two Swedish paleontologists—Nils Holmgren and Erik Jarvik—who claimed that amphibians had multiple origins among the fishes and thus were polyphyletic. Schmalhausen was uniquely qualified with his knowledge of evolutionary morphology to address this question and he did so in several papers (four titles in particular, 1957–1961), which demonstrated that amphibians were monophyletic. Others later drew similar conclusions, most notably Henryk Szarski in Poland (1961). Schmalhausen wrote numerous books. Those of greatest interest to herpetologists are "Factors of Evolution," noted earlier, and "The Origin of Terrestrial Vertebrates" (1964; English edition 1968). Hynobiid salamanders figure in a major way in the latter book.

• *References*: ["In Memoriam Ivan Ivanovich Schmalhausen (1884–1963),"] by M. S. Gilyarov and A. A. Makhotin, Zhur. Obshch. Biol., 25: 78–79, 1964; ["Academician Ivan Ivanovich Shmal'gauzen (1884–1963),"] by L. P. Tatarinov and B. A. Trofimov, Paleont. Zhur., 1964(2): 169–173, 1964 (English translation in Internatl. Geol. Rev., 7: 1682–1686, 1965); "Foreword, 1986," by D. B. Wake, p. v–xii *in* I. I. Schmalhausen, "Factors of Evolution," Univ. Chicago Press, 1986; ["Ivan Ivanovich Shmal'gauzen, 1884–1963"] by O. I. Shmal'gauzen. Nauka, Moscow, 254 pages, 1988; "Schmalhausen's Evolutionary Morphology and Its Value in Formulating Research Strategies," by D. B. Wake, Mem. Soc. Ital. Sci. Nat., 27: 129–132, 1996; Rossolimo and Dunayev, 2003 (Schmalhausen by E. I. Vorobjeva on pp. 494–517). • *Portrait and signature*: Courtesy Evgeniy Dunayev.

DICKERSON, Mary C. (1866–1923).

Mary Cynthia Dickerson, an American herpetologist, writer, nature photographer, and editor, founded the Department of Herpetology at the American Museum of Natural History and helped to launch the careers of several leading herpetologists. She was born on 7 March 1866 in Hastings, about 60 km west of Lansing in southern Michigan. Education was not a family tradition and she had to finance her own way through college, thus her attendance at the University of Michigan during 1886–1891 was intermittent. She taught high school in Michigan and Illinois to save money to complete her education at the University of Chicago during 1895–1897 (B.S. 1897), to begin graduate study there, and to attend classes in marine biology at Woods Hole. During 1897–1905 she taught biology at the Rhode Island Normal School (now Rhode Island College), a teacher's preparatory college. She resigned her administrative and teaching responsibilities in May 1905 and devoted her time to writing and publication before moving to California in September 1907 where she taught ecology for a year at Stanford University. There she also had a research collaboration with the ichthyologist, David Starr Jordan.

During her time in Rhode Island, Dickerson did the fieldwork, research, writing, and photography to produce two outstanding nature books, "Moths and Butterflies" (1901) and "The Frog Book" (1906; reprinted 1969); the latter covered the habits and life histories of the frogs and toads of the northeastern United States. In 1905 she presented an important collection of frog specimens to the American Museum, which reimbursed her for expenses. Several times during 1907–1908 she queried the museum's director and occasional writer on reptiles, Hermon C. Bumpus, about possibilities for employment, which led to her being hired. At the age of 42 she became an assistant in the Department of Woods and Forestry in November 1908, and by the following July she was also the assistant responsible for amphibians and reptiles in the newly formed Department of Ichthyology and Herpetology. In November 1909 she took on yet other duties as associate editor of the *American Museum Journal* (which in 1919 and under her editorship became *Natural History*). Thus she held two curatorships and an editorship concurrently. Yet she was still able, through determination, political skill, and some stubbornness, to develop a new department for herpetology that was established as a separate unit in February 1920 with her as curator. The museum already had herpetological collections from its inception in 1869 when it purchased the Brazilian and North American materials collected by Alexander Philipp Maximilian, prince of Wied-Neuwied, but Dickerson realized that great museums must continue to build their collections to meet the needs of research.

In developing this program, during 1916–1917 Dickerson astutely hired a number of assistants to help build the collection, among them a triumvirate of young college-trained herpetologists who later would be leaders in the discipline: Charles L. Camp from Columbia University, G. Kingsley Noble from Harvard and Columbia universities, and Karl P.

Schmidt from Cornell University. She also supported and mentored another future leader, Emmett R. Dunn, who had just graduated from Haverford College, and she sent him on his first expedition, to collect salamanders for the museum in the mountains of North Carolina.

Tragically, Dickerson's tenure as curator of the new department lasted less than a year. Some years earlier she had exhibited signs of mental disturbance brought on by overwork including hallucinations and erratic behavior. The museum's director begged her to take a vacation and finally suspended her in November 1920 to force her to rest. When she refused to stop working, police were summoned to remove her for psychiatric observation, but she returned to the museum in a deranged state in December and was then institutionalized at a hospital on Ward's Island (now the Manhattan Psychiatric Center) in the East River of New York City. Noble took over the department, temporarily at first, but Dickerson's health steadily deteriorated and she died on 8 April 1923 at the age of 57. The stated cause of death was chronic interstitial nephritis. She never married.

Dickerson wrote well on many subjects. Besides her book on anurans, which was a popular success, she authored seven herpetological papers (1911–1919), in which she discussed museum methods and described new species—an amphisbaenian and 23 lizards—from the Dominican Republic and Baja California, respectively. She developed *Natural History* into one of the finest museum magazines in the world. In addition she was responsible for conceptualizing the museum's new exhibits of amphibians and reptiles, and this sometimes required her to go into the field, which she usually did alone, to get photographs and notes from which to make the displays. Her last such trips were to Florida in 1916 in preparation for the cypress swamp forest display that was opened in 1918. As she wrote at the time, "No pictures and no reading can carry the effect . . . to the mind. It might be on another planet, so different it is from anything else on this globe . . . but nothing has prepared [me] for the influence of the Florida cypress swamp in the full sunshine of the afternoon." She pioneered many new exhibition techniques, especially those showing underwater scenes. Mary Dickerson had accomplished a lifetime's work at the museum in a scant dozen years.

• *References*: "Mary Cynthia Dickerson 1866–1923," articles by M. Slye, J. O. La Gorce, G. K. Noble, and B. Moore, Nat. Hist., 23: 506–519, 1923; "Mary Cynthia Dickerson," p. 87 *in* M. J. Bailey, American Women in Science, ABC–CLIO, Denver, Santa Barbara, and Oxford, 1994; Myers, 2000 (Dickerson on pp. 8–32, 64–80, 117–119, 169–172, 203–206). • *Portrait* (about 1912): Courtesy Charles W. Myers. • *Signature* (1909): Adler collection.

BOLKAY, Stephan J. (1887–1930).

The Hungarian herpetologist and paleontologist, István József László Bolkay (who usually published under the name "St[ephan]. J. Bolkay; see *Note*), was a specialist on European amphibians and reptiles who had a short and tragic life. He was born the son of a carpenter on 29 March 1887 in Rimaszombat (now Rimavská Sobota) about 120 km northeast of Budapest and now in Slovakia. He had an interest in nature from his childhood especially in insects and salamanders, and he conducted studies with them at his local school. He finished his secondary schooling in 1905 and proceeded to Péter Pázmány (later Budapest and now Loránd Eötvös) University where he completed his dissertation on frog tadpoles (Ph.D. 1909). This research was conducted under his mentor, the eminent zoologist Lajos Méhelÿ, who was then Curator of Herpetology at the Hungarian Natural History Museum. Bolkay began to study the systematics of European *Rana* and was the first to adequately distinguish *R. ridibunda* from *R. esculenta*, based primarily on osteology (1907, 1908). He then extended this work to a Chinese frog that had long been thought to be just a geographic race of *R. esculenta* (1909, 1911).

During the summer of 1909 Bolkay traveled to Switzerland for fieldwork. There he befriended and began to mentor the son of a Hungarian nobleman—a precocious 15-year-old boy, Géza Gyula Fejérváry—who was to become Méhelÿ's successor as curator in Budapest in 1916. That fall Bolkay became Méhelÿ's assistant at the museum and broadened his research to include fossil frogs. Because of his erratic and careless behavior, however, Bolkay developed much opposition at the museum and in 1911 he was fired. Méhelÿ tried to find him another position in Budapest, but only in 1913 did Bolkay get a permanent one, as a biology assistant at a teacher's training school where he was allowed to keep animals in terraria and conduct a modicum of research. World War I soon began and, since he was a reserve officer, Bolkay was called up in 1915. After the war, in 1919, he was able to get a curator's position at the State Museum of Bosnia and Herzegovina in Sarajevo. This was an outpost remote from the scientific centers of Europe and with a hopeless financial situation. Despite the adversity, Bolkay had an active research program that now included mammalogy. He even had to set the type and print his own papers. His supervisors forced him to publish in Croatian, which further isolated Bolkay from his international colleagues. On 17 August 1930 he shot himself and died in Sarajevo. He was 43 years old.

Despite the many misfortunes of his life, Bolkay was a gifted and productive herpetologist who had a broad perspective on his subjects including their evolution and adaptation. He authored about 65 herpetological titles (1907–1933) on a wide range of subjects, among them the systematics of European frogs, salamanders, lizards, and snakes; the Pleistocene herpetology of Hungary; comparative osteology of frogs; and the amphibians and reptiles of the Balkan Peninsula. He also described several new taxa of European amphibians and reptiles, both living and fossil.

• *References*: ["† Dr. Stjepan Bolkay,"] by R. Zaplata, Glas. Zemalj. Muz. Bosni Hercegov., 42: 191–199, 1930; "Stephan J. Bolkay," by G. J. Fejérváry, Verhandl. zool.-bot. Ges. Wien, 82: 34–51, 1932; "Die wissenschaftliche und literarische Tätigkeit von István (Stephan) Bolkay auf dem Gebeite der Zoologie," by O. G. Dely, Vertebr. Hung., 11: 11–31, 1969 (see also pp. 33–125). • *Portrait and signature* (1921): Courtesy O. G. Dely. • *Note*: The Slavic form of his name is Stephan (sometimes Stefan or, in Croatian, Stjepan) Josef Ladislaus Bolkay.

NIEDEN, Fritz (1883–1952).

The German herpetologist, Friedrich Erich Gottlieb Nieden, generally known as Fritz Nieden, was born to a Protestant family on 3 March 1883 in Elberfeld, now part of Wuppertal (North Rhine–Westphalia) about 35 km east of Düsseldorf. His father was a physician. Fritz graduated from the gymnasium in Elberfeld in 1901 and went to Bonn to begin his university training in natural sciences. After one year, he transferred to the university in Freiburg where he completed his dissertation on sexual dimorphism in the antennae of Lepidoptera (Dr.phil. 1907). In May 1907 he began work as an unpaid trainee in the Museum für Naturkunde, a part of the University of Berlin, and by April of the next year he was a salaried assistant. How he came to work in the herpetological collections is not known, but perhaps it was simply an available position. At that time these collections were curated by Gustav Tornier who specialized in African herpetology, especially the reptiles of the German colonies: German East Africa, German South-West Africa, and the two colonies in West Africa, Cameroon and Togoland. (For details on these colonies, see p. 127.)

Nieden's interests in amphibians complemented Tornier's interests in reptiles and of those of Richard Sternfeld who was informally associated with the collections from 1907 to 1913. Nieden's herpetological titles (about 25, 1907–1926) are almost entirely on materials from the four German colonies. He described many new species of frogs and caecilians, but also of African chameleons and a new teiid lizard from Venezuela. His official salaried work at the museum ended in October 1911, although he continued to work with the collections until 1915 when he was drafted into the army for service in World War I. He apparently returned to work as a volunteer in the museum after the war ended in 1918, but about 1920 he moved to Düsseldorf, not far from his birthplace. By 1936 he was retired and living in Schmalkalden (Thuringia) where he died on 16 August 1952.

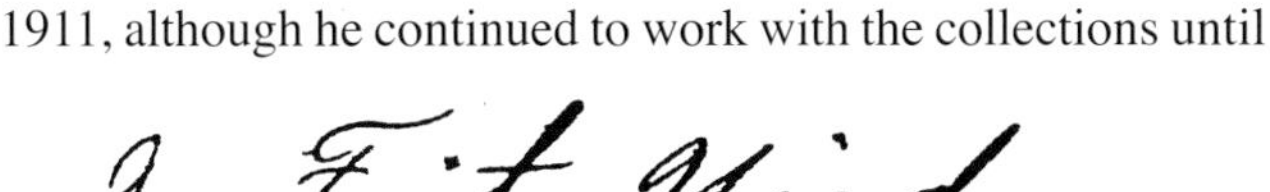

Nieden's major works include several long papers on the amphibians and reptiles of Cameroon (1910) and German East Africa (1913, 1915), except for the snakes, which were covered by Sternfeld. He published on the amphibian collections from the 1907–1908 expedition sponsored by Adolf Friedrich, Duke of Mecklenburg, to the Lakes Region of Central Africa including present-day Burundi, Rwanda, and southwestern Uganda (1912); Sternfeld covered the reptiles. Nieden also compiled three sections in the *Das Tierreich* checklists of the amphibians and reptiles of the world: caecilians (1913) and the first two parts on frogs (1923, 1926). (Ernst Ahl, who began work at the museum in 1921 and succeeded Tornier as curator in 1923, compiled the third part on frogs [1931].)

• *References*: "Fritz Nieden Dr. phil.," by M. Vasen, Das freie Wort, Schmalkalden, 1(2): 4, 22 August 1952; Personnel files (Merkblatt 2), Museum für Naturkunde, Berlin, courtesy Sabine Hackethal; Records of Stadt- und Kreisarchiv Schmalkalden, courtesy Ute Simon. • *Signature* (1914): Natural History Museum, London, courtesy A. F. Stimson.

STERNFELD, Richard (1884–1943).

The German herpetologist, Richard Sternfeld, was a specialist on the systematics of the reptiles of German colonies in Africa and the Pacific who died in the Holocaust. He was born on 8 February 1884 in Bielefeld (North Rhine–Westphalia) where his father was a merchant. He was educated at the local gymnasium and in 1903 entered the university in Freiburg to study medicine, then switched to natural science in Bonn before finishing in Freiburg (Dr.phil. 1907). His dissertation on the biology of ephemerids (mayflies) was completed under the noted evolutionary biologist, August Weismann. In the spring of 1907 Sternfeld moved to Berlin to begin his private research in the herpetology collections of the Zoological Museum at the University of Berlin that were curated by Gustav Tornier.

Like Tornier, Sternfeld focused most of his work on German colonies in Africa: German East Africa, German South-West Africa, and the two colonies in West Africa, Cameroon and Togoland. (For details on these colonies, see p. 127.) His technical papers were published in the museum's regular series, the *Mitteilungen*, and in a scientific journal. In six instances the information in them was then summarized by colony, with illustrated keys, maps, and checklists added, and published in the museum's new serial, *Die Fauna der deutschen Kolonien*. This was begun in 1909 to produce booklets for practical use in the colonies. Fritz Nieden, Sternfeld's museum colleague, authored one of the companion sections. Sternfeld also published on collections from Abyssinia, from central, southeastern, and southern Africa, and from German colonies in the South Pacific (German New Guinea, Bismarck Archipelago). He also wrote two papers on mimicry in snakes. From only six and one-half years at the museum in Berlin, Sternfeld produced 26 titles of which several were quite extensive.

In September 1913 Sternfeld began as curator of herpetology at the Natur-Museums Senckenberg in Frankfurt am Main, replacing Philipp Lehrs who had resigned after less than three years in the post. With curatorial duties, Sternfeld now had less time for research, but his geographic scope was wider: Japan, Central Africa, Pacific Islands, Australia, and South America. He was called away for military duty toward the end of World War I and served in Macedonia. As it happened, Sternfeld had little talent for curation and, following an unfortunate quarrel with the museum's director in June 1920, he was fired at the end of the year.

All told, Sternfeld produced about 35 titles on amphibians and reptiles (1908–1925). He also wrote a popular book on the reptiles and amphibians of Central Europe (1912; new edition in 1952 revised by Gerolf Steiner). He described and named several dozen species, mostly lizards and snakes including *Pogona minor, Simoselaps anomalus*, and *Tiliqua multifasciata* from Australia, the Venezuelan *Bothriopsis medusa, Paranaja multifasciata anomala* from Cameroon, and two chameleons from East Africa, *Bradypodion adolfifriderici* and *Chamaeleo schubotzi*. Sternfeld conducted no new research after leaving the museum. He worked as a zoologist in Bialowies and then labored at a munitions factory in Berlin. Because Sternfeld was a Jew, early in March 1943 he was arrested by the Nazis and murdered a short time later at an extermination camp. One of his fellow prisoners, who chanced to escape death, wrote to Robert Mertens, Sternfeld's successor as curator (in translation): "He loved his fatherland, Germany, as does every man of his origin and education, and felt himself to be entirely German."

• *References*: "Die herpetologische Sektion des Natur-Museums und Forschungs-Institutes Senckenberg," by R. Mertens, Senckenb. Biol., 48: 1–105, 1967 (Sternfeld on pp. 16–17); Rieck et al., 2001 (Sternfeld by W. Bischoff on pp. 619–621). • *Portrait*: Museum für Naturkunde, Berlin. • *Signature* (1906): Archives, Albert-Ludwigs-Universität Freiburg.

LAISTER, Alexander F. (1884–1941).

The Armenian geographer and vertebrate biologist, Alexander Filippovich Laister, who specialized on the reptiles of the Caucasus, was born on 4 April 1884 in the village of Nizniye Akhti (or Lower Akhti), now part of Razdan, about 40 km northeast of Yerevan. His father was a local judicial official. Alexander attended the gymnasium in Yerevan, and during this period made numerous field excursions into the Armenian countryside. In 1902 he went to the university in Kharkov, located in northeastern Ukraine about 400 km east of Kiev. There he attended zoology lectures by A. M. Nikolsky, then Russia's leading herpetologist who had just moved to Kharkov from St. Petersburg. After graduating from Kharkov in 1907, Laister conducted fieldwork in Armenia and made his first important herpetological discovery when he found *Vipera ursinii* at Sevan Lake. In 1909 Nikolsky invited him to be his technical assistant and help him to curate the herpetological collections that he was building at Kharkov.

Laister decided to leave Kharkov two years later to accept a position in the Caucasus Branch of the Russian Geographical Society in Tbilisi (sometimes still referred to by its Turkish

name, Tiflis), the capital of Georgia, located on the southern edge of the Caucasus mountain chain which runs along Russia's southern border with present-day Georgia and Azerbaijan. This proximity to the Caucasus gave Laister opportunities to do fieldwork in his free time. During 1931–1934 he became scientific secretary of the Institute for Information on the Caucasus of the Academy of Sciences. He spent one year in Baku, capital of Azerbaijan, located on the southwestern coast of the Caspian Sea, as professor and chair of the physical geography department of Azerbaijan Pedagogical Institute, but returned to Tbilisi in 1937 as a professor of physical geography at Tbilisi State University. He died in Tbilisi on 1 February 1941.

Virtually all of Laister's fieldwork was centered in the Caucasus including the central mountain chain and surrounding lowlands. Despite the fact that he was not employed as a biologist, he authored about ten titles in herpetology (1908–1913). His special interest was vipers and other snakes, but he published on all reptiles, mostly papers on distribution, faunal surveys, and natural history. His major work was a summary of the herpetofauna of the Yerevan region (1912). His book on the birds of Armenia (1942) was published posthumously.

• *Reference*: ["Alexandr Filippovich Laister (1884–1941) and His Role in the Investigation of the Caucasian Herpetological Fauna,"] by I. S. Darevsky, p. 4–19 *in* L. J. Borkin (ed.), Herpetological Investigations in the Caucasus. Proc. Zool. Inst. Acad. Sci., vol. 158, Leningrad, 1986 (1987). • *Portrait and signature*: Courtesy Natalia B. Ananjeva.

LANTZ, Louis A. (1886–1953).

Louis Amédée Lantz was a French amateur specialist on Palearctic salamanders and lizards. He was born in southern Alsace on 20 March 1886, in Mulhouse (Haut-Rhin), about 30 km northwest of Basel, Switzerland. He studied botany and zoology at the university in Montpellier (1903–1904) before attending a chemistry institute in Mulhouse (diploma in chemical engineering, 1907). He took a position as an industrial research chemist in Moscow, which gave him opportunities for fieldwork during his vacation periods beginning in 1908. He also kept salamanders as pets and was particularly adept at breeding and even hybridizing them. When the Russian Revolution drove him to leave, he moved to Manchester, England, in 1918, where he became director of the research laboratory for Calico Printers' Association. He retired from this post in 1950 and had intended to return to France to be associated with the museum in Paris, but he died in Basel on 3 February 1953 after a short illness.

Lantz published about 35 herpetological titles (1909–1954), primarily on salamanders and lizards. His earliest papers, published in *Blätter für Aquarien- und Terrarienkunde*, edited by Willy Wolterstorff, put him in contact with Wolterstorff, a German specialist on salamander systematics and captive care, who mentored him. This connection, in turn, led to his long-term association with Otto Cyrén, a Swedish amateur herpetologist who also happened to be a chemical engineer. Lantz and Cyrén often collaborated in their fieldwork and publications. Lantz's early papers covered the Caucasus and the Transcaspian region. Alone or with Cyrén, Lantz named many new taxa of lizards. He also reviewed the herpetological collections made by Ivan Lepechin in 1768–1769 in various Russian provinces (1922). His most important title from this period was "Les Eremias de l'Asie Occidentale" (1928, published in Tiflis [Georgia] in two parts and then reissued in 1928 as a book).

After his move to Manchester in 1918, Lantz's fieldwork shifted to the Mediterranean littoral and the Pyrenean region. So, too, did his publications, although his interests were broader still and even included papers on the North American *Ambystoma opacum* (1930, 2 papers) and new species of Japanese *Hynobius* (1931). His most surprising discovery, however, was that *Lacerta vivipara* is oviparous in the high Pyrenees Mountains (1927). Lantz's captive breeding studies allowed him to investigate sterility barriers between species and natural hybrid populations (1947), thus completing some of the earlier work by Wolterstorff who had died in 1943. Lantz continued his interest in *Triturus* and other salamandrids and, during his final years, he became convinced of the importance of genetic studies. His last papers, in fact, published in *Journal of Genetics*, dealt with background adaptations and the phenotypes and spermatogenesis of interspecific hybrids of *Triturus*.

• *References*: "L. A. Lantz," by R. Mertens, Aquar. terrar. Zeitschr. (DATZ), 6: 109–110, 1953; "L. A. Lantz," by H. W. Parker, Brit. Jour. Herpetol., 1: 174–175, 1953; "Louis A. Lantz," by H. W. Parker, Copeia, 1954: 79, 1954. • *Portrait* (1936) *and signature* (1936): Natural History Museum, London, courtesy A. F. Stimson.

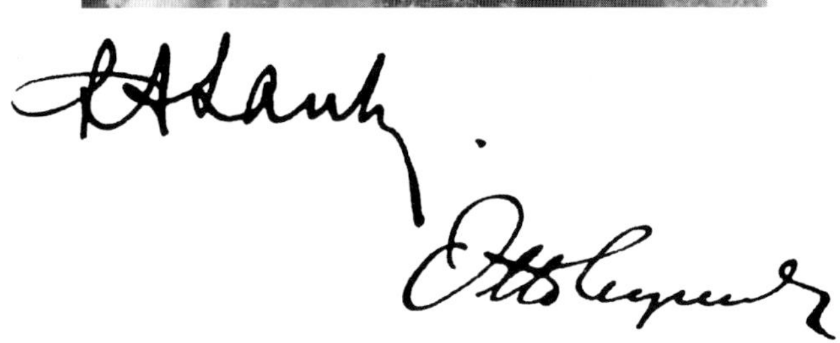

CYRÉN, Otto (1878–1946).

Carl August Otto Cyrén, a chemical engineer and amateur herpetologist, was a specialist on European amphibians and reptiles. He was born on 9 April 1878 in Göteborg, Sweden, and had an interest in reptiles as a boy. He was educated at Chalmers Technical Institute in Göteborg (graduated 1897) and the Polytechnicum in Zürich (graduated 1899). He worked in the chemical industry in Germany, Sweden, and Poland (1899–1917) and then served as ombudsman for Sveriges Kemiska Industrikontor (1917–1944), an association of Swedish chemical industries. Cyrén published extensively on chemistry and related subjects. Because of his devoted interest in herpetology, he traveled annually (except during the wars) for studies of local herpetofauna. His major surveys were in Morocco, Canary Islands, Iberian Peninsula, Madeira, and Italy, as well as the Balkans, Caucasus, and northern Iran. He often conducted his fieldwork jointly with another amateur herpetologist, Louis A. Lantz, who was also a chemical engineer by training. Cyrén's collections were divided between the Riksmuseum in Stockholm and the Göteborg Natural History Museum. He retired to Stockholm where he died on 23 September 1946.

Cyrén authored about 25 technical articles on amphibians and reptiles (1911–1948), seven of them jointly with Lantz. All of his papers deal with European and Southwest Asian herpetology and especially with lacertids. His major works dealt with the influence of climate on lizard distribution (1924), Balkan, Iberian, and Moroccan lacertids (1928, 1933, 1934), and the herpetology of the Balkans (1941). He regularly published short articles in a Swedish nature magazine, *Fauna och Flora*, founded in 1906 by Einar Lönnberg and for which Cyrén wrote 15 papers on herpetology (1909–1946). He was best known to the Swedish general public for his book, "Ormar i fantasi och verklighet" (Snakes in Fantasy and Reality; 1934).

• *References*: "Otto Cyrén," p. 168 *in* Svensk biografisk handbok 1935, P. A. Norstedts & Söner förlag, Stockholm, 1934; "Otto Cyrén 1878–1946," by H. Volsöe (=Volsøe), Fauna och Flora, 41: 241–243, 1946; "Carl August Otto Cyrén," by G[östa] N[otini], p. 155–156 *in* Svenska män och kvinnor – Biografisk uppslagsbok, vol. 2, A. Bonniers förlag, Stockholm, 1991. • *Portrait* (1936) *and signature* (1936): Natural History Museum, London, courtesy A. F. Stimson. For portrait and signature, see page 139.

FITZSIMONS, F. W. (1870–1951).

Frederick William FitzSimons (see *Note 1*), a zoologist and father of two South African herpetologists, was born on 6 August 1870 in what is now Northern Ireland (see *Note 2*). He immigrated to South Africa with his parents (see *Note 3*), to Pietermaritzburg (Natal Colony) in easternmost South Africa. (Southern Africa was then a frontier area, with the British government still establishing control over the Dutch or Afrikaner republics, and this was not resolved until the Anglo-Boer War of 1899–1902.) FitzSimons returned to Ireland for medical training and after three years there decided to devote himself to natural history. He accepted a position in 1895 or 1897 as curator of the Natal Society's museum in Pietermaritzburg where he catalogued and arranged the collections. In 1903 the museum was transferred to the control of the Government of Natal. It was during this time in Natal that FitzSimons's interest in snakes, particularly venomous ones, began.

In 1906 FitzSimons became director of the Port Elizabeth Museum, located in Port Elizabeth along the southern South African coast. There, besides almost single-handedly building and arranging the collections, he began to display living snakes in glass-fronted cages. This interest was further developed and by 1918 he founded The Snake Park, adjacent to the museum and a division of it, which was touted as a major tourist attraction, especially for passengers from ships docked in Port Elizabeth. This was the first snake park in Africa and only the second in the world after Instituto Butantan, which was founded by Vital Brazil in São Paulo in 1901. In order to provide a regular supply of snakes for his park, FitzSimons widely distributed large posters—entitled "Snake Catching. A Comparatively Easy Task"—with prices paid and photographs showing how to collect. FitzSimons was producing snake antivenoms at the park by 1920 and advertised for sale his patented "FitzSimons' First-Aid Snake Bite Outfit" as well as antisera for puff adder, cobra, mamba, and a polyvalent serum that he claimed was appropriate for "all classes of snakes." In 1925 a larger and more elaborate snake park was constructed to his plan. He retired from the museum and park in 1937 and died in Grahamstown on 25 March 1951.

FitzSimons was a prolific author. His few technical papers began in 1909 with a study on bites by the rear-fanged boomslang. He published a number of short papers and small books, often privately, on snake venom and its effect and treatment. He promoted therapeutic uses of snake venom and claimed that neurotoxins, when injected subcutaneously in non-lethal doses, reduced nervous activity and diminished reflex responses. One of his venom products (called "venene") was sold worldwide as a treatment for epilepsy.

It was his many books, however, that made him famous well beyond South Africa. His first and best-known book, "The Snakes of South Africa" (1910), was reissued in revised and expanded editions (1912, 1919, 1926). He wrote books on other vertebrate groups as well—"The Monkey-Folk of South Africa" (1911) and "The Natural History of South Africa," covering mammals and birds in six volumes (1919-1923)—but snakes remained his main focus. He authored "Snakes and the Treatment of Snake Bite" (1929 [Afrikaans edition, 1929]; reissued 1932), "Pythons and Their Ways" (1930), and "Snakes" (1932; German editions in 1934 and 1948).

FitzSimons's two sons, Vivian and Desmond, worked with their father and became herpetologists. Vivian became director of the Transvaal Museum and was author of "The Lizards of South Africa" (1943) and "Snakes of Southern Africa" (1962). His younger brother, Desmond, founded his own snake park in Durban and later directed the Clinsearch Serum Laboratories in Cape Town.

• *References*: "Frederick William Fitzsimons (1871–1938)[*sic*]," p. 125 *in* E. Rosenthal (ed.), Southern African Dictionary of National Biography, London and New York, 1966; "Frederick William FitzSimons," by D. C. FitzSimons, p. 563 *in* D. J. Potgieter (ed.), Standard Encyclopedia of Southern Africa, vol. 4, Nasou Ltd., Cape Town, 1971; "Frederick William Fitzsimons [*sic*]," by J. A. Pringle, Dict. South Afr. Biogr., 3: 299–300, 1977. • *Portrait* (about 1920): Courtesy of William R. Branch. • *Signature* (1909): Natural History Museum, London, courtesy Andrew F. Stimson. • *Note 1*: The family consistently spelled its surname "FitzSimons." • *Note 2*: Biographers differ on his place of birth. According to FitzSimons (1971), he was born in "Garvaghy" (County Down), a parish of about 5,000 inhabitants (as of 2006) located 6 km southeast of Dromore, but Pringle (1977) gives it as "Garvagh" (County Londonderry), a small town of about 1,300 persons (2001) that is 18 km south of Coleraine. • *Note 3*: The family's date of emigration is given as 1877 (Pringle 1977) and 1881 (FitzSimons 1971).

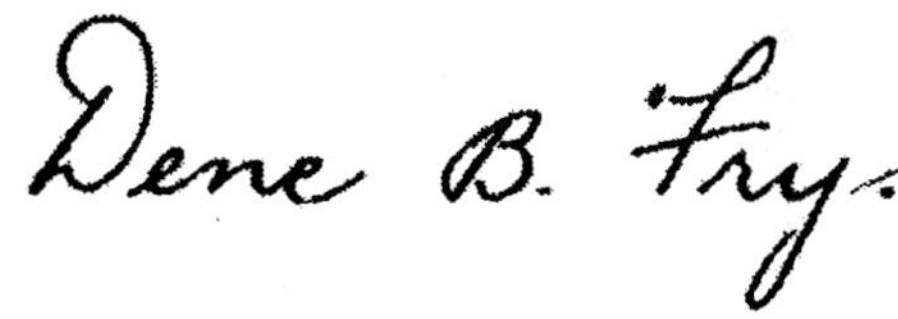

FRY, Dene B. (1893–1917).

Dene Barrett Fry, an Australian herpetologist who had just begun an outstanding career, died at the age of 23. He was born on 11 October 1893 in Lewisham, one of the inner western suburbs of Sydney. He attended school at Lindfield, Chatswood Public, and Fort Street High School, all in Sydney. It was in the rocky gullies of then-rural Lindfield that he developed his love for nature. He later attended Sydney Technical College and also was trained in drawing under Julian Ashton, an Australian artist who established his own art school, which is today an iconic Sydney institution. In 1908 Fry was one of eight boys appointed to the first group of so-called cadets who assisted the scientific staff at the Australian Museum. (Another member of this first group was J. R. Kinghorn, who later became the museum's curator of amphibians and reptiles.) In 1912, on application, Fry was promoted to junior assistant. He resigned at the end of 1913 to become a student at the University of Sydney where he had a teaching post as Demonstrator in Zoology.

In May 1915, with World War I underway, Fry enlisted in the Australian Army Medical Corps and served as a laboratory assistant aboard the hospital ship "Karoola," which saw duty off western France and in the Aegean Sea near Turkey. The Turkish campaign, which took place on the peninsula of Gallipoli (April 1915–January 1916), involved a combined British Empire and French invasion that ultimately failed, leaving 505,000 soldiers dead and 262,000 wounded; many of the latter were taken aboard hospital ships. As Fry described it in a letter home, "Gallipoli is hell personified." His duties were primarily bacteriological and microscopical, but he regarded them as being ". . . of the greatest value to a biologist." On returning to Sydney he was transferred to the infantry and, as a member of the Australian Imperial Forces, was trained at Duntroon, Australia's most prestigious military academy, in Canberra. He sailed again for Europe in August 1916, participated in several engagements in France, and was killed in action at Hermies (Pas-de-Calais), 25 km southeast of Arras in northern France, when his company rushed a German position on 9 April 1917. He is buried in nearby Beaumetz-les-Cambrai.

Fry had a short yet brilliant career. He produced seven herpetological papers (1912–1916 [1917]), an output that is all the more remarkable when one recalls that he joined the medical corps in May 1915. His interest in science never wavered, however. During a shore leave of just a few days in 1915, he went to the British Museum but had the ". . . bad luck to miss Dr. [George A.] Boulenger" who was away. Fry described new taxa of frogs, lizards, snakes, and a turtle. In his very first paper he named a new genus of microhylid frog (*Austrochaperina*) and three new and still-valid species. This paper as well as many later ones included excellent illustrations drawn by Fry. In 1913 he showed that an enigmatic sea turtle, *Chelonia* (now *Natator*) *depressa*, described by Samuel Garman in 1880, was a valid taxon, and in other papers he re-described other poorly known species. In his later papers he named 15 new taxa of which seven are valid today: the frogs *Limnodynastes interi-*

oris and *L. terraereginae* of Australia and *Aphantophryne pansa* of New Guinea; the Australian skinks, *Egernia formosa* and *Lerista picturata*; and the Australian pygopodid, *Aprasia repens,* and gecko, *Gehyra punctata*. This is exceptional productivity for a "junior assistant" whose publishing career spanned barely six years, several of them disrupted by war duty.

• *References*: ["Dene Barrett Fry,"] by H. G. Chapman, Proc. Linnean Soc. New South Wales, 43: 9–10, 1918; ["Dene Barrett Fry,"] by E. H. Gurney, Proc. Royal Soc. Queensland, 30: 3–4, 1918; "Dene Barrett Fry," p. 178 *in* Book of Remembrance of the University of Sydney in the Great War, 1914–1918, Univ. Sydney, Sydney, 1939; "Rare and Curious Specimens," by R. Strahan, Australian Mus., Sydney, (3), ii, 173 pages, 1979 (Fry on p. 53); Additional information from Harold Cogger, 2006. • *Portrait* (about 1916) *and signature* (1912): Australian Museum, courtesy Harold Cogger.

IHERING, Rodolpho von (1883–1939).

Rodolpho Theodoro Gaspar Wilhelm von Ihering, Brazilian zoologist and fish culturist, was born on 17 July 1883 in Taquara just northeast of São Leopoldo (Rio Grande do Sul) in southernmost Brazil. His early education was from his father, Hermann von Ihering (see p. 108), a German zoologist who became director of the Museu Paulista in 1893. The family thus moved to São Paulo where Rodolpho attended secondary school and graduated in 1901. He soon became an assistant to his father at the museum but went to Europe in 1905 to study in Heidelberg and again, in 1911, in Naples and Vienna. In 1916, when his father was dismissed as museum director because of anti-German sentiment in Brazil during World War I, in solidarity Rodolpho resigned his post at the museum. For the next decade he ran a small manufacturing enterprise and also published natural history literature for school children. In 1926 he joined the staff of the Instituto Biologico in São Paulo and in 1934 became head of the zoological section.

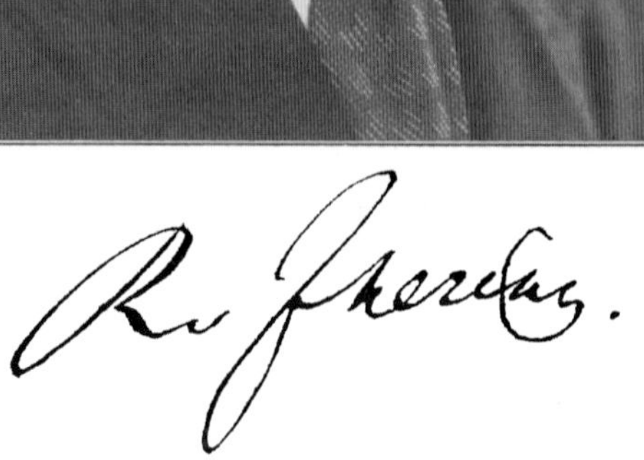

During his years at the museum, Ihering's research had dealt primarily with vertebrates and insects, but beginning in the late 1920s his interests shifted to fisheries. In 1933 he organized a government commission to improve the fisheries of the semi-arid regions in northeastern Brazil. He favored using native fishes over introduced species and pioneered the use of pituitary gland extracts to induce spawning. In 1937 he founded and became head of the national fisheries service, Serviço Nacional de Piscicultura, a position he held at the time of his death in São Paulo on 15 September 1939.

Like those of his father, Rodolpho's scientific interests were broad. Together they wrote a major catalogue of Brazilian birds (1907). Rodolpho authored several major works including the "Dicionário da Fauna do Brasil," a massive summary of the country's animals (serialized 1931–1939, 1940). His specifically herpetological output was relatively modest and consists of just three papers (all 1910 [1911]). These included the first parts of two projected works, one on the amphibians of Brazil and the other on Brazilian snakes. Neither was completed, but these initial parts are substantial contributions to Brazilian herpetology and contain descriptions of new species. His other paper is a summary of the herpetofauna (caecilians and snakes) inhabiting the floating islands of hyacinths in the Rio de La Plata.

• *References*: "Rodolpho von Ihering 1883–1939," by C. Pereira, Arq. Zool. Est. São Paulo, 1 (also Revta. Mus. Paulista, 24): 491–502, 1940; "A Permanência de Rodolpho von Ihering," ed. by M. Pinto Paiva, Fund. Brasil. Conserv. Nat., Rio de Janeiro, 212 pages, 1984. • *Portrait*: From Pereira, 1940. • *Signature*: Bibliothèque Nationale de France, Paris.

GOMES, J. Florencio (1886–1919).

João Florencio de Salles Gomes, a Brazilian expert on snakes, was born on 2 or 3 September 1886 in Tatuhy (now Tatuí), about 120 km west-northwest of São Paulo. He attended primary school locally, graduating in 1896, and then went far away for his secondary education at the Colégio São José, a venerable school run by his uncle in Salvador (Bahía). He finished in 1903 and, intending to be a physician like his father, he promptly enrolled at the Faculdade de Medicina da Bahia. He soon transferred to the Faculdade de Medicina do Rio de Janeiro (M.D. 1910). After completing a microbiology course with Oswaldo Cruz, Gomes became an assistant at the Instituto Serumtherapico Butantan, the snake research institute near São Paulo, which had been founded by Vital Brazil only nine years earlier and where the first antivenoms to pitviper bites were produced. At Butantan his primary duty was snake systematics. In addition he collaborated with Arthur Neiva in research on the bot fly, a skin parasite on humans and other mammals.

Gomes curated the snake collection in Butantan's museum, which he built up to nearly 1,850 specimens by the time of his tragically early death. He also studied and organized other major collections of snakes, notably those in the Museu Paraense (Pará), Museu Paulista (São Paulo), and Museu Rocha (Ceará) in Brazil and in the museums of Buenos Aires and La Plata in Argentina. A man of extremely modest character, Gomes was also an indefatigable worker both in the laboratory and on his numerous field excursions, yet he nevertheless succumbed to influenza during the pandemic of 1918–1919. He died in São Paulo on 29 May 1919 at the age of 32 and within the very age group that is known to be most susceptible to death from the virus.

Gomes authored four herpetological titles (1913–1918) but these are substantial contributions containing descriptions of new taxa and of poorly known species, with distributional data and highly detailed illustrations. Vital Brazil had shown that venoms varied by species, so it was critical to know which species had bitten a patient in order to select the correct antiserum to administer. Therefore the work of Gomes—to accurately identify, describe, map, and illustrate all of the species—was essential. He had planned to prepare a comprehensive work on the snakes of Brazil, and had accumulated extensive data and notes but did not live to complete it.

Afrânio do Amaral was Gomes's assistant and successor at Butantan. Amaral published two new species of snakes (1921) using his predecessor's notes and credited authorship to Gomes, but falsely claimed that these were Gomes's only unpublished works. Amaral then proceeded to publish solely as his own work several papers describing new genera and species based on other notes and unfinished manuscripts that had been left by Gomes. Friends of Gomes attempted to set the record straight in an anonymous pamphlet published in São Paulo in 1944. They reproduced facsimiles of Gomes's handwritten notes and compared them directly to relevant text pages in Amaral's papers published in 1921–1926. They demonstrated not only that Amaral had plagiarized Gomes, but that he had altered some of the original data apparently to cover his actions. This booklet was reissued in 1945 in an expanded version, but with the names of the authors added—Alcides Prado and J. B. Arantes, both staff members at Butantan. In view of these revelations, it seems appropriate to credit much of the early work of Amaral as part of the scientific legacy of Gomes, although technically authorship of the new names remains with Amaral.

• *References*: "Dr. João Florencio Gomes," by V. Brazil, Ann. Paul. Med. Cirurg., 10: 122–124, 1919; "Dr. João Florencio de Salles Gomes," by A. de E. Taunay, Revta. Mus. Paulista, 11: 563–575, 1919; "Afranio do Amaral Apropriou-se do Acervo Científico Deixado por João Florencio Gomes," by A. Prado and J. B. Arantes, privately printed, São Paulo, 32 pages, 1945. • *Portrait*: From Taunay, 1919. • *Signature* (1915): Instituto Butantan, courtesy Denise Maria Mariotti.

POWER, John H. (1884–1964).

John Hyacinth Power, a South African herpetologist and archaeologist, was born on 2 November 1884 in Waterford (Ireland), then part of the United Kingdom. Trained in Dublin, he immigrated to South Africa in 1903 to teach at Christian Brothers College, a boy's school in Kimberley, capital of the

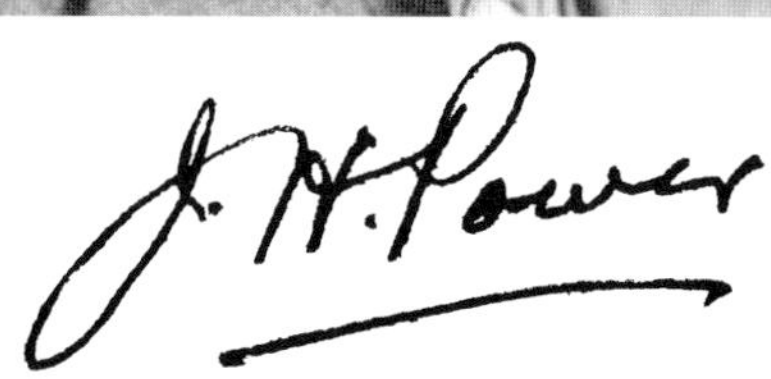

Northern Cape Province and center of one of the largest diamond-mining regions in the world. Powers's interest in herpetology began in 1905 when the herpetologist, Louis Péringuey, later director of the South African Museum in Cape Town, encouraged him to study amphibians and reptiles. Power had always been an enthusiastic collector, but now there was a focus to his work. He conducted extensive fieldwork on his own time and coincidentally collected fossils, stone implements, and other archeological material that he found. In 1920 he was named principal of the South African School of Mines (later the Northern Cape Technical College) in Kimberley. Since its founding in 1908, Power also had been informally associated with the Alexander McGregor Memorial Museum in Kimberley, the largest museum of natural history and archaeology in the province. He served as the museum's honorary curator of reptiles and amphibians. Although his early interests were in herpetology, they shifted more and more to archeology. He became director of the museum in 1948 and retired in 1958 because of ill health. He moved to Pietermaritzburg (Natal Province) in easternmost South Africa and died at his son's home in Johannesburg on 21 December 1964, a few weeks after turning 80.

Powers's herpetological publications number about 25 (1913–1944), some of them co-authored with George A. Boulenger, John Hewitt, and Walter Rose. These deal primarily with distributional records from different parts of South Africa; frog behavior and life histories (especially tadpoles); physiological adaptation in tadpoles and adult frogs; and systematic reviews of lizards in the genera *Agama, Lophosaura* (now *Bradypodion*), and *Zonurus* (now *Cordylus*). He also authored a major revision of the frog genus *Breviceps* (1926) and described new taxa of frogs, lizards, and a tortoise.

• *References*: "J. H. Power 1884–1964," by G. J. Jack (=Fock), South Afr. Archaeol. Bull., 20(78), pt. 2: 74, 1965; "John Hyacinth Power, 1884–1964: A Tribute," by B. Wilson, McGregor [Mus.] Misc., 6(2): 5–6, 1996. • *Portrait* (about 1948) *and signature* (1963): McGregor Museum, courtesy Beryl Wilson.

SLEVIN, Joseph R. (1881–1957).

The American herpetologist, Joseph Richard Slevin, studied the herpetofauna of the Galapagos Islands, Central America, East Asia, Australia, and especially that of western North America. He was born in San Francisco on 13 September 1881 to a family of naturalists. His father was an amateur ornithologist, collector of birdskins, and member of the California Academy of Sciences; one of his brothers was also an ornithologist and another was an entomologist. He attended St. Ignatius High School in the city and then St. Mary's, a tiny Jesuit college in rural Kansas during 1895–1899 but left without graduating from what was then a six-year program of study. With a desire for exploration, he went to sea, first with the United States Navy until 1903 and then with the Oceanic Steamship Company.

In 1904 the academy was planning an expedition to the Galapagos Islands and Slevin, with his family and nautical backgrounds, was ideally suited for a position as a naturalist-collector. Starting in May he was employed by the academy and trained by the Curator of Herpetology, John Van Denburgh.

The 30-year-old schooner, aptly re-named "Academy" for the voyage, departed in June 1905 and returned from the islands 17 months later. During that time, disaster struck San Francisco, with an earthquake and ensuing fire of April 1906 that destroyed the academy's building, taking with it all but 13 of its collection of 8,100 herpetological specimens. Slevin turned out to be an indefatigable collector, catching marine iguanas in numbers and toting giant tortoises on his shoulders back to camp for skinning. On one occasion the crew was loading two enormous tortoises aboard a skiff when the boat was smashed against the rocks and the animals floated off. They were found alive 18 hours later, ". . . none the worse for it," Slevin remarked in his field notes. By his return, the academy, despite its calamity, could boast having the foremost collection of Galapagos reptiles—more than 4,000 specimens in all.

Slevin became Van Denburgh's assistant and spent much of the next years collecting in the field. He visited most of the western states and Mexico. His first dozen papers were co-authored with Van Denburgh and included herpetofaunal reviews of several western states (Arizona, Utah, Nevada, Idaho) and Baja California, with a special interest in offshore islands along the Pacific Coast. They also wrote reviews of the snake genera *Thamnophis* and *Pituophis*. With Van Denburgh's early death in 1924, Slevin became curator and, with the exception of naval duty during World War I, he served in that role until his own death in San Francisco on 15 February 1957. He continued to do fieldwork after 1924 including Guatemala, Panama,

Mexico, and Australia. He published on these collections as well as others at the academy including those made by a former assistant curator, Joseph C. Thompson, in East Asia. Slevin was succeded as curator by Alan E. Leviton.

Slevin authored more than 50 herpetological titles (1913–1959; his Galapagos field notes of 1905–1906 were published in 1982). There were several larger or popular works. These include "The Amphibians of Western North America" (1928), a companion to Van Denburgh's two-volume work on the reptiles (1922), the popular "A Handbook of Reptiles and Amphibians of the Pacific States" (1934), and Slevin's posthumous "The Galápagos Islands" (1959), which details the history of their exploration. A handbook of reptiles for the general public, which he had completed by 1957, was never issued. By the time of Slevin's death, the herpetological collections at the academy, re-established in 1906, had grown from 13 specimens to more than 75,000, most of them personally handled by him during 53 years with the academy. They were the beginning of what has become one of the most extensive herpetological collections in the world.

• *References*: "Joseph R. Slevin," anonymous, Acad. News Letter (Calif. Acad. Sci.), 164: 2–3, 1953, and 207: 2–3, 1957; "Joseph Richard Slevin (1881–1957)," by V. M. Tanner, Great Basin Nat., 17: 56–58, 1957; "Joseph Richard Slevin 1881–1957," by R. C. Miller, Copeia, 1959: 84–85, 1959. • *Portrait* (1955): George S. Myers/Alan E. Leviton portrait collection, courtesy Alan E. Leviton. • *Signature* (1938): Adler collection.

SERIÉ, Pedro (1865–1951).

Pedro Serié, an Argentine herpetologist and ornithologist, was born in the department of Charente, France, on 4 January 1865. He immigrated to Argentina where he became a naturalized citizen. He was hired as an assistant at the Museo Nacional de Historia Natural in Buenos Aires in June 1891 where he worked with a succession of outstanding naturalists including Hermann Burmeister, Carlos Berg, Ángel Gallardo, and the brothers Ameghino, Florentino and Carlos, all of whom served in turn as a director of the museum. In 1920 Serié was placed in charge of the herpetological collections and in 1925 also became secretary of the museum. Concurrently, he taught at the Escuela Normal de Profesores "Mariano Acosta" beginning in 1910 and became a professor there in 1924. Through these two institutional affiliations he had broad influence in Argentina and wrote regular columns on natural history in magazines and newspapers. He was a key person in scientific circles nationally, as charter member and, later, two-term president of the Sociedad Ornitológica del Plata and editor of its journal, *El Hornero*, founding member of the Sociedad Argentina de Estudios Geográficos "Gaea," and a member of the national commission of parks and nature reserves, among many others. He died in Buenos Aires on 23 December 1951 after an extended illness.

Serié's publications are primarily on herpetology and ornithology. He authored about 35 titles on amphibians and reptiles (1914–1937) and wrote on all of the major groups including caecilians and amphisbaenians, but his primary interest was snakes. Most titles dealt with the Argentine herpetofauna with the exception of two on that of Paraguay (1915, 1916). His most substantial papers include a supplement (1915) to Berg's "Contribuciones . . . Fauna Erpetológica Argentina" (1898), reviews of the snakes (1921) and venomous snakes of Argentina (1916, 1937), a paper on feeding in serpents (1920), and another on the distributions of Argentine snakes (1936).

• *References*: "† Pedro Serié (1865–1951)," by M. A. Freiberg, Physis, 20: 527–529, 1953; Additional information from Isabel T. Mercadal de Barrio, 1991. • *Portrait*: From Freiberg, 1953. • *Signature* (1915): Natural History Museum, London, courtesy A. F. Stimson.

GHARPUREY, K. G. (1880–1956).

Khanderao Ganpatrao Gharpurey (see *Note*), author of the several editions of the book "Snakes of India and Pakistan," was born on 28 July 1880 in Wardha (Central Province, now Maharashtra State), British India. Educated at Morris College in Nagpur and Presidency College in Calcutta (B.A. 1898), he attended the medical college in Lahore (now in Pakistan) and then received medical training in the United Kingdom (L.R.C.P.&S. Edinburgh and L.R.F.P.&S. Glasgow, both 1903, with additional training in London).

Gharpurey returned to India in 1906 and joined the Indian Medical Service as a physician-surgeon attached to various army hospitals. He held a military commission from 1906 to 1934 and he was regularly seconded as a civil surgeon—the chief government doctor in charge of the hospital—in various districts of the

Bombay Presidency and for Aden, in Arabia, which at the time was administered by the Bombay government. In 1909 he saw military service during a period of unrest in Somaliland, then a British protectorate. In 1914–1918, during World War I, he served in East Africa and in Iran and retired at the rank of lieutenant colonel. In 1934 he became surgeon-general in charge of the government medical service for all of Bombay Presidency, and then he retired completely from medical duties the following year. Gharpurey died in Poona, Bombay Presidency (now Pune, Maharashtra State) on 6 March 1956.

Gharpurey's interest in snakes began on his return to India in 1906 when he became more aware of the large number of human deaths from snakebite and of the public fear of snakes in general. An incident with a patient that same year had a major impact on him. A man came to him who was violent and uncontrollable because of a snakebite that turned out to be from a non-venomous species. Gharpurey eventually calmed him by injecting distilled water and after a short rest the man walked home in peace. He soon decided to study snakes and prepared an inexpensive identification manual for use by the masses.

Gharpurey joined the Bombay Natural History Society in 1910 and became active in its affairs. He published about a dozen papers in the society's journal (1914–1935), mostly reports on collections of snakes he had made in various districts, and also articles on snakebite and some unusual specimens. His book, "Snakes of India," re-titled "Snakes of India and Pakistan" after partition, covers the entire Indian Subcontinent including Bangladesh and extends to Burma and the Andaman Islands. It first appeared in 1935 followed by four later editions (1937, 1944, 1954, 1962), but only the second edition had significant changes. These books cover snake identification and distribution with several photographic plates mostly of dead specimens, and also contain a section on snakebite treatment. Gharpurey also published a short book in Marathi (one of the major Indian languages) on the snakes of Maharashtra State (1936). This book covers identification and distribution and has numerous illustrations. It was intended for people living in the most inaccessible areas of the state.

• *References*: "Lt.-Col. K. G. Gharpurey, I.M.S. (Retd.)," by [Sálim Ali], Jour. Bombay Nat. Hist. Soc., 53: 447, 1956; Additional information from Gharpurey's son, Madhao K. Gharpurey, 1992, and grandson, Malhar W. "Bhau" Gharpurey, 2006. • *Portrait* (1947) *and signature* (1927): courtesy M. K. Gharpurey. • *Note*: Gharpurey's original forename was "Khandu," but by protocol observed in many parts of India for prominent members of society it was modified to "Khanderao."

CHABANAUD, Paul (1876–1959).

The ichthyologist and herpetologist, Paul Chabanaud, was born on 30 November 1876 in Versailles, on the southwestern edge of Paris. Interested in nature as a child, he had amassed a sizeable insect collection by the age of ten, and when he vacationed at Touraine as a teenager he collected amphibians and reptiles. He studied natural history at Poitiers for his baccalaureate and finished in 1897. His special interest was beetles and he became active in the Société Entomologique de France; he served as the society's secretary during 1904–1909. In October 1915 he volunteered his services to Louis Roule who since 1910 had been the professor in charge of reptiles and fishes at the Muséum d'Histoire Naturelle in Paris. Roule asked him to identify herpetological materials from Europe and Africa and then, in October 1919, sent him on a scientific mission to French West Africa. After traveling to Dakar (Senegal) and then Conakry (Guinea), Chabanaud set out for the interior and traveled nearly 1,200 km on foot in southern Guinea and in Liberia. He returned to Paris in June 1920 with large zoological collections.

In July 1920 Chabanaud became a preparator in a laboratory at the museum devoted to fishes and colonial products of animal origin. His research interests soon shifted to fishes of the French colonies in West Africa and in Indochina. His special interest became flatfishes—their anatomy, systematics, phylogeny, and paleontology—although he continued to publish occasionally in herpetology. He earned his state degree in 1933 at the Sorbonne (University of Paris) and also took his doctorate there with a dissertation on the neurocranium of teleost fishes (D.Sc. 1936). Chabanaud, who worked alone, was especially interested in population variation and gathered thousands of measurements, which led him to become an early user of statistical techniques. However, his regular use of strictly personal morphological terminology made his research difficult to understand and limited its impact. He died in Paris on 27 February 1959.

Most of Chabanaud's nearly 275 publications were on ichthyology. He authored about 40 in herpetology (1915–1954), all but five of them before 1926. Most of his papers dealt with systematics and distribution. These covered a wide geographic scope including North Africa from Morocco to Tunisia and West Africa as far south as Togo, but also a few papers on France, India (in-

cluding the Himalayas), Indochina, and Romania. He described about 35 new species of frogs, lizards, and snakes. His most remarkable novelty was a new genus and species of pipid frog from Guinea, *Pseudhymenochirus merlini*. He also published on the skulls of crocodilians, venomous serpents of the French colonies, skittering locomotion in frogs, and an extensive treatise on the osteology of pelomedusid turtles.

• *Reference*: "Paul Chabanaud 1876–1959," by R. P. Dollfus, Bull. Mus. Hist. Nat., ser. 2, 32: 61–79, 1960. • *Portrait and signature* (both about 1928–1929): Natural History Museum, courtesy Andrew F. Stimson.

EMELIANOV, A. A. (1878–1946).

The Soviet vertebrate zoologist, Alexander Adrianovich Emelianov, who made the most comprehensive studies of the herpetofauna of the Russian Far East, was born on 9 July 1878 in Olonki (Irkutsk Government), a small village west of Lake Baikal and about 70 km northwest of Irkutsk. As a boy he had a love for nature. He attended secondary school in Irkutsk near the southern end of Baikal, but left in 1896 because of his "political disloyalty" (his own words in translation). In 1900 he went to the Far East to work on the Ussuri Railway, a branch of the Trans-Siberian line. Beginning in 1903 he taught natural history at schools in Ussuriysk, 50 km north of Vladivostok.

From 1906 to 1910 Emelianov was a student at the Higher Pedagogical School in Tomsk, located in the Siberian Plain about 200 km northeast of Novosibirsk. Later he worked at Imperial Tomsk University as a volunteer with Nikolai F. Kashchenko who in 1898 had made collections of vertebrates in the Altai, a mountainous region that now comprises the northernmost tip of Xinjiang (China) and the adjacent parts of Russia and Mongolia. It was during this time in Tomsk that Emelianov made contact with A. M. Nikolsky, then the Soviet Union's leading herpetologist, who encouraged him to concentrate on amphibians and reptiles. From 1917 to 1923, in addition to teaching school again in the Primorye area north of Vladivostok, Emelianov collected in the Altai and also in Manchuria, a region in northeastern China that was partly under Soviet control. His collecting was done on behalf of the Zoological Museum of the Academy of Sciences in St. Petersburg and the Tomsk University Museum of Zoology.

From 1923 to 1932 Emelianov attended Far East State University in Vladivostok, graduated in 1928, and worked there until 1932. He then moved to the newly opened Far Eastern Branch of the Academy of Sciences in that city. Before 1935 the Soviet government rejected scientific degrees, but later he was awarded a candidate degree by the Presidium of the Academy of Sciences (C.Biol.Sci. 1936) on the basis of his published work. During this time he made further explorations, to the Sikhote-Alin Mountains northeast of Vladivostok, the Kamchatka Peninsula, and Sakhalin Island. When the Far Eastern Branch had to be closed in 1939, he worked at the academy's Kazakh Branch in Alma-Ata during 1939–1944 and also was head of the Department of Vertebrate Zoology at Kazakh State University. By 1940 he had completed his doctoral dissertation, entitled "Amphibians and Reptiles of the Soviet Far East," but he could not defend it until later (D.Sc. Academy of Sciences, Leningrad 1944). He declined a permanent position in Alma-Ata and returned to Vladivostok and the Far East Branch, which was re-opened in 1944. He died in Moscow on 7 January 1946. His ashes were later buried in the Ussuryisk Nature Preserve at a place called "Zmeinaya Sopka" (Snake Hill) where he had once studied pitvipers.

Emelianov's original interests were in systematics and distribution, but later in his career he studied life history and zoogeography. He conducted important studies on amphibian distribution and ecology including the first detailed studies of the hynobiid salamander, *Onychodactylus fischeri*. He was interested in the venoms of pitvipers and, jointly with a local physician, was involved in antivenom production. Emelianov authored nearly 30 herpetological works (1916–1951). His surveys of amphibians and reptiles covered multiple regions in the Russian Far East. Snakes were his special interest and he wrote many papers on their ecology, habits, and diseases. He was the first to employ skeletochronology—counting the growth layers in bones to determine age—in snakes (1929). He described two well-known species of Asian pitvipers, *Agkistrodon saxatilis* and *A. ussuriensis* (both species now in the genus *Gloydius*). Emelianov's major work was his book, "Snakes of the Far Eastern District" (1929), a review of the dozen species in the district with distribution maps, keys, lists of specimens giving basic data, and an extensive bibliography. Unfortunately, his disserta-

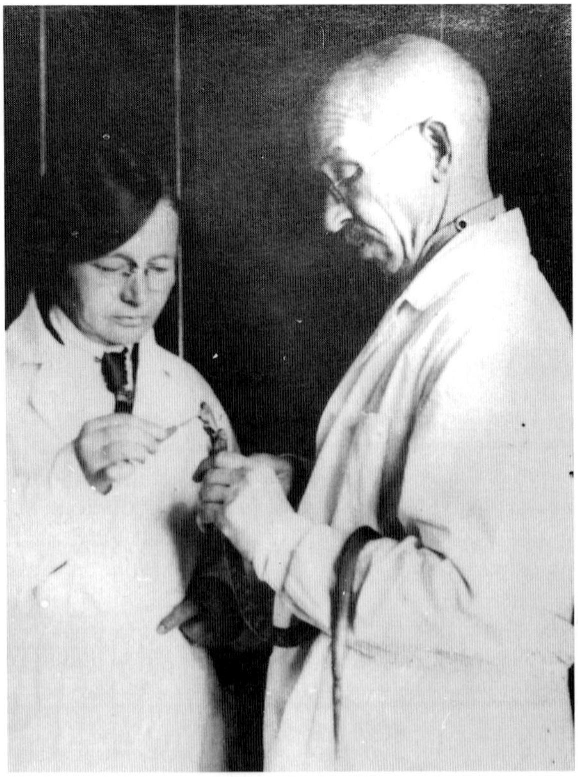

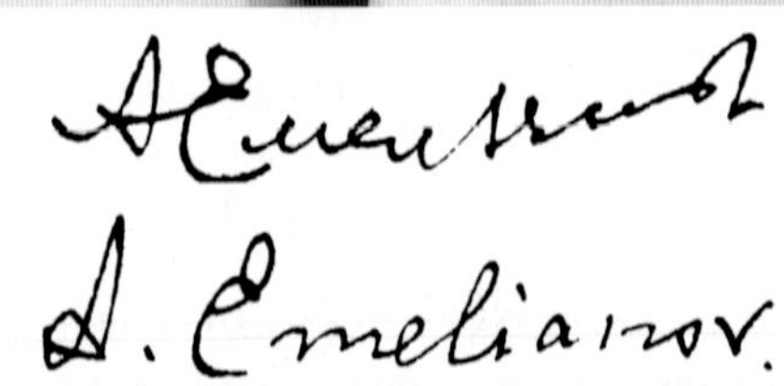

tion, which contains a wealth of data on ecology and life history, was never published because the country was still at war and he died just two years after defending it.

• *References*: ["In Memoriam Doctor of Biological Sciences A. A. Emelianov,"] by A. I. Kurentsov, Trudy Dalnevost. Fil. Komarova Akad. Nauk SSR, ser. Zool., 2(5): 119–121, 1951; "In Memoriam Doctor of Biological Sciences, Professor Alexander Adrianovich Emeljanov," by G. F. Bromley, p. 3–8 *in* L. J. Borkin (ed.), Herpetological Investigations in Siberia and the Far East. Zool. Inst. Acad. Nauk SSSR, Leningrad, 1981; Kuzmin, S. L. and I. V. Maslova, The Amphibians of the Russian Far East. Adv. Amphib. Res. Former Soviet Union, Sofia–Moscow, 8: 18–20, 2003 (Russian ed., same authors and title, Russian Acad. Sci., Moscow, pp. 25–26, 2005). • *Portrait* (1934, at right): Courtesy Nelly Filipchuk and Leo J. Borkin. • *Signatures* (Russian form, 1943; Romanized form, 1940): Courtesy Leo J. Borkin.

PROCTER, Joan B. (1897–1931).

Joan Beauchamp Procter was an English herpetologist who worked at the museum and then at the zoo in London but had a tragically short career. She was born to a wealthy family in London on 5 August 1897 and was fragile and delicate from birth. Her older sister, Chrystabel Procter, became a prominent horticulturist and they were taught by their mother to love plants and animals. As a small child Jane eschewed dolls in favor of a large Dalmatian green lizard, which traveled with her and sat beside her at meals, and a baby crocodile, which she often took to school until it caused a stir in a mathematics class. Her education at St. Paul's School for Girls (1908–1916) was often interrupted by illness and finally ended by it. While still a schoolgirl she had the habit of visiting the Natural History Museum to discuss her specimens with the curator of reptiles, George A. Boulenger, then the world's preeminent herpetologist, who encouraged her interest. When because of her poor health she had to give up her plan to attend Cambridge in 1917, Boulenger invited her to assist him. When he retired in 1920 she was appointed temporarily to supervise the department until the arrival of H. W. Parker as curator in 1923.

By this time, however, she had come to the attention of the authorities at the Zoological Society of London as someone skilled both scientifically and artistically in producing naturalistic displays. She assisted Edward G. Boulenger, director of the zoo's aquarium and son of her former mentor, in decorating the new aquarium building. Later, her skill in devising new techniques for display was amply demonstrated at the opening of the new aquarium in April 1924 in the presence of the king and queen. There had been difficulty in getting seaweeds established in time, so Procter made imitations out of thin colored sheets of rubber that were so lifelike that one distinguished biologist later remarked to the zoo director, "My dear boy, these weeds of yours will never live here, you must put in fake ones."

Procter was appointed Curator of Reptiles in November 1923, which caused a sensation in the press. She was not only a woman but also young (26), striking in appearance, and strong-willed. Much was made of her small size, which the newspapers compared to that of some of her charges including two adult Komodo monitors and a large anaconda. In fact, she would often take the male dragon with her on her daily inspection rounds, sometimes steering him by holding the tip of his tail and using it as a rudder! She held this post with distinction until her death from cancer at the age of 34 on 20 September 1931 in London. Just a few months earlier the Intercollegiate University in Chicago, an affiliate of the Orthodox American Catholic Church, awarded her an honorary doctorate (D.Sc. 1931).

Besides her scientific abilities, Procter had artistic and practical skills that made her an ideal zoo curator. The old reptile house had to be replaced and Procter designed the new one, from the general layout to the drains, electrical heating, service areas, displays, and decorations. When opened in 1927, the new building was heralded as the most advanced in technology and display of any reptile house in the world. She introduced several innovations including differential heating to provide thermal gradients, which utilized electricity rather than the inefficient and imprecise hot-water pipes traditionally used; elaborate naturalistic displays instead of nearly barren cages; and a regulated traffic flow so that visitors did not bunch up in dead-end cul-de-sacs. She also began the practice to illuminate enclosures from above and darken the visitor spaces, thus highlighting the exhibits and reducing distractions for the animals.

Procter was equally adept at handling and studying live animals. She became an expert surgeon in excising tumors and dressing sores. On one occasion, with a huge python held by 14 men, Procter calmly lanced a septic abscess that was slowly poisoning it. Despite her infirmities, which became increasingly pronounced, she would often spend evenings in the dark-

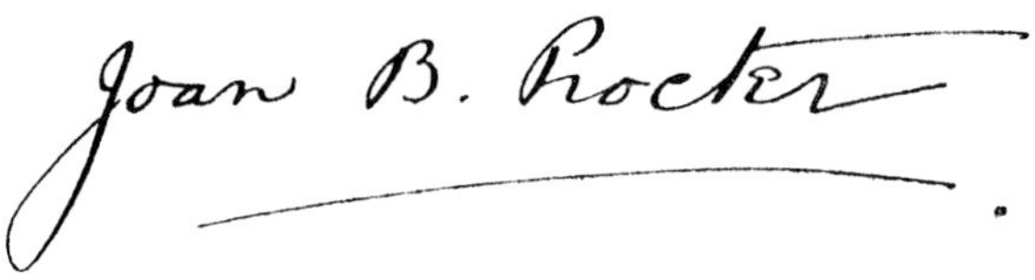

ened building to study vocalizations of the frogs and crocodilians. Toward the end of her life, prolonged by a series of major intestinal operations, she hardly passed a day without severe pain, and when she could no longer walk, Procter went to work in an "electric bath-chair." Whenever that became physically impossible she would stay in contact with her staff by telephone from her nearby home in Regent's Park. Because of her health she attempted to resign late in 1928, but her offer was declined by the zoo's council whose president, the Duke of Bedford, wrote to her: "it is most desirable that you should continue . . . and yet leave the Society in debt to you for past services admirably rendered." Her eyesight was mostly gone by mid-1930 because of the chemotherapy that she was taking, yet she could still summon great strength. Just months before she died and reportedly high on morphine, she wrestled with an escaped bear in a basement room until a keeper arrived.

Throughout her time at both the museum and the zoo, Procter maintained a diverse research program on amphibians and reptiles (about 25 papers, 1918–1929). These touched on many geographic regions including Bolivia, East Africa, India, the Indo-Australian Archipelago, and Persia. They include studies on variation in the scutellation of snakes, descriptions of new species of frogs, lizards, and snakes, and studies on behavior. Her paper on the scapulae of frogs (1921), showing how their form can be used to diagnose frog families, foreshadowed the later research of Ivor Griffiths (1963). The longest paper (1922), and perhaps her most important, was a detailed study of the development of the shell in the Pancake Tortoise (*Malacochersus tornieri*) of East Africa. Using radiographs and other techniques, she showed how the bony plates become thinned out with age, allowing the shell to become flexible so that the animal can wedge itself in rock crevices to escape predators.

• *References*: "Miss Joan Procter (1897–1931)," by E. W. MacBride, Proc. Linnean Soc. London, 1931–1932: 183–185, 1932; "Miss Joan Procter," anonymous, Jour. Bombay Nat. Hist. Soc., 35: 663–664, 1932 (reprint of obituary in The Times [London], 21 September 1931); "Joan Beauchamp Procter," by H. Bailes, Oxford Dict. Natl. Biogr., 45: 458–459, 2004; Personal files of Chrystabel Procter, Girton College archives, Cambridge, courtesy Kate Perry, and other information from Emma Cochrane, 2006. • *Portrait* (1923): Natural History Museum, London, courtesy A. F. Stimson. • *Signature* (1922): Adler collection.

RENDAHL, Hialmar (1891–1969).

The Swedish zoologist, Carl Hialmar (sometimes Hjalmar) Rendahl, studied the systematics, anatomy, and embryology of vertebrates and published on amphibians and reptiles from diverse parts of the world. He was born on 26 December 1891 in Jönköping (Småland) in southern Sweden. After completing local schools in 1910, he went to the university in Stockholm where he graduated in 1916, but he had already started working in the Department of Vertebrates at the university's Zoological Museum two years earlier. He earned his doctorate at the university (Fil.Dr. 1924) for a dissertation on the embryology of the brain in chickens, and he became a senior lecturer in zoology at the university in the same year. In 1933

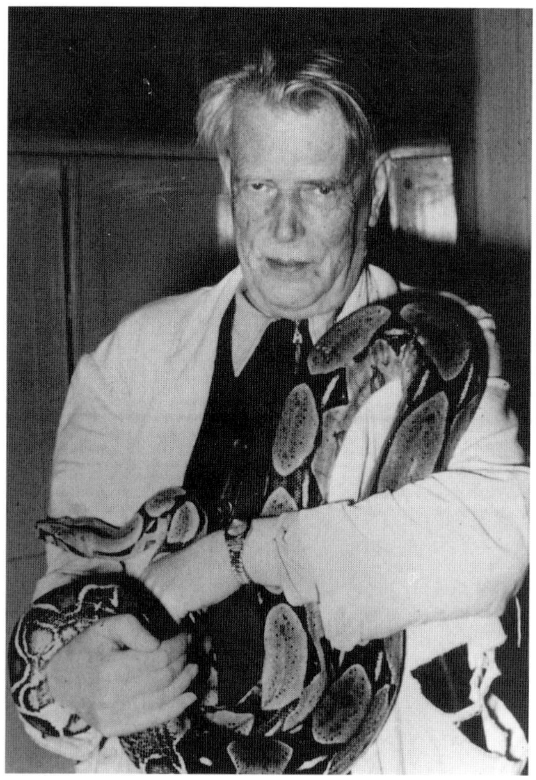

he was promoted to professor and became chief of the vertebrates department, succeeding his longtime colleague, Einar Lönnberg. Rendahl retired in 1957 and died in Stockholm on 2 May 1969.

Rendahl authored technical papers on vertebrates generally but especially on fishes and birds. He wrote popular books on birds, bird migration, and Arctic birds, as well as one on venomous snakes and snake venoms, "Giftormar och ormgifter" (1936). He also served as editor for the 22-volume Swedish translation of A. E. Brehm's series on animal life, "Djuren liv" (1929–1931). Rendahl published a dozen technical articles on amphibians and reptiles (1918–1941). These primarily focused on collections at the museum from overseas including Bolivia, Burma, Central Asia, Colombia, Ecuador, the Kuril Islands of East Asia, South Africa, and Sumatra. He described about 20 new species including one caecilian (now *Epicrionops columbianus*), but few of his other new taxa—all lizards and snakes—are recognized today. Rendahl's major herpetological work, "Zur Herpetologie der Seychellen. I. Reptilien" (1939), is based on extensive collections and behavioral observations made in 1935 by Hilmer Sundberg, an amateur naturalist and Uppsala businessman. The companion section on the amphibians of these Indian Ocean islands was never published.

• *References*: "Carl Hialmar Rendahl," p. 702 *in* Svensk biografisk handbok, P. A. Norstedts & Söner förlag, Stockholm, 1935; "Carl Hialmar Rendahl," by T. F[lensburg], p. 235–236 *in* Svenska män och kvinnor–Biografisk uppslagsbok, vol. 6. Albert Bonniers förlag, Stockholm, 1949. • *Portrait*: Erik Åhlander, courtesy Richard Wahlgren. • *Signature* (1927): Göran Nilson.

HORA, Sunder Lal (1896–1955).

Sunder Lal Hora, who was a specialist on the fishes, amphibians, and reptiles of India and the Oriental Region, was born on 22 May 1896 in Hafizabad (West Punjab), a part of British India that is now in Pakistan. He attended the Sain Das Anglo-Sanskrit School at Jullunder where he developed an early interest in science and then focused on zoology at the Government College of Punjab University in Lahore (B.Sc. 1917, M.Sc. 1919). In 1919 the British zoologist, Nelson Annandale, Director of the Zoological Survey of India, recruited him to the survey in Calcutta. Hora, Baini Prashad, C. R. Narayan Rao, and others of herpetological interest were trained by Annandale who had a policy that every staff member would spend a few months each year in the field. Hora took full advantage of the policy to know his animals in nature, which enriched his laboratory studies and publications. While in the field in the eastern Himalayas with Annandale in 1920 to study the torrential hill streams in the Manipur Valley (then in Assam, now in Manipur State), Hora first recognized the adaptive modifications of fishes and frog tadpoles to this habitat, a topic that became a major theme for his research. In 1921 he was placed in charge of the survey's collections of lower vertebrates. Rao completed his doctoral studies (D.Sc. Punjab University, 1921; D.Sc. Edinburgh University, 1928) while on staff at the survey. He was seconded to the Bengal Government as director of fisheries during 1942–1947 and then returned to the survey where he served as director until his death in Calcutta on 8 December 1955.

Hora was a tireless researcher and a prolific author who produced about 425 titles. These focused primarily on taxonomy, ecology, and zoogeography. He wrote long series of papers on fishes in the Indian Museum, the siluroid fishes of India, Burma, and Ceylon, and the game fishes of India. He also published on the ancient Hindus and their knowledge of fishes and fisheries based on Sanskrit literature and archeological records. In zoogeography he is best known for proposing a hypothesis (1937) to explain the close relationships of many fauna and flora in the Malay Peninsula to those in the Indian Subcontinent whose ancestors may have reached the Western Ghats and Sri Lanka via the Satpura Mountains in central India. As a conservationist, he served as the first Secretary-General of the Indian Board for Wildlife (1952–1955) and thus led the initial effort to establish the country's nature reserves.

Hora published a dozen papers specifically on herpetological topics (1920–1949) but there were others on stream ecology and faunal surveys that included amphibians and reptiles. His herpetological titles list several papers on tadpole morphology, especially suctorial mouthparts (1922–1933), and studies on the distributions of crocodiles, turtles, and snakes (1948–1949) as evidence for his Satpura Hypothesis. He also published a series of three papers on gekkonid, agamid, and scincid lizards in the Indian Museum (1926–1927).

• *References*: "Dr. Sunder Lal Hora (An Appreciation)," by M. Chandy, Jour. Zool. Soc. India, 7: 191–194, 1955 (1956); "Sunder Lal Hora," by E. G. Silas, Copeia, 1956: 134–136, 1956; "The Late Dr. Sunder Lal Hora (1896–1955): An Appreciation, Together With a Complete List of His Scientific Writings," by M. L. Roonwal, Rec. Indian Mus., 54: 107–137, 1956. • *Portrait* (1934): George S. Myers/Alan E. Leviton portrait collection, courtesy Alan E. Leviton. • *Signature*: From Chandy, 1956.

CUESTA TERRÓN, Carlos (1897–1948).

Carlos Cuesta Terrón, a herpetologist and fisheries researcher, was born in Mexico City on 7 May 1897. He attended schools in that city during 1903–1910 and then the Escuela Nacional Preparatoria (1910–1912), located in "San Ildefonso," an old baroque building in downtown Mexico City in which the first programs were held of what later came to be known as the Universidad Nacional Autónoma de México (UNAM). He then matriculated at another unit of UNAM, the Escuela Nacional de Medicina, where he earned his degree as medical doctor, surgeon, and midwife in 1918. Three years earlier he had come to the attention of Alfonso L. Herrera, a leading Mexican biologist whose own research interests included herpetology. He steered Cuesta Terrón into a position at the Museo Nacional de Historia Natural in 1919 where he took responsibility for the section on lower vertebrates.

Cuesta Terrón soon began his first fieldwork, a study of the fishes and reptiles of Baja California. Later he was commissioned to study the harvesting of sea turtles and sea bass in the Gulf of Mexico and the fisheries in Lake Chapala (Jalisco). After his first paper, on turtle fisheries (1919 [1920]), he reviewed the herpetofauna of Baja California in two papers (1919 [1920], 1921). He also studied the life cycles of several species of fish in Lake Chapala, the first biological field studies in Mexico directed toward the maintenance of commercial fisheries. In 1921–1922 he briefly practiced medicine in Mixquihuala (Hidalgo) before returning to government service at the Dirección de Pesquerías, the national fisheries service in Mexico City. He returned to work with Herrera at the museum during 1927–1929.

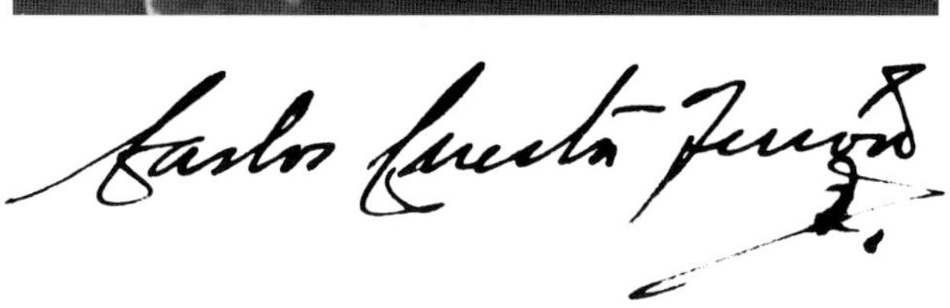

In 1929, Isaac Ochoterena, the histologist and reptilian morphologist, created the Instituto de Biología to encompass the entire biological faculty of UNAM. He invited Cuesta Terrón to become one of its founding members and to take charge of work on lower vertebrates. During this period Cuesta Terrón conducted the rest of his herpetological work. Among these were several studies on the systematics, distribution, and ecology of venomous snakes and the beaded lizard. These topics were of practical importance in a country with a largely rural population that has many species of venomous reptiles, and a country in which rattlesnakes played a major ritualistic role in the ceremonies of the Aztecs and other native peoples. In total he published 13 papers on herpetology (1919 [1920]–1932) including one on rattlesnake symbolism among the prehistoric people of Mexico (1931).

Cuesta Terrón left his position at the institute in 1932 in order to devote the rest of his life to translating important English texts into Spanish, to teaching in the medical school, and to the development of vaccines. He also continued to be consulted by the Mexican government on fisheries matters. He died in Mexico City on 21 May 1948.

• *References*: "In Memoriam Carlos Cuesta Terrón," by B. Kohlmann Cuesta, Bol. Soc. Herpetol. Mex., 1: 27–30, 1989; Additional information from Cuesta Terrón's grandson, Bert Kohlmann Cuesta, 1989–2006. • *Portrait* (1943): Courtesy Bert Kohlmann Cuesta. • *Signature* (1931): Adler collection.

WETTSTEIN-WESTERSHEIMB, Otto (1892–1967).

The herpetologist and mammalogist, Otto Wettstein-Westersheimb (see *Note*), a specialist on systematics and faunistics, was curator of one of the world's largest herpetological collections. He was born into Austria's leading family of natural scientists on 7 August 1892 in Vienna. His father and his mother's father were renowned botanists and his uncle was a prominent geologist. All were members of the nobility. As a boy, Wettstein had an interest in nature, first for insects and later, birds and mammals. However, at the age of one he had the misfortune to have an inflammation of his tympanic cavities that resulted in lifelong deafness. In 1910 he entered the University of Vienna. One of his professors there was Franz Werner, a herpetologist who in 1914 took him on a three-month expedition to the Anglo-Egyptian Sudan, a trip that had a profound influence on Wettstein's choice of career. He wrote his dissertation on the heart cavity of decapod crustaceans (Dr.phil. 1915). He was exempted from military duty during World War I because of his deafness, so in 1915 he applied for a position at the Naturhistorisches Museum where he hoped to work with birds or mammals. To his great disappointment the only opening was a volunteer position in herpetology; however, these are rich collections built up since the 1820s by Leopold Fitzinger, Franz Steindachner, and Friedrich Siebenrock. The large personal collection of his teacher, Werner, was added after Werner's death in 1939.

Wettstein's first task was to catalogue the collection of amphibians and reptiles that had been purchased in 1913 from the widow of Egid Schreiber, author of "Herpetologia Euro-

paea" (1875; second edition 1912–1913). Although most of Wettstein's publications during 1912–1920 dealt with birds and mammals, he came to enjoy his work with amphibians and reptiles and in 1920 (1921) he published his first title on them, a long paper on the herpetology of Albania (jointly with Felix Kopstein). When the curator, Siebenrock, retired in 1920, Wettstein took his position. Although he continued to publish on birds and mammals as well as insects and even plants, almost one-third of his output after 1920 was in herpetology (52 titles, 1920 [1921]–1968). One of his first tasks after succeeding Siebenrock was to reorganize the collections. This process took five years to complete and was followed by an active research period. When World War II began, however, the alcoholic collections had to be removed to underground storage in the museum's basement and to other shelters inside and outside of Vienna where they were effectively unavailable for research purposes. It was Wettstein's assistant and later his successor as curator, Josef Eiselt, who returned the collections to the museum's halls.

In 1940 Wettstein was invited to join the faculty at the University of Vienna as the professor responsible for vertebrate systematics. He made numerous field excursions throughout his career, mostly to circum-Mediterranean countries and especially to their offshore islands. His herpetological research naturally focused on the large Vienna collection. Papers thus appeared on specimens from Africa (Morocco, Senegal), Asia (Cambodia, Kazakhstan, Tibet), Central and South America (Brazil, Costa Rica, Ecuador), New Zealand, and Southwest Asia (Afghanistan, Iran, Lebanon, Palestine, Turkey) as well as from Europe including Austria and the Mediterranean islands. His major works include two herpetological sections in the never-completed series, "Handbuch der Zoologie," founded by Willy Kükenthal, which covers the biology and morphology of animals. Wettstein contributed "Rhynchocephalia" (1931) and "Crocodilia" (1937, 1954). His "Herpetologia Aegaea" (1953, supplement 1957) summarized the amphibians and reptiles of the Aegean Islands. Wettstein named numerous new taxa of amphibians and reptiles (including in the Gymnophiona and Rhynchocephalia) and especially new species and subspecies of *Lacerta*.

Wettstein retired as curator in 1945. He died in Vienna on 10 July 1967.

• *References*: "A. O. Universitätsprofessor Dr.phil. Otto Wettstein-Westersheimb †," by J. Eiselt, Ann. naturhist. Mus. Wien, 70: 1–18, 1967; "Otto Wettstein von Westersheimb †," by G. Niethammer, Zeitschr. Säugetierk., 32: 378–381, 1967; Rieck et al., 2001 (Wettstein by W. Bischoff on pp. 644–646). • *Portrait* (1936) *and signature* (1936): Natural History Museum, London, courtesy A. F. Stimson. • *Note*: His surname was sometimes also written as Wettstein von Westersheimb and as von Wettstein-Westersheimb.

LUEDERWALDT, Hermann (1865–1934).
Hermann Luederwaldt, a German entomologist and botanist who spent his career in Brazil, was born on 23 January 1865 in Gollnow (now Goleniów, Poland, about 25 km northeast of Szczecin), Pomerania, a region in northern Prussia. He attended schools in Gollnow and Stettin (now Szczecin) and immigrated to Brazil in 1897 to settle in Blumenau (Santa

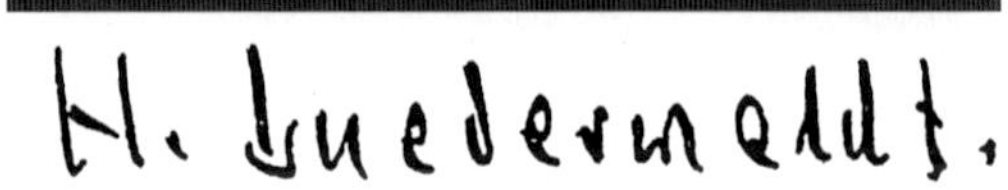

Catarina) in southern Brazil where his brother-in-law had a clinical practice. As a boy he had been interested in insects and even then harbored plans to work in a tropical region. He worked on a farm in Blumenau, which afforded him the opportunity to study nature, but his income was meager and did not allow him the necessities to build a collection. In 1905 he therefore applied to Hermann von Ihering, Director of the Museu Paulista in São Paulo, for a job. Ihering could only offer a position as gardener. Luederwaldt accepted it, for this allowed his weekends free for personal fieldwork in the area. He became a preparator in 1906 and by 1909 had published his first paper on the Brazilian fauna.

Luederwaldt spent much time at field sites—often with the museum's official collector, Ernesto Garbe—including the museum's new biological station, Estação Biologica do Alto da Serra, in the nearby mountains. In 1925 Luederwaldt was finally appointed to a curatorial post as assistant in zoology for invertebrates, a position he held until his death on 5 August 1934 in São Paulo. In the meantime he conducted numerous field excursions in the state of São Paulo including the coastal regions, and published on a wide range of subjects (vertebrates, insects, crustaceans, mollusks, plants). He authored about 65 titles on the Brazilian biota (1909–1934).

Luederwaldt published three titles on reptiles including one on the ecology of *Tupinambis,* a large teiid lizard (1921; reprinted 1972), and another that reviews the Brazilian crocodilian collection in the Museu Paulista (1926). His most important contribution was his well-illustrated monograph of Brazilian turtles, "Os Chelonios Brasileiros" (1926), which listed the Paulista collections and in which he described a new species, *Phrynops tuberculata.* There are also herpetological sections in his reviews of the mangrove swamps of the coastal island of Santos (1919) and in his papers on museum expeditions to other coastal islands, Ilha dos Alcatrazes (1923) and the large island of São Sebastião (1929).

• *Reference*: "Hermann Luederwaldt," by A. de E. Taunay, Revta. Mus. Paul., 21: 33–47, 1937. • *Portrait and signature* (1928): National Museum of Natural History, Washington, courtesy Rafael Lemaitre.

ORTENBURGER, Arthur I. (1898–1961).

Arthur Irving Ortenburger was an American vertebrate zoologist who monographed the whipsnakes and racers and studied the herpetology of Oklahoma. He was born in Detroit (Michigan) on 8 June 1898 and attended the University of Michigan in nearby Ann Arbor (A.B. 1920, M.S. 1922, Ph.D. 1925) where he was a graduate student of Alexander G. Ruthven. Prior to finishing his doctorate, Ortenburger became an assistant curator in herpetology at the American Museum of Natural History in New York (1922–1923). He moved to the University of Oklahoma in Norman in 1924 as a curator in the Stovall Museum of Science and History (now Sam Noble Oklahoma Museum of Natural History) and assistant professor of zoology. In his early years at Oklahoma he directed all fieldwork for the Oklahoma Biological Survey, located on the campus, and he built up large collections for the museum. He rose through the professorial ranks at Oklahoma and served as zoology department chair (1942–1945). On retirement in 1958, he and his wife, Roberta Deam Ortenburger (1900–1961), who collaborated in much of his herpetological fieldwork and publication, moved to Galiano Island in the Strait of Georgia off Vancouver (British Columbia) where they conducted studies on the embryos of whales. He died there on 18 November 1961.

Ortenburger authored about 25 titles in herpetology (1921–1933). While at the American Museum in the early 1920s, he worked up some of the Pacific Islands collections of the Whitney South Sea Expedition. All of his other publications were on North American amphibians and reptiles and particularly those of Oklahoma. His large collections of Oklahoma reptiles were used by one of his master's students, Robert G. Webb, to produce the book, "Reptiles of Oklahoma" (1970). Most of Ortenburger's titles deal with distribution, identification, and systematics. His *magnum opus*, based on his doctoral dissertation, was the book, "The Whip Snakes and Racers Genera *Masticophis* and *Coluber*" (1928; reprinted 2007), a masterful review of their systematics, geographic distribution, and variation in the style of generic revisions pioneered at Michigan by his major professor, Alexander Ruthven.

• *References*: "Arthur Irving Ortenburger 1898–1961," by J. T. Self, Proc. Oklahoma Acad. Sci., 42: 295, 1962; "Remembrances of the Ortenburgers," by A. N. Bragg, Biologist, 48: 10–14, 1966; "History of Oklahoma Herpetology II. Dr. Arthur Irving Ortenburger 1898–1961," by C. C. Carpenter, Bull. Oklahoma Herpetol. Soc., 2(2–3): 52–55, 1977; "The Centennial History of the Department of Zoology," by C. C. Carpenter, Publ. Dept. Zool. Univ. Oklahoma, Norman, [3], 177 pages, 1992 (Ortenburger on p. 29); Additional information from the Ortenburgers's grandson, David Ortenburger, 2007 • *Portrait*: Sam Noble Oklahoma Museum of Natural History, University of Oklahoma, courtesy J. Droke. • *Signature* (1927): Adler collection.

TAKAHASHI, Seiichi (1874–1932).

The Japanese artist and amateur ophiologist, Seiichi Takahashi, wrote and illustrated two books on the snakes of Imperial Japan. He was born on 7 March 1874 in Sendai (Miyagi), near the northeastern coast of Honshu, Japan's largest main island. Japan at that time had just emerged from the last of its medieval culture, the Edo period. It was transforming itself into a world power that would soon incorporate Formosa (now Taiwan) and Korea into the empire in 1895 and 1910, respectively, and occupy Manchuria in 1931. Takahashi was the son of a samurai of the Sendai feudal clan who became a judge in the Imperial Army. Seiichi himself became an artist and a teacher of drawing and painting, first at a middle school in Shiroishi just southwest of Sendai in 1899. He was strongly influenced there by another teacher, an Englishman who taught natural history, and Takahashi began to draw plants and animals. In about 1907 he moved to Taihoku (now Taipei) to be an art teacher at the Language School of the Formosan Government House (later Taihoku Normal School). This school functioned to teach Japanese language and culture to native Formosans from families of high position. He also was a lecturer at Government House and taught painting and natural history at the associated middle school. At the same time he drew subjects at the Formosan Medical College. In about 1925 he moved to Tanshui, just northwest of Taihoku, where he taught art at a missionary middle school.

Takahashi often went on fieldtrips to see and collect snakes. These trips included mountainous regions in Okinawa and Formosa. When he visited Makazayazaya (now Hsinmachia)—a well-known collecting site in the mountains of southern Formosa and the type locality for several taxa of snakes—he had to be accompanied by a guard of policemen because the aboriginal tribes were aggressive toward Japanese. Many specimens were brought alive to his home, where they were kept for observation and as subjects for his paintings. Takahashi continued to study reptiles throughout his life, especially snakes. He died in Taihoku on 29 March 1932 at the age of 58.

Takahashi wrote and illustrated two books on Japanese snakes. The first, "Japanese Venomous Snakes" (1922), was published in Taihoku. It consists of a series of 21 colored plates, with extensive captions, illustrating terrestrial snakes from Taiwan and Japan. A total of 20 species are illustrated including three non-venomous ones that are often confused with the others. His second book, "Terrestrial Snakes in Japan" (1930), was published in Tokyo. It has a broader taxonomic and geographic scope and includes the territory on the Asiatic mainland annexed by Japan. This book has a full text plus diagrams of head scutellation for 70 species, although a few are not identified to species or, in one instance, even to genus. There are 65 colored plates. The plates in both books are not highly detailed, but are generally lifelike, artfully drawn, and easily identified to species. However, Takahashi was not always able to apply the correct scientific names and, according to Moichirô Maki, then Japan's leading student of snakes, 24 of the 70 species were misidentified. Takahashi named three new species in his books, two of which are recognized today: a coralsnake (*Sinomicrurus hatori*) and a pitviper (*Ovophis monticola makazayazaya*), both from Taiwan. All of his alcoholic specimens and photographic negatives were donated to Taihoku Imperial (now National Taiwan) University.

• *References*: Zhao, E.-M. and K. Adler, Herpetology of China. Soc. Stud. Amphib. Rept. (Contr. Herpetol., 10), 522 pages, 1993 (Takahashi on pp. 38–39); Additional information from Takahashi's son, Makoto Takahashi, 1991. • *Portrait* (mid-1920s) *and signature* (1932)*:* Courtesy Makoto Takahashi.

MELL, Rudolf (1878–1970).

The amateur German zoologist, Rudolf Emil Mell, was an early specialist on the distribution and biology of South China reptiles, insects, and other animals. He was born on 16 February 1878 in Langenberg (now Gera-Langenberg, Thuringia), about 20 km east of Jena. Like his father, he attended a training college. During 1898–1903 Rudolf was a teacher and later, in 1903–1907, a biology instructor at a secondary school. In 1908 he sailed to China to found and run the German-Chinese Middle School in Canton (now Guangzhou), in Guangdong Province along the southern coast of China. There, in a tropical realm teeming with wildlife, he quickly indulged his interests for butterflies and snakes and later expanded them to include mammals, birds, amphibians, and other reptiles. He made extensive, personally financed expeditions throughout southeastern China, but following the end of World War I and the financial collapse in Germany, his days in China were limited. He had intended to write his masterwork, "Fauna Sinica Meridionalis," for which he amassed a huge collection of animals, but it never appeared.

He returned to Germany in 1924 and settled in Berlin, hoping to be appointed to a museum position. Although his friends arranged for an honorary doctorate for him (Dr.phil.h.c. 1924 at Rostock), his lack of proper university-level training and credentials precluded such a post, so he turned to popular writing in order to eke out a meager living. He wrote three such books based on his work in China, each with a disproportionately

large herpetological content: "Mussestunden am Tropenrande" (1947, under his pen name, Ma Lamm Kuei), "Wochenend am Wendekreis" (1955), and "Bergwaldtiere am Tropenrand" (1960). He maintained connections with the scientific establishment through memberships in several scientific societies and continued to write papers on his collections, which were donated to the Museum für Naturkunde in Berlin. He died in Berlin on 12 January 1970.

Mell's original interest in Chinese herpetology was ecological and out of necessity he also conducted taxonomic studies. He often delegated the latter work to his colleagues at the museum—Ernst Ahl, Hans Kanberg, Walter Unterstein, and Theodor Vogt—but usually reserved the snakes for himself. Mell named about 20 taxa of snakes, mostly subspecies, but only one of them is currently recognized. He was more in his element in describing behavior and analyzing distributions from a zoogeographical point of view. All told, he authored about 15 papers on amphibians and reptiles (1922–1952), some of them quite extensive.

Mell's major herpetological work was his book, "Grundzüge einer Ökologie der chinesischen Reptilien" (1929), which is an extensive study of geographic variation with respect to altitude and latitude, as well as discussions of reproduction and behavior. In it he proposed an ecomorphological law linking scale features to diet, in which he believed that snakes feeding on larger prey (endotherms) would require greater elasticity and therefore more supralabial and dorsal body scales than would ectotherm specialists. Recent research with Australian snakes has broadly supported this idea, which has been termed "Mell's Law" by Richard Shine.

• *References*: "Dr. Rudolf Mell," by G. Lederer, Entomol. Zeitschr., 67: 41–44, 1957; "Rudolf Mell," p. 59–60 *in* L. Gebhardt, Die Ornithologen Mitteleuropas, vol. 3 (Jour. Ornithol., vol. 115), 1974. • *Portrait* (about 1920): From Mell, 1960, Bergwaldtiere am Tropenrand, Stuttgart, 1960. • *Signature* (1931): Natural History Museum, London, courtesy A. F. Stimson.

De Villiers, Cornelius G. S. (1894–1978).

Cornelius Gerhardus Stephanus De Villiers was a South African comparative anatomist specializing on amphibians and reptiles, and also a writer and translator of works on music, theater, and literature. He was born on 16 December 1894 on the parental farm in Dunghye Park, a rich Afrikaaner farming region in the Caledon District (Cape Colony), South Africa, about 100 km southeast of Cape Town. His early education was at farm schools, but his intelligence was recognized and he was sent to Victoria University College in Stellenbosch (later re-named University of Stellenbosch) where he graduated (B.A. 1914, M.A. 1916) before teaching zoology in Pretoria at Transvaal University College (now University of Pretoria). After World War I he went to the Universität Zürich in Switzerland where he conducted his first research on amphibian anatomy (Ph.D. 1922), and the next year he became a professor at the university in Stellenbosch where he remained until he retired in 1959.

In the zoology department at Stellenbosch, De Villiers developed one of the leading programs in comparative anatomy especially of amphibians and reptiles. Known to all of his stu-

dents as "Doctor Con," he was a popular and dynamic lecturer. He knew six languages and regularly translated important plays and prose into his native Afrikaans. In addition, he made the first Afrikaans translations of Haydn's and Handel's musical scores, and he often lectured on music and singing. Before Stellenbosch had a drama department, De Villiers staged plays on campus. He also conducted extensive genealogical research on old South African families. He died on 25 November 1978 in Ceres (Cape Province), about 100 km northeast of Cape Town, at the home of a nephew. He never married.

De Villiers's scientific publications were almost exclusively on amphibians and reptiles (about 30 titles, 1922–1939). Most consisted of highly detailed morphological studies and were extensively illustrated with serial sections of stained material and especially of the skull. His research program involved in-depth studies of the morphology and early development of key species as an aid to understanding amphibian phylogeny. Among the frogs, he studied *Bombinator* (now *Bombina*), *Breviceps, Cacosternum, Hemisus, Hyperolius, Phrynomerus* (now *Phrynomantis*), and *Xenopus*; the salamanders, *Ambystoma* and *Rhyacotriton*; and the caecilians, *Boulengerula* and *Scolecomorphus*. He made an extensive study of *Ascaphus* (1934), in particular the so-called tail of the male, and discovered the postpubic bones, which are unique to this frog and lie within the copulatory organ. He also studied skinks of the genus *Acontias*. De Villiers was an accomplished field naturalist and collected much of his own material. As a result, he made important discoveries on the breeding habits and early life histories of frogs. He made one of the earliest observations of tadpole cannibalism under natural conditions of crowding (1929).

• *References*: "Cornelius Gerhardus Stephanus De Villiers," by J. M. De Kock, Dict. South Afr. Biogr., 5: 188–189, 1987; "Die Wêreld van Con De Villiers—'n Kultuurhistoriese Oorsig," by C. J. De Kock, Doctoral Dissertation, University of Stellenbosch, Stellenbosch, 1994. • *Portrait*: Courtesy P. le F. N. Mouton. • *Signature*: Adler collection.

Ahl, Ernst (1898–1945).

Christoph Gustav Ernst Ahl, a German ichthyologist, herpetologist, and aquarist, was born in Berlin on 1 September 1898. He attended secondary school in Berlin-Lichtenberg, but just before taking his graduation examinations in 1916, he was called to duty as a front-line soldier in an artillery regiment during World War I. After discharge in 1919, he entered Friedrich Wilhelm (now Humboldt) University in Berlin to study natural science and especially zoology. He also joined the local terrarium club "Triton" and was soon also active in another, "Salamander." Ahl received a doctorate (Dr.phil. 1921) for his systematic review of the fishes in the family Chaetodontidae.

In January 1921 he was employed as a scientific assistant in the ichthyological and herpetological collections of the university's Zoological Museum. The museum's rich collection of amphibians and reptiles, one of the largest and most important in the world, was begun in the 1820s by Hinrich Lichtenstein and enlarged by his successors, A. F. A. Wiegmann, F. H. Troschel, Wilhelm Peters, and Gustav Tornier. After Tornier's retirement in 1923, Ahl became curator for herpetology, but he continued his research on fishes. He was editor of the new German aquarium journal, *Das Aquarium*, from 1927 to 1934, in which he regularly published on fishes, amphibians, and reptiles. Ahl was also politically active and joined two Nazi organizations when Adolf Hitler became Chancellor of Germany in 1933—the Nationalsozialistische Deutsche Arbeiterpartei (NSDAP or National Socialists, the Nazi Party) and a paramilitary organization of the NSDAP, the Sturmabteilung (SA or Brownshirts)—but Ahl was expelled in 1939, apparently for his undisciplined behavior.

Ahl's work at the museum was also undisciplined. Most of his scientific publications are taxonomic in nature and he described numerous new taxa of amphibians and reptiles, especially frogs, from throughout the world. However, he did not have a thorough knowledge of the literature and his methodology was superficial and careless (for example, he often did not label types as such). A very small number of his new names are valid today—only 16% of his frogs, by one recent estimate. His most spectacular novelty, however, was a new lizard from southern China, which he named *Shinisaurus crocodilurus* (1929 [1930]) and placed in his new family Shinisauridae. It remains today the sole member of this family, named for the expedition leader, Shu-szi Sin, whose surname the Germans mistransliterated as "Shin." All told, Ahl authored some 100 titles on fishes (1921–1940) and about 65 on amphibians and reptiles (1923–1939), excluding his popular articles in *Das Aquarium*. His most extensive herpetological contributions were a review of the African frogs of the genus *Hyperolius* (1931) and volume 3 of the *Das Tierreich* checklist of the frogs of the world (1931; volumes 1–2 [1923, 1926] were authored by Fritz Nieden, a herpetologist who had been an assistant at the museum in Berlin during 1907–1915).

On 27 August 1939, five days before the outbreak of World War II, Ahl was recalled to army duty. He saw action in Poland, where he was wounded, and then was sent to North Africa. Later, he was transferred to the Balkans as an officer with a Tenth Army coastal artillery unit stationed in Herzegovina. On 14 February 1945 he was reported as missing in action in a region just west of Nevesinje, about 75 km southwest of Sarajevo and near the Adriatic coastline. He had been dismissed as curator by the museum on 31 March 1941, but whether this was because of his long absence for military service or on the basis of his performance at the museum is not known.

• *References*: "Über das Leben und Werk von Ernst Ahl," by H.-J. Paepke, Mitt. zool. Mus. Berlin, 71: 79–101, 1995; Rieck et al. 2001 (Ahl on p. 367; by Rieck on pp. 410–411). • *Portrait* (1925): George S. Myers/Alan E. Leviton portrait collection, courtesy Alan Leviton. • *Signature* (1925): Adler collection.

VELLARD, Jehan A. (1901–1996).

Jehan (born Jean) Albert Vellard was a peripatetic biologist specializing on the ethnology, herpetology, and arachnology of South America. Although ethnically French, he spent very little time in France. He was born in Béja (also called Bajah), Tunisia, about 90 km west of Tunis, on 10 June 1901 (see *Note*). Tunisia had only recently become a French colony. Vellard's father was descended from a family of Swiss mercenaries and had been sent by the French government as part of the colonizing effort. His mother's ancestors, on the other hand, were planters in the former French colony of Haiti. He developed a boyhood interest in zoology, especially in entomology.

After completing his medical training at the Université de Marseille in France (M.D. 1923), Vellard took a position at the Instituto Butantan in São Paulo, Brazil, to work with the director, Vital Brazil, on animal venoms. There Vellard conducted laboratory research on snakes, frogs, spiders, and scorpions through 1929, before he began an exploring mission to Paraguay on behalf of the Muséum d'Histoire Naturelle in Paris (1930–1932). He then moved to Recife (Pernambuco), on the Brazilian coast, where he ran a government immunology laboratory for some years before becoming director of the Museo Nacional Tihuanacu in La Paz, in Andean Bolivia (1940–1943). Although he never abandoned his zoological research, from the early 1930s Vellard's interests turned more to the ethnology of the indigenous tribes of South America. During 1943–1948 he was at the Instituto Miguel Lillo and head of the zoology department at the Universidad Nacional de Tucumán in northern Argentina.

In 1948 Vellard moved to Lima, Peru, where he founded and directed the Institut Français d'Études Andines (IFEA), a research organization run by the French ministry of foreign relations that is devoted to Andean studies in Bolivia, Colombia, Ecuador, and Peru. He also held professorships at the Universidad Nacional Mayor de San Marcos and at the Pontificia Universidad Católica del Perú and was director of the Museo de Historia Natural "Javier Prado." From this base he conducted fieldwork throughout the Andes and became particularly interested in the adaptations of humans and other animals to high elevations. He stepped down as director of IFEA in 1962 and returned to La Paz where he founded and directed the Institut Bolivien de Biologie d'Altitude. He retired from French service in 1966. Vellard returned to Argentina, living successively in Tucumán, Córdoba, and finally in Buenos Aires where he became director of the Museo Etnográfico "Juan B. Ambrosetti" of the Universidad de Buenos Aires. He died in Buenos Aires on 28 July 1996.

Vellard published an enormous body of work, mostly alone, on diverse topics: ethnology, yellow fever, curare, biogeography, conservation, and on poisonous animals (a total of more than 200 titles, among them several books including one on animal venoms, "Los Animales Venenosas de Sur América," [1947]). All of his zoological research including that on frogs was originally motivated by his interest in animal venoms. At Butantan he developed the first serums against the venom of spiders, work that was resumed in 1953 in Peru.

In addition, Vellard authored about 130 papers on frogs and snakes (1925–1991). His earliest work with snakes dealt with the nature of venoms, which he studied using biochemical and immunological techniques. Although he also published on other aspects, such as hemipenial morphology (1928) and hybrids (1929), he did not work on snake systematics while at Butantan since that subject was the domain of his erstwhile colleague, Afrânio do Amaral. Later, however,

he also wrote on geographic distribution and systematics, and described several new taxa of snakes. Vellard's first papers on frogs (1925–1929), mostly co-authored with Brazil or the bacteriologist, Arlindo de Assis, were on the cutaneous skin secretions of frogs. Only in 1946, while still in Argentina, did he begin to publish on frog systematics and distribution. He described and named numerous new taxa. He authored a series of seven long papers, each on a different genus or species group, under the general title "Estudios sobre Batracios Andinos" (1951–1960). Several of his later papers dealt with adaptations of frogs to high elevations in the Andes including his extensive studies of the frogs in Lake Titicaca on the Peruvian-Bolivian border.

• *References*: "Jehan Albert Vellard," by O. Dollfus, Bull. Inst. Franç. Andines, 25: 165–167, 1996; Personal information from Jehan Vellard, 1991. • *Portrait and signature* (1991): Adler collection. • *Note*: Vellard's date of birth is given in some sources as 17 June or 18 July 1900.

JONG, J. K. de (1895–1972).

The Dutch zoologist, Jan Kornelis de Jong, was a specialist on the reptiles of the Dutch East Indies (now Indonesia) and made detailed field studies of the Komodo Monitor (*Varanus komodoensis*). He was born on 6 April 1895 in Goedereede (South Holland) about 40 km southwest of Rotterdam. De Jong studied biology at the University of Amsterdam and received his doctorate for a dissertation on the development of the tongue sheath in snakes (Ph.D. 1925). During his time at the university he was an assistant in the Zoological Laboratory and he also worked in the Zoological Museum. In December 1927 he shipped out to the East Indies as an entomologist at the Deli Experiment Station for Tobacco in Medan (Sumatra). In 1929 he moved temporarily to the Zoological Museum in Buitenzorg (now Bogor, in Java), and from 1930 to 1935 he was an entomologist at the West Java Experiment Station, also in Bogor. Later he took a post at the Marine Biological Laboratory in Batavia (now Jakarta, in Java). After World War II De Jong was employed as head of sea fisheries at the then-new Marine Biological Laboratory in Makassar (Celebes, now Sulawesi) in 1947, and then returned to the marine laboratory in Batavia as head of sea fisheries. He retired in 1954 to the Netherlands and died in Hilversum (North Holland) about 20 km southeast of Amsterdam on 6 August 1972.

Because of the various positions he held in the East Indies, De Jong was able to report on reptile collections from various localities, both from his own collections and those of others. He described several new taxa of squamates, among them the agamid *Japalura nasuta* from Borneo and the colubrids *Cantoria annulata* and *Heurnia ventromaculata* from New Guinea. He published several papers on the Komodo monitor's anatomy and osteology and, in his last herpetological paper, an extensive report on the lizard's behavior and natural history based on his own fieldwork on Flores, Komodo, and Rintja islands in October and November 1929. All told, De Jong authored 15 titles on reptiles (1925–1937).

• *References*: "Dr. J. K. de Jong," by W. S. S. van Benthem Jutting, Bijdr. Dierk., 27: 194–195, 1939; "Jan Kornelis de Jong," by M. J. van Steenis-Kruseman, Flora Malesiana, ser. 1, 1: 266, 1950; "In Memoriam Dr. Jan Kornelis de Jong," by W. D. Margadant and A. P. M. Sanders, Vakblad Biol., 52: 342, 1972; Additional information from L. B. Holthuis, 1998. • *Portrait* (probably 1927): Amsterdam Zoo archives, courtesy Florence F. J. M. Pieters. • *Signature* (1930): Courtesy L. B. Holthuis.

MOSAUER, Walter (1905–1937).

Walter Mosauer was a specialist on functional morphology and adaptations to desert environments who provided the first accurate description of the movements and forces involved in the sidewinding locomotion of desert vipers. He was born in Vienna on 5 May 1905 and had a boyhood curiosity for animals. At the age of 15 he met the Austrian herpetologist, Franz Werner, who fostered his interests in reptiles and exploration. Mosauer attended the medical school of the University of Vienna, intending to be a dental surgeon, and during this time went on field excursions to Istria, Dalmatia, and Tunisia. It was during two trips to the Sahara in Tunisia (1926, 1928) that his interest in the locomotor patterns of desert lizards and snakes began. He was particularly fascinated with the sidewinding movements of the horned viper that had been recorded as long ago as the first century A.D. by the Roman poet, Lucan.

After graduation from medical school (M.D. 1929), Mosauer decided to attend the University of Michigan in the United States for a doctorate in zoology, probably on the advice of Werner who had made collections under contract for the university in 1927. Mosauer completed his dissertation on the myology of the trunk region in snakes (Ph.D. 1931). He studied

representatives of 51 genera and used both anatomical and X-ray techniques (published 1935). He became Instructor in Zoology at the University of California at Los Angeles (UCLA) in 1931, where he had the opportunity to study and compare the convergent anatomical and functional adaptations of reptiles in Californian deserts to those of the Sahara. Mosauer was a dedicated heliophile and in the field routinely dressed in just pants and shoes, even when in the desert. He also conducted research in Baja California and in Mexico. It was on a visit to Guerrero that he fell seriously ill. At first thought to be plant poisoning, he soon died in Mexico City from acute leukemia on 10 August 1937. He was barely 32 years old.

Despite his brief career, Mosauer made major herpetological contributions (39 papers, 1924–1940). He published several faunal studies: Guadalupe Mountains of New Mexico and Texas (1932), Tunisia (1934), and the Colorado Desert of California (1935). He also described two Mexican lizards (*Lepidophyma gaigeae, Phyllodactylus delcampoi*).

Mosauer's main interest, however, was in anatomical and behavioral adaptations to desert environments. His most important work was on sidewinding locomotion, which occurs in many snakes in deserts throughout the world. He was the first to describe it in correct mechanical terms. As Mosauer deduced, it is a "rolling" motion, in that single points along the snake's body do not slide over the sand, as in normal undulatory locomotion. Instead, each point on the body is set down on one point of sand, touching it firmly for a moment, and then is lifted off again, as in the rolling of a wheel. Therefore, only vertical forces, rather than transverse ones, are employed and there is no horizontal translation of the points along the substrate. As he discovered, sidewinding is used not only in soft sand, but as a method to achieve speed and distance as well. His work on this topic is contained in four papers (1928–1932). Mosauer also learned how some lizards and snakes "swim" beneath the sand's surface and studied their anatomical adaptations to living in such environments. At the time of his death, he was studying the tolerance of reptiles to high temperatures, research that inspired his UCLA colleague, Raymond B. Cowles, to extend this work.

• *References*: "Dr. Walter Mosauer," p. 800 *in* American Men of Science, 5th ed., Science Press, New York, 1933; "Walter Mosauer," by R. Martín del Campo, Anal. Inst. Biol. Mex., 8: 461, 1937; "Foundations of Early Modern Mexican Herpetology: An Indexed Bibliography of the Herpetological Works of Walter Mosauer, 1905–1937," by H. M. Smith and R. B. Smith, Trans. Kansas Acad. Sci., 73: 302–318, 1970 (1971). • *Portrait* (about 1935): UCLA Yearbook, "Southern Campus," p. 310, 1935, courtesy UCLA Archives. • *Signature* (1935): Adler collection.

DEVINCENZI, Garibaldi J. (1882–1943).

Garibaldi José Devincenzi was a Uruguayan physician and vertebrate zoologist who produced the first comprehensive catalogues of the lizards and snakes of his country. He was born in Maldonado (Maldonado), Uruguay, on 20 December 1882. His father was a prominent governmental and cultural leader who established a library and museum in his home department of Maldonado just east of Montevideo. From this influence young Garibaldi developed an early interest in the natural sciences. He attended schools in Montevideo (1891–1901) followed by studies at the Facultad de Medicina de Montevideo, from which he received a doctor of medicine and surgery degree (1909). Beginning in 1905 he also served

as a medical assistant at the Hospital Maciel and at the same time held a variety of teaching positions in Montevideo. He later became professor of natural history at the Facultad de Humanidades y Ciencias, Universidad de la República, where he taught botany, zoology, and geology.

In 1912 the government invited Devincenzi to be Director of the Museo Nacional de Historia Natural in Montevideo, a position once held by the zoologist and herpetologist, Carlos Berg. Devincenzi held this post with distinction for over 30 years. This was a difficult time for national support of science, yet he persevered and among other accomplishments reinitiated the museum's scientific journal, the *Anales*, in 1920. Although he was primarily a museum worker, he understood all aspects of the natural history of vertebrates.

In addition to his primary work at the museum, Devincenzi continued his medical duties at the hospital and in 1919 began to work in the emergency room. In 1926 he became a member of the national council of hygiene. He published regularly in medical journals. He retired in 1942 and died in Montevideo on 17 September the following year. Miguel Klappenbach, who later also served as director of the museum, named a frog for him (*Melanophryniscus devincenzii*) in 1968. On the 160th anniversary of the museum in 1997, Uruguay issued a commemorative $6 postage stamp in Devincenzi's honor.

Devincenzi produced the first catalogues of the vertebrates of Uruguay. His primary work was on fishes and he authored seven herpetological titles (1925–1939). The first and most extensive, "Fauna Erpétologica del Uruguay," despite its title, is restricted to lizards and snakes. It is modeled on the British Museum catalogues by George A. Boulenger and includes extensive information and illustrations based on numerous specimens examined by Devincenzi. He also published a three-part series, entitled "Ofidios del Uruguay" (1929–1931), that was later revised and enlarged as a single booklet (1939).

• *Reference*: "Garibaldi J. Devincenzi. Breve Nota Biográfica y Lista de Sus Publicaciones," by M. A. Klappenbach and M. E. Philippi, Publ. Extra, Mus. Nac. Hist. Nat., Montevideo, 39: 1–6, 1986. • *Portrait*: Courtesy of Alvaro Mones, 1986. • *Signature*: Courtesy of José Langone.

PICADO T., Clodomiro (1887–1944).

Clodomiro Emmanuel de los Angeles Picado Twight was Costa Rica's first academic biologist. He was born in San Marcos (Carazo), Nicaragua, on 17 April 1887, but his parents were from Costa Rica. They had taken temporary refuge in Nicaragua to escape the wrath of his mother's father, Enrique Twight, who did not approve of the marriage. After Twight died in 1887 and the first free elections were held in Costa Rica in 1889, the Picado family returned to Costa Rica where the father became a mathematics professor at the Colegio San Luis in San José. After Clodomiro finished his schooling in 1906, he became a professor of natural sciences at the Colegio de San Juan Gonzaga in Cartago. He soon went to Paris on scholarship to study zoology and botany at the Sorbonne, the University of Paris (diplomas in 1909 and 1912, respectively). He then studied at the Institut Pasteur and the Institut de Medecin Coloniale before receiving his doctorate at the Sorbonne (D.Sc. 1913). On returning to Costa Rica in 1914, he became director of the Laboratorio de Análisis Clínico at the Hospital San Juan de Dios in San José and over the next several years accepted professorships at several colleges in the city. At the hospital he performed research on the treatment of typhoid with a broth prepared from fungi, which many have likened as a precursor to the discovery of penicillin. His research on the ecology of the tank bromeliads is considered to be a pioneering work in tropical ecology. He also conducted research on the control of insect pests, aging, and other aspects of public health.

Picado was also interested in the physiology of envenomation by snakes, fishes, and invertebrates and directed the production of antivenoms for pitvipers. His strictly herpetological publications were few (6 titles, 1926–1933), but include reviews of the venomous snakes of Costa Rica and studies on the microornamentation of their scales. His most important work on snakes was his book, "Serpientes Venenosas de Costa Rica. Sus Venenos Seroterapia Anti-Ofídica" (1931; reprinted 1976).

Picado became director of the Instituto Nacional de Higiene in 1940. He died suddenly of a heart attack on 16 May 1944 in San José. In 1970 the Instituto Clodomiro Picado, a unit of the University of Costa Rica, was established in his honor. Because of the antisera produced by the institute, Costa Rica today has the lowest snakebite mortality rate in Latin America.

• *References*: "Vida y Obra del Doctor Clodmiro Picado T.," by M. Picado Chacón, Ed. Costa Rica, San José, 286 pages, 1964; "Clodomiro Picado Twight," p. 182 *in* G. Malavassi V. and P. R. Gutierrez, Diccionario Biográfico de Costa Rica, Univ. Autónoma Centro Amer., San José, 1992. • *Portrait* (1915): From Rev. Biol. Trop., 35: opposite p. 1, 1987. *Signature* (1937): Adler collection.

KOBA, Kazuo (1904–1981).

The Japanese herpetologist, Kazuo Koba, a specialist on the reptiles and amphibians of the Ryukyu Archipelago, was born on 25 March 1904 in Sendai (Kagoshima), on Kyushu, Japan's southernmost main island. He studied science at Tokyo College (graduated 1927) and had already visited the Ryukyus by 1926. After teaching high school for three years, he entered Tokyo Arts and Science University (graduated 1933) and continued his research there until 1936. In that year he was transferred to Manchuria, a portion of northeastern China over which Japan had completed its control in 1931. For two years there he taught elementary school in Mukden (now Shenyang, Liaoning Province). In 1938 he became curator of zoology at the newly opened Central National Museum of Manchoukuo, a Japanese puppet state in Manchuria, located in Hsinking (now Changchun, Jilin Province). While based in Manchuria, Koba made regular field excursions, mainly to study snakes. He made the first investigation (1938) of the extreme abundance of pitvipers—a partly arboreal species now known as *Gloydius shedaoensis*—on Syôryûzan Island (now Shedao or "Snake Island," Liaoning). In the southeastern part of the island he found a density of one individual per square meter!

In 1944 Koba moved to the Natural Resources Research Center in Tokyo and in 1952 joined the faculty of Kumamoto University, located on Kyushu. He soon began his studies on the herpetofauna of the Ryukyus, which extend from Kyushu to Taiwan, and began work on his doctorate (Kyushu University, D.Agr. 1961). He retired from Kumamoto in 1969 and then became the founding Director of the Japan Snake Institute, located about 100 km north of Tokyo, for three years; he was succeeded by Yoshio Sawai. Koba became a professor in the junior college associated with Ginkyo University in Kumamoto in 1973. Even at this time, when he was in his seventies, Koba conducted active research including strenuous nighttime field studies of pitvipers and early morning laboratory observations long before dawn to record egg-laying behavior. He died in Kumamoto on 10 October 1981.

Koba authored about 35 titles on amphibians and reptiles (1926–1979). His earliest were short notes on animals from various parts of Japan, and his papers naturally shifted to Manchuria after he moved there in 1936. When he went to Kumamoto in 1952, most of his research focused on the amphibians and reptiles of the nearby Ryukyus (then often called the Loo Choo Islands) and especially on the pitvipers of the genus *Trimeresurus* (now *Protobothrops*). His main interest was life history including food, eggs, egglaying, hatching, and field behavior, on which he meticulously recorded his data. His major work, based on his doctoral dissertation, is "Studies on the Snakes of the Genus *Trimeresurus* of the Amami and Tokara Islands, Japan" (1962), which also includes brief observations on the amphibians and other reptiles of these islands. In addition, Koba was interested in the biogeography of Japanese animals and published several papers on the topic, some of them with the zoologist and herpetologist, Yaichirô Okada.

• *References*: ["Dr. Kazuo Koba, 1904–1981,"] by H. Morimoto, Japan. Jour. Herpetol., 9: 103–105, 1982; ["Dr. Kazuo Koba, the Gentleman,"] by R. C. Goris and Y. Goris, Japan. Jour. Herpetol., 9: 105–106, 1982. • *Portrait* (1961): Howard K. Gloyd collection, courtesy Jonathan Campbell. • *Signatures* (Japanese form, 1973): Courtesy of Masafumi Matsui; (Romanized form, 1967): Adler collection.

ROSE, Walter (1884–1964).

Walter Rose was a dental surgeon, amateur herpetologist, and popularizer of amphibians and reptiles in South Africa. He was born in Liverpool, England, on 21 August 1884. After practicing dentistry in Ireland for 13 years, he immigrated to South Africa in 1920 and set up a practice in Cape Town. His herpetological fieldwork and writing began soon after his arrival. Rose became a prolific writer of provocative letters to the Cape Town newspapers on a wide range of subjects (especially animal welfare), which earned him a large following. He retired from dentistry in 1956 and then was associated with the South African Museum in Cape Town as Honorary Curator of Amphibians and Reptiles (1957–1960), but most of his time was spent writing on herpetology. He died in Cape Town on 13 September 1964 at the age of 80.

Rose authored about 25 titles on amphibians and reptiles (1926–1962). Most of his technical papers—on systematics, distribution, and behavior—appeared in the 1920s including

the description of *Xenopus gilli* (1927, with John Hewitt), a rare species first discovered by Rose near Cape Town. He did a great deal of fieldwork and usually sent the new species he found to others for description (Hewitt and John Power). Beginning in 1954 he published numerous semi-popular articles on amphibians and reptiles in *African Wild Life* magazine, but his professional and amateur interests merged only once, in a paper in the *South African Dental Journal* on the specialized fangs of spitting cobras (1949).

Rose is best known, however, for a series of books on South African amphibians and reptiles written for the general public. They are authoritatively researched and yet entertainingly written, and they have served to popularize these animals for more than 75 years. The first was "Veld & Vlei. An Account of South African Frogs, Toads, Lizards, Snakes, & Tortoises" (1929; the individual sections were issued in pamphlet form the same year, each with its own title, such as "Lizards of South Africa"); "The Adventures of Breviceps, The Rain Frog" (1943, with an edition in Afrikaans the same year); his most widely known book, "The Reptiles and Amphibians of Southern Africa" (1950; revised and enlarged edition 1962 [reprinted 1997]); and "Snakes—Mainly South African" (1955). His sense of humor is evident throughout. One chapter, imposingly titled "Advice for the Inexperienced on Catching and Handling Poisonous Snakes, Together with Full Instructions for Keeping Them in Captivity," is followed by a single word of text—"Don't"—to which there is a footnote acknowledging the British humor magazine, *Punch*.

• *References*: "Dr. Walter Rose," anonymous, South. Afr. Mus. Assn. Bull., 8(5): 158, 1964; "Walter Rose. (1884–1964)," by C. E. Gow, Jour. Herpetol. Assn. Rhodesia, 23–24: 58, 1965; "Dr. Walter Rose," by J. C. Greig, Cape Herpetol. News, 3: 26–28, 1980. • *Portrait:* Adler collection. • *Signature* (1936): Natural History Museum, London, courtesy A. F. Stimson.

NETTING, M. Graham (1904–1996).

Morris Graham Netting, a herpetologist, conservationist, and museum administrator, was born on 3 October 1904 in Wilkinsburg, a town that is now part of the eastern suburbs of Pittsburgh. He developed his interest in nature from visits to the town park to catch snakes. In 1922, while a student at the University of Pittsburgh (B.S. 1925), he became a volunteer in the bird section at the natural history museum of the Carnegie Institute, which administers a group of museums presented to the city in 1895 by the industrialist, Andrew Carnegie. Because of Mr. Carnegie's special interest in dinosaurs, the museum, which is today called the Carnegie Museum of Natural History, quickly established itself as one of the major natural history museums in the United States. Netting got a paid position in herpetology in 1925 at 25 cents an hour. He became a master's student at the University of Michigan (M.S. 1929); Alexander G. Ruthven, director of the Museum of Zoology, was his advisor. He also learned field techniques including long-term studies from Frank N. Blanchard of the Department of Zoology.

Netting intended to work for his doctorate but the financial climate associated with the stock market crash in October 1929 caused him to return to a paying job at the Carnegie Museum, no matter how meager the salary was. He became Assistant Curator of Herpetology in 1929 and began to build the collections, partly through purchase and also through his own collecting in eastern United States and in Panama, Trinidad, and Venezuela. By 1949 he took on additional duties as Assistant Director of the museum and became Director in 1954, thus leading the museum's renaissance after World War II.

Netting was an early conservationist and, among other efforts, co-founded the Western Pennsylvania Conservancy in 1932. As director, he established the museum's international programs to train curators and collection managers, which eventually involved students from 40 countries. His crowning achievement was to establish the Powdermill Nature Reserve, named for a Civil War–era mill that produced gunpowder, located 70 km southeast of Pittsburgh. This nearly 900-ha tract of land became the museum's biological field station, dedicated to long-term studies of natural populations of the sort that Netting had learned long before from Blanchard. Netting was awarded an honorary doctorate by Pennsylvania's Waynesburg College (D.Sc. 1950). He retired in 1975 to live at Powdermill where he carried on a long-term study of box and wood turtles. He died there on 26 August 1996.

Netting's herpetological interests were in distribution, systematics, and life history (nearly 70 titles, 1926–1969). Most of his papers dealt with the distribution of amphibians and reptiles in western Pennsylvania, West Virginia, and Virginia. The most important discovery from his West Virginia fieldwork was *Plethodon richmondi*, a new species that had masqueraded as an attenuated form of the leadback phase of *P. cinereus*. It was named for Neil D. Richmond, his eventual successor as curator in 1955 (see *Note*), and published jointly with M. B. Mittleman, who had independently discovered the same species in Ohio. Netting's fieldwork in Florida was done jointly with Coleman J. Goin, with whom he described a new frog and two salamanders including a dwarf siren (*Pseudobranchus axanthus*). They also teamed to name a new Bahamian boa (*Epicrates exsul*).

• *References*: "M. Graham Netting 1904–1996," by C. Gans, Copeia, 1997: 658–659, 1997; R. Conant, 1997 (Netting on pp. 470–471). • *Portrait* (1988): By Kraig Adler. • *Signature* (1939): Adler collection. • *Note*: Grace L. Orton joined the staff in 1945 and briefly replaced Netting as curator, but resigned in 1950.

STULL, Olive Griffith (1905–1968).

Olive Griffith Stull, a specialist on snake systematics and the pathology of chickens, was born on 10 February 1905 in Rochester, located along Lake Ontario in western New York State. She attended Smith College (A.B. 1926), an all-women's school in western Massachusetts, where the herpetologist, Emmett R. Dunn, was one of her professors. Her first paper, on geographic variation in the scutellation of snakes, was published just before she graduated. After a year at Syracuse University, not far from home, Stull transferred to the University of Michigan where she completed her graduate education (A.M. 1928, Ph.D. 1929). She was the last student to work closely with Alexander G. Ruthven before he assumed the Michigan presidency in 1929. Ruthven

had developed a new style of doctoral dissertation in herpetology—an in-depth revision of a single, well-defined genus—with his own review of *Thamnophis* in 1906. Stull chose another colubrid genus, *Pituophis*. Upon graduation from Michigan, she became a biology professor at Virginia State Teachers College in Fredericksburg for a year. In 1930–1931 she held a National Research Council fellowship at the Museum of Comparative Zoology of Harvard College, where Thomas Barbour was her sponsor. While at Michigan, Stull had planned to monograph the boas and pythons (then all placed in a single family, Boidae) and at Harvard she pursued this project in earnest. She married in 1930 (see *Note 1*) and after leaving Harvard became a homemaker, yet she still continued to publish on boids.

Stull returned to academic work as a result of World War II. Being trained as a vertebrate zoologist, she decided to do veterinary research, specifically the diseases of poultry. In 1943 she became an agent in poultry pathology of the Bureau of Animal Industry of the federal Department of Agriculture. The laboratory was located at Purdue University in West Lafayette (Indiana). Concurrently, she was also a technician, later an assistant, at Purdue and in 1953 became assistant professor in Veterinary Science, which at that time was a department in the College of Agriculture. When the School of Veterinary Medicine was established in 1959, she joined it and in 1961 was promoted to associate professor in veterinary anatomy. She taught histology and veterinary anatomy at a time when female students and faculty were rare in veterinary schools. Her research was on avian leukosis, a cancer of the white blood cells. When she contracted lymphoma in 1952, she suspected that the disease might have been transferred to her from chickens used in her research, but such a mode of transmission has never been established. She died on 15 June 1968 in Lafayette (Indiana).

Stull authored ten titles in herpetology (1926–1956), all of them on snakes. In these she named 15 taxa in the genera *Boa*, *Enygrus* (now *Candoia*), *Epicrates*, *Eryx*, *Liasis*, *Pituophis*, *Python*, and *Tropidophis* from Australia, Bolivia, Brazil, the East Indies, New Guinea and the Solomon Islands, Singapore, the United States, and the West Indies. Her major work, "Variations and Relationships in the Snakes of the Genus *Pituophis*" (1940; reprinted 2005), was not published until long after her dissertation was finished in 1929 and the text was not significantly revised in the interim. Her book received some serious criticism for her handling of localities, statistical data, and taxonomic conclusions. Later, her work on boids suffered a serious blow when virtually all of the specimens she had borrowed from museum collections were destroyed by accident. It was routine in the veterinary anatomy department at Purdue for janitors to incinerate the contents of garbage cans because animal tissues were discarded in them. Unfortunately, a substitute janitor did not know that Stull kept her preserved snakes in garbage cans filled with alcohol. Her manuscript on the boas and pythons of the world was never published (see *Note 2*).

• *References*: "My Life With Cancer," by O. S. Davis, Ladies' Home Jour., October, 81(9): 68–69, 122, 124, 126, 1964; "Prof. Olive Griffith Stull Davis," p. 1140 *in* Amer. Men Sci., 11th ed., R. R. Bowker Co., New York, 1965; "Olive Griffith Stull Davis (1905–)," p. 333 *in* M. Ogilvie and J. Harvey (eds.), Biographical Dictionary of Women in Science, vol. 1. Routledge, New York, 2000; Additional information from Ruth Ann Weiderhaft, School of Veterinary Medicine, Purdue Univ., 1999. • *Portrait* (1964): School of Veterinary Medicine, Purdue Univ., courtesy Kevin R. Doerr. • *Signatures* (1928, as Stull; 1961, as Davis): Adler collection. • *Note 1*: During 1930–1950 Stull was married to Loy E. Davis. Her herpetological publications used the surname Stull, whereas those in avian pathology used Davis, even after her divorce. • *Note 2*: The manuscript, drawings, and data sheets are in the library of the Division of Amphibians and Reptiles at the National Museum of Natural History in Washington.

Cott, Hugh B. (1900–1987).

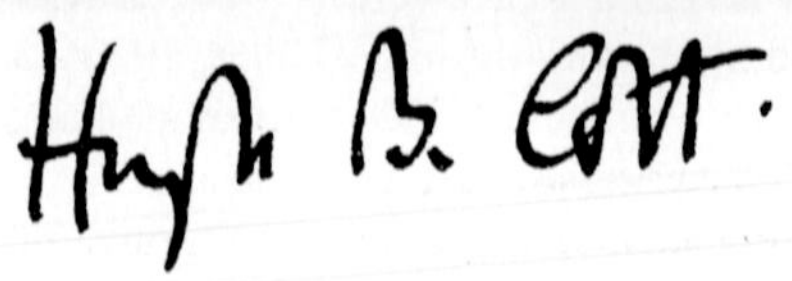

Hugh Bamford Cott was an English zoologist, expert on camouflage, conservationist, artist, and a specialist on the biology of amphibians and reptiles. He was born on 6 July 1900 in Ashby Magna (Leicester) where his father was a clergyman. After schooling at Rugby and the Royal Military College at Sandhurst, he served in the British Army in Ireland (1919–1921) before going to Selwyn College in Cambridge (B.A. 1925, M.A. 1927), originally as a priest in training. Beginning in 1928 he held lectureships in zoology, first at Bristol and then at Glasgow (where he earned a D.Sc. in 1935). He returned to Cambridge in 1938 as lecturer and Strickland Curator in the university's Museum of Zoology, a distinguished post once held by Hans Gadow. His students remember him as an enthusiastic teacher who would impress them with his exquisite chalkboard drawings, which would be faintly drawn before the lecture so that they came out perfectly! During austerity after World War II, he experimented with the edibility of exotic eggs and used the faculty and students of his college to try ostrich omelettes and other tasty concoctions. He was awarded a Sc.D. by Cambridge in 1947 on the basis of career achievements in print. Except during the war, he taught at Cambridge until his retirement in 1967.

Because of his research on animal camouflage and his book on adaptive coloration (1940), Cott's primary duties during the war were to train personnel in the most effective means to disguise people and equipment. As a British Army officer, he saw duty with the Royal Engineers in Palestine and also in the Western Desert of North Africa where he had been responsible for the camouflage of the British Crusader tanks in 1941–1942. He was chief instructor at the army's Middle East Camouflage School (1941–1943), but later was transferred to the Mountain Warfare Training Centre (1943–1944). Returning to Cambridge, Cott became dean (or chaplain) of Selwyn College (1966–1967) just before retiring, after which he was a self-employed writer and artist and conducted "Swan Hellenic Tours" safaris to East Africa. He died in Stoke Abbott (Dorset) on 18 April 1987.

Cott was a zoologist who liked to explore and to observe animals in nature. He began as a Cambridge student with trips to southeastern Brazil (1923) and the Amazon basin (1925–1926), encouraged by a fellow Selwyn College student, the herpetologist H. W. Parker. Cott's first paper reported on the behaviors of animals observed on the Ilha de Marajó at the mouth of the Amazon (1926). There he made observations on mimicry that foreshadowed his later research. In 1927 the

Zoological Society of London sponsored his expedition to the Lower Zambesi Valley. This was his first introduction to the continent that would occupy much of his academic and military life. The trip produced several herpetological papers, with new taxa of lizards and snakes, and live animals collected for the zoo's new reptile house that opened in 1927. Further observations on mimicry and on warning (or aposematic) colors were reported from several of his trips, and these topics became the central theme to his research of this period, summarized in his classic book, "Adaptive Coloration in Animals" (1940; reprinted 1966).

Following war duty his interests returned to Africa where he decided to study the biology of crocodiles and fish-eating birds. He spent the 1950s in Uganda and Northern Rhodesia (now Zambia) studying the ecology and economic status of *Crocodylus niloticus*. In 1961 he published a long monograph on this topic in which, among other results, he gave the first evidence that crocodiles swallow stones to provide ballast when in water. This work covered every aspect of the life history of this crocodile and provided the standard for later studies of African (Anthony C. Pooley), Australian (Harry Messel, Grahame J. W. Webb), and American species (John B. Thorbjarnarson).

Cott also pointed to the rapid decline of wild populations, which led him to take an active role in conservation. To protect nesting female crocodiles he arranged for the rescheduling of observation trips for tourists. He became chairman of the Crocodile Specialist Group of the International Union for the Conservation of Nature and in 1972 co-authored with Pooley a major report on the status of Africa's three species of crocodiles.

Cott's interests broadened to all of Africa's wildlife, which was the subject of his last book, "Looking at Animals. A Zoologist in Africa" (1975), and is illustrated with his own work. The final chapter, on how to photograph and draw wildlife, was clearly an effort to entice his audience into a deeper and aesthetic appreciation for all of nature.

• *References*: "Hugh Bamford Cott," p. 413 *in* Who's Who 1985, A. & C. Black, London, 1985; "Obituary: Hugh Cott," by A. Bellairs, Herpetol. Jour., 1: 206, 1987; "Dr. H. B. Cott (1900–1987)," p. 64–68 *in* Selwyn College Calendar, 1987; "Hugh B(amford) Cott 1900–1987," p. 117 *in* Contemporary Authors, vol. 122, Gale Res. Co., Detroit, 1988; Additional information from J. L. Cloudsley-Thompson, 1996–1997. • *Portrait* (1945): Selwyn College Archives, Cambridge University, courtesy Elizabeth Stratton. • *Signature* (1936): Adler collection.

PING, Chih (1886–1965).

Chih (sometimes Chi) Ping (see *Note*), the founder of modern biology in China and an early specialist on reptiles who trained a generation of Chinese herpetologists, was born on 9 April 1886 in Kaifeng (Henan Province) in east-central China. His father, a descendent of Manchus, was a public school teacher, but died when his son was 14. Ping originally studied the classics and in 1902 attended Henan College in Kaifeng. In 1904 he went to Peking (now Beijing) to study English, mathematics, and science at Imperial Capital University (now Peking University) and developed a special interest in biology and Darwinian evolution. He graduated in 1908 and decided to study in the United States. It was a propitious time to do so, because in 1908 President Theodore Roosevelt authorized funds from the indemnity imposed on China after the Boxer Rebellion of 1900 to be used to fund scholarships for Chinese students to study in America (see *Note*). At that time, China was most in need of practical skills such as agriculture and engineering and the largest number of students, including Ping, went to Cornell University in New York State. Ping entered Cornell in 1909 and graduated in agriculture (B.S. 1913) and then completed his dissertation in entomology (Ph.D. 1918). The American herpetologist, Karl P. Schmidt, was a fellow student in the same professor's laboratory. After a postdoctoral period at the Wistar Institute of Anatomy and Biology in Philadelphia, Ping returned to China in 1920.

In 1914, together with Chinese students at other American universities, Ping founded the Science Society of China, which they brought to China when they returned home. This led to transformative academic changes. At that time there were no departments of biology at Chinese universities or institutes for scientific research, but Ping soon founded both. He became a professor, in turn, at Nanking Higher Normal University (later Southeastern University, today Nanjing University), where he founded a biology department in 1921, Xiamen University, National Central University, and Fudan University. In 1922 he began the Institute of Biology of the Science Society of China in Nanking. Six years later, he established the Fan Memorial Institute of Biology (out of which grew both the Institute of Botany and the Institute of Zoology, Academia Sinica) in Beijing where he spent the last years of his life (1955–1965). In 1955 he was honored by election as an Academician of Academia Sinica. He died in Beijing on 21 February 1965.

Ping published broadly in zoology including morphology, physiology, entomology, and paleontology, and he mentored a large number of students in diverse subjects. In the 1930s his special interest was amphibians and reptiles. His students of that period included Mang-ven Chang, Tsong-han Chang, Ping-wen Fang, Hsi-fan Hsü, Tsen-hwang Shaw, and Tchong-peng Sun. In the late 1950s he mentored another herpetologist, Er-mi Zhao. Ping published about ten papers in herpetology (1927–1934). In addition to titles on sea snakes, faunal surveys, and the anatomy of gecko tongues, he authored three papers (1929–1930) on fossil and Subrecent turtles in which he described four new species and erected two new genera. He was the first to record a fossil turtle from China, a tortoise from the Eocene of Henan, his home province (1929).

• *References*: ["Brief Biography of Professor Bing Zhi,"] by X.-W. Wu et al., China Hist. Mater. Sci. Technol., 7(1): 15–29, 1986; ["Bing Zhi (1886–1965),"] by W.-Q. Zhu and Q.-H. Zhai, p. 21–35 *in* J.-Z. Tan (ed.), Biographies of Modern Biologists in China. Hunan Sci. Technol. Press, Changsha, 1988; Zhai, Q.-H. and Z.-G. Hu (eds.), Collected Works of Bing Zhi, vol. 1, Peking Univ. Press, Beijing, 2006 (Ping biography on pp. 1–19); Additional information from Ping's daughter, Qi-hui Zhai, 2006. • *Portrait* (about 1930) *and Chinese signature* (1940s): Courtesy Qi-hui Zhai. • *Romanized signature* (1918): Archives, Cornell University Library. • *Note*: Ping's original name was Ping-chih Chai. He changed it to Chih Ping in 1909 when he took the examinations for an indemnity scholarship. Many examinees changed their names to avoid embarrassment in case they did not pass. Ping never used his original name again, except during the war with Japan (1937–1945) to hide his identity. The modern Pinyin version of his name is Zhi Bing.

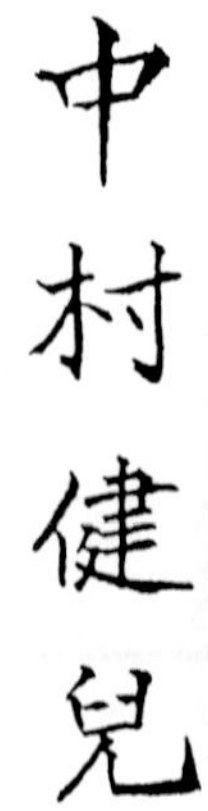

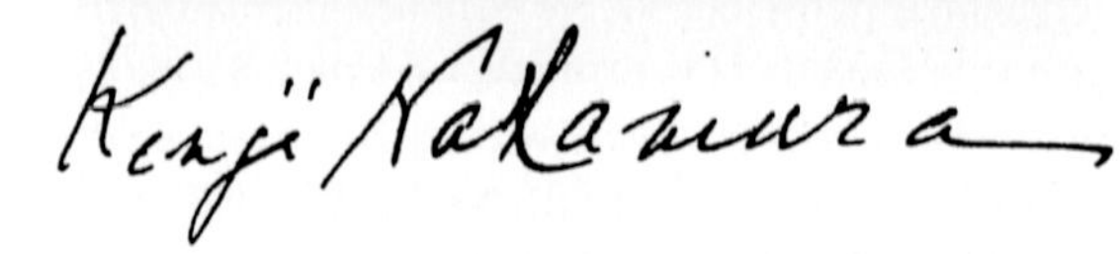

NAKAMURA, Kenji (1903–1984).

The Japanese cytologist and morphologist, Kenji Nakamura, who studied the chromosomes, morphology, and systematics of reptiles, was born to a noble family in Tokyo on 23 November 1903. His father, a baron, was a general in the Imperial Army and once served as minister of the Imperial Household. Kenji studied zoology at Kyoto Imperial University and finished in 1927. He continued for graduate study during 1927–1929, after which he became a lecturer and worked for his doctorate (D.Sc., 1935) under Taku Komai who had been a student of the American geneticist and future Nobel laureate, Thomas H. Morgan. In July 1937 Nakamura was called up for military duty as an officer in the First Telegraph Regiment and served, intermittently, throughout World War II until August 1946. He then returned to the university (renamed Kyoto University in 1947) as an assistant professor and was promoted to professor in 1949. Because of the university's age limit, he retired in 1967, but then served as a professor at Osaka Physical Education University (1967–1980). He died in Kyoto on 30 July 1984.

Nakamura studied a range of topics including sex chromosomes in orthopteran insects and fertilization and cleavage in sea urchins. His herpetological work (about 25 titles, 1927–1988) covered three major areas. His first interest was reptilian chromosomes (12 papers, 1927–1937). He was an accomplished cytologist and personally owned the best microscopes. He published on techniques including methods he developed in the late 1920s for fixing and staining the chromosomes of turtles, lizards, and snakes. As a result, he established 48 as the fundamental number of chromosomes in reptiles and he discovered the ZZ-ZW method of sex determination in reptiles in which the female's sex chromosomes are heteromorphic. His next studies were on snake morphology and he published papers on the skeleton, reproductive organs, and blood vessels of *Typhlops* (4 papers, 1938–1941). He discovered the nuchal poison glands in *Natrix tigrina* (1935, 1936) that are used in defensive behaviors. When he first dissected the glands, the nuchal secretions squirted into his eyes. He experienced acute pain and had to be rushed to the hospital. Recently, it was discovered that these snakes obtain these steroids by consuming toads and sequestering their toxins. Edmond Malnate (1960) used the presence of this gland as one character to resurrect the generic name *Rhabdophis* for this and 14 other species of Asian natricine snakes.

Nakamura was also interested in the systematics and distribution of the Japanese herpetofauna, beginning with his work on Taiwanese turtles (2 papers, 1934). However, he is best known to the general public for his two books, "Snakes and Their Benefits and Problems for Man" (1947) and "Japanese Reptiles and Amphibians in Colour" (1963, co-authored with Shun-Ichi Uéno, with excellent color drawings of each species by Takaji Matsui). In 1960 a publisher in Tokyo decided to produce a series of textbooks entitled "Systematic Zoology," with a volume about each class of animals. Nakamura was invited to write the section on reptiles and produced a manuscript of more than 2,000 pages, in large part based on Alfred S. Romer's "Osteology of the Reptiles" (1956), but this was left incomplete and outdated at his death in 1984.

Two herpetologists at Kyoto University—Tsutomu Hikida and Masafumi Matsui—massively revised and expanded Nakamura's manuscript. This part of the reptile section was published in 1988, with Nakamura as senior author. (The second part of the section on reptiles was authored by Matsui alone and published in 1992.) This text is the most comprehensive reference in Japanese covering the morphology, systematics, and distribution of the reptiles of the world.

• *References*: ["Dr. Kenji Nakamura, 1903–1984,"] by H. Fukada, Japan. Jour. Herpetol., 10: 120–122, 1984; H. Sato and N. Kobayashi (eds.), ["Memorials for the Late Dr. K. Nakamura on the First Anniversary of His Death,"] privately publ., Kyoto, 32 pages, 1985; ["Mr. Kenji Nakamura,"] by M. Ueno, Syst. Zool. Month. Rep., Nakayama Shoten, Tokyo, 19: 1–2, 1988; Additional information from Hajime Fukada, 1990. • *Portrait* (about 1975): courtesy Hajime Fukada. • *Signatures* (Japanese form): courtesy Hajime Fukada; (Romanized form, 1956): courtesy Nakamura's son, Yasuo Nakamura.

MORI, Tamezô (1884–1962).

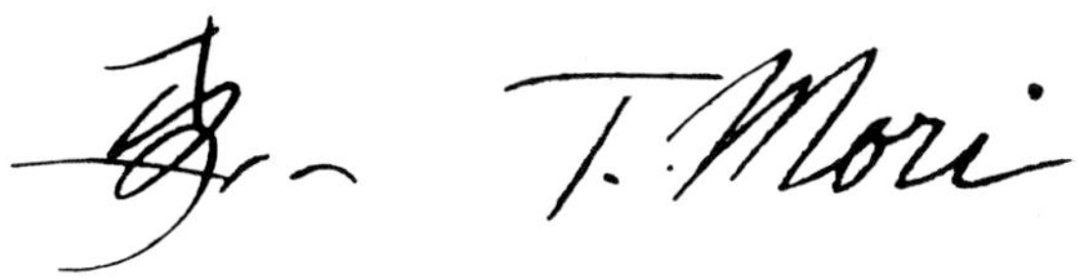

The Japanese vertebrate biologist, Tamezô Mori, who was born on 1 June 1884 in Himeji (Hyôgo), was a specialist on the fishes, amphibians, and reptiles of Korea and Manchuria. He entered the natural history course in the preparatory center for teachers associated with Tokyo Imperial University (now University of Tokyo; graduated 1904), after which he taught at high schools in Japan. Following Japan's annexation of Korea in 1910, Mori became a teacher at Keijo Higher Primary School in Keijo (now Seoul), a school for native Koreans. He also began to conduct fieldwork to study the vertebrates of Korea and made study trips to the British Museum (Nat. Hist.) in England and to Stanford University in the United States during 1921–1922. In 1925 he became a professor in the preparatory department of Keijo Imperial University (now, after several mergers, Seoul National University).

Earlier, Japan had taken control of China's Liaotung Peninsula (Liaodong Bandao) and in 1931 captured the rest of Manchuria, which together with the part already under its control was re-named Manchukuo. During 1933 Mori, as zoologist, joined the First Manchukuo Expedition, led by Shigeyasu Tokunaga. The zoologist and herpetologist, Yaichirô Okada, wrote the section on amphibians and reptiles (1935). Until this time, Mori could travel freely throughout Korea and adjacent China, even as far south as Shanghai, but with the outbreak of war with China in 1937 followed by World War II, he had to confine most of his later fieldwork to Korea. He received his doctorate at Kyoto Imperial University (now Kyoto University; D.Sc. 1936) under the supervision of Taku Komai for a dissertation on the freshwater fishes of East Asia. After the end of the war, Mori was repatriated late in 1945 and Keijo Imperial University was closed. He became president of Hyogo Prefectural Medical University in 1946, then professor at Hyogo University of Agriculture in 1949 (in 1951, also vice president), and in 1957 professor at Mukogawa Women's University. He died on 18 July 1962.

Mori published a number of books on the animals and plants of East Asia including "A Hand-List of the Manchurian and Eastern Mongolian Vertebrata" (1927 [1928]) and other books on freshwater fishes of East Asia and of Manchuria, butterflies of Korea, and plants of Korea. He authored six papers on the amphibians and reptiles (1927–1929) including reports on Quelpart (now Cheju) Island off the southern tip of the Korean Peninsula and Tsushima Island between Korea and Japan. He also named two new taxa of salamanders (*Hynobius*) of which *H. quelpaertensis* of southwestern Korea is currently recognized.

• *References*: ["Dr. Tamezo Mori,"] anonymous, Japan. Jour. Ichthyol., 6, insert preceding p. 75, 1957; Hasegawa, 1967 (Mori on pp. 85–86). • *Portrait:* Anonymous, 1957. • *Signatures* (Japanese form, 1927; Romanized form, 1928): courtesy Tsutomu Hikida.

BRONGERSMA, L. D. (1907–1994).

The Dutch vertebrate zoologist, Leo Daniël Brongersma, made significant contributions to Indo-Australian herpetology, exploration of New Guinea, snake anatomy, and sea turtle biology. He was born on 17 May 1907 in Bloemendaal (North Holland), near Haarlem. He attended local schools and in 1925 entered the University of Amsterdam where his interest in herpetology started. He passed his examinations (comparable to B.Sc. and M.Sc.) in 1929 and 1931. By 1928 he had become an assistant in the Zoological Museum in Amsterdam where he began his study of the amphibians and reptiles of the Dutch East Indies (now Indonesia). Before receiving his doctorate (Dr.Sci. 1934, under the supervision of Johan E. W. Ihle), he became Curator of Reptiles and Amphibians at the Rijksmuseum van Natuurlijke Historie (now Nationaal Natuurhistorisch Museum) in

Leiden. He held this position from 1932 (except for an early interruption because of the world financial crisis of the early 1930s) until 1958 when he became director of the museum. He was succeeded as curator by Marinus Hoogmoed.

Brongersma's first tropical field work came in 1945–1946 when he visited Port Dickson, in Malaya opposite Sumatra, and Buitenzorg (now Bogor), Java. Beginning in 1952 he made several trips to Dutch (that is, western) New Guinea. The most extensive of these was the Brongersma-led expedition in 1959 to the Mountains of the Stars near the border of Papua New Guinea. This was a broad-based study of the anthropology, ethnology, geology, and biology of this previously unknown region that often had to be accessed by helicopters. For his leadership of this important expedition, which included 15 scientific personnel as well as a medical staff, marines, and a police detachment, Brongersma was honored by Queen Juliana of the Netherlands and made an officer in the Order of Orange and Nassau.

During 1962–1964 Brongersma participated in zoological and geological surveys of the region surrounding the bay of Ria de Arosa in northwestern Spain. In addition to his primary work at the museum, he held a volunteer faculty position at the University of Leiden, beginning in 1938; in 1964 he became professor of systematic zoology. Brongersma retired as museum director in 1972, but continued his research there. He died in Leiden on 24 July 1994 at the age of 87.

Brongersma's first and most consistent scientific interest was the systematics of the amphibians and reptiles of the Dutch overseas possessions—the East Indies, Surinam in northern South America, and islands in the Lesser Antilles—but he also published papers on New Caledonia, western China, Colombia, Venezuela, Malaya, El Salvador, and the British West Indies. A second line of research concerned the osteology and soft anatomy of snakes, in particular the trachea and lungs and their associated blood vessels (17 papers, 1938–1961). Garth Underwood, in his classic review of snake classification (1967), incorporated this research in his book and emphasized its significance for snake systematics. Finally, beginning in 1960, most of Brongersma's research became focused on sea turtles in the North Atlantic including their movements and strandings on beaches.

Brongersma authored nearly 130 herpetological titles (1927–1983) including his two major works: "Contributions to Indo-Australian Herpetology" (1934), which served to update Nelly de Rooij's "The Reptiles of the Indo-Australian Archipelago" (1915–1917), and "European Atlantic Turtles" (1972), a summary of the literature, both scientific and popular, on the subject. Brongersma described numerous new frogs, lizards, and snakes, mostly from the East Indies and Surinam. He defined the snake family Tropidophiidae (1951), based partly on the anatomy of the lungs. Perhaps his most spectacular novelty was a new python, with boldly marked labial scales, from the mountains of New Guinea, which he named *Liasis boeleni* (1953).

• *References*: "Notes on the Scientific Career of Professor Dr. L. D. Brongersma, Compiled at the Occasion of His Sixty-Fifth Birthday," by H. Boschma, Zool. Med. Leiden, 47: viii–xxii, 1972; "L. D. Brongersma 1907–1994," by M. S. Hoogmoed, Copeia, 1995: 513–516, 1995; "In Memoriam Prof. Dr Leo Daniel Brongersma (1907–1994)," by M. S. Hoogmoed, Zool. Med. Leiden, 69: 177–201, 1995. • *Portrait* (1982) *and signature* (1931): Courtesy Marinus S. Hoogmoed.

Eisentraut, Martin (1902–1994).

Martin Bruno Eisentraut was a mammalogist and explorer of the Bolivian Chaco and the mountains of Cameroon, and also a specialist on the lacertid lizards of western Mediterranean islands. He was born in the village of Groß-Töpfer (Thuringia), Germany, on 21 October 1902. Early in childhood he learned from his mother to love animals, especially birds, and he studied taxidermy at his local school. In 1921 he entered the university in Halle to study zoology and wrote a dissertation on orthopteran chromosomes (Dr.phil. 1925) under the geneticist, Valentin Haecker. He became an assistant in the Museum für Naturkunde at the university in Berlin in 1925, first in herpetology and later also in mammalogy. He joined the German Army at the outbreak of World War II and carried out work to control malaria and spotted fever in captured lands.

After the war he returned to the museum and during 1947–1950 was curator of herpetology. Together with Kurt Deckert and Heinz Wermuth, Eisentraut was responsible for moving the herpetological collection from wartime storage in the museum's cellars back to the halls. The cellars were damp and many of the jar labels, which in most cases were the only easy means to identify the contents, had fallen off. So Eisentraut

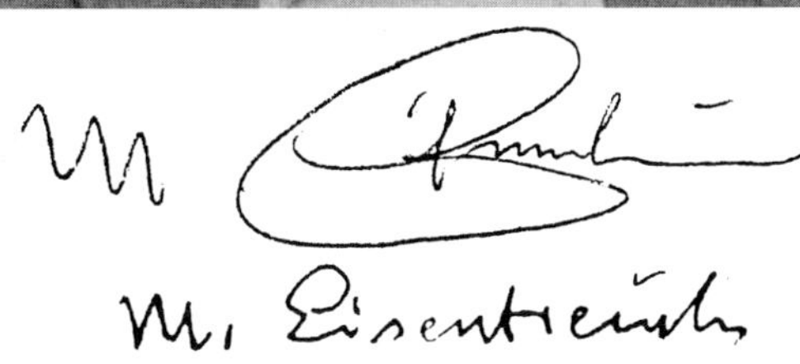

and his colleagues had the daunting task to determine the identities by checking against literature descriptions of the specimens. This was critical in a collection of such great historical importance with many type specimens. Although most of Eisentraut's research was on mammals, one of his doctoral students, Wermuth, went on to a herpetological career.

Eisentraut lived in West Berlin and worked at the museum, which after the war happened to be in the Soviet-controlled sector in the east. This situation created difficulties, not the least of which was being paid in East German marks but dwelling in high-priced West Berlin. In 1950 he moved to a new post as curator of mammals at the Staatlichen Museum für Naturkunde in Stuttgart where he worked until becoming director of the Zoologisches Forschungsinstitut und Museum Alexander Koenig in Bonn in 1957. Except for 1969–1971, he held this post until retirement in 1977. He died in Bonn on 5 July 1994.

Eisentraut was interested in broad questions of evolution, physiological ecology, faunistics, and zoogeography. His research regularly took him into the field. He first went to the Balearic Islands (including Ibiza and Formentera, grouped by Eisentraut as the Pityusic Islands) and Columbretes Islands in the western Mediterranean Sea (1928–1930). He then went to the Gran Chaco region of Bolivia (1930–1931), followed by many trips to Cameroon and the offshore island of Fernando Poo (now Bioko) in West Africa (1938–1973). His return trip to Bolivia after 50 years (1979) was summarized in a book (1983).

He authored nearly 240 titles of which 43 were on his primary research subject, bats. Of 16 titles on amphibians and reptiles (1927–1981), his primary research topic was the evolution of lizards. He named numerous geographic varieties of lacertids from the western Mediterranean islands, but few are recognized today. All of this work was summarized in his long monograph, "Die Eidechsen der spanischen Mittelmeer Inseln" (1949; a book version was issued in 1950), and in two later papers (1950, 1954). He was particularly interested in the evolution of melanism in island lizards and conducted translocation experiments in which he released melanistic and amelanistic lizards on other islands including some tiny islets that were originally lizard-free. It should be noted that these releases were done in 1930, many decades before experimental biogeography developed as a discipline. His last paper on this subject was published in 1981, together with his museum colleague, Wolfgang Böhme.

Eisentraut also wrote papers on egg-laying behavior in two species of Bolivian frogs of the genus *Leptodactylus* (1932), melanism in lizards of the genus *Cnemidophorus* on an island off the Venezuelan coast (1933), and nesting behavior of sea turtles on Fernando Poo (1964). His extensive studies in Cameroon—especially on Mt. Fako (or Mt. Cameroon), at 4,095 m the highest mountain in western or central Africa and on which he conducted studies of faunal zonation—and Fernando Poo were summarized in books (1963 and 1973, respectively), both of which have significant herpetological content.

• *References*: "Martin Eisentraut—Philosopher, Scientist, and Explorer: A Tribute," by W. Böhme, Bat Res. News, 25: 13–14, 1984; "Leben und Werk von Martin Eisentraut (1902–1994)," by W. Böhme and R. Hutterer, Bonn. Zool. Beitr., 48: 367–382, 1999; Rieck et al., 2001 (Eisentraut on pp. 368–369; by W. Böhme on pp. 440–442). • *Portrait* (1982): Courtesy Wolfgang Böhme. • *Signatures* (informal version above; formal version, 1984, below): courtesy Wolfgang Böhme.

Burt, Charles E. (1904–1963).

Charles Earle Burt, an American specialist on lizards, wrote widely on amphibians and reptiles of the Western Hemisphere and the Pacific region. He was born in Neodesha (Kansas) on 12 August 1904 and educated at Kansas State College (now University; B.S. 1926, M.S. 1927). At the University of Michigan he became a student of Frank N. Blanchard in the Department of Zoology. He had met him during the summer of 1927 at Michigan's biological station where Blanchard taught. Burt also had informal associations with the herpetologists at the university's Museum of Zoology, notably Helen T. Gaige and Alexander G. Ruthven. Before receiving his doctorate (Ph.D. 1930), he had moved to the American Museum of Natural History in New York as an assistant curator in herpetology (1929–1930) and began to study its rich collections of lizards from South America and the Pacific. He was not reappointed after one year, despite his prodigious output, which included numerous misidentifications of specimens.

Burt briefly taught at Trinity University in Waxahachie (Texas) during 1930–1931, but then moved to a professorship at the Methodist-affiliated Southwestern College in Winfield (Kansas) where he remained until 1944. During 1932–1936 he

worked under contract with the United States National Museum in Washington to collect amphibians and reptiles throughout the American Southeast. All of his fieldwork and many of his publications were done in collaboration with his wife, May Danheim Burt (1898–1988). In 1939 they started a small biological supply company in Winfield and in 1944 moved it to Topeka (Kansas). This firm, Quivira Specialties Company, sold biological specimens (living and preserved), books, reprints, equipment, and other related materials to museums, schools, universities, and to individuals. The presence of breeding colonies of European *Lacerta* and *Podarcis* in the Topeka area today is directly traceable to lizards that escaped from the company's premises. The firm continued until shortly after his death from cancer at the age of 58, in Topeka, on 13 July 1963.

Charles Burt authored about 65 herpetological titles (1927–1954), covering a wide range of subjects including distribution, systematics, behavior, and ecology, but his taxonomic focus was lizards. His earliest major work was "The Lizards of Kansas" (1928). The well-known book, "A Study of the Teiid Lizards of the Genus *Cnemidophorus*" (1931), resulted from his doctoral work and covered a group of lizards that ranges from near the Canadian border to Argentina. He was senior author of major reviews of the South American lizards in the collections of the National Museum (1930, with May Burt), the American Museum (1931, with May Burt), and those at Stanford University in California (1942, with George Myers). In 1932, with his wife, he published a summary of the Pacific Island herpetological collections made by the Whitney South Sea Expedition of the American Museum. In 1933, again with his spouse, he published a checklist of the lizards of South America. Many of his papers contained useful keys to species, but doubtless the most widely used was "A Key to the Lizards of the United States and Canada (1935 [1936]), written in the tradition of the key to snakes (1925) by his doctoral adviser, Frank Blanchard.

• *References*: "Dr. Charles E. Burt 1904–1963," by W. W. Tanner, Herpetol. Newsl. (accompanying Herpetologica, Sept. 1963), p. [4]; "Charles E. Burt," by W. W. Tanner, Copeia, 1963: 722–723, 1963; "Update on *Lacerta* in Topeka, Kansas," by J. E. Gubanyi, Kansas Herpetol. Soc. Newsl., 118: 13–14, 1999; Myers, 2000 (Burt on pp. 43, 47–48, 60, 152). • *Portrait* (1950) *and signature* (1940): Adler collection.

TANNER, Vasco M. (1892–1989).

The American entomologist and herpetologist, Vasco Myron Tanner, was born in Payson, about 15 km south of Provo (Utah Territory) on 29 October 1892. Utah became a state only four years later and was then still the American frontier. As a child, Tanner wore moccasins and spent time with Ute Indian boys with whom he made bows and arrows to hunt ground squirrels. At this early age he came to love plants and animals. In 1912 he graduated from North Sanpete High School in Mount Pleasant, in central Utah, then studied zoology at Brigham Young University (A.B. 1915) and paleontology at the University of Utah (M.A. 1916). From 1916 to 1924 he was head of the biology department at Dixie Junior College and High School in St. George, located in the southwest corner of Utah. He went to Stanford University for his doctoral work in insect morphology

(Ph.D. 1925), under the guidance of Gordon F. Ferris. At Stanford, his future research orientation was influenced by a close association with David Starr Jordan, the noted ichthyologist who had been president of the university.

On returning to Utah in 1925, Tanner became chairman of the newly established Department of Zoology and Entomology at Brigham Young University. He held this position for 33 years. He directed the work of the department, which included field surveys of Utah, Nevada, and the surrounding states, effectively the entire Great Basin region. In 1939 he founded the journal *Great Basin Naturalist* and served as its editor through 31 volumes. Also active in civic affairs, he was elected to the Utah House of Representatives in 1962. He retired from his university work in 1970 and died in Provo on 25 April 1989 at the age of 96.

Most of Tanner's publications are on entomology, mostly results of faunal surveys and descriptions of new taxa. His wife once wrote: "His first love was a new species . . . At first, it was hard to believe that I was not essential to his happiness." He also authored about 30 titles in herpetology (1927–1965), one with his younger brother and herpetologist, Wilmer W. Tanner, who joined his department as a faculty member in 1949. Most of these papers were distributional and systematic in nature including a review of Utah amphibians (1931), a description of the Utah race of *Leptotyphlops humilis* (1938), and a major review of the spadefoot toads (1939). His most important work dealt with a faraway region he never visited. During World War II several of his students served in the United States armed forces in the Pacific region (1943–1946) and made large collections of amphibians and reptiles, some 2,300 specimens, all of which were sent to Tanner for study. Despite the title of this series of eight papers—"Pacific Island Herpetology" (1948–1953), which covers the Marianas, Marquesas, Philippines, and Solomons—it also includes Australia and New Guinea as well as Korea and Selangor, a state on the west coast of peninsular Malaysia. In this series he described new taxa of squamates in the genera *Emoia* and *Typhlops* and, in a related paper with Walter C. Brown (1949), a new species of *Pseudogekko* from the Solomons.

• *References*: "Vasco M. Tanner," by C. L. Hayward et al., Great Basin Nat., 30: 181–218, 1970; "Vasco Myron Tanner: 1892–1989," by H. D. Smith and H. H. Frost, Jour. Mammal., 72: 430–432, 1991; Tanner family records, courtesy Wilmer W. Tanner, 1991. • *Portrait* (about 1965): Courtesy Wilmer W. Tanner. • *Signature* (1941): Adler collection.

WOODBURY, Angus M. (1886–1964).

Angus Munn Woodbury was an American naturalist and ecologist who published on the herpetology of Utah. His parents were early settlers in Pine Valley, located just north of St. George (Utah Territory), an isolated pioneer settlement in the southwestern corner of Utah. He was born there on 11 July 1886, ten years before Utah became a state, and went to school at Brigham Young Academy in Provo. Woodbury attended Brigham Young University starting in 1906, but withdrew to become a forest ranger in Dixie National Forest (1908–1920) and teach at Dixie Junior College and High School in St. George (1920–1926), before finishing at Brigham Young (B.S. in zoology 1927). He then took his master's at the University of Utah (M.S. 1928) before proceeding to the University of California at Berkeley where he earned his doctorate for a dissertation on the biotic relationships of Zion Canyon in southwestern Utah (Ph.D. 1931). During the summers of 1925–1933 he became the first park naturalist at Zion National Park, the state's original national park, and it was here that his interest in ecology began.

Woodbury joined the zoology faculty at the University of Utah in 1931 and retired in 1952. He later served as director of ecological research at Dugway Proving Ground, a remote 3,250-ha United States Army chemical testing facility 140 km southwest of Salt Lake City (1952–1956), and also for the Colorado River Project (1957–1964), a federal Bureau of Reclamation project to control the flow of water in the upper Colorado River. Woodbury's work on the latter, which he published in *Science* (1961), provided a key argument for the protection of Rainbow Bridge National Monument, located in southeastern Utah. Rainbow Bridge is the world's largest natural stone bridge. While still working on this project, Woodbury, together with his wife, died as the result of an automobile accident just north of Loveland (Colorado), on 1 August 1964.

Woodbury published on a wide array of subjects including ecology, distribution, systematics, behavior, conservation, and field techniques—especially about the amphibians and reptiles of Utah. His herpetological papers number about 35 titles (1928–1960). Woodbury described three new taxa of squamates from Utah including a rattlesnake, *Crotalus viridis concolor* (1929). His major work was "A Descriptive Catalogue of the Reptiles of Utah" (1931). He conducted extensive studies on the life history and ecology of the tortoise, *Gopherus agassizii* (1940 and 1948, both papers with Ross Hardy), and also published several articles on snake dens (1950–1954).

• *References*: "Angus Munn Woodbury 1886–1964," by V. M. Tanner, Great Basin Nat., 25: 81–88, 1965; "Register of the Papers of Angus Munn Woodbury," Spec. Coll. Dept., Univ. Utah Libraries, collection 176, 73 pages, 1979. • *Portrait* (about 1930) *and signature* (1941): Adler collection.

RADOVANOVIĆ, Milutin (1900–1968).

Milutin Milorad Radovanović (see *Note*), a Yugoslavian herpetologist and evolutionary biologist, was born on 15 March 1900 at the parental farm in Tabanović, near Šabac, about 75 km west of Belgrade. Serbia at this time was an independent kingdom that had only recently emerged from over four centuries of Turkish rule. In 1918, after the disintegration of the Austro-Hungarian Empire following World War I, it united with Croatia and Slovenia and in 1929 this union was officially named Yugoslavia. These and future periods of political change had significant impacts on Radovanović's life and academic career. He first went to school in Tabanović but, because of World War I, he could not attend secondary school at the gymnasium in Šabac until after a long hiatus. He finally graduated in 1922 and then went to Germany to study natural sciences in Jena where, except for one semester at the university in Innsbruck, Austria, he later graduated with a doctor's degree (Thüringischen Landes-Universität, 1928). His dissertation, on the poison apparatus of snakes, was completed under the supervision of the evolutionary biologist, Ludwig Plate.

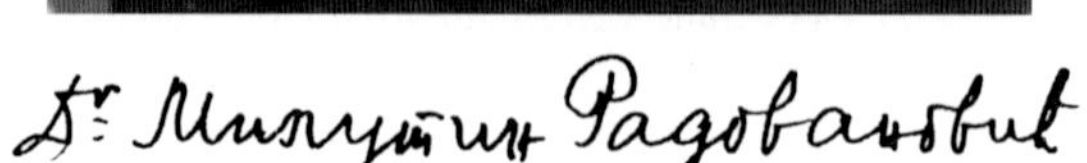

Radovanović returned to Yugoslavia where he taught at some secondary schools and then became an assistant at the university in Ljubljana (Slovenia) and later, in 1932, curator at the Zemaljski Museum in Sarajevo (Bosnia-Hercegovina). Before World War II, he worked in Belgrade at the Museum of Serbian Lands and then returned to Sarajevo as the first professor in the new Faculty of Agriculture. During the war, in which Yugoslavia declared neutrality but was later invaded and forced to join the Axis powers, he was associate professor in the Faculty of Agriculture and Forestry at the University of Belgrade. By the end of the war the political climate was changing. A republic was declared and a Soviet-style constitution instituted in 1945. Radovanović was demoted to assistant professor in the Faculty of Philosophy but still continued his academic work despite this and other misfortunes. He wrote a zoology textbook (1946; revised edition 1947) and another on evolution (1955; revised edition 1959). He resumed his scientific research and in 1956 became full professor in the Faculty of Natural Sciences and Mathematics in Belgrade. Two other books followed (1963, 1965) on morphology, phylogeny, and systematics. The relaxing political situation created other opportunities for him including travel abroad. Ironically, having survived wars and often-violent political upheaval in his homeland, he died overseas. On a trip to do fieldwork in southern Africa, he perished in an airplane accident on 20 April 1968 at the Ondekaremba International Airport, located about 45 km northeast of Windhoek, Namibia.

Radovanović's two areas of research were trichopteran insects (the aquatic caddisflies) and amphibians and reptiles. Of his 74 papers and 8 books, 30 titles are on herpetology (1928–1970). His earliest work, beginning with his doctoral dissertation, was on the soft anatomy and osteology of the head of snakes; this resulted in two major reviews (1935, 1937). From 1937, however, his herpetological research shifted to faunal, taxonomic, and evolutionary studies including work on the zoogeography and speciation of island lizards in the Adriatic Sea. He named 17 new taxa of newts (*Triturus*) and lizards (*Podarcis*) from Yugoslavia and the Adriatic islands. His major herpetological work was a book on Yugoslavian amphibians and reptiles, "Vodozemci i Gmizavci Naše Zemlje" (1951), which includes keys, distribution maps, a zoogeographical review, comments on ecology, and an extensive German summary.

• *References*: ["Dr. Milutin Radovanović (1900–1968),"] by M. Pljakić, p. 504–505 *in* H. Rot (ed.), Biologija Enciklopedijski Leksikon, Interpres, Belgrade, 1973; ["Milutin M. Radovanović (1900–1968),"] by G. V. Džukić and I. R. Savić, p. 289–345 *in* Lives and Work of Serbian Scientists, vol. 7, Serbian Acad. Sci. Arts, Belgrade, 2001. • *Portrait* (1941) *and signature*: Courtesy Georg V. Džukić and David M. Damkaer. • *Note*: In German publications, his name is spelled "Radowanowitsch."

CONANT, Isabelle Hunt (1901–1976).
CONANT, Roger (1909–2003).

Isabelle dePeyster Hunt Conant, an animal photographer and artist, and Roger Conant, a herpetologist and curator of reptiles and director at the Toledo (Ohio) and Philadelphia zoos, teamed to create the most widely used herpetological book ever published, a field guide to the amphibians and reptiles of eastern and central North America.

Isabelle Conant was born to a wealthy family in Staten Island (New York) on 17 May 1901. She was raised by a governess and attended private school. When the family's finances were ruined and her father died during her teenage years, she had to find a paying job and became a reporter for a daily newspaper, the *Staten Island Advance*. In 1936 she joined the staff of the just-opened Staten Island Zoo as business manager, publicity director, and, later, photographer. Isabelle married the zoo's charismatic curator of reptiles, Carl Kauffeld, but after they divorced she moved to the venerable and much larger Philadelphia Zoological Garden in 1942.

Isabelle was hired by Roger Conant, the zoo's curator of reptiles, public relations counsel, and membership secretary. She soon became the zoo's official photographer. With her photographic skills and Roger's talent for publicity, the Philadelphia Zoo became one of the best-known zoos in the world. Isabelle was a fearless photographer, often entering the cages of animals to get close-up shots that the newspaper photographers refused to take. In April 1947 Roger and Isabelle were married and thereafter she supported all of his herpetological work, provided photographs for his publications, and served as cook and assistant during his fieldwork throughout the United States and Mexico.

Roger Conant was born in Mamaroneck (New York) on 6 May 1909. His interest in reptiles, especially snakes, began at the age of 12 when he captured some water snakes at a Boy Scout camp. He attended the University of Pennsylvania (1927–1929), but had to withdraw because of his widowed mother's meager finances. He had wanted to be a zoo director anyway and in 1929, at the age of 19, he successfully applied for the post of curator of reptiles at the small Toledo Zoo, which was a fortuitous choice. Toledo is just 70 km south of Ann Arbor and the University of Michigan, which then had the largest graduate training program in herpetology in the United States. Conant became an unofficial student there and was advised by Helen Gaige, Frank Blanchard, and their graduate students including Howard K. Gloyd who became Roger's lifelong colleague. Conant designed the zoo's first reptile house and also developed a research program, unusual for small zoos at that time. He made collections throughout the state and produced a comprehensive book, "The Reptiles of Ohio" (1938; revised edition 1951), that was called "a landmark in the development of herpetology in the Eastern United States" by Karl P. Schmidt.

By 1935, however, Conant had returned to Philadelphia and to his new position at the zoo. He became director of the zoo in 1967. Although he still curated the herpetological collection, pioneered some of the earliest captive breeding programs, and designed a new reptile house, his duties were far broader. He edited the zoo's publications for 28 years, wrote and conducted a weekly radio show for 33 years, and made frequent television appearances, all to promote the zoo. He had extensive interests in animals, had a commanding presence as a speaker, and became a prominent spokesman for zoos.

In 1951, following preliminary correspondence, the ornithologist, Roger Tory Peterson, visited the zoo to ask Conant to write the volume on amphibians and reptiles in the field guide series that bears the Peterson name. Because of Roger's prior publications—both scientific and popular—and Isabelle's photographic skills, they were the ideal team for the task. They set out to illustrate and map every species and subspecies found in eastern North America. To photograph the animals, the Conants at one time or another temporarily possessed living examples of almost every one of them. Isabelle's technique required that the specimens be photographed in black and white. She then brown-toned the prints in her darkroom before coloring each one with watercolor dyes by hand, using the live specimen as her guide. Since the dyes were permanent, any error required her to start over again. The accuracy and artistry of this method can only be fully appreciated by seeing the original colored illustrations because the four-color printing methods of the day were mediocre. Assembling the highly detailed distribution maps and writing the text were Roger's tasks. The result, "A Field Guide to Reptiles and Amphibians of the United States and Canada East of the 100th Meridian," was published in 1958 and was an instant success.

A second edition was issued in 1975 but covered more territory to the west since Robert Stebbins's companion volume on western species (1966) extended only to the eastern borders of the Rocky Mountains. The new edition was appropriately retitled "Eastern and Central North America." A third edition, the last to have Roger's active participation, was issued in 1991 with the assistance of Joseph T. Collins. Together, these three editions have sold more than 500,000 copies. In 1998, a "third edition, expanded" appeared. Conant issued a public statement at the time, lamenting what he called a mere "face lifting" perpetrated by the publisher to boost sales. There was no new text, but color photographs were added, and with the volume expanded by nearly 170 pages and the use of thicker paper, the book was less a field guide than were the earlier editions. He publicly wondered if Peterson would have been horrified. Clearly Conant was.

Isabelle did not live to see either of the third editions published. Because of her declining health, Roger had retired early from the zoo in 1973 and they moved to Albuquerque (New Mexico), where he had an adjunct professorship at the University of New Mexico. Earlier, the University of Colorado had awarded him an honorary doctorate (D.Sc. 1971), which recognized his professional accomplishments. Without his zoo administrative tasks, Roger had more time for herpetology. Shortly after Isabelle's death—in Albuquerque on 3 November 1976—Roger visited Howard Gloyd, a friend from his Ohio days who by then was living in Tucson (Arizona). Conant decided to re-enter a project that they had started in 1932, to monograph the pitvipers of the genus *Agkistrodon*, which had been put aside when Gloyd moved to a job in Chicago and Conant returned to Philadelphia. Gloyd was terminally ill by 1976 and died in 1978, so it was Conant who completed the book, together with the aid of Gloyd's widow, Kathryn J. Gloyd, whom Conant married in 1979. He continued to publish until his very last year when at the age of 94 he died in Albuquerque on 19 December 2003. Roger's scientific collections including those made in Ohio, as well as Isabelle's photographic negatives, are deposited at the American Museum of Natural History in New York.

Roger Conant's scientific career spanned eight decades. He authored about 210 papers, reviews, and books covering the systematics, life histories, and distribution of amphibians and reptiles (1929–2003). He described about two dozen new snakes (in the genera *Agkistrodon, Alsophis, Elaphe, Eunectes, Natrix,* and *Thamnophis*) and a turtle (*Apalone*). He had a lifetime association with the Boy Scouts of America and authored four editions of their merit badge pamphlet on reptiles (1944–1976). Because of these pamphlets and the several editions of the "Field Guide," Conant carried on an active correspondence with numerous budding herpetologists who were mentored by him. He chaired a national committee that began the standardization of common names for the amphibians and reptiles of the United States and Canada (published 1956). His major herpetological works were his book on the reptiles of Ohio and the various editions of the "Field Guide," noted earlier; another book, "What Snake Is That?" (1939, with William Bridges); a monograph of the watersnakes of Mexico (1969); and a review of the gartersnakes of Mexico's transvolcanic belt (2003). His most extensive and largest work, "Snakes of the *Agkistrodon* Complex: A Monographic Review" (1990), covered the 15 American and Asian species then assigned to four genera. This massive tome represents the culmination of Conant's career and, at the same time, memorializes his 60-year-long project with Howard Gloyd, his longtime colleague.

• *References on Isabelle Conant*: "Herpetology's Loss," anonymous, Herpetol. Rev., 7: 180, 1976 (reprinted in Copeia, 1977: 209–210, 1977); "Isabelle Hunt Conant (1901–1976)," by E. V. Malnate, Herpetol. Rev., 8: 9, 1977. • *References on Roger Conant*: "The Remarkable Career of Roger Conant," by K. Adler, p. 16–23 *in* J. B. Murphy et al. (eds.), Captive Management and Conservation of Amphibians and Reptiles. Soc. Study Amphib. Rept., Contr. Herpetol., vol. 11, 1994; Conant, R., A Field Guide to the Life and Times of Roger Conant. Selva, Provo, Utah, xix, 498 pages, 1997; "In Memoriam: Roger Conant (1909–2003) with Reflections by Some of Roger's Many Friends and Colleagues," by K. Adler et al., Herpetol. Rev., 35: 101–107, 2004. • *Portrait* (middle to late 1960s) *and signatures* (Isabelle Conant, 1970; Roger Conant, 1975): Adler collection.

MAHENDRA, Beni Charan (1904–1995).

The Indian vertebrate zoologist and anatomist, Beni Charan Mahendra, studied the biology of the house gecko and the systematics of Indian snakes. He was born on 11 July 1904 and attended St. John's College, an institution founded by English missionaries in Agra (Uttar Pradesh), the historic capital of the Mughal emperors and site of the Taj Mahal in north-central India. In 1927, when St. John's became affiliated with Agra University (now Dr. Bhim Rao Ambedkar

University), Mahendra began teaching herpetology to master's students. Bombay University (now University of Mumbai) awarded him a D.Sc. in 1942 on the basis of his published papers. In 1944 he moved to Birla College in Pilani (Rajasthan) as Head of the Department of Zoology. Mahendra returned to St. John's College in 1949 as Head of its Department of Zoology and later became Dean of the Faculty of Science at Agra University. In 1954 he founded the Academy of Zoology, an international organization headquartered in Agra. He became the first president and edited its journal, *Annals of Zoology*, from 1955 until its cessation in 1991. He was elected a fellow of the Indian Academy of Sciences in 1938. Mahendra died in Agra in 1995.

From about 1950 Mahendra began a quest for the ancient Hindu views of life and decided to read the ancient texts in their original Sanskrit, a classical language of South Asia he had learned as a boy. He became increasingly disillusioned with purely Western views of science that disregarded studies of nature by the ancient Indian scholars as well as the traditions of Indian students. It became his goal "to bring the hoary East and the modern West together," which he attempted to do in his book, "Biology of the Individual Human Organism with the Original Bhagwadgita" (1987 [1990]). Appropriately, the book was dedicated to Aristotle, the Greek philosopher, and his Eastern counterpart, Vyâs, the original guru of the Hindu Dharma.

Mahendra authored about 40 titles on herpetological subjects (1929–1984). He published eight papers (1935–1953) on the life history, anatomy, reproduction, and development of the Indian house gecko (*Hemidactylus flaviviridis*); numerous papers on snake anatomy, systematics, and phylogeny; and others on the vertebral column of frogs and the tails of vertebrates. His major herpetological work was a book, "Handbook of the Snakes of India, Ceylon, Burma, Bangladesh, and Pakistan" (1984). Unlike Malcolm Smith's volume on snakes (1943) in the "Fauna of British India" series, Mahendra's did not include the Indo-Chinese region. Mahendra's book covered 298 species and he erected two new genera of natricines. He had conceived this book in the late 1920s and apparently began writing the first draft about 1943 when, as he wrote in 1984, "the project was rudely interrupted" by publication of Smith's volume. Mahendra had not published on snakes since 1938 and his handbook seems not to have been significantly revised since about 1970.

• *Source*: Indian Academy of Sciences, courtesy G. Madhavan, 2006. • *Portrait*: Sant Prakash and Indraneil Das. • *Signature* (1988): Adler collection.

FANG, Ping-wen (1903–1944).

The ichthyologist and herpetologist, Ping-wen Fang (see *Note*), who was a specialist on the systematics and morphology of turtles, salamanders, and the loaches of the family Homalopteridae, was born in 1903 in the village of Nuotian (She District, Anhui Province) in eastern China. In 1926 he graduated from Dongnan University in Nanking (now Nanjing, Jiangsu Province), after which he became a zoologist at the Metropolitan Museum of Natural History, a unit of Academia Sinica in Nanking. There he was mentored by a leading zoologist, Chih Ping, and began his research on lower vertebrates. In 1935 Fang went to Paris to study fishes at the Muséum National d'Histoire Naturelle under the direction of the ichthyologist, Louis Roule. Fang worked at the museum throughout World War II and, ironically, it was during the Liberation of Paris that he was killed as the result of an aerial bombardment in the fifth district of the city on 26 August 1944. He was 41 years old.

Fang authored five herpetological papers. He published three on turtles including detailed studies of collections from Kwangsi (now Guangxi Zhuang Autonomous Region) in southernmost China (1930); the turtles (and snakes) of the Nanking region (1931, with Tsong-han Chang, another of Ping's students); and a third on specimens from eastern and northeastern China including sea turtles (1934). In the last, Fang described as new *Chinemys megalocephala* from Nanking, which probably represents a large-headed morph of *C. reevesii*. Megacephaly is well known in large specimens of many turtle species and is correlated with a molluscan diet. Fang's other herpetological research dealt with amphibians, in which he had the collaboration of Mang-ven Chang, who also worked with Ping and preceded Fang by one year to study in Paris. Fang and Chang authored two important papers together. The first was an extensive and well-illustrated review of the amphibians of the Nanking region (1931). In the other (1932) they demonstrated the validity of *Tylototriton* (now *Echinotriton*) *asperrimus* and described a new species, *T. kweichowensis*, from moderate elevations in Kweichow (now Guizhou) Province in southern China.

• *Reference*: ["Fang Bingwen,"] p. 245 *in* T. Zhou (ed.), Atlas for Identification of Turtles and Tortoises, China Agric. Publ. House, Beijing, 2004. • *Portrait* (1936): Zhou, 2004. • *Signature* (1929): American Museum of Natural History, New York, courtesy Charles W. Myers. • *Note*: The modern Pinyin, or phonetic, spelling of Fang's name is Bingwen Fang.

MAKINO, Sajiro (1906–1989).

The Japanese cytogeneticist, Sajiro Makino, made early contributions to knowledge of amphibian and reptilian chromosomes, and discovered the first triploid reptile. He was born on 21 June 1906 in the little town of Narita (Chiba), 80 km east of Tokyo, today the site of Tokyo's international airport. He was interested in biology as a schoolboy and at age 12 won a prize for breeding leghorn chickens. This led him in 1924 to study zoology with the Faculty of Agriculture at Hokkaido Imperial University in Sapporo, on Japan's northernmost main island. He graduated in 1930 but not before meeting a cytologist, Kan Oguma, who became his mentor. Makino joined the staff of the Zoological Institute at the university that same year, became assistant professor in 1935, and earned his doctorate under Oguma (D.Sc. 1940). During World War II and its aftermath, when osmic acid, the most important cytological fixative of the period, was not available, Makino developed new squash techniques and discovered the water pretreatment method for improving his preparations. After the war, he succeeded Oguma as professor at the university, which was renamed Hokkaido University in 1947. He worked on the chromosomes of amphibians and reptiles primarily in the 1930s and 1940s. In 1950 he shifted his attention to cancer cells and their chromosomal peculiarities and in 1959 to the relationship between chromosomal aberrations and specific human diseases.

Makino had an enormous output and published more than 400 papers and 16 books. Among the latter was his classic, "Human Chromosomes" (1975). He is best known to vertebrate zoologists for his book, "An Atlas of the Chromosome Numbers in Animals" (Japanese edition 1950; enlarged English edition with the assistance of Warren H. Leonard [see *Note*]; revised edition 1956). These two editions list chromosome numbers for 3,317 and 4,850 species and built on the earlier compilations for vertebrates published by Oguma and Makino (1931, 1932, 1937). Makino's influence was also spread through his training of more than 80 students who earned doctorates under his supervision, among them several who specialized on the chromosomes of amphibians and reptiles including Masahiro Ito, Sei-ichi Kohno, Motomichi Sasaki, and Takeshi Seto. Makino retired in 1970 and died at his home in Sapporo on 6 August 1989.

Makino's early interest was in amphibian chromosomes. He later studied those of some reptiles, but largely left this area to his colleague in Kyoto, Kenji Nakamura. Makino published 30 titles on strictly herpetological topics (1930–1970) including several papers on the breeding habits and ecology of local salamanders. His chromosomal studies focused on hynobiids, salamandrids, bufonids, ranids, and rhacophorids, and also included some snakes, agamid and gekkonid lizards, and sea turtles. His main interest was to determine chromosome number and gross morphology including differences between the sexes. He discovered the unusually low chromosome number of 40 in *Hynobius retardatus* (no other hynobiid is known to have fewer than 56). In 1949 Makino and his colleague, Ei Momma, reported on a gecko from the Ogasawara (or Bonin) Islands—said to be *Gehyra variegata ogasawarasimae* but now known to be *Lepidodactylus lugubris*—that had an unusually large number of chromosomes. This was the first report of triploidy in a reptile.

• *References*: ["In Memory of Professor Sajiro Makino,"] by T. Okigaki, Tiss. Cult., 15: 450–452, 1989; "Sajiro Makino (1906–1989)," by H. Hori, Zool. Sci., Tokyo, 6: 1045–1046, 1989; "Dr. Sajiro Makino (1906–1989)," by M. Sasaki and S. Matsui, Cancer Genet. Cytogenet., 44: 139–141, 1990; Additional information from Takeshi Seto, 1992 and 2006. • *Portrait* (1976) *and signatures* (Japanese form, 1976; Romanized form, 1975): Courtesy Takeshi Seto. • *Note*: Leonard, a plant geneticist, came to Japan as a young army officer with the American occupation forces. Makino always acknowledged Leonard's key support of his research and for obtaining chemicals that were otherwise unavailable to him in post-World War II Japan.

HELLMICH, Walter (1906–1974).

Karl George Walter Hellmich was a German herpetologist who came to specialize on the systematics and biogeography of reptiles, especially of South America. He was born on 26 February 1906 in Oschatz (Saxony), about 55 km east of Leipzig. After graduating from secondary school in 1925, he entered the university in Munich and wrote his doctoral dissertation on developmental physiology and regeneration in amphibians (Dr.phil. 1929; published 1930). In May 1929 he began as an assistant at the Zoologischen Staatssammlung

München, but left in September to become a research fellow at Yale University in the United States and then at the Universidad de Chile. After returning from South America in 1931, he came to the museum to assist Lorenz Müller in the herpetological section and brought with him the huge collections that he and his Munich colleague, Wilhelm Goetsch, had made in Chile, which included about 2,000 lizards. In 1942 Hellmich became a curator in the herpetological section. During World War II he served in the army in an administrative capacity in Munich, and from May 1945 to April 1946 was held as a prisoner of war in an American-run camp. In June 1948 he returned to the museum as head of the herpetological section, in succession to Müller. Müller had been chief curator of zoology since 1928, and by now was heavily involved in reconstruction of the museum, which had been severely damaged in a bombing raid in April 1944. In 1959 Hellmich became director of the museum and five years later was made a professor at the University of Munich.

From the early 1960s Hellmich's attention turned to the Himalayas. He had long had an interest in alpine environments in Europe and South America. The German Research Council began mapping the southern slopes of Mount Everest in the 1950s and in 1959 the Nepal Research Centre was established to undertake a comprehensive study of the region with offices in Munich and Kathmandu. Hellmich became managing director of the center and editor of its publications. In 1965, when the original scheme was coming to an end, Hellmich personally discussed its continuation in a meeting with King Mahendra in Nepal. After retiring from the museum in 1971, Hellmich devoted himself to the center's activities. He died on 29 July 1974 at his home in Garmisch-Partenkirchen, located at the foot of the Bavarian Alps about 90 km south of Munich.

Hellmich's herpetological publications number about 85 (1928–1970). The majority of them are on the lizards and snakes of South America, especially lizards of the genus *Liolaemus* in Chile and Argentina. Many of these papers were co-authored with Müller. Other papers by them dealt with Angola, China, Lebanon, and Spain. Hellmich also co-authored several articles on the herpetofauna of Turkey with Muhtar Başoğlu. New taxa were named in many of Hellmich's papers. He was interested in geographic and elevational variation and in broad biogeographic and evolutionary questions. His major herpetological works include the two herpetological reports from the German expeditions during 1925–1932 to the Gran Chaco, a vast prairie region in the interior of south-central South America. The report on the amphibians, turtles, and crocodilians was co-authored with Müller (1936) and that on lizards (1960) was by Hellmich alone. He also wrote a field guide to the amphibians and reptiles of Europe (1956; English edition 1962).

• *Reference*: Rieck et al., 2001 (Hellmich by D. Fuchs on pp. 474–477). • *Portrait* (1963): Carnegie Museum, courtesy M. Graham Netting. • *Signature* (1956): Adler collection.

ALLEN, Ross (1908–1981).

The showman and naturalist, Ensil Ross Allen, who was born in Pittsburgh (Pennsylvania) on 2 January 1908, supported and popularized the study of reptiles. By the age of 12 he was collecting snakes and other animals and he had started his own museum. He was a powerful swimmer and canoeist, which became important in reaching otherwise inaccessible places to find reptiles and to subdue large reptiles such as alligators underwater. The family lived in Ohio for some years and then moved to Winter Haven (Florida) in 1924. He attended college briefly in 1927, but dropped out to pursue taxidermy and tend to his collection of live reptiles.

Florida at that time was becoming a tourist destination. In the 1920s, two entrepreneurs from Ocala in north-central Florida recognized the development potential of Silver Springs, one of the world's largest artesian formations. These springs, located just east of Ocala, produce more than two billion liters of crystal-clear water daily and reveal an underwater environment that can be readily seen from the glass-bottomed boats that were invented there. The developers were looking for individuals to establish tourist attractions at Silver Springs. In 1929 Allen set up a taxidermy shop there, organized a display of snakes, alligators, and other native wildlife, and later an Indian village with Seminoles who were recruited by him from the Everglades. This entire facility became known as the Florida (later, Ross Allen's) Reptile Institute. After tourists watched Indians wrestle alligators and Allen milk rattlesnakes for their venom, they could purchase souvenir live reptiles and reptile products—such as alligator soup and rattlesnake meat "with supreme sauce"—to take home. The extracted venoms were made available to Wyeth Inc. of Philadelphia to produce antivenins. Production was increased at the time of World War II to provide protection

for troops being trained in the southern states where they were exposed to pitvipers. During those war years, Allen personally milked nearly 74,000 snakes.

Allen was charismatic and athletic, and he had a flair for showmanship that made him a celebrity. He was a stand-in for Johnny Weissmuller, the actor who played Tarzan in Hollywood movies in the 1930s, whenever underwater scenes with large reptiles were scripted. In 1965 Allen sold his institute to the American Broadcasting Company with an agreement to remain as director for ten years. The institute had been a popular tourist attraction for four decades until 1971 when attendance dropped after Walt Disney World was opened near Orlando. The institute closed in 1975. Allen then worked at the Sarasota Jungle Gardens and the Saint Augustine Alligator Farm until 1979 when he moved to Lake City, located in northern Florida and, importantly, along one of the major automobile traffic streams leading to and from Disney World from the north. There, Allen planned to open Ross Allen's Alligator Town, but he died, in Gainesville (Florida) on 17 May 1981. The exhibit opened the next month.

Ross Allen has sometimes been dismissed as only a showman, but he had a deep interest in reptiles. He once remarked, ". . . I love herpetology and want to do everything I can to further it." In 1938 he began publishing technical articles on amphibians and reptiles. After the war, he realized he could not do science and also run the institute, so in 1949 he hired Wilfred T. Neill, a noted herpetologist, to direct a newly established research division at the institute. Together, they published many papers on the Floridian herpetofauna including life histories, new taxa, distribution, and a spectrum of natural history notes. Their booklet, "Keep Them Alive!" (1950, with new editions in 1954 and 1959), became a standard reference for reptile keepers. The institute even established its own research journal (1950–1959). Allen had visited British Honduras (now Belize) in 1937 to collect. During 1957–1960, he returned several times with Neill and together they made extensive collections throughout the country, and they reported on them in ten papers. Neill left the institute in 1962 and Allen did not continue scientific research.

Allen influenced generations of budding herpetologists through his activities at the institute, publications, and public appearances. He also encouraged youngsters to take a serious interest in reptiles by providing personal tours of the institute, carrying on correspondence with them, and by sending them live specimens. The institute sold live reptiles to zoos and other public displays as well as to individuals. It was not unusual for Allen to include a few unordered specimens in a shipment to a teenager. Sometimes there were other surprises. One 15-year-old had paid for a juvenile iguana but received a full-grown adult instead, with an explanatory note from Allen to the young man's nervous mother.

• *References*: "Ross Allen's Reptile Institute," by W. T. Neill, Ross Allen's Rept. Inst., Silver Springs, Florida, 32 pages, [1950]; "Ensil Ross Allen 1908–1981," by R. Conant, Herpetol. Rev., 12: 99, 1981; "Ross Allen (1908–1981)," by S. Grenard, Rept. Amphib. Mag., May/June: 11–13, 1990; Conant, 1997 (Allen on p. 448). • *Portrait and signature* (1948): Adler collection.

BOGERT, Charles M. (1908–1992).

Charles Mitchill Bogert made diverse contributions to the systematics and biology of amphibians and reptiles, and succeeded G. K. Noble as curator at the American Museum of Natural History in New York. He was born on 4 June 1908 on a cattle ranch owned by his parents in Mesa (Colorado), but when it failed the family moved to southern California. Bogert was already interested in natural history by this time and even brought some live lizards with him. He attended Los Angeles schools and in 1926, while still in high school, he met Laurence M. Klauber, the amateur specialist on rattlesnakes. Klauber sent him reprints, corresponded with him regularly, and whenever he came to Los Angeles would take Bogert to dinner to talk about reptiles. It was Klauber who inspired him to become a professional herpetologist. Bogert attended Pasadena Junior College for two years before transferring to the University of California at Los Angeles (B.A. 1934, M.A. 1936). At UCLA he had the good fortune to study with two faculty members who worked with desert reptiles—Raymond B. Cowles and Walter Mosauer—and the former became Bogert's advisor for his master's thesis on the Patch-nosed Snakes (*Salvadora*) which was published in 1939. Bogert was unable to work on a doctorate, which he always regretted, but financial circumstances brought on by the Great Depression prevented it.

Instead, Bogert was hired in December 1936 as an assistant to Noble at the American Museum. Noble had founded the museum's Department of Experimental Biology in 1934 and

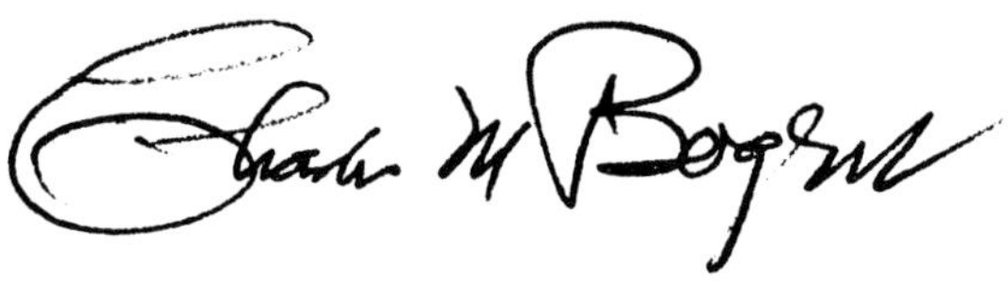

needed someone to handle the day-to-day curating in herpetology. He could hardly have chosen more wisely. Like Noble, Bogert had a broad range of interests and combined field and laboratory studies in innovative ways. Bogert's first assignment was to work up the herpetological collections from the museum's 1925 expedition to Angola. This project introduced him to some of the main themes of his life's work: African snakes, elapids, and venom delivery mechanisms. Because of his prior work in the American Southwest, Bogert also had a natural interest in Mexico and his fieldwork there began in 1939.

Bogert was promoted to Assistant Curator of Herpetology in 1937 and by 1940 he was settling into his post when, in December of that year, Noble suddenly died at the age of 46. Bogert, just 32, was put in charge of the herpetology department. He was promoted to full curator in 1943. From 1941 until 1954 he was the sole herpetological curator at the museum, except for about three years in the 1940s when James A. Oliver assisted him. Despite his duties to curate one of the world's major collections, Bogert had a substantial research program and made seminal contributions to several fields. In 1966 UCLA, his *alma mater*, honored him with an LL.D. degree. After early retirement in 1968, Bogert toyed with the idea of moving to Mexico and his beloved Oaxaca, but he settled instead in Santa Fe (New Mexico). He continued to write and work on a revision of the patch-nosed snakes, but a series of strokes beginning in April 1988 and arthritis debilitated him. Frustrated with his seriously declining health, he ended his own life on 10 April 1992 at his home in Santa Fe.

Bogert authored about 65 scientific titles on reptiles and amphibians (1930–1991) plus 150 reviews and popular publications beginning in 1927. These covered a broad range of topics. In 1938, Bogert's former adviser, Cowles, established a field station in the Coachella Valley just southeast of Palm Springs (California) in order to study the adaptation of reptiles to hot environments. Bogert collaborated in this research, which resulted in their classic paper, "A Preliminary Study of the Thermal Requirements of Desert Reptiles" (1944; reprinted 1974). This work, together with independent research by a Soviet physiologist, Alexei M. Sergeev, ushered in the study of behavioral thermoregulation in reptiles. Cowles, Bogert, and Bogert's museum colleague, the paleontologist Edwin H. Colbert, conducted research on alligators that provided the first evidence for the role of body size in thermoregulation (1946). Bogert performed independent work on thermoregulation in Mexican lizards (1949) and Appalachian salamanders (1952).

Concurrent with his investigations on thermoregulation, Bogert began studies on African snakes including taxonomic and anatomical work on colubrids (1940) and investigations of the venom-injecting and venom-spraying mechanisms of the elapids of Africa and Asia (1943). This, in turn, led to his interest in the living descendents of that group of lizards thought to be ancestral to snakes, resulting in two of his major works. The first, with the anatomist Samuel B. McDowell, Jr. (1954), concerning the Bornean lizard *Lanthanotus*, dealt with the classification of lizards and the origin of snakes. His other classic, with the Mexican herpetologist Rafael Martín del Campo (1956; reprinted 1993), covered the closely related genus *Heloderma*, the only lizards with venom and an injection apparatus. Bogert's interest in Mexican herpetology led to about 15 titles (1939–1969). These include a major review of the herpetofauna of Sonora (1945, with Oliver) and the descriptions of two new genera (*Cryophis*, *Exiliboa*) and many new species and subspecies of salamanders, turtles, lizards, and snakes including a box turtle (*Terrapene nelsoni klauberi*), a fringe-toed lizard (*Uma exsul*), and a dwarf boa (*E. placata*).

Bogert also made significant contributions to the knowledge of frog behavior. He conducted one of the earliest studies of homing behavior in frogs (1947) and introduced toe-clipping as a simple marking device for individual recognition. His studies on frog vocalizations, however, were of far greater importance. Peter Paul Kellogg and Arthur A. Allen of Cornell University's Laboratory of Ornithology were the first to publish the recordings of frog calls ("Voices of the Night," 1948). These were originally made direct-to-disc and on the sound tracks of motion picture film, and provided the species-specific call of each frog with little discussion of function. Bogert's work on this topic, however, was a decided advance. He was one of the first zoologists to make use of portable tape recorders and a new instrument—the sound spectrograph—to analyze calls. Armed with this new technology, he was able to address many issues of biological significance including how calls are produced, the effects of body size and temperature, and the diversity of calls and their role as isolating mechanisms. He introduced several experimental innovations, such as the playback paradigm in which frogs respond, positively or negatively, to tape-recorded calls. He published a phonograph record of the calls of North American frogs (1958, with explanatory booklet, reissued in 1998 as a compact disc; the single track of Panamanian frogs was supplied by Kellogg) and a seminal review of the subject, "The Influence of Sound on the Behavior of Amphibians and Reptiles" (1960, with an accompanying demonstration record; index published 1970).

• *References*: "Charles Mitchill Bogert June 4, 1908–April 10, 1992," by R. Conant, Herpetol. Rev., 23: 102–105, 1992; "Biographical Sketch and Bibliography of Charles Mitchill Bogert, 1908–1992," by C. W. Myers and R. G. Zweifel, Herpetologica, 49: 133–146, 1993; "Charles M. Bogert 4 June 1908–10 April 1992," by C. Gans, Copeia, 1993: 264–266, 1993; Conant, 1997 (Bogert on pp. 453–454); Myers, 2000 (Bogert on pp. 54–60). • *Portrait* (about 1950) *and signature* (1962): Adler collection.

GRANT, Chapman (1887–1983).

Chapman Grant, a grandson of former President Ulysses S Grant and founder of the journal *Herpetologica* and The Herpetologists' League, was an amateur herpetologist who published extensively on North American and West Indian amphibians and reptiles. He was born on the family farm near Salem Center, a hamlet that is now part of North Salem just north of New York City on 22 March 1887. His father was the youngest of three sons of the 18th President of the United States, but his parents disdained their family's prominence. Chapman Grant, who moved with his family to southern California at the age of one, once remarked, ". . . not a lot was made of the family heritage. In California, it didn't matter who was your grandfather." This independence and a tendency to question authority, often in a brusque manner, continued throughout his 95 years of life. It was a peripatetic life of constant moves from place to place and career to career, but the common thread throughout was herpetology. He attended private schools in California, then a postgraduate year at a school in New Jersey before attending Williams College in Massachusetts where he studied biology and geology (B.A. 1910). During 1910–1913 he held a variety of jobs in New York and California and volunteered on an expedition to Mexico sponsored by the American Museum of Natural History (1910). He also briefly held positions at museums and aquaria and developed some skill in designing the latter.

In October 1913, however, Grant was commissioned a second lieutenant in the United States Army cavalry, thus beginning a 23-year period in the military. This took him to many posts—in Georgia, Texas, New York, Arizona, New Jersey, Oklahoma, Minnesota, and a short tour to France in 1919 by which time he reached the rank of major—and these gave him opportunities to collect. In September 1930 he was posted to San Juan (Puerto Rico) and, although he had published on reptiles as early as 1911, his herpetological research blossomed there. By 1932 he was in Indiana, then Kentucky, and in December 1935 he retired from the army because of ill health at the rank of major. Grant returned to California in 1935 and became a real estate operator in San Diego. In 1953 he moved to Escondido where he owned several ranches and avocado orchards. There he lived until his death on 5 January 1983.

Beginning in 1910, Grant authored at least 220 scientific articles on a wide variety of subjects; about 60 percent of them were on herpetology. These covered a range of topics but especially systematics. His most productive period began in 1931 after arriving in Puerto Rico where he discovered and named 18 new taxa of lizards, snakes, and frogs. After a 1938 trip to Jamaica and the Cayman Islands, he published major papers on those herpetofaunas (1941). His West Indian collections are primarily in the museums at the University of Michigan and Harvard University, but those from Middle America are at the University of Illinois. After 1960 his herpetological publications became more irregular. The last was a 1971 note in *Science*. This was a critique of an earlier paper in that journal on chemical cues in newborn snakes in which, true to form, he challenged an authority on the topic—but wrongly so.

For many years Grant had published in *Copeia* and other outlets. He became increasingly annoyed with the pronouncements of editors who, he believed, were frustrating his attempts to publish his work as he saw fit. Accordingly, he founded the journal *Herpetologica* in 1936 as an alternative in which the editors exercised very gentle oversight of content. Grant and Walter L. Necker, a herpetologist then working at the Chicago Academy of Sciences, served as editors. Grant ran the business and subscription side of the operation and Necker the printing and mailing of the journal. The academy had its own hot-type printing press and, with the benevolence of the academy's director, the herpetologist Howard K.

Gloyd, Necker used it to print the journal. When Necker was called for military duty in 1943, Grant took over complete control in San Diego and then brought the journal to Escondido when he moved there in 1953.

In 1946 Grant established The Herpetologists' League, a society intended to provide support and visibility for the journal, yet he continued to control the editorial and financial aspects of the journal. During this early period, when subscriptions were low, Grant personally made up the deficits, but his lax editorial standards were criticized. He defended himself in print on numerous occasions and noted that *Herpetologica* had been "established because of a hatred for the omniscience and omnipotence of editors in general." Two earlier attempts had been made by Grant to divest himself of the journal—to Ross Allen's Reptile Institute and to Tulane University—but both were aborted when Grant balked at the last minute. In 1960 *Herpetologica* was transferred to Brigham Young University and the editorship of Wilmer W. Tanner, and it finally came under the full control of the league. Grant continued to publish in *Herpetologica*, but as its editorial standards improved he became disillusioned with it when it took on what he saw as discriminatory practices. His last paper in this journal appeared in April 1961.

• *References*: "Major Chapman Grant (1888 [*sic*]–1983)," by W. W. Tanner, Herpetologica, 39: 330–331, 1983; "Chapman Grant, *Herpetologica*, and The Herpetologists' League," by H. M. Smith, Herpetologica, 42: 1–32, 1986. • *Portrait* (1947, with painting of President Grant on wall): San Diego Historical Society (Union-Tribune coll. 3875). • *Signatures* (two forms, 1934 above, 1971 below): Adler collection.

WELLER, W. H. (1913–1931).

There is no herpetologist of taxonomic note who had such a tragically short life as the American salamander specialist, Worth Hamilton Weller. He was born in Cincinnati (Ohio) on 28 May 1913. His interest in natural history was ignited by walks in woodlands and fields near his home to find snakes and salamanders together with a boyhood friend, Karl H. Maslowski, a future professional wildlife photographer. In 1928 Ralph Dury, then director of the museum of the Cincinnati Society of Natural History, established a junior affiliate for ". . . young men who are interested in collecting and studying objects of natural history." Weller and Maslowski soon signed up and by January 1930 Weller had published his first paper, on the plethodontid salamander *Aneides aeneus*, in the first issue of the *Proceedings of the Junior Society of Natural Sciences*. When Weller was only 16 years old, he was already business manager and curator of herpetology of the Junior Society and by July of 1930 was also its editor.

This group of teenagers wrote the *Proceedings*. The typesetting and printing were also done by them, on an old-fashioned, hot-type printing press in the museum's basement. The Junior Society had meetings twice a month and sponsored field trips. Weller quickly became the leader of this group and showed a propensity for scientific work that belied his years. For his first paper he had even consulted with Thomas Barbour, herpetologist at Harvard College, and with the nation's leading student of salamanders, Emmett R. Dunn of Haverford College in Pennsylvania. Thus began a steady correspondence with Dunn who later agreed to accept Weller as his student.

Numerous additional papers followed that were based on Weller's study of specimens in the museum's collection and the results of his own field collections in Ohio, Kentucky, Tennessee, and North Carolina. In 1930 he described a new salamander from Kentucky and named it for his mentor; today it is known as *Gyrinophilus porphyriticus duryi*. Salamanders quickly became Weller's specialty, although he did not neglect reptiles or other amphibians. Two more species were described as new but these turned out to be synonyms; inexperience had its cost. His most spectacular discovery, however, was yet to come.

In August 1930, Dury, Weller, and other members of the Junior Society surveyed the herpetofauna of Grandfather Mountain in the Blue Ridge Mountains of western North Carolina. Today the rocky summit of this mountain is easily accessible by paved road, but in 1930 it was a strenuous and dangerous climb. In the spruce forest above 1,500 m Weller found a small plethodontid that he did not recognize. Further study at the museum convinced him that it was new. A few days after his graduation from Walnut Hills High School in June 1931, Dury led a second field trip to Grandfather Mountain so that Weller could get fresh material that would help him to define his new species.

The group arrived on the afternoon of 22 June. As they set up a tent camp a severe thunderstorm was brewing but Weller, always impetuous, rushed away to collect. By eleven o'clock he had failed to return, and Dury notified local authorities. An intensive search continued for four days before Weller's body was found lodged among boulders of rain-swollen Stonestack Creek. A sack was found with his body that contained specimens of his new species. It is presumed that after collecting the animals, darkness had fallen and in his attempts to return to camp, Weller had slipped off a cliff to his death. He had just turned 18 the previous month. The new species was put into the hands of Weller's colleague, Charles F. Walker of the Ohio State Museum, who named it *Plethodon welleri* (1931).

During his very short career Weller published 12 papers on herpetology (1930–1932; most of them reprinted 1965). The *Proceedings* of the Junior Society ceased publication in 1932 and the society itself lingered on for a few more years. Weller had been the driving force behind the organization, but he had not prepared it for the inevitable succession in leadership.

• *References*: "Foreword" by K. Adler *in* Herpetological Papers from the *Proceedings of the Junior Society of Natural Sciences* (1930–1932)," Ohio Herpetol. Soc., Columbus, 1965; "Worth Hamilton Weller," by K. H. Maslowski, Cincinnati Mus. Nat. Hist. Quart., 21(4): 6–13, 1988. • *Portrait* (1931): Courtesy of Weller's brother, Arthur D. Weller, Arlington (Virginia). • *Signature* (1930): Cincinnati Museum of Natural History, courtesy John W. Ferner.

STUART, L. C. (1907–1983).

The American herpetologist, Laurence Cooper Stuart, an explorer, biogeographer, and a specialist on Guatemalan herpetology, was born on 9 July 1907 in Du Bois, located in northwestern Pennsylvania. He grew up in Canton in northeastern Ohio. He wanted to study forestry and went to the University of Michigan in 1926 where he befriended a fellow undergraduate, Norman Hartweg, and together they came into contact with the herpetologist, Alexander G. Ruthven, in the Museum of Zoology. Stuart earned three degrees at Michigan (B.S. 1930, M.S. 1931, Ph.D. 1933), the latter under the daily supervision of Ruthven's associate, Helen T. Gaige, because by 1929 Ruthven had become university president. In 1931 Stuart married Ruthven's daughter, Katherine. Stuart joined the Michigan faculty in 1934 as an instructor and rose to full professor by 1958. During much of this period he was appointed in the Laboratory of Vertebrate Biology and the Institute of Human Biology, and also had an office in the nearby Museum of Zoology where he was close to the herpetology collections. He generally worked alone and never had graduate students. His frequent expeditions to Guatemala began in 1933 at a time when travel around that country could be exceedingly difficult and many areas had not been explored at all. In 1956 he built a home in Panajachel on Lake Atitlán to which he retired in 1969. He died on 27 May 1983 in Cleveland (Ohio) where his sister lived and where he had gone for hip surgery.

Stuart authored 62 herpetological titles (1930–1971), nearly all of them on Guatemala. Beginning with the British and French biological explorations of Central America in the late 19th Century, collectors had regularly visited Guatemala, but Stuart was the first to specialize on the country's herpetofauna. He named nearly 40 new species of salamanders, frogs, lizards, and snakes from Guatemala, nearly all of which are currently recognized. He also conducted numerous faunal and biogeographic studies. He published major systematic reviews of the colubrid snake genus *Dryadophis* (1941) and the Guatemalan lizards of the genus *Anolis* (1955). He also authored classic faunal reviews of the savanna region of Petén (1935); Alta Verapaz, a mountainous region and adjacent lowlands in central Guatemala (1948, 1950); and the Guatemalan Plateau (1951). His Guatemalan work culminated with his checklist of the herpetofauna of Guatemala (1963).

Stuart's work outside Guatemala was limited to descriptions of a dozen new snakes and a lizard from Colombia, Venezuela, the West Indies, and especially Mexico; a study of lizard microhabitats in the Middle Pahvant Valley of Utah; and a book about the land vertebrates on islands in eastern Lake Michigan, co-authored by Robert T. Hatt and others.

• *References*: "Dr. Laurence Cooper Stuart," p. 1101 *in* American Men of Science, vol. 2. Science Press, Lancaster, Pennsylvania, 1955; "Laurence Cooper Stuart," by W. E. Duellman, Copeia, 1983: 1116–1118, 1983; "L. C. Stuart 1907–1983," by J. Villa, Brenesia, San José, 22: [i-ii], 1984 (1985). • *Portrait* (1936): Howard K. Gloyd collection, courtesy Jonathan A. Campbell. • *Signature* (1941): Adler collection.

HARTWEG, Norman (1904–1964).

Norman Edouard Hartweg, who was curator of herpetology at the University of Michigan and a specialist on turtles and Mexican herpetology, was a principal founder of the Organization for Tropical Studies. He was born on 20 August 1904 in Warren, located in northwestern Pennsylvania. His interest in animals arose from his love of the outdoors. Not wishing to go to college after high school, he became a pipe layer for a year. On the recommendation of a family friend, he then volunteered for work at the Museum of Comparative Zoology at Harvard College. For several years he assisted Samuel Garman who studied fishes and reptiles, especially turtles. Garman had been a field assistant for Edward D. Cope and later a student of Louis Agassiz. Hartweg then went to the University of Michigan (A.B. 1930, M.S. 1931, Ph.D. 1934) and its Museum of Zoology. Technically, he was a student of the herpetologist, Alexander G. Ruthven, but since Ruthven had become university president in 1929, Ruthven's associate, Helen T. Gaige, provided Hartweg's day-to-day supervision. On completing his doctorate, Hartweg became an assistant curator and in 1946 succeeded Gaige as curator-in-charge. By 1961 he was assistant director of the museum, but his life was cut short by pancreatic cancer and he died at the age of 59 on 16 February 1964 in Ann Arbor. Donald Tinkle succeeded him as curator.

Hartweg was devoted to the museum, his students, and his profession. He had many doctoral students who went on to notable careers, among them Fred Cagle, Herndon Dowling, William Duellman, Richard Etheridge, James Mosimann, Grace Orton, and James Peters. He served as the herpetological editor of the journal, *Copeia* (1950–1955). Perhaps his most important professional contribution was his involvement in the founding of the Organization for Tropical Studies (OTS), a research and educational program in Costa Rica run by a consortium of universities in the United States and the Universidad de Costa Rica. Several universities had initiated efforts to train students and faculty in tropical biology by establishing field stations and courses. In 1957 the University of Michigan, because of its longstanding interest in Mexican biology, attempted to establish a field station in Chiapas or Oaxaca in southern Mexico, based on specific sites evaluated by Hartweg's Michigan colleague, L. C. Stuart. After years of negotiations involving several trips to Mexico by Hartweg, Mexican authorities could not make a firm commitment and Hartweg thus had to look elsewhere.

The first formal steps to establish OTS were taken by Hartweg when, in December 1962, he invited representatives from nine institutions to meet in Miami early in 1963 during which OTS was established. Hartweg was elected the first president, wrote the by-laws, and submitted the charter to the State of Florida for approval. Fortunately, Jay M. Savage of the University of Southern California had already established a summer field course in Costa Rica in 1961, and he graciously allowed this to be incorporated into OTS in 1963. Throughout this key organizational phase of OTS, Hartweg was critically ill. During the final months of his life, OTS was his main preoccupation day and night. To conserve his time and energy, he often worked overnight at the museum and slept on the concrete floor.

Hartweg's research program was always secondary to his other duties. He authored about 30 titles (1931–1963), primarily on turtles and Mexican herpetology. Among them is a four-part series on the herpetology of the Isthmus of Tehuantepec (1937–1940), co-authored with a Michigan graduate student, James Oliver. Hartweg described 19 new taxa from Mexico and Texas, nearly all of which are recognized today including the frogs

Plectrohyla matudai and *P. sagorum*, the turtles *Kinosternon creaseri* and *Trachemys gaigeae*, a lizard, *Sceloporus smithi*, and the pitviper, *Porthidium dunni*. His ongoing study of the turtle family Kinosternidae and his book on the herpetology of the Great Lakes Basin were never published. Hartweg was one of the first to document overwintering in turtle hatchlings, a phenomenon in which the newly hatched turtles remain underground in the nest and emerge in the spring (1944, 1946).

• *References*: "Dr. Norman Hartweg," p. 480 *in* American Men of Science, vol. 2. Science Press, Lancaster, Pennsylvania, 1955; "The Organization for Tropical Studies," by D. E. Stone, p. 143–187 *in* F. Almeda and C. M. Pringle (eds.), Tropical Rainforests. Calif. Acad. Sci., San Francisco, 1988; Conant, 1997 (Hartweg on pp. 464–466); Additional information from Thomas E. Moore, 1997, Margaret (Mrs. Norman) Hartweg, 1999, and Arnold G. Kluge, 2002. • *Portrait* (1962): Carnegie Museum, Pittsburgh, courtesy M. Graham Netting. • *Signature* (1955): Adler collection.

WALKER, Charles F. (1904–1979).

Charles Frederic Walker, a naturalist, ornithologist, and herpetologist, was a specialist on the systematics of New World amphibians and for three decades curator of herpetology at the University of Michigan. He was born on 27 December 1904 in Columbus where he attended Ohio State University (A.B. 1930, M.S. 1931). Beginning in 1927 as an undergraduate student, he became an assistant in the Department of Natural History at the Ohio State Museum, which was then adjacent to the university campus. Walker's original interest

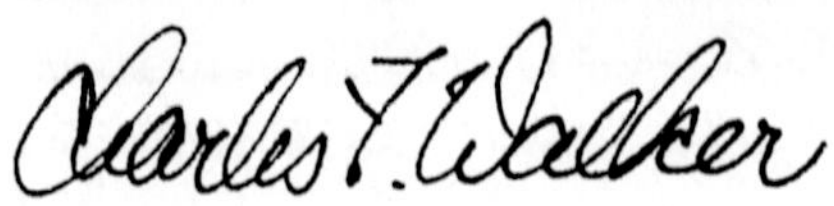

was birds, on which he published his first ten papers, but by 1928 his focus turned to the amphibians of Ohio, the subject of his master's thesis. For his doctorate he attended the University of Michigan (Ph.D. 1935) in Ann Arbor. His official advisor was the herpetologist, Alexander G. Ruthven, who since 1929 had been president of the university. Thus, like other Ruthven students of that period, Walker was actually supervised by Ruthven's able associate, Helen T. Gaige.

During 1935–1938 Walker was employed by the Soil Conservation Service of the United States Department of Agriculture. In 1938 he became an associate professor of zoology at Ohio State, not at the university's main campus but at the F. T. Stone Laboratory, the country's oldest freshwater biological field station located on Gibraltar Island in western Lake Erie. Walker served in the United States Army in 1942–1943. In 1947 he returned to Michigan as associate curator in the Museum of Zoology and was the colleague of Norman Hartweg who had become curator-in-charge of the herpetological collections the year before and with whom he long enjoyed a long friendship that had begun during their days as fellow graduate students at Michigan. Walker succeeded Hartweg as curator-in-charge upon the latter's death in 1964 and retired in 1975. Ronald Nussbaum was his successor as curator. Walker died in Ann Arbor on 16 February 1979. He never married.

Walker had extensive field experience and a broad knowledge of zoology, botany, and geology, which benefited his work with amphibians. In addition to morphology, he emphasized life history details and evaluated distribution in terms of Pleistocene glaciology. His interest in birds caused him to focus on the vocalizations of frogs. He was thus able to resurrect *Pseudacris brachyphona* from synonymy (1932) and identify a new species, *P. brimleyi* (1933, with B. B. Brandt). Walker was the first to recognize the existence of two call types in *Hyla versicolor*, as early as 1931, which are now known to represent sibling species. This work on anurans led to Walker's major work, "The Amphibians of Ohio. Part I. The Frogs and Toads" (1946; reprinted 1967), a book that summarizes the systematics, distribution, and natural history of the 13 species found in the state. He never published his companion report on Ohio salamanders, but a volume on that topic was issued in 1989, authored by a large number of specialists, many of them associates or former students of Walker's and under the editorship of Ralph A. Pfingsten and Floyd L. Downs.

On returning to Michigan, Walker's interests shifted to the Neotropics. His work on the osteology and myology of the Mexican Burrowing Toad (*Rhinophrynus dorsalis*) firmly established its classification in a family unto itself (1938). He described ten new species from Ecuador, Mexico, and Venezuela, among them an Ecuadorian frog (*Syncope antenori*) and a Mexican lizard (*Lepidophyma micropholis*) and snake (*Thamnophis mendax*). His first new species, however, was from North Carolina: *Plethodon welleri*, a salamander named for his teenage friend, Worth Hamilton Weller, who died in a fall on Grandfather Mountain after collecting the type specimens.

Walker authored 18 herpetological titles (1931–1978), but his personal research program was secondary to his commitment to curation of the museum's large collection of amphibians and reptiles and to his students. He was the consummate curator, exhibiting extreme attention to accuracy of identification, cross-indexing, and data entry in the bound catalogues. He supervised ten doctoral students, many of whom went on to hold prominent positions. Among them are N. Bayard Green who was trained at Ohio State and at Michigan, Kraig Adler, Paul S. Martin, George B. Rabb, Priscilla Starrett, Thomas M. Uzzell, Henry M. Wilbur, and George R. Zug.

• *References*: "Dr. Charles Frederic Walker," p. 1179 *in* American Men of Science, vol. 2. Science Press, Lancaster, Pennsylvania, 1955; "Charles F. Walker," by D. W. Tinkle, Copeia, 1979: 563–564, 1979; "Dedication to Charles F. Walker," by K. Adler, p. xiv–xvii *in* R. A. Pfingsten and F. L. Downs (eds.), Salamanders of Ohio. Bull. Ohio Biol. Surv., new ser., vol. 7, no. 2, 1989. • *Portrait* (1963): Carnegie Museum, Pittsburgh, courtesy M. Graham Netting. • *Signature* (1951): Adler collection.

Young, C. C. (1897–1979).

Chung-chien Young (see *Note*), the founder of vertebrate paleontology in China, had a special interest in fossil reptiles and described some of the most spectacular dinosaurs unearthed in that country. He was born on 1 June 1897 in Hwahsien (now Huaxian, Shaanxi Province), just southwest of Xi'an in north-central China where his father was a schoolteacher. He studied geology and zoology at National Peking University (now Peking University) under the guidance of the German-American geologist, A. W. Grabau. Young graduated in 1923 and went to Germany for his doctoral work later that year. He completed his dissertation in vertebrate paleontology at the university in Munich (Dr.phil. 1927) under the direction

of the paleontologist, Ferdinand Broili. Karl von Frisch, the zoologist and future Nobel laureate, was a member of Young's doctoral committee. On returning to China in 1927, he became head of the Cenozoic Research Laboratory in Peking, which he and others later developed into the Institute of Vertebrate Paleontology and Paleoanthropology, a unit of Academia Sinica. He held professorships at several universities and served as director of the Museum of Natural History in Beijing. During the reign of the infamous "Gang of Four" in the 1970s, Young was persecuted despite his advanced age, yet still continued to work on his collections of Triassic fauna from Xinjiang in northwestern China. His career was honored by his election as Academician of Academia Sinica. He died in Beijing on 15 January 1979.

Young was China's first scientist to specialize in vertebrate paleontology and his research covered all vertebrate groups. He was particularly interested in reptiles and also published on amphibians (in total about 90 titles, 1931–1979). The nature of his work is reflected in the titles of some papers: New dinosaurs from Inner Mongolia, fossil frog from the Miocene of Shantung, new Triassic and Cretaceous reptiles in China, fossil crocodiles in China, new sauropods from China, first occurrence of fossil salamanders from the Upper Miocene of Shantung, and a new fossil turtle from Kansu. He had several students who also worked with reptiles, the most prominent being Hsiang-k'uei Yeh who specializes on turtles.

• *References*: ["To the Memory of the Famous Paleontologist Chung-chien Young,"] anonymous, Verteb. PalAsiatica, 17(2): 97–98, 1979; Additional information from Hsiang-k'uei Yeh, 1991. • *Portrait* (1956) *and signatures*: Courtesy Hsiang-k'uei Yeh. • *Note*: The Pinyin (phonetic) form of Young's name is Zhongjian Yang.

CHANG, Mang-ven L. Y. (1903–1993).

The Chinese vertebrate zoologist, Mang-ven Ling Yu Chang (see *Note 1*), specialized on Chinese amphibians and reptiles. He was born on 5 July 1903 (see *Note 2*) in Ningpo (now Ningbo City, Zhejiang) and entered Southeastern (now Nanjing) University in Nanking (now Nanjing). After graduation (B.S. 1926), he was invited by Chih Ping to become a professor in the School of Agriculture at Peking University where he taught anatomy and zoology. Ping later asked him to become a researcher and his personal secretary at the Science Society of China, headquartered in Nanking. Ping, one of China's leading zoologists, worked with amphibians and reptiles and introduced Chang to them. In 1934 Chang went to France on a scholarship to work at the University of Paris under the direction of Louis Roule, professor in charge of reptiles and fishes at the Muséum d'Histoire Naturelle. At the museum Chang worked with Fernand Angel, who also mentored him. After receiving his doctorate in 1936 and some travel in Europe, Chang returned to China to help lead the country's modernization in biological research and teaching.

In 1937 Chang became a professor at National Chekiang (now Zhejiang) University in Hangchow (now Hangzhou). He moved to Fudan University in 1943 as a professor, first at its Chungking (now Chongqing) campus and in 1951 as chairman of biology at the Shanghai campus. However, during the Anti-Rightist Movement that primarily targeted intellectuals, in the fall of 1957 Chang was accused of right-wing political activity. All of his belongings were taken from his home on three occasions. He was removed from his academic post and sent to Harbin in extreme northeastern China and to Heilongjiang University. His first wife, one daughter, and young son all died there owing to the harsh conditions. He was rehabilitated in 1962 and was transferred to Harbin Normal College (now University). Because his own health had deteriorated, the authorities allowed him to return to Shanghai in 1972. In 1980 Chang returned to duty as a professor of biology at East China Normal University. Besides his many books and his activity in several scientific societies, Chang had wide influence through his training of a large number of students who went on to productive careers of their own, among them Hui-qing Gu, Bu-qing Hu, Zheng-yi Huang, Zhi-xun Li, Ken-tang Zhao, and Yu Zong. Chang died in Shanghai on 2 October 1993.

Chang published several general texts—the history of biology (1956), comparative anatomy of vertebrates (in two volumes, 1986 and 1988), and vertebrate zoology (1987). He authored about 25 titles in herpetology (1932–1998). These include reports on collections from various provinces (Szechwan [now Sichuan], Kiangsi [Guangxi Zhuang], Chekiang [Zhejiang]); descriptions of salamander larvae; the ribs of *Tylototriton*; the Chinese Giant Salamander (*Andrias davidianus*); and taxonomic studies of particular species. He described the salamandrids *Tylototriton chinhaiensis* and *T. kweichowensis* as well as the genus *Paramesotriton*. He also named *Rana daunchina*. His most elaborate publications, which helped to establish Chang as one of China's most eminent herpetologists, are three books. The first, derived from his doctoral dissertation, is "Contribution à l'Étude Morphologique, Biologique et Systématique des Amphibiens Urodèles de la Chine" (1936; reprinted 1968), the first complete monograph

of China's salamanders. During his internal exile in Heilongjiang, Chang wrote "Reptile Fauna of Heilongjiang Province" (1961), but since it was never officially published and was printed in a very small edition, it is not widely known. Posthumously, together with Zong, his former student, and Ji-fan Ma, Chang produced the first reptile volume in the "Fauna Sinica" series, which covered the general account for reptiles and the sections on turtles and crocodilians (1998).

• *References*: ["Brief Biography of Professor Mengwen Zhang,"] by L. P. Ye, p. 1–12 *in* Y.-W. Qian, E.-M. Zhao, and K.-T. Zhao (eds.), Animal Science Research. A Volume Issued to Celebrate the 90th Birthday of Mangven Chang, China Forestry Press, Beijing, 1991 (1992); "Mangven Chang (1903–1993): Pioneer Herpetologist of China," by P.-C. Wang, Herpetol. Rev., 25: 95, 1994; • *Portrait*: Courtesy Er-mi Zhao. • *Signatures* (Chinese and Romanized forms, both 1936): Adler collection. • *Note 1*: Chang's modern Pinyin (phonetic) name is Mengwen Zhang. Most of his publications, however, use the traditional Wade-Giles spelling and, unusual for Chinese authors, also include his pen name, Ling Yu, normally used only within one's family and for literary works. Ling-ven is the modern transliteration of his pen name. • *Note 2*: The birth date of 7 May given in Wang (1994) is incorrect (*fide* P.-C. Wang, 2006).

MONARD, Albert (1886–1952).

The Swiss zoologist, Albert Monard, who specialized on the herpetology of sub-Saharan Africa, was born on 2 September 1886 in Les Ponts-de-Martel, near the French border 15 km west of Neuchâtel. As a child he was fascinated by nature. This interest was encouraged by a teacher at his secondary school in nearby La Chaux-de-Fonds. His family had very modest financial means so he could not attend the Université de Neuchâtel until much later. There, he worked on his doctorate with the amphibian biologist, Otto Fuhrmann, and studied the rich benthic fauna of Lake Neuchâtel for his dissertation (D.Sc. 1917), especially the copepod crustaceans on which he worked for many years.

Monard returned to La Chaux-de-Fonds where he spent the rest of his life, except for the final months before his death on 27 September 1952 when he returned to his place of birth. In La Chaux-de-Fonds he became a teacher in 1919 at his former secondary school where he later established a museum of natural history. The museum eventually outgrew these quarters, and in 1928 it was moved to the town's post office where it occupied the second floor. By then Monard's interests had shifted to Africa. He visited Africa on four privately financed trips. His first two expeditions (1928–1929, 1932–1933) were to Angola, then a province of Portugal, where he collected throughout the countryside. He went to Portuguese Guinea (now Guinea–Bissau) in 1937–1938 and to French Cameroun (now Cameroon) in 1946–1947, the latter trip in company with Villy Aellen, a herpetologist at the university in Neuchâtel. Monard brought back large collections of animals. He personally studied the vertebrates (except fishes), scorpions, and copepods, and he described many new taxa of animals, among them about 25 new frogs, lizards, and snakes. All told, he authored eight herpetological papers (1931–1951) including major reviews, with keys, of the herpetofauna of Angola (1937, 2 papers), the first since the summary published in 1895 by the Portuguese zoologist, J. V. Barbosa du Bocage.

Monard stopped teaching in 1932 in order to become full-time curator of the museum he had founded, the Musée d'Histoire Naturelle de la Chaux-de-Fonds. He retired from this post early in 1952 because of the age limit and died a few months later.

• *Reference*: "Albert Monard 2 Septembre 1886–27 Septembre 1952," by J. G. Baer and V. Aellen, Bull. Soc. Neuchatel. Sci. Nat., 76: 171–175, 1953. • *Portrait and signature*: Adler collection.

HEDIGER, Heini (1908–1992).

The Swiss herpetologist, behaviorist, and zoo director, Heini Karl Peter Hediger, is recognized as the father of zoo biology, the multidisciplinary study of captive animals. He was born in Basel on 30 November 1908. Hediger was interested in animals as a child and went to the Universität Basel to study zoology (graduated 1929). Because of his interest in wild animals and their behavior, he wanted to be a zoo director, but his early career was academic. During 1929–1931 he joined an ethnographic expedition to the South Pacific where he surveyed the herpetofauna of New Britain and other islands in the Bismarck Archipelago that are now part of Papua New Guinea. On his return to Basel he became a curator at the Naturhistorisches Museum where Jean Roux long had been curator of herpetology. Hediger completed his doctoral dissertation at Basel (Ph.D. 1932), based on his collections from New Britain, and this became his major herpetological publication, "Beitrag zur Herpetologie und Zoogeographie Neu Britanniens" (1934). In it, he reviewed 52 species of amphibians and reptiles. Later, Hediger joined expeditions to Morocco in 1933 and 1937 and reported on the herpetological collections (1935, 1937). He described new taxa of frogs, lizards, and snakes from the South Pacific. During this period he also produced what became the standard guide to the snakes of Central Europe (French edition 1936; German edition 1937). Hediger succeeded Roux as curator early in 1937.

Despite such work expected of a museum curator, Hediger's real interest was animal behavior. Indeed, one of his earliest papers (1932) was about the "flying" behavior of snakes, which he had observed in the Bismarcks in 1930. Therefore, in September1937 he switched careers and became director of a new zoo in Bern, then moved to the zoo in Basel as director (1944–1953), and finally in 1954 to the directorship of Zoo Zürich. Throughout his zoo career he focused on animal behavior and studied topics such as territorial marking, defense, feeding, escape, fear, trauma, and sleep. From 1954 to 1973 he also served as a professor at the Universität Zurich where he supervised 17 doctoral students. He published a seminal book on the psychology and behavior of captive animals in zoos and circuses (English edition 1955; German edition 1961). He pioneered new ways to exhibit animals, taking into account his intimate knowledge of their natural behaviors. His innovations, however, eventually brought him into conflict with the zoo's trustees who fired him in 1958, but after a public outcry he was reinstated and the trustees themselves were replaced. He continued to publish on amphibians and reptiles including papers on brooding behavior in pythons, breathing in frogs and turtles, male combat in monitor lizards, and feeding behavior in the alligator snapping turtle, which he once called the "most perfect death-trap in the animal kingdom." Hediger retired from Zoo Zürich in 1973 and wrote books about his work in zoos and on animal psychology. He died in Bern on 29 August 1992.

• *References*: "Heini Hediger—A Pioneer in the Science of Animal Behavior," by G. B. Rabb, Zool. Gart. (N.F.), 63(3): 163–167, 1993; "Heini Hediger (1908–1992)," by R. E. Honegger, Copeia, 1993: 584–585, 1993; Conant, 1997 (Hediger on pp. 491–492); Rieck et al., 2001 (Hediger by R. E. Honegger on pp. 473–474). • *Portrait* (1970): Zoo Zürich, courtesy René E. Honegger. • *Signature* (1934): Adler collection.

HAAS, C. P. J. de (1895–1949).

The Dutch agriculturist, Carel Pieter Johannes de Haas, made significant contributions to our knowledge of the snake fauna of the Dutch East Indies (now Indonesia). He was born on 20 May 1895 in Breda (South Holland), 45 km southeast of Rotterdam. As a boy he had strong interests in natural history, especially insects, and went to the College of Agriculture in Wageningen. Since his main focus then was in tropical cultivated plants, he transferred to the Colonial Agriculture School in Deventer. He graduated in 1914 and became a planter in Java. By 1942 he had spent 15 years on various plantations in the Priangan Mountains in central West Java when the Japanese invasion occurred. He had served in the Dutch Auxiliary Forces and thus was interned, first at Tjimahi, Java, and finally in Singapore until he was liberated in 1945. His family had been interned in Bandung, Java, and he was not reunited with them until 1946 in the Netherlands. Four years later, put-

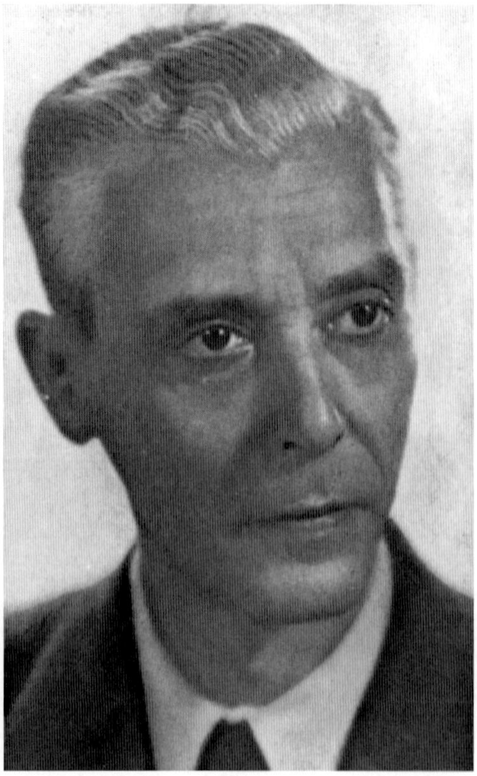

ting aside the trauma of the war years, he was able to return to the Priangan Mountains. He had planned to take charge of his former estate at Tjikadjang, but he found the factory ruined and the plantations abandoned. He therefore became an agricultural advisor at Garut, West Java, and cheerfully looked forward to his new assignment when he and his wife were killed in an automobile accident on 25 September 1949, near Trogong (now Tarogong), near Garut.

When he first arrived in Java, Haas collected beetles and butterflies as dried specimens. Since specimens were regularly destroyed by insect pests, he switched to keeping living birds and reptiles. Soon he developed a particular interest in snakes, and he sent specimens to museums in the Netherlands and to the Institut Pasteur in Bandung for identification. He became interested in geographical distribution and relative abundance of snake species and made his own preserved collection for reference, which was ultimately deposited at the Rijksmuseum in Leiden. Later, because of his difficulty in assigning the correct scientific names, he became interested in taxonomic problems. When he visited the Netherlands, he spent most of his time at the museums in Amsterdam and Leiden and, when museum systematists would not prepare a competent checklist of snakes for the area, he did so himself.

Haas authored about 15 titles on snakes (1933–1950), most of them short notes on individual species. His major papers concerned the biology and distribution of snakes in West Java (1941) and a review of the watersnakes of the Malay Peninsula (1949). His most important work, issued posthumously, was his 115-page "Checklist of the Snakes of the Indo-Australian Archipelago" (1950; reprinted 1972) with a distributional list organized by island.

• *Reference*: "Carel Pieter Johannes de Haas," by M. A. Lieftinck, Treubia, Bogor, 20: 507–510, 1950. • *Portrait* (1947) *and signature*: From Lieftinck, 1950.

BRAGG, Arthur N. (1897–1968).

The American herpetologist, Arthur Norris Bragg, who specialized on the biology of anurans, was born in Pittsfield (Maine) on 18 December 1897. His father, a farmer, died when Bragg was six, and to support the family his mother moved from place to place in Maine to find work. This disrupted Arthur's schooling and his socialization, but gave him opportunities to find frogs and turtles in the field before he was ten. He graduated from high school at the ripe age of 21 and matriculated at Maine's Bates College (B.S., 1924). After a year of graduate school at Johns Hopkins University in Baltimore, he taught at Marquette University in Milwaukee (1925–1933) before returning for further graduate study at Boston University (A.M. 1934). He then proceeded to the University of Oklahoma in Norman for his doctoral work on the mitotic cycle in embryonic toads (Ph.D. 1937), completed under the supervision of the cytologist, Aute Richards.

Thus began Bragg's interest in toad biology, which remained the focus of his research program. He joined the zool-

ogy staff in 1934 as an instructor and became full professor by 1953; he also had an appointment as herpetologist for the Oklahoma Biological Survey, which is headquartered on the university campus. Bragg rarely attended national meetings, perhaps partly because of his progressive deafness beginning in the early 1940s, thus few of his contemporaries knew him personally. Through Bragg's quiet efforts, a 32-ha tract of bottomland forest located near Norman was deeded to the university in 1957 as the Oliver Wildlife Preserve. Less than two months after retiring from the university, he suffered a coronary and died on 27 August 1968 in Norman.

Bragg published about 175 titles in herpetology (1933–1970), most of them on the biology of North American toads of the genera *Bufo*, *Scaphiopus*, and *Spea*. He studied every aspect of their life cycles including their breeding biology, early embryology, egg laying, hatching, tadpoles, behavior, ecology, distribution, and economic value. He had a special interest in cannibalism among spadefoot tadpoles. His knowledge of the life cycles of toads was so extensive that his students often referred to him as "Bufo Bragg" or "Mister Toad." Indeed, he claimed he could think like a toad. He reported (1960) being bitten by a specimen of a toad-eating snake (*Heterodon nasicus*) after handling some frogs, and he allowed the snake to ingest his thumb for about 15 min. He described the sensations, wounds, bleeding, and inability to close his hand. The swellings reached his wrist, but began to recede about 24 h later. This was the first recognition that this snake is mildly venomous.

Bragg described two new taxa of toads. One of them was a dwarf race of *Bufo americanus* from Oklahoma, which he named for his colleague, the ecologist Charles C. Smith, Sr., who had first introduced him to the toads of Oklahoma and was a frequent field companion. Besides his many short papers, Bragg published two larger works: "Researches on the Amphibia of Oklahoma" (1950), the result of a symposium that he edited and for which he wrote nine of the ten chapters, and his semi-autobiographical book on spadefoot toads, "Gnomes of the Night" (1965). His manuscript on "Frogs and Toads of Oklahoma" was never published.

• *References*: "Arthur Norris Bragg," by J. H. Black and C. C. Carpenter, Copeia, 1969: 419–420, 1969; "Arthur Norris Bragg 1897–1968," by H. P. Brown, Proc. Oklahoma Acad. Sci., 49: 223, 1970; "Arthur Norris Bragg (1897–1968) Biography and Bibliography," by C. J. McCoy [Jr.] and J. H. Black, Smithsonian Herpetol. Inform. Serv., 27: [i], 1–6, 1974. • *Portrait*: Sam Noble Oklahoma Museum of Natural History, University of Oklahoma, courtesy J. Droke. • *Signature* (1956): Adler collection.

FREIBERG, Marcos A. (1911–1990).

Marcos Abraham Freiberg (see *Note*), an Argentine zoologist who specialized on reptiles, amphibians, and birds, was born in Buenos Aires on 23 April 1911 to Jewish immigrants. His father was from Germany and his mother from Russia. Marcos became interested in reptiles as a teenager. He attended university at the Instituto Nacional del Profesorado Secundario in Buenos Aires and graduated in 1931 in natural sciences. Several of his teachers were on the staff of the Museo Nacional de Historia Natural (now Museo Argentino de Ciencias Naturales "Bernardino Rivadavia") including Pedro Serié who invited him to assist with the herpetological collections. Freiberg was on the museum's paid staff from 1934 to 1938. After these years he moved to the Museo de Entre Ríos in Paraná, about 350 km northwest of Buenos Aires, where by 1939 he was assistant director and chief of the museum's zoology department. He then relocated to a place not far from the Paraguayan border in northern Argentina to a national laboratory devoted to insect pests in the town of Presidencia Roque Sáenz Peña (Chaco), 900 km from Buenos Aires. He also had temporary work assignments in other venues. Although he could do field work in these rural locations, the financial conditions for his family were difficult and he was also remote from Buenos Aires, the country's scientific center.

Freiberg decided to enroll in dentistry school at the Universidad Nacional de Buenos Aires (D.D.S. 1942). He practiced dentistry in the city and taught biology at some local colleges, and he continued his work in herpetology and in conservation. In 1961 he was elected the first president of the Asociación Latino-Americana de Ictiólogos y Herpetólogos and in 1974 was awarded a Fulbright grant to study museum collections in the United States. In 1984 he and his wife immigrated to the United States to be near their children and they settled in San Francisco. Freiberg translated medical texts into Spanish and authored the "Bilingual Dictionary of Dental Terms" (1990). He died at his home in San Francisco on 19 August 1990.

In addition to Freiberg's technical papers on insects and birds, those on amphibians and reptiles number about 30 (1935–1974). Most are national or provincial checklists, distributional reports, and descriptions of new taxa of reptiles. Among his taxa recognized today are a lizard (*Stenocercus doellojuradoi*), a crocodilian (*Caiman latirostris chacoensis*), a snake (*Leptotyphlops australis*), and two sideneck turtles (*Acanthochelys pallidipectoris* and *Podocnemis argentinensis*), the latter a Cretaceous species. Freiberg wrote several books for the general public starting with "Vida de Batracios y Reptiles Sudamericanos" (1954). Later, he authored a series of books in "El Mundo," a series published in Buenos Aires: snakes (1970), turtles (1971), birds (1973), amphibians (1975), prehistoric animals (1976), and lizards (1977). He wrote a textbook on the ecology of Argentina (1982 and also produced three popular books in English: two on the reptiles of South America—turtles (1981) and snakes (1982)—and another on venomous animals (1984, with Jerry G. Walls).

• *References*: "Marcos A. Freiberg," anonymous, Ciencia Investig., Buenos Aires, 24(6): "332" (=281), 1968; "En Homenaje Dr. Marcos A. Freiberg," by J. M. Gallardo, Bol. Asoc. Herpetol. Argentina, 6(1): 1, 1990 (1991); "Recordando a Marcos Freiberg a un Año de Su Desaparicion," by J. C. Chebez, Bol. Asoc. Herpetol. Argentina, 7(2): 28–30, 1991; "Marcos A. Freiberg (1911–1990) El Dolor de Su Partida," by J. C. Chebez, Nuestra Aves, Buenos Aires, 9(24): 11–12, 1991; Additional information from Freiberg's son, Edward Freiberg, and grandsons, Daniel and Michael Kreiman, 2006. • *Portrait* (1984): By Kraig Adler. • *Signatures* (left, 1964; right, 1974): Adler collection. • *Note*: Born Abraham Freiberg, "Marcos" was added during his childhood because of discrimination against Jews.

GISLÉN, Torsten (1893–1954).

Torsten Richard Emanuel Gislén, who specialized on marine and freshwater biology especially echinoderms, was also a student of Swedish herpetology. He was born on 21 June 1893 in Kalmar (Småland Province), on the east coast of southern Sweden where his father was the county treasurer. He was interested in wildlife in his youth including amphibians and reptiles. Gislen's family moved to nearby Jönköping where he finished high school. He studied at Uppsala University (graduated 1916) and completed his doctoral dissertation there on echinoderms in 1924. After serving in posts at the zoological station in Kristineberg and the zoological department of Uppsala University, he became a professor of zoology at Lund University in 1932 where he remained for the rest of his career. Gislén was primarily a field biologist with interests in natural history, ecology, and zoogeography. He made excursions throughout Scandinavia and overseas including California, Hawaii, Japan, and Yugoslavia. It was on an outing that he died of a sudden heart attack on 26 August 1954 in Hällekis, near a hill overlooking the country's largest lake, Vänern, located in southern Sweden. He had intended to bring his students there the next day to climb the hill.

Gislén had a secondary interest in amphibians and reptiles on which he authored about 25 titles (1935–1959). These cover natural history, distribution, behavior, and systematics. His major work was finished after his death by one of his doctoral students, Hans Kauri, who was a refugee from Estonia. Their book, "Zoogeography of the Swedish Amphibians and Reptiles with Notes on Their Growth and Ecology" (1959), summarized data that Gislén had assembled for 20 years. It covers the complete natural history of each of the 12 species of amphibians and six of reptiles with distribution maps for each one. They discussed the climatic and human factors possibly responsible for the continued southward shift of faunal limits since the beginning of the Iron Age.

• *References*: "Torsten Richard Emanuel Gislén," by B. T[unbla]d, p. 66–67 *in* Svenska män och kvinnor–Biografisk uppslagsbok, vol. 3. A. Bonniers förlag, Stockholm, 1946; "Torsten Richard Emanuel Gislén," anonymous, Kungl. Fysiogr. Sällsk. i Lund Förh., 24(17): 61•–65•, 1954. • *Portrait and signature*: From Anonymous, 1954.

TWITTY, Victor C. (1901–1967).

Victor Chandler Twitty was a specialist on the experimental embryology, systematics, and behavior of salamanders. He was born on 5 November 1901 near Loogootee in southwestern Indiana. After he finished high school, his family moved to Indianapolis where he attended a local college, Butler (B.S. 1925). At Yale University in Connecticut he joined the laboratory of Ross G. Harrison, one of the world's leading embryologists and the inventor of tissue culture. The 1920s were exciting times for embryology. There were numerous discoveries, such as that of the embryonic organizer by the German embryologist, Hans Spemann, who worked primarily with salamanders. The Harrison group's organism of choice was the spotted salamander—what embryologists then called *Amblystoma punctatum* and is now *Ambystoma maculatum*—and

Twitty's dissertation was on the transplantation of eyes and limbs between that species and *A. tigrinum* (Ph.D. 1929). After further work at Yale, he went to Berlin in 1931 to work with a former student of Spemann's, Otto Mangold, who was then at a Kaiser Wilhelm (now Max Planck) Institute.

In 1932 Twitty joined the faculty of Stanford University in California where he spent the rest of his career. He was biology department chairman for 15 years beginning in 1949. Twitty had a distinguished career, honored by his election to the National Academy of Sciences in 1950. His career was remarkable in that his research was in two vastly different fields of study: experimental embryology and animal behavior. In California his animals of choice were newts of the genus *Taricha*. His embryological work dealt with growth regulation of eyes and limbs, pigment cell activity during development, the diffusible nature of inducer substances, and the first evidence of a potent neurotoxin in newt tissues (named tarichatoxin, a non-protein anti-predator toxin, which was later found to be chemically identical to tetrodotoxin, first identified in the Japanese puffer fish). He collected his newts locally and became interested in their natural history and systematics. He described two taxa recognized today: *T. rivularis* and *T. torosa sierrae*.

Twitty was also interested in the genetic relationships of newts and thus conducted hybridization experiments. Beginning in 1953 he released large numbers of hybrid larvae (20,000 in that year alone) into the wild at a ranch in the mountains of Sonoma County along the coast north of San Francisco, some 180 km from Stanford. This study raised the question whether, once mature, the newts would return to breed along the same segment of stream where they were released as larvae. Rather than wait for his answer, he decided to study the resident population of *T. rivularis* by marking, displacing, and recapturing adults. Between 1953 and 1966 Twitty and his students toe-clipped and moved more than 20,000 adults, some as much as 10 km, and found that they homed to their individual-specific segments of the stream. Furthermore, the newts took a beeline home, even over mountainous terrain; only one animal over those years ever entered another stream. Since random wandering was thus ruled out, many experiments followed to determine the sensory basis for homing. One set of studies involved removing the eyes of adults captured along a stream and displacing them. The animals appeared at their original capture sites the next year, suggesting that some sensory cue or cues other than vision was involved. Twitty wrote: "Of the countless displaced newts that I have handled, I think none has made such an impact on me as the first one of these blinded animals to be recaptured. As I examined its empty eye-sockets and emaciated body [blinded newts often do not feed], and then looked downstream toward the heavy forest and rugged terrain it had traversed in coming home, my respect for its accomplishment came as near awe and reverence as can be inspired by lowly organisms."

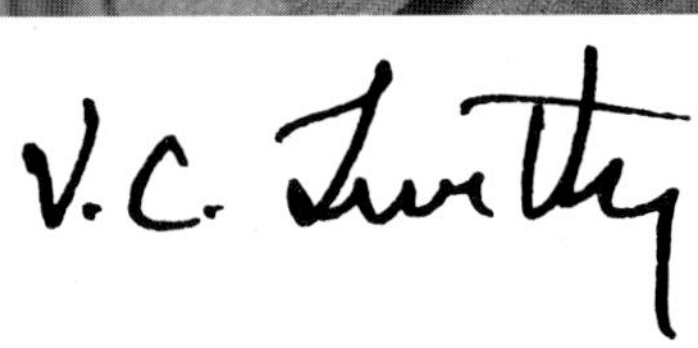

Unfortunately, Twitty would not follow up these tantalizing results. On the morning of 22 March 1967 his body was found in his Stanford laboratory, a suicide from "self-inflicted cyanide poisoning" according to a news release by his university. He was to have retired that June but had been seriously ill with cancer. Some of Twitty's marked newts continued to return to their breeding spots even into the late 1990s. By then, the multisensory basis of newt navigation—including their use of the earth's magnetic field as both a compass and a map—had been established with a different species of newt (*Notophthalmus viridescens*) by John B. Phillips and his associates then at Indiana University.

• *References*: V. C. Twitty, Of Scientists and Salamanders, W. H. Freeman Co., San Francisco, xi, 178 pages, 1966; "Victor Chandler Twitty November 5, 1901–March 22, 1967," by D. E. Kelly, Copeia, 1967: 887–888, 1967; "Victor Chandler Twitty November 5, 1901–March 22, 1967," by N. K. Wessells, Biogr. Mem. Natl. Acad. Sci., 74: 333–346, 1998; "Victor Chandler Twitty," by J. E. Lesch, Amer. Natl. Biogr., 22: 67–69, 1999. • *Portrait* (1951): Department of Special Collections, Stanford University Libraries. • *Signature* (1944): Adler collection.

HIBBARD, Claude W. (1905–1973).

The first articulated skeleton of a dinosaur was recovered from a New Jersey clay pit in 1858. Shortly thereafter the first dinosaur fossils in western North America were discovered and set off a scramble to locate, describe, and name them and those of other vertebrates. This led to a fierce competition between two American paleontologists—Edward D. Cope and O. C. Marsh—for sole control of the best fossils, and it broke out into a feud in 1872 that lasted until Cope's death in 1897. The fossils receiving greatest attention were the large animals—the so-called megafauna—such as dinosaurs, rhinoc-

eroses, horses, camels, huge crocodiles, and giant tortoises. By comparison, the microfauna received less attention, especially because it consisted of mostly disarticulated vertebrae and other bits of bone. Part of the problem was simply the difficulty of finding such tiny, isolated bones in the deposits and of unearthing them in large and representative quantities. In 1928 Claude William Hibbard, then an undergraduate student, devised a simple technique for sifting tiny bones of amphibians, reptiles, and other small animals from the matrix. With its further development by him and others, enormous collections of microfossils were recovered which led to a more complete understanding of faunal succession and paleoclimatic changes in the American West.

Hibbard was born on 21 March 1905 on the family farm in Toronto, an agricultural community in southeastern Kansas. He entered the University of Kansas in 1926 (B.A. 1933, M.A. 1934) and beginning in 1928 became an assistant in the Museum of Vertebrate Paleontology. After finishing his master's degree, he served in the National Park Service in Kentucky for a year before returning to Kansas as Assistant Curator of Fossil Vertebrates. While retaining his post at Kansas, he entered the University of Michigan as a doctoral student in 1936 under the direction of the mammalogist and biogeographer, Lee R. Dice. After completing his formal course work, Hibbard returned to Kansas to finish writing his dissertation (Ph.D. 1941). During his curatorship there he conducted summer field trips to western Kansas to collect fossils, a pattern that continued almost without interruption throughout his life. In 1946 he moved back to Michigan as Associate Curator of Fossil Vertebrates and became curator in 1949. He remained at Michigan for the rest of his career and died on 9 October 1973 in Ann Arbor.

Hibbard was primarily a mammalogist, but he also published extensively on fossil amphibians and reptiles. His major scientific accomplishments were the development of the faunal sequence for the Plio-Pleistocene rocks of the High Plains and his evidence for climatic equability in Pleistocene faunas. He authored more than 150 titles (1933–1975) with several on Recent amphibians and reptiles. Most of his herpetological work is contained in papers on entire vertebrate faunas.

His most lasting contribution, however, was the development and application of his technique for collecting microfossils. During the summer of 1928 he was able to join a fossil-collecting trip to northwestern Kansas by agreeing to be the cook and to set up camp. In those days, the overburden—the topsoil above a fossil-bearing layer—was removed with a plow, scraper, and a team of horses. Having been a farm boy, Hibbard also drove the horses. His dream had been to find a fossil rhinoceros or camel, but he was handed a pair of tweezers and told to look for small pieces of bone. He tried it for a few hours, crawling around on his knees or on his belly in the blazing sun, but then thought of a better way.

The next day he returned with a square piece of window screening fixed in a wooden frame. He took sacks of fossil-containing sands and his new device to a buffalo wallow—depressions in the plains created by a buffalo when it rubs its body on the ground and then fill with water. He put a shovel full of sand with fossils into the basket-like sieve, then lifted and lowered it in the water. The fine sand passed through the screen, leaving the fossil bones on the screen's surface. By the end of the day he had enough fossils to fill a small matchbox. His supervisor, however, was not impressed. As he told Hibbard, with that one small box "you have enough small fossils for all of the museums in the world." Hibbard was told to start digging for horses and camels, but he decided that when he had a chance to collect on his own he would focus on microfossils and perfect his technique. This technique permitted the study of Pliocene and Pleistocene amphibians and reptiles and has been adopted worldwide for vertebrate microfossils. Over the following decades Hibbard and his students were able to develop the best-documented sequence for demonstrating the evolution, faunal migration, and climatic change for the late Cenozoic anywhere in the world.

• *References*: "Claude W. Hibbard, 1905–1973," by H. A. Semken, Jr., and R. J. Zakrzewski, Jour. Mammal., 56: 275–279, 1975; "Studies on Cenozoic Paleontology and Stratigraphy," ed. by G. R. Smith and N. E. Friedland, Mus. Paleontol., Univ., Michigan, Ann Arbor, 1975 (Hibbard on pp. v-vi, 129–138); "Memorial to Claude William Hibbard 1905–1973," by J. A. Wilson, Mem. Geol. Soc. Amer., 5: 1–8, 1977; "Claude William Hibbard," p. 272 *in* Who Was Who in America, vol. 7. Marquis, Chicago, 1981. • *Portrait and signature*: Courtesy Gerald R. Smith.

CARVALHO, Antenor Leitão de (1910–1985).

Antenor Leitão de Carvalho, a Brazilian specialist on frogs and fishes, was born in Barreira do Piquete, about 180 km northeast of São Paulo on 15 April 1910. It was in this small

town in the coastal mountains, midway between São Paulo and Rio de Janeiro, where he developed his love for wildlife. Following his local schooling, he joined the merchant marine as a pilot (1927–1932). When his ships docked in Rio de Janeiro, he would visit the Museu Nacional to volunteer his assistance. In 1933 he became a staff technician and field collector for Alipio de Miranda-Ribeiro, then the museum's senior zoologist and Brazil's foremost ichthyologist who also happened to study frogs. Carvalho developed similar interests and soon went on his first major expedition, which was to northeastern Brazil. His first publication on frogs appeared in 1937 but at that time the museum's publication funds were exhausted and he, like Miranda-Ribeiro, had to publish in *O Campo*, a Brazilian farm journal. Miranda-Ribeiro died in 1939 and although he intended Carvalho to be his successor as curator of herpetology, Carvalho had to enter an open competition established by the museum's director before he could assume this role in 1941. Beginning in 1957 Carvalho also held administrative posts at the museum including chief of the department of zoology and, later, of vertebrates as well as vice-director of the museum.

In 1947 Carvalho was awarded a Guggenheim Foundation fellowship for advanced study in the United States, but the Brazilian government would not permit him to accept it until 1951 when he went to California to work with George S. Myers at Stanford University for two years. Myers, an ichthyologist and herpetologist who had spent time in Brazil (1942–1944), had been associated with Carvalho at the Museu Nacional and became Carvalho's other mentor. This period with Myers represented the only academic training that he ever had. In addition, Carvalho had a long collaborative association with Joseph R. Bailey, a snake specialist at Duke University in North Carolina who spent several years in Brazil (1940–1942, 1953–1954). Jointly with Eugênio Izecksohn, Carvalho taught the herpetology course to postgraduate students at the Universidade Federal Rural do Rio de Janeiro during 1975–1979. He retired from the museum in 1979 and died in Rio de Janeiro on 11 December 1985.

First and foremost Carvalho was a field naturalist and participated in numerous expeditions to many parts of Brazil. Among the areas visited were the Amazon basin, Mato Grosso, and the coastal forests. Most of these trips occurred between 1936 and 1955.

In addition to his work on fishes and insects, Carvalho authored about two dozen titles on amphibians and reptiles (1937–1965). These are primarily on frogs but there are also papers describing new species of reptiles and reviews of Brazil's caimans. Jointly with Myers (1945) he confirmed that a plethodontid salamander supposedly from Uruguay and described by Marcos Jiménez de la Espada in 1875, was based on a mislabeled specimen from California. Carvalho's work on frogs was broad-ranging and included ecology, zoogeography, descriptions of new genera and species, natural history, and morphology of tadpoles and adults (his last paper, with Ivor Griffiths, concerned the use of larval characteristics in anuran phyletics). He had intended to write a monograph on the American microhylids, but only a synopsis appeared (1954) in which he reviewed the genera and established six new ones. Carvalho died with many diverse projects left unfinished. One of them was a new species of turtle he discovered in the state of Maranhão in the early 1960s. It was re-discovered and named *Trachemys adiutrix* by P. E. Vanzolini in 1995.

• *References*: "A Obra Científica de Antenor Leitão de Carvalho (1910–1985)," by H. Nomura, Revta. Bras. Zool., 10: 545–552, 1993 (1994); "Antenor Leitão de Carvalho," by P. E. Vanzolini, Herpetol. League Commun., 1(2): 1, [1995]; Additional information from Ulisses Caramaschi, 2006. • *Portrait* (1943–1944): George S. Myers/Alan E. Leviton portrait collection, courtesy Alan Leviton. • *Signature*: Courtesy José Pombal Júnior.

ELKAN, Edward (1895–1983).

Edward Rudolph Elkan was a physician, pathologist, amateur herpetologist, and specialist on the diseases of lower vertebrates. Born Rudolf Eduard Elkan in Hamburg, Germany, on 21 February 1895, he had a boyhood fascination with amphibians and reptiles, but his scientific interest was whetted when he got a microscope at age 16. After boarding school he entered the army and lost sight in one eye as a result of service in World War I. He studied pathology in medical school in Freiburg (M.D. 1921) and at the Hamburg Institute of Tropical Medicine before working as a physician in Palestine (1925–1930), which was then under British mandate. This gave him an opportunity for first-hand work with tropical diseases and, coincidentally, introduced him to exotic reptiles. On returning to Hamburg he set up a private practice, but because he was a Jew and Adolf Hitler was rising to power, he decided to immigrate to England in 1933. He quickly achieved the necessary certifications (L.R.C.P.&S. Edinburgh and

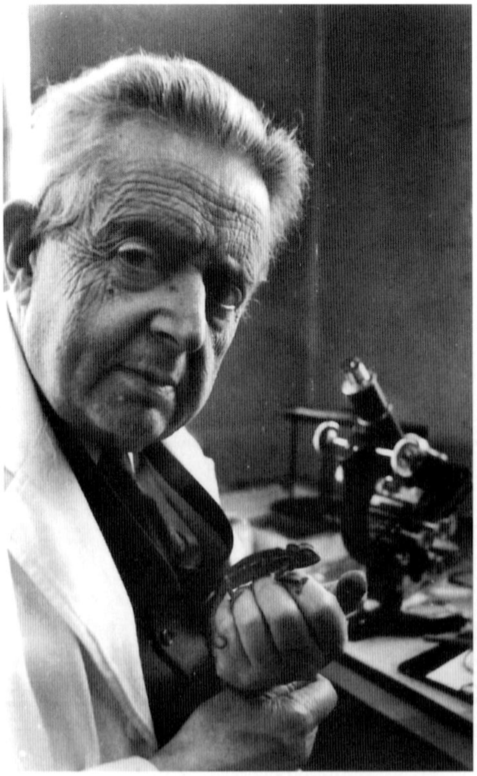

L.R.F.P.&S. Glasgow, both 1934) and then opened a practice in gynecology and obstetrics in Pinner (Middlesex) on the western edge of London.

In 1937 Elkan began to use *Xenopus* frogs to detect early pregnancy in his female patients, which was based on research by the British endocrinologist Lancelot Hogben (1930, 1931) who was then working in Cape Town. Elkan published a seminal paper in *British Medical Journal* (1938) promoting the standardized use of *Xenopus* in human pregnancy tests, describing how the tests should be conducted and discussing their reliability—all based on his own research. Originally he imported his frogs from South Africa but soon decided to breed them, guided by the work of British zoologist Edward J. Bles (1901, 1905) who first described the life history of *X. laevis*. Elkan shifted his interests to procedures for breeding (previously thought to be difficult) and maintaining healthy colonies, which brought him back to his early interest in pathology. He published on these topics and extended his interest to the biology and diseases of other amphibians and reptiles. At the same time a German specialist on the diseases of fishes, Heinz H. Reichenbach-Klinke, published three slender volumes on the diseases of fishes, amphibians, and reptiles (1957, 1961, 1963) that referred to work published by Elkan. When an English translation was contemplated, Elkan was the obvious choice to do it. However, the resulting book, "The Principal Diseases of Lower Vertebrates," which was issued in a single volume (1965; reprinted), was not merely a translation but was enlarged by Elkan based on his own researches. This was the first book in the English language on the subject of diseases of amphibians and reptiles and their treatment.

Elkan always did his own post mortem examinations, cut and stained tissue sections from specimens, and made his own drawings and photographs, all as a hobby adjunct to his medical practice. At first the work was conducted at his medical offices, but after World War II when the British Ministry of Health opened a laboratory at Watford (1945) with a large breeding colony of *Xenopus*, Elkan was asked to supervise it. (Since the 1960s *Xenopus* is no longer used for human pregnancy tests and has now become a model for developmental biology studies.) He conducted his research there until the laboratory was closed in 1972 when he moved it to the Mount Vernon Hospital in Northwood. In 1974 Elkan and Reichenbach-Klinke produced an enlarged version of their book under the title "Color Atlas of the Diseases of Fish, Amphibians and Reptiles," which remains a standard reference in this rapidly expanding field.

Elkan's major accomplishment was to establish the field of comparative pathology and medicine of amphibians and reptiles and to encourage veterinarians to take a serious interest in it. He published a wealth of descriptions of diseases and showed that individual animals could be treated and survive. His enormous collections of microscope slides, drawings, photomicrographs, and reference publications are now at the Royal College of Surgeons in London under the curatorship of his disciple, John E. Cooper.

Elkan retired from his practice in 1980. He died at the age of 88 at his home in Pinner on 4 July 1983.

• *References*: "E Elkan," by J. E. Cooper and M. H. Bennett, Brit. Med. Jour., 287: 1071, 1983; "Dr. Edward Elkan 1895–1983," by J. E. Cooper, Herpetol. Rev., 14: 105, 1983; "'The Guiding Hand—Foundations of Lower Vertebrate Pathology' The First Edward Elkan Memorial Lecture," by J. E. Cooper, Herpetopathologia, Angers, 1: 1–4, 1989; Additional information from Maya (Mrs. E.) Elkan, 1990. • *Portrait* (1973) *and signature*: Courtesy Maya Elkan.

SZARSKI, Henryk (1912–2002).

Henryk Maria Szarski, a vertebrate evolutionary biologist who specialized on amphibians, was born on 13 September 1912 in Kraków in southern Poland. He was a member of an old mercantile family that sold teas but did not follow in the trade when, on testing by his father, he could not differentiate the kinds of tea. Instead, he studied biological sciences at Jagiellonian University in Kraków, graduating in 1935, and continued there under the supervision of the comparative anatomist, Zygmunt Grodziński, to do his dissertation on the digestive system of oligochaete worms (Ph.D. 1937). He joined Grodziński's department and switched his research to the vascular system of lymphatic organs in amphibians with his first papers appearing in 1937.

Szarski's research was interrupted when the German Army invaded Poland in September 1939 to begin World War II. In November the Nazis arrested most of the faculty at Jagiellonian University as part of a larger effort to exterminate the Polish intellectual elite, many of which were also Jewish. Most were sent to Dachau and other camps. Szarski was not yet a professor but was of Jewish origin, so he went into hid-

ing in Kraków and later to the village of Zawoja at the foot of Babia Gora, a peak in the Carpathian Mountains near the Czech border where he worked at a fish hatchery.

Only in 1946 could Szarski complete his habilitation. Two years later he became a professor at Nikolaus Copernicus University in Toruń, located about 180 km northwest of Warsaw. As dean in 1950–1952, he organized a faculty of biology and earth sciences and served as rector during 1956–1959. His research shifted to comparative studies of skin vascularization in amphibians and included species of special evolutionary importance such as the New Zealand frog, *Leiopelma*.

In 1967 Szarski returned to his *alma mater* and succeeded Grodziński as head of the university's Department of Comparative Anatomy, a post he held until his retirement in 1982. Besides his research, throughout his career Szarski spent much of his time popularizing science and for several decades wrote articles for the journal *Wszechświat* (Universe) on interesting scientific discoveries. His many textbooks served generations of Polish faculty and students including one on the history of vertebrates (1967, with editions as late as 1998) and another concerning the mechanisms of evolution (1972, later editions through 1986). He became a major propagator in the socialist countries of modern evolutionary ideas in the post-Lysenko era. He died in Kraków on 16 January 2002.

Although his main research encompassed general evolutionary questions about vertebrates, amphibians became his primary objects for study. His later research focused on two main topics: amphibian monophyly and cell size in vertebrates. During the 1930s and 1940s, the monophyly of all amphibians, living and extinct, was challenged by some paleontologists, especially the Swedes Nils Holmgren and Erik Jarvik, who believed that even the living amphibians had multiple origins among ancient fish groups. Szarski, in a classic paper in *The Quarterly Review of Biology* (1962; published more extensively in Polish as a short book in 1961), presented an extensive, tightly argued defense of monophyly based on morphological, developmental, and paleontological evidence. He also drew from his own work on respiratory patterns and the interpretations of the Russian evolutionary morphologist, Ivan I. Schmalhausen (1957), and the American physiological ecologist, Raymond B. Cowles (1958), regarding the probable environment in which the first tetrapods arose. Szarski argued not only that all Amphibia had a single origin among the fishes, but that all living amphibians are monophyletic. He did not, however, use the term Lissamphibia, which was proposed to include the living groups of amphibians by the anatomist, Hans Gadow (1901), nor did he reference the unique amphibian tooth structure found only among the lissamphibians, as first noted by the British herpetologist, H. W. Parker (1956).

Szarski's other major research program was the evolutionary differentiation of cell dimensions in vertebrates and his recognition that DNA content and tissue and blood cell size in amphibians, especially salamanders, was unusually large among vertebrates. A number of papers by him and his students, notably Grażyna Czopek, Juliusz Czopek, and Jacek M. Szymura, provided extensive comparative data on liver, erythrocyte, and other cell types. Szarski believed that large cell size was an adaptive change due to metabolic rate and genome dimensions.

• *References*: "Henryk Szarski," p. 429–430 *in* J. Adamczewski, Mała Encyklopedia Krakowa, Wydawnictwo "Wanda," Kraków, 1996; "Henryk Szarski (1912–2002)," by J. Rafiński, Kosmos, Warsaw, 51(2): insert between pp. 122 and 123, 2002; "Henryk Szarski 1912–2002," by W. M. Kilarski, Copeia, 2002: 889–890, 2002. • *Portrait*: From Kilarski, 2002. • *Signature* (1994): Courtesy of Zbigniew Szyndlar.

SERGEEV, Alexei M. (1912–1943).

The Soviet embryologist and physiologist, Alexei Mikhailovich Sergeev (sometimes Sergeyev or Sergeeff), was one of the discoverers of behavioral thermoregulation in reptiles. He was born on 7 November 1912 in Moscow and graduated from high school in 1931. At Moscow State University he was a student of the zoologist and ecologist, Alexander N. Formozov, and the evolutionary morphologist, Boris S. Matveev. Sergeev's first interest was birds, but on a field expedition to the Caspian Sea in 1934 he shifted to reptiles. After receiving his diploma in 1936, Sergeev proceeded to complete his candidate's thesis and it was converted to a doctorate after he defended it in 1940. He then became a lecturer at the university where he taught evolutionary morphology of vertebrates.

With World War II underway, Sergeev quickly joined the army in June 1941; his last note home, postmarked Voronezh, a city located about 500 km south of Moscow, arrived in September 1943. German forces earlier had encircled and captured his army division during a battle near Kiev in Ukraine, but the Soviet Army rescued him. However, soldiers who were lucky

enough to survive captivity, having not died in defense of the motherland, were then arrested. They were subjected to trials to determine their political loyalty and even to find evidence of cowardice or treason. This was a humiliating procedure for those, like Sergeev, who had faced death on the battlefield. Many did not have the fortitude to endure offensive questioning or were slandered by fellow soldiers in order to avoid their own punishment. Sergeev was originally sentenced to die by firing squad, but instead was given ten years of prison and then five of internal exile. He received the harshest penalty of his group since, knowing German, he could speak with the guards and was therefore suspected of collaboration with the enemy. Sergeev was sent to the Usolsk camp near Berezniki, about 150 km north of Perm and along the Kama River at the western edge of the Ural Mountains. This camp was part of the vast Soviet slave-labor Gulag where, ill from dystrophy and in despair, he hanged himself sometime late in 1943. In 1965 his widow appealed for his postmortem rehabilitation, but was rejected by the Soviet government.

Sergeev authored many works during his short academic career, of which about a dozen were on amphibians and reptiles (1937–1943). He made important contributions on the evolution of embryonic adaptations of reptiles to terrestrial environments (including descriptions of the stages of development of the amnion and embryonic blood circulation); the ecological physiology of amphibians and reptiles; and the origin of viviparity in reptiles. He developed a retrospective method to back-calculate the earlier body lengths of lizards and turtles based on measurements of growth rings in their bones and shell scutes. Much of his research was summarized in a book on the evolution of embryonic adaptations in reptiles (1943). Sergeev's most important discovery was published in 1939, based on field studies that he conducted in the deserts of Turkmenistan in 1938. He reported that agamid and lacertid lizards as well as tortoises could, through behavioral means, regulate their body temperatures independent of the environment. This work was performed independently of that done in the United States by Raymond B. Cowles and his student, Charles M. Bogert, who discovered the same phenomenon in iguanid lizards (published 1939–1944).

• *References*: ["In Memory of Aleksey Michailovich Sergeev—Distinguished Zoologist (1912–1943),"] by E. A. Baburina et al., Bull. Moscow Soc. Invest. Nature, sect. Biol., 87(4): 118–122, 1982; Rossolimo and Dunayev, 2003 (Sergeev on pp. 329–343, 534); Additional information from Evgeniy Dunayev, 2006, and Boris Vassiliev, 2007. • *Portrait* (1940–1941): Courtesy Evgeniy Dunayev. • *Signature* (1927–1928): Courtesy Boris Vassiliev and Ella Smirina.

CAGLE, Fred R. (1915–1968).

Fred Ray Cagle, who studied the life histories and systematics of North American amphibians and reptiles, especially turtles, was born on 9 October 1915 at Marion in southern Illinois. He was interested in reptiles as a boy and attended Southern Illinois Normal University (now Southern Illinois University) in nearby Carbondale (B.Ed. 1937). For graduate work he went to the University of Michigan where he began to study hom-

ing behavior in turtles with Frank Blanchard. With Blanchard's premature death in September 1937, the turtle systematist Norman Hartweg became his supervisor (M.S. 1938, Ph.D. 1943). Cagle conducted most of his fieldwork in southern Illinois, allowing him simultaneously to become a faculty member at his *alma mater* in Carbondale. During World War II he served in the Pacific and took advantage of the opportunity to do field research in the Mariana Islands.

In 1946 Cagle accepted a professorship at Tulane University in New Orleans where he spent the rest of his career. He became chair of the Department of Zoology in 1952, coordinator of all research at the university in 1959, and was promoted to vice president in 1963. In the latter, he had responsibility for institutional development, public and governmental relations, alumni activities, and oversight of research and the research campuses. Cagle also served in numerous national and international positions where he was an effective leader for the biological community. He continued his herpetological activity through editorships of *Copeia, Biological Abstracts*, and *American Midland Naturalist*. His greatest legacy to herpetology, however, was the training of a large number of notable master's and doctoral students including Allan H. Chaney, James L. Dobie, Robert E. Gordon, Richard M. Johnson, C. Robert Shoop, Donald W. Tinkle, and Robert G. Webb, an undergraduate student, Richard Etheridge, and two local herpetologists, J. Whitfield Gibbons and Ernest A. Liner. Cagle's sudden death at the age of 52—on 8 August 1968 in New Orleans—was attributed to an aneurysm.

Before his publications turned to public policy concerning biology, Cagle authored more than 40 research titles on the systematics, behavior, and life histories of amphibians and reptiles (1937–1955). His major works include a study of homing behavior and migration in turtles (1944), based on his doctoral dissertation, the life history of *Pseudemys* (now *Trachemys*) *scripta troostii* (1950), and descriptions of four new taxa of *Graptemys* (1953, 1954). He also wrote the chapter on reptiles for the popular university-level textbook, "Vertebrates of the United States" (1957; second edition 1968), co-authored with W. Frank Blair, Albert P. Blair, Pierce Brodkorb, and George A. Moore. Cagle's most influential publications, however, were two papers outlining how life history studies of reptiles (1953) and amphibians (1956) should be conducted, which were used by numerous researchers in the United States and overseas as patterns for their own work.

• *References*: "Dr. Fred R(ay) Cagle," p. 161 *in* American Men of Science, vol. 2. Science Press, Lancaster, Pennsylvania, 1955; "Fred Ray Cagle (1915–1968)," by J. T. Collins, Herpetol. Rev., 1(6): 6, 1969; "Fred Ray Cagle 1915–1968," by D. W. Tinkle, Copeia, 1969: 220, 1969; Conant, 1997 (Cagle on pp. 454–456). • *Portrait* (1964): Courtesy Josephine (Mrs. Fred) Cagle. • *Signature* (1946): Adler collection.

OLIVER, James A. (1914–1981).

The American herpetologist, James Arthur Oliver, who became curator and director of several scientific and cultural institutions, was born on 1 January 1914 in Caruthersville, along the Mississippi River in southeastern Missouri. When he was two, his family moved to Saint Louis where his father became United States District Attorney. James brought his first live snake home when he was four. As a teenager he was a regular visitor to the reptile house at the Saint Louis Zoo. The curator, R. Marlin Perkins, often took him along on collecting trips. Wanting to be a herpetologist, Oliver enrolled at the University of Texas at Austin in 1932, but two years later he transferred to the University of Michigan (B.A. 1936, M.A. 1937, Ph.D. 1942). His doctoral chairman, Alexander G. Ruthven, was president of the university so Oliver's daily supervision was provided by Ruthven's associates in the Museum of Zoology, Helen T. Gaige and Norman Hartweg. In 1942, as assistant curator, he joined Charles M. Bogert in the Department of Herpetology of the American Museum of Natural History in New York. He served in the United States Navy during 1943–1945 and then returned to the museum in 1946. He was promoted to associate curator in 1947. Because of deteriorating museum finances he resigned in 1948 and joined the biology faculty at the University of Florida.

In 1951 Oliver returned to New York as Curator of Reptiles at the Bronx Zoo, a distinguished post first held by Raymond Ditmars. During this period he oversaw the design and construction of the new reptile house. The new building had naturalistic displays and one result was the first successful captive breeding of king cobras (published by Oliver in 1956). He took on additional duties as Assistant Director in April 1958 and Director of the zoo two months later. He returned to the American Museum in September 1959, now as Director, and guided it through a major period of growth. More than a dozen exhibition halls were planned and opened during his tenure. After retirement from the museum in 1969, he became Director of the New York Aquarium, a facility located in Brooklyn and, like the Bronx Zoo, operated by the New York Zoological Society (now the Wildlife Conservation Society). Oliver retired from the aquarium in 1976. He died in New York City on 2 December 1981.

Because of his succession of administrative posts after 1958, Oliver's scientific work occurred earlier in his career (about 30 titles, 1937–1959). His first research was on Mexico, beginning with a four-part series on the herpetology of the Isthmus of Tehuantepec (1937–1940, with Hartweg). During the summers of 1938–1939, he participated in a fresh-water fish survey of New Hampshire and also collected amphibians and reptiles. This work led to the first summary of the state's herpetofauna (1939 [1940?], co-authored with two of his graduate school colleagues, Joseph Bailey and Grace Orton). Oliver's papers in the late 1930s and 1940s were mostly on behavior and systematics (he named 20 taxa of lizards and snakes), but his most significant works were a monograph on the herpetofauna of Sonora, Mexico (1945, jointly with Bogert) and a revision of the Neotropical genus *Leptophis*, then known as *Thalerophis* (1948). His 1948 paper on the anoles of Bimini (Bahamas), in which he studied resource partitioning, introduced the ecological dimension of perch height and led, in turn, to the ecomorph concept of Ernest Williams and Stanley Rand. In 1953, Oliver co-authored the first review of the amphibians and reptiles of the Hawaiian Islands with Charles E. Shaw of the San Diego Zoo.

Oliver's best-known works, however, were two books. The first of these, "The Natural History of North American Amphibians and Reptiles" (1955), was highly influential in shifting attention from a focus on species to a focus on the habits of these animals. Topics included economic value, occurrence, locomotion, food and feeding, reproduction, and growth, size, and longevity. In his other book, "Snakes in Fact and Fiction" (1958; revised edition 1963), he separated myths and superstitions from truth and reality.

• *References*: "Dr. James Arthur Oliver," p. 839 *in* American Men of Science, vol. 2. Science Press, Lancaster, Pennsylvania, 1955; "New Director," by J. O'Reilly, Museum Memo [Amer. Mus. Nat. Hist.], 9(1): [1–3], 1959; "James Arthur Oliver," p. 23–25 *in* Current Biography, January 1966; "Dr. Oliver Retires as Aquarium Director," anonymous, Anim. Kingdom, 79(3): 1, 1976; Myers, 2000 (Oliver on pp. 56–57, 59). • *Portrait* (1961): Carnegie Museum, courtesy M. Graham Netting. • *Signature* (1966): Adler collection.

TIHEN, Joseph A. (1918–2002).

Joseph Anton Tihen, who studied the systematics and morphology of fossil and extant amphibians and reptiles, was born on 20 November 1918 in Harper (Kansas), about 75 km southwest of Wichita. As a teenager he came to know Claude Hibbard, a vertebrate paleontologist at the University of Kansas. Hibbard used Harper as a base for his collecting trips for fossils during the summers and Tihen was invited on several of them. Tihen later attended the University of Kansas where he took courses from the herpetologist, Edward H. Taylor, and was also influenced by Taylor's graduate student, Hobart M. Smith. After graduation (A.B. 1940), Tihen stayed at Kansas for a year of graduate work, but then transferred to the University of Rochester in New York State. There he became a student of the salamander specialist, Sherman C. Bishop, and was further influenced by Smith who had just joined the Rochester faculty. Tihen's dissertation was on gerrhonotine anguid lizards (Ph.D. 1945). He also conducted studies on the morphology of scolecophidians—*Liotyphlops* and other blindsnakes—and initiated work on the osteology of salamanders. He published parts of his doctoral research, but in 1944 he became part of a genetics team at the university's medical center that studied the effects of radiation on mouse populations as part of the top-secret Manhattan Project, which built the first atomic bombs.

Tihen taught at Tulane University in New Orleans during 1946–1947 and then returned to Kansas to assist in his father's real estate and insurance business. He resumed publishing on gerrhonotines and named nine new Mexican taxa now placed in the genera *Abronia*, *Gerrhonotus*, and *Mesaspis* and a new genus (*Coloptychon*) based on a Panamanian species. In 1957–1958, while at the University of Florida Medical School, he wrote up the results of the mouse genetics project and published his important research on the osteology of ambystomatid salamanders. This included his establishment of the subfamilies Dicamptodontinae and Rhyachotritoninae; these were later raised to family status. In 1958 he joined the faculty of the University of Illinois where one of his colleagues again was Smith, who had moved there from Rochester. At Illinois his research turned to bufonid frogs, both fossil and living, and he named two new genera of African bufonids. Finally, in 1960 Tihen became Professor of Biology at the University of Notre Dame in northern Indiana where he remained for the rest of his career. Charles J. Chantell was his doctoral student and Neclâ Özeti, the Turkish herpetologist, studied with him there. He authored major works on bufonids and named numerous new fossil species. He also named new fossil salamanders including new species of *Cryptobranchus* and *Ambystoma*, all collected during field trips in 1962–1964 from Miocene and Pliocene deposits in central United States.

Tihen was a quiet man who had great strength of his convictions. At a national meeting in 1964 he lectured on the zoogeographic implications of his work on the Tertiary herpetofaunal record, which he attributed in part to plate tectonics. He then withstood withering, although probably expected, criticism for invoking a mechanism that did not start to become established theory until the late 1960s. Tihen retired from Notre Dame in 1989 and moved to Fort Myers (Florida) where he died on 12 November 2002.

• *References*: "Dr. Joseph Anton Tihen," p. 1137 *in* Amer. Men Sci., vol. 2. Science Press, Lancaster, Pennsylvania, 1955; "Joseph A. Tihen," by A. Grobman and M. M. Stewart, Copeia, 2002: 1152–1153, 2002; Additional information from Frances (Mrs. Joseph) Tihen, 2004. • *Portrait* (1963): Carnegie Museum Pittsburgh, courtesy M. Graham Netting. • *Signature* (1998): Courtesy Frances Tihen.

BATTERSBY, James C. (1901–1993).

For 45 years (1916–1961) James Clarence Battersby was custodian of the amphibian and reptile collections at the British Museum (Natural History) in London and as such he had day-to-day charge of one of the largest and most important herpetological collections in the world. He was born in Chelsea, London, on 11 April 1901 and attended several schools, mostly in London, but never had university training. He was the son of a police constable who worked at the museum and, at the age of 15, James was hired as a "boy attendant" to assist George A. Boulenger, then the most renown systematic herpetologist in the world who was within four years of retirement. Boulenger had a global knowledge of amphibians and reptiles and very disciplined work and curatorial habits, which Battersby absorbed and emulated for the rest of his life.

Battersby was promoted to attendant in 1920, but without an honors university degree he could not apply for the scientific grades at the museum. Thus, in addition to his custodial duties over the ensuing years, it was his duty to introduce a series of curators to the departmental traditions—Joan Procter, H. W. Parker, and Alice G. C. Grandison—for a curatorial post he could never hope to occupy. Battersby's main tasks were specimen identification, cataloguing, processing loan requests, and assisting visitors. Since the alcoholic specimens were kept in a separate building, fetching and returning them was a time-consuming task. Battersby always put service before his personal research and the herpetological community honored his service in 1959 when the American Society of Ichthyologists and Herpetologists elected him an honorary foreign member. The fact that his mentor, Boulenger, had been elected the first such foreign member must have made the honor all the more meaningful to him.

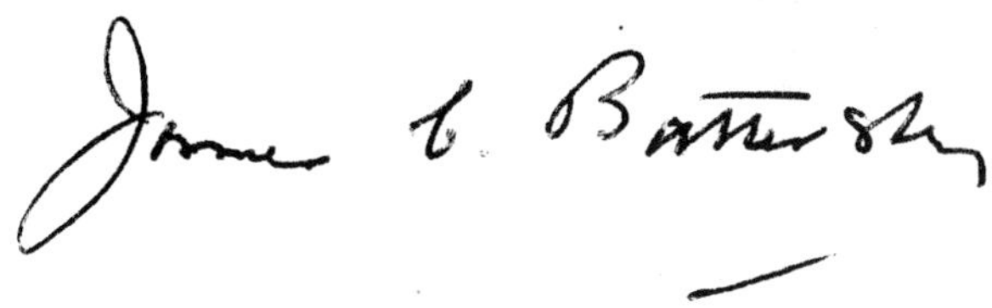

During World War II, when the alcoholic specimens had to be moved out of London, Battersby had the Herculean task of storing them in caves in the chalk hills of Surrey south of London. He also served on the museum's fire brigade during the London Blitz when bombs fell near the museum. Eventually he was sent for army duty in North Africa and Italy for which he was later decorated with the British Empire Medal. From 1946 to 1953 he served as the compiler for the Amphibia and Reptilia sections of *Zoological Record* and for Reptilia separately until 1960.

Given his heavy responsibilities for the collections, Battersby had little time for personal research, but he did produce about ten titles (1938–1971), the last being a warm remembrance of Boulenger. His scientific papers cover diverse geographic areas (West Indies, East Africa [where he did his only field work, in 1951], Middle East, and Nepal). Included in these are descriptions of five new taxa of lizards and snakes, and a new genus of geckos from Arabia.

Battersby retired from the museum in 1961 and moved to Cranleigh (Surrey) where he died from pneumonia on 3 December 1993.

• *References*: "James Clarence Battersby 1901–1993," by A. G. C. Grandison, Herpetol. Rev., 25: 44, 1994; "James Clarence Battersby 1901–1993," by A. G. C. Grandison, Copeia, 1994: 838–840, 1994. • *Portrait* (1958): George S. Myers/Alan E. Leviton portrait collection, courtesy Alan Leviton. • *Signature* (1968): Adler collection.

DING, Han-bo (1912–2003).

The Chinese zoologist and developmental biologist, Han-bo Ding (see *Note*), a specialist on amphibians and reptiles, was born on 3 November 1912 in Gutian County, about 145 km northwest of Fuzhou, Fujian Province. Ding's father was a Methodist pastor who attended college in Boston and returned to China to become a professor and later a leader in the Chinese church. The family moved to Fuzhou in 1919, just after his father returned from the United States. Han-bo Ding attended Anglo-Chinese College in Fuzhou and then went to study medicine at Yenching (now part of Peking) University. There he contracted tuberculosis and while spending a year in a sanatorium he decided to switch to biology (B.S. 1936). He went on to earn a master's degree under the American biologist, Alice M. Boring, a specialist on amphibians who taught at Yenching and trained many other herpetologists including Tso-kan Chang and Cheng-chao Liu.

After completing his research (M.S. 1939), Ding returned to his native province and became a professor at Fukien Christian University in Fuzhou. In 1947 he went to the United States where he earned a doctorate at Ohio State University with a dissertation on amphibian embryonic development and nucleo-cytoplasmic relationships (Ph.D. 1949). After a postdoctoral year at Ohio State, he returned to China and his post at Fukien Christian University. In 1951 the university was merged with two other colleges to become Fujian Teachers (or Normal) University where he was professor of biology.

Later he held other posts including chairman of biology (1957–1979), director of the Developmental Biology Research Laboratory (1979–1984), and vice president of the university (1979–1983). He retired as a professor in 1986 and died in Fuzhou on 7 March 2003.

Ding's research with amphibians began under Boring's tutelage. He studied hermaphroditism, Bidder's organ, seasonal reproductive changes, and hybridization. These topics interested him throughout his career. He published nearly 100 titles, about half of them dealing with amphibians and reptiles (1938–1991). Ding authored faunal reviews of amphibians and reptiles from various parts of Fujian Province, life history studies, use of male anurans for detection of human pregnancy, distributional records, and studies on fertilization, hatching, and development. In 1979 he and his student, Ming-chang Ts'ai (now Ming-zhang Cai), described a diminutive frog, *Rana minima*, which Ding had first found in 1958 in water-filled ditches near Lingshi Monastery about 65 km south of Fuzhou. With others, Ding wrote two identification manuals of Chinese amphibians (1959) and reptiles (1962). With his university colleague, Ji Cheng (now Ji Zheng), he authored a survey of Fujian turtles and tortoises (1965). Ding's major works, however, were a book on the snakes of Fujian (1974, with Cheng, although their names do not appear on the title page of the first printing because personal authorship was forbidden in China during the mid-1970s) and another on the venomous snakes of Fujian, snakebites, and treatment of bites (1978, with several co-authors). These are excellent field guides, with keys, diagrams showing the diagnostic scutellation, and photographs of specimens.

• *References*: "Han Bo Ding, Ph.D. (1912–)," p. 199 *in* M. B. Ogilvie and C. J. Choquette, A Dame Full of Vim and Vigor. A Biography of Alice Middleton Boring. Harwood Acad. Publ., London, 1999; "An American's Appreciation of Professor Ding Hanbo," by K. Adler, p. 26–28 *in* Y.-Z. Ding and Y.-L. Ding (eds.), A Memorial to Professor Ding Hanbo. Fujian Normal Univ., Fuzhou, 2005; Additional information from Ding's son, You-zhen Ding, 2006. • *Portrait* (1959): Courtesy You-zhen Ding. • *Signatures* (Chinese form, 1991; Romanized form, 1995): Adler collection. • *Note*: Ding spelled his name Han-po Ting in publications through 1982, according to the Wade-Giles transliteration system, but thereafter used the Pinyin (phonetic) form.

EL-TOUBI, M. R. (1909–1999).

The vertebrate comparative anatomist, Mohamed Rashad El-Toubi (see *Note*), was born on 15 August 1909 in Tantâ (Gharbeyya), Egypt, about 90 km north of Cairo. He studied zoology (B.Sc. 1932, M.Sc. 1936) at the Egyptian University (now Cairo University), located in Giza in the western suburbs of Cairo. (From 1938 to 1952 the university was called Fouad I University during the reign of Fouad's son, King Farouk.) El-Toubi became a demonstrator in zoology at the university in 1932 and five years later went to the United Kingdom for his doctorate at University College London, a unit of the University of London (Ph.D. 1940). There he studied with the fish paleontologist, D. M. S. Watson, but also spent two terms at Oxford University with the anatomist, E. S. Goodrich. El-Toubi returned to Giza as lecturer in zoology and became a professor of zoology in 1952. He served as vice-dean of the university's science faculty during 1952–

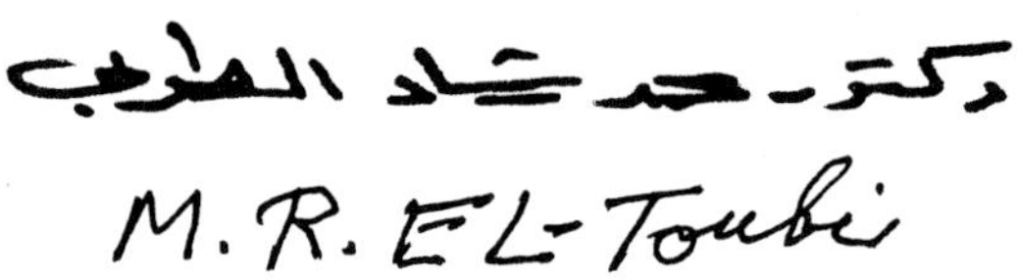

1955 and as counselor to the Ministry of Scientific Research in Cairo the following year. He took leave from Cairo to hold zoology professorships at Tripoli University in Libya (1968–1969) and at Riyadh University in Saudi Arabia (1969–1974). He retired from his professorship in Cairo in 1974. El-Toubi's career was honored with the Science Prize for 1989, presented by the Egyptian Academy of Science and Technology. He died in Cairo on 20 September 1999.

El-Toubi was a prolific author and translator of key biological texts from English into Arabic. He authored ten books on a variety of subjects (1964–1987) including bees, marine life, animals, and human embryology. Among the dozen English books he translated are "Reptiles and Amphibia" by Maurice Burton and Robert Burton (English edition 1974, Arabic 1977) and "The Vertebrate Body" by Alfred S. Romer (English edition 1962, Arabic 1971). He wrote about 40 scientific publications, primarily on the anatomy and development of the skull in vertebrates. Of these, 27 are on reptiles (1938–1973): both lizards (agamids including *Uromastyx*, gekkonids, lacertids, scincids) and snakes (boids, colubrids, viperids). These include the development of the skull in *Chalcides ocellatus* (1959, 2 papers) and in *Ptyodactylus hasselquistii* (1961, 3 papers) as well as a three-part series on the osteology of Egyptian lacertids (1966–1969). He also authored papers on the origin and phylogenetic relationships of snakes based on their skull development (1965, 2 papers) and the development of the skull in *Malpolon monspessulana* (1973, 2 papers). Together with Abdel-Hamid Khalil, one of his former master's students, El-Toubi discovered the peculiar caudal ribs of geckos (1950), which had not theretofore been reported in living lizards.

• *Sources*: Information from M. R. El-Toubi, 1992; Wendy Butler, University College London, 2006; and Adel Ibrahim, Suez Canal University, 2006–2007. • *Portrait* (about 1990): Adler collection. *Signatures* (Arabic form): Courtesy El-Toubi's daughter, Salwa M. R. El-Toubi, and Adel Ibrahim; (1992, Romanized form): Adler collection. • *Note*: El-Toubi used the then-standard English spelling of his name on his papers, but the modern form is Mohammad Rashad At-Toubi.

PERKINS, C. B. (1888–1955).

Clarence Basil Perkins (see *Note*) was a specialist on American snakes who developed the reptile house at the San Diego Zoo into one of the premier programs of its kind in the world. He was born in Denver (Colorado) on 11 November 1888 and had a boyhood interest in herpetology especially for snakes. He attended Lawrenceville School, an elite boarding school then only for boys near Princeton (New Jersey). He went on to Princeton University where he kept snakes and turtles in his dormitory room. His graduation was delayed by one year so he could complete a program in herpetology (Litt.B. 1913). After graduating he returned to Denver to work as a banker, but soon joined the United States Navy Air Corps as a lieutenant. During World War I he regularly visited San Diego and its zoo, which first exhibited reptiles in 1922 under the volunteer curatorship of a local amateur herpetologist, Laurence M. Klauber.

After the war Perkins returned to Denver and a career as stockbroker and banker, and continued his interest in snakes. At this time Klauber was beginning his work on rattlesnakes and advertised widely his interest in obtaining specimens, especially large samples from a specific locality and short time period. This would permit him to better understand population variation in scutellation and other characteristics. On 17 October 1930 he received a fateful letter from Perkins: "I am sending you today, by express, one hundred and fifty-three of these snakes [*Crotalus viridis*]." And they were being sent alive! Perkins had collected them from a single locality, an underground prairie dog "town" used by the snakes as a hibernaculum near Platteville (Colorado). Eventually, he collected 861 specimens from this site, providing Klauber with what is probably the largest sample of snakes of a single species from a single locality ever amassed. This became the most important sample on which Klauber based his seven-part series entitled "A Statistical Study of the Rattlesnakes" (1936–1940).

The connection with Klauber soon led to Perkins's hiring by the zoo in 1931 as supervisor of reptiles. His first major task was to design and build a new, modern reptile house, a concrete structure with radiant heating for all cages, which opened in 1936. With a new facility, he was able to display a much larger variety of species and to keep them healthy longer. At that time, most zoos simply replaced animals that died and made little effort to maximize their longevity or to achieve breeding in captivity. Perkins, on the other hand, kept his animals in more naturalistic and environmentally appropriate surroundings. The animals not only lived longer, they tended to display more natural behaviors and reproduced, and he published some of his findings.

Altogether, Perkins authored 18 papers (1938–1955). During 1947–1955, he published his annual surveys of longevities in captivity, the first-ever for collections in the United States. The first listed 22 taxa at the San Diego Zoo, 15 of them snakes, but in his last there were 66 taxa, all snakes, in 14 American collections. His major works were "Snakes of San Diego County" (1938) and "A Key to Snakes of the United States" (1940; second edition 1949), the latter based on the key by Frank Blanchard (1925). Both booklets were widely used. He published a few other notes, but many of his observations appeared in Klauber's book on rattlesnakes (1956) including notes on longevity, captive breeding, mating behavior, and temperament, among other topics. Perkins was the first to note the genetic basis for albinism in snakes (in the gophersnake, *Pituophis catenifer*; reported in Klauber 1956) and he developed an albino strain of this species. He publicized amphibians and reptiles and the zoo through his regular appearances on a San Diego radio program, "Animal Kingdom." He also took time to encourage serious amateurs including locals such as Sheldon Campbell and Charles E. Shaw. Later, Shaw was hired as Perkins's assistant and in January 1954 succeeded him as curator. Shaw carried on Perkins's annual longevity lists through 1962. Perkins died on 28 April 1955 in San Diego.

• *References*: "C. B. Perkins, Herpetologist," by C. E. Shaw, Zoonooz (Zool. Soc. San Diego), April: 15, 1955; "C. B. Perkins 1889–1955," by W. B. Wade, Copeia, 1955: 323–324, 1955; Princeton University Alumni Records (box 389), courtesy Rosalba D. Varallo. • *Portrait* (1950): Zoological Society of San Diego, courtesy Marjorie Shaw. • *Signature* (1934): Adler collection. • *Note*: It has long been rumored that C. B. Perkins was the brother of the herpetologist, Richard Marlin Perkins (1905–1986), director of several American zoos and a national television personality. Marlin Perkins had two brothers, neither of whom was named Clarence Basil. (R. Marlin Perkins Papers, Western Historical Manuscript Collection, University of Missouri, Saint Louis, courtesy Jennifer Sparks).

SHAW, Charles E. (1918–1971).

Charles Edward Shaw was a specialist on North American lizards and snakes who became Curator of Reptiles at the San Diego Zoo. He was born on 31 July 1918 in San Diego. He grew up in pre-war San Diego when it was still a small city surrounded by wilderness abounding in wildlife. Shaw attended a junior high school that fortuitously was located adjacent to the zoo. By the age of 15, he had become a protégé of C. B. Perkins, the Supervisor (later Curator) of Reptiles, who put him to work scrubbing the shells of giant tortoises among other chores. Through Perkins he met Laurence M. Klauber, the amateur herpetologist and rattlesnake expert who was a consultant at the zoo. In 1938 Perkins put Shaw on the zoo's part-time payroll and Klauber gave him work in the private museum at his home, preserving and cataloging specimens and counting scales. Shaw also volunteered in the herpetological collections at the San Diego Natural History Museum located adjacent to the zoo. He attended San Diego State College (now University; A.B. 1943). On graduation, he served

in the United States Marine Corps and was stationed in Honolulu where he collected reptiles in his free time. When his military service ended in 1945, he was re-hired by the zoo as Perkins's full-time assistant and in 1954 he succeeded him as curator. Shaw assumed the additional duties of Assistant Director of the zoo and Superintendent of the Children's Zoo in 1958 and continued in all three posts until his untimely death from cancer at the age of 52 on 30 May 1971 (see *Note*).

Shaw authored about 35 titles on reptiles (1939–1974) and Klauber's influence is evident from the technical nature of his early papers. He described four new taxa of lizards from Baja California and Arizona based on Klauber's collection including a legless lizard (*Anniella geronimensis*) and a chuckwalla (*Sauromalus klauberi*). Shaw monographed the genus *Sauromalus* in 1945 while he was still in military service. Once he came to work full-time at the zoo, his papers mostly dealt with living animals in the reptile house: male combat in colubrid, crotalid, and elapid snakes; breeding in xantusiid lizards; diets; captive reproduction; and longevity records. He won several national awards for being the first to breed certain endangered species of reptiles in captivity including the Galapagos tortoise, which also earned him an award from the Charles Darwin Foundation. He continued publishing the annual longevity lists for snakes begun by Perkins and in 1950 authored a widely used booklet on the lizards of San Diego County, a companion to Perkins's 1938 pamphlet on the county's snakes. In 1953 he and James A. Oliver, then reptile curator at the Bronx Zoo in New York, teamed to publish the first herpetology of the Hawaiian Islands, largely based on Shaw's fieldwork there during the war.

Like Perkins, as reptile curator at the zoo, Shaw became a local celebrity through his regular appearances on the San Diego television show "Zoorama." When Klauber died in 1968, having only revised the first three chapters for the second edition of his rattlesnake book, Shaw helped to edit the rest of that monumental work. Shaw's only book of his own, "Snakes of the American West" (1974), was issued posthumously, co-authored with his boyhood friend and another Perkins protégé, Sheldon Campbell, amateur herpetologist and San Diego stockbroker.

• *References*: "Charles Edward Shaw July 31, 1918–May 30, 1971," by S. Campbell, Zoonooz (Zool. Soc. San Diego), July: 17, 1971; "Charles Edward Shaw * 31. Juli 1918, † 30. Mai 1971," by R. E. Honegger, Salamandra, Frankfurt am Main, 7: 47–48, 1971; Conant, 1997 (Shaw on pp. 476–477). • *Portrait* (1967): Zoological Society of San Diego, courtesy Marjorie Shaw. • *Signature* (1949): Adler collection. • *Note*: Shaw's death date has sometimes been published as 1 or 2 June, in error.

KAWAMURA, Toshijiro (1906–2003).

The embryologist, Toshijiro Kawamura, who used frogs and salamanders as his experimental models, made important contributions to the knowledge of parthenogenesis, interspecific hybridization, and other genetic phenomena. He was born on 7 March 1906 in Otsu (Shiga), 30 km east of Kyoto. After graduating from Hiroshima Higher Normal School in 1928, he taught at Nagano Normal School, and in 1930 went to Hiroshima University of Literature and Science (renamed Hiroshima University after World War II) to study biology (graduated 1933). He then became an assistant and, in 1937, a lecturer at the university while he worked on his doctoral dissertation on artificial parthenogenesis in frogs (D.Sc. 1941). Fortuitously, he was briefly a lecturer at Kyoto Imperial University in 1945, when Hiroshima University was almost completely destroyed by the atomic bombing of the city by the United States on 6 August. Kawamura spent the rest of his career at Hiroshima and was instrumental in the re-building of the biological facilities. He became professor in 1951, director of the university library in 1963, and served from July 1966 to January 1969 as president of the university.

In 1965, the University of Michigan in the United States established an Amphibian Research Facility under the directorship of George W. Nace to develop genetic strains of amphibians for research use. It had been planned that Kawamura would join the Michigan facility after his retirement and bring with him the many strains he had developed in his own laboratory, but he was unable to do so when he became university president. Instead, in 1967 the Japanese government and Hiroshima University designated Kawamura's laboratory as a special research center, the Laboratory for Amphibian Biology. The laboratory grew enormously, from three professionals and a small facility in 1967 to 15 staff occupying 4,000 square meters of laboratory and greenhouse space by 1994. The laboratory originally was directed by Ryozo Yagiu and then by Hiroshi Nagahama. From 1972, Midori Nishioka, Kawamura's longtime associate, held the post until the laboratory was closed in March 1995. The center published its own annual

serial, the *Scientific Report of the Laboratory for Amphibian Biology, Hiroshima University*, which extended to 13 volumes (1972–1994). The laboratory developed some 3,000 special strains of amphibians for use by biologists in Japan and throughout the world, and it maintained close ties with the Michigan facility. Research at the Hiroshima center, which was differentiated into four research divisions, was very broad and included studies on embryology, genetics and gene expression, physiology, biochemistry, bioacoustics, ecology, and evolution. Today, these studies are continued at the university's Institute of Amphibian Biology, opened in April 1995 as part of the Faculty of Science. The institute is directed by Yoshio Yaoita, formerly of the University of Tokyo, whose research focuses on programmed cell death (apoptosis) during amphibian metamorphosis.

Kawamura's own publications on amphibians numbered nearly 50 (1939–1990) and dealt with artificial parthenogenesis in frogs, hybridization studies in frogs and salamanders, triploid and tetraploid frogs, sexual isolation between species, nucleo-cytoplasmic hybrids in frogs, and hybridization among European ranid frogs. He retired as professor early in 1969 but continued active research. He died in Hiroshima on 27 January 2003.

• *References*: ["Laboratory for Amphibian Biology,"] by M. Nishioka, Bull. Biol. Sci. Hiroshima Univ., 57: 3–7, 1991; "Toshijiro Kawamura (1906–2003)," by M. Nishioka, Zool. Sci., Tokyo, 21: 499–501, 2004. • *Portrait* (1969): Courtesy Toshijiro Kawamura. • *Signatures* (Japanese form, about 1973; Romanized form, 1988): Adler collection.

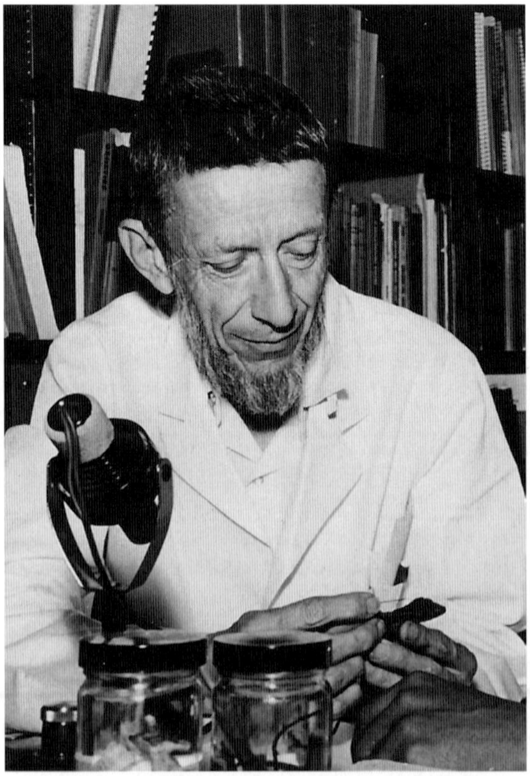

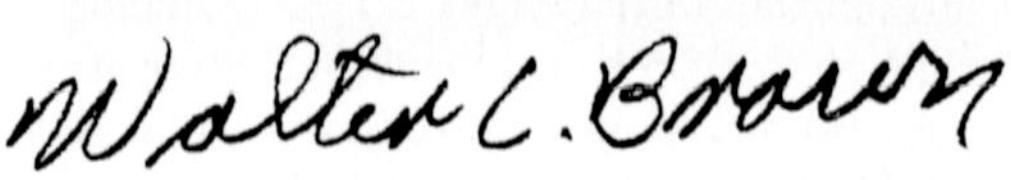

BROWN, Walter C. (1913–2002).

The American herpetologist, Walter Creighton Brown, who became a specialist on the amphibians and reptiles of the South Pacific and the Philippines, was born on 18 August 1913 at Butte in southwestern Montana. He was raised on a ranch and attended a one-room schoolhouse. When he was ten, his family moved to Tacoma (Washington State) where he graduated from high school and the College (now University) of Puget Sound (A.B. 1935, M.A. 1938). James R. Slater, his college biology professor, introduced him to herpetology and Brown published his first papers with him. After teaching high school biology in Tacoma, he enlisted in the army in 1942 and served in the South Pacific and the Philippines, experiences that determined the geographic focus of his future research career. During 1946–1947 Brown studied at the University of Rochester in western New York State under Sherman C. Bishop, with whom he described a plethodontid salamander (*Desmognathus aeneus*). At Stanford University Brown was a student of George S. Myers and wrote his dissertation on the frogs of the Solomon Islands (Ph.D. 1950). After serving on the faculty of Northwestern University in Illinois during 1951–1953, Brown moved to Menlo College in Atherton (California), not far from the Stanford campus, where he spent the rest of his career.

Brown served as a Fulbright Visiting Professor during 1954–1955 at Silliman University in Dumaguete City, the primary university city of the southern Philippines and located on Negros Island. Previously, his research had focused on the reptiles and amphibians of South Pacific islands, but from this point on his work dealt principally with the Philippines. In 1954 he met Angel C. Alcala who had just finished his bachelor's degree, and Alcala became his principal research collaborator for the rest of Brown's career. They conducted comprehensive surveys of the herpetofauna of the Philippines, and Brown returned again in 1958, 1963, and 1991. Brown retired in 1978 and continued his association with the California Academy of Sciences in San Francisco where most of his specimens are deposited including 27,000 from the Philippines. He died on 15 July 2002 at a retirement home in Murphys (California), after a long illness.

Brown authored nearly 100 titles on amphibians and reptiles (1939–1998) and described numerous frogs, lizards, and snakes from the South Pacific (Fiji, Marianas, New Britain, New Guinea, Palau, Solomons) and the Philippines. Brown and Alcala jointly named 36 new Philippine species. They also conducted studies on biogeography and ecology, especially of the lizards. Brown's major works are "The Amphibians of the Solomon Islands" (1952), two parts of a review of "Philippine Lizards" (Gekkonidae, 1978, and Scincidae, 1980, both with Alcala), a review of the skinks of the genus *Emoia* (1991), and a color-illustrated field guide, "Philippine Amphibians" (1998, with Alcala).

• *References*: "Walter Creighton Brown," p. 823 *in* American Men & Women of Science, 21st ed., vol. 1, Thomson Gale, Farmington Hills, Michigan, 2003; "A Life of Service to Science. A Tribute to Walter Creighton Brown 1913–2002," by A. C. Alcala, Silliman Jour., 45: 258–262, 2004. • *Portrait*: Senckenberg Museum, Frankfurt am Main, courtesy Gunther Köhler. • *Signature*: Adler collection.

LAURENT, Raymond F. (1917–2005).

Raymond Ferdinand Louis-Philippe Laurent, a Belgian-born specialist on the herpetology of Africa and South America, was born on 16 May 1917 in Wasmes (Hainaut) near the French border. His parents moved to Brussels when he was two and gave him some newts when he was seven, thus beginning his hobby and, later, a career. He received his schooling at the Athénée Royal in nearby Ixelles (1928–1934) and then attended the Université Libre de Bruxelles (Licencié 1938, D.Sc. 1940). Paul Brien supervised his doctoral work on the osteology and systematics of African frogs, and Laurent also had substantial advice from H. W. Parker of the British Museum. Unfit for military service because of pleurisy, Laurent attempted to flee Belgium after the German invasion in May 1940 but he was forced to return home. He soon began his research at the Musée du Congo Belge (now Musée Royal d'Afrique Centrale) in Tervuren on the outskirts of Brussels where he remained throughout and after the war.

The Belgian Congo (now Zaïre)—Belgium's most important colony—was then moving toward independence. In an effort to allow Belgian institutions to prolong their work in Central Africa, in 1947 the government established the Institut pour la Recherche Scientifique en Afrique Centrale in the Congo and Ruanda-Urundi. It was headquartered in Leopoldville (now Kinshasa) and had five regional centers. In 1949 Laurent joined the center in Uvira on the northwest edge of Lake Tanganyika and then moved to the center in Astrida (now Butare) in Ruanda (now Rwanda) in 1951 before spending 1955–1956 on the River Zaïre at Mabali (now Mbandaka) and again at Uvira. In 1957 he became a professor at Université Officielle du Congo Belge et du Ruanda-Urundi in Elisabethville (now Lubumbashi), capital of Katanga Province. He could hardly have chosen a more dangerous place to be when, on declaration of Zaïre's independence in June 1960, Katanga seceded. The atrocities that followed (including massacres and even cannibalism) have been well recorded. Thus, in 1961, Laurent moved to the United States and the Museum of Comparative Zoology at Harvard University to which he brought his collections and continued his African research.

In 1967 Laurent moved to the Fundación Miguel Lillo in San Miguel de Tucumán in northern Argentina where he held several positions (including director of the Instituto de Zoología from 1970 and head of the herpetology program from 1983) and also became professor at the Universidad Nacional de Tucumán (1967–1975). His active research shifted more to Argentina and South America. He continued to work his African collections but these were eventually sent to the museum in Tervuren in about 2000. Laurent retired from the institute in 1996 and died in Tucumán on 3 February 2005. He was thus present during the so-called "Dirty War" of 1976–1983 when the government kidnapped and killed many of its opponents and thousands of Argentines simply disappeared. Ironically, despite the adverse places in which he lived, Laurent's worst professional loss occurred when much of his personal library was destroyed during a flood at his home in the United States.

Despite the many disruptions during Laurent's life he was one of the most productive herpetologists of his era. He published about 215 titles (1940–1997), nearly all of them on herpetological topics. His primary research was on the amphibians and reptiles of Sub-Saharan Africa and included major systematic reviews of frogs (*Afrixalus* and *Hyperolius*, 1950; *Ptychadena*, 1954; *Hemisus*, 1972; *Leptopelis*, 1973; *Cryptothylax*, *Phlyctimantis*, and *Kassina*, 1976), turtles (*Pelusios*, 1965), and—jointly with his compatriot and specialist on the Belgian Congo, Gaston-François de Witte—the lizards (*Scelotes*, 1943) and colubrid snakes (*Calamelaps*, *Miodon*, and *Aparallactus*, 1947). Laurent published long monographs covering the amphibians and reptiles of Angola (3 parts, 1950–1964), the turtles and snakes of the Great Lakes region of Central Africa (1956, 1960), the herpetofauna of the Somali Republic (3 papers in 1965, 2 of them with Carl Gans) and the amphibians of Virunga (formerly Albert) National Park (1972; the reptile section by de Witte was published in 1941). He described many new species as well as the frog families Hyperoliidae, Mantellidae, and Phrynobatrachidae. In his important review of the genus *Atractaspis* (1950), Laurent showed that mole vipers did not belong in the family Viperidae. He continued to publish on African herpetology through 1988.

Beginning in 1967, Laurent's work turned mostly to Argentina and South America, especially in numerous papers on frogs (*Gastrotheca*, 1967–1986; *Pleurodema*, 1969–1975; and *Telmatobius*, 1971–1990) and lizards (*Liolaemus*, 1982–1997). He wrote on relationships between the herpetofaunas of Africa and South America (1979) and authored a long and masterful chapter on the systematics of the Lissamphibia in the amphibian volume of Pierre-Paul Grassé's series, "Traité de Zoologie" (1986). Laurent was also responsible for training a significant number of the current generation of Argentine herpetologists.

• *References*: "Raymond Laurent," by M. M. Stewart and M. Halloy, Copeia, 2002: 245–247, 2002; "Raymond F. Laurent (1917–2005)," by E. O. Lavilla and G. J. Scrocchi, Cuad. Herpetol., Tucumán, 19: 65–67, 2005; Additional information from Monique Halloy, 2006. • *Portrait* (1979): Natural History Museum, London, courtesy A. F. Stimson. • *Signature* (1963): Adler collection.

EISELT, Josef (1912–2001).

The Austrian systematic zoologist, Josef Eiselt, was a specialist on the herpetofauna of Southwest Asia and on copepod crustaceans. He was born in Vienna on 3 May 1912 and educated at local schools. After first considering innkeeping as a career, he enrolled at the University of Vienna in 1933 to study natural history and physics. Later, he joined the university's Institute of Zoology where he studied with the Dutch-born anatomist, Jan Versluys. Eiselt's dissertation (Dr.phil. 1939) was on the comparative anatomy of the middle ear of frogs (published 1941). During his time as a student he volunteered in the Herpetological Collection at the Natural History Museum, curated by Otto Wettstein who was also a professor in the institute. Upon finishing his doctorate, Eiselt was an assistant at the institute before being drafted into the German Army (Austria had been annexed into the Third Reich in 1938).

After the war, Eiselt was briefly held by the British and released in the fall of 1945. He then assumed a variety of positions including work as a secondary schoolteacher (1949–1951), but he also volunteered to help rebuild the institute and the collection. In September 1952 he became curator of the herpetological collection, one of the most historically important in the world. These collections were in safekeeping during the war, but Eiselt, almost by himself, transferred the 100,000-specimen alcoholic collection and 3,000 skeletons and stuffed specimens to new quarters. He became head of the museum's department of vertebrates in 1972 and retired five years later. He maintained an active research program at the museum until his death in Vienna on 25 July 2001. During 1979–1985 he served as the first president of Societas Europaea Herpetologica where his calm presence and diplomatic skills built a common purpose among the divergent national interests on the society's council. Eiselt died in Vienna on 25 July 2001.

Before and just after the war, when travel was difficult, Eiselt confined his field expeditions to the Mediterranean, but in 1962–1977 these shifted to Northeast Africa and especially Southwest Asia. He visited Nubia (between Aswan, Egypt, and Wadi Halfa in northern Sudan), Turkey, Iraq, Iran, and Afghanistan. After retirement he went to Turkey again and, at the age of 81, Yemen. He made a total of 15 trips to Turkey. Early in his career, research focused on existing museum materials—mainly on salamanders, lizards, and the Austrian herpetofauna. After the Nubian expedition, he emphasized the systematics and distribution of the amphibians and reptiles of Southwest Asia, especially of Turkey and the Transcaucasian region, with a focus on *Lacerta* during his last 15 years. His most frequent field companions were Josef F. Schmidtler, an amateur German herpetologist, and a Viennese photographer, Inge Adametz; Eiselt co-authored new taxa of frogs, lizards, and a snake with the former. In addition to papers on copepods, he published about 45 on herpetological topics (1940–1996). Perhaps his most surprising discovery was the presence of a hynobiid salamander in Iran, far to the west of any previous record (jointly described in 1970 with Hans M. Steiner as a new species, *Batrachuperus persicus*).

• *References*: "Zum Gedenken an Dr. Josef Eiselt (1912–2001)," by J. F. Schmidtler, Salamandra, 38: 1–4, 2002; "Dr. Josef Eiselt† 3 May 1912–25 July 2001," by F. Tiedemann et al., Amphibia-Reptilia, 23: 125–128, 2002; "Wirklicher Hofrat Dr. Josef Eiselt 1912–2001†," by F. Tiedemann et al., Herpetozoa, 14: 99–105, 2002; "Dr. Josef Eiselt (3. Mai 1912–25. Juli 2001)," by J. F. Schmidtler, Die Eidechse, 13(1): 27, 2002. • *Portrait* (1961): Natural History Museum, London, courtesy A. F. Stimson. • *Signature* (1989): Adler collection.

TWEEDIE, M. W. F. (1907–1993).

Michael Willmer Forbes Tweedie was a British zoologist in Singapore who specialized on Malayan crabs, freshwater fishes, and reptiles. He was born in Playden, near Rye (East Sussex) on 2 September 1907. After his education at Tonbridge School (1921–1925) and at Cambridge University (B.A. 1928, M.A. 1929), he was employed as a paleontologist

by the Caribbean Petroleum Company in Venezuela (1929–1931). These were the Depression years and, when he eventually lost his job, he returned to Cambridge to study fossil ostracods. In need of employment, he applied for and won a position as curator at the Raffles Museum in Singapore (1932). His work there was interrupted during the war when he received a commission in the Royal Air Force to work on camouflage and to be a liaison with the Malay-speaking people. He was taken prisoner by the Japanese in Java in 1942 and interned until the end of the war. In 1946 Tweedie became director of the Raffles Museum. He retired in 1957 and returned to England where he died in Rye on 25 March 1993.

Tweedie was a life-long naturalist. He had studied a range of natural sciences at Cambridge and specialized in geology. The Malayan fauna inspired his interest in reptiles. He became an excellent field naturalist and made collections throughout Southeast Asia and the Malay Peninsula and even on remote Christmas Island in the Indian Ocean. His publications are similarly broad in scope with books on the poisonous animals of Malaya (1941), birds of the Malay Peninsula (1960), a popular book on dinosaurs (1977), and mammals of Malaysia (1978). His technical papers, however, are mostly on Stone Age archaeology, crustaceans, freshwater fishes, and reptiles. Tweedie's papers on reptiles are few (18 papers, 1940–1959 [1960]), the most notable being the descriptions of a new lizard (*Larutia trifasciata*) and a snake (*Enhydris pahangensis*) and his discovery of a hitherto unrecorded breeding ground for *Dermochelys coriacea* on the coast of Trengganu, Malaya. He is best known, however, for his book, "Snakes of Malaya" (three editions, the first in 1953 [reprinted 1954], second 1957 [reprinted 1961], and third 1983), which became and still is the standard reference on the topic. In 1954 (reprinted 1965, 1970, and 1988) Tweedie and J. L. Harrison, a professor of zoology at the University of Singapore, co-authored "Malayan Animal Life," which has a short section on amphibians and reptiles.

• *References*: "On Michael Willmer Forbes Tweedie," by P. K. L. Ng and C. M. Yang, Raffles Bull., Zool., 37: 160–167, 1989; "Tribute M W F Tweedie 1907–1993," by B. L. Lim, Malayan Nat., 47: 18, 1993 (1994); "Michael Willmer Forbes Tweedie (1907–1993)," by P. K. L. Ng, Crustaceana, 68: 111–115, 1995; Additional information from M. W. F. Tweedie, 1990. • *Portrait* (1974) *and signature* (1990): Adler collection.

HU, Shu-qin (1914–1992).

The Chinese herpetologist, Shu-qin Hu (see *Note*), who specialized on the systematics of frogs, was born on 13 February 1914 in Changshu (now Changzhou, Jiangsu Province), about 55 km northwest of Shanghai. In 1937 Hu graduated from Dongwu University in Soochow (now Suzhou, Jiangsu), just west of Shanghai, where she studied biology. Later, she worked with the prominent Chinese zoologist, Cheng-chao Liu, who was a professor at Soochow University. After the outbreak of the Second Sino-Japanese War in July 1937 and the Japanese capture of Shanghai in November, many Chinese moved to the country's interior and beyond Japanese-controlled territory. Liu and Hu went to Chengtu (now Chengdu, Sichuan Province), the central agricultural basin of China, where Liu took a position at West China Union University and Hu worked in succession at Huaxi University, Yenching University (evacuated from Beijing temporarily during the war), and Sichuan Medical Institute. They married in 1941. In 1960 she moved to the Sichuan Biological Research Institute (now Chengdu Institute of Biology, a division of Academia Sinica). Hu died in Chengdu on 10 December 1992.

Hu's research program was intimately tied to that of her husband's and together they developed the Chengdu Institute into the largest herpetological center in China. Altogether she co-authored about 40 titles (1940–1993), primarily on amphibians. During the mid-1970s when personal authorship was not permitted, it was a general practice to credit publications to work units, such as the Sichuan Biological Research Institute. Thus Hu's name does not appear on some title pages. When new species were being described, however, the names of Hu and her colleagues were inserted after the new Latin names, as required by the International Code of Zoological Nomenclature. Despite wartime conditions, she participated in extensive field surveys especially during the period 1938–1949, in the provinces of Gansu, Sichuan, and Sikang (formally merged into Sichuan in 1951). Surveys to more distant provinces occurred after 1950 including Guangdong (Hainan Island), Guangxi, Guizhou, and Yunnan.

Based on these collections, Liu, Hu, and their colleagues published many papers reporting new distribution data, numerous new taxa, and life history information. Hu described 50 currently recognized species of Chinese amphibians and reptiles, 44 of them salamanders and frogs in nine families including two salamandrids, nine megophryids, 16 ranids, and ten rhacophorids. Most of these new taxa were co-

authored with Liu, but some others with her Chengdu colleagues, Liang Fei, Guan-fu Wu, Fu-hua Yang, and Chang-yuan Ye. The new lizards and snakes were all co-authored with Er-mi Zhao, with whom Hu also joined in erecting the hynobiid genus *Liua* in honor of Liu.

Hu co-authored several books. She participated in writing three identification guides (1959, "Atlas of Chinese Animals—Amphibians," with Liu and Han-bo Ding; 1962, "Atlas of Chinese Animals—Reptiles," with Ding, Bu-qing Hu, and Zhu-jian Huang; and a combined second edition in 1987, "Atlas of Chinese Animals—Amphibians and Reptiles," with Zhao, Yi-sheng Wang, Yao-ming Jiang, and Qing-yun Huang). Hu's masterwork was the book, "Chinese Tailless Amphibians" (1961, with Liu), which contains 28 plates of detailed watercolor drawings of tadpoles and adult frogs. She was, in fact, senior author of "Systematic Key to the Amphibians of China" (1977, credited to the Sichuan Biological Research Institute). Hu was also the senior author and editor of an extensive report on the amphibians and reptiles of Xizang, or Tibet (1987, with Zhao, Jiang, Fei, Ye, Qi-xiong Hu, Qing-yun Huang, Yong-zhao Huang, and Wan-shu Tian).

Hu co-authored three books posthumously. Two of them were published in 1993: "Latin–Chinese–English Names of Amphibians and Reptiles" (with Zhao, Jiang, Fei, Qing-yun Huang, and Ye), which has worldwide coverage, and "Rare and Economic Amphibians of China" (with Ye and Fei). In 2006 she co-authored the first volume on amphibians in the "Fauna Sinica" series (with Fei as lead author).

• *References*: ["Obituary,"] by Committee on Funeral Arrangements for Shu-qin Hu, Chengdu, one page, 1992; Additional information from Shu-qin Hu, 1991, and Er-mi Zhao, 2006. • *Portrait and signatures* (Chinese form, 1991; Romanized form, 1986): Courtesy Shu-qin Hu. • *Note*: Until the mid-1980s, Hu spelled her name Shu-chin (sometimes Shu-ching) Hu in publications, according to the Wade-Giles transliteration system, but thereafter used the Pinyin (phonetic) form.

BARBOUR, Roger W. (1919–1993).

Roger William Barbour was a vertebrate zoologist, naturalist, and wildlife photographer who studied the amphibians and reptiles of his native Kentucky and wrote and illustrated several major herpetological books. He was born on 5 April 1919 to a farming family in Morehead, located in northeastern Kentucky where he spent much of his youth roaming the nearby hills and woodlands. He entered Morehead State Teachers College (now Morehead State University) at the age of 15 (B.S. 1938). At Morehead he was introduced to herpetology and ornithology by his zoology professor, Wilfred A. Welter, after whom he would later name a Kentucky plethodontid (*Desmognathus welteri*). He went to Cornell University for his master's degree (M.S. 1939) to study towees with the ornithologist, Arthur A. Allen. He then began his doctorate there in herpetology with Albert H. Wright, but World War II soon intervened.

While waiting for his induction notice, Barbour taught at colleges in Kentucky and then again after serving in the army (1945–1946) and before returning to Cornell. By then Wright's laboratory was full, so Barbour worked with Wright's colleague, the mammalogist William J. Hamilton, Jr., with whom

he completed his dissertation about the mammals, amphibians, and reptiles of Big Black Mountain in the coal country of southeastern Kentucky (Ph.D. 1949). He spent 1949–1950 teaching nature study in Wheeling (West Virginia) before taking a faculty position at the University of Kentucky in Lexington in 1950, where he spent the rest of his career. He retired in 1984 and, after a long illness, died on 26 August 1993 in Lexington.

Barbour was primarily a herpetologist and mammalogist, but he also published on other wildlife. Among his non-herpetological books are titles on the bats of America, Kentucky wildflowers and ferns, birds of Kentucky, Kentucky trees and shrubs, mammals of Kentucky, and the darters of America. His titles on amphibians and reptiles number about 40 (1940–1994) including five books. The papers focus on the systematics, distribution, life history, food habits, behavior, and ecology of Kentucky species. He is best remembered, however, for his herpetological books, "Amphibians and Reptiles of Kentucky" (1971), "Turtles of the United States" (1972), "Snakes of Eastern North America" (1989), "Turtles of the World" (1989), and "Turtles of the United States and Canada" (1994). The last four were co-authored by his former doctoral student, Carl H. Ernst, and the last one also with another turtle biologist, Jeffrey E. Lovich, a former student of Ernst's. Barbour was an outstanding wildlife photographer and contributed most of the color and black-and-white photographs in all of his 11 books.

• *References*: "Roger William Barbour 1919–," by W. H. Davis, Jour. Tenn. Acad. Sci., 66: 123–124, 1991; "Roger William Barbour 5 April 1919–26 Aug. 1993," by C. H. Ernst, Copeia, 1994: 569–572, 1994. • *Portrait and signature* (1981): Special Collections, Camden-Carroll Library, Morehead State University.

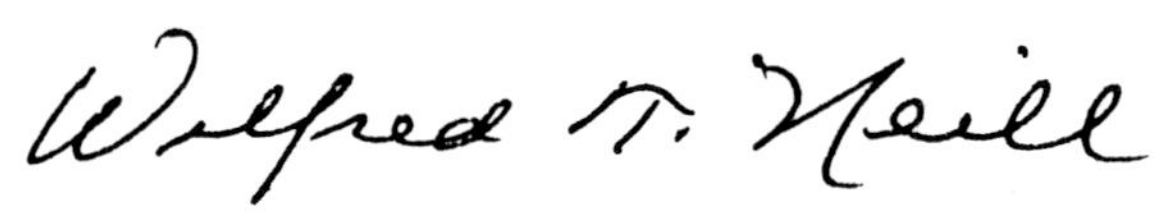

NEILL, Wilfred T. (1922–2001).

The herpetologist and field naturalist, Wilfred Trammell Neill, Jr., who was born on 12 January 1922 in Augusta (Georgia) along the South Carolina border, studied the herpetofaunas of southeastern United States and British Honduras (now Belize). He kept live reptiles as a boy and published his first paper—on skinks in Georgia—before graduating from the University of Georgia (B.S. 1941). He then taught zoology and German briefly at Augusta Junior College in his hometown before joining the Army Air Corps during World War II (1942–1945) when he saw duty in the Pacific. After discharge, he taught again in Augusta and by 1949 was a zoology professor. That same year Ross Allen hired him to become founding director of the Research Division of Ross Allen's Reptile Institute in Silver Springs, located in north-central Florida. The institute, begun in 1929 by Allen, was a live animal exhibit for tourists and a place for the extraction of snake venom for production of antivenom. Allen had a serious interest in herpetology, but little time to pursue it scientifically given his management duties for the institute. He had known Neill since 1932 when the latter started purchasing live reptiles from him.

By the time Neill was hired he already had 30 papers to his credit, but now, being able to devote most of his time to herpetological research, his productivity soared. He also edited the institute's publications, including its own research journal (published from 1950 to 1959). Neill attempted to develop the Research Division into an academic department by inviting visiting students to study specimens and receive training, and by instituting a seminar program jointly with the University of Florida in nearby Gainesville. He continued to do fieldwork throughout the Southeast and accompanied Allen on several expeditions to British Honduras in 1957–1960.

Neill resigned his position at the institute in 1962 and moved to New Port Richey, near the Gulf of Mexico just north of Tampa, to care for his aging parents. In January 1964 he went to the Florida State Museum in Gainesville as associate curator of the collections. Some preserved specimens formerly in the Allen-Neill collection at Silver Springs were donated to the museum at this time; others went to the national museum in Washington, but many were simply lost. Neill briefly became a graduate student at the University of Florida, but at the age of 43 he could not abide the course schedule that was standard for the other new students half his age. In the fall of that year he abruptly left Gainesville and returned to New Port Richey. He spent the next decade writing books, but a near-fatal rattlesnake bite in 1978 caused his health to deteriorate. This was his 41st snake bite. By then he had developed a severe allergy to snake venom and would break out in hives if he was anywhere near a Cottonmouth (*Agkistrodon piscivorous*). He died in a retirement home in Lakeland, just east of Tampa, on 19 February 2001.

Neill published extensively on amphibians and reptiles (about 185 titles, 1940–1974, half of them popular articles in *Florida Wildlife* magazine) and, from 1952, on Florida's Seminole Indians and archeology. He was an energetic field collector who also carefully observed behavior, ecology, and other aspects of natural history. Among his many insightful contributions are papers on the herpetofaunas in bromeliads, the role of crawfish in the ecology of amphibians and reptiles, algae on turtles, salamander vocalizations, and caudal luring in juvenile snakes. He also produced major papers on the natural history of *Farancia erytrogramma* and *Macrochelys temminckii*. He discovered and described more than a dozen new taxa of frogs, salamanders, snakes, and a turtle, including the one-toed *Amphiuma pholeter*. He and Allen authored ten papers from their expeditions to British Honduras (1959–1965). Neill's major publications in herpetology were his summary of the occurrences of amphibians and reptiles in saltwater environments (1958), "The Last of the Ruling Reptiles" (1971), still considered by many to be the best summary of crocodilian natural history, and "Reptiles and Amphibians in the Service of Man" (1974), concerning the many ways in which these animals are used for research, teaching, medicine, and commercial purposes. He published several other books about biogeography, Indonesia, and archeology.

• *References*: "Wilfred T. Neill – Florida's Premier Herpetologist," by D. L. Auth, Gainesville Herpetol. Soc. Newsl., 7(8): 9–15, 1991 (reprinted in Liner et al., 1993); "Biographical Sketch and Bibliography of Wilfred T. Neill," by E. Liner et al., Smithson. Herpetol. Informat. Serv., 95: 1–24, 1993; "Wilfred Trammell Neill Jr.," by R. Pinney, New York Herpetol. Soc. Newsl., 189: 3, 2001; Additional information from Ray E. Ashton, Jr., 1990, and from Neill's son, W. Trammell Neill, III, 1993 and 2001. • *Portrait* (about 1945): Courtesy Trammell Neill. • *Signature* (1957): Adler collection.

BELLAIRS, Angus d'A. (1918–1990).

The British physician and anatomist, Angus d'Albini Bellairs, was a specialist on the structure and development of reptiles who also popularized these animals to the general public. He was born in Bournemouth (Dorset) along the southern coast of England on 11 January 1918 (see *Note*) and attended Stowe, an elite boy's school in the Buckinghamshire countryside (1931–1935). He studied zoology and medicine at Queen's College, Cambridge (B.A. 1939) where one of his teachers was the naturalist, Hugh Cott, but with war looming Bellairs decided to get his medical qualifications (M.R.C.S., L.R.C.P., University College Hospital, London, 1942). That same year he was called to duty in the Royal Army Medical Corps and served in Tunisia, Egypt, and Italy. He collected reptiles wherever he went and it was understood that any soldier reporting sick, even malingerers, would get a day's sick leave if they arrived with a live reptile. In 1944 he was transferred to a unit in Burma (and later to India), where he would sometimes go collecting with a subordinate toting a submachine gun.

On returning to civilian life he did not want to do clinical medicine, so he took a post in 1949 at London Hospital Medical College in the anatomy department headed by J. Dixon Boyd. Boyd had been one of his teachers at Cambridge and with whom he had published on the lachrymal apparatus of squamates (1947). Medical anatomy departments in those days had a stronger tradition in comparative work than they do today. When Boyd returned to Cambridge in 1951, he took Bellairs along. However, Bellairs did not like the heavier teaching duties and, besides, he missed London. So in 1953 he returned and became reader at St. Mary's Hospital Medical School. In 1970 the University of London awarded Bellairs a personal chair as Professor of Vertebrate Anatomy. He died in London on 26 September 1990 at the age of 72.

Bellairs was a gentleman academic who had loved reptiles from his boyhood. He often paraphrased T. E. Lawrence by saying: "Nothing reptilian is alien to me." He authored about 75 herpetological titles (1942–1989) including several books. Most of his research was on the structure and development of the skull and its associated organs, especially the nose. He discovered the functions of Jacobson's organ in lizards and the nasal excrescence of sexually mature male gharials. Jointly with a fellow physician-herpetologist, Malcolm A. Smith, he studied the head glands and egg-teeth of squamates. With Garth Underwood he wrote a seminal paper on the origin of snakes. In the area of teratology he discovered in snakes a human-like type of cleft palate, and he studied cyclopia in turtles. He and Carl Gans did a study of the orbitosphenoid bone in amphisbaenians that had implications for the classification of those reptiles. With his student, Susan V. Bryant, he conducted research on regeneration and autotomy of lizard tails.

Bellairs wrote or edited eight books. With Gans he helped plan the classic series, "Biology of the Reptilia," and co-edited volume 1 (1969). He also co-edited "Morphology and Biology of Reptiles" (1976) with C. Barry Cox. It was his writing for the educated public, however, where Bellairs provided such important service. His little book "Reptiles" (1957) went through several revised and enlarged editions (1968, 1970, 1975, plus several reprintings), the last co-written with

John Attridge. With Richard Carrington he wrote the popular "The World of Reptiles" (1966), but it was his two-volume "The Life of Reptiles" (1969, with several overseas editions) that was particularly successful because of its detailed and scholarly treatment of reptilian biology. With H. W. Parker, herpetologist at the British Museum, he wrote a book on amphibians and reptiles (1972) that appeared in German, French, and Spanish. It was based on Bellairs's "The Life of Reptiles" and a manuscript on the lives of amphibians left unpublished by Parker on his death in 1968. Bellairs joined with J. F. Deryk Frazer to prepare the fifth edition (1973) of Malcolm Smith's "British Amphibians and Reptiles," which had first appeared in 1951. Finally, on the occasion of the First World Congress of Herpetology, held in Canterbury, England, in 1989—for which Bellairs served as honorary president—he published his only novel, "The Isle of Sea Lizards." Ostensibly it was a thriller about mosasaurs still living at a remote island, but in reality it was about his long-held views on academic politics, biological research, and the funding of research.

• *References*: "Professor Angus d'Albini Bellairs–An Appreciation," by M. W. J. Ferguson, p. 3–10 *in* M. W. J. Ferguson (ed.), The Structure, Development and Evolution of Reptiles," Symp. Zool. Soc. London no. 52, 1984; "Professor Angus d'Albini Bellairs, 1918–1990," by M. W. J. Ferguson, Herpetol. Jour., 1: 582–583, 1991; "Angus d'Albini Bellairs 1918–1990," by S. V. Bryant, Amer. Zool., 31: 279–280, 1991; "Angus d'Albini Bellairs–An Appreciation," by C. Gans, Copeia, 1991: 869–872, 1991; "In Memoriam A. d'A. Bellairs," by A. S. Breathnach, Jour. Anat., 176: 211–213, 1991. • *Portrait:* Courtesy Ruth (Mrs. Angus) Bellairs. • *Signature* (1969): Adler collection. • *Note*: Bellairs's school, Stowe, records his birth as 19 January.

LOWE, Charles H. (1920–2002).

Charles Herbert Lowe, Jr., herpetologist, field naturalist, and a leading specialist on the ecology of the Sonoran Desert, was born in Los Angeles (California) on 16 April 1920. He was interested in the desert and reptiles as a youngster and wanted to be a professional herpetologist. He was encouraged by Laurence M. Klauber who took him on local field excursions. Lowe went to the University of California at Los Angeles (UCLA) where he soon met Raymond B. Cowles, a founder of reptilian physiological ecology, who would become Lowe's future doctoral advisor. In the summer of 1942, Lowe participated in his first expedition to Sonora (Mexico) as an assistant to Charles Bogert of the American Museum who was a former Cowles student. He published his first paper on reptiles at the end of 1942 and graduated the next year (A.B. 1943). He served in World War II as a junior naval officer (1944–1946) before returning to UCLA for his doctoral research (Ph.D. 1950). During 1947–1950, he and his fellow UCLA students, Kenneth Norris and Richard Zweifel, worked for the United States Atomic Energy Commission in White Sands (New Mexico). They collected data on ambient radioactivity along a transect that tracked the path of the atomic cloud as it drifted north from "Ground Zero" in 1945 and they collected reptiles along the way.

On finishing his doctorate, Lowe assumed a position as Instructor in Zoology at the University of Arizona, located in the Sonoran Desert in the southeastern part of the state. There he developed an active field-based research program on the systematics and ecology of the Sonoran herpetofauna. He trained many notable doctoral students including Daniel D. Beck, Robert L. Bezy, Charles J. Cole, Stephen R. Goldberg, and John W. Wright. Lowe built a large and comprehensive museum collection of amphibians and reptiles from the southwestern United States and adjacent Mexico and of which he was highly protective. A tragic fire in 1994 destroyed his home and many of his records, leaving a number of his long-term projects unpublished. He retired in 1995 and, after a long period of declining health, died in Tucson on 13 September 2002.

Lowe made multiple contributions to the ecology of Sonoran vertebrates. He authored nearly 140 titles (1942–2006) including two important books, "The Vertebrates of Arizona" (1964), which includes detailed descriptions of the state's natural environments, and "The Venomous Reptiles of Arizona" (1986, with Cecil Schwalbe and Terry B. Johnson). Lowe named 20 new taxa of frogs, salamanders, lizards, and snakes. It was Lowe, together with his colleague, Zweifel, and several of Lowe's students—Wright, Cole, and Bezy—who showed in the 1960s that many whiptail lizard species are clones of hybrid origin and are all-female. Lowe's breadth of interest is reflected in the titles of some of his papers: aggressive behavior in male sidewinders, isolating mechanisms in sympatric populations of southwestern anurans, background color-matching in amphibians and reptiles, evolution of parthenogenetic species of *Cnemidophorus*, chromosomes and karyotypes of sceloporine iguanid lizards, natural free-running period in vertebrate animal populations, and a new plethodontid salamander from Sonora, Mexico. Lowe also prepared the first edition of a widely used map of the biotic communities of the Southwest (1968), which has been revised several times, most recently in 1980 by Lowe and David E. Brown. Through his research, training of students, his books and other publications, and development of a major museum collection of amphibians and reptiles, Lowe became recognized as the pre-eminent reptile ecologist of the American Southwest.

• *References*: "Prof. Charles Herbert Lowe, Jr.," p. 696 *in* American Men of Science, vol. 2. Science Press, Lancaster, Pennsylvania, 1955; "Charles Herbert Lowe, Jr., Noted Desert Ecologist, Herpetologist, and Professor, 1920–2002," by [P. C. Rosen], Sonoran Herpetol., 15: 110, 2002; "Charles Herbert Lowe, Jr. 1920–2002," by P. Rosen, Copeia, 2004: 961–972, 2004. • *Portrait* (1989): Courtesy Wilmer W. Tanner. • *Signature* (1949): American Museum of Natural History, courtesy Charles W. Myers.

ROMER, J. D. (1920–1982).

John Dudley Romer, an English naturalist, provided the first comprehensive review of Hong Kong's amphibians and reptiles. He was born on 9 September 1920 in Richmond (Surrey) where he development a boyhood interest in reptiles from family trips into the countryside. He was educated at Richmond County School (1931–1936) and at Kingston-on-Thames Technical College (1937–1940). In 1941 he was called up for service in the Royal Army Ordnance Corps. During 1942–1945 he served in India (Assam) where he took on pest control duties (rats and mosquitoes) along the front lines. Following the war, in 1946 he went to the British Crown Colony of Hong Kong to advise on rodent control as an anti-plague measure. After a year back in London he returned to Hong Kong, eventually becoming the colony's pest control officer until his retirement. He spent one year in Port Harcourt, Nigeria (1951–1952), to consult on the preservation of food stocks in tropical climates. As pest control officer for Hong Kong, he advised the government on all matters relating to prevention or control of pests of public health or medical importance and supervised a staff of about 500 persons.

At heart Romer was a field naturalist. In 1949 he co-founded the Hong Kong Natural History Society. After his repeated urgings in the 1950s, the Botanic Gardens established a small zoo. When the zoo became popular, the gardens were renamed the Hong Kong Zoological and Botanical Gardens and Romer served it as honorary curator of mammals (1957–1979). He conducted numerous field trips to all parts of the colony to survey its fauna especially amphibians, reptiles, and mammals, and became the primary recipient of unusual animals from local people. In 1958, a supposedly dead Chinese cobra brought in for identification bit Romer on a finger, which required skin grafts and he needed months of recovery. Since Hong Kong is a major transshipment point for cargo it was one of Romer's tasks to determine whether odd animals brought in were native or escapees.

For his many services to the colony Romer was decorated by Queen Elizabeth II with the Imperial Service Order (I.S.O.) in a ceremony at Buckingham Palace in 1980. Malcolm Smith named a tiny rhacophorid frog in his honor (*Philautus romeri*) in 1953. Romer retired to England in 1980, eventually settling in the Isle of Wight where he died from cancer at the hospital in Ryde on 15 March 1982.

Romer published numerous titles on mammals and pest control. His first herpetological papers appeared while he was in Assam during the war. All told, he authored about 45 papers on amphibians and reptiles (1944–1981) including descriptions of two new frogs (*Amolops hongkongensis* and *Megophrys brachykolos*). Most of Romer's papers deal with distribution, behavior, and pattern and color variants but especially records of species newly recorded from Hong Kong. His primary publications, however, are annotated checklists with illustrated keys to the colony's herpetofauna: snakes (1961, revised 1970, 1979), venomous snakes (1959, revised 1965, 1972, 1975, 1977, with a 1983 revision by Anthony Bogadek), lizards (1975), turtles (1978), and amphibians (1979).

Collectively these checklists comprise the first herpetology of Hong Kong, which has an especially rich herpetofauna (105 of China's 661 species as of 1993). Unfortunately, Romer's ambition to write a comprehensive book on the topic was not realized because of his early death. Fortunately, Bogadek, an Italian priest and teacher at St. Louis School in Hong Kong and amateur naturalist, started working with Romer in 1975 and it was Bogadek and his students, Stephen Karsen and Michael Lau, who provided the first such book (1986 [Chinese edition 1988]; revised 1998 [1999]).

• *References*: "John Romer and the Natural History Society," by J. Fearon, Mem. Hong Kong Nat. Hist. Soc., 18: 5–6, 1988; Romer family records, courtesy Raymonde (Mrs. John) Romer, Hugh D. Romer, Paul Romer, and Anthony Bogadek. • *Portrait* (1979): Courtesy G. N. Richardson. • *Signature* (1979): Adler collection.

MINTON, Sherman A., Jr. (1919–1999). MINTON, Madge Rutherford (1920–2004).

The Mintons, who became one of the most widely known husband-and-wife teams in herpetology through their popular books on reptiles, were both born and raised in rural southern Indiana. Sherman Anthony Minton, Jr., physician-herpetologist, was born on 24 February 1919 in New Albany, located along the Ohio River across from Louisville (Kentucky). His father was a lawyer who became prominent in the political and judicial life of Indiana and the United States. His mother had grown up on a farm and encouraged her son's early fascination with amphibians and reptiles. By the time Sherman Jr. was 11, he had already started collections of living and preserved specimens. By high school he had learned their Latin names. He wanted to study herpetology in college, but his father insisted he go into law or medicine. Sherman reluctantly chose the latter and enrolled at Indiana University in Bloomington (B.S. 1939). By this time, his father had been elected to the United States Senate from Indiana (1934–1940) and later an old friend from the Senate, Harry S Truman, now President, appointed him to the United States Supreme Court in 1949.

Madge Alice Shortridge Rutherford, pioneer women's airforce pilot and her husband's collaborator, was born on 22 March 1920 in Greensburg, about 75 km southeast of Indianapolis, but grew up in nearby Saint Paul. Her father was a teacher and eventually moved the family to Indianapolis where he taught zoology at Arsenal Technical High School. Madge was interested in many subjects including natural history and snakes, and she went to college in Bloomington, two years behind Sherman. From her opening remark to him in November 1937—"I understand you collect snakes"—they became fast friends, but they did not marry until October 1943 because of World War II. Madge had been fascinated by airplanes since childhood and wanted to be a pilot. In 1938 Butler University in Indianapolis began an Advanced Civilian Pilot Training Program, sponsored by the federal Civil Aeronautics Authority (CAA), whereby selected students, mostly men, could learn to fly and earn a pilot's license. She transferred to Butler in 1938 and entered the flight program, leaving Sherman in Bloomington. Madge graduated among the top ten trainees who were guaranteed admission to an advanced aeronautics program, but she was refused on account of her sex. That same day, 7 August 1940, she wrote to Eleanor Roosevelt, wife of President Franklin D. Roosevelt, explaining her plight. Nine days later, a letter arrived from the CAA authorizing her admittance to the advanced program.

Madge graduated from Butler (B.A. 1941) and, with the outbreak of war in December 1941, was invited to apply to the newly formed Women Airforce Service Pilots (WASP). The WASPs served, not as combat fliers, but to ferry aircraft from factories to domestic air bases, and to train test pilots and flight instructors. They also towed targets for artillery practice. Although their service was in the United States, 38 of them died in the line of duty. Discrimination against these women continued, however, and when Madge arrived at the training center in Texas, the base commander, who apparently had not been informed that women would be arriving, exploded: "What in the hell are you doing here? Get off my air base!" Undaunted, Madge finished her training and in August 1943 was transferred to Long Beach (California), which became her base of operations. Sherman, who had finished at Indiana University Medical School (M.D. 1942) and was commissioned as a medical officer in the United States Navy, had been serving since January 1943 as an intern at the nearby San Diego Naval Hospital. They married in October. Sherman saw duty aboard the U.S.S. "Brooks" in the Pacific including the horrific invasion of Japanese-held Luzon Island in the Philippines during which his ship was hit by a Kamikaze plane. Madge continued flying for the WASPs until August 1944 when she learned she was pregnant and resigned to return home to Indiana.

After Sherman was discharged from the navy in 1946, he decided that he did not want to practice medicine any longer. Instead, he wanted to do research and teach medical microbiology and continue herpetology as an avocation. Accordingly, the Mintons lived in Ann Arbor during 1946–1947 where Sherman took courses in microbiology at the University of Michigan and the herpetology course taught by Norman Hartweg and Charles Walker. In September 1947 Sherman became Assistant Professor

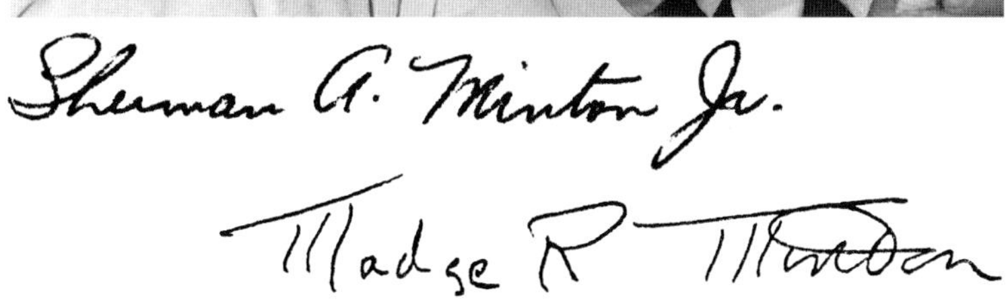

of Microbiology and Immunology at the medical school from which he had graduated only five years before. Here he remained until his retirement in 1984, except for the period 1958–1962 when he and his family resided in Karachi, Pakistan.

The United States Government asked Indiana University to assist Pakistan (then West Pakistan) to establish the Basic Medical Science Institute (now the Jinnah Postgraduate Medical Center). Sherman was the staff microbiologist. As a boy, he had been fascinated by Rudyard Kipling's stories of British India in his "Jungle Books," which had numerous herpetological references, and had long been inspired to see Kipling's India. It was an opportunity to experience a totally different herpetofauna from those of their travels in the United States and Mexico, and it led to some of Sherman's most significant publications. During their time in Pakistan, the Mintons traveled more than 69,000 km and collected some 1,500 herpetological specimens, aided by their three daughters. On returning to Indiana, Sherman resumed his plan to monograph the amphibians and reptiles of Indiana, a project suggested to him by Hartweg and Walker during his Michigan days. The Mintons spent the rest of their lives in Indianapolis where Sherman died on 15 June 1999 and Madge on 7 November 2004.

Sherman Minton's research interests were diverse, both in terms of subject matter and of geography. He authored more than 200 titles on herpetology (1944–2002): papers on the herpetofauna of the Big Bend region of West Texas and of Mexico and Central America; sea snakes of Southeast Asia and the South Pacific; toxic bites by rear-fanged snakes; biochemical properties of snake venoms; and relationships of groups of snakes based on immunology. Minton also published numerous articles for amateur and professional snake keepers. He discovered that he could distinguish two kinds of Jefferson salamander, a lead-colored form (*Ambystoma jeffersonianum*) and a blue-spotted form that he recognized as a separate species (*A. laterale*). It was not until the ecological and cytological work of Thomas Uzzell ten years later that the genetic differences between these species confirmed Minton's findings. From his research in the early 1950s, Minton showed that the original antivenoms developed by Wyeth Laboratories were not effective against the major North American pitviper venoms, which led Wyeth to significantly improve their crotalid antivenom in a way that went unchanged for nearly half a century. He was the first to note that the venom of *Crotalus scutulatus* was ten times more toxic in the Big Bend region than it was in snakes from the central part of the range of the species. In the 1980s, he became a vociferous opponent of the quack theory for electric shock as a treatment for snakebite. He also served various scientific societies in responsible capacities, two of them as president (International Society of Toxinology, Society for the Study of Reptiles and Amphibians).

Sherman wrote numerous books on herpetology and snake venoms. In 1962 the United States Navy issued "Poisonous Snakes of the World," an anonymous, error-filled compilation from the literature, which reported among other things the absence of cottonmouths in Florida and the presence of sea snakes in landlocked Afghanistan! The Navy, realizing its blunder, soon withdrew the book from circulation. Minton, however, recognizing the great value of having an accurate manual on poisonous snakes for military personnel, together with Herndon G. Dowling and Findlay E. Russell volunteered to revise the book (published 1968). Minton's major scientific herpetological treatise was "A Contribution to the Herpetology of West Pakistan" (1966). He produced two books on snake poisoning, "Snake Venoms and Envenomation" (1971, as editor) and "Venom Diseases" (1974), which also covers invertebrates and other vertebrates. Beginning in 1944, Minton had begun to gather information about Indiana's amphibians and reptiles, which became a lifelong project. This culminated in his book, "Amphibians and Reptiles of Indiana" (1972) and, posthumously, an expanded, color-illustrated edition (2001). In 1992, with American military operations underway in Southwest Asia, he collaborated with Alan E. Leviton, Steven C. Anderson, and Kraig Adler to produce a handbook with illustrated keys to the amphibians and reptiles of the central portion of the Middle East.

Madge had always supported Sherman's work at the medical school and on field trips. She did his library research and cared for his captive snakes, even milking the venomous ones for his research. They co-authored half a dozen papers, but it was their joint work on two books that made them household names throughout the world. "Venomous Reptiles" (1969; revised and expanded edition 1980) and "Giant Reptiles" (1973) were very popular, especially the latter. "Giant Reptiles" drew heavily upon their own work, particularly observations made in Pakistan. Ironically, although trained as a physician Sherman became best known as a herpetologist.

• *References on both*: "Madge and Sherman Minton," by M. M. Stewart, Copeia, 2000: 304–309, 2000; "Life, Love, and Reptiles," autobiography by S. A. Minton, Jr., Krieger Publ. Co., Malabar, Florida, xvii, 217 pages, 2001; Additional information from their daughter, April M. Kieser, 2006. • *References on Sherman Minton*: "In Remembrance of Sherman A. Minton, Jr. 1919–1999," by H. B. Bechtel, Herpetol. Rev., 30: 202–204, 1999; S. A. Minton, 2001 (biography by K. Adler on pp. ix–xvi). "Sherman A. Minton, Jr. 1919–1999," by D. R. Karns, Copeia, 2001: 891–894, 2001. • *References on Madge Minton*: "The Flygirl," by M. F. Greene, Life Mag., June: 80–86, 1999; "Madge Rutherford Minton 1920–2004," anonymous, Herpetol. Rev., 36: 2, 2005. • *Portrait*: Courtesy April Minton Kieser. • *Signatures* (Sherman Minton, 1955): American Museum of Natural History, courtesy C. W. Myers; (Madge Minton, 1999): Courtesy April Minton Kieser.

WORRELL, Eric (1924–1987).

Eric Frederic Arthur Worrell, a naturalist, author, snake venom extractor, and Australia's "Snake Man," popularized reptiles in his country. He was born on 27 October 1924 in Granville (New South Wales), a Sydney suburb. As a young boy he caught frogs and snakes in the city parks. At the age of 11 he decided to be a naturalist and set up his own zoo where visitors paid a small fee. An indifferent student, he left Sydney High School after eighth grade "and took on bush work." He was drafted into the army in 1940 (see *Note*) and sent to Darwin (Northern Territory) to help install artillery and was present during the Japanese bombing of the city in February 1942. After the war he lived with Aborigines in the territory to study its wildlife and then returned to Sydney where he taught about animal care at Sydney Technical College. In 1948, fulfilling his boyhood dream to establish a public display of wildlife, he es-

tablished the Ocean Beach Aquarium in Umina along the coast just north of Sydney. As one of its programs, the staff began to collect snake venoms for the Commonwealth Serum Laboratories located in Parkville (Victoria) where antivenoms were produced (only that for tiger snakes was then available). In the early 1950s there was a public outcry because of a spate of human deaths from the taipan. Worrell personally collected the taipan venom and became the main supplier before taipan antivenom had been produced and at a time when all recorded bites were fatal. Taipan antivenom became available in 1955.

In 1958 Worrell's facility was moved further north to its present site in Gosford and re-named the Australian Reptile Park. Here it flourished as a tourist attraction, reaching a zenith of 250,000 visitors in 1970, but changing traffic patterns later caused attendance to dwindle. Financial difficulties and Worrell's health problems led to a reorganization in 1986 and his effective retirement. Worrell calculated that during his career at the park he was bitten about once a year, or one in every 3,000 snakes he milked. Many of these bites were serious and required hospitalization, yet ironically it was a heart attack in the yard of his home just adjacent to the park that suddenly ended his life on 12 July 1987. Queen Elizabeth II had honored Worrell for his public contributions when she personally invested him with the Member of the British Empire (M.B.E.) decoration during her 1970 visit to Sydney.

Eric Worrell had a reputation far beyond Sydney because of his many books on snakes and other reptiles, which have served to popularize these animals. He regularly wrote articles for magazines and newspapers and prepared numerous documentaries for television. During the 1950s and 1960s, he was the best-known naturalist in the country. His first book appeared in 1953, a slim volume entitled "Dangerous Snakes of Australia"; it went through many editions and was expanded to include New Guinea. This was followed by his most widely popular book, "Song of the Snake" (1958), an account of his experience as a naturalist-writer in the Northern Territory, Queensland, and Tasmania. His most important title, "Reptiles of Australia" (1963; second edition 1970), was the most comprehensive book on the subject up to that time and went through several printings. It was the standard reference on the topic until the appearance of Harold Cogger's book in 1975. Worrell's only other strictly herpetological book was "Australian Snakes, Crocodiles, Tortoises, Turtles, Lizards" (1966), which is mainly a picture book.

Worrell also contributed to the technical literature (20 papers, 1945–1963). He named several taxa of squamates including the elapid genera *Austrelaps, Cryptophis, Drysdalia, Parasutra,* and *Sutra* and the species *Glyphodon dunmalli, Parasutra dwyeri*, and *Simoselaps minimus*.

• *References*: "Vale, Eric Worrell," by D. B. Millar, Herpetofauna, 17(2): 32, 1987; "Eric Worrell 1924–1987," p. 480 *in* H. May and S. M. Trosky (eds.), Contemporary Authors, Gale Res. Co., Detroit, 1988; Additional information from John Weigel, 1997, and Harold Cogger, 2006. • *Portrait* (1985): Courtesy Robert W. Murphy. • *Signature* (1974): Courtesy John Weigel. • *Note*: Technically, Worrell served the Allied Works Council, an organization that carried out duties required by the Allies within Australia during World War II.

HOOFIEN, Jacob H. (1913–1997).

Jacob (Yaakov in Hebrew) Haim Hoofien, an amateur Israeli herpetologist, was born in Jaffa (now Tel-Aviv), Palestine, on 14 October 1913 (see *Note*). His parents were immigrants from the Netherlands and he received his education partly there (Amsterdam 1919; Leiden 1931–1936) and partly in Palestine (Jaffa 1920–1930). He returned to Palestine in 1937 to work for a public contractor in Haifa until 1939 when he moved to the Anglo-Palestine Bank (now Bank Leumi Le-Israel). His father was a senior executive at the bank and Jacob worked there for the rest of his career. Because Hoofien loved nature, he preferred being in the field to the luxurious offices in the bank and became active in herpetology in 1944. A man of broad knowledge, when once asked how his fascination with reptiles began, he quoted the Rubaiyat of Omar Khayyám of 11th Century Persia (closing line of quatrain 60): "Who is the Captive, pray, and who the Captor?" He made two major expeditions outside his country: to Jerash, Transjordan (now Jordan) with Georg Haas in 1945, and to the Horn of Africa with Heinrich Mendelssohn and Oskar Theodor in 1958. Hoofien became one of the founders of the Society for the Protection of Nature and regularly went into the field until 1976 when his health made this impractical. He died after a long illness on 21 June 1997 in Kfar Shmaryahu, Israel (see *Note*).

Because of his extensive reading and close associations with leading herpetologists including Haas, Mendelssohn, and Louis Lantz, among others, Hoofien had an extensive knowledge of reptile systematics and ecology, especially the taxonomy and zoogeography of the reptiles of Israel. He published about 30 herpetological titles (1945–1995). These include papers on distribution, geographic variability, behavior, and a four-part series on the herpetofauna of Mount Hebron (1967–1973), which included the description of a new gecko (*Cyrtodactylus amictopholis*). There are also papers on the reptiles of Somaliland and the Dahlak Archipelago off the Red Sea coast of Eritrea. His major contribution to herpetology, however, was his book, "Reptiles of Israel. A Guide and a Key" (1956; reprinted several times), written in Hebrew jointly with Alexander Barash. This is a comprehensive field manual with descriptions and keys, and it is fully illustrated with drawings and photographs.

• *References*: "Yaakov H. Hoofien 1913–1997," by E. Kochva, Israel Jour. Zool., 44: 1–2, 1998; "Jacob Hoofien (1913–1997), Herpetologist," by J. Wahrman and Y. L. Werner, Ecol. Environ., 5(1): 49–50, 1998; Additional information from Jacob H. Hoofien, 1991. • *Portrait* (1978): Adler collection. • *Signatures* (Hebrew form, 1954): Courtesy Yehudah Werner; (Romanized form, 1989): Adler collection. • *Note*: These birth and death dates have been confirmed by Hoofien's son, Danny Hoofien (2006), although other dates have been given in the literature.

WERMUTH, Heinz (1918–2002).

Heinz Wermuth, who was interested in the systematics and conservation of crocodilians and turtles, was born on 5 January 1918 in Berlin, Germany, the son of a master baker. He had a pet turtle by the age of ten and through frequent, sometimes daily, visits to the zoo became quite knowledgeable about reptiles and their scientific names. He graduated from the Friedrich-Werdersches Gymnasium in 1936 and entered the university in Berlin as a zoology student. He soon gravitated to the university's Museum für Naturkunde and its herpetological collection, but in August 1939 the curator, Ernst Ahl, was recalled to army duty just days before the start of World War II. Kurt Deckert, who had completed his doctorate in 1938, replaced Ahl. Wermuth volunteered to assist Deckert and his associate, Martin Eisentraut, but eventually all three of them were called for military service. Wermuth was drafted into a German infantry unit in 1939 and was wounded in action on the eastern front in 1943. He was held as a prisoner of war by American forces for a year beginning in April 1945.

Wermuth returned to the museum in November 1946 and during 1948–1950 helped with the daunting task of moving the large herpetological collection from the museum's damp cellars, where it had been stored for safekeeping during the war, to the museum's halls. He also resumed his academic studies at the university in 1947 under Eisentraut, who was now curator of herpetology, and completed his dissertation on a statistical analysis of variation in *Anguis fragilis* (Dr.rer.nat. 1950), which became his first scientific paper (1950 [1951]).

By August 1950, Wermuth had become staff scientific assistant for the herpetological collection and in 1952 was promoted to curator and head after Eisentraut's departure for a museum post in Stuttgart. In 1961 Wermuth had been nominated for a professorship at the university, but his life was soon to become unbearably complicated. Wermuth lived in West Berlin, but the museum was in the Soviet-controlled sector in East Berlin. At great personal risk, he smuggled fake passports among his work papers on his daily trips to the museum, which allowed many of his colleagues there to escape to West Berlin. When the Berlin Wall was erected in August 1961, his daily treks across the city became more and more difficult. The East German government demanded that he move his household to East Berlin, but he refused. When his role in smuggling passports was discovered, the museum fired him and he was banned as an enemy of the state and subject to immediate arrest in East Germany.

Fortunately, in 1962 a special curatorship was created for Wermuth in Stuttgart and he moved there to the Staatlichen Museum für Naturkunde. This museum had been damaged during the war and Wermuth spent much of his time refurbishing the herpetological collection. In 1972, he was made head of zoology at the museum and he retired 11 years later. He was honored in 1997 by his adopted state of Baden-Württemberg by being appointed professor, a title that had eluded him in Berlin. Wermuth died at his home on 28 December 2002 in Freiberg, on the Neckar River northeast of Ludwigsburg and 10 km north of Stuttgart.

Besides his research, Wermuth had many other interests. Beginning in 1949 he became a regular author in terrarium and aquarium journals and served on their editorial boards. In

1958, the German-language checklists of the animals of the world, *Das Tierreich*, which had been suspended during the war, were revived and Wermuth became editor. Amphibians and reptiles led in this rebirth and Wermuth wrote a sample account by covering a small family of lizards, the Helodermatidae (1958). Under his editorship, 16 additional herpetological parts appeared through 1997 including five compiled by Wermuth himself. In the late 1970s, Wermuth and Wolf Herre initiated the series "Handbuch der Reptilien und Amphibien Europas" and the first volume appeared in 1981 under the editorship of Herre's former student, Wolfgang Böhme.

Wermuth was also busy in conservation activities especially for crocodilians and turtles. In June 1956 he circulated a petition to leading herpetologists worldwide, entitled "The Urgency for International Protection of Crocodiles Throughout the World." It was signed by 124 persons and Wermuth submitted it to the International Union for the Protection of Nature at its meeting in Brussels the next year, but protection was slow to come. Even the famed missionary doctor and Nobel laureate, Albert Schweitzer, wrote to him from Lambaréné in west-central Africa to offer his support. In 1972 Wermuth visited reptile leather wholesalers in Paris and was shocked to find that they processed about 250,000 crocodile skins each year. This led him to produce an illustrated key to crocodile species and their hides (1978, jointly with Karlheinz Fuchs, a tannery engineer in Frankfurt am Main) for the use of government agencies charged with enforcing conservation laws, and also for the tanning industry itself with which he maintained a mutually productive relationship.

Wermuth's research program, which began with his doctoral work on variation in *Anguis*, was fundamentally changed in 1949 when he first met Robert Mertens who became his mentor and colleague. Wermuth soon focused on the systematics of crocodilians, turtles, and lizards. All told, he authored about 150 titles (1949–1996) including those in the popular journals. His major works, besides the *Das Tierreich* checklists, include a major review of crocodilians (1953); a checklist of turtles, crocodilians, and tuatara, "Schildkröten Krokodile Brückeneschsen" (1955, jointly with Mertens); and an expanded, illustrated version of the 1955 checklist with keys (1961, with Mertens; reprinted 1996 with an afterword by Wermuth and revisionary commentary by Fritz Jürgen Obst). He also co-authored with Mertens the third edition of the standard checklist of European amphibians and reptiles (1960; the earlier editions of 1928 and 1940 were by Mertens and Lorenz Müller). He also authored a color-illustrated pocket guide to the amphibians and reptiles of Germany (1957). Wermuth named some new taxa of turtles and lizards, among them one honoring his mentor—*Testudo hermanni robertmertensi* of the western Mediterranean.

• *References*: "Heinz Wermuth 1918–2002," by U. Friederich, Salamandra, 39: 1–4, 2003; "Heinz Wermuth (1918–2002)," by R. E. Honegger, Elaphe, 11(2): 11–12, 2003; "Dr. Heinz Wermuth 1918–2002 †," by F. Tiedemann, Herpetozoa, 16: 79–80, 2003; "Heinz Wermuth," by A. Schlüter, Amphibia-Reptilia, 26: 1–2, 2005; Additional information from Wermuth's son, Johannes Wermuth, 2006. • *Portrait* (1963): Carnegie Museum, Pittsburgh, courtesy M. Graham Netting. • *Signature* (1956): Adler collection.

MILSTEAD, William W. (1928–1974).

William Wright Milstead, a herpetologist who studied ecology and systematics especially that of lizards and turtles, was also an accomplished biology educator and administrator. He was born on 12 June 1928 in Abilene, in central Texas, but the family moved to Houston where he finished high school. While in school, his interest in reptiles was stimulated by collecting snakes in the Houston area and in the Big Thicket of East Texas with two other budding herpetologists: Richard Etheridge, a student at another local high school, and later with Walter Auffenberg who briefly worked in Houston. Milstead first attended Baylor University and was associated with the Strecker Museum as a student curator, but then transferred to the University of Houston (B.S. 1949). His graduate work was done at the University of Texas in Austin (M.A. 1951, Ph.D. 1953) where he completed his doctoral dissertation on competition in natural populations of whiptail lizards under the direction of W. Frank Blair.

During 1953–1955, Milstead was in Brazil as Visiting Professor of Ecology at the Universidade Federal do Rio Grande do Sul in Porto Alegre after which he joined the biology faculty of Sul Ross State College in Abilene (1955–1958). After a year at Texas Technological College (now University) in the West Texas town of Lubbock, he moved to the University of Missouri in Kansas City in 1959 where he spent the rest of his career. During 1971–1973, he was on leave in Austin as Coordinator for Environmental Programs of the International Biological Program. Milstead

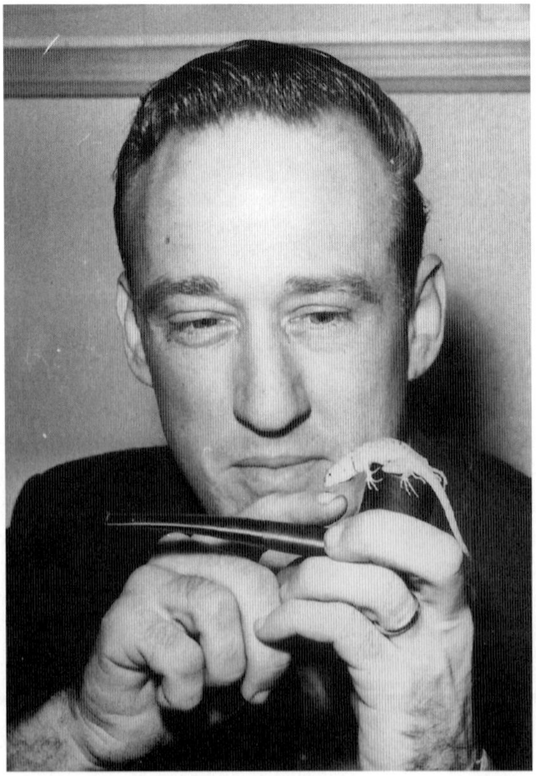

regularly consulted overseas concerning the improvement of high school and college biology teaching and visited Brazil, East Pakistan (now Bangladesh), India, and Uruguay for this purpose. He became chairman of his department in Kansas City at a very difficult time of transition when a local university merged with his own. Milstead suffered from diabetes since childhood and died from its complications, in Kansas City, on 29 November 1974. He was only 46.

Milstead authored more than 40 titles on amphibians and reptiles (1945–1977), primarily on ecology, systematics, geographic variation, and distribution. He was particularly interested in competitive relationships of natural populations and in biogeography. An interest in Pleistocene turtles led to his description of a new tortoise (*Testudo wilsoni*) from Texas and several later papers on box turtle systematics and evolution including his monograph, "Studies on the Evolution of Box Turtles (Genus *Terrapene*)" (1969). In 1965 he organized a major national symposium in Kansas City, which was published in book form under his editorship with the title "Lizard Ecology" (1967). One of his last papers (1974, with A. Stanley Rand and Margaret M. Stewart) was a provocative study of the existence of polymorphism in the convergent "cricket frog" ecomorph on five continents.

• *References*: "William Wright Milstead," p. 4303 *in* American Men and Women of Science, vol. 4. R. R. Bowker, New York, 1972; "[William W. Milstead]," anonymous, Herpetol. Rev., 5: 104, 1974; Additional information from Milstead's son, W. Bryan Milstead, 1999, and Archives, University of Missouri–Kansas City, 1999. • *Portrait* (1961): Courtesy Bryan Milstead and Harlan D. Walley. • *Signature* (1969): Adler collection.

GUIBÉ, Jean (1910–1999).

The French zoologist, Jean Marius René Guibé, who specialized on the frogs and snakes of Madagascar, was born in Paris on 18 February 1910. His family was from Normandy where his father was a surgeon. After obtaining his certificate in natural sciences from the Faculté des Sciences at Caen in Normandy in 1930, he studied medicine for a year, but then returned to natural sciences for his degree in 1934. By this time he already had established relationships with several researchers at the Muséum d'Histoire Naturelle in Paris including the comparative anatomist, Jacques Millot. After receiving his doctorate in 1939, also at Caen, and serving in the military during the war, Guibé joined the museum in 1945 as an assistant to Léon Bertin who put him to work on reptiles and amphibians even though Guibé's previous research had been with invertebrates. He conducted field excursions in Algeria (1949) and Madagascar (1950–1951). In 1957 he replaced Bertin and thus became part of a distinguished lineage of professors at the museum who, beginning with Count de Lacepède in 1795, had responsibility jointly for reptiles and fishes. Guibé was the last, however, to have this dual role. When he retired in 1975 on medical grounds because of recurrent violent migraines, the post was split and his successor, Édouard-Raoul Brygoo, became the first professor solely in charge of reptiles and amphibians. It is said that Guibé was taciturn but not without humor. He was a meticulous researcher with a talent for drawing. He died in Caen on 4 May 1999.

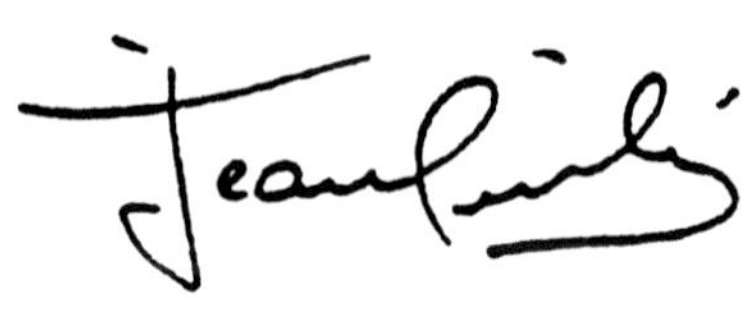

Guibé published more than 150 titles, the majority of them on amphibians and reptiles (1946–1978). His special interests were systematics and anatomy and also behavior and biogeography. He focused on frogs and snakes of Madagascar and West Africa, regions where France long has had political interests. He often collaborated with his museum colleagues, especially Fernand Angel, Maxime Lamotte, and Millot on frog systematics and with Jean Anthony on the anatomy of snakes, particularly boids. He studied sexual dimorphism in reptiles including the peculiar rostral extension in the Madagascan vinesnakes (*Langaha*), but why it should have such a different form in females and males is still not known. Guibé named about 60 new taxa of squamates and frogs including several new genera. He authored several larger works, among them the lists of amphibian and lizard types at the museum ([1950] and 1954, respectively), "Les Serpents de Madagascar" (1958), and several popular books on amphibians and reptiles in French and German. In 1970 he contributed 22 chapters on anatomy and behavior to the two reptile volumes in Pierre-Paul Grassé's series, "Traité de Zoologie." It was Grassé who had originally recommended him for a post at the museum. Guibé's final work, and his most extensive scientific contribution, was "Les Batraciens de Madagascar" (1978); of the 128 species of frogs monographed, he had named 40.

• *References*: Jaussaud and Brygoo, 2004 (Guibé on pp. 266–267); "Professeur Jean Guibé (1910–1999): Biographie et Bibliographie," by R. Bour, Alytes, 24: 133–146, 2006. • *Portrait*: George S. Myers/Alan E. Leviton portrait collection, courtesy Alan Leviton. • *Signature* (1958): Natural History Museum, London, courtesy Andrew F. Stimson.

SAINT GIRONS, Hubert (1926–2000).

The naturalist and histologist, Hubert Jean-Marie Jacques Saint Girons, who specialized on the reproductive biology, behavior, and ecology of snakes especially vipers, was born in Paris on 13 May 1926. He received his diploma from the École Régionale d'Agriculture in Rennes in 1946 and his doctorate at the Sorbonne, the University of Paris, in 1951 for a thesis on the reproductive biology and ecology of European vipers. He then became a research assistant at the Centre National de la Recherche Scientifique (CNRS), France's largest public research organization, and worked at the Laboratoire d'Evolution des Etres Organisés at the Sorbonne. In 1960 Saint Girons moved to the Laboratoire d'Ecologie of the Muséum National d'Histoire Naturelle, located at Brunoy in the southern suburbs of Paris, but in 1969 he returned to his former laboratory at the Sorbonne, directed by Pierre-Paul Grassé. CNRS, in fact, supported Saint Girons's research program and salary throughout his career. In 1964 he became Directeur de Recherche, the highest research position within CNRS and equivalent to full professor.

Since the late 1940s, Saint Girons had maintained a house at Bohallard, near Nozay, about 40 km north of Nantes in the Loire countryside some 330 km southwest of Paris. It was here that he conducted many of his field studies on vipers, spending nearly half of each year at Bohallard and the other half at his Paris laboratory on Boulevard Raspail. In much of his research he had the collaboration of his wife, Marie-Charlotte Saint Girons, not only in Bohallard but on his overseas visits to work on snakes in Morocco and tuatara in New Zealand. He also had a close collaboration with the histologist, Manfred Gabe, his colleague at the Sorbonne, and with Donald Bradshaw at the University of Western Australia in Perth. Saint Girons died during the night of 17–18 April 2000 in Nantes.

Saint Girons authored nearly 200 titles; nearly all are about reptiles (1946–1998) and primarily on snakes. He was among the first to conduct both field and laboratory studies of snake biology, and this led to an interest in ecophysiology and studies on the energy sources that the animals stored prior to reproduction. He used outdoor enclosures in order to study snakes in semi-controlled but relatively natural conditions that did not require radiotelemetry. From his first to last papers, vipers were the main subjects of his life's work, but he also studied the reproductive biology and ecology of tropical reptiles in Cambodia and New Caledonia. He and his wife had a passion for French Morocco where they sometimes worked up to an elevation of 3,100 m in the High Atlas Mountains. With Gabe, he published major studies on the tuatara and on the cytology of the nasal cavity and Jacobson's Organ in reptiles. Some other topics of interest were the seminal receptacles in chameleons, ecology of seasnakes, sexual cycles in lizards, comparative morphology of the hypophysis in squamates, circadian activity rhythms in reptiles, external nasal glands in skinks (with Bradshaw), thermoreceptors in snakes, and the biogeography of reptiles. His major faunal work was his book, "Les Serpents du Cambodge" (1972), which summarized his work in Cambodia.

• *References*: "In Memoriam Hubert Saint Girons (1926–2000)," by J. Lescure, Bull. Soc. Herpétol. France, 98: 57-60, 2001; "Vale—Hubert Saint Girons (1926–2000)," by D. Bradshaw et al., Herpetol. Rev., 32: 9–10, 2001; "Hubert Saint Girons, 1926–2001," by M. Lambert, Herpetol. Bull. (Brit. Herpetol. Soc.), 78: 2, 2001. • *Portrait and signature* (1988): Adler collection.

KIRTISINGHE, P. (1903–1981).

Parakrama Kirtisinghe, a Ceylonese herpetologist and conchologist, was born 12 February 1903 at Hikkaduwa, a small coastal village about 90 km south of Colombo. As a boy, the eldest of seven brothers, Hikkaduwa's now-famous golden beaches and living coral reef teemed with marine life and there he began his love for nature. After completing his secondary education at Ananda College in Colombo, he proceeded to England to study medicine at University College London, but his childhood interest in animals caused him to switch to zoology (B.Sc. 1929). He was a university teacher throughout his career, beginning in zoology at the University of Ceylon in Colombo. Kirtisinghe returned to University College to complete a master's degree (1941), and sometime after 1960 he moved to Aquinas University College, also in Colombo, as professor and head of zoology. The University of Sri Jayewardenepura awarded him an honorary doctorate (D.Sc. 1981) in recognition of his contributions to Sri Lankan zoology. He died later that year, in Colombo, on 15 June.

Kirtisinghe's interest in conchology began as a boy. He amassed a vast private collection and subsequently authored a comprehensive guidebook, "Sea Shells of Sri Lanka" (1978). His interest in frogs began much later, probably in the 1930s, and was intensified from 1939 when, because of wartime fuel rationing, it became too difficult to drive an automobile to the seashore at Hikkaduwa. Thus, his fieldwork became more local and inland. He published two papers on frogs in 1946 including the description of a new species of ranid, today called *Nannophrys marmorata*, but his major contribution was his book, "The Amphibia of Ceylon," published personally in 1957. This became the standard reference on Sri Lankan amphibians (30 species of frogs and 2 caecilians) for nearly half a century. It was based on his own field and laboratory work including detailed information on tadpoles and life histories, and from direct comparisons to Sri Lankan and Indian specimens at the British Museum (Nat. Hist.) made during his visits to London in 1954 and 1955. Kirtisinghe published two additional papers (1957, 1958) on previously undescribed tadpoles; these were his last contributions to herpetology.

• *References*: De Silva, 1996 (Kirtisinghe on p. 20); Additional information from members of the Kirtisinghe family, 1993. • *Portrait* (about 1941): Courtesy Anslem de Silva. • *Signature* (1956): Natural History Museum, London, courtesy A. F. Stimson.

KAURI, Hans (1906–1999).

The zoologist, Hans Kauri, was a specialist on harvest-spiders, dragonflies, and horseflies, and also worked on the amphibians and reptiles of Scandinavia and the Baltic region. He was born on 30 May 1906 in Konguta, in the Tartumaa area of Estonia, which was then part of the Russian Empire. During World War II, he earned a master's degree at Tartu University in Estonia with a thesis on Odonata. He was finishing his doctoral dissertation on harvest-spiders in 1944 when suddenly he had to leave his homeland. Although Estonia declared its independence in 1918, the pact between Nazi Germany and the Soviet Union in 1939 led to Estonia's annexation by the latter in 1940. The German Army, however, occupied the country until 1944. When the Soviets then asserted control, they deported most academics to Siberia and replaced them with Russians. Kauri had to leave so quickly that he could not even take the manuscript of his nearly completed dissertation with him.

Kauri went to Sweden and became a museum assistant at Lund University. He later completed a totally new doctoral dissertation there, on speciation in European frogs and its relationship to climate (Fil.Dr. 1959). He became an associate professor of zoology at Lund and in 1963 moved to the University of Bergen in Norway as full professor. During 1969–1974, Kauri directed an investigation of the fauna of Hardangervidda, the high mountain plateau of southern Norway, sponsored by the International Biological Program. He retired in 1976 and continued to work at the university's museum until 1998. Kauri died on 30 January 1999 in Stockholm.

Kauri published about 120 titles on a wide range of subjects including the systematics, speciation, ecology, zoogeography, and environmental physiology of vertebrates and invertebrates. He authored 15 titles in herpetology (1946–1976). His first paper reviewed the distributions of Estonian amphibians and reptiles and was followed by papers on distribution of the green toad, systematics of European green frogs, and effects of light on development of tadpoles. Kauri's major works include "Zoogeography of the Swedish Amphibians and Reptiles with Notes on Their Growth and Ecology" (1959), a book initiated by his professor at Lund, Torsten Gislén, but completed by Kauri after Gislen's death in 1954. It includes detailed distribution maps for each species. Kauri's

doctoral dissertation on speciation in European *Rana* was published the same year. Kauri also wrote the herpetological volume (1962; several reprints in the 1960s and 1970s) in the Swedish series on animals of the world, "Djurens Värld." This is a well-illustrated review of the anatomy, biology, and distribution of amphibians and reptiles. Kauri also named the northernmost race of the smooth newt (*Triturus vulgaris borealis*), which reaches to nearly 65° North Latitude in Sweden where it occurs only in a neotenous form.

• *References*: "Professor Fil. Dr. Hans Kauri 80," by O. A. Sæther, Rahvuslik Kontakt, Stockholm, 1986(2): 29–35, 1986; "Hans Kauri," by O. A. Sæther, Internavis Univ. Bergen, 1999(4): 6, 1999; Additional information from Ole A. Sæther, 2006. • *Portrait* (1976) *and signature* (1955): Courtesy Ole A. Sæther.

ORCÉS V., Gustavo (1903–1999).

Gustavo Edmundo Orcés Villagómez, an Ecuadorian vertebrate zoologist, was born on 31 July 1903 in Quito. Orphaned at an early age, he was raised by his mother's relatives who sparked his interest in animals by giving him books on the subject. He attended schools in Quito (Escuela Padre Borja and Colegio San Gabriel) and upon attaining legal age he managed his family's hacienda, San Xavier, located near Machachi, about 30 km south-southwest of Quito. Throughout his youth Orcés was often out in the countryside, in the forests as well as the high Andean páramos. In 1936 his self-training in scientific zoology began when he was asked to catalogue the large bird collection of the Instituto Nacional Mejía, a secondary school in Quito. When the Escuela Politécnica Nacional (EPN) reopened in 1946, he became associated with its biology department and, coincidentally, with a visiting French paleontologist, Robert Hoffstetter, who taught in the department and who mentored him in research. During this time Orcés worked on his doctorate at the Universidad Central del Ecuador in Quito, which he completed in 1958 with a dissertation on marine fishes of Ecuador. He served as Director of the Instituto Nacional de Caza y Pesca in 1950. He was also mentored by another foreign visitor—the American herpetologist James A. Peters of the national museum in Washington—who worked in Quito during 1958–1959 and several years in the 1960s.

Orcés spent his career in the biology department of EPN and served for 42 years as director of the department. He published on all major vertebrate groups but mostly in collaboration with overseas scholars or with his Ecuadorian colleagues and students. His primary interests were in taxonomy and distribution. He named 11 new species of fishes and reptiles including two spectacular lizards, a leaf-nosed anole (1956, with Peters) and a giant phenacosaur (1996, with Ernest E. Williams and others), and two colubrids of the genus *Sibynomorphus* (1989, with Ana Almendáriz). He published about a dozen papers on herpetology (1948–2004), which covered many groups including venomous snakes, turtles, and amphisbaenians. Orcés developed a large personal herpetological collection, which today is mostly at the national museum in Washington; smaller collections are at some other museums in the United States.

Orcés was responsible for training a large number of the current generation of zoologists at Ecuadorian universities and conservation organizations. For his far-reaching influence on science in Ecuador, he received many awards including an

honorary doctorate from EPN (1994). The university's Museo de Historia Natural was renamed in his honor in 2000. In 1989, the Fundación Herpetologica Gustavo Orcés was founded in Quito to promote the conservation of amphibians and reptiles in Ecuador. Orcés retired from EPN in 1990 at the age of 87 and died in Quito on 22 March 1999.

• *References*: "Científico de la Fauna Ecuatoriana," by F. Ortiz-Crespo, Diario Hoy, Quito, 27 March 1999; "Dr. Gustavo Orcés Villagómez," by L. Albuja V., Politécnica (Escuela Politécnica Nacional), Quito, 25(1): 7–10, 2004. • *Portrait*: From Albuja V., 2004. • *Signature*: Courtesy Ana Almendáriz.

AUFFENBERG, Walter (1928–2004).

Walter Auffenberg, Jr., who was born in Detroit (Michigan) on 6 February 1928, became a specialist on fossil and living amphibians and reptiles—especially salamanders, snakes, tortoises, and monitor lizards. His parents encouraged his boyhood infatuation with reptiles, thinking it was a passing fancy, and even allowed him to hitchhike to Florida and Mexico to collect. He graduated from high school at the age of 16 and, with no plans for college, went to Deland in east-central Florida to tend some orange groves that his parents had purchased for income during their retirement years. Early in 1945, at the end of World War II, he enlisted in the navy to become a paramedic and was stationed in Corpus Christi (Texas). Two years later, he briefly worked for a Houston oil company as a draftsman until his supervisor found out about

his wish to become a herpetologist and urged him to go to college. Auffenberg promptly resigned and returned to Deland to attend the local college (Stetson University, B.S. 1951). For his graduate work he moved to nearby Gainesville and the University of Florida (M.S. 1953, Ph.D. 1956), which would become his academic home for most of his life.

Before finishing his graduate degrees, however, he was a curator at the Charleston Museum in South Carolina (1954–1955) and then Curator of Paleontology at the Florida State Museum (now Florida Museum of Natural History). In 1959 he followed his doctoral advisor, Arnold B. Grobman, to the University of Colorado to become Associate Director of the Biological Sciences Curriculum Study. BSCS was responsible for producing a series of innovative textbooks that revolutionized the teaching of high school biology in the United States and overseas. Auffenberg returned to the museum in 1963 as Chairman of the Department of Natural Sciences, a post he held for a decade. He became Curator of Herpetology (1973–1984) and from 1973 also held a joint appointment as Professor of Zoology at the university. He retired in 1991. In December 1995 he suffered a debilitating stroke but continued his work with the help of his son, Kurt Auffenberg. He died in Gainesville on 17 January 2004.

Auffenberg authored more than 130 titles about amphibians and reptiles (1948–2004) on a disparate variety of topics. His first major interest was fossil amphibians and reptiles, a subject on which he continued to publish as late as 1988. His major contributions were a review of the fossil snakes of Florida (1963), derived from his doctoral dissertation, and a checklist of the fossil land tortoises of the world (1974). He was interested in the body musculature of limbless or nearly limbless amphibians and reptiles including snakes, the lizard *Ophisaurus*, and salamanders such as *Siren* and *Amphiuma*. He made numerous studies on chelonians, both fossil and extant species. His primary works include papers on the fossil *Terrapene* (1958) and tortoises (1963) of Florida and a study on the ecology of *Gopherus berlandieri* (1969, with his doctoral student, William G. Weaver, Jr.), plus numerous shorter papers on systematics, behavior, and ecology. At one point, the Auffenbergs had a colony of no less than 32 *Geochelone radiata* living in their backyard for easy observation (together with several 2-m-long alligators in their sinkhole ponds). Auffenberg also conducted several major works on reptile systematics and evolution including one on geographic variation in *Coluber constrictor* (1955) and another on the osteology and phylogenetic relationships of tortoises of the genus *Gopherus* (1976). By the end of his career, his geographic focus was Pakistan and together with his son, Kurt, and other collaborators, he published several papers.

Despite all of these many studies, however, the research for which Auffenberg will most be remembered is his work on monitor lizards. Ironically, it began with tortoises. He had proposed to do a long-term study of the Aldabra tortoise. He needed to go for a full year, together with his wife and young family, but the British government would only permit a month or two because they did not want the responsibility for a family living in such a remote place for so long. Coincidentally, in 1968 F. Wayne King of the New York Zoological Society conceived the idea for a long-term study of the Komodo monitor and asked for Auffenberg's advice. The result was that the Auffenbergs volunteered for the task, thus beginning his interest in these lizards that would result in a trilogy of autecological studies. Auffenberg spent a total of 13 months on Komodo and surrounding islands (1969–1972) studying this largest of living lizards. He and his family lived in canvas-covered, A-frame-like structures within the home ranges of these top carnivores that are known to disembowel adult water buffalo and even to kill humans. On one occasion, a large rogue male tracked his 15-year-old son, Kurt, and Auffenberg's graduate student, Putra Sastrawan, for several kilometers after entering their tent and tearing apart articles of their clothing. Auffenberg's book, "The Behavioral Ecology of the Komodo Monitor," appeared in 1981, but by this time he was already doing field work on two other monitors with decidedly different food habits and lifestyles.

In 1973, Auffenberg began a study of the Bengal monitor, a highly mobile, largely insectivorous species. This study—which required 13 years of intermittent fieldwork from Pakistan to Java as well as laboratory work—was published as a book in 1994. In the meantime, he had begun yet another monitor project, on a species then known from just three museum specimens after 130 years and thought by some to be extinct. Gray's monitor of the Philippines, which he re-discovered, is the most frugivorous varanid, a relatively sedentary species that Auffenberg radiotracked high in fruit-laden trees during some 22 months of fieldwork from 1976 to 1983. His book on this species appeared in 1988. Despite these substantial accomplishments, Auffenberg's stroke prevented him from pursuing what he often called his "dream study"—the monitor relative, *Lanthanotus* of Borneo.

• *References*: "Introduction: The Past to Now," by K. Auffenberg and W. Auffenberg, p. 1–19 *in* J. B. Murphy et al. (eds.), Komodo Dragons Biology and Conservation. Smithsonian Inst. Press, Washington, 2002; "A Conversation with Walter Auffenberg," by W. Card, Reptiles Mag., June: 38–43, 2004; "Walter Auffenberg (1928–2004)," by R. Franz, Herpetol. Rev., 35: 215–216, 2004; "In Memory of Walter Auffenberg," by P. A. Meylan and L. H. Ogren, Chelon. Res. Monogr., 3: 11–17, 2006; Additional information from Auffenberg's son, Kurt Auffenberg, 2005–2006. • *Portrait* (about 1976) *and signature* (1994)*:* Courtesy Kurt Auffenberg.

REICHENBACH-KLINKE, H.-H. (1914–1995).

Heinz-Hermann Reichenbach-Klinke (born Klinke; see *Note 1*) was a specialist on the diseases of fishes, and wrote the first texts devoted exclusively to the diseases of amphibians and reptiles and their treatment. He was born on 14 August 1914 in Fürstenwalde (Brandenburg), along the Spree River 50 km southeast of Berlin. After receiving his doctoral degree at the university in Berlin (Dr.rer.nat. 1938) for a thesis on the anatomy of beetle intestines, he was conscripted into the German Army. Following World War II, he became a scientific assistant in the Zoological Institute of the Technical School (now University) in Braunschweig and achieved his habilitation in 1952. His primary research interest was in applied aspects of the diseases of fishes, particularly trematodes, fungi, and other parasites and their effects on the host, and in environmental factors leading to infections. Besides his research interests, he studied fish diseases as a hobbyist and in August 1948 became the edi-

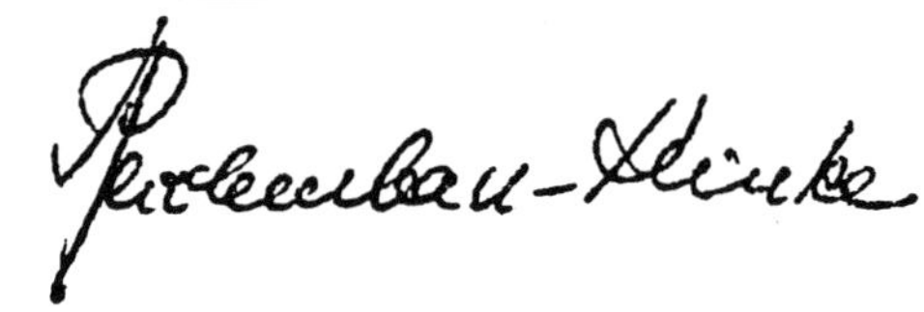

tor of *Wochenschrift für Aquarien- und Terrarienkunde*, one of the principal magazines for hobbyists, which in 1951 transformed into *Die Aquarien- und Terrarien-Zeitschrift* (or DATZ; see *Note 2*). These periodicals covered amphibians and reptiles as well as fishes, thus it was natural that Reichenbach-Klinke's interests broadened. He began to publish technical papers on amphibian diseases in 1948, mostly dealing with skin diseases, bone degeneration, and nasal parasites.

In 1957 Reichenbach-Klinke moved to the Bayerische Biologische Versuchsanstalt (now the Bayerisches Landesamt für Wasserwirtschaft), the fisheries research institute of the University of Munich, Germany's largest university. This research facility, begun in 1881, is devoted to the study of fish diseases, fish handling, and wastewater issues related to fisheries. When he joined the institute, he developed a special unit devoted to the diseases of amphibians and reptiles and recognized the importance of linking the institute to the university's veterinary sciences faculty of which he became a professor of zoology in 1962. He retired in 1979 and died in Munich on 10 July 1995.

Reichenbach-Klinke authored many papers on the diseases of fishes and several texts on the subject including "Krankheiten der Aquarienfische" (1957), "Krankheiten und Schadigungen der Fische" (1966), and "Fish Pathology" (1973). Although other authors had published papers on specific diseases of amphibians and reptiles or devoted chapters to the subject in larger works, Reichenbach-Klinke was the first to devote an entire book, actually two books, to the topic: "Krankheiten der Amphibien" (1961) and "Krankheiten der Reptilien" (1963). Both cover infectious diseases, tumors, environmental factors, a variety of syndromes (such as paralysis, blindness, bone damage, and developmental abnormalities), and various treatments including surgery, healing of wounds, and regeneration.

In order to make these books more generally available, an English edition was planned. The obvious choice to do the translation was Edward Elkan, a German physician and pathologist who had immigrated to England in 1933. Elkan was one of the pioneers in using *Xenopus* frogs to detect early pregnancy in women, beginning in 1937, and when he decided to breed frogs in England rather than to continue importing them from South Africa, he developed an interest in their diseases and cures. Elkan had been in contact with Reichenbach-Klinke when the latter was writing his books on amphibians and reptiles. Elkan also added to the content of the English edition based on his own experience. The result was their co-authored book, entitled "The Principal Diseases of Lower Vertebrates" (1965), the first on the subject in the English language. When this volume went out of print in the early 1970s, it was reprinted in three separate volumes, one on each of the three classes. In 1974, Elkan and Reichenbach-Klinke co-authored "Color Atlas of the Diseases of Fish, Amphibians and Reptiles" as a companion to their earlier books. These volumes represent a standard reference on the topic yet today.

• *References*: "Prof. Dr. Heinz Hermann Reichenbach-Klinke †," by R. Hoffmann and W. Körting, DATZ, 48: 684, 1995; Additional information from Reichenbach-Klinke's sons, Klaus and Matthias Reichenbach-Klinke, 2006. • *Portrait and signature*: Courtesy Matthias Reichenbach-Klinke. • *Note 1*: Heinz-Hermann Klinke's father died in World War I. When Klinke was adopted by his stepfather, it was decided to add his stepfather's surname to his own. • *Note 2*: The first three volumes of DATZ were titled *Deutsche Aquarien- und Terrarienzeitschrift*.

ALEKPEROV, A. M. (1912–1981).

The vertebrate zoologist, Abdulla Mustafaevich Alekperov, who wrote the first herpetology of Azerbaijan, was born on 15 March 1912 in Shemakha, about 110 km west of Baku, in the eastern Caucasus region of what was then the Russian Empire. He completed his secondary education at a technical school in Baku in 1929 and then entered the Azerbaijan Pedagogical Institute in that city to study chemistry and biology. After graduation in 1932, he taught at a secondary school until 1934 when he went to Leningrad (now St. Petersburg) State University. He completed his candidate thesis in 1938 on the lower vertebrates of the southern Taiga Subregion of the Leningrad area and then returned to teach at the pedagogical institute in Baku. During World War II he served in the Soviet Army, and it was after the war in 1945 that his fieldwork with the reptiles of Azerbaijan began. Alekperov had very broad research interests including faunistics, ecology, systematics, and conservation of fishes, amphibians, and reptiles. In 1958 he published his doctoral dissertation, which he defended in 1959 at the Zoological Institute of the Academy of Sciences in Leningrad where he worked with the Soviet herpetologist, Sergius A. Chernov. This was a comprehensive study of the amphibians and reptiles of Azerbaijan, the result of many years of fieldwork and study.

In the meantime, Alekperov was appointed a professor at Azerbaijan State University in Baku and from 1951 held several administrative posts including dean of the biology faculty, chair of the zoology department, and pro-rector of the university. In 1967 he was made full professor of vertebrate zoology and held this post until his death in Baku on 16 October 1981.

Alekperov authored about 90 publications; many are on amphibians and reptiles, especially venomous snakes. He contributed the herpetological section to the book "Animal Life of Azerbaijan" (1951). His major work in herpetology, however, was his book, "Amphibians and Reptiles of Azerbaijan" (1978), an update of his doctoral dissertation, which covers nine species of amphibians and 54 of reptiles. In addition, there is an extensive section on fossil amphibians and reptiles (especially turtles), with the description of new taxa. The book covers systematics, distribution (with maps), ecology, and food habits.

• *Reference*: "Abdulla Mustafaevitch Alekperov 1912–1981," by N. B. Ananjewa (=Ananjeva) and I. S. Darevsky, Amphibia–Reptilia, 4: 318–319, 1983. • *Portrait and signature*: Courtesy Natalia B. Ananjeva.

GORDON, Robert E. (1925–1996).

Robert Edward Gordon, a vertebrate zoologist, ecologist, and specialist on the salamanders of eastern North America, was for three decades a science administrator at the University of Notre Dame and a leader of national biological organizations. Although he was born in New York City on 20 June 1925, his family moved to Atlanta (Georgia) when he was a child and it was there that his interest in animals began. During World War II, he served in the United States Army Medical Corps (1944–1946). He studied biology at Emory University in Atlanta (A.B. 1949) and went on to work with the celebrated ecologist, Eugene P. Odum, at the University of Georgia (M.S. 1950). Gordon's master's research on the Green Salamander (*Aneides aeneus*) was begun in 1948 at the Highlands Biological Station in western North Carolina. Thus began his long association with this field station, which became the base of operations for him and his students and associates until the early 1970s. Gordon proceeded to Tulane University in New Orleans for his doctorate (Ph.D. 1956) with Fred R. Cagle. Gordon's dissertation on the Green Anole (*Anolis carolinensis*) was never published but has been regularly cited for the wealth of life history information he amassed on this widespread species.

After four years on the faculty of Northeast Louisiana State College (now University of Louisiana at Monroe), Gordon moved to northern Indiana in 1958 where he spent the rest of his career at Notre Dame. He served the university successively as head of the biology department (beginning when he was 39), associate dean of science, and then, during 1971–1989, as its first Vice President for Advanced Studies with responsibility for the graduate school and all sponsored research. A man with great leadership skills and a mesmerizing personality, Gordon was the highest-ranking layman in the administration of this Catholic institution. On his retirement, Notre Dame awarded him an honorary doctorate (LL.D.) in recognition of his key role in transforming its postgraduate programs. In addition to these academic posts, Gordon served as editor of the *American Midland Naturalist*, a biological journal published by Notre Dame. Nationally, Gordon became a leader in several organizations, serving as president of the American Institute of Biological Sciences, president of the trustees of *Biological Abstracts*, chairman of the Council of Graduate Schools, and executive secretary of the Council of Biology Editors. He was long active in the Society for the Study of Amphibians and Reptiles and served as president in 1971. Gordon returned to Georgia when he retired in 1990 and lived in La Grange where he died on 2 June 1996.

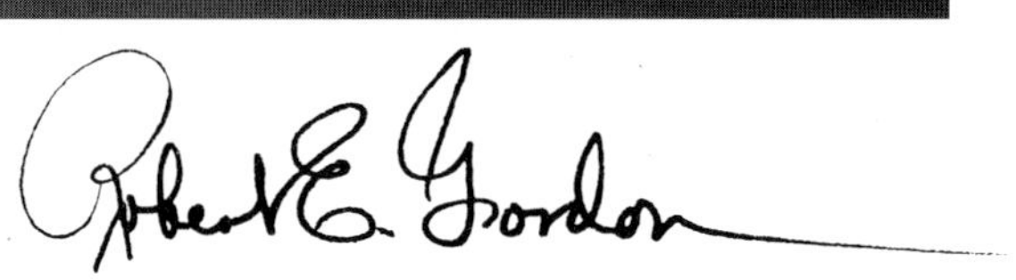

Because of Gordon's swift rise in university administration, most of his research publications on amphibians and reptiles were produced early in his career (15 titles, 1949–1962). These cover the life history, distribution, behavior, and ecology of species native to eastern United States; about half of them are on plethodontid salamanders. His major research contribution covered the life history and ecology of the green salamander (1952). He also published an influential paper on the relative abundance, microhabitats, and temporal behavior of plethodontids (1962), co-authored with one of his doctoral students, James A. MacMahon, and a beginning graduate student at the University of Southern California, David B. Wake, who worked with him at Highlands. Gordon was a longtime collector of postage stamps with images of amphibians and reptiles and wrote a regular column for *Biophilately*. He authored several publications on the subject including a handbook issued by the American Topical Association (1977). Gordon also influenced herpetology through his training of graduate students (including Ronald Gratz, Julian Harrison, III, James MacMahon, David Snyder, and Kerwin Werner) and by mentoring numerous budding amphibian biologists (among them Kraig Adler, Margaret Stewart, and David Wake).

• *References*: "Robert Edward Gordon," anonymous, p. 289 *in* American Men & Women of Science, 19th ed., R. R. Bowker, New Providence, New Jersey, 1994; "Robert Edward Gordon (1925–1996)," by K. Adler, Herpetol. Rev., 27: 109–110, 1996; "Biological Sciences," by P. B. Carter, Notre Dame Mag., Autumn 1996: 69–70, 1996. • *Portrait* (1988): Courtesy Catherine (Mrs. Robert) Gordon. • *Signature* (1960): Adler collection.

SCHWARTZ, Albert (1923–1992).

Albert Schwartz, Jr., the preeminent herpetological taxonomist of the West Indies, was born on 13 September 1923 in Cincinnati, along the Ohio River in southwestern Ohio. His father was a surgeon, but both of his parents had died by the time he graduated from college (University of Cincinnati, B.S. 1944), leaving him, their only child, with a sizeable trust fund. He enlisted in the army on 10 August 1945, four days before the Japanese surrender. A chance trip to Florida during college introduced him to the subtropics, so he decided to attend the University of Miami for his master's work (M.S. 1946). His doctorate from the University of Michigan (Ph.D. 1952) was on the mammals of southern

Florida, supervised by William H. Burt, which involved much fieldwork in Florida. At Michigan he met an undergraduate student and budding herpetologist, William E. Duellman, who accompanied him on field trips to the Appalachian Mountains and to the western United States. These experiences began Schwartz's interest in herpetology. He became Curator of Vertebrates at the Charleston Museum in South Carolina in 1952 and studied the regional herpetofauna. It was a field trip to Cuba in the winter of 1954, however, that began his nearly four-decades-long work on the amphibians and reptiles of the West Indies.

In 1956 Schwartz joined the biology faculty at Pennsylvania's Albright College and from this base made several additional trips to Cuba. He was a hard-driving, obsessive collector for whom knowledge of animals in the field—their natural colors, habits, and habitats—was all-important. The president of the college required that all faculty attend graduation exercises, so Schwartz would have his field vehicle fully packed and parked just outside the building and would depart for Cuba within minutes after the ceremonies were over. Albright, however, was a long distance from Cuba. In 1959 his father's brother died, the last of six brothers all of whom became dentists and physicians. Schwartz again came into a large inheritance that allowed him to escape from academic work and focus on systematic research. So in 1960 Schwartz resigned from the faculty and moved to Miami. At first he supported himself entirely from his own funds. He turned his house and garage into a research facility, complete with museum, library, and laboratory. Using his trust fund, he could support large field parties, sometimes as many as a half dozen persons or more. He soon realized, however, that his personal funds were limited, so in 1967 he accepted a position at nearby Miami-Dade Community College. Being entirely a teaching institution, no school time was allocated for research and no graduate students were available to assist him. Nevertheless, he was able to organize his teaching efficiently and leave time for fieldwork, research, and writing. He retired in 1988.

While in South Carolina, Schwartz had described new taxa of salamanders, a frog, turtles, and a snake, but his later work in the West Indies resulted in the identification and naming of 86 species and 279 subspecies of frogs, lizards, and snakes. After the Cuban Revolution ended in 1959, he reluctantly turned his attention elsewhere, first to Hispaniola and then to other places. He eventually collected on more than 130 West Indian islands. Schwartz produced 217 titles in herpetology (1949–1991) of which 195 were on the West Indies. He continued to work on the Cuban herpetofauna through collaboration with Orlando Garrido in Havana with whom he co-authored 15 papers (1967–1985). Schwartz invited many students to join him on his field excursions and thereby inspired them to develop careers in biology and herpetology. Among them were Julian Harrison, Dennis Paulson, and Richard Thomas from his Charleston days, and George Zug from those at Albright College. Schwartz and his associates collected more than 80,000 specimens of West Indian amphibians and reptiles. These are deposited in the University of Kansas Museum of Natural History, except for those from Cuba, which are at the American Museum in New York.

Schwartz published several major works on amphibians and reptiles. He co-authored a major summary of the amphibians and reptiles of southern Florida (1958, with Duellman) and a checklist of West Indian amphibians and reptiles (1975, with Thomas; supplement in 1978, with Thomas and Lewis Ober). By the mid-1980s, chronic arthritis in his knees made fieldwork nearly impossible for him. Schwartz's associates continued to collect for him and he confined his work to home. Beginning in 1982, Robert W. Henderson of the Milwaukee Public Museum became his major collaborator. They produced two guides to Hispaniolan amphibians and reptiles (both 1984; Spanish edition with Sixto Incháustegui), a complementary guide to the West Indies exclusive of Hispaniola (1985), a revised West Indian checklist (1988), and their magisterial "Amphibians and Reptiles of the West Indies: Descriptions, Distributions, and Natural History" (1991).

By the late 1980s, Schwartz could no longer go up the ladder to retrieve bottles of specimens in his private museum. Much of his writing by this time was on West Indian butterflies. By 1990, crippled with arthritis, he became homebound. After surgery following a fall at his home, Schwartz died in a Miami hospital on 18 October 1992. He never married.

• *References*: "Albert Schwartz (1923–1992)," by R. W. Henderson, Carib. Jour. Sci. 29: 1–3, 1993; "Albert Schwartz 13 Sept. 1923–18 Oct. 1992," by W. E. Duellman et al., Copeia, 1993: 927–932, 1993; "Remembrances of Albert Schwartz," by W. E. Duellman et al., p. 15–27 *in* R. Powell and R. W. Henderson (eds.), Contributions to West Indian Herpetology: A Tribute to Albert Schwartz. Soc. Study Amphib. Reptiles, Ithaca, New York, 1995; Conant, 1997 (Schwartz on pp. 475–476). • *Portrait* (early 1970s) *and signature* (1953): Adler collection.

BOKERMANN, Werner C. A. (1929–1995).

Werner Carlos Augusto Bokermann, a Brazilian specialist on frogs and birds, was born on 4 July 1929 in Botucatu, a provincial town about 70 km northwest of the state capital, São Paulo. After finishing secondary school in Botucatu in 1947, he went to São Paulo seeking employment and became an assistant in the Departamento de Zoologia of the Secretaria da Agricultura de São Paulo. In 1950 the department head, noted ornithologist Hélio Ferraz de Almeida Camargo, arranged for Bokermann to work in the herpetological section, the domain of P. E. Vanzolini, an expert on South American reptiles. An arrangement was reached by which Bokermann would study frogs and curate the frog collections. He flourished in this environment and soon published his first paper on frogs and eventually authored 83 titles on the subject (1950–1993).

Bokermann's research was primarily taxonomic and he described 70 new species, primarily in the families Hylidae and Leptodactylidae, but also in the Bufonidae, Dendrobatidae, and Microhylidae. He co-authored one of them with the elder of Brazilian frog study, Bertha Lutz (1963), and 11 others with his protégé, Ivan Sazima (1973–1982 [1983]). Bokermann continued to work with and describe the collections in the department of zoology through 1956 but thereafter described new species from his private collection kept at his home. The latter was built up from his extensive fieldwork throughout the country and from exchanges with museums. During 1967–1968 Bokermann held a Guggenheim Foundation fellowship to study frogs in museum collections in Colombia, Ecuador, and the United States.

A major administrative change occurred at Bokermann's home institution in 1969 by which the collections from the department of zoology were transferred to a new Museu de Zoologia at the Universidade de São Paulo. He, however, was assigned to the Instituto de Biociências at the university. Concurrently, he took a position at the Fundação Parque Zoológico de São Paulo to work in the zoo's division of birds of which he eventually became unit head in 1984. After 1970 Bokermann's attention turned more and more to his work at the zoo and to ornithology, although he was also involved with the zoo's amphibian collection and its museum. He was particularly interested in nest building and conservation of birds. In 1977 Bokermann took leave from the institute to enroll at the Faculdade Farias Brito in Guarulhos (São Paulo) from which he graduated in 1978. In 1980–1981 he conducted postgraduate study at the Universidade de São Paulo and in 1991, at the age of 61, defended his doctor's dissertation on the biology of the macuco, a species of bird. Most of his publications after 1977 were on birds.

Bokermann's research on frogs was characterized by careful attention to details of all aspects of their natural history and morphology. He paid particular attention to descriptions of tadpoles and was one of the first to incorporate information on their vocalizations through the use of audiospectrograms in the descriptions of new species. In addition to Sazima, he helped to train many of the next generation of Brazilian herpetologists including Adão Cardoso, Célio F. Haddad, and José Pombal Júnior. He regularly participated in examinations for masters and doctoral students in herpetology and ornithology at São Paulo area universities. He amassed a large private collection of frogs, which numbered nearly 40,000 specimens, and generously made it available to other scientists. The collection is now in the Museu de Zoologia, Universidade de São Paulo. Colleagues in North and South America honored Bokermann by naming no less than ten frog species after him. He died in São Paulo on 1 April 1995 (see *Note*).

• *References*: "In Memoriam: Werner C. A. Bokermann," by A. S. R. dos Santos, Bol. Centro Estud. Ornitol., 12: 2–19, 1995; "In Memoriam: Werner C. A. Bokermann," by H. M. F. Alvarenga, Ararajuba (now Revta. Bras. Ornitol.), 3: 101–102, 1995. • *Portrait* (1968): Courtesy William E. Duellman. • *Signature* (1968): Adler collection. • *Note*: Santos (1995) gives the date of death incorrectly as 1 May.

UNDERWOOD, Garth (1919–2002).

Garth Leon Underwood, a British specialist on the classification and evolution of snakes and lizards, was born on 16 July 1919 in Isleworth (Middlesex), West London. Both of his parents were artists, yet it was his visits to the zoo in Regent's Park and countryside trips with his father that introduced him to nature. In 1938 he went to University College London, but soon was conscripted into the British Army. His first duty was in France (1940); he almost lost his life aboard the troop ship H.M.T. "Lancastria" when it was sunk. Later, he was stationed in India (1943–1946) where he did fieldwork and began his serious interest in reptiles. One of his army duties in India was to help develop the system that broadcast the sounds of tank battles into the jungles to confuse the enemy. After the

war, he finished his university work (B.Sc. 1948) and promptly immigrated to Jamaica to take a position as demonstrator at the University College of the West Indies (now University of the West Indies) in what was then the British West Indies. Underwood rose through the ranks to senior lecturer by 1960, the same year he was awarded his D.Sc. by the University of London on the basis of his published work. He did a great deal of fieldwork in Jamaica, and also on other islands in the eastern Caribbean, from St. Kitts south to Trinidad. After serving as professor and dean at the agricultural branch of his university, which is located in Trinidad, he returned to London in 1964 as a research fellow at the British Museum (Nat. Hist.). He then accepted a post at Sir John Cass College (renamed City of London Polytechnic and now London Metropolitan University) during 1967–1984, after which he retired and became a research fellow again at the museum. He died in London on 15 October 2002 at the age of 83.

Underwood's herpetological publications began in 1945 while he was stationed in India with a note on egg-laying behavior in an agamid lizard. Throughout his career, his chief research interests were the evolution and classification of snakes and lizards. In this work he championed the use of the widest possible array of characters, and he introduced several novel ones. Among his earliest papers are those on the systematic position of *Loxocemus* and others on the classification and phylogeny of geckos and pygopodids. Inspired by Gordon L. Walls's book, "The Vertebrate Eye and Its Adaptive Radiation" (1942; reprinted 1963), Underwood's paper on the evolution of reptilian retinas (1951) began one of several lines of evidence for a reevaluation of the evolution of snakes. In that same year, he co-authored with Angus Bellairs a seminal paper that marshaled evidence for a fossorial origin of snakes. Underwood's classic book, "A Contribution to the Classification of Snakes" (1967), fully displayed his scope and original thinking. (As he later wrote, "How a man thinks matters more in the long run than what he thinks.") Besides a long list of anatomical and osteological characters, he discussed visual cells, chromosomes, and the biochemistry of bile salts. To this list he continued to add, most recently his study of the rictal glands in the corner of the mouth of many snakes (2002). Other titles display the richness of his interests: simplification and degeneration in the course of evolution, parallel evolution, affinities of burrowing asps, classification of pythons, and affinities of vipers.

While he was in Jamaica, Underwood developed an interest in the local reptile fauna, especially in the anoline lizards, which brought him into collaboration with the American herpetologist, Ernest Williams. They co-authored a major review of the anoles of Jamaica (1959). Underwood published many other papers on these and other West Indian reptiles in which he described several new taxa. His largest publication from this period was an identification manual, "Reptiles of the Eastern Caribbean" (1962 [1963]), which was published in a loose-leaf format so that supplements could be added (one in 1964). His most unusual discovery from the West Indies was a new species, *Clelia errabunda* from St. Lucia (1993), based on museum specimens collected in the 1880s. This snake has not been seen alive for over a century; in 1995, he wrote a popular booklet on the subject, "A Tale of Old Serpents." Most of Underwood's collections are at the Natural History Museum in London and at Harvard University.

• *References*: "Garth Underwood (1919–2002): A Vision of Reptile Systematics," by R. S. Thorpe, Herpetol. Rev., 34: 6–8, 2003; "Garth Leon Underwood 1919–2002," by C. J. McCarthy, Copeia, 2003: 222–225, 2003; "Garth Underwood," by M. M. Stewart, Copeia, 2003: 415–418, 2003. • *Portrait* (1988): By Kraig Adler. • *Signature* (1969): Adler collection.

WILLIAMS, Ernest E. (1914–1998).

The herpetologist and evolutionary biologist, Ernest Edward Williams, was an authority on the morphology of reptiles and the taxonomy and ecology of lizards of the genus *Anolis*. He was born on 7 January 1914 in Easton, located on the eastern border of Pennsylvania. As a child he became acquainted with the local fauna especially the salamanders. His scientific interest in reptiles began at the local college, Lafayette (B.S. 1933), where he studied with the anatomist, Beverly W. Kunkel, a specialist on turtle morphology. After working in the family-owned store during the Great Depression, Williams entered graduate school at Columbia University in New York City in 1940 where he was a student of William K. Gregory who had previously mentored two other herpetologists, Charles L. Camp and G. Kingsley Noble. During 1942–1945 Williams served in the United States Army in Europe and returned to Columbia in 1946, the year Gregory retired. George G. Simpson then chaired his doctoral committee, but Gregory continued to influence his dissertation research on the neck skeleton of turtles (Ph.D. 1949).

In 1949 Williams became an instructor at Harvard University and taught the laboratories associated with Alfred S. Romer's comparative anatomy course. (Romer had also been a Gregory student at Columbia.) Williams became Assistant Professor of Biology in 1952 and rose through the ranks to professor by 1970. After the retirement of Arthur Loveridge in 1957, he assumed additional duties as Curator of Reptiles and Amphibians in the Museum of Comparative Zoology (MCZ). Williams retired as curator in 1980 and as a professor in 1984, although he continued his research at the MCZ. In September 1997 he returned to Pennsylvania where he died on 1 September 1998 in Mount Bethel, 25 km north of his hometown, Easton. He never married.

Williams had an influential career as a researcher on reptiles and amphibians, as curator of one of the great museum collections, and as a supervisor of doctoral students. His research during the 1950s, inspired by Kunkel, was on turtle anatomy and systematics. His first paper (1950) composed his doctoral work and included a new classification of turtles. His other papers on turtles included several insightful discussions of morphology, which often had systematic implications; rediscovery of the Australian pleurodire, *Pseudemydura umbrina*; fossil Cuban *Testudo* and the evolution of tortoises; and the first recognition of *Geochelone carbonaria* and *G. denticulata* as distinct species. His major work of this period was the classic book, "Revision of the African Tortoises and Turtles of the Suborder Cryptodira" (1957, with Loveridge).

Beginning in the late 1950s, Williams's research shifted to West Indian lizards of the genus *Anolis*. He had made two collecting trips to Jamaica in the 1950s and, given the large West Indian collections at the MCZ amassed by a former director, the herpetologist Thomas Barbour, Williams recognized the potential of this highly speciose genus of lizards for evolutionary and ecological studies. He soon collaborated with a British herpetologist, Garth Underwood, who was then teaching in Jamaica. Williams began publishing on the systematics of anoles, and urged his doctoral students to study an array of biological questions using these lizards as models. The result was the development of a graduate research program that for three decades studied the systematics, ecology, evolution, and behavior of anoles and developed some of the key concepts in evolutionary ecology including biogeography, ecology of colonization, cytological and biochemical studies, resource partitioning (most notably the ecomorph concept), and the roles of thermoregulation and behavior, all within an evolutionary context. Among his students of this period were George Gorman, Paul Hertz, Raymond Huey, Ross Kiester, Stanley Rand, Jonathan (now Joan) Roughgarden, and Thomas Schoener. His other students, Carl Gans and Allen Greer, did not study anoles. Williams also influenced Robin Andrews, David Crews, Thomas Jenssen, James Lazell, Jr., Thomas Parsons, Rodolfo Ruibal, Robert Trivers, and P. E. Vanzolini. The MCZ's focus on West Indian anoles continues today with the recent appointment of Jonathan Losos.

Although his main research focus then were the anoles of the West Indies (later expanded to include South and Central America), Williams made significant contributions to other topics. He authored a critical study of the development of tetrapod vertebrae (1959), which rejected Hans Gadow's

longstanding ideas (1896) of their derivation from embryonic "arcualia." Together with Parsons, a student of Romer's, Williams studied the unique tooth structure of amphibians (1962), following earlier work by H. W. Parker (1956). They were able to support the studies of Ivan Schmalhausen and Henryk Szarski who had argued for the monophyly of living amphibians, the Lissamphibia. Williams also published on other subjects: Solomon Islands elapid snakes; breakdown of polymorphism in a salamander; the skull of a fossil Argentinian snake; convergence in the digital pads of scincid lizards; and a computational approach for comparison and identification of species in difficult groups. Williams's 1972 paper on the evolution of ecomorphs anticipated the use of phylogenetic approaches to understanding evolutionary phenomena by two decades. Williams authored nearly 100 titles in herpetology (1950–1999) during his career.

Despite his research and students, Williams did not neglect his curatorial responsibilities. Under Barbour, with his personal wealth and the assistance of Loveridge, the MCZ developed a herpetological collection, particularly that of the West Indies, which was unrivalled among American museums in terms of number of species. It sometimes was said that Barbour simply wanted a pair of every species, but Williams recognized that to do modern evolutionary studies large series of specimens of both sexes, of various size classes, and from many localities were required. Thus, the MCZ's herpetological collections changed in concept and Williams was instrumental in this transition, which had occurred earlier at some other museums in the United States. During the first half of his 23-year curatorship, Williams had the unpaid volunteer assistance of Benjamin Shreve who had begun working at the MCZ in the early 1930s. Shreve, an amateur herpetologist who was a member of a distinguished Boston family, published many new species. During Williams's tenure as curator, the museum's rich herpetological holdings grew from 72,550 to 303,925 specimens, but the numbers alone are deceptive because of the numerous types, rare species, and skeletal material that are included.

• *References*: "Introduction," by P. E. Vanzolini, p. vii–x *in* A. G. J. Rhodin and K. Miyata (eds.), Advances in Herpetology and Evolutionary Biology. Essays in Honor of Ernest E. Williams. Mus. Comp. Zool., Cambridge, Massachusetts, 1983; R. Powell and R. E. Henderson (eds.), Contributions to West Indian Herpetology. A Tribute to Albert Schwartz. Soc. Study Amphib. Reptiles, Ithaca, New York, 1996 (Williams on pp. 37–45); "Ernest Edward Williams 1914–1998," by P. E. Vanzolini, Copeia, 1999: 1141–1142, 1999; "Ernest Edward Williams 1914–1998," by C. Gans, Herpetol. Rev., 31: 10–11, 2000. • *Portrait* (1962): Carnegie Museum, courtesy M. Graham Netting. • *Signature* (1958): Adler collection.

GALLARDO, José M. (1925–1994).

José María Alfonso Félix Gallardo, an Argentine zoologist and herpetologist, was born in Buenos Aires on 1 August 1925 (see *Note*). He liked nature from an early age, influenced by vacations at his family's country estate in Bella Vista and by his grandfather, Ángel Gallardo, who had served as director of the Museo Argentino de Ciencias Naturales "Bernardino Rivadavia" in Buenos Aires. That institution would become José's academic home for nearly 50 years. He attended secondary school at Colegio Champagnat de Buenos Aires and began work at the museum in 1946 while he was a student of natural sciences at the Universidad de Buenos Aires (Licentiate 1950). His first interest, like that of his grandfather, was in entomology, but by 1950 he turned to herpetology. During 1959–1960, he visited several museums in the United States and Germany to study their collections of amphibians and reptiles from South America. In the early 1960s, he worked on collections in the Departamento de Zoologia of the Secretaria da Agricultura de São Paulo in Brazil.

In 1962 Gallardo became an investigator with support from the Consejo Nacional de Investigaciones Científicas y Técnicas (CONICET), and in 1971 was appointed director of the museum, a position he held for the rest of his life. Concurrently, he served as professor of vertebrates at the Universidad de Buenos Aires and also lectured at other institutions in the city. Throughout this period he was working on his doctoral dissertation—a study of the amphibians and lizards of Argentina, which addressed their taxonomy, ecology, zoogeography, and behavior—at the Universidad de Buenos Aires (D.Sc. 1983). He became involved in conservation, speaking out in particular about the modification of ecosystems through human activities. He died in Buenos Aires on 12 October 1994.

Gallardo published on all major vertebrate groups, but most of his works were on amphibians and reptiles (about 95 titles, 1951–1993). A majority of these were on frogs and lizards, but he also studied turtles, amphisbaenians, and snakes. His taxonomic papers were comprehensive studies of individual species or entire genera, and they included descriptions of new taxa such as *Melanophryniscus*, a well-known genus of frogs (1961), and numerous new species. His papers were often enriched with information about natural history, ecology, behavior, and zoogeography. Gallardo's major works on herpetology include popular books on the amphibians (1974) and reptiles (1977) of the Buenos Aires region, a guide for the identification of Argentine amphibians (1987), a monograph on the ecology and behavior of the frogs of Argentina co-authored with Esperanza V. de Olmedo (1992), and a book on cultural aspects of amphibians and reptiles (1994).

• *References*: "Investigadores del Museo," anonymous, Bol. Mus. Argent. Cienc. Nat. "Bernardino Rivadavia," 2(3): 6–9, 1992; "Remembranzas de José María Gallardo," by J. C. Chebez, Nuestras Aves, 12(30): 18–19, 1994; "Homenaje al Dr. Jose Maria Gallardo," by W. Volkheimer, Mus. Argen. Cienc. Nat. "Bernardino Rivadavia," Extra, nuev. ser., 130: 1–7, 1996. • *Portrait*: From Volkheimer, 1996. • *Signature* (1966): Courtesy Jorge D. Williams. • *Note*: Volkheimer (1996) gave the birth date incorrectly as 1 July (see Á. Gallardo, Memorias de Ángel Gallardo. Ed. Elefante Blanco, Buenos Aires, 644 pages, 2003).

FUKADA, Hajime (1913–2002).

The Japanese herpetologist, Hajime Fukada, who was born on 23 August 1913 in Kyoto, became his country's foremost authority on snake life histories. His father was a professor at Kyoto Imperial University (after 1947, Kyoto University) and, when Hajime was 11 years old, helped him catch a rat snake to maintain as a pet, the first of many snakes and amphibians his parents allowed him to keep. In 1932, while still in high school, Hajime made a fortunate acquaintance with Moichirô Maki, then Japan's leading student of snakes, who mentored him. Fukada finished high school in 1934 and en-

tered Kyoto Imperial University to study applied entomology (graduated 1937). He became a graduate student with Taku Komai to work on the genetics of parasitic wasps. Komai also had been Maki's professor.

In 1938 Fukada was conscripted into the Imperial Army. He had been a horseback rider as a high school student, so he was assigned to a cavalry regiment. He went into training for the meteorological corps and then joined its teaching staff. By July 1941 he was posted to a meteorological station at Heito in southern Formosa (now Pingtung, Taiwan), where he collected and kept many snakes in his officer's quarters. He was sent to Manila in January 1942, but by December was demobilized and in 1943 became a zoology professor at Ryojun High School in Port Arthur (now Lüshun, Liaoning) in Japanese-occupied China.

After the war, Fukada was repatriated to Kyoto and, in despair over having trained young men for war, refused to become a professor. Instead, he was employed as an entomologist at a hydroponic farm operated by the American occupation forces in Otsu (Shiga) where he was allowed to keep snakes in the laboratory and on holidays to run his first mark-recapture studies in the field. In 1954 he became an instructor at Kyoto Gakugei University (after 1966, Kyoto University of Education), where he began his comprehensive mark-recapture studies in the paddy fields west of Tambabashi in the southern part of Kyoto. Five years later, because Komai had retired in 1946, Fukada continued his graduate work with Kenji Nakamura, another of Komai's former students, at Kyoto University. Fukada earned his doctorate (D.Sc. 1961) for a dissertation on life history studies of snakes. During 1955 to 1969, he had amassed 4,769 mark-and-recapture records. He retired his professorship in 1977 because of the age limit, but moved his snake colony home where it was kept until 1988. Fukada taught briefly at Kyoto University of Foreign Languages (1983–1988). He died in Kyoto on 12 March 2002.

Fukada followed the advice of his mentor, Moichirô Maki, to study snake life histories along the lines of research pioneered in the United States by Frank N. Blanchard of the University of Michigan. Maki sent Fukada copies of Blanchard's papers, which Fukada committed to memory! His studies focused on snake reproduction, growth, longevity, and estimates of population size. He also studied the development of the nuchal poison gland of *Rhabdophis tigrinus*, discovered by Nakamura. All told, Fukada authored about 135 titles in herpetology (1951–1997), most of them on snakes. In addition to books for children on snakes and frogs (1970) and on turtles (1972), he wrote two technical books. The first was "Crocodilians of the World" (1966) with Wataru Kimura, director of the Atagawa Tropical Garden and Crocodile Farm. The other reviewed his life's studies on snake biology, entitled "Snake Life History in Kyoto" (1992). Fukada was a quiet and modest person who was always helpful to colleagues. He was one of the founders of the Herpetological Society of Japan and served as its president during 1977–1989. In 1986, Emperor Hirohito honored him with the Order of Merit of the Rising Sun.

• *References*: ["Obituaries of the Past President, Prof. Hajime Fukada,"] by M. Matsui et al., Bull. Herpetol. Soc. Japan, 2002: 99–129, 2002; "Obituary: Prof. Hajime Fukada, 1913–2002," by M. Matsui, Curr. Herpetol., Kyoto, 21: 61–66, 2002; Additional information from Hajime Fukada, 1990. • *Portrait* (mid-1960s): From Matsui, 2002 (both titles). • *Signature* (Japanese form, 2002): Courtesy Tae (Mrs. Hajime) Fukada; (Romanized form, 1986): Adler collection.

Špinar, Zdenek V. (1916–1995).

Although they are mostly rare elsewhere, fossil frogs are abundant in the carbonaceous deposits of Bohemia. There, Zdeněk Vlastimil Špinar, who became the leading student of the Tertiary frogs of Central Europe, was born on 4 April 1916, in Čáslav. He had a boyhood interest in nature especially birds, and attended local schools. In 1935 he went to Prague to study biology and chemistry at Charles University, one of the ancient universities of Europe. Although he finished his studies in 1939 just before the Nazis closed the Czech universities, he could graduate with his doctor's degree only after their re-opening (RNDr. 1945). During World War II he was forced to work building roads and serving as an administrator for the chemical industry. After receiving his degree, he became a professor in the Geological-Paleontological Institute of Charles University.

Špinar's earliest publications, beginning 1940, were on invertebrate fossils, but later, influenced by his teacher, the paleontologist Josef Augusta, he decided to focus on fossil amphibians. His first paper on the subject (1952), based on his doctoral studies, was on Moravian discosauriscids. Although these had long been regarded as reptiles, Špinar discovered larval specimens with gills, which demonstrated that they were amphibians. Soon after, he began his studies on fossil frogs. He discovered a rich site in northern Bohemia where the frogs were exceptionally well preserved (including fossilized soft parts), in Oligocene-Miocene deposits. Špinar and his students

performed all of the excavations and preparations themselves. He published a long series of papers on the subject, culminating in his book, "Tertiary Frogs from Central Europe" (1972). This has descriptions of species based on 1,100 adults and 80 tadpoles, mostly of the families Palaeobatrachidae and Pelobatidae, but also some records of Ranidae and Bombinatoridae. His treatment of osteological variation is the most extensive for any fossil frogs. Because of the extraordinarily good conditions for fossilization, the ossicles of the middle ear and various soft parts (organs, glands, nerves, and even pigments) were preserved. Špinar also published on fossil frogs from Britain, Germany, India, Israel, Libya, and Mongolia.

Špinar undertook advanced training at Charles University (a candidatus degree, C.Sc. 1958; Dr.Sc. 1968). He published several textbooks including "Principles of Invertebrate Paleontology" (1960), "Systematic Invertebrate Paleontology" (1966, with a team of authors), and "Paleontology of Vertebrates" (1984). Being written in Czech and thus accessible, these had a profound effect on the development of paleontological studies in the country. His most widely known work, however, was to continue a project of his teacher, Josef Augusta. Špinar collaborated with two artists—Zdeněk Burian and, later, Lubomir Dédek—to prepare a series of 97 illustrations of reconstructed fossil animals. Using this artwork, several popular books, most especially Spinar's "Life Before Man" (1973), and scientific films for the general public were produced. His work in all its phases was one of the major reasons for the renaissance of paleontological studies in Czechoslovakia in the post-war years. A man with an ebullient personality and great generosity to colleagues and students, Špinar died on 14 August 1995, at his summer house in northern Bohemia near the town of Česká Kamenice, north of Prague, and very close to one of his Tertiary-frog localities.

• *References*: "The Seventieth Anniversary of Prof. RNDr. Zdeněk V. Špinar, DrSc.," by V. Pokorný, Acta Univ. Carolin., Geol., 2: 133–135, 1986; "Czech Palaeontologist Zdeněk V. Špinar Died," by Z. Roček, Jour. Czech Geol. Soc., 40: 94, 1995 (reprinted, Russ. Jour. Herpetol., 3: 103–104, 1996). • *Portrait* (1976) *and signature* (1989): Adler collection.

DELY, Olivér Gy. (1927–2003).

Olivér György Dely, a Hungarian herpetologist and editor, curated the national herpetological collection and authored the most comprehensive summary of his country's herpetofauna. He was born on 27 April 1927 in Nagyszalonta (now Salonta, in westernmost Romania), an ethnically Hungarian town in Transylvania that was part of the Kingdom of Romania at the time of his birth. He attended local schools, but this was interrupted by service in World War II, after which he finished his secondary education in Budapest in 1946. He attended Péter Pázmány (later Budapest and now Loránd Eötvös) University during 1946–1950 and then began work in the herpetology collection of the Hungarian Natural History Museum in Budapest. This is a collection with a distinguished history under the eminent zoologist, Lajos Méhelÿ, and his successor, the evolutionary biologist, Géza Gyula Fejérváry, who died prematurely in

1932. The latter's widow, Aranka Mária Fejérváry-Lángh, a paleontologist and specialist on frog osteology, succeeded him and in 1952 Dely replaced her as curator.

Dely was thus in charge of the collection at the time of the Hungarian Revolution (October–November 1956), which was suppressed primarily by Soviet troops. On 5 November, an incendiary bomb shot by cannon from nearby Gellért Hill penetrated the roof of the museum. The alcoholic collections were on the top floor. These immediately exploded and caused a fire. Most of the zoological department and part of its library were destroyed. The entire herpetological collection of more than 40,000 specimens (including types) and the manuscript for Dely's dissertation were lost.

Undaunted, Dely began to rebuild the collection. He was finally able to complete his graduate degree in 1964 (Hungarian Academy of Sciences, a candidate of biological sciences degree, C.Sc., 1965) based on his research on the taxonomy and ecology of *Triturus alpestris*. Dely organized several expeditions in order to obtain new materials for the museum and for exchange purposes: Egypt (1957), Algerian Sahara (1973), and North Korea (1978). Unfortunately, during the Algerian trip he suffered a shotgun accident to his left leg. The healing took two years, yet his leg continued to break again throughout his life and seriously restricted his work. Dely founded the museum's journal, *Vertebrata Hungarica*, in 1959 and served as its first editor. In 1981 he organized the First Herpetological Conference of the Socialist Countries in Budapest, to which selected scientists from Western countries were invited, and edited the volume of proceedings. Although this was a small gathering, it was a key event in making East-West connections that facilitated planning for the First World Congress of Herpetology in 1989. Still painfully suffering from his 1973 accident, Dely died in Budapest on 19 November 2003.

Dely published more than 80 technical titles in herpetology (1952–1998). These cover a wide range of subjects. His major interest was morphological variability at the population level to study microevolution of co-existing and isolated populations. Most of his studies focused on the Hungarian herpetofauna, but he also published on a South American amphisbaenid, Egyptian turtles, and Mongolian lacertids. He wrote the herpetologies of three Hungarian national parks. His major works, however, were the herpetological sections of "Fauna Hungariae" (Amphibia, 1967; Reptilia, 1978). These are well-illustrated accounts in Hungarian with detailed descriptions, distribution, and literature, but no maps.

• *References*: "In Memoriam Dr. Olivér György Dely (1927–2003)," by Z. Korsós and J. Vörös, Ann. Hist.-Nat. Mus. Natl. Hungar., 96: 5–17, 2004; "In Memoriam Dr. Olivér György Dely (1927–2003)," by Z. Korsós, Biblioth. Herpetol., 5(2): 7–9, 2005. • *Portrait*: Hungarian Natural History Museum, courtesy Zoltán Korsós. • *Signature* (1989): Adler collection.

CHERCHI, Maria A. (1927–1985).

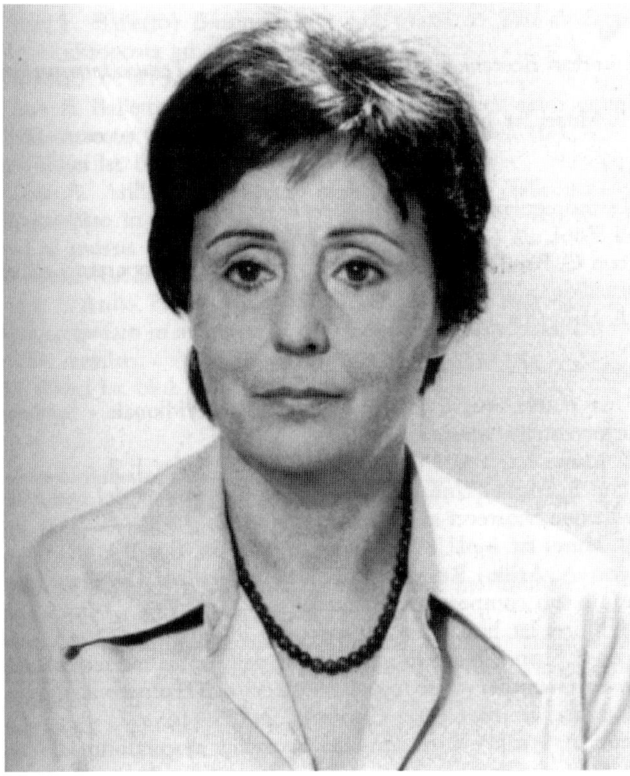

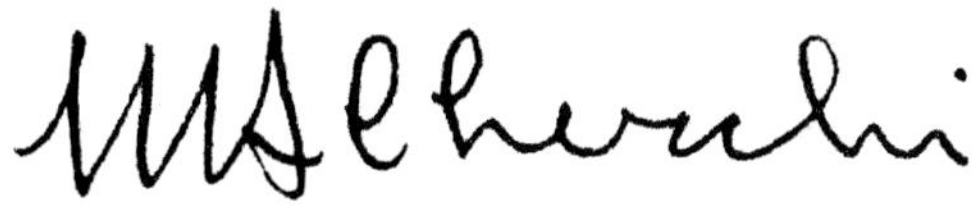

Maria Adelaide Cherchi (see *Note*), an Italian specialist on amphibians and reptiles of arid regions and their physiology, was born in Sassari (Sardinia) on 27 June 1927. She graduated in natural sciences from the Università degli Studi di Genova (1951), received her doctorate (Ph.D. 1959) under the direction of the zoologist, Giuseppe Scortecci, and then became a professor there in 1963, a post she held throughout her life. She taught courses both in the Faculty of Sciences and in the School of Medicine and Surgery. Cherchi was at home in the laboratory and in the field. Even after several surgeries in 1982, she participated in two long research excursions to Spain in 1983 and 1984. After a lengthy illness, she died on 18 July 1985.

Cherchi's interests spanned a broad range of topics including systematics, faunistics, and physiology. Although her first paper was on butterfly behavior (1951), she quickly turned to amphibians and reptiles as research subjects and authored about 60 herpetological titles (1952–1987 [1988]). The university's museum has large herpetological collections from North Africa and Arabia, principally made by Scortecci, and Cherchi's taxonomic papers dealt mostly with Somalia and South Arabia. From these areas (and the Andaman Islands) she described several new taxa of frogs and lizards. Her most extensive faunistic work was a review of the amphibians of the Arabian Peninsula (1985, jointly with Emilio Balletto and John Gasperetti). She often used computer-based multivariate analyses for her biometrical studies of taxonomy and faunistics.

From the early 1950s, Cherchi also developed an interest in physiological adaptation to extreme environments including caves and deserts, and was particularly interested in thermoregulation. She studied this phenomenon in salamanders, lizards, and turtles. Among other discoveries, Cherchi found that tortoises can survive long periods of anoxia (first paper in 1963) and, by shifting to an anoxic metabolism that produces less heat, these animals can better survive the midday sun of Mediterranean summers. In seeking the biochemical basis for this result, she studied how anoxic animals reverse certain steps in the Krebs citric acid cycle (1982).

• *References*: "Maria Adelaide Cherchi (1927–1985)," by A. Arillo and E. Balletto, Boll. Zool., 53: 119–121, 1986; "Maria Adelaide Cherchi (27 June 1927–18 July 1985)," by E. Balletto, Amphibia-Reptilia, 7: 96–99, 1986. • *Portrait*: From Balletto, 1986. • *Signature* (1951): Courtesy Emilio Balletto. • *Note*: Cherchi's married name, Coddé, sometimes appears on her papers.

RAND, A. Stanley (1932–2005).

Austin Stanley Rand, a specialist on the ecology and behavior of Neotropical amphibians and reptiles, was a mentor to numerous students of tropical biology. He was born on 29 September 1932 in Seneca Falls, located in central New York State 60 km north of Ithaca where his father, a Canadian ornithologist, had just completed his doctorate at Cornell University. His mother, whose home was in Seneca Falls, was also a biologist and had an interest in snakes (see *Note*), so it is little surprise that young Stan published so early. His note on chorus frogs in Quebec appeared in 1944 when he was 12. His father was a zoologist at the National Museum of Canada in Ottawa and in 1947 became Curator of Birds at the Chicago (later, Field) Museum of Natural History. Beginning in 1950, the younger Rand became an assistant at the museum and worked for the herpetologist, Karl P. Schmidt, whom his father succeeded as Chief Curator of Zoology in 1955. After Stan graduated from Indiana's DePauw University (B.A. 1955), he served two years in the army before entering Harvard University for graduate study with Ernest Williams and beginning his work on the ecology of West Indian anole lizards (Ph.D. 1961). Rand spent the next year studying anoles in Jamaica before he moved to São Paulo (Brazil) and the Departamento de Zoologia of the Secretaria da Agricultura where P. E. Vanzolini curated the herpetological collections. For most of the next two years Rand was at the museum's station in the cloud forest at Boracéia studying frogs and in Belém, near the mouth of the Amazon River, researching the ecology of lizards.

In 1964, while he was still in Brazil, Rand received a telephone call from Martin Moynihan, director of a field laboratory operated by the Smithsonian Institution on Barro Colorado Island (BCI) in the Canal Zone of Panama, to offer him a position as biologist there. The connection was so poor that Moynihan could not hear Rand's acceptance, even when he shouted it out! Once just a big hill, BCI was created in 1910 when a valley was flooded to become Lake Gatún during the building of the Panama Canal. Smithsonian scientists soon joined in a biological inventory of the Canal Zone and in 1923 the zone's governor declared BCI a biological reserve. Thomas Barbour, the Harvard herpetologist, headed a group that developed BCI for research purposes. He hired the first superintendent (James Zetek, an entomologist working for the Panama Canal Company), used his personal wealth to buy out the banana growers, and supervised construction of the first laboratory. Moynihan was appointed director in 1957 and decided that a permanent staff was needed for the station to realize its full potential as a place for visiting researchers to conduct studies and for students to be trained in tropical biology. The growing numbers of staff and research facilities on BCI and the marine laboratories on both coasts were consolidated and re-named the Smithsonian Tropical Research Institute (STRI) in 1966.

Moynihan could hardly have chosen STRI's first staff herpetologist more wisely. Rand had a highly productive research career there for 33 years, during which he was Coordinator of the Environmental Sciences Program (1973–1979), responsible for monitoring tropical marine and terrestrial habitats, and Assistant Director (1974–1979). He became Senior Biologist in 1979 and continued to do his research and to advise students. He retired in 1997 and moved to Alexandria (Virginia) on the western outskirts of Washington. He died on 14 November 2005 at a hospital in Baltimore (Maryland).

Rand's research interests in amphibians and reptiles were broad. His early work at the Field Museum and at Harvard was in systematics. He named a new salamander and a snake from Central America, West Indian anoles, and chameleons from East Africa. At Harvard, he joined Williams's long line of students who studied *Anolis*. Rand developed several key ideas, often together with Williams, including the ecomorph concept (to explain how anoles of different species had adapted to almost identical niches on different islands) and the residency effect (to explain the outcomes of dominance interactions). They used information theory to understand redundancy of signals in communication. At STRI, his work shifted to behavioral and physiological studies of communally nesting green iguanas and the behavioral ecology and reproduction of Panamanian anoles, crocodilians, and frogs.

In 1977, with Ronald Heyer, Rand co-authored his first paper on *Physalaemus pustulosus*, the Túngara Frog, a nondescript but abundant leptodactylid, which initiated his most important research program. He had been gathering data on its communication system since the mid-1960s and in 1978, with the arrival of Michael J. Ryan, then a Cornell graduate student, Rand began a long collaboration that lasted until his death. They investigated the significance of vocal diversity; mate recognition and its relationship to speciation; anti-predator systems including studies of dramatic predation events by bats (with a mammalogist, Merlin D. Tuttle); and female choice, sexual selection, and sensory exploitation. Rand and Ryan conducted a 5,000-km transect from Mexico to Panama, collecting audio recordings and tissue samples from throughout the range of this species, and later made collections in South America. In collaboration with a physiological psychologist, Walter Wilczynski, they studied the neural mechanisms that underlie mate choice, and with him and David Cannatella, still-ongoing studies of phylogenetics and molecular neurobiology.

Rand authored about 145 titles on amphibians and reptiles through 2006. He co-edited two books: "Iguanas of the World: Behavior, Ecology, and Conservation" (1982, with Gordon M. Burghardt; reprinted [1995]) and "The Ecology of a Tropical Forest: Seasonal Rhythms and Long-Term Changes" (1982, with Egbert G. Leigh, Jr., and Donald M. Windsor; Spanish edition 1990). Possibly even more important than Rand's publications, however, was his influence on more than 70 students—undergraduate, graduate, and postdoctoral—that were advised, inspired, and constructively critiqued by him on projects involving plants and vertebrates of all classes. He taught field courses for the Organization for Tropical Studies in Costa Rica (1978–1991), but most of his teaching was done in the laboratory and in the field in Panama. Although STRI is not a degree-granting institution, it may reasonably be said that Stanley Rand had as important a role in educating students on the biology of amphibians and reptiles as any university professor.

• *References*: "A. Stanley Rand (1932–2005)," by M. J. Ryan, Herpetol. Rev., 37: 134–137, 2006 (reprinted in Copeia, 2006: 573–577, 2006); "In Celebration of A. Stanley Rand," by W. R. Heyer, Phyllomedusa, São Paulo, 5: 8–10, 2006. • *Portrait* (1997): By Kraig Adler. • *Signature* (1993): Courtesy Patricia (Mrs. Stanley) Rand, 2006. • *Note*: Rand's mother, Rheua Vaughn Medden (1905–1985), studied snakes with Albert H. Wright at Cornell University (A.B. 1927, M.S. 1928). She published several papers on the subject (1927–1931) including an eight-part historical review on rattlesnakes from the works of early travelers in North America.

GORHAM, Stanley W. (1917–1984).

Stanley Warren Gorham, a Canadian specialist on amphibians, was born on 22 November 1917 on the family farm in Brown's Flat (Kings County) in the Saint John River Valley of New Brunswick. Gorhams have lived in this valley since 1786 and were among the Loyalist families who moved north after the American Revolution. As a schoolboy Stanley would carry frogs and snakes to class in his pockets. He never completed high school and, like most men of his generation, soon went to work as a woodsman. Gorham also worked for the railroad until World War II when, in 1939, he enlisted in the marine branch of the Royal Canadian Air Force and ferried supplies to northern bases. When in ports he would visit museums and meet curators, which set his eventual career path. Because he lacked a high school diploma, after the war he apprenticed himself to a taxidermist to learn a trade that might give him access to a museum position. In 1953 he became a technician at the National Museum of Canada in Ottawa. His initial work was taxidermy but with the support of the museum's herpetology curator, J. Sherman Bleakney, Gorham began his formal scientific work. He participated with museum teams in numerous field surveys throughout Canada including the Arctic.

In 1965, however, an opportunity to return to New Brunswick came with an offer to become both assistant curator of natural sciences and taxidermist at the New Brunswick Museum in Saint John. Gorham was promoted to curator of vertebrate zoology and, later, acting head of the Department of Natural Sciences. He made several overseas trips that included Fiji and Australia (1966), the Falkland Islands in the South Atlantic (1967), Brazil (1970), and the Soviet Union (1978). To recognize his scientific contributions, the University of New Brunswick awarded him an honorary doctorate (D.Sc. 1975). Gorham retired from the museum in 1982 but he already had been having recurrent fevers from psittacosis (parrot fever) that had been contracted during his Falklands trip. This caused serious heart problems and it was during his second open-heart surgery that he died in Halifax (Nova Scotia) on 23 March 1984.

Gorham was primarily a field naturalist and many of his 36 herpetological titles (1955–1991) dealt with distributional records and natural history. For his province he authored the

booklet, "The Amphibians and Reptiles of New Brunswick" (1970). He had a worldwide view of amphibians, however, and from about 1947 he began work on a global checklist. This was no easy task, given that he did not have a research post until 1965 and that some museum curators actively discouraged him. Despite his shy and gentle demeanor, Gorham was a man of great patience and determination who, once he developed an idea or a pattern, stuck to it, as he did in his personal life. (For the last 25 years of his life he always wore a blue shirt, blue jacket, and a blue tie!) He produced a series of papers comparing the number of Canadian amphibian species to those in other countries (1957, 1962, 1963). He also wrote papers on amphibians and reptiles of distant places, in particular those of the Fiji Islands (6 papers, 1965–1977).

Gorham compiled two sections on amphibians for the *Das Tierreich* checklist series (caecilians 1962; primitive frogs 1966). His *magnum opus*, however, first conceived during his museum visits during the war some 30 years earlier, was his "Checklist of World Amphibians" (1974). Even when a larger, more comprehensive checklist appeared in 1985, authored by a team of 59 specialists from 20 countries, Gorham's single-authored, simpler list still had the advantage of including synonyms and subspecies and remained a useful reference for many years.

• *References*: "Portrait of a Curator. The N.B. Museum's Stan Gorham," by H. McLeod, Atlantic Advocate, Fredericton (New Brunswick), January: 14–17, 1981; "Former Museum Curator Dies," anonymous, Evening Times-Globe, Saint John, 26 March: 14, 1984; "A Tribute to Stanley Warren Gorham, 1917–1984," by F. R. Cook, Canadian Field-Nat., 105: 592–596, 1991. • *Portrait*: New Brunswick Museum, courtesy Donald F. McAlpine. • *Signature* (1967): Adler collection.

Álvarez del Toro, Miguel (1917–1996).

Miguel Álvarez del Toro, a vertebrate biologist and conservationist, was born in Colima (Colima) in western Mexico on 23 August 1917. As a boy, he liked to collect birds, insects, and other animals near his rural home, but in 1931 his family moved to Mexico City where he completed high school at the Colegio Morelos in 1936. He had been strongly influenced as a teenager by reading a manual on taxidermy and learned to prepare his own specimens. These experiences were put to use, first in 1938–1939 as a collector of bird specimens for the Academy of Natural Sciences in Philadelphia, and then in 1939 when he became a taxidermist at the Museo de la Flora y Fauna, a hunting and fishing museum in Mexico City. He was promoted to subdirector of the museum in 1940.

In 1942 Álvarez moved to southernmost Mexico to be general zoologist at the Departamento de Viveros Tropicales y Museo de Historia Natural (now Instituto de Historia Natural) in Tuxtla Gutiérrez, the capital of Chiapas, which is located in the central valley of the state. With the premature death of the museum's director in 1944, Álvarez took charge of the museum and of the associated zoo. In 1980 the zoo was moved to a tropical forest, El Zapotal, just south of the city, where Chiapan species of both invertebrates and vertebrates are exhibited in their natural habitats. In March and April 1982, however, the museum and zoo were very nearly destroyed when a nearby and long-dormant volcano, Tzitzunal (also called Chichón or Chichonal), erupted for 15 days, spewing heavy ash everywhere within a radius of 20 km. For three days there was practically no daylight. In places the ground was too hot to walk on and many animals were killed at the zoo, but rainstorms and strong winds soon cooled the land and washed the vegetation clean. Álvarez's life's work survived.

Álvarez was a self-trained zoologist and never attended a university. He began his own research and conservation programs in Tuxtla Gutiérrez and was instrumental in the establishment of most of the nature reserves in Chiapas. His studies on the ecology and behavior of the Mexican crocodilians, begun in 1944, led to their protected status. Álvarez earned several national and international awards for his conservation activities, and two universities conferred honorary doctorates on him. He died in Tuxtla Gutiérrez on 2 August 1996, at the age of nearly 79, while still serving as director of the museum and zoo.

Álvarez was a prolific writer for both the lay public and the scientific community. His publications, which include nine books, cover a wide range of topics: vertebrates (especially birds), ecology, conservation, and captive breeding. He authored nearly 20 titles on herpetology (1956–1983) and described several new taxa of Chiapan lizards, jointly with Hobart M. Smith and including a new xantusiid, *Lepidophyma lipetzi* (1977). His major herpetological book was "Los Reptiles de Chiapas" (1960), a comprehensive review that includes much natural history information based mostly on his own observations. Second and third editions appeared in 1972

(1973) and 1982; he had completed a fourth edition at the time of his death but it was not published. These represent the most important contributions on the reptiles of southern Mexico by a Mexican national. Álvarez's other major work on reptiles was his book, "Los Crocodylia de Mexico" (1974), that was based on his nearly 30 years of work in the field and at the zoo including the first description of parental care in the three Mexican species.

• *References*: "Miguel Álvarez del Toro. Curriculum Vitae de un Autodidacta," by B. Álvarez Rincón, Gob. Est. Chiapas, Tuxtla Gutiérrez, 23 pages, 1996; "Miguel Álvarez del Toro (1917–1996)," by G. Casas Andreu, Bol. Soc. Herpetol. Mex., 7: 27–30, 1996; "Biographical Sketch of Miguel Álvarez del Toro (Don Miguel): 1917–1996," by O. A. Flores-Villela and W. L. Hodges, Herpetol. Rev., 30: 69–70, 1999. • *Portrait*: From Álvarez Rincón, 1996. • *Signature* (1961): Adler collection.

BRAME, Arden H., Jr. II (1934–2004).

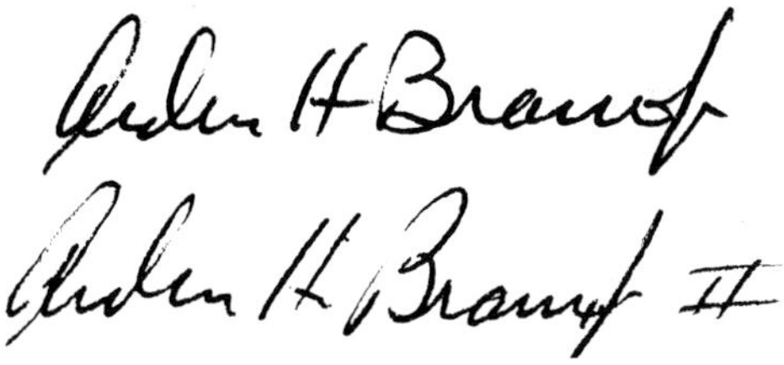

Arden Howell Brame, Jr. II (see *Note*), a specialist on plethodontid salamanders, was born on 19 March 1934 in Beverly Hills, a city surrounded by Los Angeles. His interest in salamanders began in 1952 when he found his first live specimen, after which he began correspondence with German salamander specialists—Friedrich Pölz and, later, Günther E. Freytag—who mentored him. Brame had a large collection of living animals including an adult Asiatic giant salamander, which he kept in his family's bathtub!

Brame attended the University of California at Los Angeles during the mid-1950s, where Bayard H. Brattstrom, a graduate student with Raymond Cowles, mentored him. Brame transferred to the University of Southern California (USC) and graduated in botany (B.A. 1957). With the arrival of Jay Savage as a professor at USC in 1956, Brame's interests in salamanders became more scientifically focused and he became Savage's student (M.S. 1967). Beginning in 1960, Brame also worked part-time at the Los Angeles County Museum of Natural History (LACM) and, in the absence of a curator during 1961–1965, was in charge of its then-small herpetological collections. During 1957–1963, Brame was a member of the United States Army Ready Reserve and, during six months of active duty at Fort Ord in 1958, he was able to spend weekends collecting animals throughout the length of California. When he went on maneuvers at Fort Ord, he found specimens of *Batrachoseps*, which he brought back to camp in his canteen! He discovered that moisture-loving salamanders could be found and were even abundant during long periods without rain because of the almost-continuous summer fogs.

In 1965 Brame became supervisor of the Eaton Canyon Nature Center in Altadena, on the northern outskirts of Los Angeles. This is one of several such centers operated by the Los Angeles County Department of Parks and Recreation. Late in 1968 he moved back to the LACM as collections manager in herpetology, but in 1970 returned to his former position at the nature center. This post was terminated in 1978 in a reorganization of the parks department. Brame became a professional genealogist in 1973 and published often on the subject. He was active in local conservation programs and served as president of the Pasadena Audubon Society (1971–1976). Brame was a strong supporter of the Society for the Study of Amphibians and Reptiles in the 1960s and served as its president in 1973. On 19 August 2004 he died at a care center in Pico Rivera, a Los Angeles suburb, of hepatitis that had been contracted many years earlier from a blood transfusion during heart surgery at a time when blood supplies were not regularly screened.

Brame's papers on salamanders (40 titles, 1956–2002 [2003]) were primarily on plethodontids. In the mid-1950s, he decided to produce a checklist of the world's Caudata, living and fossil, for the serial *Das Tierreich*. This never appeared, but the plan focused his work on the topic. In 1959 he participated in a USC-sponsored expedition to Costa Rica for five months, which inspired him to study Neotropical salamanders. Beginning in 1960, Brame published a steady stream of papers, nearly half of them co-authored with another of Savage's graduate students, David B. Wake. Numerous new species were described from eight countries extending from Mexico to Peru. Among them were 26 species of *Bolitoglossa* and seven of *Oedipina*. Brame also named four species of *Batrachoseps* from arid regions of southern California and a new *Plethodon* from Oregon, the last co-authored with Richard Highton. All told, Brame named 38 species of plethodontids, all still recognized and representing fully ten percent of the living members of this large family.

• *References*: "Arden Howell Brame, Jr.," anonymous, Pasadena Area Hist., 1972: 168–169, 1972; "Arden H. Brame, Jr., II," p. 37 *in* M. K. Meyer and P. W. Filby (eds.), Who's Who in Genealogy & Heraldry, vol. 1, Gale Res. Co., Detroit, 1981; "In Memorium [*sic*]: Arden Brame," by M. Long, The Wrentit (Pasadena Audubon Soc.), 52(5): 3, 2004; Additional information from Richard Breithaupt, Jr., 2006–2007. • *Portrait*: Courtesy Richard Breithaupt, Jr. • *Signatures* (above, 1973; below 1974): Adler collection. • *Note*: In 1973, Brame discovered he had an older half-brother living in Colorado who shared the same father and had the same name as his. Early in 1974, he had his name changed legally to Arden H. Brame, Jr. II.

STEWART, Margaret M. (1927–2006).

Margaret McBride Stewart was an ecologist who studied the amphibians of eastern United States, Malawi, Jamaica, and Puerto Rico. She was born on 6 February 1927 on the family farm in rural Guilford County (North Carolina). Her first experiences with wild animals were on the farm where she found salamanders in the family's springhouse and rat snakes in the henhouse. After graduating from a country high school 15 km from Greensboro, she attended the University of North Carolina at Greensboro (A.B. 1948) and, while doing some part-time biology teaching in the area, completed her master's at the University of North Carolina at Chapel Hill (M.A. 1951). Her scientific interest in amphibians began during the summer of 1950 at Highlands Biological Station in western North Carolina. There she met Robert E. Gordon, a doctoral student of Fred Cagle's at Tulane University, who was conducting research. He introduced her to salamander biology. Both her master's thesis and later her doctorate at Cornell University concerned salamander life history. At Cornell, her mentor was the mammalogist, William J. Hamilton, Jr., although he was not her official chairman. He suggested that she study amphibian life histories using the outline developed by Cagle for reptiles.

After graduation from Cornell (Ph.D. 1956), Stewart was offered a position as Assistant Professor of Biology at the New York State College for Teachers in Albany (later, State University of New York at Albany and now University at Albany) where she spent her entire career. She taught a variety of courses, but her favorite was "Field Biology," a natural history of plants and animals that sometimes drew as many as 150 students a year. Outdoors was where Stewart most wanted to be and she used every opportunity in her teaching and research to be there. She conducted many field studies in New York including the Allegany Indian Reservation in the western part of the state, the Adirondacks in the north, and the Albany Pine Bush, a glacial remnant community around the city, but much of her work was done overseas.

During 1963–1964 she studied the amphibians of newly independent Malawi (formerly Nyasaland) in southeastern Africa where she contracted malaria and tended to her husband when he was bitten by a mole viper (*Atractaspis*). Beginning in the summer of 1966 and including her sabbatical leave of 1972–1973, she worked in Jamaica where she was introduced to the leptodactylid frogs of the genus *Eleutherodactylus*. She first visited Puerto Rico in 1969 and during the summers from 1978 and including two sabbaticals she worked there, primarily studying the ubiquitous Coqui (*E. coqui*). This species became the primary focus of her ecological and behavioral work for some 20 years, although her fieldwork with it ended in 1989 because of Hurricane Hugo.

At Albany, Stewart was promoted through the ranks and became Distinguished Teaching Professor in 1977. She was active in numerous scientific and conservation organizations and was elected president of two of them (Society for the Study of Amphibians and Reptiles, American Society of Ichthyologists and Herpetologists). As president of SSAR in 1979, she was the first woman to head a major herpetological society. The University of Puerto Rico at Mayagüez awarded her an honorary doctorate (D.Sc. 1996) in recognition for her long-term studies of the coqui frog. She retired in 1997, but continued some teaching and founded a new master's program at her university devoted to the conservation of biodiversity. Stewart's long and productive career became a model for aspiring herpetologists especially women. She died on 2 August 2006 at her home on the Helderberg Plateau overlooking Albany.

Stewart authored about 65 titles in herpetology (1956–2004), half of them on the biology of frogs. Her papers include studies on development in salamanders (*Ambystoma*, *Eurycea*), food habits of sympatric species of ranid frogs, ontogenetic and sexual differences in the dorsal patterns of the mink frog, and the herpetofauna of the Nyika Plateau of northern Malawi. Her work on the coqui frog was often done in collaboration with colleagues and her doctoral students. Among these studies was the first demonstration of internal fertilization in an oviparous frog, habitat partitioning and its physiological basis, population densities and their relation to retreat sites, vocalizations and defense of retreat sites, and the dramatic parachuting behaviors of this species. Her work on Malawian amphibians (55 species of frogs and one caecilian) was published in book form (1967).

• *References*: "Biologist Meg Stewart Combines an Enthusiasm for Scholarship and Field Work," by C. Ricci, Univ. Albany Mag., 4(1): 3–5, 1990; "Margaret McBride Stewart," by W. S. Brown and A. R. Breisch, Copeia, 2005: 701–708, 2005; "Margaret M. Stewart (1927–2006)," by W. S. Brown and A. R. Breisch, Herpetol. Rev., 37: 396–398, 2006 (2007). • *Portrait* (1960s): Adler collection. • *Signature* (2002): Courtesy Sheila E. Tuttle and William S. Brown.

LEBEDKINA, Natalia S. (1918–2000).

The evolutionary morphologist, Natalia Sergeevna Lebedkina, who specialized on fishes and amphibians, was born on 6 June 1918 in Vinnitsa, Ukraine, about 200 km southwest of Kiev. She was descended from nobility on her mother's side of the family. Her father was a physician and comparative morphologist. The family often moved from place to place because of his work assignments, first to the Crimea, then Minsk in Byelorussia (now Belarus), and finally to Leningrad (now St. Petersburg) where he became a professor at the Lesgaft Institute of Natural Sciences. Natalia finished secondary school in Leningrad in 1935 and then entered Leningrad University, and she began graduate studies at the Lesgaft Institute in 1939.

Her studies were interrupted during World War II by the German invasion of the Soviet Union and by the siege of Leningrad, which began in September 1941. Until the siege was ended nearly 900 days later, an estimated 1.1 million Leningraders died out of an original population of 2.5 million from the German bombardment, exposure to the elements, disease, and starvation. Lebedkina's father died there in March 1942. Food rationing was severe, yet one of the assistants of Nikolai Vavilov—the country's leading geneticist who was then in prison for opposing Trofim Lysenko's anti-Mendelian views—starved to death in Leningrad surrounded by edible seeds, in order that the country's seed bank would be preserved for future generations.

Natalia and the rest of her family were evacuated to Yaroslavl Province in 1942 where she worked as a hospital nurse. In 1944 she moved to Moscow and became an assistant at a fisheries institute and in 1946 resumed graduate work at Moscow State University under the zoologist and morphologist, Boris S. Matveev. She completed her candidate's thesis in 1950, on a morpho-functional analysis of the skulls of rabbits and voles.

In April 1951 Lebedkina was arrested for "anti-Soviet activity"—for denouncing a political comment told at a small party—and sentenced to five years imprisonment, first in Moscow and later at a labor camp in Kuibyshev (now Samara) Province. After Joseph Stalin's death in March 1953, she was released in a general amnesty. Having been a political prisoner, she could not take an academic post, so late in 1953 she accepted a job in a factory that manufactured plaster anatomical casts for use in schools and medical colleges. In 1954 she was a researcher at the Institute for Scientific Information. She was rehabilitated in 1955, and in April 1956 became a junior investigator with the evolutionary morphologist, Ivan Schmalhausen. He had fallen victim to Lysenkoism in 1948 and had lost all of his academic posts, but was permitted to continue working in the Institute of Evolutionary Morphology in Moscow. Lebedkina continued to work at the institute after Schmalhausen's death in 1963. In 1966 she was invited to attend the Nobel symposium on the evolution and phylogeny of lower tetrapods, but Soviet authorities denied her an exit permit because of her earlier political activities. She earned a doctorate in zoology (D.Sc., Academy of Sciences, Leningrad 1976) and retired in 1981, but continued to work at the institute. She died in Moscow on 7 June 2000.

Lebedkina authored about 30 titles on skull development in fishes and amphibians, a subject of central interest in the Schmalhausen laboratory, which focused on the vertebrate transition from water to land. About half of her publications were on amphibians (1957–1995). Much of her work was with two hynobiid salamanders, *Ranodon sibiricus* and *Salamandrella keyserlingii*, for which she worked out the developmental staging tables (1964). Her *magnum opus* was the book, "Evolution of the Amphibian Skull" (1979; English edition 2004 with staging tables added). This was the result of 20 years of careful study of the structure, development, and homologies of the cranial elements and had implications for the establishment of cranial structure in the earliest land vertebrates.

• *References*: Rossolimo and Dunayev, 2003 (Lebedkina by S. V. Smirnov on pp. 186–203; also p. 533); "Foreword," by S. V. Smirnov, p. ix-x *in* N. S. Lebedkina, Evolution of [the] Amphibian Skull, Advances in Amphibian Research in the Former Soviet Union, vol. 9, 2004; Additional information from Evgeniy Dunayev, 2006, and Leo J. Borkin, 2007. • *Portrait*: Courtesy Evgeniy Dunayev. • *Signature*: Courtesy Sergius Kuzmin.

ESTES, Richard (1932–1990).

The American paleoherpetologist, Richard Dean Estes, was a specialist on the evolution and systematics of fossil salamanders, frogs, and lizards and the paleoecology of lower vertebrate faunas. He was born on 9 May 1932 in San Rafael, located just north of San Francisco (California). His parents were teachers interested in the natural and earth sciences. Estes attended the College of Marin, a local junior college, intending to study ceramics, but he chanced to take a paleontology course and after just three lectures had decided on his career. He transferred to the University of California at Berkeley (A.B. 1956, A.M. 1957, Ph.D. 1960) where he completed a dissertation on the Cretaceous lower vertebrates from the Lance Formation of Wyoming. He officially studied with Donald E. Savage, but also came under the influence of Charles L. Camp who he regarded as his "spiritual mentor." He accepted a faculty position in biology at Boston University in 1960. On arriving at the Museum of Comparative Zoology at nearby Harvard University, he spotted a man in the hallway dressed in a gray suit with a black tie and 1930s glasses. Estes asked directions to the office of the renown paleontologist, Alfred S. Romer. The man quietly led him to the office and turned out to be Romer himself! He offered Estes an unpaid position as Associate in Vertebrate Paleontology at the museum and became his mentor. In 1974, Estes returned to California to become Professor of Zoology at San Diego State University.

Estes was a dedicated teacher and mentor in his own right, at both Boston and San Diego State. Among those who completed masters or doctoral degrees with him on herpetological topics are Stephen Brown, Daisy Folk, Charles Meszoely, Mark Norell, Richard Schatzinger, and Robert Sullivan. In addition, Ana María Báez and Borja Sanchíz, from Argentina and Spain, respectively, studied under his direction. Estes was editor of the *Journal of Vertebrate Paleontology* (1983–1987) and in 1990 received the Romer–Simpson Medal of the Society of Vertebrate Paleontology, the youngest person ever to receive this prestigious award. Estes continued to work at San Diego State until his death, on 28 December 1990 in San Diego, after a yearlong illness.

Estes authored about 85 titles on fossil and living amphibians and reptiles (1957–1997). The breadth of his work is reflected in the titles of some papers: a fossil gerrhosaur from the Miocene of Kenya, fossil salamanders and salamander origins, Prosirenidae–a new family of fossil salamanders, studies on the fossil snake *Dinilysia patagonica*, first fossil record of caecilian amphibians, early fossil record of frogs, and ontogenetic variation in the molariform teeth of lizards. His major published works include his doctoral dissertation (1964); two volumes in the "Handbuch der Paläoherpetologie" series ("Gymnophiona, Caudata" in 1981; "Sauria Terrestria, Amphisbaenia" in 1983), which have world-wide coverage; and the book, "Phylogenetic Relationships of the Lizard Families" (1988, co-edited with Gregory Pregill), which has become a standard reference on the topic.

• *References*: "Professor Richard Estes, 1932–1990," anonymous, News Bull. Soc. Vert. Paleontol., 151: 77–78, 1991; "Romer–Simpson Medal, 1990, Lawrence, Kansas," response by R. Estes, Jour. Vert. Paleontol., 11: 141–142, 1991. • *Portrait* (1978) *and signature* (1973): Courtesy Richard Etheridge.

McCOY, Clarence J. (1935–1993).

Clarence John McCoy, Jr., generally known as Jack McCoy, studied the herpetofauna of the United States, Mexico, and Belize. He was born on 25 July 1935 in Lubbock, north-central Texas. His father was an oilfield engineer. The McCoy family moved to Tulsa (Oklahoma) when young Jack was five, where he developed an interest in the outdoors, especially hunting and fishing. However, it was his Tulsa high school science teacher, the herpetologist and ornithologist Edith R. Force (see *Note*), who interested him in reptiles. McCoy went to Oklahoma State University to study wildlife ecology and ended up studying lizard systematics (B.S. 1957, M.S. 1960). For his doctorate he became a student of T. Paul Maslin at the University of Colorado where he wrote a dissertation on the natural history and ecology of a bisexual species of whiptail lizard (Ph.D. 1965). During his time at Colorado, McCoy participated in numerous field trips to Mexico, which became the main geographic focus of his career.

In 1964, prior to completing his doctorate, McCoy took a position in the Section of Amphibians and Reptiles at the Carnegie Museum in Pittsburgh (Pennsylvania). This museum had a reputation in herpetology because of the work of two herpetologists: M. Graham Netting, former section curator and later museum director, and Neil D. Richmond, section curator. McCoy became assistant curator in 1965, associate curator in 1970, and curator-in-charge in 1972 when Richmond became Curator of Environmental Studies. McCoy continued to do extensive fieldwork in Mexico and also in Belize, and he made shorter excursions to southern South America and Algeria. He built the collection aggressively, particularly with respect to Middle America. During his nearly 30-year tenure at the museum, the herpetological collections more than doubled to about 180,000 specimens. He built the freshwater turtle collection into one of the world's largest. Besides his generous service to the herpetological community as curator of one of the largest collections in the United States, McCoy trained many students in curatorial methods, especially students from overseas. He also served the major herpetological societies in various responsible positions and was president of two of them (Society for the Study of Amphibians and Reptiles, Herpetologists' League). McCoy died of a heart attack at his home in Penn Hills, a Pittsburgh suburb, on 7 July 1993. He was not quite 58 years old.

McCoy's research program was broad and covered systematics, distribution, biogeography, life history, and reproductive biology. He authored more than 160 titles on amphibians and reptiles (1958–2001) and named a dozen new taxa of lizards, turtles, and a salamander including two species of *Graptemys* from the Gulf Coast of the United States and an *Amphisbaena* from Argentina. He was interested in geographic variation in reproductive parameters, publishing important studies on

ovarian cycles and clutch size in lizards (1966, with Gerard A. Hoddenbach) and sex-determining temperatures in turtles (1982, with James J. Bull and Richard C. Vogt). His major works include a short textbook, "Vertebrates" (1968), an illustrated key to the amphibians and reptiles of Mexico (1979, with Gustavo Casas Andreu), "Amphibians and Reptiles in Pennsylvania" (1982), which includes distribution maps for each species, and a monograph on the ecogeography of the Mexican Bolson Tortoise (*Gopherus flavomarginatus*), coedited with David J. Morafka (1988). McCoy's last book, published posthumously and co-authored with Arthur C. Hulse and McCoy's former museum assistant and successor, Ellen J. Censky, was "Amphibians and Reptiles of Pennsylvania and the Northeast" (2001), which would have benefited from McCoy's keen editorial skills.

• *References*: "Clareance [*sic*] J. McCoy Jr. 25 Jul 1935–6 Jul 1993," by J. K. McCoy et al., Bol. Soc. Herpetol. Mex., 6: 15–24, 1994; "Clarence J. McCoy 1935–1993," by M. G. Netting; C. Gans, Copeia, 1994: 840–842, 1994; "Biographical Sketch of C. J. McCoy: 1935–1993," by J. K. McCoy et al., Herpetologica, 50: 379–382, 1994; "In Remembrance of C. J. McCoy," ed. by S. D. Busack and E. J. Censky, Herpetol. Rev., 26: 65–66, 1995. • *Portrait* (1982): Courtesy Ellen J. Censky. • *Signatures* (above, 1960; below, 1985): Adler collection. • *Note*: Edith Rhoda Force Kassing (1890–1966) had been a student of Frank N. Blanchard at the University of Michigan Biological Station in the 1920s and published nine papers on reptiles during 1925–1936. She founded the Tulsa Audubon Society in 1935.

OREJAS-MIRANDA, Braulio R. (1933–1985).

Braulio Rubí Orejas-Miranda, a Uruguayan herpetologist, conservationist, and educator, was born in Montevideo on 19 February 1933. After receiving his teaching degree in 1952, he taught at several secondary schools in the city. He also served as a contract professor at universities in Venezuela (Universidad Central in Caracas, 1960–1961, during which time he also directed the city's zoological gardens) and in Colombia (Universidad del Valle in Calí, 1964; Instituto National de Ciencias in Bogotá, 1970). In 1969 he received the degree of licentiate in biological sciences from the Universidad de la República in Montevideo and then did postgraduate study at the Universidade de São Paulo in Brazil (1971–1973).

His association with the Museo Nacional de Historia Natural in Montevideo began in 1954 when Orejas-Miranda became a part-time preparator in botany. The museum's curator of herpetology, Miguel A. Klappenbach, chose him to participate in the museum's expeditions to the Mato Grosso of Brazil (1955) and the Río Caura of central Venezuela (1957). He soon began his research in herpetology but with Klappenbach's promotion to museum director in 1970, Orejas-Miranda took an increasing role for the herpetological collections.

Concurrent with these educational and research activities, Orejas-Miranda began his involvement with national and international programs in science policy and conservation. During 1967–1970 he had national responsibilities for primary and secondary school science education, but then he moved to Washington and the Organization of American States where, as a member of the Basic Sciences Unit (1971–

1982), he supervised a broad range of research projects throughout the Americas. In 1982 he returned to Montevideo and UNESCO's Regional Office for Science and Technology for Latin America and the Caribbean and its "Man and the Biosphere" program. By this time, most of his publications dealt with renewable resources and conservation. He died suddenly in Montevideo of a heart attack on 8 May 1985. He was 52 years old.

Orejas-Miranda wrote 45 publications and nearly all of his research contributions were on herpetology (24 titles, 1958–1977). His technical papers deal with snakes, particularly those of the family Leptotyphlopidae in which he described ten new species. He co-authored a semi-popular book on the amphibians and reptiles of Uruguay (1969) with Klappenbach. His most elaborate scientific work, however, and one of the classic contributions to Latin American herpetology, was the snake volume of the "Catalogue of the Neotropical Squamata" (1970; reprinted with addenda, 1986), co-authored with James A. Peters and Roberto Donoso-Barros. This two-volume book covers all Latin American lizards and snakes south of Mexico and contains keys and checklists in parallel Spanish and English texts.

• *References*: "Lic. Braulio R. Orejas-Miranda (1933–1985). Una Bio-Bibliografía," by A. Mones, Bol. Soc. Zool. Uruguay, ser. 2, 3: 64–66, 1985; "In Memoriam [Dr. Braulio Orejas-Miranda]," by E. Gudynas, Herpetol. Rev., 16: 98–99, 1985. • *Portrait*: Courtesy Alvaro Mones. • *Signature* (1967): From Mones, 1985.

HILLENIUS, Dick (1927–1987).

Dirk Hillenius, generally known as Dick Hillenius, a Dutch herpetologist, poet, and essayist, specialized on chameleons. He was born in Amsterdam on 29 May 1927. As a boy his interests were in music and biology. He attended the University of Amsterdam and in 1951 became head of the public displays in the university's Zoological Museum. In 1954 he became curator of the museum's herpetological section. At the same time, he began his doctoral research on differentiation within the genus *Chamaeleo* and received his degree under Hendrik Engel (Ph.D. 1959).

Although chameleons became Hillenius's primary research focus, he also published essays and poetry beginning in 1961. Among his many titles are (in translation): "Against Vegetarianism" (1961), "The Mechanics of Romanticism" (1969), and "A Small Machine to Fight Consistency" (1975). As these titles suggest, Hillenius had a contrarian's point of view. He also had an infectious enthusiasm that made his writing popular to a wide Dutch audience. He often worked amphibians and reptiles into his writing, particularly chameleons, toads, and frogs, but also treated Darwinian theory in a light-hearted way. He was a regular and popular guest on Dutch television. At the same time, he was a serious curator of a modest-sized herpetological collection with important historical materials, especially from the former Dutch East Indies (now Indonesia) and Surinam, which had been studied by former staff members, L. D. Brongersma, J. K. de Jong, P. N. van Kampen, and Nelly de Rooij. The museum also obtained the important collections of amphibians made by Rose M. A. Blommers-Schlösser and specimens from Hillenius's own collecting, especially chameleons. He died in Amsterdam on 4 May 1987.

Hillenius authored about 15 papers on herpetology (1959–1988). Although he also published on the distribution of *Salamandra* and the body-form of frogs, most of his papers were on chameleons including their systematics, osteology, and behavior. His approach to systematics was phenetic. He named *Calumma guibei* (1959) from Madagascar and *Chamaeleo intermedius* from the Miocene of Kenya (1978). He also published reviews of the chameleons of Arabia (1966 and 1984, the latter with John Gasperetti) and North Africa (1978).

• *References*: "In Memoriam: Dick Hillenius," by W. Bergmans, Lacerta, 45: 128, 1987; "Dick Hillenius (29 May 1927–4 May 1987)," by J. W. Arntzen and A. Zuiderwijk, Amphibia-Reptilia, 8: 305–306, 1987; "Herinneringen aan D. Hillenius," collected papers in Hollands Maandbl., 1987(5–6): 1–30, 1987; "Morphological and Allozyme Studies of Midwife Toads (Genus *Alytes*), Including the Descriptions of Two New Taxa from Spain," by J. W. Arntzen and M. García-París, Contr. Zool., 65: 5–34, 1995 (Hillenius on pp. 27–28). • *Portrait*: Foundation for the Production and Translation of Dutch Literature, Amsterdam. • *Signature* (1965): Natural History Museum, London.

SZCZERBAK, Nikolai N. (1927–1998).

The Ukrainian vertebrate zoologist, Nikolai Nikolaevich Szczerbak (or Shcherbak), who studied the systematics, ecology, and conservation of Palearctic amphibians and reptiles, was born in Kiev on 31 October 1927. As a schoolboy, his first encounter with nature was in the Kiev suburbs. After the outbreak of World War II, his family was evacuated to eastern Russia where he observed the fauna of the dry steppes, or grasslands, along the Volga River. On returning to Kiev in 1944, his father, who was an engineer, insisted that he attend the Kiev Polytechnic Institute, but Nikolai had no interest and switched to Kiev State University to study biology. In 1948, soon after returning from the Carpathian Mountains of western Ukraine on his first expedition, he was arrested. He had been slandered by someone who claimed that Szczerbak had overheard, but did not report, some counterrevolutionary remarks. He was sentenced to seven years of hard labor for "betrayal of his motherland." No appeal was permitted. He served his sentence at a camp in Mordova, a heavily wooded region in the Middle Volga Valley where he cut trees. He contracted tuberculosis and, following his recovery, he became a technician in the camp infirmary. After Joseph Stalin died in March 1953, Szczerbak requested that his case be reviewed and he was released in December 1954. He was kept under surveillance until 1992 when he was finally and fully exonerated.

Szczerbak resumed his studies at the university in 1955 and graduated two years later. It was during this period that he met Sergius A. Chernov, a Ukrainian who was curator of the country's largest herpetological collection, at the Zoological Institute in Leningrad (now St. Petersburg). Szczerbak, whose main interest previously had been birds, soon focused on herpetology and thereafter regarded Chernov as his teacher. He monographed the herpetofauna of the Crimea as the thesis for his candidate degree at the Academy of Sciences in Kiev (C.Biol.Sci. 1962), which was defended in Leningrad.

Since 1957, Szczerbak was on the staff of the Zoological Institute of the Academy of Sciences in Kiev and in 1966 he was appointed director of its Zoological Museum, a post he held for life. Previously, the museum consisted of dusty cabinets lining a few hallways, but he built the collection into a respectable institution for research and public display. Many new specimens were collected for the museum on Szczerbak's numerous expeditions, which took him to the Crimea, the Carpathians, Caucasus, Middle Asia, Siberia, and the Far East as far as the Kuril Islands between the Kamchatka Peninsula and Japan, all Soviet territory. Field trips outside the country were virtually impossible, especially given his status as a former political prisoner, but through a little sleight of hand he even visited his long-dreamed-of Mongolia. Szczerbak discovered that part of the Tuva region of Siberia was only accessible by cutting through northwestern Mongolia. Permission could be granted at the Tere-Khol' border station. Then he and his team would conveniently get "lost" in Mongolia, collecting all the time, before returning to the motherland!

During this period Szczerbak was working on his doctoral dissertation, a review of the taxonomically complex lacertid genus *Eremias*. Much of the research was completed at the Zoological Institute in Leningrad. The degree, however, was awarded by the Academy of Sciences in Kiev (Dr.Biol.Sci. 1971). He became a professor at Kiev State University in 1975, but long before that he had begun training students and by the end of his life some two dozen of them had completed their candidate's or doctor's degree under his supervision. Among them were Valery K. Eremtschenko of Kirghizstan; Mikhail Tertishnikov of Russia; Takhir Sattorov of Tadjikistan; and Irina B. Dotsenko, Mikhail L. Golubev, Evgeny M. Pisanetz, and Anatoly A. Tokar of Ukraine. In the 1980s Szczerbak had become active in efforts to conserve the Ukrainian flora and fauna. He was elected a full member of the Ukrainian Academy of Sciences in 1993. In 1980 he was diagnosed with cirrhosis of the liver, believed to be the result of his treatments for tuberculosis at the prison camp in which unapproved and later banned drugs were used. He died of liver cancer on 27 January 1998 in Kiev.

Szczerbak authored more than 250 titles. The vast majority of them were on amphibians and reptiles (1959–2003). He described new taxa—many lizards and a toad—from Afghanistan, Iran, Kirghizstan, Pakistan, Turkmenistan, Ukraine, and Vietnam. He wrote or co-authored a number of major works. His first book, "Herpetologia Taurica" (1966), the herpetofauna of the Crimea, was based on his candidate's thesis and his second, on the lacertid lizards of the Palearctic desert (1974), on his doctoral dissertation. In 1975, he issued a short book on African desert lacertids, a complement to his 1974 book, but based solely on the literature. In 1977, Andrei P. Bannikov, Ilya S. Darevsky, and several other colleagues including Szczerbak authored the second edition of an illustrated guide and key to the amphibians and reptiles of the Soviet Union. "Amphibians and Reptiles of the Ukrainian Carpathians" (1980), by Szczerbak and Mikhail I. Szczerban', a herpetologist from the Carpathian Mountains, is based on collections from the Carpathians made by Szczerbak beginning in 1948. His next book, "Gecko Fauna of the USSR and Contiguous Regions" (1986; English edition 1996) is the result of his collaboration with a former student, Mikhail Golubev. Szczerbak's collaboration with another student, Valery Eremtschenko, led to a book on the ablepharine skinks of the same region (1986). In September 1989, the 7th All-Union Herpetological Conference was held in Kiev and Szczerbak edited the volume of abstracts, entitled "Problems of Herpetology" (1989). Szczerbak's last book published during his lifetime (1993) was a volume of papers by himself and 15 of his students and former students on the biology of the widely distributed lacertid of the Russian Steppes, *Eremias arguta*. His last book, finished and edited by Golubev, is a color-illustrated guide to the reptiles of the eastern Palearctic, effectively the former Soviet Union, with keys, distribution maps, and short accounts of natural history (2003); it covers 54 genera and 161 species.

• *References*: "Nikolai Nikolaevich Szscerbak [*sic*] (1927–1998)," by I. S. Darevsky, Russian Jour. Herpetol., 5: 167–168, 1998; "Nikolai Nikolaevich Szczerbak," by K. Adler, Gekko, Niskayuna (New York), 1: 36–37, 1999; ["Nikolai Nikolaevich Szczerbak (1927–1998),"] by T. I. Kotenko, Vest. Zool., 36(5): 93–95, 2002; Additional information from Michael L. Golubev, 2000. • *Portrait* (about 1987) *and signature* (1989): Adler collection.

STORR, Glen M. (1921–1990).
Glen Milton Storr was a vertebrate zoologist who specialized on the taxonomy of Western Australian reptiles and the faunistics of Australian birds. He was born in Adelaide (South Australia) on 22 December 1921, the son of a master plumber and traffic inspector. As a boy he had an interest in birds. He attended city schools and after graduating from high school (1938) he joined the South Australian Lands Department in 1939 as a cadet assistant to land surveyors. This work took him into the field where his interest in natural history, especially birds, was rekindled. During World War II, he enlisted in the Australian Imperial Forces in 1942 and saw active duty in Queensland and New Guinea (1943–1945), which provided more field opportunities. Starting in 1946, he was a licensed surveyor for the lands department but in 1952 decided to switch to a career in biology. In 1953 he moved to Perth and enrolled at the University of Western Australia (B.Sc. 1958, Ph.D. 1960) where he wrote his dissertation on the nutrition of the quokka, a small kangaroo, under the supervision of A. R. Main. While doing his fieldwork as a graduate student, Storr took an active interest in reptiles and decided to shift his research program to study their systematics. He believed that basic alpha-level taxonomy must be completed in order to provide a sound basis for biological research.

In May 1962, Storr became assistant curator in the Department of Vertebrate Zoology at the Western Australian Museum, with responsibilities for all vertebrates except mammals. He brought with him, as his "dowry" he often said, his personal collection of 5,000 reptiles collected as a student. He was promoted to curator the next year and by 1965 his title changed to Curator of Ornithology and Herpetology. He never applied to be a senior curator because as a bachelor he felt he was paid enough already. He continued to conduct active fieldwork until 1970 when his lung function became seriously impaired owing to his chain smoking. In 1971 he nearly died and was unable to walk more than a short distance without being short of breath. Thereafter he redoubled his efforts to complete his research. Fortunately, his colleagues—Ronald E. Johnstone and Lawrence A. Smith—were able to do the fieldwork and collaborate with him on his papers and a series of field guides. Other staff members, John Dell and Richard How, as well as many local amateurs who Storr mentored, also collected for him. Storr retired from his curatorship in 1986 and continued his research at the museum.

Western Australia is Australia's largest state and comprises one-third of the country's territory. Previously the only person to study its reptiles was Storr's predecessor as curator, Ludwig Glauert, but his publications did not reveal the enormous diversity of reptiles. Storr's fieldwork, together with that of Johnstone, Smith, the American ecologist Eric Pianka, and others, radically changed our understanding of that diversity. The reptile collection at the museum in Perth grew from 12,000 specimens in 1962, when Storr became curator, to 100,000 in 1990. He authored about 120 herpetological titles (1960–1999) and by 2000, of the 802 recognized species of Australian squamates, he had named an astonishing 172 of them (21.5%). Even in some very large genera he had described more than half of the species (*Ctenotus*, 54 of 92 species; *Lerista*, 42 of 79). Although most of his work was at the species level, he named the agamid genera *Caimanops* and *Pogona* and the scincid *Ctenotus*. By the time of his death in Perth on 26 June 1990, he had published or written a series of field guides to the reptiles of Western Australia that will be a lasting monument to his original goal to settle the taxonomy of the state's reptiles. The reptile volumes, all co-authored with Smith and Johnstone, are: "Lizards of Western Australia" ("I. Skinks," 1981, revised 1999; "II. Dragons and Monitors," 1983; "III. Geckos and Pygopods," 1990) and "Snakes of Western Australia" (1986, revised 2002). The planned volume on turtles and crocodiles was never published.

• *References*: "Dr. G. M. Storr," by R. E. Johnstone, West. Austr. Nat., 18: 139–147, 1990; "Glen Milton Storr 22 December 1921–26 June 1990," by P. F. Berry, Rec. West. Austr. Mus., 15: 1–12, 1991. • *Portrait* (1986): Western Australian Museum. • *Signature* (1962): Adler collection.

KLAPPENBACH, Miguel A. (1920–2000).

Miguel Angel Klappenbach Childe, a Uruguayan herpetologist, malacologist, and historian of science, was born on 4 May 1920 in Dolores (Soriano), Uruguay. His family lived in many regions of the country because of his father's job as a bank official and these locations fostered his interest in natural history. He followed his father into banking but also became a teacher of natural history at secondary schools. Eventually he came to live in Artigas where he taught school and was a civil employee at the government's insurance bank.

In 1955 he moved to Montevideo to become an assistant in zoology and then curator of the herpetology and malacology collections at the Museo Nacional de Historia Natural. He soon organized two major expeditions—to the Mato Grosso of Brazil (1955) and to the Río Caura of central Venezuela (1957)—where he led large teams of zoologists. He held a Guggenheim Foundation fellowship during 1967–1968, which allowed him to conduct research at various natural history museums in eastern United States. On his return to Montevideo in 1968 he received the degree of licentiate in biological sciences from the Facultad de Humanidades y Ciencias, Universidad de la República.

In 1970 Klappenbach became director of the natural history museum, a post once held by Carlos Berg and Garibaldi Devincenzi, both of whom also published on amphibians and reptiles. Klappenbach served as director until 1984. During and after this period, he held a number of important national and international leadership positions in Uruguay and other countries. Concurrent with his museum post, he was dean of the Facultad de Humanidades y Ciencias (1974–1984) and it was during his deanship that professional degree programs in anthropology, geology, and biological oceanography were initiated. After retiring as director and dean in 1984, he continued his research at the museum until his death in Montevideo on 25 January 2000.

Klappenbach published about 75 titles during his career, 11 of them on herpetology (1960–1992). These include descriptions of new species of frogs and reptiles, historical reviews of the herpetofauna of Uruguay, and conservation of the country's amphibians and reptiles. His major works are a semi-popular and well-illustrated guide to the amphibians and reptiles of Uruguay (1969), written together with Braulio R. Orejas-Miranda who wrote the section on reptiles, and a checklist with synonymies and distribution of the amphibian fauna of Uruguay (1992), co-authored by José A. Langone. One of his final works was a masterful study (1997), with Alvaro Mones, of Félix de Azara (1742–1821), the Spanish naturalist who explored the Río de la Plata region and wrote the earliest summary of the reptiles of Paraguay (1801).

• *Reference*: "Miguel Angel Klappenbach 1920–2000 Ensayo Bio-Bibliográfico," by A. Mones, Publ. Extra, Mus. Nac. Hist. Nat. Anthropol., 51: 1–17, 2001. • *Portrait* (about 1970): Courtesy Alvaro Mones. • *Signature* (1962): Courtesy Jorge Williams.

MORESCALCHI, Alessandro (1939–1997).

Alessandro Morescalchi, an Italian cytogeneticist, was a specialist on amphibian chromosomes and their implications for understanding the evolution of the Amphibia. He was born on

16 August 1939 in Viareggio (Lucca) on the Tuscan coast just northwest of Pisa. He received his doctorate at the university in Pisa (Ph.D. 1961) and the next year accepted a professorship in histology and embryology at the Università Studi degli di Napoli where he began his first work on amphibian chromosomes. He became a full professor at Naples in 1975 and ten years later moved to the Università Studi degli di Genova as Chair of Histology and Embryology and, from 1989, also Director of the Institute of Comparative Anatomy. Although most of his research focused on amphibian (and occasionally reptilian) chromosomes, in his last years, as part of the international "Antarctica Project," he made major contributions to the karyology of notothenioid fishes. He died in Genoa of cancer on 16 March 1997.

Morescalchi published numerous titles on herpetological subjects over the period 1963–1996. His studies on amphibian chromosomes confirmed the monophyly of the Lissamphibia (1971) and showed that karyological evolution in the three orders of amphibians followed the same path, from bimodal and asymmetrical chromosome sets to unimodal and symmetrical ones (1971). He studied the evolution of genome size (1975) and repetitive DNA sequence families in salamanders (1974) and noted the parallelism among large genomes, large amounts of repetitive DNA, and paedomorphosis (1975). The results of his studies reveal interesting correlations between variations in certain DNA fractions and some cellular and metabolic parameters that may play a significant role in amphibian phylogeny and their natural history (1975).

Morescalchi sponsored several doctoral students, among them Teresa Capriglione, Gaetano Odierna, and Ettore Olmo of herpetological interest. His daughter, Maria Alessandra Morescalchi, also studies amphibian chromosomes.

• *References*: "Alessandro Morescalchi 16 August 1939–16 March 1997," by E. Olmo, Chrom. Res., 5: 281–282, 1997; "Alessandro Morescalchi 16 Agosto 1939–16 Marzo 1997," by E. Olmo, Italian Jour. Zool., 64: 391–392, 1997 (1998). • *Portrait*: From Olmo, 1997, p. 391. • *Signature* (1966): Adler collection.

ÁLVAREZ, Ticul (1935–2001).

José Ticul Álvarez Solórzano, a Mexican mammalogist, herpetologist, and archeozoologist, was born in Mexico City on 26 February 1935. His father was an ichthyologist. Ticul completed college at the Escuela Nacional de Ciencias Biológicas of the Instituto Politécnico Nacional in Mexico City (IPN; Licentiate 1959) and then pursued graduate studies at the University of Kansas in the United States (M.A. 1962) under the direction of the mammalogist, E. Raymond Hall. From 1963 Álvarez was a faculty member in the Departamento de Zoología of the IPN where he remained for the rest of his career. He founded the herpetological collection in his department, which has since become one of the largest and most important museum collections of amphibians and reptiles in Mexico. He also founded the Laboratorio de Paleozoología (today re-named the Labora-

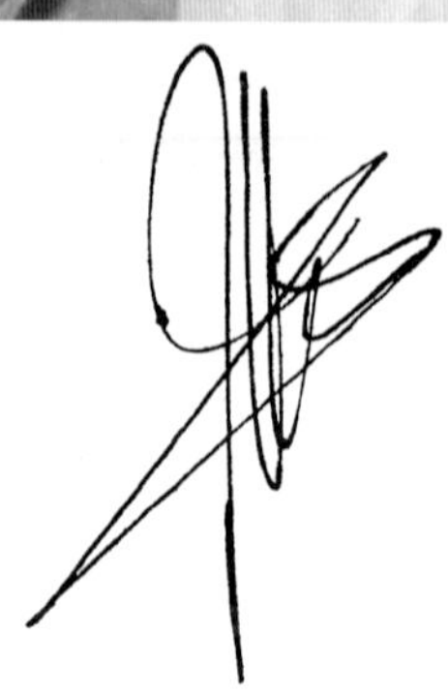

torio de Arqueozoología "M. en C. Ticul Álvarez Solórzano" at the Instituto Nacional de Antropología e Historia in Mexico City). This recognizes his important contributions to the zoology of prehistoric Mexico, a subject on which he published extensively. During the 1970s, he was a technical advisor to the wildlife branch of the government's department of agriculture during which he influenced policy regarding protection of Mexico's native fauna. In 1980, after a long struggle with diabetes, he lost his eyesight, but his professional work continued. He served as a professor at the IPN until he retired in 1989. He died in Mexico City on 16 October 2001.

Álvarez authored about 25 papers devoted specifically to amphibians and reptiles (1964–2002) plus several others on terrestrial vertebrates in general. His primary interests were in taxonomy, morphology, distribution, and biogeography. He described a new species of salamander (*Pseudoeurycea*) and two new lizards (*Lepidophyma*) from southern Mexico, wrote several papers on geographic variation, distribution, and appendicular skeleton of a two-legged amphisbaenian (*Bipes*), and reported on the bony remains of amphibians, reptiles, and other vertebrates from numerous archeological excavations made throughout Mexico. His major work, which includes a section on amphibians and reptiles, was the volume "Fauna" in the series "Atlas Cultural de México" (1987, co-authored with Manuel González E.), which summarizes Mexico's prehistoric fauna.

• *References*: "Las Contribuciones en Herpetología de Ticul Álvarez Solórzano," by O. Flores V. and H. M. Smith, p. 35–68 *in* J. Arroyo-Cabrales and O. J. Polaco (coords.), Homenage a Ticul Álvarez. Inst. Nac. Antropol. Hist., Colec. Científ., Mexico, 1997; "Ticul Álvarez Solórzano (1935–2001)," by S. Murillo and O. Flores V., Bol. Soc. Herpetól. Mex., 10: 65–69, 2002; "José Ticul Álvarez Solórzano: 1935–2001," by S. T. Álvarez-Castañeda, Jour. Mammal., 84: 762–770, 2003. • *Portrait*: From Álvarez-Castañeda, 2003. • *Signature* (1970): Adler collection.

Đào, Văn Tiến (1920–1995).

Văn Tiến Đào, a Vietnamese vertebrate biologist and patriot, was born on 23 August 1920 in Nam Dinh (Nam Dinh), about 75 km southeast of Hanoi. His father was a schoolteacher. At that time Vietnam was a French colony and in 1940, with the cooperation of the collaborationist Vichy government, Japanese troops occupied the country. Near the end of World War II, Japan allowed the Vietnamese to proclaim independence (March 1945), but not to establish a viable regime. During the war, Đào attended the French-run Université Indochinoise in Hanoi, specializing in zoology (B.S. 1944, M.S. 1946). He was a student of Boris Noyer, director of the Laboratoire de Biologie et de Médecine Expérimentales. They conducted physiological and pharmacological research in which frogs and turtles were sometimes used. René Bourret, who spent his career studying the herpetofauna of Indochina, had retired from the university in 1940, yet he collaborated with Noyer and Đào in some of this research.

When the French attempted to regain control of Vietnam after the war, conflict broke out with the Vietnamese in December 1946, and Đào served as a soldier in the guerrilla movement against the French. In 1951 he and some other scientists established a pedagogical institute in his home province. The war ended in July 1954, and by 1956 Đào was back in Hanoi where he taught at Université Indochinoise, later renamed Hanoi State University. Most of his research dealt with mammals on which he published about 85 titles (1960–1994) including his major monograph, "Scientific Results of Mammal Surveys in North Vietnam" (1985). He also described ten new taxa of Vietnamese mammals. Đào was promoted to professor in 1980 and retired in 1987. He died in Hanoi on 3 May 1995.

Đào's herpetological work was incidental to his research on mammals. He published only nine papers on amphibians and reptiles (1965–1982), but these ushered in the modern era of Vietnamese herpetology. Unlike the works of Bourret, Đào's were in Vietnamese. Most importantly, he authored a series of five papers (1977–1982) that together represent the first checklist and key to all of the amphibians and reptiles of the entire country. From this modest beginning and aided by collaborations with overseas herpetologists, in particular those of the former Soviet Union, the current high level of herpetological research activity in Vietnam has emerged.

• *References*: ["Memorial of Professor Đào Văn Tiến,"] by G. V. Kuznetsov, p. 283–285 *in* L. P. Korzun and M. V. Kalyakin (eds.), Materials of Zoological and Botanical Studies in Fan Si Pan Summit Area (North Vietnam)," Russian-Vietnamese Tropical Centre, Moscow–Hanoi, 1998; "Herpetological Exploration, Research, and Conservation in Vietnam," by K. Adler, p. 33–56 *in* V. S. Nguyen, T. C. Ho, and Q. T. Nguyen, Herpetofauna of Vietnam, Ed. Chimaira, Frankfurt am Main, 2007 (Đào on pp. 46–48); Additional information from Vu Khoi Le and Đào's son, Hai Đào, 2006. • *Portrait*: Courtesy of Hai Đào and Quang Truong Nguyen. • *Signature*: Courtesy Vu Khoi Le.

Gasperetti, John (1920–2001).

The American surveyor and oilfield engineer, John Gasperetti, who devoted his life to studying the fauna, especially the her-

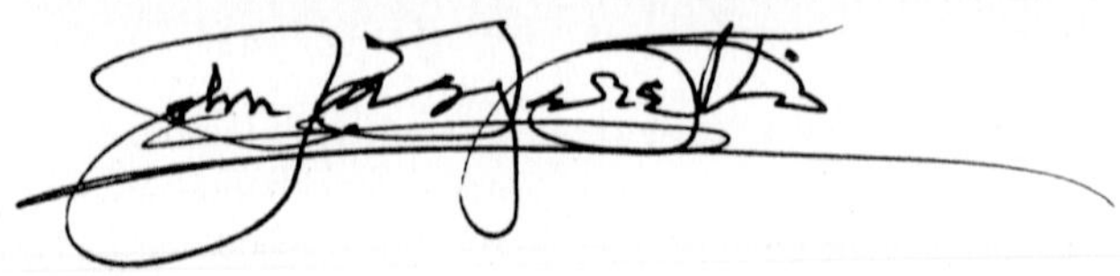

petofauna, of Arabia, was born in Walsenberg in south-central Colorado on 24 October 1920. In 1938 he graduated from Huerfano County High School in Walsenburg. During the Great Depression, college was not an option for him, so he joined the federal Civilian Conservation Corps where he learned the basics of surveying and later worked for construction and engineering firms. With America's entry into World War II in December 1941, Gasperetti attempted to enlist, but the uniformed services rejected him because of poor sight in one eye. Instead, the War Manpower Commission sent him to Saudi Arabia in 1944 to work in the oilfields run by the Arabian American Oil Company, thereby introducing him to Arabia and its wildlife. From then until 1986, he worked in many places in Southwest Asia—Afghanistan, Iraq, Kuwait, Lebanon, and Saudi Arabia—under contract with private and government-owned oil companies and Saudi ministries including the Getty Oil Company. He developed a particular friendship with J. Paul Getty, Sr., and traveled with him widely in the Middle East. Getty chose him to be his special representative at a January 1961 conference to determine the boundaries of the Kuwait–Saudi Arabia Neutral Zone. Throughout these years Gasperetti collected specimens, explored the environment, and studied his collections. He began to publish the results in 1967.

In 1986, Gasperetti gave up surveying and engineering to work for the Meteorology and Environmental Protection Administration (MEPA), a division of the Saudi Ministry of Defense and Aviation, located in Jeddah on the Red Sea. This post was more aligned with his interests in wildlife and gave him additional time to prepare his major works for publication. However, his contract with MEPA ended in October 1994. He was about to sign a new contract with them when he suffered a stroke in May 1996 that left him partly disabled. He and his wife, Patricia R. Gasperetti, moved to Egypt in December 1996 to work on marine turtles in the Red Sea, but after a few months another stroke forced his retirement. He died in Cairo on 10 December 2001.

Gasperetti authored ten papers on amphibians and reptiles (1967–1994) plus some on other wildlife and a book, "The Wildlife of Arabia" (1981, with his wife). Decades of his personal collecting, visits to museums in London and elsewhere, and careful study of these materials as well as his intimate knowledge of the Arabian environment led to a series of preliminary papers and five detailed monographs in the "Fauna of Saudi Arabia" series. The first monograph, on the chameleons of Arabia, appeared in 1984 and was co-authored with the Dutch chameleon specialist, Dick Hillenius. The next year, Gasperetti, together with the Italian herpetologists, Emilio Balletto and Maria Cherchi, published "Amphibians of the Arabian Peninsula." In 1988, Gasperetti alone authored his largest monograph, a 280-page treatise on the snakes. "Turtles of Arabia" was issued in 1993, authored by the Gasperettis and the British, American, and Australian herpetologists, Andrew F. Stimson, Jeffrey D. Miller, and J. Perran Ross, respectively. Finally, in 1994, Gasperetti's last publication, a general contribution to Arabian herpetology, was issued, co-authored with the Swiss herpetologist, Beat Schätti. These monographs included literature reviews, extensive synonymies, distribution maps, illustrations, and the other extreme detail and documentation that became Gasperetti's trademark. Unfortunately, his manuscript on the remaining lizards was not finished; it is being completed for publication by Miller.

• *References*: "Middle Eastern Herpetology Loses John Gasperetti, 1920–2001," by Y. L. Werner, Zool. Middle East, 27: 5–6, 2002; Additional information from Patricia (Mrs. John) Gasperetti, 2006. • *Portrait* (1980): By Ingrid El-Khalidi, photographed by Gabriel Mikhail. • *Signature* (1990): Adler collection.

RUIZ-CARRANZA, Pedro M. (1932–1998).

Pedro Miguel Ruiz-Carranza, a Colombian biologist and specialist on frogs, was born in Bogotá on 3 February 1932. He spent his childhood at the parental farm in Togüí (Boyacá) on the western slopes of the eastern range of the Andes, about 155 km north-northeast of Bogotá. This is a region of unsurpassed landscapes and opportunities to learn about the natural world, but also one that became embroiled in social unrest and violence that eventually claimed the life of his father in 1959. Ruiz attended primary school in Togüí and began his secondary education in nearby Tunja at the Jesuit-run Colegio José Joaquín Ortíz, but because of increasing violence in the area, he had to transfer to the Colegio Hispanoamericano in Bogotá where he graduated in 1949. He then studied medicine at the Sorbonne (University of Paris) and in 1954 returned to Bogotá for additional work at the Universidad Nacional de Colombia. There, he took courses of study in systematic botany at the Instituto de Ciencias Naturales (ICN; completed in 1957) and graduated in bacteriology at the Facultad de Medicina (1959).

In the meantime, Ruiz became a faculty member at the university, first in medicine (1958–1965), then in the biology department (1965–1975), and finally in the ICN, where he became chief of zoology (1975–1982) and then director of the ICN–Museo de Historia Natural (1982–1986, 1994–1996). He was largely responsible for building the museum's collection of amphibians and reptiles into Colombia's largest. His own herpetological work only began in 1965 and was largely self-taught. By 1968 he had already organized and run, in Bogotá, an international symposium on the evolution of frogs. In 1983 he created an extensive, multi-institutional study of the country's fauna—Programa Fauna de Colombia—and served as its national coordinator and driving force during 1985–1993. He trained a large number of the current generation of Colombian herpetologists. Ruiz retired as professor in 1996 but actively continued his research. He died in Bogotá on 12 September 1998.

Ruiz authored about 60 herpetological papers (1976–1998), nearly all of them on frogs. This work began as the result of collaboration with a faculty colleague, Jorge I. Hernández-Camacho. Their first paper (1976) was a comprehensive study of diminutive *Atelopus*-like frogs that inhabit the high Andean páramos of Colombia and Ecuador, which they placed in their new genus *Osornophryne*. On the basis of their detailed study of these unusual frogs, which display inguinal amplexus and have peculiar skeletal and morphological features, they confirmed that the atelopodids properly belong in the Bufonidae. Subsequently, Ruiz produced numerous papers on Colombian frogs in which dozens of new species were described, mostly in the families Centrolenidae, Dendrobatidae, and Hylidae, and in the genus *Eleutherodactylus*. Most of these were co-authored with his colleagues at ICN, especially Hernández and María Cristina Ardila-R., the latter briefly his successor in charge of amphibians at the museum, and the American herpetologist, John D. Lynch, who has been curator of these herpetological collections since 2000. Ruiz wrote other papers on various aspects of anuran natural history including behavior, reproduction, tadpoles, and morphology.

• *Reference*: "Pedro Miguel Ruiz-Carranza," by M. C. Ardila-R., Revta. Acad. Colomb. Cienc. Exact. Fis. Nat., 23(Supl. Espec.): 13–39, 1999 (2000). • *Portrait and signature*: From Ardila, 1999 (2000).

RASMUSSEN, Jens B. (1947–2005).

Jens Bødtker Rasmussen, a Danish specialist on African snakes, was born in Fredriksberg, near Copenhagen, on 16 April 1947. Even as a young student he had an interest in snakes and kept them in terraria. He received his higher education at the University of Copenhagen (Cand.Scient. 1975 and Lic.Scient. 1977, the latter roughly equivalent to Ph.D., under the supervision of Frits Bræstrup). He also received his Dr.Sc. degree there in 1994 for a thesis on African rear-fanged snakes, long after having established himself as a leading authority on African snakes. Rasmussen was appointed Curator of Reptiles and Amphibians in the university's Zoological Museum in 1977. This is a venerable and important museum whose early curators included J. T. Reinhardt and Christian Lütken. The collection of the Royal Natural History Museum was merged into it in 1864. Rasmussen was also a teacher at the university and, at the time of his relatively early death from cancer, was an associate professor. He died on 3 May 2005 in Copenhagen at the age of 58.

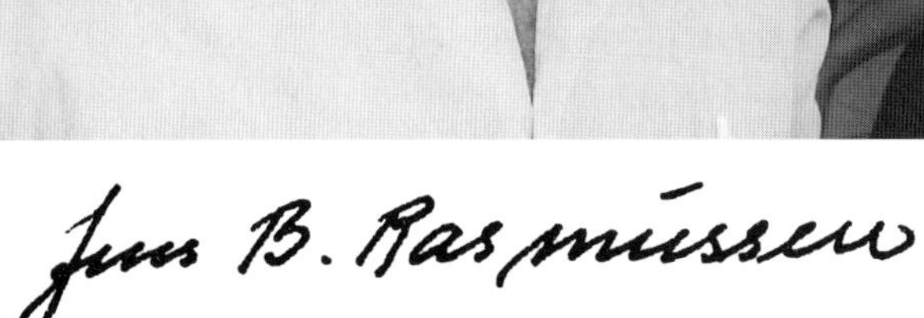

Rasmussen's research focus was the systematics, phylogeny, and biogeography of African snakes, although he also published on some Asian taxa. He authored about 60 titles on amphibians and reptiles (1969–2005), mostly on snakes, plus 38 herpetological entries in the Danish national encyclopedia (1994–2001). He published major reviews of snakes in the genera *Adenorhinos, Atheris, Atractaspis, Boiga, Causus, Crotaphopeltis, Coluber, Dipsadoboa* (including *Chamaetortus*), *Geodipsas, Lycodonomorphus, Macroprotodon, Micrelaps, Psammodynastes, Psammophis, Pseudoboodon,* and *Telescopus*. These studies included several new taxa and detailed distributional data, and they used a wide variety of morphological characteristics. His largest work was a catalogue of the snakes of Ethiopia co-authored with his English colleague, Malcolm J. Largen (1993), former director of the museum in Addis Ababa. This monograph reviews, illustrates, and maps the 88 species known to inhabit that country, nine of which are endemic. This is a remarkable total in itself, but they identified 11 additional species likely to occur in Ethiopia. It had been Rasmussen's goal to produce an atlas of African snakes, complete with an analysis of biogeographic patterns and conservation priorities. This project is to be carried on by his colleagues at the museum.

• *References*: "Mindeord over Jens Bødtker Rasmussen," by H. Bringsøe, Nordisk Herpetol. Foren., 48(4): 127–128, 2005; "Obituary for Jens Bødtker Rasmussen (1947–2005)," by W. Böhme, Salamandra, 41: 161–165, 2005. • *Portrait* (1997): By Kraig Adler. • *Signature* (1967): Courtesy Richard Wahlgren.

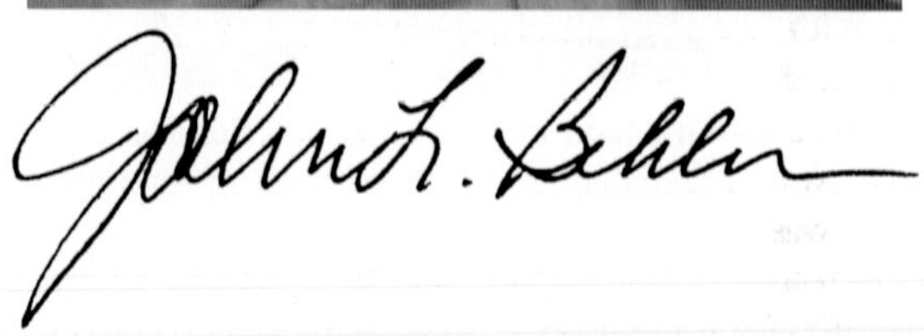

BEHLER, John L. (1943–2006).

The New York Zoological Society (now Wildlife Conservation Society), which operates the Bronx Zoo in New York City, has had a long line of curators of reptiles who influenced herpetology in the United States and abroad, beginning with Raymond L. Ditmars. John Luther Behler, Jr., who recently held that post, became an author of prominent books on amphibians and reptiles and played a key role in the conservation of reptiles—especially of turtles and crocodilians. He was born in Allentown, in eastern Pennsylvania, on 28 June 1943. Although he had a boyhood interest in snakes, frogs, and butterflies, he was expected to follow the family tradition of a career in dentistry until his father recognized his real passion and urged him to study zoology. He received an associate's degree at Mitchell College in Connecticut (A.S. 1964), a bachelor's at the University of Miami (B.S. 1966), and a master's at East Stroudsburg State College (now University) near his hometown (M.Ed. 1968). He taught at Hobart and William Smith Colleges in the Finger Lakes Region of central New York State (1968–1970). An academic life did not suit him, however, so he became an intern at the Bronx Zoo in 1970, Assistant Curator of Reptiles in 1973, and Curator three years later, succeeding F. Wayne King.

Behler had long had an interest in turtles and in conservation. He had worked as a summer naturalist for the National Park Service in 1968 and 1969 and banded ducks for the New York State Department of Environmental Conservation while teaching at Hobart and William Smith. At the zoo, conservation of reptiles became his priority and he quickly used the zoo's prestige and resources to have an impact in the state, the country, and overseas. He was instrumental in the creation of the Bog Brook Unique Area, a 55-ha fen and wildlife management area located near Brewster just north of New York City that was purchased by the state in 1981 specifically to protect bog, spotted, and wood turtles. At the time of his death, on 31 January 2006 at his home in nearby Amawalk (New York), Behler was still involved in local conservation efforts as co-leader of a cooperative project between the zoo and the National Park Service to conduct herpetological surveys in parks throughout the northeastern United States.

Behler also had an interest in conserving endangered reptiles throughout the world and no species was then considered more threatened than *Alligator sinensis*, of which fewer than 100 were believed to exist in the wild in China. The zoo had maintained a conservation interest in this species since the 1950s and unsuccessful efforts were later made to breed them in the Reptile House. Behler decided to try a different approach. In 1976, two pairs of alligators—one had been in captivity for 37 years and the other for 19, both without breeding—were transferred to special enclosures at the Rockefeller Wildlife Refuge in the coastal marshes of southwestern Louisiana. The first young was produced in 1977 and by 1980 two dozen had hatched. Behler maintained the studbook for this species, the first ever established for a crocodilian, in order to create a registry to safeguard genetic diversity, and he was still the species coordinator at the time of his death.

In 1991 Behler became the chairman of the Tortoise and Freshwater Turtle Specialist Group of the International Union for the Conservation of Nature. He soon drew attention to the impending crisis caused by China's trade in chelonians for their use as food and to prepare traditional medicines. Turtles and tortoises were being imported from as far away as Indonesia and even the United States. Chinese were using specially trained dogs to find specimens, thus capturing a large fraction of the breeding population and pushing several species to the brink of extinction. Behler argued for the opening of turtle-rearing farms in China to supply the growing commercial market. Today it is estimated that up to 80 percent of all turtles sold at market are captive bred.

As a result of his international efforts, Behler traveled overseas regularly, even at the end of his life when the symptoms of congestive heart failure were becoming apparent. Although he visited regularly throughout Latin America, Australia, and Africa, most of his trips were to Madagascar (7 visits, 1990–2001) and to China, India, Japan, and Southeast Asia (15 visits, 1978–2003), often into rugged field sites. He published about 50 titles in herpetology, mostly popular in nature and beginning in 1970. He also authored or co-authored seven books on amphibians and reptiles including the widely used "Audubon Society Field Guide to North American Reptiles and Amphibians" (1979, jointly with King). Smaller guides followed in 1988, 1989, and 1999, with specialty books on crocodilians (1998), snakes (2002), and frogs (2005), the last two written with his wife, Deborah A. Behler. Another book, on the reptiles and amphibians of New York State that was co-authored with five colleagues, was published posthumously (2007).

• *References*: "John L. Behler, Protector of Reptiles, Is Dead at 62," by J. Pearce, New York Times, sect. B, p. 34, 5 February 2006; "John Luther Behler (1943–2006)," by P. Brazaitis and B. [=W.] Holmstrom, Herpetol. Rev., 37: 137–138, 2006. • *Portrait and signature* (2006): Wildlife Conservation Society, courtesy William Holmstrom.

DILMUKHAMEDOV, Marat (1940–1994).

The comparative morphologist, Marat Eltokovich Dilmukhamedov, studied the integument and digestive tracts of amphibians and reptiles. He was born on 9 April 1940 in Tashkent, capital of Uzbek SSR, then part of the Soviet Union. After World War II, Dilmukhamedov's family—his father was a professor of history and his mother a physician—moved to Alma-Ata (now Almaty), then the capital of neighboring Kazakh SSR. He attended Kazakh State University in Alma-Ata where he earned a diploma in biology (graduated 1964). He then began his graduate studies there with the morphologist, Bronislav A. Dombrovsky, studying the internal anatomy of fishes. After completing his military service, Dilmukhamedov returned to post-graduate school to complete his candidate's thesis on the comparative morphology of the digestive tract of some reptiles (C.Biol.Sci. 1975). The next year, he became an assistant in the zoology department at the university. During 1980–1982 he taught biology at the university in Tuléar (now Toliara), on the southwestern coast of Madagascar where he was able to make collections for his future research. He returned to the university in Alma-Ata and in 1983 became a lecturer in the department where he taught and supervised students. He died in Almaty on 21 March 1994.

Dilmukhamedov studied the morphology of amphibians and reptiles and authored about 15 papers on the subject (1972–1994). He was interested in the alimentary canal of reptiles especially those of lizards and turtles, and the systematic implications of these structures. He also studied the skin, especially the secretory and sensory organs. Together with Natalia Ananjeva in St. Petersburg and a colleague in Alma-Ata, Tatjana N. Matveeva (now Dujsebayeva), he co-authored an important study of skin sensory organs in iguanian lizards (1991) including representatives of both genera of Madagascan iguanids. Building on earlier work by Giuseppe Scortecci, Ernest Williams, and others, they found cutaneous receptors of several types (cells with single or multiple hairs and cells without hairs) using histology and scanning electron microscopy. He also studied the morphology of a hynobiid salamander (*Salamandrella keyserlingii*), focusing on the morphology and development of the digestive tract and the structure of the skin including its melanin pigmentation (3 papers in 1994).

• *Reference*: "Obituary Marat Dilmukhamedov (1940–1994)," by T. N. Duisebaeva (=Dujsebayeva) and N. B. Anan'eva (=Ananjeva), Russian Jour. Herpetol., 1: 185–186, 1994 (1995). • *Portrait* (1985) *and signature*: Courtesy of Tatjana Dujsebayeva.

KUBYKIN, Rudolf A. (1937–2001).

The vertebrate zoologist, Rudolf Alexandrovich Kubykin, studied the distribution, ecology, and conservation of amphibians and reptiles in Kazakhstan. He was born on 16 August 1937 in the village of Koturkul (Szczuzhinskii District, Kokchetavskaya Province) in the northernmost part of Kazakh SSR, then part of the Soviet Union. After secondary school and service in the Soviet Army (1956–1959), he moved to Alma-Ata, capital of present-day Kazakhstan. Beginning in 1965 he attended Kazakh State University in the evenings to study biology (graduated 1971). In the meantime, he had been employed at the Institute of Zoology of the Academy of Sciences in Alma-Ata where he worked for the rest of his life. He started in the ornithology laboratory in 1969, and also collected some amphibians and reptiles. That same year, Ilya Darevsky of the Zoological Institute in Leningrad (now St. Petersburg) invited Kubykin to participate in joint surveys in different parts of Kazakhstan and by 1972 herpetology became his main research focus. Kubykin's fieldwork was almost entirely in Kazakhstan and included the Mangyshlak Peninsula in the western part of the country; the Kyzylkum Desert in the south; the desert region between Lake Balkhash and the Ili River; the Dzungarian Alatau Mountains in the southeast; and the Zaysan Depression and Lake Alakol in the east.

From the early 1980s, together with his colleague, Zoya Brushko, at the institute in Alma-Ata, Kubykin began a study of population numbers in a commercially exploited tortoise (*Agrionemys horsfieldii*), for which he was able to provide a basis for its use as a sustainable resource. During 1996–1999, he collaborated with Sergius Kuzmin from Moscow and two German herpetologists, Burkhard Thiesmeier and Hartmut Greven, to conduct a study of a highly endangered hynobiid salamander (*Ranodon sibiricus*). Their data on distribution and long-term population changes provided the basis for inclusion of this species in the "IUCN Red List." Kubykin also became active in international efforts to conserve amphibians and reptiles throughout Middle Asia. Besides his conservation studies, Kubykin researched the reproductive ecology of agamid and lacertid lizards, fossil amphibians and reptiles in Kazakhstan, the behavior of *R. sibiricus*, and the role of frosts and rain on breeding in toads of the *Bufo viridis* complex. All told, he published about 65 titles (1973–2005).

Together with Brushko, Kubykin was active in building the herpetological collections at the institute in Alma-Ata and together they co-authored a catalogue of these collections (1988), which are the largest of their kind in Kazakhstan. Kubykin conducted active fieldwork to the very end of his life, which ended in Alma-Ata on 1 August 2001 when he died from heart failure.

• *Reference*: "Rudolf Aleksandrovich Kubykin (August 16, 1937–August 1, 2001)," by Z. K. Brushko et al., Russian Jour. Herpetol., 9: 258–260, 2002. • *Portrait*: From Brushko et al., 2002. • *Signature* (1994): Courtesy Tatjana Dujsebayeva.

GIBBONS, John R. H. (1946–1986).

John Richard Hutchinson Gibbons, discoverer of the Fiji Crested Iguana, was born in Dunmow (Essex), England, on 3 June 1946. As a boy he was interested in butterflies, newts, and other animals. He attended Bloxham School in Banbury, Mid Essex Technical College, and Sheffield University (B.Sc. 1968) before spending a year studying chimpanzees at Gombe Stream in Tanzania with Jane Goodall. After graduate work at the University of Manitoba (M.Sc. 1971) and employment by the Fisheries Research Board of Canada, he moved to Australia for his doctoral studies on the ecology and behavior of agamid lizards (Ph.D. University of Adelaide 1978, but his writing was completed at Monash University in Victoria). From 1978 until his untimely death at the age of 40, he was a lecturer in biology at the University of the South Pacific in Suva, Fiji.

Gibbons was a zoologist with primary interests in ecology, behavior, and evolution. Although he published important theoretical work on sympatric speciation in insects (1979), Gibbons was more at home in the field. Fiji—a group of about 320 islands lying about 4,450 km southwest of Hawaii—presented him with just such an opportunity. Even though Fiji had been visited by several exploring expeditions during the 19th Century, they did not visit the smaller islands comprehensively. Gibbons started to explore the outlying islands shortly after his arrival in Fiji. On 4 January 1979 he visited Yaduataba, an uninhabited 70-ha islet lying between Fiji's two main islands, and found a large iguanid lizard quite unlike the more widespread *Brachylophus fasciatus*. This new animal, with females reaching a meter in total length, has a well-developed dorsal crest and several other distinguishing features. He soon found more on several nearby islets. Gibbons named it *B. vitiensis* (1981) and speculated that its ancestors rafted to Fiji from South

America some six million years ago before the Isthmus of Panama was formed and cut the Caribbean off from the Pacific Ocean. His rafting hypothesis for *Brachylophus* was supported by his finding that these iguanas have an exceptionally long incubation period (6–9 months).

Discovery of such a spectacular animal in the small nation of Fiji made headline news. Images of the animal soon appeared on the national telephone book cover and on T-shirts. However, populations were small and its habitat was threatened by large numbers of free-ranging goats put on the islands to fatten for slaughter. Being an ardent conservationist, Gibbons soon petitioned the government to remove the goats and set Yaduataba aside as a preserve—an unpopular move in a country that must import goat meat to satisfy the huge public demand. This was accomplished, but at the ceremony in August 1980 declaring Yaduataba the country's first wildlife sanctuary, Gibbons was not permitted to attend because his strong advocacy had alienated some government officials. Fortunately, he lived to see his iguana featured on the new Fiji 40-cent postage stamp issued in August 1985.

Gibbons continued to explore the Fiji island group and made other discoveries including two new species of skinks (1986) and a new gecko (1988) from Tonga. His other papers are on behavior, life history, and biogeography. In total, he authored 14 herpetological titles (1979–1988). Gibbons's career ended tragically on 15 November 1986 in a boating accident off the island of Lakeba, about 250 km east of Suva, in the Lau Group of Fiji. His wife and two young children perished with him when their small boat capsized as it passed through a reef and then was swept along its seaward side.

• *References*: "Dr John Gibbons Adventurous Scientist," anonymous, Times (London), p. 18, 3 December 1986; "Dr. John R. H. Gibbons," by W. Kenchington, South Pacific Jour. Nat. Sci., 8: 126–136, 1986. • *Portrait and signature* (1985): Adler collection.

Camarillo R., José L. (1955–2002).

José Luis Camarillo Rangel, a Mexican herpetologist, was born on 17 April 1955 (see *Note*) in Mexico City. He attended college at the Escuela Nacional de Estudios Profesionales–Iztacala (ENEP, now Facultad de Estudios Superiores Iztacala), a regional campus of the Universidad Nacional Autónoma de México (UNAM) located in Tlalnepantla on the northern outskirts of Mexico City. There, inspired by the teaching of Oscar Sánchez Herrera, he got interested in herpetology by keeping live reptiles as pets. He completed a thesis for his degree (Licentiate 1981) that compared the elevational distributions of amphibians and reptiles at localities in the Mexican states of Morelos and Mexico. His master's degree was earned at the main campus of UNAM in Mexico City (M.C., 1987) with Zeferino Uribe Peña. At the time of his premature death, he was a doctoral student of Gustavo Casas Andreu at UNAM, working on a dissertation concerning the biogeography and ecology of the Mexican herpetofauna.

Camarillo was a member of the modern generation of Mexican herpetologists who had a broad knowledge of amphibians and reptiles and of biology. He published over a wide range of topics: systematics, distribution (geographic and elevational), cytogenetics, evolution, reproductive biology, parasites, and conservation. At ENEP, where he taught from 1980, he founded the college's museum collection for herpetology. This began as the result of his privately funded fieldwork, mostly in central

Mexico. His collecting later extended east to Puebla and Veracruz and south to Guerrero and Oaxaca. He authored a major review of the state of Oaxaca's herpetofauna and its conservation in 1996. His primary area of interest was lizard biology. Together with others, he published a study on the population cytogenetics and evolution of the *Sceloporus grammicus* complex (1991) and co-authored major reviews of the genera *Diploglossus* (1994, with Jonathan Campbell) and *Lepidophyma* (2002, with Robert Bezy), and he described new species of each. He was responsible for the Spanish version of the first comprehensive list of the scientific and common names for the amphibians and reptiles of Mexico (1994, together with Ernest Liner). In total, he was the author or co-author of about 45 herpetological titles (1983–2004). Camarillo often published in international journals and collaborated with leading herpetologists in the United States on topics of fundamental biological importance, which gave his research an unusually broad impact in the field. His last paper, in *American Naturalist*, dealt with testing species boundaries in an ancient species complex using *Lepidophyma* as the model group.

Camarillo died on 28 May 2002 in Mexico City at the age of 47. He died by suicide after a long period of depression, self-isolation, and suffering from a kidney disease. Despite this tragic end, he had risen from his humble native Indian heritage to become one of the best-trained and most productive herpetologists in Mexico.

• *Reference*: "José Luis Camarillo Rangel (1955–2002)," by F. Mendoza Quijano, Bol. Soc. Herpetol. Mex., 10: 23–24, 2002. • *Portrait*: Courtesy Gustavo Casas Andreu. • *Signature* (1987): Courtesy Hobart M. Smith. • *Note*: Although Mendoza Quijano (2002) gave 27 April as Camarillo's birth date, 17 April is confirmed by Camarillo's family (*fide* Edmundo Perez Ramos, 2006).

SLOWINSKI, Joseph B. (1962–2001).

Joseph Bruno Slowinski, an authority on the systematics and evolution of elapid snakes, died tragically from the bite of one, a krait, in the northern mountains of Myanmar (Burma). He was born on 15 November 1962 in New York City and raised in Kansas City (Missouri) where he developed an interest in fossils. During a fossil-hunting expedition to South Dakota in July 1979, his interest in a paleontological career shifted to one in herpetology when he became fascinated with rattlesnakes. One of them had bitten him on an earlier field trip when he was 15. Slowinski attended the University of Kansas (B.S. 1984) and during his college years he spent much time in the field. He often caught venomous snakes and was bitten three times by copperheads. His doctoral research at the University of Miami (Ph.D. 1991) with Jay M. Savage was on coralsnake phylogeny. After postdoctoral appointments in Washington (D.C.) and then in Louisiana where he learned molecular techniques and further refined his use of morphological characteristics, he was hired as Assistant Curator of Herpetology at the California Academy of Sciences in San Francisco (1997).

Slowinski's first paper (1985) was on an aberrantly patterned rattlesnake. His other early research covered diets and species richness of a leaf-litter herpetofauna in Costa Rica. From 1989, however, his work shifted almost exclusively to venomous snakes, in particular to elapid snakes, their evolution and biology, and to their harmless mimics. He employed a range of molecular techniques including amino acid sequencing and nuclear and mitochondrial genes, to study elapid phylogeny and relationships and, in turn, to investigate general questions of phylogenetics. Through 2006, he had published about 30 herpetological titles, plus some others on speciation and molecular clocks. He also authored an introductory text on genetics in 1998.

Much of Slowinski's work was laboratory-based, but his real joy was fieldwork, which he carried out with great exuberance. He liked to explore remote places and to find new species, which he once described as the "ultimate discovery." He had field experiences throughout the Neotropics during graduate school including a full year in Peru (1990–1991), but in 1997 he made the first of a dozen trips to Myanmar. This is one of the least-explored and most biologically rich countries in the world and it possesses cobras, kraits, and other elapids. On that very first trip Slowinski discovered a new cobra (*Naja mandalayensis*), one of the species that spits its venom. Later, one of them spat in his eyes, which caused him excruciating pain. The venom had to be washed out quickly in order to prevent permanent blindness. A cobra bit him during a later journey to Myanmar as he was putting it in his collecting bag. The fang penetrated deeply but no venom was injected, a so-called blank or dry bite. Since Slowinski was allergic to antivenin, there would have been no point to using antivenin in any case. His expeditions to Myanmar were highly successful. From 1997 through 2001, nearly 3,000 specimens including at least 18 new species were sent to the museum in San Francisco for study.

His final trip to Myanmar, during the monsoon season, began in early September 2001. It was projected to last six weeks and take his team over 300 km on foot along leech-infested trails into northern Myanmar and the Himalayan foothills near the Chinese border. At seven o'clock on the morning of 11 September, Slowinski was bitten when his team was in the village of Rat Baw, at the bottom of a fog-filled valley five days' march north of Putao. He had reached into a snake bag to pick up what he thought was a non-venomous mimic. In fact, it was the model, a krait (*Bungarus multicinctus*). He knew immediately from the scale pattern what it was and told his team members not to worry because the animal was a juvenile and it had barely scraped his skin. By late morning, he began to feel the effects of neurotoxic poisoning. Slowinski explained in great detail to his team what his future symptoms would be and what care they would need to provide for him. Soon he began to lose motor functions and his speech became slurred.

Myanmar forbids the use of satellite cell phones, so his Burmese colleagues had to run nearly 15 km to the nearest radiotelephone. A helicopter was sent, but twice it had to be called back in mid-flight because of bad weather. When his diaphragm stopped working, team members took turns to provide mouth-to-mouth respiration. Although Slowinski could not speak, through prior arrangement he signaled to them by wiggling his fingers or toes—once for yes, twice for no—since his brain continued to function for a while and he was still aware of his surroundings. By early morning on the following day, 12 September 2001, his pulse was very weak and he did not respond to voices. Ultimately, his heart stopped by early afternoon, 30 hours after he had been bitten, even though his team continued cardiopulmonary resuscitation for several more hours. He was 38 years old. Because of the severe weather, only the next day was a helicopter able to land to remove his body.

• *References*: "Bitten," by R. Sullivan, New York Times Mag., 30 December: 34, 2001; "Joseph Bruno Slowinski 1962–2001," by M. A. Donnelly and B. I. Crother, Copeia, 2003: 424–428, 2003. • *Portrait* (1999): California Academy of Sciences, courtesy Dong Lin. • *Signature* (1992): American Museum of Natural History, courtesy Charles W. Myers.

THIS CURRENCY, ISSUED IN 1993, HONORS THE CAREER OF Vital Brazil (1865–1950), founder of the Instituto Butantan and developer of the first polyvalent antiserum for pitvipers. (For his biography, see page 129.) The bank note exhibits a portrait of Brazil and the head of a stylized Brazilian pitviper (*Bothrops*). A demonstration of venom extraction and the distinctive domed enclosures used to house snakes at Butantan, the world's first snake park, are also depicted. Vital Brazil promoted conservation of the snake-eating Mussurana (*Clelia clelia*), shown here consuming a pitviper. (Adler collection.)

LITERATURE CITED AND FURTHER READING

These titles include biographical information about herpetologists. Often, herpetology was just one of a person's interests especially for those who lived before 1900. References to major historical works about ornithologists, ichthyologists, entomologists, and other specialists are listed if they contain information of particular importance to herpetology.

This list supplements that in volume 1 (pages 134–135); those titles specifically referred to in volume 2 are repeated here. There has been a trend in recent years to include an historical introduction in regional works on herpetology. Thus, only a sampling of these chapters can be referenced below.

ADLER, K. 1990. Cheng-chao Liu from an American Perspective, p. 26–30. *In* E.-M. Zhao (ed.), From Water Onto Land. China Forestry Press, Beijing.

ADLER, K. 2003. Salamander Classification and Reproductive Biology. An Historical Overview, p. 1–29. *In* D. M. Sever (ed.), Reproductive Biology and Phylogeny of Urodela. Science Publ., Enfield (New Hampshire).

ADLER, K. 2007. Herpetological Exploration, Research, and Conservation in Vietnam, p. 33–56. *In* V. S. Nguyen, T. C. Ho, and Q. T. Nguyen, Herpetofauna of Vietnam. Ed. Chimaira, Frankfurt am Main.

ALESSANDRINI, A. and A. Ceregato. 2007. Ulisse Aldrovandi: Natura Picta. Ed. Compositori, Bologna, 670 pages (herpetology by M. Delfino, pp. 103–105, pls. 442–462).

ANANJEVA, N. B. 1998. Department of Herpetology, Zoological Institute, Russian Academy of Sciences, St. Petersburg, Russia: History and Current Research. Herpetol. Rev., 29: 136–140.

ANANJEVA, N. B. 2005. History and Anniversary Dates of Russian Herpetology in St. Petersburg, p. 5–10. *In* N. Ananjeva and O. Tsinenko (eds.), Herpetologia Petropolitana. Russian Jour. Herpetol., vol. 12 (supplement).

ANDERSON, S. C. 1999. The Lizards of Iran. Soc. Study Amphib. Reptiles, Ithaca (New York), Contr. Herpetol., 15: vii, 1–442 (history, pp. 15–41).

ANDREONE, F. and E. Gavetti. 1998 (1999). Gli Studi Erpetologici in Piemonte e in Valle d'Aosta, p. 17–36. *In* F. Andreone and R. Sindaco (eds.), Erpetologia del Piemonte e delle Valle d'Aosta Atlante degli Anfibi e dei Rettili. Monogr. Mus. Reg. Sci. Nat. Torino, vol. 26.

ARDILA-ROBAYO, M. C. 2003. Desarrollo de la Herpetología en Colombia. Aportes al Conocimiento Histórico de los Estudios sobre Anfibios. Revta. Acad. Colomb. Cienc., 27: 233–244.

BARBAGLI, F. and C. Violani. 2004. Breve Storia dell' Erpetologia Lombarda, p. 13–33. *In* F. Bernini et al. (eds.), Atlante degli Anfibi e dei Rettili della Lombardia. Monogr. Pianura, no. 5, Cremona.

BAUCHOT, M. L., J. Daget, and R. Bauchot. 1990. L'Ichtyologie en France au Debút du XIXe Siècle. L'Histoire Naturelle des Poissons de Cuvier et Valenciennes. Bull. Mus. Natl. Hist. Nat., ser. 4, 12(sect. A, 1, suppl.): 3–142.

BARTHOLOMEW, B., R. Wahlgren, and A. M. Bauer. 2001. Recent Literature on the History and Bibliography of Herpetology: 1. Newsl.-Bull. Internatl. Soc. Hist. Biblio. Herpetol. (now Biblio. Herpetol.), 3(1): 14–23 (annotated listing of 182 references).

BAUER, A. M. (ed.) 2004. Studies in the History of Herpetological Exploration. Bonner Zool. Beitr., 52(3–4): (1), 181–335.

BAUER, A. M. and K. Adler. 2001. The Dating and Correct Citation of A. F. A. Wiegmann's "Amphibien" Section of Meyen's *Reise um die Erde*, with a Bibliography of Wiegmann's Herpetological Publications. Arch. Nat. Hist., 28: 313–326.

BAUER, A. M., R. Günther, and M. Klipfel. 1995. The Herpetological Contributions of Wilhelm C. H. Peters (1815–1883), p. 5–87. *In* Wilhelm C. H. Peters, Herpetological Contributions. Soc. Study Amphib. Reptiles, Ithaca (New York), Facs. Repr. Herpetol., 714 pages.

BAUER, A. M. and R. A. Sadlier. 2000. The Herpetofauna of New Caledonia. Soc. Study Amphib. Reptiles, Ithaca (New York), Contr. Herpetol., 17: xii, 1–310 (history, pp. 55–65).

BELTRÁN, E., A. Jáuregui de Cervantes, and R. C. Arvea. 1990 (1991). Alfredo Dugès. Gob. Est. Guanajuato, Guanajuato, 261 pages (reprinted 1997).

BIBLIOTHECA HERPETOLOGICA. 1999 *et seq*. A journal published by the International Society for the History and Bibliography of Herpetology. During 1999–2004, the journal was called *Newsletter and Bulletin of the ISHBH*. Published in Lund, Sweden.

BOEWE, C. (ed.). 2003. Profiles of Rafinesque. Univ. Tennessee Press, Knoxville, xli, 411 pages.

BONDESON, J. 1990. Ormen i magen. Sydsvenska medicinhist. sällsk. årsskrift, 27: 141–174.

BOOS, H. E. A. 2001. The Snakes of Trinidad and Tobago. Texas A&M Univ. Press, College Station, xvi, 270 pages (history, pp. 11–34).

BORKIN, L. J. and E. I. Kolchinsky (eds.). 2000–2001. Russian-German Links in Biology and Medicine: 300-year Experience of Interactions. St. Petersburg Assoc. Sci. Scholars, St. Petersburg, vol. 1, 203 pages; vol. 2, 182 pages (in Russian).

BRANCH, W. R. 1999 (2001). Reptile Systematic Studies in Southern Africa: A Brief History and Overview. Trans. Royal Soc. South Africa, 54: 137–156.

BRANCH, W. R. and A. M. Bauer. 2005. The Life and Herpetological Contributions of Andrew Smith, p. 1–19. *In* Herpetological Contributions of Sir Andrew Smith. Soc. Study Amphib. Reptiles, Villanova (Pennsylvania), Facs. Repr. Herpetol.

BRASSLER, H. 1989–1994. Bibliographie herpetologischer Bibliographien I–III. Cour. Forsch.-Inst. Senckenberg, 116: 1–214; 140: 1–129; 167: 1–138.

BROADLEY, D. G. 1976. The Status of Herpetology in Southern Africa. Zool. Africana, 11: 233–240.

BROADLEY, D. G. 1996. A History of the Herpetological Association of Africa. African Herp News, 25: 18–21.

BRUNO, S. 1989. I Grandi Protagonisti dell'Ofidiologia, p. 19–40. *In* Umanesimo della Pietra—Verde, Ann. Gruppo Umanesimo della Pietra, no. 4, Martina Franca, Italy.

CARD, W. and J. B. Murphy. 2000. Lineages and Histories of Zoo Herpetologists in the United States. Soc. Study Amphib. Rept., Herpetol. Circ., 27: 1–49.

CHANNING, A. 1999 (2001). Historical Overview of Amphibian Systematics in Southern Africa. Trans. Royal Soc. South Africa, 54: 121–135.

COGGER, H. G. 1993. History of the Discovery of the Reptilia, p. 92–97. *In* C. J. Glasby et al. (eds.), Fauna of Australia, vol. 2A Amphibia and Reptilia. Austr. Govt. Publ. Serv., Canberra.

COLE, F. J. 1934. Historical Introduction, p. xiii–xxxi. *In* E. T. B. Francis, The Anatomy of the Salamander. Oxford Univ. Press, Oxford (reprinted 2002).

COLE, F. J. 1944. A History of Comparative Anatomy from Aristotle to the Eighteenth Century. Macmillan & Co., London, viii, 524 pages.

CONANT, R. 1997. A Field Guide to the Life and Times of Roger Conant. Selva (Canyonlands Publ. Group), Provo, xix, 498 pages (with biographies of 24 herpetologists, pp. 448–498).

COOK, F. R. 1992. The Once and Future National Reference Collection in Herpetology. Bull. Canadian Assn. Herpetol., 6: 2–9.

CUVIER, G. 1995. Historical Portrait of the Progress of Ichthyology, from Its Origins to Our Own Time. English ed. with commentary by T. W. Pietsch. Johns Hopkins Univ. Press, Baltimore, xxiv, 366 pages.

DAMKAER, D. M. 2002. The Copepodologist's Cabinet A Biographical and Bibliographical History. Amer. Philos. Soc., Philadelphia, xix, 300 pages (vol. 2 now with publisher).

DAS, I. 1998. Herpetological Bibliography of Indonesia. Krieger Publ. Co., Malabar (Florida), 92 pages.

DAS, I. 2002. An Introduction to the Amphibians and Reptiles of Tropical Asia. Nat. Hist. Books (Borneo), Kota Kinabalu, v, 207 pages (history, pp. 17–26).

DAS, I. and Y. Norsham. 2007. Status of Knowledge of the Malaysian Herpetofauna, p. 31–81. *In* L. Chua et al. (eds.), Proceedings of the Seminar on the Status of Biological Diversity in Malaysia and Workshop on Threat Assessment of Plant Species in Malaysia. Forest Res. Inst. Malaysia, Kepong (history, pp. 34–42).

DAS, I., B. Dattagupta, and N. C. Gayen. 1998. History and Catalogue of Reptile Types in the Collection of the Zoological Survey of India. Jour. South Asian Nat. Hist., 3: 121–172.

DATHE, W. and R. M. González López. 2002. Johann Christoph Gundlach (1810–1896). Un Naturalista en Cuba. Basilisken-Press, Marburg, 1–245 pages.

DE SILVA, A. 1995. An Introduction to the Herpetofauna of Sri Lanka, part 2: Historical Aspects. Lyriocephalus, Peradeniya, 2: 3–16.

DE SILVA, A. 1996. The Herpetofauna of Sri Lanka: A Brief Review. Publ. by author, Peradeniya, 99 pages (history, pp. 12–27, pls. 1–10).

DISI, A. M. 2002. The Herpetofauna of Jordan. Publ. by author, Amman, 288 pages (history, pp. 15–20).

DISI, A. M., D. Modrý, P. Necăs, and L. Rifai. 2001. Amphibians and Reptiles of the Hashemite Kingdom of Jordan. Ed. Chimaira, Frankfurt am Main, 408 pages (history, pp. 28–30).

DRIESSEN-VAN HET REVE, J. J. 2006. De Kunstkamera van Peter de Grote. De Hollandse inbreng, gereconstrueerd uit brieven van Albert Seba en Johann Daniel Schumacher uit de jaren 1711–1752. Verloren, Hilversum, 350 pages.

DUELLMAN, W. E. and K. Adler. In press. The Evolution of Amphibian Systematics: An Historical Perspective. *In* H. Heatwole and M. J. Tyler (eds.), Amphibian Biology, vol. 7. Surrey Beatty, Chipping Norton (Australia).

ESTRADA, A. R. and R. Ruibal. 1999. A Review of Cuban Herpetology, p. 31–62. *In* B. I. Crother (ed.), Caribbean Amphibians and Reptiles. Academic Press, San Diego.

EVENHUIS, N. L. 1997. Litteratura Taxonomica Dipterorum (1758–1930). Backhuys Publ., Leiden, vol. 1, ix, 1–426 pages; vol. 2, iv, 427–871 pages.

FRANSEN, C. H. J. M., L. B. Holthuis, and J. P. H. M. Adema. 1997. Type-Catalogue of the Decapod Crustacea in the Collections of the Nationaal Natuurhistorisch Museum, with Appendices of Pre-1900 Collectors and Material. Zool. Verhandel., 311: xvi, 1–344.

FUKADA, H. 2002. A Brief History of the Herpetological Society of Japan. Bull. Herpetol. Soc. Japan, 2002(1): 13–17 (in Japanese).

GARANIN, V. I. and A. G. Bakiev. 2002. Bibliography of Amphibians and Reptiles of the Volga Basin (18th–20th Centuries). Inst. Ecology, Acad. Sci., Tolyatti, 127 pages (in Russian).

GESELLSCHAFT FÜR NATURSCHUTZ UND ORNITHOLOGIE (eds.). 1995. Maximilian Prinz zu Wied. Jäger, Reisender, Naturforscher. Fauna Flora Rheinland-Pfalz, 17: 1–354 (herpetology in parts 18–19, 1996).

GROTE, A. (ed.). 1994. Macrocosmos in Microcosmo. Die Welt in der Stube. Zur Geschichte des Sammelns 1450 bis 1800. Berliner Schriften zur Museumskunde 10. Leske + Budrich, Opladen, 967 pages.

GOODE, G. B. 1901. A Memorial of George Brown Goode, Together with a Selection of His Papers on Museums and on the History of Science in America. Ann. Rept. Board Regents Smithsonian Inst. for 1897 (Rept. U.S. Natl. Mus., 2): xii, 1–515.

GRUBER, U. 1992. Die Sektion Herpetologie (Lurch und Kriechtiere) der zoologischen Staatssammlung München. Spixiana, suppl., 17: 126–132.

HAINES, S. 2000. Slithy Toves. Illustrated Classic Books at the University of Kansas in Pictures and Conversations. Soc. Study Amphib. Reptiles, Ithaca (New York), Contr. Herpetol., 16: vii, 1–180.

HARPER, F. 1940. Some Works of Bartram, Daudin, Latreille, and Sonnini, and Their Bearing Upon North American Herpetological Nomenclature. Amer. Midl. Nat., 23: 692–723.

HASEGAWA, H. 1967. Materials on the Lives and Contributions of the Deceased Japanese Entomologists Since the Meiji Era. Kontyû, 35(suppl.): 1–98 (in Japanese).

HENDERSON, R. W. and R. Powell (eds.) 2003. Islands and the Sea. Essays on Herpetological Exploration in the West Indies. Soc. Study Amphib. Reptiles, Ithaca (New York), Contr. Herpetol., 20: viii, 1–304 (history, pp. 9–20).

IKOMA, Y. 1973. Collected Works on the Giant Salamander (*Megalobatrachus japonicus*) in Japan. Tsuyama Mus. Sci. Educat., Tsuyama City, vi, 480 pages (in Japanese).

JAUSSAUD, P. and É.-R. Brygoo. 2004. Du Jardin au Muséum en 516 Biographies. Publ. Mus. Natl. Hist. Nat., Paris, 630 pages.

KLUGE, A. G. 1996. The Early Collections and Curators, Division of Reptiles and Amphibians, Museum of Zoology, University of Michigan. Herpetol. Rev., 27: 176–178.

KORSÓS, Z. and C. Horváth. 1992. Some Aspects of the History of Hungarian Herpetology, p. 35–44. *In* Z. Korsós and I. Kiss (eds.), Proc. 6th Ordin. Gen. Meet., Soc. Europ. Herpetol., Budapest.

KUZMIN, S. L. 1999. The Amphibians of the Former Soviet Union. Pensoft Publ., Sofia-Moscow, 538 pages (history, pp. 8–18; Russian edition by Kuzmin and I. V. Maslova, KMK Publ., Moscow, 2005 [history, pp. 18–32]).

KUZMIN, S. L. and I. V. Maslova. 2003. The Amphibians of the Russian Far East. Adv. Res. Former Soviet Union, 8: xii, 1–464 (history, pp. 11–25).

LAMBERT, M. R. K. 1997. The British Herpetological Society: The First 50 Years, 1947–1997. Herpetol. Jour., 7: 129–141.

LAURENT, R.-F. 1965. Contribution à l'Histoire de l'Herpétologie Congolaise et Bibliographie Générale. Acad. Roy. Sci. d'Outre-Mer, ser. Sci. Nat. Méd., 16: 1–55.

LEE, J. C. 1996. The Amphibians and Reptiles of the Yucatán Peninsula. Comstock Publ. Assoc., Cornell Univ. Press, Ithaca (New York), xii, 500 pages (history, pp. 16–29).

LESCURE, J. 2002. La Naissance de l'Herpétologie. Bull. Soc. Herpétol. France, 101: 5–27.

LESCURE, J. and B. Le Garff. 2006. L'Étymologie des Noms d'Amphibiens et de Reptiles d'Europe. Éd. Belin, Paris, 207 pages (biographies, pp. 143–186).

LEVITON, A. E. and M. L. Aldrich. 2000. India: A Case Study of Natural History in a Colonial Setting, p. 51–80. *In* M. T. Ghiselin and A. E. Leviton (eds.), Cultures and Institutions of Natural History. Mem. California Acad. Sci., vol. 25.

LITTLEJOHN, M. 1993. 40 Years On: Tangled Threads. Newsl. Australian Soc. Herpetol., 35: 12–16.

MEARNS, B. and R. Mearns. 1998. The Bird Collectors. Academic Press, San Diego, xviii, 472 pages.

MENZ, H. 2000. Martin Heinrich Rathke (1793–1860) Ein Embryologe des 19. Jahrhunderts. Basilisken-Presse, Marburg, 280 pages.

MCCRANIE, J. R. and L. D. Wilson. 2002. The Amphibians of Honduras. Soc. Study Amphib. Reptiles, Ithaca (New York), Contr. Herpetol., 19: x, 1–625 (history, pp. 27–33).

MORIARTY, J. J. and B. Bartholomew. 2007. Society for the Study of Amphibians and Reptiles: A Fifty Year History 1958 to 2007. Soc. Study Amphib. Reptiles, Herpetol. Circ., 36: 1–64.

MURPHY, J. B. 2005. Chameleons: Johann von Fischer and Other Perspectives. Soc. Study Amphib. Reptiles, Herpetol. Circ., 33: 1–123.

MÜSCH, I. 2000. Geheiligte Naturwissenschaft: Die Kupfer-Bibel des Johann Jakob Scheuchzer. Vandenhoeck & Ruprecht, Göttingen, 286 pages.

MÜSCH, I. et al. 2001. [Introductory Essays,] p. 7–39. *In* Albertus Seba, Cabinet of Natural Curiosities. Taschen, Cologne, 587 pages (includes color reproductions of all 449 plates from Seba's "Thesaurus").

MYERS, C. W. 2000. A History of Herpetology at the American Museum of Natural History. Bull. Amer. Mus. Nat. Hist., 252: 1–232.

NAKAMURA, K., T. Hikida, and M. Matsui. 1988. Systematic Zoology. Vol. 9, no. 2B. Reptilia A. Nakayama Shoten, Tokyo, 3, 308 pages (history, pp. 4–20).

OBST, F. J. 2002. Herpetology and Herpetological Type Specimens at the Museum für Tierkunde Dresden with a Bibliography of Herpetological Contributions. Faun. Abhandl. Staatl. Mus. Tierkunde Dresden, 23: 3–34.

OLIVEIRA, J. L. de. 1980–1981 (1983). Cronologia do Instituto Butantan (1888–1981). Mem. Inst. Butantan, 44–45: 11–79.

ORLOVA, V. F. with E. A. Dunayev. 1991. Herpetological Collections, p. 137–149. *In* O. L. Rossolimo (ed.), Two Centuries in the Collections of the Zoological Museum of Moscow State University. Moscow State Univ., Moscow (in Russian).

PELLEGRIN, J. 1938. La Chaire de Zoologie (Reptiles et Poissons) du Muséum National d'Histoire Naturelle. Bull. Mus. Natl. Hist. Nat., ser. 2, 10: 314–327.

PIETSCH, T. W. and W. D. Anderson (eds.). 1997. Collection Building in Ichthyology and Herpetology. Amer. Soc. Ichthyol. Herpetol., Spec. Publ., 3: xiii, 1–593 pages.

POWELL, R. and R. W. Henderson (eds.) 1996. Contributions to West Indian Herpetology. A Tribute to Albert Schwartz. Soc. Study Amphib. Reptiles, Contr. Herpetol., 12: 1–457 (history, pp. 29–50).

RAND, A. S. 2001. A History of Frog Call Studies 405 B.C. to 1980, p. 8–19. *In* M. J. Ryan, Anuran Communication. Smithsonian Inst. Press, Washington.

RIECK, W., G. Hallmann, and W. Bischoff (eds.). 2001. Die Geschichte der Herpetologie und Terrarienkunde im deutschsprachigen Raum. Mertensiella, 12: 1–759.

RODRÍGUEZ-ROBLES, J. A., D. A. Good, and D. B. Wake. 2003. Brief History of Herpetology in the Museum of Vertebrate Zoology, University of California, Berkeley, with a List of Type Specimens of Recent Amphibians and Reptiles. Univ. California Publ. Zool., 131: xv, 1–96 (history, pp. 1–13, 77–96).

ROSSOLIMO, O. L. and E. A. Dunayev (eds.). 2003. Moscow Herpetologists. KMK Publ., Moscow, 580 pages (in Russian with English summaries).

ROUX, J. 1924. Notice Historique sur la Collection Erpétologie du Museum d'Histoire Naturelle de Bâle. Verhandl. Naturf. Gesell. Basel, 35: 60–70.

SALEH, M. A. 1997. Amphibians and Reptiles of Egypt. Egyptian Environ. Affairs Agency, Publ. Natl. Biodivers. Unit, 6: viii, III, 1–234 (history, pp. 1–3).

SAVAGE, J. M. 2002. The Amphibians and Reptiles of Costa Rica. Univ. Chicago Press, Chicago, xx, 934 pages (history, pp. 34–54).

SCHMIDTLER, J. F. 2007. Die Wurzeln einer bayrischen Herpetofaunistik im 18. und beginnenden 19. Jahrhundert. Zeitschr. Feldherpetol., 14: 93–119.

SEKRETÄR. 2001 *et seq*. A journal published by the working group Literatur und Geschichte der Herpetologie und Terrarienkunde of the Deutschen Gesellschaft für Herpetologie und Terrarienkunde. During 2001–2002, the journal was called *Beiträge zur Geschichte der Herpetologie und Terrarienkunde*. Published in Rheinbach, Germany.

SINDACO, R., G. Doria, E. Razzetti, and F. Bernini. 2006. Atlante degli Anfibi e dei Rettili d'Italia / Atlas of Italian Amphibians and Reptiles. Ed. Polistampa, Florence, 789 pages (history, pp. 18–95).

SMITH, H. M. and P. David. 1999. George Shaw and the Herpetology Volume in His Series, General Zoology, p. 5–96. *In* G. Shaw, General Zoology, Volume III. Amphibians and Reptiles. Soc. Study Amphib. Reptiles, Ithaca (New York), Facs. Repr. Herpetol.

VIJAYARAGHAVAN, B. 2005. Snake Studies: India. Chennai Snake Park Trust, Chennai, 30 pages.

WILLIAMS, E. E. 1999. Over 300 Years of Collecting in the Caribbean, p. 1–30. *In* B. I. Crother (ed.), Caribbean Amphibians and Reptiles. Academic Press, San Diego.

ZHAO, E.-M. and K. Adler. 1993. Herpetology of China. Soc. Study Amphib. Reptiles, Oxford (Ohio), Contr. Herpetol., 10: 1–522 (history, pp. 15–66).

INDEX TO BIOGRAPHIES

This index is comprehensive for volumes 1 and 2. Page numbers in **boldface** refer to the biographies of the 437 individuals featured in these volumes. Where the same page number is repeated for a single entry, the individual mentioned is in more than one account on that page. Alternate or additional names are given for some individuals. Full names have been provided so far as known. Because volume 1 is out of print, the text has been put on the web for public access. Visit www.ssarherps.org and click on "Publications."

A

B

C

D

E

F

G

H

I

J

K

L

R

T

U

V

W

Y

Z

Index of Authors in Taxonomic Herpetology
Second Edition

by John S. Applegarth

As a contribution to the history of herpetology, this index seeks to include all authors who have participated in the naming of one or more taxa, at the level of genus or below, within the living families of amphibians and reptiles. Authors are identified by their complete names, birth and death dates, countries of residence, and orders of the taxa they proposed. Also included in this index are authors who have had a taxon within the living families of amphibians and reptiles named in their honor *and* who have authored at least one contribution to herpetology. Originally this index was begun as an adjunct to my index of available generic names in the living families of amphibians and reptiles, so it does not include many distinguished herpetologists who, for whatever reasons, never named any taxa and no taxa have been named for them. An index of all herpetological authors would be a far larger project. In the future I hope to revise both this index and my index of available generic names.

Names.—Last (family) names and initials of first (given) names are in boldface print. However in Myanmar people do not have inherited family names, so their given names are treated here in the same way as family names. In China, Korea and Vietnam the inherited family name is the first word of the entire name. Many family names can be preceded by a particle (such as the French preposition "de"). In this index generally names are alphabetized by the form (particle first or particle last) that would best facilitate locating publications by that author. Alternative forms may have cross-reference entries that direct the reader (with the word "*see*") to the primary entry for that author. Accent marks are included but ignored in the alphabetizing of names. If the mother's family name is given after the father's family name, the two surnames may be linked by a hyphen (mainly for the benefit of non-Hispanic readers), and together they are alphabetized as a unit but only the father's family name is entirely in boldface. Equivalent names and alternative transliterations are in parentheses and are preceded by an equal sign (=). Diminutives and nicknames that have appeared in publications are placed within parentheses and quotation marks. Non-academic titles and personal relationships to other authors in this list are included and placed in parentheses. The titles are not capitalized in order to distinguish them from names. Names or parts of names that authors have seldom used are included here for the benefit of historians and librarians and are placed within square brackets. Many of the living authors prefer to be known by certain parts of their complete name, and some avoid or dislike other parts. Please respect the feelings of these people and address them using the forms they prefer.

Dates.—Solid dots separate the categories of information—who, when, where, and what. After the first dot are the author's date of birth and, if applicable, their date of death. Birth and death dates are written symmetrically to facilitate reading only the years. If there is only one date (or year) then that refers to the author's birth and the person is presumed to be living. Any year that has not been found is represented by a question mark. A date (or any other information) that is questionable is placed within parentheses and preceded by a question mark. If neither birth nor death year are known *and* the author published over 80 years ago, then the known years of their publications are given in parentheses with dashes and question marks on the outside of the parentheses on both sides.

Dates based on old calendar systems are given within quotation marks and parentheses. In general, to convert old Julian calendar dates to modern Gegorian calendar dates, add 10 days to all dates from 15 October 1582 to 28 February 1700 inclusive, add 11 days to all dates from 1 March 1700 to 28 February 1800, and add 12 days to all dates from 1 March 1800 to 28 February 1900. An exception is Sweden, which tried to make a gradual change beginning in 1700, but was out of phase with both systems so it returned to the Julian system in 1712, and then changed to Gregorian in 1753 (see the entry for Linnaeus). Before Japan adopted the Gregorian calendar on 1 January 1873, their system of dates was complex and older dates are given here in quotation marks and parentheses.

Countries.—After the second dot are short names for the countries or regions where each author was born, resided, and died. The names of multiple places of residence are separated by a slash mark (/) and are given in chronological order. If a country or region substantially changed its name during the life of a person, then that is treated as though there had been multiple places of residence.

Orders.—After the third dot in each author's entry are the names of the herpetological orders, in alphabetical order, in which taxa were proposed by, named for, or named by others within publications of that author. *If* the first letter of an order is capitalized, then this is a code indicating that it represents living taxa within that order. If the first letter of an order is *not* capitalized and there is an asterisk (*) in front of that order, then this indicates that it represents only fossil or extinct taxa. If the name of an order is both capitalized and preceded by an asterisk, then it represents both living and fossil taxa.

If one or more taxa in an order were named *for* an author to honor that person *and* that author did not personally name any taxa in that order, then parentheses are placed around that order. If the proposed taxa were published by someone else within the work of an author *and* that author did not otherwise participate in naming any taxa in that order, then the order representing those taxa is within square brackets. Therefore if an order is within both parentheses and square brackets, it means that one or more taxa within that order were named for that author and were published in works by that author—not necessarily the same taxa—*and* that the author never named any taxa in that order. The square brackets can be either inside or outside of the parentheses; the sequence has no significance other than when the outer marks enclose a larger number of orders. For example [(Anura), Sauria].

***Acknowledgments* (1989 edition).**—I am indebted to the many people who have helped in many ways. Kraig Adler provided author information and editorial guidance. I thank everyone who answered my data requests, especially the following people who took time to gather biographical information about others: T. Alvarez-S., N. B. Ananyeva, M. Atatür, S. C. Ayala, R. Barbault, J. Bons, B. Brattstrom, R. Conant, A. J. Coventry, R. I. Crombie, Y. Develey, S. K. Dutta, R. Etheridge, R. Formas, D. R. Frost, E. S. Gaffney, J. M. Gallardo, H. W. Greene, K. Haker, H. S. Harris, H. Heatwole, T. Hikida, J. J. Kirk, E. Kramer, M. A. Labra, B. Lanza, R. Lawson, P. R. Manzani, C. J. McCoy, I. Mercadal, K. Miyazaki, M. Penna-V., H. Pieper, G. K. Pregill, J. C. Rage, P. M. Ruiz-C., H. Saint Girons, M. A. Serna-D., T. Seto, N. N. Shcherbak, J. E. Simmons, H. M. Smith, A. F. Stimson, F. H. Test, B. W. Thomas, F. Tiedemann, H. G. Tunner, M. J. Tyler, A. Veloso-M., Y. Werner, E. E. Williams, L. D. Wilson, and especially E.-m. Zhao (who gathered data for over 60 Chinese authors!). For local assistance I am grateful to James Kezer, Susan Crawford, and the staff of the University of Oregon libraries (especially R. H. Felsing, P. D. Morrison, and B. K. Wycoff). I owe special thanks to the Oregon Herpetological Society for a grant-in-aid toward postage expenses.

***Acknowledgments* (2007 edition).**—Again I thank Kraig Adler (Cornell University) who again helped me with an enormous amount of biographical information. Dawn Potter (Cornell University) cheerfully researched many technical questions. Timothy Perry (Cornell University) facilitated the final formatting of the text and inclusion of special accent marks. For assistance with Asian languages I am indebted to Robert Felsing (University of Oregon). Decades ago Hobart Smith entrusted me with Walter Necker's bibliography of herpetology (the original card file) and that file continued to be helpful. I give sincere thanks to everyone who answered my e-mailed requests, especially the following people who took time to send information about others: Andrey Bakiev, Breck Bartholomew, Chris Bell, Federico Bolaños, Jason Brown, Luis Canseco, Sergio Carvalho-e-Silva, Lawan Chanhome, Alan Channing, Patrick Couper, Atherton de Villiers, Steve Donnellan, Bill Duellman, Bruce Erickson, Oscar Flores, Ansel Fong, Justin Gerlach, Ariovaldo Giaretta, Lee Grismer, Ren Hirayama, Koji Iizuka, Hinrich Kaiser, Rod Kennett, Jörn Köhler, Dirk Megirian, Robert Morrison, Le Fras Mouton, John Murphy, Francisco Ortega, Ioana Petrisor, Martin Pickersgill, Neville Pledge, Viatcheslav Ratnikov, Miguel Rodrigues, Vincent Russo, Susan Schick, Glenn Shea, Rick Shine, Matthew Spriggs, Roy Swain, Tan Fui-Lian, Andrew Turner, Mike Tyler, Álvaro Velásquez, and Harold Voris.

Corrections.—I extend my apologies to anyone who was overlooked or incorrectly cited. In anticipation of another (third) edition of this index, you can send corrections to me at 1915 Terresa Avenue, Eugene, Oregon 97408, USA (or e-mail them to actinemys@earthlink.net). Thank you.

TAXONOMIC AUTHORS

A

Aagaard, **C**arl **J**ohan [Ove Mønster] • 1882–? • (? Denmark) / Siam • (Serpentes)

Abalos, **J**orge **W**ashington • 1915–1979 • Argentina • Serpentes

Abbott, **C**harles **C**onrad • 4 June 1843–1919 July 27 • USA • Sauria

Abbott, **W**illiam **L**ouis • 23 February 1860–1936 April 2 • USA • (Anura, Sauria)

Abdala, **C**ristian **S**[imón] • ? • Argentina • Sauria

Abe, **A**ugusto **S**hinya • 23 November 1946 • Brazil • Sauria

Abe, **Y**oshio • 3 January 1891–1960 April 22 • Japan • Caudata

Acevedo, **M**anuel • ? • Guatemala • Anura, Sauria

Acharji, **M. N.** (= Mahendranath) • ? • India • Serpentes

Achaval [-Elena], **F**ederico • 27 July 1941 • Uruguay • (Anura)

Acosta, **J. A.** • ? • (? Argentina) • Sauria

Acosta, **J**uan **C**arlos • ? • Argentina • Sauria

Acosta [-Galvis], **A**ndrés **R**[ymel] • 9 September 1970 • Colombia • Anura, Caudata

Adams, **A**ndrew **L**eith • 1827–1882 August • Scotland / England / Ireland • *testudines

Adams, **L**everett **A**[llen] • 23 September 1877–1976 August • USA • *caudata

Adams, **M.** • ? • Australia • Anura

Adler, **K**raig [Kerr] • 6 December 1940 • USA • Anura, Caudata, (Sauria, Serpentes), Testudines

Adriani, **G**iuseppe • 20 September 1949 • Italy • Sauria

Adrover, **R**afael • ? • Spain • Anura

Aellen, **V**illy • 4 December 1926 • Switzerland • Caudata

Agassiz, [Jean] **L**ouis [Rodolphe] • (? 26 *or* 28) May 1807–1873 December 14 • Switzerland / USA • Anura, Caudata, *Crocodilia, Sauria, Serpentes, Testudines

Agnolin, **F**ederico **L**[isandro] • 18 October 1982 • Argentina • *anura, *testudines

Aguayo-Vedia, **C**[iidar] **R**odrigo • 26 November 1973 • Bolivia • Anura

Aguiar, **O**dair, Jr. • ? • Brazil • Anura

Aguilar [-Puntriano], **C**ésar [Augusto] • 5 October 1971 • Perú • Anura, Serpentes

Ahl, [Christoph Gustav] **E**rnst • 1 September 1898–1945 February 14 • Germany / Yugoslavia • Anura, Gymnophiona, Sauria, Serpentes, *Testudines

Ahmed, **M**ohammed **F**iroz • 1 May 1975 • India • Anura

Aichinger, **M**anfred • ? • Austria • Anura

Akhmedov, **M. I.** • ? • USSR • Sauria

Akker, **S**cott [Alexander] • 27 November 1970 • England • Anura

Alayo-Dalmau, **P**astor • 1 June 1915–2001 February 2 • Cuba • (Sauria)

Alberch, **P**ere • 2 November 1954–1998 March 13 • Spain / USA / Spain • (Anura)

Albino, **A**driana **M**[aria] • 5 June 1960 • Argentina • *serpentes

Alcala, **A**ngel **C**[hua] • 1 March 1929 • Philippines • Anura, Sauria, (Serpentes)
Alcala, **E. L.** • ? • Philippines • Anura, Sauria
Alcock, **A**lfred **W**illiam (lieutenant colonel) • 23 June 1859–1933 March 24 • India / England / Scotland / India / England • Gymnophiona, Sauria, Serpentes
Alcover, **J**osep **A**ntoni • ? • Spain • *anura
Aldrovandi, **U**lisse • 11 September 1522–1605 (? May 4 *or* November 10) • Italy • (Sauria)
Alefaio, **S**emese • ? • Tuvalu (South Pacific) • Sauria
Alefaio, **T**ataua • ? • Tuvalu (South Pacific) • Sauria
Alekperov, **A**[bdulla] **M**[ustafaevich] • 15 March 1912–1981 October 16 • Russia / USSR (Azerbaijan) • *testudines
Alekseev (= Alexejew), **A**[leksei] **K**[arpovich] • ? • Russia • *sauria
Aleman G., **C**ésar • ? • Venezuela • (Anura), Serpentes
Alexander, **A**[lexander] **A**llan • 19 July 1928–1999 October 28 • USA • Sauria
Alfaro [Gonzáles], **A**nastasio • 16 February 1865–1951 January 20 • Costa Rica • (Caudata, Sauria)
Allen, **E**[nsil] **R**oss • 2 January 1908–1981 May 17 • USA • (Anura), Caudata, Serpentes
Allen, **G**lover **M**[orrill] • 8 February 1879–1942 February 14 • USA • (Anura, Sauria), Serpentes
Allen, **J**oel **A**saph • 19 July 1838–1921 August 29 • USA • (Serpentes)
Allen, **M**orrow **J.** • 24 November 1909–1988 October 22 • USA • Sauria, (Serpentes)
Allison, **A**llen • 12 May 1950 • USA / Papua New Guinea / USA • Anura, Sauria, Serpentes
Alluaud, **C**harles [A.] • 4 May 1861–1949 December 8 • France • (Anura, Sauria, Serpentes)
Almeida, **A**ntonio de **P**adua [Leite Serra de] • 10 August 1968 • Brazil • Anura
Almeida-Reinoso, **D**iego [Patricio] • 27 March 1974 • Ecuador • Anura
Almendáriz, **A**na • ? • Ecuador • Anura, Sauria
Alonso-Andrés, **L**uis • ? • Spain • *testudines
Alonso-Santiago, **L.** • ? • Spain • *testudines
Altamirano, **F**ernando • ?–? • México • ([Caudata], Sauria)
Altig, **R**onald [Gail] ("Ronn") • 8 June 1941 • USA • Anura, Serpentes
Altobello, **G**iuseppe • ? • Italy • Caudata
Alvarado-Díaz, **J**avier • ? • México • Serpentes
Álvarez, **A**ntenor • 2 May 1864–1948 July 28 • Argentina • (Serpentes)
Alvarez [de Avanza], **B**lanca **B**[eatriz] • ? • Argentina • Anura, Sauria, Serpentes
Álvarez del **T**oro, **M**iguel (don) • 23 August 1917–1996 August 2 • México • (Caudata), Sauria
Álvarez S[olórzano], [José] **T**icul • 26 February 1935–2001 October 16 • México • Caudata, Sauria
Álvarez: *see also* **Gonzáles-Á**lvarez, **L**ucindo
Alves, **A**na **C**laudia **R**eis • ? • Brazil • Anura
Amaral, **A**frânio [Pompílio Bastos] do • 1 December 1894–1982 November 29 • Brazil • Sauria, Serpentes
Amat, **F**èlix • 14 October 1967 • Spain • Caudata
Ambrosetti, **J**uan **B**autista • 22 August 1865–1917 May 28 • Argentina • *crocodilia, *sauria
Ameghino, **F**lorentino (= Fiorino) • 18 September 1854–1911 August 6 • Italy / Argentina • (Anura), *anura, (*crocodilia), *sauria, *testudines
Amiet, **J**ean-**L**ouis • ? • Cameroons • Anura, (Sauria)
Ammerman, **L**oren **K.** • 23 March 1965 • USA • Serpentes
Ammon, **L**udwig [Johann Georg Friedrich] von • 14 December 1850–1922 July 26 • Germany • *testudines
Amthauer, **E.** • ? • ? • Testudines
Ananyeva (= Ananjeva), **N**atalia **B**[orisovna] ("Natasha") (*wife of* Nikolai L. Orlov) • 5 January 1946 • USSR / Russia • Sauria, [Serpentes]
Anderson, **A**tholl **J**[ohn] • 5 April 1943 • New Zealand / Australia • (*crocodilia)
Anderson, **J**ames **D**[onald, Jr.] • 16 August 1930–1976 November 20 • USA • Caudata, Sauria, Serpentes
Anderson, **J**ennifer **A.** • ? • USA • Caudata
Anderson, **J**eromie **A.** • ? • (? USA) / Pakistan • Sauria
Anderson, **J**ohn • 4 October 1833–1900 August 15 • Scotland / British India / England • Anura, Caudata, Sauria, Serpentes, Testudines
Anderson, **R**obert **A.** • ? • USA • Sauria
Anderson, **S**teven **C**[lement] • 7 September 1936 • USA • Sauria, Serpentes
Andersson, **L**ars **G**abriel • 22 February 1868–1951 February 13 • Sweden • Anura, Caudata, Sauria, Serpentes
Andrade, **G**ilda **V**[asconcellos de] • 22 April 1960 • Brazil • Anura, Sauria
Andreas, **B**ritta • ? • Germany • Testudines
Andrén, **C**laes • ? • Sweden • Anura, Sauria, Serpentes
Andreone, **F**ranco • 23 May 1961 • Italy • Anura, Caudata, Sauria, Serpentes
Andrews, **C**harles **W**illiam • 1866–1924 May 25 • England • *crocodilia, *serpentes, *testudines
Andrews, **E**[dward] **W**yllys, IV • 11 December 1916–1971 July 3 • USA • Serpentes
Andrews, **E**than **A**llen • 10 September 1859–1956 • USA • (Anura)
Andrzejowski, **A**ntoni • 1785–1868 • (? Poland) • Caudata, Sauria, Serpentes
Angel, **F**ernand • 2 February 1881–1950 July 13 • France • Anura, (Gymnophiona), Sauria, Serpentes
Angulo [-Arce], **A**riadne • 22 June 1969 • Perú / Canada • Anura
Annandale, [Thomas] **N**elson • 15 June 1876–1924 April 10 • Scotland / British India • Anura, Gymnophiona, Sauria, Serpentes, Testudines
Anstis, **M**arion • ? • Australia • Anura
Aoki, **R**iosuke • 8 September 1954 • Japan • *crocodilia
Aparicio [-Effen], **J**ames • ? • Bolivia • Anura
Aplin, **K**enneth **P**[eter] ("Ken") • 9 October 1958 • England / Australia • Anura, Sauria, Serpentes
Applegarth, **J**ohn **S**[tirling] • 3 December 1937 • USA / Colombia / USA • Sauria
Arambourg, [Louis Joseph] **C**amille • 3 February 1885–1969 November 19 • France • (*crocodilia, *sauria), *testudines
Araújo, **A**lexandre **F**[ernandes] **B**amberg de • ? • Brazil • Sauria
Araya, **K**unio • ? • Japan • (Anura)

Archer, **M**ichael ("Mike") • ? • Australia • Testudines
Archey, (sir) **G**ilbert [Edward] • 9 August 1890–1974 • England / New Zealand • (Anura)
Ardila-R[obayo], **M**aría **C**ristina • 15 February 1947 • Colombia • Anura
Arends-R., **A**lexis • ? • Venezuela • Anura, Sauria
Argôlo, **A**ntônio **J**orge **S**uzart • 1957 • Brazil • Serpentes
Arikan, **H**üseyin • ? • Turkey • Anura, Sauria
Aristotle • BC 384–322 BC • Greece • (Serpentes)
Arizábal-Arriaga, **W**ilfredo • 7 February 1967 • Perú • Sauria
Armstrong, **B**arry **L**. • ? • USA • Serpentes
Armstrong-Ziegler, **J**udy **G**ail • ? • USA • *serpentes
Arndt, **R**udolf **G**erhard [Ernst] • 28 October 1941 • Germany / USA • Serpentes
Arnold, **D**ouglas **L**. • ? • USA • Sauria
Arnold, **E**dwin **N**icholas ("Nick") • 16 October 1940 • England • (Anura, Caudata), Sauria, Serpentes, (Testudines)
Arnoult, **J**[acques] • 25 February 1914–1995 August 10 • France / Madagascar / France • (Anura, Sauria)
Arntzen, **J**an **W**. • ? • England • Anura
Arosemena, **F**ernando **A**. • ? • Panamá • Sauria
Arribas [-Amo], **O**scar **J**. • 13 September 1965 • Spain • Sauria
Arrieta, **D**iego • ? • Uruguay • Anura
Arroyo [-Sánchez], **S**andy **B**[ibiana] • 27 September 1974 • Colombia • Anura
Arthaber, **G**ustav [Adolph] (edler) von • 21 October 1864–1943 April 29 • Austria • *testudines
Ashe, **J**ames [O. P.] • 1925–2004 • Cyprus / South Africa / Kenya / USA / Kenya • Serpentes
Astre, **G**aston [Prosper] • 16 April 1896–1975 February • France • *crocodilia
Ataev, **C**hary **A**taevich • 6 October 1936 • USSR • Sauria
Atatür, **M**ehmet [Kutsay] • 15 February 1947 • Turkey • Anura, Caudata
Audouin (*not* Audoin), **J**ean-**V**ictor • 27 April 1797–1841 November 9 • France • Anura, Sauria
Audubon, **J**ohn **J**ames • 26 April 1785–1851 January 27 • Haiti / France / USA / Scotland / USA • (Sauria)
Auffenberg, **W**alter [Jr.] ("Walt") • 6 February 1928–2004 January 17 • USA • *anura, *caudata, Sauria, *Serpentes, *testudines
Augé, **M**arc • 21 April 1950 • France • *sauria, *serpentes
Auliya, **M**ark [Andre] • 16 March 1963 • South Africa / Germany • Serpentes
Austin, **R**ichard **M**[aurice], Jr. • ? • USA • Caudata
Auth, **D**avid **L**[eslie] • 7 July 1945 • USA • [Sauria], Serpentes
Avci, **A**ziz • ? • Turkey • Serpentes
Averianov, **A**lexander **O**[legovich] (= Alexandr Aver'yanov, "Sasha") • 4 October 1965 • USSR / Russia • *sauria, *serpentes, *testudines
Avery, **D**avid **F**[ranklin] • 30 January 1939 • USA • Sauria
Ávila, **L**uciano **J**avier • 1965 • Argentina • Sauria
Ávila, **R**obson **W**aldemar • ? • Brazil • Serpentes
Ávila-Pires, **T**eresa **C**[ristina] **S**[auer] de • 20 June 1955 • Brazil • Sauria, Serpentes
Awan Khwi Shein • ? • Burma / Myanmar • Anura, [Sauria]
Axtell, **C**arol **A**. • ? • USA • Sauria
Axtell, **R**alph **W**[illiam] • 20 April 1928 • USA • Sauria
Ayala, **S**tephen **C**[harles] • 13 June 1942 • USA / Ecuador / Colombia / USA • Sauria
Ayarzagüena [-Sanz], **J**osé • ? • Venezuela • Anura, Serpentes
Aylmer, **G**uy • ? • Sierra Leone • (Anura)
Aymard, **A**uguste • 5 December 1808–1889 June 26 • France • *anura

B

Babbitt, **L**ewis **H**all • 29 October 1902–1972 September 28 • USA • Serpentes
Babcock, **H**arold **L**[ester] • 30 May 1886–1953 January 21 • USA • Testudines
Babocsay, **G**ergely • 10 February 1969 • Hungary / Israel / Hungary • Serpentes
Bach, **S**tefanie • ? • Germany • Sauria
Bachmayer, **F**riedrich • 10 September 1913–1989 July 25 • Austria • *serpentes, *testudines
Bacon, **J**ames **P**[atterson], Jr. • 22 August 1939–1986 April 4 • USA • Sauria
Bacqué, **A**lfred • ?–? • Paraguay • Serpentes
Badam, **G**. **L**. • ? • India • *testudines
Badham: *see* **Caughley**, **J**udith **A**. **B**adham
Badmaeva, **V**. **I**. • ? • USSR • Sauria
Báez, **A**na **M**aría • ? • Argentina • *anura
Báez, **E**. **C**. • ? • Argentina • Serpentes
Bagnara, **J**oseph **T**[homas] • 26 July 1929 • USA • Anura
Baha El Din, **M**indy (*wife of* Sherif M. Baha El Din) • 1958 • USA / Egypt • (Sauria)
Baha El Din, **S**herif **M**. (*husband of* Mindy Baha El Din) • ? • Egypt • Anura, Sauria
Bahir, **M**ohomed **M**. • ? • Sri Lanka • Sauria
Baig, **K**halid **J**aved • ? • Pakistan • Sauria
Baikie, **W**illiam **B**alfour • 27 August 1825–1864 December 12 • Scotland / Nigeria • Crocodilia, (Sauria, Testudines)
Bailey, **A**. • ? • (? Australia) • [Sauria]
Bailey, **J**ohn **W**endell (lieutenant colonel) • 9 January 1895–1967 • USA • Sauria
Bailey, **J**oseph **R**[andle] • 17 September 1913–1998 September 18 • USA • (Anura), Serpentes
Bailey, **V**ernon [Orlando] • 21 June 1864–1942 April 20 • USA • (Sauria)
Bailón, **S**alvador • 26 August 1956 • Spain / France • *serpentes
Bain, **R**aoul **H**[arley] • 18 October 1970 • Canada / USA • Anura
Baird, **S**pencer **F**[ullerton] • 3 February 1823–1887 August 19 • USA • Anura, Caudata, Sauria, Serpentes, Testudines
Bakde, **R**amesh • ? • India • Anura, Sauria
Baker, **J**ames **K**enneth • ? • USA • Caudata
Bakiev, **A**[ndrey] **G**[ennadyevich] • 11 September 1959 • USSR / Russia • Serpentes
Bakradze, **M**ichael **A**leksandrovich • 20 June 1939 • USSR • Sauria
Baldo, [Juan] **D**iego • ? • Argentina • Anura
Balfour, **A**ndrew • 21 March 1873–1931 January 30 • Scotland / Sudan / England • (Anura)
Balletto, **E**milio • 19 October 1943 • Italy • Anura
Balli, **A**ntonio • ? • Italy • Serpentes
Ballinger, **R**oyce **E**[ugene] • 21 February 1942 • USA • Sauria

Ballion, Ernst **E**. von • 1816–1901 • Russia • Caudata
Balouet, **J**ean-**C**hristophe ("Chris") • 12 November 1956 • France • *crocodilia
Baloutch, **M**. • ? • Iran • Sauria
Bamberg: *see* **Araújo**, **A**lexandre **F**[ernandes] **B**amberg de
Bangoura, **M**ohamed **A**lhassane • ? • Guinea • Anura
Bangs, **O**utram • 12 January 1863–1932 September 22 • USA • (Testudines)
Banta, **B**enjamin **H**[arrison] • 2 January 1927–1999 June 26 • USA • Sauria, Serpentes
Barabanov, **A**ndrei **V**. • ? • USSR / Russia • Sauria
Baran, **İ**brahim • 15 January 1940 • Turkey • Anura, Caudata, Sauria, Serpentes, Testudines
Barbadillo, **L**[uis] **J**[avier] • ? • Spain • Anura
Barbault, **R**obert • 24 January 1943 • France • Sauria
Barbour, **R**oger **W**[illiam] • 5 April 1919–1993 August 26 • USA • Caudata, (Testudines)
Barbour, **T**homas • 19 August 1884–1946 January 8 • USA • Anura, (Caudata), Gymnophiona, *Sauria, Serpentes, (*serpentes), Testudines
Barbosa Rodrigues: *see* **Rodrigues**, **J**oão **B**arbosa
Barboza du Bocage: *see* **Bocage**, **J**[osé] **V**[icente] **B**arboza du
Barker, **D**avid **G**. • 6 January 1952 • USA • Sauria, Serpentes
Barker, **T**racy **M**. • ? • USA • (Serpentes)
Barla, **M**aria **J**ulia • ? • Argentina • Anura
Barnes, **B**en • 1903 • Germany • *serpentes
Barnes, **D**aniel **H**enry • 25 April 1785–1828 October 27 • USA • Caudata
Barnett, **B**rian [Francis] • 18 May 1943 • Australia • (Sauria)
Barrio, **A**velino • 10 August 1920–1979 June 30 • Spain / Argentina • Anura, Serpentes
Barrio-Amorós, **C**ésar **L**[uis] • 3 April 1971 • Spain / Venezuela • Anura, Caudata, Sauria, Serpentes
Barron, **P**. **A**. **R**. • ? • (? England) / Siam • (Serpentes)
Barry, **J**ohn **C**hase • 21 January 1946 • USA • [Sauria]
Bartell, **R**oger • ? • Australia • (Sauria)
Bartenjev, **A**. • ? • USSR • Sauria
Bartholomew, **G**eorge **A**[delbert, Jr.] • 1 June 1919–2006 October 2 • USA • (Serpentes)
Bartlett, **E**dward • (*about* 1836)–1908 April • England / Sarawak / England • Sauria, Testudines
Barton, **B**enjamin **S**mith • 10 February 1766–1815 December 19 • USA • Caudata
Bartram, **W**illiam • 9 (? February *or* April) 1739–1823 July 22 • USA • (Anura, Testudines)
Baskin, **J**onathan **N**[oel] • 25 July 1939 • USA • Sauria
Bas [-Lopez], **S**antiago • ? • Spain • Serpentes
Başoğlu, **M**uhtar • 6 April 1913–1981 February 21 • Turkey • Caudata, (Sauria), Serpentes
Basso, **N**éstor **G**[uillermo] • ? • Argentina • Anura
Bastos, **R**ogério **P**[ereira] • ? • Brazil • Anura
Bataller-Calatayud, **J**osé **R**amón • 10 August 1890 • Spain • *crocodilia, *testudines
Bate, **D**orothea **M**[inola] **A**[lice] ("Dorothy") • 1879–1951 • (? England / Palestine) • *testudines
Battersby, **J**[ames] **C**[larence] • 11 April 1901–1993 December 3 • England • (Anura, Gymnophiona), Sauria, Serpentes
Batuwita, **S**udesh • ? • Sri Lanka • Sauria
Bauer, Aaron M[atthew] • 27 June 1961 • USA • Sauria
Baumann, **F**ranz • 25 December 1885–1961 May 2 • Switzerland • Anura, Sauria, Serpentes
Baur, **G**eorge (= Georg Hermann Carl Ludwig) • 4 January 1859–1898 June 25 • Germany / USA • *sauria, *Testudines
Bavay, **A**rthur [René Jean Baptiste] • 28 April 1840–1923 • France • Sauria, Serpentes
Baxter, **G**eorge **T**. • 19 March 1919–2005 March 20 • USA • (Anura)
Baylor, **E**dward **R**[andall] • 21 January 1914–1994 February 23 • USA • Anura
Beargie, **K**athleen • 10 September 1924 • USA • Sauria
Beauvois: *see* **Palisot** (baron de Beauvois), **A**mbroise-**M**arie-**F**rançois-**J**oseph
Beçak, **W**illy • 1932 • Brazil • Anura
Beccari, **N**ello • 11 January 1883–1957 March 20 • Italy • (Anura, Sauria)
Bechstein, **J**ohann **M**atthäus • 11 July 1757–1822 February 23 • Germany • Caudata, Serpentes, Testudines
Beck, **R**ollo **H**oward • 26 August 1870–1950 November 22 • USA • (Sauria, Testudines)
Beddard, **F**rank **E**[vers] • 19 June 1858–1925 July 14 • England • Anura
Beddome, **R**[ichard] **H**[enry] (colonel) • 11 May 1830–1911 February 23 • England / British India / England • Anura, Gymnophiona, Sauria, Serpentes
Bedford, **S**tuart **H**[ugo] • ? • Australia / New Zealand • *crocodilia
Bedot, **M**aurice • 1859–1927 August • Switzerland • (Serpentes)
Bedriaga, **J**acques de (= Yakov Vladimirovich Bedryaga) • 1854–1906 • Russia / Germany / France / Italy • Anura, Caudata, Sauria, Serpentes, Testudines
Beebe, [Charles] **W**illiam • 29 July 1877–1962 June 4 • USA / Trinidad • (Anura, Sauria)
Beerli, **P**eter • ? • Switzerland / USA • Anura
Behrndt, **A**. **C**. • ? • Australia • Sauria
Beireis (*not* Bereis), **G**ottfried **C**hristoph • 28 February 1730–1809 September 18 • Germany • Anura
Bélanger, **C**harles-**P**aulus • 29 May 1805–1881 November 18 • France / Martinique • [Anura, Sauria, Serpentes, (Testudines)]
Belding, **L**yman • 12 June 1829–1917 November 22 • USA • (Anura, Sauria)
Bell, **B**en (= Benjamin D.) • ? • New Zealand • Anura
Bell, **C**hristopher **J**[ackson] ("Chris") • 12 March 1966 • USA • *crocodilia
Bell, **E**dwin **L**[ewis, II] • 13 May 1926 • USA • Sauria
Bell, **L**. **N**eil • ? • USA • (Caudata), Serpentes
Bell, **T**homas • 11 October 1792–1880 March 13 • England • Anura, Caudata, Sauria, Serpentes, *Testudines
Belluomini, **H**élio **E**merson • ? • Brazil • Serpentes
Benavides, **E**dgar • ? • Chile • Anura
Bendiscioli, **G**iuseppe • ?–? • Italy • Serpentes
Beneden, **P**ierre **J**oseph van • 19 December 1809–1894 January 8 • Belgium • *testudines
Bennett, **D**aniel • ? • Scotland • (Serpentes)
Bennett, **E**dward **T**urner • 6 January 1797–1836 August 21 • England • Crocodilia, (Sauria), Testudines

Bennett, **G**eorge • 31 January 1804–1893 September 29 • England / Australia • (Anura, [*sauria])

Bequaert, **J**oseph **C**harles [Corneille] • 24 May 1886–1982 January 19 • Belgium / USA • (Anura, Sauria, Serpentes)

Bereis: *see* **Beireis**, **G**ottfried **C**hristoph

Berg, **C**arlos (= Frederick Wilhelm Carl) • ("21 March") 2 April 1843–1902 January 19 • Russia / Argentina • Anura, Serpentes

Berg, **D**ietrich **E**. • 1932 • Germany • *crocodilia

Berg, **J**ohannes • ?–? • Germany • (Sauria)

Bergallo, **H**elena **G**odoy • ? • Brazil • Sauria

Berger, **T**homas **J**[oseph] • ? • USA • Anura

Bergna, **S**. **A**. • ? • Argentina • Serpentes

Bergounioux, [Edmond André] **F**[rédéric]-**M**[arie] (père) • 15 October 1900 • France • *testudines

Berman, **D**avid **S**. • 10 January 1940 • USA • *sauria

Bernal-Carlo, **A**manda • ? • Colombia / USA • Serpentes

Bernarde, **P**aulo **S**. • ? • Brazil • Anura

Bernardes, **A**[line] **T**[ristão] • ? • Brazil • Anura

Berry, **C**harles **T**[hompson] • 6 November 1906 • USA • *testudines

Berry, **J**ames **F**[rederick] • 22 December 1947 • USA • Testudines

Berry, **P**. **Y**. (= Pek-yeong) • ? • Malaysia • Anura

Berthelot, **S**abin • 4 April 1794–1880 November 10 • France / Canary Islands • [Sauria]

Berthold, **A**[rnold] **A**[dolph] • 26 February 1803–1861 February 3 • Germany • Anura, Caudata, Gymnophiona, Sauria, Serpentes, Testudines

Bertin, **L**éon • 8 April 1896–1956 February 6 • France • (Anura)

Bertolotto, **C**arolina **E**lena **V**iña • ? • Brazil • Sauria

Besharse, **J**oseph **C**[ulp] • 21 January 1944 • USA • Caudata

Beshkov, **V**ladimir **A**. • ? • Bulgaria • (Caudata)

Bethencourt **F**erreira: *see* **Ferreira**, **J**ulio **B**ethencourt

Betta: *see* **De Betta**, [Francesco] **E**doardo

Beurden: *see* **van Beurden**, **E**ric **K**.

Beutler, **A**xel • 1949 • Germany • Sauria

Beyer, **G**eorge **E**[ugene] • 9 September 1861–1926 June 2 • Germany / USA • (Caudata)

Bezy, **R**obert **L**[ee] • 26 September 1941 • USA • Sauria

Bhaduri, **J**eetendra **L**al • ?–? • India • Anura

Bhatta, **G**opalakrishna • ? • India • Gymnophiona

Bhupathy, **S**[ubramanian] • ? • India • Sauria

Bianco: *see* **Manera de Bianco**, **T**eresa

Bianconi, **G**iovanni **G**iuseppe • 31 July 1809–1878 October 18 • Italy • Anura, Sauria, Serpentes

Bibron, **G**abriel • 1806–1848 March 27 • France • Anura, Caudata, Crocodilia, Gymnophiona, Sauria, Serpentes, *Testudines

Biedermann [-Imhoof], **W**. **G**. **A**dolf • ?–(1862–1876)–? • Switzerland • *testudines

Bielz, **E**duard **A**lbert • 4 February 1827–1898 May 27 • Hungary • [Sauria]

Bigelow, **J**acob • 27 February 1786–1879 January 10 • USA • Serpentes

Biju, **S**[athyabhama] **D**[as] • ? • India • Anura

Bird, **R**oland **T**[haxter] • 29 December 1899–1978 January 24 • USA • *crocodilia

Bischoff, **W**olfgang • ? • Germany • Sauria

Bishop, **S**herman **C**[hauncey] • 18 November 1887–1951 May 28 • USA • Caudata

Biswas, **S**adabrata • ? • India • Sauria, Serpentes

Black, **D**ennis • ? • Papua New Guinea • Sauria

Blainville, **H**[enri]-**M**[arie] **D**ucrotay de • 12 September 1777–1850 May 1 • France • *crocodilia, Sauria, Serpentes, Testudines

Blair, **A**lbert **P**[atrick] (*brother of* W. F. Blair) • 3 November 1913–2004 December 3 • USA • Anura

Blair, **W**[illiam] **F**ranklin ("Frank", *brother of* A. P. Blair) • 25 June 1912–1984 February 9 • USA • (Anura, Serpentes)

Blake, **A**nthony **J**ohn **D**yson • ? • Australia • Anura

Blanc, **C**harles **P**[ierre] • ? • Madagascar / France • (Anura), Sauria

Blanchard, [Charles] **É**mile • 7 March 1819–1900 February 11 • France • Caudata, Sauria

Blanchard, **F**rank **N**[elson] • 19 December 1888–1937 September 21 • USA • (Anura, Sauria), Serpentes

Bland-Sutton, (sir) **J**ohn • 21 April 1855–1936 December 20 • England • (Anura)

Blaney, **R**ichard **M**. • 10 March 1942 • USA • Serpentes

Blanford, **W**[illiam] **T**[homas] • 7 October 1832–1905 June 23 • England / British India / England • Anura, Sauria, Serpentes, Testudines

Blasius, **J**ohann **H**einrich • 7 October 1809–1870 May 26 • Germany • Anura, (Caudata)

Blatchley, **W**[illis] **S**[tanley] • 6 October 1859–1940 May 28 • USA • Caudata

Bleeker, **P**ieter • 10 July 1819–1878 January 23 (*not* 24) • Netherlands / Dutch East Indies / Netherlands • Anura, Gymnophiona, Sauria, Serpentes, Testudines

Blochmann, **F**riedrich [Johann Wilhelm] • 21 January 1858–1931 September 22 • Germany • (Sauria)

Blomberg, **R**olf [David] • 11 November 1912 • Sweden • (Anura, Serpentes)

Blomefield: *see* **Jenyns**, **L**eonard

Blommers-Schlösser, **R**ose **M**[arie] **A**[ntoinette] • 30 January 1944 • Netherlands • Anura

Blotto [-Acuña], **B**oris **L**[eonardo] • 28 July 1979 • Argentina • Anura

Bloxam, **Q**uentin **M**. **C**. • ? • England • Sauria

Blumenbach, **J**ohann **F**riedrich • 11 May 1752–1840 January 22 • Germany • Anura, Caudata, Serpentes, Testudines

Blyth, **E**dward • 23 December 1810–1873 December 27 • England / British India / England • Anura, Sauria, Serpentes, Testudines

Bocage, **J**[osé] **V**[icente] **B**arboza du • 2 May 1823–1907 November 3 • Madeira / Portugal • Anura, Caudata, Gymnophiona, Sauria, Serpentes

Bocourt, **M**arie-**F**irmin • 19 April 1819–1904 February 3 • France / Central America / France • (Anura, Caudata), Crocodilia, Sauria, Serpentes, Testudines

Bocquentin-Villanueva, **J**ean [-Claude Yves] • ? • Venezuela • *crocodilia, *testudines

Boda, **A**nton • ? • Hungary • *testudines

Boddaert, **P**ieter (= Petrus) • 1730–(? 1795 *or* 1796) May 6 • Netherlands • Anura, (Sauria, Serpentes), Testudines

Bodenheimer, **F**riedrich **S**imon • 1897–1959 • Israel • Caudata, Testudines
Boeadi • ? • Indonesia • Anura
Boehme: *see* **Böhme**, **W**olfgang
Boettger (= Böttger), **C**aesar **R**udolf (*nephew of* O. Boettger) • 20 May 1888–1976 September 8 • Germany • Sauria
Boettger (= Böttger), **O**skar (*uncle of* C. R. Boettger) • 31 March 1844–1910 September 25 • Germany • Anura, Caudata, Gymnophiona, *Sauria, Serpentes, Testudines
Bogachev, **V**[ladimir] **V**[ladimirovich] • ("7 February") 19 February 1881–1965 December 11 • Russia / USSR • *testudines
Bogadek, **A**nthony (father) • 15 March 1931 • Italy / Hong Kong • (Sauria)
Bogert, **C**harles **M**[itchill] • 4 June 1908–1992 April 10 • USA • Anura, Caudata, Sauria, (*sauria), Serpentes, Testudines
Bohaska, **D**avid **J**. • ? • USA • [*testudines]
Bohlin, [Anders] **B**irger • 26 March 1898–1990 • Sweden • *testudines
Bohls, **J**ohann [Friedrich Wilhelm] • 1863–? • Germany • (Sauria), Testudines
Böhme (= Boehme), **W**olfgang • 21 November 1944 • Germany • Anura, Caudata, *Sauria, Serpentes
Bohn, **P**éter • ? • Hungary • *testudines
Boie, **F**riedrich (*brother of* H. Boie) • 4 June 1789–1870 March 3 • Germany • Gymnophiona, Sauria, Serpentes
Boie, **H**einrich (*brother of* F. Boie) • 4 May 1794–1827 September 4 • Germany / Netherlands / Java • Anura, Caudata, Serpentes, (Testudines)
Bois-Reymond, **C**laude du • 21 December 1855–1925 • Germany / China • (Anura)
Boistel, **R**enaud • 1973 • France • Anura
Böker, **H**ans • 14 November 1886–1939 April 23 • México / Germany • (Sauria)
Bokermann, **W**erner **C**[arl] **A**[ugust] • 4 July 1929–1995 (? April *or* May) 1 • Brazil • Anura, (Gymnophiona)
Bolaños [-Vives], **F**ederico • 3 August 1961 • Costa Rica • Caudata
Bolkay, **S**tephan **J**oseph [Ladislaws] (= Istvan József Bolkay) • 29 March 1887–1930 August 17 • Hungary / Yugoslavia • *Anura, Caudata, *Sauria, *Serpentes
Bologna, **M**arco • ? • Italy • Caudata
Bölsche, **W**ilheim • 1843–1893 • Germany • Rhynchocephalia
Bonaccorso, **E**lisa • 12 September 1976 • Venezuela / USA • Anura
Bonaparte, **C**harles **L**ucien [Jules Laurent] (= Carlo Luciano, prince) • 24 May 1803–1857 July 29 • France / England / Italy / USA / Italy / France • Anura, Caudata, Sauria, Serpentes, Testudines
Bonett, **R**onald **M**. • ? • USA • Anura, Caudata
Bonnaterre, **P**ierre-**J**oseph (l'abbé) • 24 November 1751–1804 September (? 20 *or* 21) • France • Anura, Caudata, Crocodilia, Sauria, Serpentes, Testudines
Bonpland, **A**imé [-Jacques Alexandre] ("Goujaud") • 28 April 1773–1858 March 11 • France / Argentina / Brazil • [Serpentes]
Bons, **J**acques • 1933 • Morocco / France • Anura, Sauria, Serpentes
Borges-Nojosa, **D**iva **M**aria • 1 April 1963 • Brazil • Anura, Sauria
Boring, **A**lice **M**[iddleton] • 22 February 1883–1955 September 18 • USA / China / USA • Anura, (Caudata)
Borkin, **L**[eo] **Y**[akovlevich] (= Jakovlevich) • 31 August 1949 • USSR / Russia • [Serpentes]
Börner, **A**chim-**R**üdiger • 1 April 1955 • Germany • Sauria
Bornschein, **M**arcos **R**[icardo] • ? • Brazil • Anura
Borre, **A**lfred **P**reud'homme de • 1833–1905 • Belgium • Crocodilia, Sauria
Borroughs: *see* **Burrows, Robert Jonathan**
Borsuk-Białynicka, **M**agdalena • ? • Poland • *anura, *sauria
Bory de Saint-Vincent, **J**ean-**B**aptiste **G**eneviève **M**arcellin (baron de) • 6 July 1778–1846 December 22 • France • Anura, Caudata, Crocodilia, Sauria, Serpentes, Testudines
Bosc [d'Antic], **L**ouis **A**ugustin **G**uillaume • 29 January 1759–1828 July 10 • France / USA / France • (Anura), Sauria, (Testudines)
Boscá [y Casanoves], **E**duardo • 11 February 1843–1924 [*not* 1928] May 28 • Spain • Anura, Caudata, Sauria, Serpentes
Bosch, **H**erman **A**. **J**. in den • ? • Netherlands • Serpentes
Boschma, **H**[ilbrand] • 22 April 1893–1976 July 22 • Netherlands • (Sauria, Serpentes)
Bossu, **J**ean-**B**ernard • 29 September 1720–1792 May 4 • France • [Serpentes]
Bossuyt, **F**ranky • ? • Belgium • Anura
Bostanchi, **H**amid • 3 March 1964 • Iran • Serpentes
Bostic, **D**ennis **L**. • 30 September 1937–1975 March 19 • USA • Sauria, (Serpentes)
Botta, **P**aolo **E**milio (= Paul Émile) • 6 December 1802–1870 March 29 • Italy / France • (Serpentes)
Böttger: *see* **Boettger**, **C**aesar **R**udolf *and* **O**skar
Boubée, **N**érée • 12 May 1806–1863 August • France • Anura
Bouët, **G**eorges [Théodore Louis] • 1869–1957 • France • (Sauria, Serpentes)
Boulenger, **E**dward **G**eorge (*son of* G. A. Boulenger) • 8 May 1888–1946 April 30 • England • Anura, Caudata, Sauria
Boulenger, **G**eorge **A**lbert (*born* Georges, *father of* E. G. Boulenger) • 19 October 1858–1937 November 23 • Belgium / England / Belgium / France • Anura, Caudata, Crocodilia, Gymnophiona, Sauria, Serpentes, *Testudines
Boundy, **J**eff (= John Jeffrey) • 12 April 1959 • USA • Serpentes
Bour, **R**oger [Henri] • 9 July 1947 • France • Testudines
Bourgat, **R**obert • ? • Madagascar / Togo • (Anura)
Bourne, **G**odfrey **R**[oderick] • 20 March 1943 • Guyana / USA • Anura
Bourret, **R**ené [Léon] • 28 January 1884–1957 July 28 • France / French Indochina / France • Anura, Caudata, Sauria, Serpentes, Testudines
Boutan, **L**ouis [Marie Auguste] • 6 March 1859–1934 April 6 • France • Sauria
Bowdich, **T**homas **E**dward • 20 June 1791–1824 January 10 • England • [Sauria, Serpentes]
Boycott, **R**ichard **C**[harlton] • 5 January 1952 • South Africa / Swaziland • Anura
Boyle, **L**awrence **J**. • ? • USA • Sauria
Braack, **H**arold **H**. • ? • South Africa • Anura

Braconnier, Séraphin • ?–? • France • (Anura), Sauria, (Serpentes)
Bradshaw, S[tephen] **D.** • ? • Australia • (Sauria)
Brady, **M**aurice **K**[irby] • 1904–? • USA • Serpentes
Braestrup, **F**[rits] **W**[impffen] • 20 September 1906–1999 February 9 • Denmark • (Gymnophiona, Serpentes)
Bragg, **A**rthur **N**[orris] • 18 December 1897–1968 August 27 • USA • Anura
Brain, **C**[harles] **K**[imberlin] ("Bob") • 1931 • Rhodesia / South Africa • Sauria, Serpentes
Braithwaite, **R**ichard **W.** • ? • Australia • Sauria
Bräm, **H**einrich • ? • Switzerland • *testudines
Bramble, **D**ennis **M**[arley] • ? • USA • Testudines
Brame, **A**rden **H**[owell], Jr. II • 19 March 1934–2004 August 19 • USA • *Caudata
Braña [-Vigil], **F**lorentino • ? • Spain • Serpentes
Branch, **W**illiam **R**[oy] ("Bill") • 12 May 1946 • England / South Africa • Anura, Sauria, Serpentes
Brandão, **R**euber **A**lbuquerque • ? • Brazil • Anura
Brandon, **R**onald **A**[rthur] • 3 December 1933 • USA • Caudata, Sauria
Brandoni de Gasparini, **Z**ulma [**N**elida] • 15 May 1944 • Argentina • *crocodilia, (*testudines)
Brandt, **B**artholomew **B**randner • 30 July 1898–1983 May • USA • Anura
Brandt, **J**ohann **F**riedrich (= Fedor Fedorovich) • ("13 May") 25 May 1802–1879 July 15 ("July 3") • Germany / Russia • Serpentes, Testudines
Branner, **J**ohn **C**asper • 4 July 1850–1922 March 1 • USA • (Anura)
Brasileiro, **C**inthia **A.** • ? • Brazil • Anura
Brattstrom, **B**ayard **H**[olmes] • 3 July 1929 • USA • *anura, *sauria, *serpentes, *testudines
Brauer, **A**ugust [**B**ernhard] • 3 April 1863–1917 September 10 • Germany • (Sauria, Serpentes)
Braun, **C**[ristina] **A**[ssuncão] **S**[irangelo] (*wife of* P. C. Braun) • ? • Brazil • Anura
Braun, **M**ax (= Maximilian Gustav Christian Carl) • (? 13 *or* 30) September 1850–1930 February 22 • Germany • Sauria
Braun, **P**[edro] **C**[anisio] (*husband of* C. A. S. Braun) • ?–? • Brazil • Anura
Bravard, [Pierre-Joseph] **A**uguste • 18 June 1803–1861 March 20 • France / Argentina • *crocodilia, *testudines
Bravo, **T**elésforo • ? • Canary Islands • *sauria
Brazenor, **C**harles **W**alter • 2 April 1897–1979 April • England / Australia • Serpentes
Brazil [Mineiro da Campanha], **V**ital • 28 April 1865–1950 May 8 • Brazil • (Serpentes)
Breder, **C**harles **M**arcus, Jr. • 25 June 1897–1983 October 28 • USA • (Anura)
Brelih, **S**avo • ? • Yugoslavia • Sauria
Bréthes, **J**ean (= Juan) • 24 February 1871–1928 July 2 • France / Argentina • Serpentes
Breuil, **M**ichel • ? • France • Caudata
Breukelen: *see* **van Breukelen**, **F**rank
Briceño-Rossi, **A**ntonio **L**eocadio • ? • Venezuela • Serpentes
Brickell, **J**ohn, • (? 1749)–(? 1809 *or* 1810) • USA • Serpentes
Brieva [-Vasquez], **L**ila **M**[aria del] • 28 February 1958 • Chile • Anura
Brimley, **C**lement **S**amuel (*brother of* H. H. Brimley) • 18 December 1863–1946 July 23 • England / USA • (Anura), Caudata, Testudines
Brimley, **H**erbert **H**utchinson (*brother of* C. S. Brimley) • 7 March 1861–1946 April 4 • England / USA • (Caudata)
Brinkerink, **J**ohann **P.** • ? • Netherlands • *testudines
Brinkman, **D**onald **B.** • ? • Canada • *crocodilia
Britto-Pereira, **M**ônica **C**ox de • ? • Brazil • Anura, Gymnophiona
Brizzi, **R**ossana • 30 April 1948 • Italy • Sauria
Broadley, **D**onald **G**[eorge] • 29 May 1932 • England / Rhodesia / Zimbabwe • Anura, Sauria, Serpentes, Testudines
Brocchi, **P**aul [-Louis-Antoine] • 2 May 1838–1898 August • France • Anura, Caudata, (Sauria), Serpentes
Brochu, **C**hristopher **A**[ndrew] ("Chris") • 12 August 1967 • USA • *crocodilia
Brock, **V**ernon **E**[ugene] • 24 June 1912–1971 January • USA • Serpentes
Brode, **J**ohn **M**[artin] • 31 May 1933 • USA • Caudata
Brodie, **E**dmund **D**[arrell], Jr. • 29 June 1941 • USA • Anura, Caudata, Sauria, Serpentes
Broghammer, **S**tefan • ? • Germany • (Serpentes)
Broili, **F**erdinand • 11 April 1874–1946 April 30 • Germany • (Caudata)
Broin: *see* **Lapparent de Broin**, **F**rance de
Bromme, **T**raugott • 1802–1866 September 4 • Germany • [Anura, Serpentes]
Brongersma, **L**[eo] **D**[aniël] • 17 May 1907–1994 July 24 • Netherlands • Anura, *Sauria, Serpentes
Brongniart, **A**lexandre • 5 February 1770–1847 October 7 • France • (*crocodilia, Sauria, Testudines)
Bronn, **H**einrich **G**eorg • 3 March 1800–1862 July 5 • Germany • *testudines
Brooker, **M**ichael **G.** • ? • Australia • (Sauria)
Brookes, **J**oshua • 24 November 1761–1833 January 10 • England • Caudata, (Sauria)
Broom, **R**obert • 30 November 1866–1951 April 6 • Scotland / South Africa • *rhynchocephalia, Sauria, (Serpentes)
Brown, **A**rthur **E**rwin • 14 August 1850–1910 October 29 • USA • Anura, Serpentes
Brown, **B**arnum • 12 February 1873–1963 February 5 • USA • (*crocodilia, Serpentes)
Brown, **B**ryce **C**[ardigan] (*husband of* Lillian M. Brown) • 7 May 1920 • USA • Anura, Sauria, Serpentes
Brown, **J**ason **L**[ee] • 27 June 1980 • USA • Anura
Brown, **J**ill **R.** (*wife of* Lauren E. Brown) • 20 February 1943 • USA • Anura
Brown, **L**auren **E**[vans] (*husband of* Jill R. Brown) • 4 September 1939 • USA • Anura
Brown, **L**illian **M.** (*wife of* Bryce C. Brown) • 1 August ? • USA • (Anura)
Brown, **R**afe **M**[arion] • ? • USA • Anura, Sauria
Brown, **T**ed (= Teddy Lee) • 3 April 1943 • USA • (Sauria)
Brown, **W**alter **C**[reighton] • 18 August 1913–2002 July 15 • USA / Philippines / USA • Anura, Caudata, Sauria, Serpentes
Bruce, **R**ichard **C**[onrad] • 2 June 1936 • USA • [Caudata]
Brüggemann, **F**riedrich • 1850–1878 April 6 • Germany • (Sauria)

Bruguière, **J**ean-**G**uillaume • 1750–1799 October 1 • France • Serpentes, Testudines
Brun, **C**aroline • ? • USA • Anura
Brunner, **G**eorg • ? • Germany • *caudata, *sauria, *serpentes
Bruno, **S**ilvio • 1941 • Italy • Caudata
Brushko, **Z**oya **K**arpovna • 25 March 1931 • USSR • Sauria
Brygoo, **É**douard **R**aoul • 22 April 1920 • France / Madagascar / France • (Anura), Sauria, (Testudines)
Bryson, **R**obert **W**[illiam], Jr. • ? • USA • Serpentes
Buchholz, **K**arl **F**[reiderich] • 1 February 1911–1967 July 1 • Germany • Sauria, Serpentes
Buchholz, **R**einhold **W**ilhelm • 2 October 1837–1876 April • Germany • Anura, Sauria, Serpentes
Buck, **E**mil [August] • 20 April 1840–1899 December 17 • Germany • Sauria
Buckeridge, **J**ohn [St James] **S**tewart • 6 December 1949 • New Zealand / Australia • *testudines
Buckland, **W**illiam • 12 March 1784–1856 August 24 • England • *crocodilia
Buckley, **L**arry **J**. • ? • USA • Sauria
Budak, **A**bidin • 9 November 1943 • Turkey • Sauria, Testudines
Budde, **M**ichael • ? • Germany • Testudines
Buden, **D**onald **W**[illiam] • 1943 • USA • (Sauria), Serpentes
Budgett, **J**ohn **S**amuel • 16 June 1872–1904 January 19 • England • Anura
Buffetaut, **E**ric [Charles Nicolas] • 19 November 1950 • France • *crocodilia
Buffon, **G**eorges-**L**ouis **L**eclerc (= Georges-Louis le Clerc, comte de Buffon) • 7 September 1707–1788 April 16 • France • [Crocodilia, Serpentes]
Bui Ngok **T**hanh • ? • Vietnam • Sauria, Serpentes
Bujes, **C**lóvis [de Souza] • 2 August 1964 • Brazil • Sauria
Buller, (sir) **W**alter **L**awry • 9 October 1838–1906 July 19 • New Zealand / England • Rhynchocephalia, Sauria
Bullini, **L**uciano • 11 July 1941 • Italy • Anura, Caudata
Bullock, **D**. **J**. • ? • Scotland • Sauria
Bulmer, **R**alph **N**[eville] **H**[ermon] • 3 April 1928–1988 July 18 • England / New Guinea / New Zealand • (Anura)
Bumzahem (*changed to* C. Boyd G. Campbell), **C**arlos **B**. • 27 July 1934 • USA • Anura, Caudata, Sauria
Burbidge, **A**ndrew **A**. • ? • Australia • (Sauria)
Burbrink, **F**rank **T**[homas] • ? • USA • Serpentes
Burchell, **W**illiam **J**ohn • 23 July 1781–1863 March 23 • England / South Africa / Brazil / England • (Sauria), Serpentes
Burden, **W**[illiam] **D**ouglas • 24 September 1898–1978 November 14 • USA • (Sauria)
Buresch, **I**wan (= Ivan) • 1885–1980 • Bulgaria • Caudata, (Sauria), Serpentes
Bürger, [Heinrich] **O**tto [Wilhelm] • 4 May 1865–1945 January 18 • Germany / Chile / Germany • (Anura, [Sauria])
Burger, **M**arius • 16 October 1966 • South Africa • Anura
Burger, **W**[illiam] **L**eslie [Jr.] • 23 May 1925–1988 April 1 • USA / Japan / USA • Anura, Caudata, Sauria, Serpentes
Burley, **F**. **W**illiam • 20 August 1941 • USA • Sauria
Burmeister, [Karl] **H**ermann [Konrad] (= Carlos Germán Conrado) • 15 January 1807–1892 May 2 • Germany / Argentina • Anura, *crocodilia, Sauria, (Serpentes)
Burns, **D**ouglas **M**[urray] • 7 February 1935–1976 November • USA / Thailand • Caudata
Burns, **N**eville • ? • Australia • (Sauria)
Burrowes [-Gomez], **P**atricia **A**. • 8 April 1961 • USA / Colombia / USA / Puerto Rico • Anura
Burrows (*not* Borroughs), **R**obert **J**onathan • 29 January 1958 • USA • Anura
Burt, **C**harles **E**[arle] (*husband of* May Danheim Burt) • 12 August 1904–1963 July 13 • USA • Sauria
Burt, **M**ay **D**anheim (*wife of* Charles E. Burt) • 15 January 1898–1988 April 22 • USA • Sauria
Burton, **E**dward • ?–(1836)–? • England • Sauria
Burton, **T**homas **C**harles • ? • Australia • Anura
Busack, **S**tephen **D**[ana] • 3 October 1944 • USA • Anura, Sauria
Buscalioni, **A**ngela **D**[elgado] • 8 December 1958 • Morocco / Spain • *crocodilia
Bush, **F**rancis **M**[arion] • 5 September 1933 • USA • Caudata
Busse, **K**laus • ? • Germany • Anura
Bustamante [-Rosero], **M**artín **R**[afael] • ? • Ecuador • Anura
Bustard, **H**. **R**obert • 1938 • Australia / Scotland • Sauria
Bustos-Obregón, **E**duardo • ? • Chile • [Anura]
Butler, **A**mos **W**[illiam] • 1 October 1860–1937 August 5 • USA • (Serpentes)
Butler, **A**[rthur] **L**[ennox] • 1873–1939 • Malaya / Sudan • (Anura, Sauria), Serpentes
Butler, **W**illiam **H**enry ("Harry") • 25 March 1930 • Australia • (Sauria, Serpentes)
Bwong, **B**eryl **A**[koth] • 19 May 1975 • Kenya • Anura

C

Cabrera, **M**ario **R**[oberto] • 29 July 1954 • Argentina • Sauria
Cádiz, **A**ntonio • ? • Cuba • Anura
Cadle, **J**ohn **E**[verett] • 7 December 1953 • USA • Anura, Sauria, Serpentes
Cagle, **F**red **R**[ay] • 9 October 1915–1968 August 8 • USA • Testudines
Cai Chunmo • 8 February 1941 • China • Caudata
Caicedo: *see* **Rancés** Caicedo, José
Calabresi, **E**nrica • 1891–1944 • Italy • Anura, Sauria
Caldwell, **D**avid **K**eller • 6 August 1928–1990 June 1 • USA • Testudines
Caldwell, **J**analee **P**[aige] • 20 July 1942 • USA • Anura, Sauria, (Serpentes)
Călinescu, **R**aul [I.] • 27 December 1901–1970 August 30 • Romania • Anura, Caudata, Testudines
Call, **R**[ichard] **E**llsworth • 13 May 1856–1917 • USA • Serpentes
Callison, **G**eorge • 1 June 1940 • USA • *rhynchocephalia
Camarillo-Rangel, **J**osé **L**[uis] • 27 April 1955–2002 May 28 • México • Sauria
Camerano, **L**orenzo (= Laurentio) • 9 April 1856–1917 November 22 • Italy • Anura, Sauria, Serpentes
Camp, **C**arlos **D**[uane] • 15 April 1955 • USA • Caudata
Camp, **C**harles **L**ewis • 12 March 1893–1975 August 14 • USA • *Anura, Caudata, Sauria
Campbell, **J**onathan **A**[twood] • 13 May 1947 • USA • Anura, Caudata, Gymnophiona, Sauria, Serpentes

Campbell: *see also* **Bumzahem**, Carlos **B**.
Campden-Main, Simon **M**[ichael] • 7 October 1944 • England / Canada / USA • Serpentes
Campo: *see* **Martín del Campo** [y Sánchez], **R**afael
Canedo, **C**larissa • ? • Brazil • Anura
Cann, [Robert] **J**ohn • 19 March 1951 • Australia • (Serpentes), Testudines
Cannatella, **D**avid **C**[harles] • 25 December 1954 • USA • Anura
Canning, **K**[atharine] **L**aura • 9 September 1969 • England • Sauria
Canseco-Márquez, **L**uis • 30 January 1971 • México • Anura, Caudata, Serpentes
Cantor, **T**heodor **E**dvard (= Theodore Edward, esquire) • 6 February 1809–1860 March 26 • Denmark / Germany / British India / *died at sea* • Anura, Sauria, Serpentes, Testudines
Capellini, **G**iovanni • 23 August 1833–1922 May 28 • Italy • *crocodilia, *testudines
Capetillo [-Arcos], **J**uana [Adelina] • 13 May 1949–2006 • Chile • Anura
Capolongo, **D**omenico • 4 October 1938 • Italy • Sauria, Serpentes, (Testudines)
Capula, **M**assimo • 13 August 1956 • Italy • Anura, Caudata
Capurro [-S.], **L**uis **F**. • ? • Chile • Anura
Cara, **A**lberto • 1847–? • Italy • Sauria
Caramaschi, **U**lisses • ? • Brazil • Anura, Gymnophiona, Serpentes
Carcerelli, **L**uiz **C**laudio • ? • Brazil • Anura
Cardoso, **A**dão **J**[osé] • 13 January 1951–1997 February 20 • Brazil • Anura
Čarevskij: *see* **Tsarevsky**, **S**ergey **F**edorovich
Carfì, **S**alvatore • 3 June 1939 • Italy • Sauria
Carié, **P**aul • ? • Mascarene Islands • (*serpentes)
Carlleyle, **A**. **C**. **L**. • ?–(1869)–? • (? England) / British India • Sauria, Serpentes
Carmely, **H**illa • ? • Israel • [Sauria, Serpentes]
Carnaval, **A**na **C**arolina **O**. **Q**. • 29 April 1972 • USA • Anura
Carpenter, **C**harles **C**[ongden] ("Chuck") • 2 June 1921 • USA • (Sauria)
Carpenter, **G**[eoffrey] **D**[ouglass] **H**ale • 26 October 1882–1953 January 30 • England / Uganda / England • (Sauria, Serpentes)
Carr, **A**rchie (= Archibald Fairly, Jr.) • 16 June 1909–1987 May 21 • USA • (Anura), Caudata, (*Sauria), Testudines, (*testudines)
Carranza, **S**alvador • 20 April 1968 • Spain • Caudata, Sauria
Carrillo de **E**spinoza, **N**elly • 2 November 1932 • Perú • (Anura, Sauria), Serpentes
Carrington da Costa: *see* **Costa**, **J**oão **C**arrington [Simões] da
Carrizo, **G**ustavo **R**. • ? • Argentina • Anura
Carruthers, **V**incent **C**raig • 16 June 1942 • South Africa • Anura
Carter, **H**enry [John] • 1813–1895 • British India • (Sauria), *testudines
Carvalho, **A**ntenor **L**[eitão] de • 15 April 1910–1985 December 11 • Brazil • Anura, Crocodilia, (Sauria), Serpentes
Carvalho-e-Silva, **A**na **M**[aria] **P**[aulino] **T**elles de (*born* Ana Telles, *wife of* Sergio Carvalho-e-Silva *from* 1983) • 25 May 1961 • Brazil • Anura
Carvalho-e-Silva, **S**ergio **P**otsch de (*husband of* Ana Carvalho-e-Silva *from* 1983) • 22 May 1955 • Brazil • Anura, (Serpentes)
Casamiquela, **R**odolfo **M**[agín] • 11 December 1932 • Argentina • *anura
Cascon, **P**aulo • ? • Brazil / USA / Brazil • Anura, Gymnophiona
Case, **E**[rmine] **C**[owles] • 11 September 1871–1953 • USA • *crocodilia
Case, **J**udd **A**. • ? • USA • *sauria
Case, **T**ed **J**[oseph] • 19 July 1947 • USA • Sauria
Castañeda [-Menendez], **F**ranklin **E**[nrique] • 26 December 1972 • Honduras / Costa Rica • Anura, Serpentes
Castanho, **L**uciano **M**endes • 9 August 1968 • Brazil • Anura
Castaño, **C**arlos **J**ulio • ? • Colombia • Anura
Castelnau: *see* **Laport**, **F**rançois [Louis Nompar de Caumont, comte de Castelnau]
Castoe, **T**odd **A**[dam] • 20 November 1977 • USA • Sauria
Castro, **L**. **P**. • ? • Argentina • Sauria
Castro [-Herrera], **F**ernando • 17 May 1952 • Colombia • Anura, Sauria
Castroviejo, **J**avier • ? • Spain • Sauria
Catenazzi, **A**lessandro • 21 June 1975 • Switzerland / Perú / USA • (Anura)
Catesby, **M**ark • (? 31 Dec 1682 *or* 24 March 1683 *or* 3 April 1683)–1749 December 23 • England / Carolina / England • (Anura, Serpentes), Testudines
Cattaneo, **A**ugusto • ? • Italy • Serpentes
Cattoi, **N**oemi **V**. • ? • Argentina • *testudines
Caughley, **J**udith **A**. **B**adham • ? • Australia • Sauria
Caup: *see* **Kaup**, **J**ohann **J**akob
Cautley, (sir) **P**roby **T**homas • 3 January 1802–1871 January 25 • England / British India / England • *crocodilia, *testudines
Cei, **J**osé **M**[iguel] • 23 March 1918–2007 January 8 • Italy / Argentina / Portugal / Argentina • Anura, Sauria, (Serpentes)
Censky, **E**llen **J**[oan] • 20 September 1955 • USA • Sauria
Cépède: *see* **Lacépède**, **B**ernard **G**ermain **É**tienne [de la Ville sur Illon] (comte de)
Cerdas [-Fallas], **L**uis [Gonzalo] • 24 June 1941 • Costa Rica • Serpentes
Černov: *see* **Chernov**, **S**ergius **A**leksandrovich
Céspedez, **J**orge **A**bel • ? • Argentina • Anura, Sauria
Cetti, **F**rancesco • 9 August 1726–1778 November 20 • Italy • (Sauria, Serpentes)
Chabanaud, **P**aul • 30 November 1876–1959 February 27 • France • Anura, Sauria, Serpentes
Chacón, **A**ndrés • ? • Venezuela • Anura
Chamberlain, **D**ee **A**nn • ? • USA • Caudata
Chan-ard, **T**anya • ? • Thailand • Anura, (Serpentes)
Chanda, **S**hyamal **K**umar • ? • India • Anura
Chaney, **A**llan **H**[arold] • 11 December 1923 • USA • (Sauria)
Chang **M**ang-ven (= Zhang Mengwen, *formerly* Chang Ling-yu) • 5 July 1903–1993 October 2 • China / France / China • Anura, Caudata, Serpentes
Chang **T**so-**k**an • ? • China • Anura
Chang **T**song-**h**an • ? • China • Testudines

Chang: *see also* **Zhang F**uji, **Zhang M**inghua, *and* **Zhang Y**uxia

Chanhome, **L**awan • 24 June 1964 • Thailand • Sauria, Serpentes

Chani, **J**osé **M**[aría] • 10 December 1943 • Argentina • *sauria

Channing, **A**lan [Eric] • 26 September 1948 • England / South Africa • Anura

Chantell, **C**harles **J**[oseph] • 19 May 1931 • USA • *anura, *caudata

Chao: *see* **Zhao E**rmi *and* **Zhao K**entang

Chaparro [-Auza], **J**uan **C**arlos • 4 April 1973 • Perú • Anura, Sauria

Chapin, **J**ames **P**aul • 9 July 1889–1964 April 5 • USA • (Anura, Sauria, Testudines)

Chapman, **A**ndrew • ? • Australia • (Sauria)

Charas, **M**oïse (*or* Moyse) • 2 April 1619–1698 January 21 • France / England / Netherlands / Spain / France • (Serpentes)

Cheke, **A**nthony **S**. • ? • England • Sauria

Chelazzi, **G**uido • ? • Italy • (Serpentes)

Chen Fuguan • 3 August 1929 • China • Serpentes

Chen Gongxin • 1936 • China • *testudines

Chen Huojie • ? • China • Anura

Chen L.-s. • ? • China • Anura

Chen Shyh-**h**wang • ? • Taiwan • Sauria

Chen Szu-**l**ung • ? • Taiwan • Anura, Sauria

Chen Yue-**s**hung • ? • Taiwan • Anura

Cheng Hsien-**y**u • 1953 • Taiwan • Sauria

Cheng: *see also* **Zheng J**i

Cherchi, **M**[aria] **A**[delaide] • 27 June 1927–1985 July 18 • Italy • Anura, Sauria

Cherepanov, **G**ennady **O**legovich • ? • USSR • *crocodilia

Cherlin, **V**ladimir **A**[lexandrovich] • 6 March 1951 • USSR • Serpentes

Chermock, **R**alph **L**[ucien] • 25 August 1918–1977 November • USA • (Caudata)

Chernov, **S**[ergius] **A**[leksandrovich] • 28 July 1903–1964 January 2 • Russia / USSR • Anura, *caudata, Sauria, Serpentes, (*testudines)

Chiappe, **L**uis **M**. • 18 June 1962 • USA • *crocodilia

Chigira, **Y**oshinori • ? • Japan • (Serpentes)

Chippindale, **P**aul **T**[homas] • 15 August 1963 • USA • Caudata, Serpentes

Chirio, **L**aurent [Vincent] • 3 November 1957 • France • Sauria

Chiszar, **D**avid [Alfred] • 21 October 1944 • USA • Anura, Sauria, Serpentes

Chitoku, **T**sutomu • ? • Japan • *testudines

Chkhikvadze (= Čkhikvadze *or* Tchikvadze), **V**iacheslav **M**[ikhailovich] • 4 January 1940 • USSR / Georgia • *caudata, *sauria, *Testudines

Chopra, **S**hiv **R**aj **K**umar • 1931 • India • *crocodilia

Chosatzky: *see* **Khosatzky**, **L**eo (= Lev) **I**[saakovich]

Chou Wen-**h**ao • ? • Taiwan • Anura, Sauria

Chow Ming-**c**hen (= Zhou Mingzhen) • 9 November 1918 • China • *testudines

Chrapliwy, **P**eter **S**[tanley] ("Pete") • 20 October 1923 • USA • Sauria, Serpentes

Christie, **M**iguel **I**. • ? • Argentina / USA • Sauria

Christoph, **H**ugo **T**heodor (= Hugo Fedorovich) • 4 April 1831–1894 • (? Russia) • Serpentes

Christov, **L**. • ? • Bulgaria • *testudines

Chumakov: *see* **Tschumakov**, **I**van **S**ergeevich

Churchill, **G**reg • ? • Australia • (Serpentes)

Cifelli, **R**ichard **L**[awrence] • ? • USA • *sauria

Ciobanu, **M**ihai • ? • Romania • *testudines

Cirer, **A**ntonia **M**. • ? • Spain • Sauria

Cisneros-Heredia, **D**iego **F**[rancisco] • ? • Ecuador • Anura

Cisternas, **R**afael • ?–? • Spain • (Anura)

Clark, **E**rica **D**. (*wife of* Richard J. Clark) • ? • Norway • (Sauria)

Clark, **H**erbert **C**harles • 11 November 1877–1960 November 8 • USA • (Gymnophiona, Serpentes)

Clark, **H**ubert **L**yman • 9 January 1870–1947 July 31 • USA • (Serpentes)

Clark, **J**ohn • 26 July 1909–1967 • USA • *testudines

Clark, **R**ichard **J**. (*husband of* Erica D. Clark) • ? • Norway • (Sauria)

Clark (*not* Clarke), **W**illiam **B**ullock • 15 December 1860–1917 July 27 • USA • *crocodilia, *testudines

Clarke, **B**arry **T**[homas] • 6 June 1951 • England • Anura

Clausen, **R**obert **T**[heodore] • 26 December 1911–1981 December 31 • USA • Caudata

Clay, **B**rian **T**. • ? • Australia • (Sauria)

Clay, **W**illiam **M**[arion] ("Bill") • 3 October 1906–1983 April 1 • USA • Serpentes

Clayton, **D**avid **R**. • ? • USA • Caudata

Clench, **W**illiam **J**ames • 24 October 1897–1984 February 22 • USA • (Sauria)

Clergue [-Gazeau], **M**onique [Andrée] • 20 November 1931 • France • Caudata

Cliff, **F**rank **S**[amuel] • 3 April 1928–2000 October 1 • USA • Serpentes

Clift, **W**illiam • 14 February 1775–1849 June 20 • England • Crocodilia, (Sauria)

Cloquet, [Joseph] **H**ippolyte • 10 March 1787–1843 • France • Caudata, Sauria, Serpentes

Clos, **L**ynne **M**[obley] • 2 September 1955 • USA • *sauria

Cobo: *see* **Serra-Cobo**, **J**ordi

Cochran, **D**oris **M**[able] • 18 May 1898–1968 May 22 • USA • Anura, (Caudata), Sauria, Serpentes

Cocroft, **R**eginald **B**[ifield, III] • 5 January 1960 • USA • Anura

Cocteau, **J**ean-**T**héodore • 15 March 1798–1838 May 13 • France • Anura, Sauria, Serpentes, Testudines

Cocude-Michel, **M**arguerite • ? • France • *rhynchocephalia

Cogger, **H**arold **G**[eorge] ("Hal") • 1935 • Australia • Anura, Sauria, Serpentes

Colbert, **E**dwin **H**[arris] ("Ned") • 28 September 1905–2001 November 15 • USA • *crocodilia

Cole, **C**harles **J**[ames] • ? • USA • Sauria

Cole, **L**eon **J**[acob] • 1 June 1877–1948 February 17 • USA • Sauria, Serpentes

Colenso, **W**illiam (reverend) • 17 November 1811–1899 February 10 • England / New Zealand • Rhynchocephalia, Sauria

Colgan, **D**onald **J**. • ? • Australia • Sauria

Collett, **R**obert • 2 December 1842–1913 January 27 • Norway • (Anura, Sauria)

Colli, **G**uarino **R**inaldi • 22 November 1962 • Brazil / USA / Brazil • Sauria

Collin de Plancy, Victor [Émile Marie Joseph] • 1853–(? 1922 *or* 1923) • France • (Anura)

Collins, **R**[obert E] **L**ee • 22 October 1898–1952 April 22 • USA • *testudines

Coloma [-Roman], **L**uis **A**[urelio] • 4 March 1962 • Ecuador / USA / Ecuador • Anura, Sauria

Colombelli, **B**runella • 25 April 1952 • Switzerland • Anura

Combaz, **J**. • ? • (? Belgium) • Anura

Comeau, **N**oël-**M**. • ?–1976 • Canada • Caudata

Conant, **I**sabelle [de Peyster] **H**unt (*wife of* Carl Kauffeld *from* ? *to* ?, *and wife of* Roger Conant *from* 1947 *to* 1976) • 17 May 1901–1976 November 3 • USA • (Serpentes)

Conant, **R**oger (*husband of* Isabelle Hunt Conant *from* 1947 *to* 1976) • 6 May 1909–2003 December 19 • USA • (Caudata), Serpentes, Testudines

Condon, **H**erbert **T**homas • 27 February 1912–1978 January 12 • Australia • Anura, Sauria

Constable, **J**ohn **D**avidson • ? • (? USA) • Sauria, Serpentes

Cook, **H**arold **J**ames • 31 July 1887–1962 September 29 • USA • *caudata, (*sauria)

Cook, **R**obert • ? • Australia • (Testudines)

Cooper, **B**yrum **W**. • ? • USA • Anura

Cooper, **E**dwin **L**[owell] • 23 December 1936 • USA • (Gymnophiona)

Cooper, **J**[ames] **G**[raham] • 19 June 1830–1902 July 19 • USA • Caudata, (Sauria), [Serpentes,] Testudines

Cope, **E**dward **D**rinker • 28 July 1840–1897 April 12 • USA • *Anura, Caudata, *Crocodilia, Gymnophiona, *Sauria, *Serpentes, *Testudines

Copland, **S**tephen **J**[ohn] • ?–1981 August 19 • Australia • Anura, Sauria

Corben, **C**[hris] **J**. • ? • Australia • Anura

Cordeiro, **C**[armem] **L**[úcia dos] **S**[antos] • ? • Brazil • Serpentes

Cordes, **J**ames **E**[dwin] • 5 March 1955 • USA • Sauria

Córdova [-Santa Gadea], **J**esús **H**[umberto] • ? • Perú • (Anura)

Cornalia, **E**milio [Baldassare Maria] • (? 25 August 1824 *or* 24 August 1825)–1882 June 8 • Italy • Anura, Testudines

Corredor E., **V**ladimir • ? • Colombia • Sauria

Corsi, **M**. • ? • Italy • Sauria

Cosgriff, **J**ohn **W**[illiam, Jr.] • 10 November 1931–1985 April 28 • USA • (*serpentes)

Costa, **E**lizabeth **M**aria **M**amede da • ? • Brazil • Sauria

Costa, **G**abriel **C**. • ? • Brazil • Sauria

Costa, **J**oão **C**arrington [Simões] da • 1891 • Portugal • *testudines

Costa, **O**ronzo **G**abriele • 26 August 1787–1867 November 7 • Italy • Caudata, *crocodilia, *rhynchocephalia, Sauria

Cott, **H**ugh **B**[amford] • 6 July 1900–1987 April 18 • England / British East Africa / England • (Anura), Sauria, Serpentes

Coues, **E**lliott • 9 September 1842–1899 December 25 • USA • Serpentes

Couper, **P**atrick **J**. • ? • New Zealand / Australia • Sauria, Serpentes

Courtice, **G**illian **P**. • ? • Australia • Anura

Coutinho: *see* **Silva Coutinho**, **J**oão **M**artino (major) da

Couturier, **G**ustavo **A**. • ? • Argentina • Serpentes

Couvering: *see* **Van Couvering**, **J**udith **A**nne **H**arris

Covacevich, **J**eanette [A.] • 1945 • Australia • Sauria, Serpentes

Coventry, **A**[lbert] **J**ohn • 24 April 1936 • Australia • Sauria

Cowles, **R**aymond **B**[ridgman] ("Ray") • 1 December 1896–1975 December 7 • South Africa / USA • (Sauria, Serpentes)

Cragin, **F**rancis **W**hittemore ("Frank") • 4 September 1858–1937 June 15 • USA • Sauria, *testudines

Cranbrook, **G**athorne **G**athorne-**H**ardy (5th earl of, *formerly* lord Medway) • 20 June 1933 • England / Malaya / England • Sauria

Crawford, **A**ndrew **J**[ackson] • ? • USA / Panamá • Anura

Creaser, **E**dwin **P**[hilip] • 6 February 1907–1981 March • USA • (Testudines)

Crenshaw, **J**ohn **W**[alden], Jr. • 17 May 1923–2004 April 23 • USA • Caudata

Crespo, **E**duardo **G**. • ? • Portugal • Anura, (Caudata)

Cretzschmar, **P**hilipp **J**akob • 11 June 1786–1845 May 4 • Germany • [Crocodilia]

Creveld, ? • ?–(1809)–? • (? Netherlands *or* Germany) • Sauria

Crombie, **R**onald **I**[an] • 28 January 1949 • USA • Anura, Sauria

Crouzel, **F**ernand [Jacques] • 1913 • France • *testudines

Crump, **M**artha **L**[ynn] • 23 August 1946 • USA • Anura

Crutsinger, **G**regory **M**[itchell] • 23 July 1980 • USA • Anura

Cruz, **C**arlos **A**lberto **G**onçalves da • 17 July 1944 • Brazil • Anura

Cruz, **F**élix **B**[enjamin] • ? • Argentina / USA • Sauria, Serpentes

Cruz-Díaz, **G**ustavo **A**[dolfo] • ? • Honduras • Caudata, Sauria, Serpentes

Csiki, **E**rnö (= Ernestus) • 22 October 1875–1954 July 10 • Hungary • (Anura), Serpentes

Csiki, **Z**oltán • ? • Romania • *anura

Cuc: *see* **Ho Thu Cuc**

Cuentas, **D**aniel **A**lfredo • ? • (? Colombia) • Anura

Cuesta-Terrón, **C**arlos • 7 May 1897–1948 May 21 • México • Sauria, Serpentes

Cuevas, **C**ésar **C**. • ? • Chile • Anura

Cullom, **S**helby **J**. • ? • USA • Sauria

Cunha, **H**élio **J**. da • ? • Brazil • Sauria

Cunha, **O**svaldo **R**odrigues da • ? • Brazil • Sauria, Serpentes

Cunningham, **M**ichael **J**. • ? • Australia / South Africa • Anura

Curcio, **F**elipe **F**. • ? • Brazil • Sauria

Curio, **E**berhard • ? • Germany • Sauria

Curry-Lindahl, **K**ai • 10 May 1917 • Sweden • (Serpentes)

Curtiss, **A**nthony • ? • Haiti • (Sauria)

Cuvier, **G**eorges [Jean-Léopold-Nicolas-Frédéric Dagobert] (baron) • 23 August 1769–1832 May 13 • France / Germany / France • Anura, *Caudata, Crocodilia, Gymnophiona, Sauria, Serpentes, Testudines, (*testudines)

Cyrén, [Carl August] **O**tto • 9 April 1878–1946 September 23 • Sweden • (Caudata), Sauria, (Serpentes)

Czarewskii: *see* **Tsarevsky**, **S**ergey **F**edorovich

Czechura, **G**regory **V**[incent] • 11 January 1953 • Australia • Sauria

Czernov: *see* **Chernov**, **S**ergius **A**leksandrovich

D

da Costa: *see* **Costa**, **E**lizabeth **M**aria **M**amede da

da Costa Prudente: *see* **Prudente**, Ana Lúcia da Costa
da Cruz: *see* **Cruz**, Carlos Alberto Gonçalves da
da Cunha: *see* **Cunha**, **Hélio J.** da *and* **Osvaldo Rodrigues** da
Dahl, Chris • ? • Papua New Guinea • Anura
Dahl, Georg • 23 November 1905–1979 • Sweden / Colombia • (Testudines)
Dalman, Johan Wilhelm (= Vilhelm) • 4 November 1787–1828 July 12 • Sweden • Serpentes
Daltry, Jennifer C. • ? • England • Anura, Serpentes
Daly, John W[illiam] • 8 June 1933 • USA • Anura
Dam, **G. P. F.** van • ? • South Africa • Sauria
Dammerman, Karel Willem • 4 July 1885–1951 November 19 • Netherlands / Dutch East Indies (Java) / Netherlands • (Anura)
Danforth, Stuart T[aylor] • 23 September 1900–1938 November 25 • USA / Puerto Rico • (Sauria, Serpentes)
Daniel, (hermano, = Daniel Julián González Patiño) • 28 January 1909–1988 July 29 • Colombia • (Anura, Serpentes)
Daniel, J[ivanayakam] C[yril] • 9 July 1927 • India • Anura, (Gymnophiona, Sauria)
Danielyan, Felix D[anielanovich] • 1938 • USSR • Sauria
Danilevski, **B.** • ?–? • ? • (Sauria)
Dao Van Tien • 23 August 1920–1995 May 3 • Vietnam • (Serpentes), Testudines
Darevsky, I[lya] S[ergeevich] • 18 December 1924 • USSR / Russia • (Anura), *caudata, *Sauria, Serpentes, (*testudines)
Darling, Donald M. • 22 March 1925–2006 May 20 • USA • Caudata
Darlington, Philip J[ackson], Jr. • 14 November 1904–1983 December 16 • USA • (Anura, Sauria, Serpentes)
Darwin, Charles [Robert] • 12 February 1809–1882 April 19 • England • ([Anura], Crocodilia, [Sauria], Serpentes, Testudines)
Das, Indraneil • 18 December 1964 • England / Malaysia • Anura, Sauria, Serpentes
da Silva: *see* **Silva**, **Hélio R**[icardo] da *and* **Silva**, **Nelson Jorge** da
Dasmann, Marlene M. • ? • USA • Sauria
Dattatri, [Chandra] S[hekar] • ? • India • Serpentes
Daubenton, Louis Jean-Marie • 29 May 1716–1800 January 1 • France • Caudata
Daudin, François-Marie • (? 24 *or* 25) March 1774–1804 • France • Anura, Caudata, Crocodilia, Gymnophiona, Sauria, Serpentes, Testudines
Daugherty, Charles H[ines] • ? • USA / New Zealand • Anura, Sauria
Davenport, Tim R. B. (= Timothy Richard Bentley Davenport) • 6 February 1964 • England / Tanzania • Anura
David, [Jean-Pierre] Armand (l'abbé *or* père) • 7 September 1826–1900 November 10 • France / Italy / China / France • Anura, Caudata, (Serpentes), [Testudines]
David, Patrick • 21 January 1959 • France • Anura, Sauria, Serpentes
Davidge, Christine • ? • Australia • (Sauria)
Davies, Margaret • 8 November 1944 • Australia • Anura
Davis, D[elbert] Dwight • 30 December 1908–1965 February 6 • USA • Anura
Davis, John • 25 January 1761–1847 January 14 • USA • Serpentes
Davis, Nathan Smith, Jr. • 5 September 1858–1920 December 21 • USA • Anura
Davis, William B[ennoni] • 14 March 1902–1995 June 19 • USA • Anura, Sauria
Davis: *see also* **Stull**, **Olive Griffith** (= mrs. Loy Erman Davis)
Dawood, Abeda • ? • South Africa • Anura
De Betta, [Francesco] Edoardo (nobile) • 6 June 1822–1896 November 4 • Italy • Anura, Sauria, Serpentes, (Testudines)
De Blainville: *see* **Blainville**, **H**[enri]-**M**[arie] **Ducrotay** de
de Borre: *see* **Borre**, **Alfred Preud'homme** de
de Broin: *see* **Lapparent de Broin**, **France** de
Decary, Raymond • 1891–1973 • (? France) / Madagascar • (Anura, Sauria)
Deckert, Kurt • 24 March 1907–1987 August 1 • Germany • Anura
Deckert, Richard F. • (? 5 *or* 15) December 1878–1971 January • USA • (Serpentes)
de Espinoza: *see* **Carrillo** de **Espinoza**, **Nelly**
De Filippi, Filippo • 20 April 1814–1867 February 9 • Italy • Sauria, ([Serpentes])
Degen, Edward • ?–? • (? England / Uganda) • (Serpentes)
Degenhardt, William G[eorge] • 16 April 1926 • USA • Sauria
Degerbøl, Magnus • 1895–1977 • Denmark • Serpentes
de Grijs: *see* **Grijs**, **Pedro** de
Dehne, **A.** • ?–(1856)–? • Germany • Sauria, Serpentes
de Jeude: *see* **Lidth de Jeude**, **T**[heodorus] **W**[illem] van
de Jong: *see* **Jong**, **Jan Kornelis** de
De Kay, James E[llsworth] • 12 October 1792–1851 November 21 • Portugal / USA • Caudata, *crocodilia, (Serpentes)
de Lacépède: *see* **Lacépède**, **Bernard Germain Étienne** [de la Ville sur Illon] (comte de)
De la Fuente, Marcelo • ? • Argentina • *testudines
de Lapparent de Broin: *see* **Lapparent de Broin**, **France** de
De la Riva, Ignacio [José] • 8 January 1960 • Spain • Anura
De la Torre-Loranca, Miguel Angel • ? • México • Serpentes
del Campo: *see* **Martín del Campo** [y Sánchez], **Rafael**
Delean, Steven • ? • Australia • (Sauria)
de Lema: *see* **Lema**, **Thales** de
Delfortrie, Eugène [-Edmond] • 28 March 1816–1885 September 3 • France • *testudines
Dell, John • ? • Australia • (Sauria)
Delorme, Magali • ? • France • Anura
Del Prato, Alberto • ?–? • Italy • Sauria
Delwig, **W.** • ? • (? Azerbaijan) • Anura
Dely, Olivér György • 27 April 1927–2003 November 19 • Hungary • Caudata
de Massary: *see* **Massary**, **Jean-Christophe** de
Demathieu, Georges R[aymond] • 7 February 1920 • France • *crocodilia
Demegillo, Arnold D. • ? • Philippines • Serpentes
Demidov, Anatoly [Nikolaevich] (= Anatole de Démidoff, prince of San Donato) • 24 March (? 1812 *or* 1813)–1870 April 29 • Italy / Russia / France • [Sauria, Serpentes]
Dempsey, Paula • ? • Australia • Sauria
Denburgh: *see* **Van Denburgh**, **John**
Dendy, Arthur • 20 January 1865–1925 March 24 • England / Australia / New Zealand / South Africa / England • (Sauria)

Deng Qixiang • 24 October 1932 • China • Caudata
Dennis, **A**ndrew **J**. • ? • Australia • Anura
Dennis, **D**avid **M**[ichael] • 30 November 1939 • USA • Anura, Caudata
Dennys, **N**[icholas] **B**[elfield] • *about* 1840–1900 • China / Singapore • (Anura, Serpentes)
Dent, **J**ames **N**orman • 10 May 1916 • USA • Anura
Denton, **R**obert **K**[enneth], Jr. • 6 October 1953 • USA • *sauria
Denzer, **W**olfgang • ? • (? Germany / England) • Sauria
Depéret, **C**harles [Jean Julien] • 25 June 1854–1929 May (? 17 *or* 18) • France • *anura, (*crocodilia), *sauria, *serpentes, *testudines
de Plancy: *see* **Collin de Plancy**, **V**ictor [Émile Marie Joseph]
Deppe, **W**ilhelm • ?–(1830)–? • Germany • Sauria
de Queiroz, **K**evin • 4 June 1956 • USA • Sauria
Deraniyagala, **P**[aulus] **E**[dward] **P**[ieris] • 8 May 1900–1973 December 1 • Ceylon • *Crocodilia, Sauria, Serpentes, *Testudines
Deriugin (= Derjugin), **K**[onstantin] M[ikhailovich] • ("27 January") 8 February 1878–1938 December 27 • Russia / USSR • (Caudata)
Derleyn, **P**. **P**. • ? • (? Belgium / Burundi) • Serpentes
de Rochebrune: *see* **Rochebrune**, **A**lphonse [Amédée] Trémeau de
de Rooij: *see* **Rooij**, "**N**elly" de (= Petronella Johanna de Rooij, mrs. Breijer-de Rooij)
de Sá, **R**afael **O**[mar] • 4 March 1961 • Uruguay / USA • Anura
Desjardins, **J**ulien [-François] • 27 July 1799–1840 August 18 • France • Sauria
Des Lauriers, **J**ames **R**[obert] • 21 May 1939 • USA • *testudines
de Schauensee: *see* **Meyer de Schauensee**, **R**odolphe
De Sola, [Clarence] **R**alph • 26 July 1908–1993 May 8 • USA • Testudines
Despax, **R**aymond • ?–? • France • Anura, Sauria, Serpentes
Despott, **G**iuseppe • (? 24 *or* 26) July 1878–1936 September 22 • Malta • (Sauria)
Dessauer, **H**erbert **C**[lay] • 30 December 1921 • USA • Sauria
De Stefano, **G**iuseppe • ?–? • Italy • *caudata, *crocodilia, *sauria, *serpentes, *testudines
Deuve, **J**ean (*pseudonym* = Michel Caply) • 1918 • France / Laos / Japan • Serpentes
De Villiers, **A**therton **L**[ilford] • 27 August 1955 • South Africa • Anura
De Villiers, **C**[ornelius] **G**[erhardus] **S**[tephanus] ("Con") • 16 December 1894–1978 November 25 • South Africa • (Testudines)
Devincenzi, **G**aribaldi **J**[osé] • 20 December 1882–1943 September 17 • Uruguay • (Anura)
De Vis, **C**[harles] **W**[alter] • 9 May 1829–1915 April 30 • England / Australia • Anura, *crocodilia, *Sauria, Serpentes, *testudines
DeWeese, **J**ames **E**[dward] • ? • USA • Anura
de Witte: *see* **Witte**, **G**aston-**F**rançois de
De Zigno, **A**chille (barone) • 14 January 1813–1892 January 14 • Italy • *crocodilia, *serpentes
Dias, **E**duardo **J**osé dos **R**eis • ? • Brazil • Sauria
Dias-Filho, **M**anoel **M**artins • ? • Brazil • Sauria
Diaz, **N**elson **F**. • ? • Chile • Anura, Sauria
Díaz [-Beltrán], **L**uis **M**[anuel] • ? • Cuba • Anura, Sauria, Serpentes
Díaz-Gómez, **J**uan **M**anuel • ? • Argentina • Sauria
Di Bernardo, **M**arcos • 4 November 1963–2006 June 16 • Brazil • Anura, Serpentes
Dice, **L**ee **R**aymond • 15 July 1887–1977 January 31 • USA • (Sauria)
Dickerson, **M**ary **C**ynthia • 7 March 1866–1923 April 8 • USA • Sauria
Diesmos, **A**rvin **C**[antor] • ? • Philippines • Anura
Dijk: *see* **Van Dijk**, [David] **E**duard ("Eddie") *and* **P**eter **P**aul
Ding Hanbo (= Ting Han-po) • 3 November 1912–2003 March 7 • China / USA / China • Anura
Dinnik, **N**ikolai **Y**akovlevich • ?–? • Russia • (Serpentes)
Dirksen, **L**utz • 11 October 1965 • Germany • Serpentes
Distant, **W**illiam **L**ucas • 12 November 1845–1922 February 4 • England • (Sauria, Serpentes)
Di Tada, **I**smael **E**[duardo] • ? • Argentina • Anura, (Sauria)
Ditmars, **R**aymond **L**[ee] • (? 20 *or* 22) June 1876–1942 May 12 • USA • (Sauria), Serpentes
Dixo, **M**arianna [Botelho de Oliveira] • ? • Brazil • Anura, Sauria
Dixon, **J**ames **R**[ay] • 1 August 1928 • USA • Anura, Caudata, Sauria, Serpentes
Djao: *see* **Zhao Ermi**
Doan, **T**iffany **M**[arie] • ? • USA • Sauria
Dobie, **J**ames **L**[ogie] • ? • USA • *testudines
Doi, **H**ironobu • ? • Japan • Sauria
Dole, **J**im **W**[alter] • 28 May 1935 • USA • Anura
Dollo, **L**ouis [-Antoine-Marie-Joseph] • 27 December 1857–1931 April 29 • Belgium • (Anura), *crocodilia, *sauria, Serpentes, *testudines
Domergue, **C**harles **A**. • ? • France / Madagascar • (Anura), Sauria, Serpentes, Testudines
Donadio, **O**scar **E**nrique • 8 March 1954 • Argentina • *sauria
do Nascimento: *see* **Nascimento**, **F**rancisco **P**aiva do
Dong Qian • 6 February 1927 • China • [Anura]
Dong Zhiming • ? • China • *caudata, *crocodilia, *sauria
Donndorff, **J**ohann **A**ugust • 23 March 1754–1837 November 22 • Germany • Sauria, Serpentes, Testudines
Donnellan, **S**tephen **C**. • 18 November 1954 • Australia • Anura, Sauria, Serpentes
Donnelly, **M**aureen **A**[nn] • 10 August 1954 • USA • Anura, Sauria, Serpentes
Donnezan, **A**lbert • 14 October 1846–1914 • France • *testudines
Donoso Barros, **R**oberto • 5 October 1922–1975 August 2 • Chile • Anura, Sauria, Serpentes, (Testudines)
D'Orbigny: *see* **Orbigny**, **A**lcide [Dessalines] de *and* **C**harles [Victor Dessalines] de
Doria, **G**iacomo (marchese) • 1 November 1840–1913 September 19 • Italy • Anura, Sauria, Serpentes
Dorson, **E**dward **E**. • ? • USA • Sauria
dos Reis: *see* **Reis**, **S**érgio **F**[urtado] dos
dos Santos-Costa: *see* **Santos-C**osta, **M**aria **C**ristina dos
Dotsenko, **I**rina **B**. • ? • USSR / Ukraine • Serpentes
Douglas, **A**thol **M**ardon • ? • Australia • (Sauria)
Douglass, **E**arl • 28 October 1862–1931 January 13 • USA • *sauria, *serpentes, (*testudines)

Doumergue, **F**rançois • 11 May 1858–1939 December 23 • France / Algeria • Sauria
Dowling, **H**erndon **G**[lenn, Jr.] • 2 April 1921 • USA • Serpentes
Downs, **F**loyd **L**[eslie] • ? • USA • Serpentes
Drewes, **R**obert **C**[lifton] • 14 February 1942 • USA • Anura, Serpentes
Drewry, **G**eorge **E**[arl] • 1935 • USA • Anura
Dring, **J**ulian **C**[hristopher] **M**[ark] • ? • England • Anura, Sauria
Dryden, **G**ilbert **L**ittleton • 23 July 1938 • USA • Serpentes
Dubinin, **V**[sevolod] **B**[orisovich] • 11 January 1913–1958 May 7 • Russia / USSR • Sauria
Dubois, **A**lain [Camille Paul] (*husband of* Annemarie Ohler) • ? • France • Anura, Caudata
Dubois, **E**ugène • 28 January 1858–1940 December 16 • Netherlands • *crocodilia
du Bois-Reymond: *see* **Bois-Reymond**, **C**laude du
Duda, **P. L.** [V. K. G.] • ? • India • Sauria
Dudley, **A**lexander **P.** • ? • Australia • (Sauria)
Duellman, **W**illiam **E**[dward] (*husband of* Linda Trueb) • 6 September 1930 • USA • Anura, Caudata, Sauria, Serpentes
Duerden, **J**ames **E**dwin • 1869–1937 September 4 • England / USA / Ireland / Jamaica / USA / South Africa / England • (Serpentes), Testudines
Duerigen: *see* **Dürigen**, **B**runo
Duff-MacKay, **A**lexander ("Alex") • 1939–2003 March 6 • Kenya • Anura
Dugès, **A**lfredo [Augusto Delsescautz] (= Alfred Auguste Delsescautz) • 16 April 1826–1910 January 7 • France / México • Anura, Caudata, Sauria, Serpentes, Testudines
Dujardin, **F**élix • 5 April 1801–1860 April 8 • France • *crocodilia
Dumas, **P**hilip **C**[onrad] • 9 April 1923–1992 October 17 • USA • (Sauria)
Duméril, **A**ndré **M**arie **C**onstant (*father of* A. H. A. Duméril) • 1 January 1774–1860 August 14 • France • Anura, Caudata, Crocodilia, Gymnophiona, Sauria, Serpentes, *Testudines
Duméril, **A**uguste **H**enri **A**ndré (*son of* A. M. C. Duméril) • 30 November 1812–1870 November 12 • France • Caudata, Crocodilia, Gymnophiona, Sauria, Serpentes, Testudines
Dunayev, **E**vgeniy **A**[natolyevich] • ? • Russia • Sauria
Duncan, **R**obert • ? • USA • Caudata
Duncker, **G**eorg • 1870–1953 • Germany • (Serpentes)
Dundee, **H**arold **A.** • 23 August 1924 • USA • (Anura), Caudata
Dunger, **G**erald **T.** • ? • Nigeria • Sauria, Serpentes
Dunkle, **D**avid **H**osbrook • 9 September 1911 • Canada / USA • *sauria, (Serpentes)
Dunlap (*and* Dunlap-Pianka), **H**elen **L**ouise (*wife of* Eric R. Pianka *from* 1965 *to* 1980) • 9 April 1940 • USA • (Sauria)
Dunn, [Alta] **M**erle **T**[aylor] (*wife of* Emmett Reid Dunn *from* 1930 *to* 1956) • ? • USA • (Anura), Sauria
Dunn, **E**mmett **R**eid (*husband of* Merle T. Dunn *from* 1930 *to* 1956) • 21 November 1894–1956 February 13 • USA • Anura, Caudata, (*caudata), Gymnophiona, Sauria, Serpentes, (*serpentes), Testudines
Du Pasquier, **L**ouis • 17 March 1941 • France / Switzerland • Anura
Duperrey, **L**ouis-**I**sidore • 21 October 1786–1865 August 25 • France • [Anura, Sauria, Serpentes]
Durant, **P**edro • ? • Venezuela • Anura
Dürigen (= Duerigen), **B**runo • 1853–1930 • Germany • Anura, Caudata, Sauria, Serpentes, Testudines
Dury, **R**alph • 23 July 1899–1984 June • USA • (Caudata)
du Toit: *see* **Toit**, **C**ornelius **A**lbertus du
Dutta, **S**ushil **K**[umar] • 23 July 1952 • India • Anura
Duval, **J**ulian **J.** • ? • USA • Sauria
Duvernoy, **G**[eorges] **L**[ouis] • 6 August 1777–1855 March 1 • France • Sauria, Serpentes
Dvigubsky, **J**ean (= Ivan Alekseevich Dwigubskij) • 30 December 1771–1839 January 11 • Russia • Serpentes
Dwyer: *see* **Kaiser**, **C**hristine **M**[ary]
Dybowski, **B**enedykt [Ivanovich] • ("30 April") 12 May 1833–1930 January (? 30 *or* 31) • Russia / Poland / Siberia / Poland • (Anura), Caudata

E

Eason, **P**erri **K**[aye] • 26 September 1958 • USA • Sauria
Eberhardt, **H**ans-**J**oachim • 6 July 1920–1980 • Germany • Caudata
Eberle, **W**[ayne] **G**ary • 31 March 1944 • USA • Serpentes
Echelle, **A**lice **F.** • ? • USA • Sauria
Echelle, **A**nthony **A**[llan] • 9 September 1940 • USA • Sauria
Echternacht, **A**rthur **C**[harles] • 3 September 1939 • USA • Sauria
Edeling, **A. C. J.** • ?–1895 • East Indies • Anura, Sauria, Serpentes
Edgren, **R**ichard **A**[rthur], Jr. • 28 May 1925 • USA • Serpentes
Edwards, **E**uan • ? • Madagascar • (Serpentes)
Edwards, **G**eorge • 3 April 1693–1773 July 23 • England • (Sauria, Serpentes)
Edwards, **J**ohn **E.** • ? • Australia • (Serpentes)
Edwards, **S**tephen **R**[ichard] • 12 January 1942 • USA / Switzerland • Anura
Edwards: *see also* **Milne-Edwards**, **H**enri (*originally* Henri-Milne Edwards)
Efimov (= Yefimov), **M**[ikhail] **B**[orisovich] • 1947 • USSR / Russia • *crocodilia
Ehmann, **H**arry [F. W.] • ? • Australia • Sauria
Ehrenberg, **C**hristian **G**ottfried • 19 April 1795–1876 June 27 • Germany • (Anura), Sauria
Ehrl, **A**ndreas • ? • Germany • Sauria, Serpentes
Eibl-Eibesfeldt, **I**renäus • 15 June 1928 • Austria / Germany • Sauria, (Serpentes)
Eichwald, [Carl] **E**duard [Ivanovich] von • 4 July 1795–1876 November 10 • Latvia / Russia • Anura, Sauria, Serpentes
Eigenmann, Carl **H**[enry] • 9 March 1863–1927 April 24 • Germany / USA • Caudata, (Sauria)
Eimer, [Gustav Heinrich] **T**heodor • 22 February 1843–1898 May 29 • Switzerland / Germany • Sauria
Eiselt, **J**osef • 3 May 1912–2001 July 25 • Austria • Anura, Caudata, (Gymnophiona), Sauria, Serpentes, (Testudines)
Eisen, **G**ustav (= Gustavus Augustus) • 2 August 1847–1940 October 29 • Sweden / USA • (Serpentes)
Eisentraut, **M**artin [Bruno] • 21 October 1902–1994 July 5 • Germany • Sauria

Elbel, **R**obert **E**[dwin] • 8 July 1925–2005 December 1 • USA / Thailand • Anura, Sauria, Serpentes
El Din: *see* **Baha El Din**, **M**indy *and* **S**herif **M**.
Elias, **P**aul • ? • USA / England • Caudata
Elliot, **D**aniel **G**iraud • 7 March 1835–1915 December 22 • USA • [Serpentes]
Elliott, (sir) **W**alter • 16 January 1803–1887 March 1 • Scotland / British India / Scotland • (Sauria), Serpentes
Elpatyevsky (= Elpatjewskij), **V**ladimir **S**[ergeevich] • ?–? • Russia • Sauria, Serpentes
Emelianov, **A**leksander **A**drianovich • ("27 June") 9 July 1878–1946 January 7 • Russia / USSR • (Anura), Serpentes
Emin Pasha: *see* **Schnitzer**, **E**duard (*alias* Emin Pasha)
Emlen, **J**ohn **T**[hompson], Jr. • 28 December 1908–1997 November 8 • USA • Anura, Sauria
Emmons, **E**benezer • 16 May 1799–1863 October 1 • USA • *crocodilia
Endo, **R**iuji • 1891–? • Japan • *rhynchocephalia
Engels, **W**illiam **L**ouis • 28 October 1905–1980 October • USA • Serpentes
Entz, **G**éza • 29 March 1842–1919 December 4 • Hungary • (Anura)
Eremchenko (= Jeriomtschenko), **V**alery **K**[onstantinovich] • 27 June 1949 • USSR / Kyrgyzstan • Sauria
Ergueta-Sandoval, [Sonia] **P**atricia • ? • Bolivia • Anura
Erickson, **B**ruce **R**[onald] • 29 September 1930 • USA • *crocodilia, *testudines
Erman, **A**dolphe (= Georg Adolf) • 12 May 1806–1877 July 12 • Germany • [(Caudata)]
Ernst, **C**arl **H**[enry] • 28 September 1938 • USA • Testudines
Ernst, **R**affael • 28 July 1975 • Germany • Anura, Serpentes
Ersch, **J**ohann **S**amuel • 23 June 1766–1828 January 16 • Germany • [Serpentes]
Eschscholtz, **J**ohann **F**riedrich von • 1 November 1793–1831 May 7 • Russia (Estonia) • Caudata, Sauria, Testudines
Eshelman, **R**alph **E**[llsworth] • 1947 • USA • *sauria
Espada: *see* **Jiménez** de la **E**spada, **M**arcos
Espinal, **M**ario **R**. • ? • Honduras • Caudata, (Serpentes)
Espinoza, **R**obert **E**. • 14 January 1967 • USA • Sauria
Espinoza: *see also* **Carrillo** de **E**spinoza, **N**elly
Essex, **R**obert • ?–? • England / South Africa / England • Sauria
Estes, **R**ichard [Dean] • 9 May 1932–1990 December 28 • USA • *anura, *caudata, *crocodilia, *gymnophiona, *sauria
Estrada [-Acosta], **A**lberto **R**. • 3 September 1953 • Cuba / Puerto Rico • Anura, Sauria, Serpentes
Eterovick, **P**aula **C**abral • 1 September 1973 • Brazil • Anura
Etheridge, **R**ichard **E**[mmett] • 16 September 1929 • USA • (Anura), *Sauria, (Serpentes)
Evans, **G**eorge **H**enry (colonel) • 28 November 1863–1948 June 1 • British India • Serpentes
Evans, **H**oward **E**[dward] ("Howie") • 22 September 1922 • USA • Anura
Evans, **S**usan **E**. • ? • England • *anura
Evermann, **B**arton **W**arren • 24 October 1853–1932 September 27 • USA • (Sauria)
Eversmann, **E**duard [Friedrich Aleksandrovich] • ("12 January") 23 January 1794–1860 April 14 • Germany / Russia • Sauria
Eydoux, [Joseph] **F**ortuné [Théodore] • 1802–1841 • France • Sauria, Serpentes

F

Fabiani, **R**amiro • 15 May 1879–1954 April 29 • Italy • *crocodilia
Fahr, **A**enny • 1883–1964 • Germany • (Sauria)
Faivovich, **J**ulian • 14 January 1975 • Argentina / USA • Anura
Falanruw, **M**arjorie **V**. **C**. • ? • Guam • Sauria
Falconer, **H**ugh • 29 February 1808–1865 July 31 • Scotland / British India / England • *crocodilia, *sauria, *testudines, (Testudines)
Falk, **K**urt • ? • Angola • Sauria
Falla, (sir) **R**obert **A**lexander • 21 July 1901–1979 February 23 • New Zealand • (Sauria)
Fan Tsang-how • ? • China • Sauria, Serpentes, Testudines
Fang Junjiu • 25 August 1934 • China • [Serpentes]
Fang Pingwen (= Bingwen) • 1903–1944 August 26 • China / France • Caudata, Testudines
Fang Rongsheng • 25 July 1932 • China • Anura
Fatio, [Paul] **V**ictor • 28 November 1838–1906 March 19 • Switzerland • Anura, Caudata, Sauria, Serpentes
Fattizzo, **T**iziano • ? • Italy • Testudines
Fauvel, **A**lbert-**A**uguste • 7 November 1851–1909 November 3 • France • Crocodilia
Fayrer, (sir) **J**oseph (first baronet) • 6 December 1824–1907 May 21 • England / British India / England • Serpentes
Fea, **L**eonardo • 1852–1903 • Italy • (Anura, Sauria, Serpentes)
Febvre, **S**éverine • 13 April 1982 • France / South Africa / France • Anura
Fedchenko, **A**[leksei] **P**[avlovich] • ("7 February") 19 February 1844–1873 September 15 ("September 3") • Russia / Mont Blanc in the Alps • [Sauria]
Feener, **D**onald **H**., Jr. • ? • USA • Anura
Fehlmann, **H**[erman] **A**dair • 31 January 1917–2005 August 4 • USA • Sauria
Fei Liang • 3 July 1936 • China • Anura, Caudata
Feichtinger, **W**olfgang • ? • Germany • Anura
Feio, **R**enato **N**[eves] • ? • Brazil • Anura
Fejérváry [von Komlós-Keresztes], **G**éza **J**yula (= Géza Gyula Imre, báró) • 25 June 1894–1932 June 2 • Hungary • (Anura), *anura, Caudata, Sauria, (*testudines)
Feltrim, **A**line **C**osta • ? • Brazil • Sauria
Ferguson, **D**enzel **E**[dward] • 28 August 1929–1998 December 13 • USA • Caudata
Ferguson, **G**ary **W**[right] • 15 April 1941 • USA • Sauria
Ferguson, **W**illiam • 1820–1887 July 31 • England / Ceylon • Anura, Sauria, Serpentes
Fernádez: *see* **Fernández-S**áfadez, **I**ngrid (*not* Fernádez)
Fernandes, **R**onaldo • 27 October 1966 • Brazil / USA / Brazil • Serpentes
Fernández, **K**ati • ? • Argentina • (Anura)
Fernández-Sáfadez, **I**ngrid (*not* Fernádez) • ? • Bolivia • Sauria
Fernando, **S**amantha **S**uranjan • ? • Sri Lanka • Anura
Ferner, **J**ohn **W**[illiam] • 14 April 1945 • USA • Sauria, Serpentes
Ferrarezzi, **H**erbert • ? • Brazil • Serpentes
Ferrari-Pérez, **F**ernando • ?–1927 • México • (Sauria), [Serpentes]
Ferraris, **C**arl **J**., Jr. • ? • USA • Anura

Ferreira, **J**ulio **B**ethencourt • ?–? • Portugal • Anura, Sauria, Testudines
Ferreira, **T**alita **G**ancev • ? • Brazil • Serpentes
Ferreira, **V**anda **L**ucia • ? • Brazil • Serpentes
Fetcho, **J**oseph **R**[obert] ("Joe") • 28 June 1957 • USA • Serpentes
Filho, **C**arlos **M. M. R.** • ? • Brazil • Sauria
Filhol, [Antoine Pierre] **H**enri • 11 May 1843–1902 April 28 • France • (*caudata), *crocodilia, *sauria, *serpentes
Filippi: *see* **De Filippi**, **F**ilippo
Finn, **F**rank • 1868–1932 October 1 • England • Sauria, Serpentes
Firschein, **I**rwin **L**ester • ? • USA • Anura, (Caudata)
Fischberg, **M**ichail • 2 June 1918–1988 June 26 • Russia / Switzerland / England / Switzerland • Anura
Fischer, **E**berhard • ? • Germany • (Gymnophiona)
Fischer, **J**[ohann] **G**[ustav] • 1 March 1819–1889 January 27 • Germany • Anura, (Caudata), Gymnophiona, Sauria, Serpentes, Testudines
Fischer von Waldheim, [Johann] **G**otthelf (= Grigory Ivanovich)• 13 October 1771–1853 October 18 • Germany / Russia • Caudata, Sauria, Serpentes
Fisher, **A**laric • ? • Australia • Sauria
Fisher, **A**lbert **K**enrick • 21 March 1856–1948 June 12 • USA • (Anura)
Fisher, **R**obert **N.** • ? • USA • Sauria
Fisk, **G**eorge **H. R.** • ?–? • (? South Africa) • (Serpentes, Testudines)
Fitch, **A**lison **J.** • ? • Australia • Sauria
Fitch, **H**enry **S**[heldon] • 25 December 1909 • USA / Central America / USA • Sauria, (Serpentes)
Fitzgerald (= Fitz Gerald), **E**[dward] **A**[rthur] (major) • 10 May 1871–1931 January 2 • England • ([Sauria])
Fitzinger, **L**eopold [Joseph Franz Johann] • 13 April 1802–1884 September 20 • Austria • Anura, Caudata, Gymnophiona, *rhynchocephalia, Sauria, Serpentes, *Testudines
FitzSimons, **F**[rederick] **W**[illiam] (*father of* V. F. M. FitzSimons) • 6 August 1870–1951 March 25 • Ireland / South Africa • (Sauria)
FitzSimons, **V**[ivian] **F**[rederick] **M**[aynard] (*son of* F. W. FitzSimons) • 7 February 1901–1975 August 1 • South Africa • Anura, Sauria, Serpentes, Testudines
Fleay, **D**avid [Howells] • 6 January 1907–1993 August 7 • Australia • (Anura)
Fleischmann, **F**riedrich **L**udwig (= Friderico Ludovico) • 1806–1886 • Germany • Serpentes
Fleming, **J**ohn (reverend) • 10 January 1785–1857 November 18 • Scotland • Caudata, Sauria, Serpentes, Testudines
Fletcher, **J**[oseph] **J**[ames] • 1850–1926 May 15 • New Zealand / Australia / England / Australia • Anura
Flores, **G**lenn • ? • USA • Anura
Flores-Villela, **O**scar [Alberto]• 10 January 1954 • México • Sauria
Flower, **S**tanley **S**myth (major) • 1 August 1871–1946 February 3 • England / Siam / Egypt / England • (Anura), Sauria, ([Serpentes], Testudines)
Flury, **A**lvin **G**odfrey • 1 November 1920 • USA • Serpentes
Folkerts, **G**eorge **W**[illiam] • 26 November 1938 • USA • (Caudata), Testudines
Fong-G[rillo], **A**nsel • 16 August 1971 • Cuba • Anura, Sauria
Forbes, **H**enry **O**gg • 30 January 1851–1932 October 27 • Scotland / Portugal / East Indies / New Zealand / England • (Sauria)
Forcart, **L**othar [Hendrich Emil Wilhelm] • 10 December 1902–1990 • Switzerland • Anura, (Gymnophiona, Serpentes)
Ford, **J**ulian **R**[alph] • 3 November 1932–1987 January 31 • Australia • Sauria
Ford, **L**inda **S**[herill] • ? • USA • Anura, Serpentes
Formas [-Cortés], **J**[osé] **R**amón • 25 December 1939 • Chile • Anura
Forskål (= Forsskål), **P**er (= Petrus *or* Peter) • 11 January 1732–1763 July 11 • Finland / Sweden / Germany / Sweden / Arabia • Sauria, Serpentes, Testudines
Forster, **J**ohann **R**einhold • 22 October 1729–1798 December 9 • Germany / England • Serpentes
Foster, **M**ercedes **S.** • 16 August 1942 • USA • Anura, Serpentes
Foucart, **A**ntoine • ? • France • Sauria
Foufopoulos, **J**ohannes • 21 November 1967 • Greece / USA • Anura
Fouquette, **M**[artin] **J**[ohn], Jr. ("Jack") • 4 June 1930 • USA • Anura, Serpentes
Fourtau, **R**ené • ?–? • France • *crocodilia
Fowler, **D**anny **C.** • ? • Dominican Republic • Anura, (Sauria)
Fowler, **H**enry **W**eed • 23 March 1878–1965 June 2 • USA • Anura, Caudata, Sauria, Testudines
Fowler, **J**ames **A**[bbott] ("Jim") • 15 July 1916 • USA • Caudata
Fowler, **S**amuel **P**age • 22 April 1800–1888 • USA • (Anura)
Fox, **R**ichard **C**[arr] • 3 October 1933 • USA / Canada • *sauria
Fox, [Rufus] **W**ade, Jr. • 2 June 1920–1964 September 20 • USA • Serpentes
Fox, **S**tanley **F**[orrest] • 21 March 1946 • USA • Sauria
Fraas, **O**scar [Friedrich] von • 17 January 1821–1897 November 22 • Germany • *serpentes
França, **M**arco **A**urélio **G**allo de • 6 May 1981 • Brazil • *testudines
Franco, **F**rancisco **L**uis • ? • Brazil • Serpentes
Frank: *see* **van Frank**, **R**ichard
Franz, [L.] **R**ichard [Jr.] • 12 July 1941 • USA • Anura, Sauria
Franzen, **M**ichael • ? • Germany • Caudata, Serpentes
Fraser, **D**ouglas **F**[yfe] • 17 June 1941 • USA • Serpentes
Fraser, **N. C.** • ? • England • *rhynchocephalia
Frauenfeld, **G**eorg (ritter) von • 3 June 1807–1873 October 8 • Austria • (Anura)
Freeman, **H**oward **L**[eslie] • ? • USA • Caudata
Freiberg, **M**arcos **A**[braham] • 23 April 1911–1990 • Argentina / USA • (Anura), Crocodilia, Sauria, Serpentes, *Testudines
Freire, **E**liza **M**aria **X**avier • ? • Brazil • Anura, Sauria, Serpentes
Freitas, **M**arco **A**ntônio de (*husband of* Thais Figueiredo Santos Silva) • ? • Brazil • Sauria
Fréminville, **C**hristophe-Paulin de La Poix de • 23 January 1787–1848 January 12 • France • (Serpentes)
Frey, **E**berhard • ? • Germany • *crocodilia
Freycinet, **L**ouis **C**laude de **S**aulces de • (? 7 *or* 8) August 1779–1842 August 18 • France • (Anura, [Sauria])

Freyer, **H**einrich • 2 July 1802–1867 August 21 • Slovenia • (Caudata), Serpentes

Freytag, **G**ünther **E**[rich] • 14 January 1918–1989 August 20 • Germany • Caudata

Friant, **M**adeleine • ? • France • *anura

Fries, **C**arl, Jr. • 30 September 1910 • USA / México • *sauria

Fries, **I**sabelle [Scott] • 5 December 1947 • USA • Serpentes

Fritts, **T**homas **H**[arold] • 23 April 1945 • USA • Anura, Sauria, Serpentes

Fritz, **U**we • 31 October 1963 • Germany • Testudines

Fritze, **A**dolf • ?–(1894)–? • Germany • Serpentes

Frogner, **K**arl **J**[ohn] • 7 June 1941 • USA • Anura

Frost, **C**harles • ?–1915 September 18 (*age* 62) • Australia • (Anura), Sauria, Serpentes

Frost, **D**arrel **R**[ichmond] • 12 June 1951 • USA • Anura, Sauria

Frost, **J**ohn **S**[tephen] • 22 May 1945 • USA • Anura

Fry, **D**ene **B**[arrett] • 11 October 1893–1917 April 9 • Australia / France • Anura, Sauria, Serpentes, Testudines

Fu, **J**inzhong • 4 February 1966 • China / Canada • Caudata

Fuchs, **E**rnst • 18 May 1910–2000 August 4 • Austria • *testudines

Fuchs, **K**arlheinz • ? • Germany • Crocodilia

Fuenmayor: *see* **Rivas** Fuenmayor, **G**ilson

Fuente: *see* **De la Fuente**, **M**arcelo

Fuentes: *see* **Jiménez** Fuentes, **E**miliano

Fuentes-Ramos, Oswaldo • ? • Venezuela • Anura, Caudata

Fuhn, **I**on **E**duard • ("23 January") 5 February 1916–1987 August 31 • Romania • Caudata, Sauria

Fuhrmann, **O**tto • 1 April 1871–1945 January 26 • Switzerland • (Anura), Gymnophiona, (Serpentes)

Fülleborn, **F**riedrich **G**[eorg Hans Heinrich] • 13 September 1866–1933 September 9 • Germany • (Anura, Sauria)

Fuller, **P**hillip **J**ohn • ? • Australia • (Sauria)

Funk, **W**. **C**hris • ? • USA • Anura

Funkhouser, **A**nne [Black] (*wife of* John W. Funkhouser) • 10 October 1929 • USA • Anura

Funkhouser, **J**ohn **W**[illiam] (*husband of* Anne Funkhouser) • 28 August 1926–1974 December 3 • USA • Anura

G

Gabadage, **D**. • ? • Sri Lanka • Anura

Gabunia, **L**eo **K**[alistratovich] • ? • USSR / Russia • [*crocodilia], *testudines

Gachet, [Antoine-] **H**ippolyte • 15 November 1798–1842 November 22 • France • Serpentes

Gadow, **H**ans [Friedrich] • 8 March 1855–1928 May 16 • Germany / England • Anura, (Caudata), Sauria, *testudines

Gaffney, **E**ugene **S**[pencer] ("Gene") • 12 August 1942 • USA • (*testudines)

Gaiduchenko, **L**. **L**. • ? • USSR • *testudines

Gaige, **F**rederick **M**cMahon ("Dick", *husband of* Helen T. Gaige) • 3 July 1890–1976 October 20 • USA • (Anura, Sauria, Serpentes)

Gaige, **H**elen [Beulah] **T**[hompson] (*wife of* Frederick McMahon Gaige) • 24 November 1890–1976 October 24 • USA • Anura, Caudata, Sauria, Serpentes, (Testudines)

Gaimard, [Joseph-] **P**aul • 31 January 1793–1858 December 10 • France • Sauria, (Serpentes), Testudines

Gainsbury, **A**lison **M**. • ? • Brazil • Sauria

Galbreath, **E**dwin **C**[arter] • 18 March 1913–1989 January 20 • USA • (*sauria, *serpentes), *testudines

Gàlgano, **M**ario • 27 September 1907–1985 August 3 • Italy • (Anura)

Gali, **F**rank • ? • USA • Sauria

Gallagher, **D**aniel **S**tephen, Jr. • ? • USA • Sauria

Gallardo, **J**osé **M**aría [Alfonso Félix] • 1 August 1925–1994 October 12 • Argentina • Anura, Sauria

Galleguillos-G[onzález], **R**icardo • ? • Chile • Anura

Galvan, **M**ark **A**[llan] • 31 May 1965 • USA • Sauria

Gamero, **M**aria **L**ourdes **D**iaz de • ? • Venezuela • *testudines

Gancev: *see* **Ferreira**, **T**alita **G**ancev

Gans, **C**arl • 7 September 1923 • Germany / USA • (Anura), Sauria, Serpentes

Gao Keqin (= **Kequin**) • 1955 • China / USA / China • *anura, *caudata, *sauria

Garanin, **V**[alerian] **I**[vanovich] • ? • Russia • Serpentes

Garcia, **P**aulo **C**[hristiano de] **A**[nchietta] • ? • Brazil • Anura

García-Lopez, **J**osé **M**[iguel] • 17 February 1970 • Puerto Rico / USA • Anura

García-París, **M**ario • ? • Spain / USA • Anura, Caudata

García-Pérez, **J**uan **E**lías • ? • Venezuela • Sauria, Serpentes

García [-Piedrahita], **E**varisto • 17 November 1845–1921 June 19 • Colombia • Serpentes

Garcia-Porta, **J**oan • ? • Spain • *serpentes

García [-R.], **J**uan **C**arlos • ? • Colombia • Anura

Garda, **A**drian **A**. • ? • Brazil • Sauria

Garden, **A**lexander • January 1730–1791 April 15 • USA • Caudata

Gardner, **A**ndrew **S**. • ? • Oman • Sauria

Gardner, **J**ames **D**. • ? • Canada • *caudata

Garín, **C**arlos • ? • Chile • Sauria

Garman, **S**amuel [Walton] (*brother of* W. H. Garman) • 5 June 1843–1927 September 30 • USA • Anura, Sauria, Serpentes, Testudines

Garman, [William] **H**arrison (*brother of* S. W. Garman) • 27 December 1856–1944 August 7 • USA • Caudata

Garnier, **J**[ohn] **H**[utchinson] • 1823–1898 February 1 • Scotland / Ireland / France / Canada • Caudata

Garrido, **O**rlando **H**. • ? • Cuba • Sauria, Serpentes

Garrido-R., **A**. • ? • Venezuela • Sauria

Gartside, **D**onald **F**[orsyth] • ? • Australia • Anura

Gasc, **J**ean-**P**ierre • ? • France • Gymnophiona, Serpentes

Gasparini: *see* **Brandoni de Gasparini**, **Z**ulma [**N**elida]

Gasperetti, **J**ohn • 24 October 1920–2001 December 10 • USA / Saudi Arabia / Egypt • (Sauria, Serpentes)

Gasser, **F**rançois • ? • France • Caudata

Gaucher, **P**hilippe • ? • French Guiana • Anura

Gaulke, **M**aren • ? • Germany • Sauria, Serpentes

Gauthier, **J**acques **A**rmand • 7 June 1948 • USA • *sauria

Gay, **C**laude • 18 March 1800–1873 November 29 • France / Chile / France • (Anura), Sauria

Gayda, **H**enrik **St**. • ? • (? Germany) • Anura, Caudata, (Serpentes)

Gayet, **M**ireille • ? • France • *caudata

Gehlbach, **F**rederick **R**[enner] • 5 July 1935 • USA • Caudata

Geisenheyner, **L**udwig • 1841–1926 • Germany • Sauria

Gené, [Carlo] **G**iuseppe (= Josepho) • 9 December 1800–1847 July 14 • Italy / Sardinia / Italy • Anura, Caudata, Sauria
Geniez, **P**hilippe • ? • France • Sauria
Genta, **P**aolo • ? • Italy • Testudines
Geoffroy Saint-Hilaire, **É**tienne [-François] • 15 April 1772–1844 June 19 • France / Egypt / France • Crocodilia, Sauria, Serpentes, Testudines
Georges, **A**rthur • 25 October 1953 • Australia • Testudines
Georgi, **J**ohann **G**ottlieb • 31 December 1729–1802 October 27 • Germany / Russia • [Serpentes]
Gerlach, **J**ustin • 26 July 1970 • Seychelles / England / Seychelles • Anura, Sauria
Gervais, [François Louis] **P**aul • 26 September 1816–1879 February 10 • France • (*anura), Caudata, *crocodilia, [*rhynchocephalia], *Sauria, Serpentes, *Testudines
Gesner, **C**onrad (= Cunrat Gessner) • 26 March 1516–1565 December 13 • Switzerland • (Caudata)
Giard, **A**lfred [Mathieu] • 8 August 1846–1908 August 8 • France • (Serpentes)
Giaretta, **A**riovaldo **A**[ntônio] • 17 April 1966 • Brazil • Anura
Gibbes, **L**ewis **R**eeve • 14 August 1810–1894 November 21 • USA • Caudata
Gibbons, **J**[ames] **W**hitfield ("Whit") • 5 October 1939 • USA • (Testudines)
Gibbons, **J**ohn **R**[ichard] **H**[utchinson] • 3 June 1946–1986 November 15 • England / Australia / Fiji • Sauria, (*sauria)
Giebel, **C**[hristoph] **G**[ottfried Andreas] • 13 September 1820–1881 November 14 • Germany • *caudata, *crocodilia, Sauria, Testudines
Giglioli, **E**nrico (= Henry Hillyer) • 13 June 1845–1909 December 16 • England / Italy • (Caudata), Sauria
Gillam, **M**ike **W**. • ? • Australia • Sauria
Gillespie, **J**ohn • ? • USA • Serpentes
Gilliams, **J**acob • 1784–1868 February 4 • USA • Caudata, Sauria
Gilmore, **C**harles **W**[hitney] • 11 March 1874–1945 December 27 • USA • (*anura), *caudata, *crocodilia, *rhynchocephalia, *sauria, *serpentes, *Testudines
Ginés, (hermano, = Pablo Mandazen-Soto) • 26 June 1912 • Spain / Venezuela • (Anura)
Gingerich, **P**hilip **D**[erstine] • 23 March 1946 • USA • *crocodilia
Ginsburg, **L**éonard • 1927 • France • *crocodilia, (*testudines)
Girard, **C**harles [Frédéric] • 9 March 1822–1895 January 29 • Alsace / Switzerland / USA / France • Anura, Caudata, Sauria, Serpentes, Testudines
Giraudo, **A**lejandro **R**[aul] • ? • Argentina • Serpentes
Giri, **V**arad [Bhagwan] • 8 July 1970 • India • Gymnophiona
Gislén, **T**orsten [Richard Emanuel] • 21 June 1893–1954 August 26 • Sweden • [Caudata]
Gistel, **J**ohannes [von Nepomuk Franz Xaver] • 11 August 1809–1873 • Germany • Anura, Caudata, Sauria, Serpentes, Testudines
Glass, **B**ryan **P**[ettigrew] • 21 August 1919 • USA • Anura, Testudines
Glauert, **L**udwig • 5 May 1879–1963 February 1 • England / Australia • (Anura), Sauria, Serpentes, Testudines
Glaw, **F**rank • ? • Germany • Anura, Sauria, Serpentes
Gleadow, **F**rank • ?–? • British India • (Sauria)
Gloor, **H**ans [Jacob] • 8 February 1917 • Switzerland • Anura
Glos, **J**ulian • 23 March 1970 • Germany • Anura
Gloyd, **H**oward **K**[ay] • 12 February 1902–1978 August 7 • USA • Serpentes
Glückselig, **A**ugust **M**aria • 1908–1870 • Prague • Sauria
Gluesenkamp, **A**ndrew **G**[allagher] • ? • USA • Anura
Gmelin, **C**arl **C**hristian • 18 March 1762–1837 July 26 • Germany • Serpentes
Gmelin, **J**ohann **F**riedrich • 8 August 1748–1804 November 1 • Germany • Crocodilia, Sauria, Serpentes, Testudines
Gmelin, **S**amuel [Gottlieb] • 4 July 1744–1774 July 27 • Germany / Russia • Testudines
Goberdhan-Vigle, **D**eborah **C**. **I**. • ? • USA • Anura
Godman, **F**rederick **D**u**C**ane • 15 January 1834–1919 February 19 • England • ([Anura], Sauria, Serpentes)
Goeldi: *see* **Göldi**, **E**mil **A**ugust (= Emilio Augusto Goeldi)
Goff, **C**arlos **C**lyde • 2 October 1905–1939 January 13 • USA • Serpentes
Goin, **C**oleman **J**[ett] (*husband of* Olive B. Goin) • 25 February 1911–1986 May 12 • USA • Anura, (*anura), *Caudata, Serpentes, Testudines
Goin, **O**live [Lynda] **B**[own] (*wife of* Coleman J. Goin) • 2 December 1912–2000 June 2 • USA • Anura
Goldfuss, [Georg] **A**ugust • 18 April 1782–1848 October 2 • Germany • *caudata, *rhynchocephalia, Sauria, Serpentes
Göldi, **E**mil **A**ugust (= Emilio Augusto Goeldi) • 28 August 1859–1917 July 5 • Switzerland / Brazil / Switzerland • Anura, Testudines
Golliez, **H**enri • 9 December 1861–1913 October 28 • Switzerland • *testudines
Golubev, **M**ichael **L**[eonidovich] • 30 July 1947 • USSR / USA • Sauria
Gomes, **J**oão **F**lorêncio [de Salles] • (? 2 *or* 3) September 1886–1919 May 29 • Brazil • Serpentes
Gomez: *see* **Burrowes G**omez, **P**atricia **A**.
Gómez: *see* **Díaz G**ómez, **J**uan **M**anuel
Gonçalves, **J**[osé] **M**[oura] • ? • Brazil • Serpentes
Gonçalves: *see also* **Cruz**, **C**arlos **A**lberto **G**onçalves da
Gonzáles-Álvarez, **L**ucindo • ? • Bolivia • Anura, Sauria, Serpentes
Gonzáles-Sponga, **M**anuel **A**. • ? • Venezuela • Serpentes
Good, **D**avid **A**[ndrew] • 6 August 1956 • USA • Sauria
Goode, **G**[eorge] **B**rown • 13 February 1851–1896 September 6 • USA • (Sauria)
Goode, **J**ohn • ? • Australia • (Testudines)
Gordo, **M**arcelo • ? • Brazil • Anura
Gordon, **K**enneth **L**[lewellyn] • 7 June 1899–1983 October 29 • USA • (Caudata)
Gorham, **S**tanley **W**[arren] ("Stan") • 22 November 1917–1984 March 23 • Canada • Anura
Gorman, **J**oe (= Joseph B., Jr.) • 25 April 1918–1999 April 18 • USA • Anura, Caudata
Gorzula, **S**tefan **J**. • ? • Venezuela • Anura, Serpentes
Gosse, **P**hilip **H**enry • 6 April 1810–1888 August 23 • England • Anura, Sauria
Gough, **L**ewis **H**enry • ?–? • South Africa • (Anura), Serpentes
Gouvêa, **E**lio • ? • Brazil • Anura
Gow, **G**raeme **F**[rancis] • ? • Australia • Sauria, Serpentes
Gower, **D**avid **J**[ohn] • 11 June 1969 • England • Gymnophiona

Gradwell, **N**orman • ? • Australia • (Anura)
Grafe, **T**[homas] **U**lmar • 13 November 1959 • Germany / India / Germany / USA / Germany / Brunei • Anura
Graham, **E**ugene **D.**, Jr. • ? • USA • Sauria
Grandidier, **A**lfred • 20 December 1836–1921 September 13 • France / Madagascar / France • Anura, *Crocodilia, Sauria, Serpentes, Testudines, (*testudines)
Grandison, **A**lice **G**[eorgie] **C**[ruickshank] • 25 March 1927 • England • Anura, (Gymnophiona), Sauria
Granger, **W**alter [Willis] • 7 November 1872–1941 September 6 • USA • (*crocodilia, *sauria, Testudines), *testudines
Grant, **C**hapman (major) • 22 March 1887–1983 January 5 • USA • Sauria, Serpentes, (Testudines)
Grant, **T**aran • 18 March 1972 • Canada / Colombia / USA • Anura, Sauria, Serpentes
Gravenhorst, **J**ohann **L**udwig **C**hristian [Carl] • 14 November 1777–1857 January 14 • Germany • Anura, Caudata, Sauria, Testudines
Graves, **L.** • ?–? • (? Belgium) • Crocodilia
Gray, **F**rancis **C**alley • 19 September 1790–1856 December 29 • USA • Serpentes
Gray, **H**eather **M**[arie] • ? • Canada • Anura
Gray, **J**ohn **E**dward • 12 February 1800–1875 March 7 • England • Anura, Caudata, *Crocodilia, Rhynchocephalia, Sauria, Serpentes, *Testudines
Gray, **P**eter • 23 May 1953 • USA • Anura
Graybeal, **A**nna • 26 August 1965 • USA • Anura
Greeff, **R**ichard • 14 March 1829–1892 August 30 • Germany • (Sauria)
Green, **D**avid **M**[artin] • 20 January 1953 • Canada • Anura
Green, **J**acob • 26 July 1790–1841 February 1 • USA • Caudata, Sauria
Green, **N**[orman] **B**ayard • 15 November 1905–1988 May 23 • USA • Caudata
Green, **R**obert **H.** ("Bob") • ? • Australia • (Sauria)
Greenbaum, **E**li • ? • USA • Caudata
Greene, **H**arry **W**[alter] • 26 September 1945 • USA • *sauria
Greer, **A**llen **E**[ddy], Jr. • ? • USA / Australia • Sauria, Serpentes
Gregson, **R**. **A**. **M**. • ? • Australia • Sauria
Gressitt, **J**[udson] **L**insley • 16 June 1914–1982 April 26 • USA • Anura, Sauria, Serpentes
Grey, (sir) **G**eorge • 14 April 1812–1898 September 19 • Portugal / England / Australia / New Zealand / South Africa / New Zealand / England • [Anura, (Sauria), Serpentes, Testudines]
Griffin, **L**awrence **E**dmonds • 10 September 1874–1949 September 12 • USA • Sauria, Serpentes
Griffith, **E**dward • 1790–1858 January 8 • England • [Caudata, Sauria, Testudines]
Griffith, **H**ugh [William] • ? • Canada • Sauria
Griffiths, **I**vor • ? • England / Tanganyika • Anura
Grigg, **G**ordon [Clifford] • 15 August 1942 • Australia • Anura
Grijs (= Grys), **P**edro de • ?–1940 February 3 • Germany • Anura, Sauria, (Serpentes)
Grillo, **H**amilton [César Zanardi] • ? • Brazil • Anura
Grismer, **J**esse **L**[eland] (*son of* L. Lee Grismer)• ? • USA • Sauria, Serpentes
Grismer, **L**[arry] **L**ee (*father of* Jesse L. Grismer) • 19 November 1955 • USA • Anura, Sauria, Serpentes
Gritis, **P**aul **A**[lexander] • 22 December 1952 • USA • Anura
Grobman, **A**rnold **B**[rams] • 28 April 1918 • USA • Caudata, Serpentes
Groessens-van Dyck, **M**arie-**C**laire • ? • Belgium • *testudines
Grohmann, **F**rancesco **S**averio • ?–(1832)–? • Sicily • Sauria
Gronow, **L**orenz **T**heodor (= Laurens Theodorus Gronovius) • 1 June 1730–1777 August 8 • Netherlands • (Sauria, Serpentes)
Grosjean, **S**téphane • ? • France • Anura
Grosse, **W**olf-**R**üdiger • ? • Germany • Anura
Grossmann, **P**eter • ? • Germany • Testudines
Grossmann, **W**olfgang • ? • Germany • Sauria
Grubant, **V**. **N**. • ? • USSR • Serpentes
Gruber, **H**ans **J**ürgen • ? • Germany • Sauria
Gruber, **J**ohann **G**ottfried • 29 November 1774–1851 August 7 • Germany • [Serpentes]
Gruber, **U**lrich • 7 May 1932 • Germany • Sauria, Serpentes, Testudines
Grys: *see* **Grijs**, **P**edro de
Guayasamin [-Ernst], **J**uan **M**[anuel] • 20 December 1974 • Ecuador / USA • Anura
Gubin, **Y**urii **M.** • ? • Russia • *anura
Gudynas, **E**duardo • ? • Uruguay • Sauria
Gueldenstaedt: *see* **Güldenstädt**, **J**ohann **A**nton von
Guenther: *see* **Günther**, **A**lbert [Carl Ludwig Gotthilf]
Guérin-Méneville, **F**elix-**É**douard • 12 October 1799–1874 January 26 • France • [Anura], Crocodilia, Gymnophiona, Sauria, Serpentes, Testudines
Guibé, **J**ean [Marius René] • 18 February 1910–1999 May 4 • France • Anura, Sauria, Serpentes
Guichenot, [Antoine] **A**lphonse [Théodore] • 31 July 1809–1876 February 17 • France / Algeria / France • Sauria, Serpentes
Guicking, **D**aniela • ? • Germany • Testudines
Guillette, **L**ouis **J**[oseph], Jr. • 29 August 1954 • USA • Sauria
Guimarães-Neto, **A**ristides **S**[algado] • ? • Brazil • Anura
Güldenstädt, **J**ohann **A**nton von • 26 April 1745–1781 March 23 • Latvia / Russia • Serpentes
Gulia, **G**iovanni • 11 February 1864–1918 January 22 • Malta / Egypt • *Sauria
Gumprecht, **A**ndreas • ? • Germany • Serpentes
Gundlach, **J**uan (= Johannes Christopher) • 17 July 1810–1896 March 15 • Germany / Cuba • (Anura), Sauria, Serpentes
Günther, **A**lbert [Carl Ludwig Gotthilf] • 3 October 1830–1914 February 1 • Germany / England • Anura, Caudata, Crocodilia, Gymnophiona, (Rhynchocephalia), *Sauria, Serpentes, *Testudines
Günther, **R**ainer • 11 September 1941 • Germany • Anura, Sauria
Gürich, **G**eorg [Julius Ernst] • 25 September 1859–1938 August 16 • Germany • *crocodilia
Gururaja, **K**[otambylu] **V**[asudevacharya] • ? • India • Anura
Gutberlet, **R**onald **L**[ouis], Jr. • ? • USA • Sauria, Serpentes
Gutiérrez-Mayén, [María] **G**uadalupe • 12 December 1960 • México • Anura, Caudata, Serpentes
Guttman, **S**heldon **I.** • 9 March 1941 • USA • Caudata
Gyi, **K**o **K**o: *see* **Ko Ko Gyi**

Gyldenstolpe, **N**ils **C**arl **G**ustav **F**ersen (greve) • 30 September 1886–1961 April 10 • Sweden • (Anura, Sauria), Serpentes

H

Ha Dinh Duc • ? • Vietnam • Testudines
Haacke, **W**ulf **D**ietrich • 15 December 1936 • South Africa • Sauria, Serpentes
Haas, **A**lexander • 17 September 1964 • Germany • Anura
Haas, **G**eorg • 19 January 1905–1981 September 13 • Austria / Germany / Israel • Sauria, Serpentes, *testudines
Haas, **W**infred • ? • Germany • Anura
Haase, **E**rich • ?–? • Germany / Siam • (Sauria)
Haast, **W**illiam **E**[dward] ("Bill") • 30 December 1910 • USA • (Serpentes)
Haddad, **C**élio **F**[ernando] **B**[aptista] • 5 December 1959 • Brazil • Anura
Hahn, **D**onald **E**[dgar] • 2 September 1939 • USA • Sauria, Serpentes
Haile, **N**eville **S**eymour • 5 June 1928–2004 June 20 • England / Malaysia / England • Anura
Hairston, **N**elson **G**[eorge, Sr.] • 16 October 1917 • USA • Caudata
Haldeman, **S**amuel **S**tehman • 12 August 1812–1880 September 10 • USA • (Caudata, Serpentes)
Hall, **C**harles **W**illiam • 8 February 1922 • USA • Serpentes
Hall, **D**avid **H.** • ? • USA • Anura
Hall, **W**illiam **P**urington, III • ? • USA • Sauria
Hallermann, **J**akob • ? • Germany • Sauria, Serpentes
Halliday, **T**[imothy] **R**[ichard] ("Tim") • 1945 • England • (Anura)
Hallmann, **G**erhard • ? • Germany • (Sauria)
Hallowell, **E**dward • 14 September 1808–1860 February 20 • USA • Anura, Caudata, Sauria, Serpentes, Testudines
Halstead [Tarlo], **L**[ambert] **B**[everly] • 13 June 1933–1991 April 30 • England • *crocodilia, *testudines
Haly, **A**myrald • ?–(1887 *and* 1888)–? • Ceylon • Sauria, Serpentes
Han Defen • 1933 • China • *testudines
Hanitsch, [Karl] **R**ichard • 22 December 1860–1940 August 11 • Germany / England • (Anura, Serpentes)
Hanken, **J**ames • 14 July 1952 • USA • Caudata
Hanlon, **M**ark (= Timothy Marcus Stephen) • ? • Australia • (Sauria)
Hansen, **R**obert **W**[illiam] • 21 April 1954 • USA • Caudata
Hardwicke, **T**homas (major-general) • (? 1755 *or* 1756)–1835 (? March *or* May) 3 • England / British India / England • Sauria, Serpentes, (Testudines)
Hardy, **D**avid **L.** • ? • USA • (Serpentes)
Hardy, **G**raham **S**tuart • 25 April 1947 • New Zealand • Sauria
Hardy, **J**erry **D**avid, Jr. • 1929 • USA • Anura, Sauria
Hardy, **L**aurence **M**[cNeil] ("Mac") • 24 February 1939 • USA • Anura, Serpentes
Harlan, **R**ichard • 19 September 1796–1843 September 30 • USA • Anura, Caudata, *crocodilia, Sauria, Serpentes, Testudines
Harold, **G**regory • ? • Australia • (Sauria)
Harper, **F**rancis • 17 November 1886–1972 November 17 • USA • Anura, Sauria
Harris, **D**ennis **M**[artin] • 1950 • USA • Sauria
Harris, **H**erbert **S**[tanley], Jr. • 30 June 1942 • USA • Serpentes
Harrison, **D**avid **L.** • ? • England • *anura, *serpentes
Harrison, **J**ulian **R**avenel, III • 23 August 1934 • USA • Anura, Caudata
Harrison, **L**auncelot • 13 July 1880–1928 February 20 • Australia • Anura, (Sauria)
Harrison, **T**erry • ? • USA • *anura
Hartert, **E**rnst [Johann Otto] • 29 October 1859–1933 November (? 11 *or* 14) • Germany / England / Germany • Sauria
Hartman, **D**eron • ? • USA • Serpentes
Hartmann, **M**ax (= Maximilian) • 7 July 1876–1962 October 11 • Germany • Sauria
Hartmeyer, **R**obert [Hermann] • 19 May 1874–1923 October 13 • Germany • [Anura, Sauria, Serpentes]
Hartweg, **N**orman [Edouard] • 20 August 1904–1964 February 16 • USA • Anura, (Caudata), Sauria, Serpentes, Testudines
Harvey, **C**hris • ? • Australia • Sauria
Harvey, **M**ichael **B**[rown] • 11 January 1966 • USA • Anura, Sauria, Serpentes
Hasbún [-Rose], **C**arlos **R**oberto • 17 May 1963 • El Salvador / England / El Salvador • Sauria
Hasselquist, **F**rederik • 3 January 1722–1752 February 9 • Sweden • (Sauria)
Hasselt, **A**[lexander] **W**[illem] **M**[ichiel] van • 9 August 1814–1902 September 16 • Netherlands • Serpentes
Hasselt, **J**ohan **C**oenraad van • 24 June 1797–1823 September 8 • Netherlands / Dutch East Indies (Java) • Anura, Gymnophiona, Sauria, Serpentes, (Testudines)
Hassler, **W**illiam **G**[rey] ("Bill") • 1 September 1906–1979 September 29 • USA • Anura, Sauria
Hatt, **R**obert **T**[orrens] • 17 July 1902–1989 September 16 • USA • [*sauria]
Haughton, **S**[idney] **H**[enry] • 7 May 1888–1982 May 24 • England / South Africa • *anura, *testudines
Häupl, **M**ichael • ? • Austria • Sauria
Hauschild, **A**ndree • ? • ? • [Sauria]
Haworth, **A**drian **H**ardy • 19 May (? 1767 *or* 1768)–1833 August 24 • England • Serpentes
Hay, **J**ennifer **M**[ary] • ? • New Zealand / USA • Anura
Hay, **O**liver **P**erry (*father of* W. P. Hay) • 22 May 1846–1930 November 2 • USA • Caudata, Serpentes, *Testudines
Hay, **W**illiam **P**erry (*son of* O. P. Hay) • 8 December 1871–1947 January 26 • USA • Testudines
Hayden, **F**erdinand **V**andiveer • 7 September 1829–1887 December 22 • USA • (Serpentes)
Haynes, **D**avid [Polk] • 13 January 1942 • USA • Testudines
Heang: *see* **Kiew Bong Heang**
Heatwole, **A**udry [Ann Yoder] (*wife of* Harold F. Heatwole) • 22 September 1932 • USA / Puerto Rico / Australia / USA • Anura
Heatwole, **H**arold **F**[ranklin] (*husband of* Audry Heatwole) • 2 December 1934 • USA / Puerto Rico / Australia / USA • Anura, Sauria, Serpentes
Hebrard, **J**ames **J**ack • 1944 • USA / Kenya • Sauria
Hechenbleikner, **H**erbert • ? • USA • Sauria
Hecht, **G**ünther • 1902 • Germany • Sauria, Serpentes

Hecht, **M**ax **K**[nobler] (*husband of* Bessie Louise Matalas *from* 1947) • 15 February 1925–2002 March 17 • USA • *anura, (Sauria), *sauria, *serpentes
Heck, **J**ohann **G**eorg • ?–1857 • (? Germany) • [Caudata]
Hedges, **S**[tephen] **B**lair • ? • USA • Anura, Sauria, Serpentes
Hediger, **H**eini [Karl Peter] • 30 November 1908–1992 August 29 • Switzerland • Anura, Sauria, Serpentes
Heifetz, **W**illiam • ? • USA • Sauria
Heilprin, **A**ngelo • 31 March 1853–1907 July 17 • Hungary / USA • (Anura), Serpentes
Heinze, **A**lbert **A.** • ? • USA • Caudata
Helfenberger, **N**otker • ? • Switzerland • Sauria, Serpentes
Heller, **E**dmund • 21 May 1875–1939 July 18 • USA • Sauria, (Serpentes)
Heller, **F**lorian • 1905 • Germany • *serpentes
Hellmich, [Karl George] **W**alter • 26 February 1906–1974 July 29 • Germany / Chile / Germany • Anura, Caudata, (Gymnophiona), Sauria, (Serpentes), Testudines
Hemmer, **H**elmut • 7 December 1940 • Germany • Anura
Hemprich, **F**ridericus **G**uilelmus (= Friedrich Wilhelm) • 24 June 1796–1825 June 30 • Poland / Germany / Eritrea • Sauria, Serpentes
Henderson, **J**ohn **R**obertson • 21 May 1863–1925 October 26 • England / British India • Testudines
Henderson, **R**obert **W**[illiam] • 14 October 1945 • USA • Sauria, Serpentes
Hendricks, **F**red **S**[amuel] • 4 October 1946 • USA • Sauria, Serpentes
Hendrickson, **J**ohn **R**[oscoe] • 26 August 1921–2002 September 6 • USA / Malaysia / USA • Anura
Henkel, **F**riedrich **W**ilhelm ("Willi") • 17 June 1949 • Germany • Sauria
Henle, **K**laus • 5 January 1955 • Germany • Anura, Sauria, Serpentes
Hennig, [Emil Hans] **W**illi • 20 April 1913–1976 November 5 • Germany • Sauria, Serpentes
Henrici, **A**my **C.** • ? • USA • *anura
Hensel, **R**einhold [Friedrich] • 1826–1881 • Germany • Anura, Serpentes
Henshaw, **H**enry **W**[etherbee] • 3 March 1850–1930 August 1 • USA • Anura, (Sauria, Serpentes)
Hensley, **M**[arvin] **M**ax • 5 April 1922–2000 February 19 • USA • Serpentes
Henzl, **M**artin **J.** • ? • Austria • Anura
Heppich, **S**usanna • 3 December 1953 • Austria • Anura
Hermann, **J**ohannes (= Jean) • 31 December 1738–1800 October (? 4 *or* 8) • France • Sauria, Serpentes, Testudines
Hernández, **E**fraín • ? • Spain • Sauria
Hernández, **F**rancisco • 1517–1587 January 28 • Spain • (Sauria)
Hernández-**C**amacho, **J**orge **I.** • 17 January 1935 • Colombia • Anura, Sauria
Hernández-**P**acheco, **E**duardo • 1872–? • Spain • *testudines
Hernández **S**ampelayo: *see* **Sampelayo**, **P**rimitivo **H**ernández
Héron-Royer, **L**ouis **F**rançois • ?–1891 December • France • Anura
Herre, **A**lbert **W**[illiam] **C**[hristian] **T**[heodore] • 16 September 1868–1962 January 16 • USA • (Anura, Sauria)
Herre, [Karl] **W**olf • 3 May 1909 • Germany • *Caudata
Herrera [-Lopez], **A**lfonso [Luis] • 1868–1942 • México • (Sauria, Serpentes, Testudines)
Herrmann, **H**ans-**W**erner • 6 December 1961 • Germany / USA • Anura, Sauria, Serpentes
Hersey, **F**rederick **K.** • ? • Australia • (Sauria)
Hershkovitz, **P**hilip • 12 October 1909–1997 February 15 • USA • Gymnophiona
Hertz, **P**aul **E**[ric] • 8 July 1951 • USA • Sauria
Hesse, **P**aul • 3 February 1857–1938 February 26 • Germany • (Sauria)
Heude, **P**ierre-**M**arie (père) • 25 June 1836–1902 January 2 • France / China • Testudines
Heuglin, [Martin] **T**heodor (hofrath) von • 20 March 1824–1876 November 5 • Germany • (Anura)
Hewitt, **J**ohn • 23 December 1880–1961 August 4 • England / Sarawak / South Africa • Anura, [Crocodilia], Sauria, Testudines
Heyden, **C**arl [Heinrich Georg] von • 20 January 1793–1866 January 7 • Germany • Anura, Sauria
Heyer, **W**[illiam] **R**onald • 29 August 1941 • USA • Anura, Sauria
Heymons, **R**ichard [Friedrich Wilhelm Carl] • 29 May 1867–1943 December 1 • Germany • (Anura)
Hibbard, **C**laude **W**[illiam] • 21 March 1905–1973 October 9 • USA • (*anura, *caudata), *sauria, (*serpentes), *testudines
Hickman, **J**ohn **L.** • ? • Australia • (Sauria)
Hidalgo, **H**ugo • 1951 • USA • Sauria
Higginbottom, **J**ohn • 1788–1876 • England • Caudata
Higgins, **S. B.** • ?–(1873)–? • USA • Serpentes
Highton, **R**ichard • 24 December 1927 • USA • Caudata, Serpentes
Hikida, **T**sutomu • 9 March 1951 • Japan • Sauria
Hilgendorf, **F**ranz **M**artin • 5 December 1839–1904 July 5 • Germany / Japan / Germany • *sauria, Serpentes
Hill, **J**ames **P**eter • 21 February 1873–1954 May 24 • Scotland / Australia / England • (Anura)
Hillenius, **D**irk ("Dick") • 29 May 1927–1987 May 4 • Netherlands • (Anura), *Sauria
Hillis, **D**avid **M**[ark] • 21 December 1958 • USA • Anura, Caudata, Sauria, Serpentes
Hilton, **J**ohn **W**illiam • 9 September 1904–1983 November 27 • USA • (Serpentes, Testudines)
Hilton, **W**illiam **A**[twood] • 27 June 1878–1970 August • USA • Caudata
Hinckley, **M**ary **H.** • 6 April 1845–1944 • USA • Anura
Hinkel, **H**arald • ? • Germany • Gymnophiona
Hirayama, **R**en • 3 November 1956 • Japan • *testudines
Hirsch, **K**arl **F**[ranz] • 20 March 1921–1996 June 1 • Germany / USA • *crocodilia
Hitz, **R**obert • 7 March 1951 • ? • [Sauria]
Hla Tun • ? • Burma / Myanmar • [Anura, Sauria], Serpentes
Hô Thu Cúc • ? • Vietnam • Anura, Sauria, Serpentes
Hoard, **R**obert **S.** • ? • USA • Serpentes
Hocking, **P**eter • ? • Perú • Serpentes
Hödl, **W**alter • ? • Austria • Anura
Hodrová, **M**arcela (= mrs. Marvanová) • ? • Czechoslovakia • *caudata
Hoegg, **S**imone [Isabel] • 21 July 1977 • Germany • Anura

Hoernes, **R**udolf • 7 October 1850–1912 August 20 • Austria • *testudines

Hoeven, **J**an van der • 9 February 1801–1868 March 10 • Netherlands • *Caudata, Sauria

Hoevers, **L**eo **G**. • 24 October 1938 • Netherlands / Guyana / Belize / Kenya / Sudan • Serpentes

Hoffman, **A**braham **C**arel • 1903 • South Africa • Anura, Serpentes

Hoffman, **R**ichard **L**[awrence] • 25 September 1927 • USA • Caudata, Sauria

Hoffmann, **J**oseph • ? • Luxemburg • Anura

Hoffstetter, **R**obert [Julien] • 11 July 1908–1999 December 29 • France • *caudata, *sauria, *serpentes

Hofmann, **A**dolf • 17 January 1853–1913 September 9 • Czech (Žebrák / Styria / Prague) • *crocodilia, (*sauria)

Hoge, **A**lphonse **R**[ichard] (*husband of* Silvia Alma Romano -Hoge) • 5 September 1912–1982 December 25 • Brazil / Belgium / Brazil • Serpentes, (Testudines)

Hogg, **J**ohn • 1800–1869 September 16 • England • Caudata

Holbrook, **J**ohn **E**[dwards] • 30 December 1794–1871 September 8 • USA • Anura, Caudata, Sauria, Serpentes, Testudines

Holl, **F**riedrich • ?–(1820 *to* 1834)–? • Germany • *caudata

Holland, **R**ichard **L**auritz • 26 March 1939 • USA • Serpentes

Holland, **W**illiam **J**acob • 16 August 1848–1932 December 13 • Jamaica / USA • *crocodilia, (Serpentes, *testudines)

Hollingsworth, **B**radford **D**[amian] • ? • USA • Anura, Sauria

Holm, **P**eter **A**[lfred] • ? • USA • Sauria, Serpentes

Holman, **J A**lan ("Al") • 24 September 1931–2006 August 12 • USA • *anura, *Caudata, *Sauria, *serpentes, *testudines

Holmer, **L**ars **M**agnus (= Laurentius Magnus) • 1764–1800 • Sweden • Sauria

Holsinger, **J**ohn **R**obert • 6 April 1934 • USA • Caudata

Holt, **T**imothy **R**. • ? • Australia • *crocodilia

Hombron, [Jacques Bernard] • 15 January 1798–1852 September 16 • France • Sauria, Serpentes

Home, (sir) **E**verard • 6 May 1756–1832 August 31 • England • (Testudines)

Honda, **M**asanao • ? • Japan • Sauria

Honnorat-[Bastide], **É**douard **F**. • ?–? • France • (Anura)

Hoofien, **J**acob (= Yaakov) **H**[aim] • 23 October 1913–1997 June 21 • Israel • Sauria

Hoogmoed, **M**arinus **S**[teven] • 19 March 1942 • Netherlands • Anura, Gymnophiona, Sauria, Serpentes

Hooijer, **D**irk **A**lbert ("Dick") • ? • Netherlands • *testudines

Hooley, **R**eginald **W**alter • 5 September 1865–1923 May 5 • England • *testudines

Horn, **H**ans **G**eorg • ? • Germany • Sauria

Hornaday, **W**illiam **T**emple • 1 December 1854–1937 March 6 • USA • Crocodilia

Horner, **P**aul **G**[eorge] • 3 June 1948 • Australia • Sauria, Serpentes

Hornstedt, **C**laës **F**redrik • 1758–1809 May • Sweden • Serpentes

Horowitz, **S**amuel **B**oris • 26 August 1927 • USA • Sauria

Horst, **R**utger • ?–(1883)–? • Netherlands • Anura

Horton, **D**[avid] **R**[obert] • 4 April 1945 • Australia • Sauria

Horváth, **G**éza • 23 November 1847–1937 September 8 • Hungary • (Sauria)

Horwitz, **P**ierre • ? • Australia • Anura

Hoser, **R**aymond **T**[errence] • 8 February 1962 • England / Australia • Sauria, Serpentes

Hoskin, **C**onrad **J**[ules]. • 28 January 1976 • South Africa / Australia • Anura / Sauria

Hosmer, **W**illiam ("Bill") • ? • Australia • Anura, Sauria

Hossini, **S**aida (= Saïda *or* Saòda) • ? • France / Morocco • *anura

Hotz, **H**ansjürg • ? • Switzerland / USA / Switzerland • Anura

Hou Lianhai • ? • China • *sauria

Houck, **L**ynne **D**[iane] • ? • USA • [Caudata]

Houston, **T**[erry] **F**[rancis] • ? • Australia • Sauria

Houttuyn, **M**artinus (= Maarten Willem) • 1720–1798 • Netherlands • Caudata, Sauria, (Serpentes)

Howell, **K**im **M**[onroe] • 15 September 1945 • USA / Tanzania • Anura

Howell, **T**homas **R**[aymond] • 17 June 1924–2004 December 14 • USA • Serpentes

Hoy, **P**hilo **R**omayne (? *or* Romeyn) • 3 November 1816–1892 December 9 • USA • (Testudines)

Hoyos [-Hoyos], **J**ulio **M**ario • ? • Colombia • Caudata

Hoyt, **D**ale **L**[yon] • ? • USA • Anura

Hsu Chunhua • ? • China • *testudines

Hsu (*or* Hsü), **H**si-fan • 9 March 1906–1990 February 2 • China / USA • Anura, Testudines

Htun Win • ?–? • Burma / Myanmar • ([Anura, Sauria]), Serpentes

Hu Buqing • 4 April 1915–1979 September 30 • China • Serpentes

Hu Qixiong • 25 April 1945 • China • Anura, Caudata

Hu Shuqin (= Shu-chin, *wife of* Liu Ch'eng-chao) • 13 February 1914–1992 December 10 • China • Anura, Caudata, Sauria, Serpentes

Hua, **S**téphane • ? • France • *crocodilia

Huang Kangcai • 23 December 1935 • China • Anura, Sauria

Huang Meihua • 25 November 1930 • China • Serpentes

Huang Qingyun • 18 June 1938 • China • Sauria, Serpentes

Huang Wanbo • ? • China • *crocodilia

Huang Yongzhao • 24 June 1939 • China • Anura, Serpentes

Huang Zhengyi • 3 November 1933 • China • Anura, Caudata, Serpentes

Huang Zhujian • 1932 • China • [Anura], Serpentes

Hubrecht, **A**[mbrosius] **A**[rnold] **W**[illem] • 1853–1915 • Netherlands • Sauria, Serpentes, Testudines

Hubricht, **L**eslie • 11 January 1908–2005 September 19 • USA • (Caudata)

Huene, **F**riedrich von • 22 March 1875–1969 April 4 • Germany • *crocodilia, *rhynchocephalia

Huerta-Ortega, **S**ara **M**. • ? • México • (Serpentes)

Huey, **R**aymond **B**[runson] • 14 September 1944 • USA • Sauria

Hughes, **R**ichard **C**hester • 21 July 1900–1985 May • USA • Caudata

Hugi, **F**ranz **J**oseph • 23 January 1796–1855 March 25 • Switzerland • (*testudines)

Huheey, **J**ames **E**[dward] • 2 August 1935 • USA • Serpentes

Hulke, **J**ohn **W**hitaker • 6 November 1830–1895 February 19 • England • *crocodilia

Hulse, **A**rthur **C**[harles] • 20 December 1945 • USA • Sauria

Hulselmans, Jan **L. J.** • ? • Belgium • Anura

Humbert, **A**loïs • 1829–1887 • Switzerland • *testudines

Humboldt, [Friedrich Wilhelm Heinrich] **A**lexander von • 14 September 1769–1859 May 6 • Germany • (Anura, Caudata), Serpentes

Hummel, **K**arl [A.] • 14 October 1889–1945 April 7 • Germany • *testudines

Hummelinck: *see* Wagenaar Hummelinck, P[ieter]

Humphrey, **F**rances **L.** (*wife of* Frederick A. Shannon *from* 1964 *to* 1965) • ? • USA • Anura, Serpentes

Humphries, **R**obert **B.** • ? • Australia • (Sauria)

Hunsaker, **D**on, II • 6 April 1930 • USA • (Sauria)

Hunt, **L**awrence **E.** • ? • USA • Sauria

Hurter, **J**ulius • 2 July 1842–1917 November • Switzerland / USA • (Anura), Testudines

Husband, **G**rant **A.** • ? • Australia • (Sauria)

Hussain, **S. A.** • ? • India • (Anura)

Hutchinson, **M**ark **N**[orman] • 31 August 1954 • Australia • *Sauria

Hutchison, **J**[ohn] **H**oward • 29 December 1939 • USA • (*testudines)

Hutterer, **R**ainer • ? • Germany • *Sauria

Hutton, **A**ngus **F.** • ? • (? England) / India / Australia • (Serpentes)

Hutton, **F**rederick **W**ollaston • 16 November 1836–1905 October 29 • England / New Zealand • Sauria

Huxley, **T**homas **H**enry • 4 May 1825–1895 June 29 • England • Crocodilia

I

Ibáñez-D[íaz], **R**oberto • 9 September 1958 • Panamá • Anura, Caudata, Sauria

Ibarra-Vidal, **H**éctor [Guillermo] • ? • Chile • Anura

Ibisch, **P**ierre **L**[eonhard] • 11 September 1967 • Germany • (Anura), Sauria

Ibrahim, **A**del **A**[bdush-Shakour Assayyed] • 9 March 1960 • Egypt • (Serpentes)

Ihering (= Jhering), **H**[ermann] von (*father of* Rodolpho von Ihering) • 9 October 1850–1930 February 24 • Germany / Brazil / Germany • (Anura, Sauria, Serpentes)

Ihering, **R**odolpho [Theodoro Gaspar Wilhelm] von (*son of* H. von Ihering) • 17 July 1883–1939 September 15 • Brazil • Gymnophiona, Serpentes, Testudines

Ilgaz, **Ç**etin • ? • Turkey • Sauria, Serpentes

Incháustegui-M[iranda], **S**ixto **J.** • ? • Dominican Republic • Sauria, (Serpentes)

in den Bosch, **H**erman **A. J.**: *see* **Bosch**, **H**erman **A. J.** in den

Ineich, **I**van [René] • 3 November 1957 • France / French Polynesia / France • Sauria, Serpentes

Inger, **R**obert **F**[rederick] ("Bob", *husband* of Tan Fui-Lian) • 10 September 1920 • USA • Anura, Caudata, Sauria, Serpentes

Ingoldby, **C. M.** • ?–? • British India • Sauria, Serpentes

Ingram, **G**len **J**[oseph] • 22 April 1951 • Australia • Anura, Sauria, Serpentes

Ingram, **W**illiam, III • 1945 • USA • Sauria

Ingrasci, **M**atthew **J.** • ? • USA • Sauria

Ionides, **C**[onstantine] **J**[ohn] **P**[hilip] • 1900–1968 September 22 • England / Tanganyika • (Sauria, Serpentes)

Irwin, **S**teve (= Stephen Robert) • 22 February 1962–2006 September 4 • Australia • (Testudines)

Isenschmidt, **M**oritz • ?–? • Switzerland • Anura

Ishikawa, **C**hiyomatsu • ("8 January") 30 January 1860–1935 January 17 • Japan / Germany / Japan • (Anura), Caudata

Iskandar, **D**joko **T**[jahjono] • 1950 • Indonesia • Anura, (Sauria)

Isle (= l'Isle) [du Dreneuf], **A**rthur de la • ?–? • France • *anura, Caudata

Ito, **A**kiko • ? • Japan • Serpentes

Ito, **R**iu • ? • Japan • Anura

Iturra-C[onstant], [Iris] **P**atricia • 22 September 1947 • Chile • Anura

Ivanov, **M**artin • ? • Czech Republic • *serpentes

Iverson, **J**ohn **B**[urton, III] • 4 October 1949 • USA • Sauria, (Serpentes), Testudines

Ives, **J**ames **E**dmund • 19 September 1865–1943 January 2 • England / USA • Sauria

Iwan: *see* **Lepekhin**, **I**van [Ivanovich] (= Iwan Lepechin)

Izecksohn, **E**ugênio • ? • Brazil • Anura

J

Jackman, **T**odd **R**[ichard] • 20 April 1965 • USA • Caudata, Sauria

Jackson, **D**ale **R**obert • 1949 • USA • *testudines

Jackson, **H**erbert **W**illiam • 5 January 1911–2002 May 19 • USA • (Caudata)

Jackson, **J**ames **F**rederick • 11 November 1943 • USA • Sauria

Jacobs, **H**ans **J.** • ? • Germany • Sauria

Jacobs, **J**eremy **F**[ischer] • 3 May 1955 • USA • Caudata, Sauria

Jacobsen, **N**iels **H. G.** • ? • South Africa • Sauria, Serpentes

Jacquin, **J**oseph Franz (freiherr) von • 7 February 1766–1839 December 9 • Austria • Sauria

Jacquinot, **H**onoré • 1 August 1815–1887 • France • Sauria, Serpentes

Jaeger, **R**obert **G**[ordon] • 16 December 1937 • USA • [Caudata]

Jaekel, **O**tto [Max Johannes] • 21 February 1863–1929 March 6 • Poland / Germany • *rhynchocephalia

Jäger, **G**eorg [Friedrich] von • 25 December 1785–1866 September 10 • Germany • (Serpentes)

Jain, **S**ohan **L**all • 1931 • India • *testudines

James, **C**ameron • ? • Australia • Sauria

James, **E**dwin • 27 August 1797–1861 October 28 • USA • [Anura, Caudata, Sauria, Serpentes]

Jameson, **D**avid **L**[ee] • 3 June 1927 • USA • Anura

Jan, **G**iorgio (= Georg) • 21 December 1791–1866 May 8 • Austria / Italy • Sauria, Serpentes

Janensch, **W**erner • 10 November 1878–1969 October 20 • Germany • *rhynchocephalia

Jarocki, **F**eliks **P**aweł • 14 January 1790–1865 March 25 • Poland • Caudata, Serpentes, Testudines

Jaslow, **A**lan **P**[hilip] • ? • USA • Anura

Jeitteles, **L**udwig **H**einrich • 1830–1883 • Austria • Sauria

Jenkins, **R**ob **W. G.** • ? • Australia • (Sauria)

Jensen, **A**dolf **S**everin • 1866–1953 • Denmark • Serpentes

Jenyns, Leonard (*changed to* Leonard Blomefield *in 1871*) • 25 May 1800–1893 September 1 • England • [Caudata]
Jepsen, Glenn Lowell • 4 March 1903–1974 October • USA • (*caudata), *sauria
Jerdon, T[homas] C[laverhill] • 12 October 1811–1872 June 12 • England / British India / England • Anura, Sauria, Serpentes, Testudines
Jesu, Riccardo • ? • Italy • Anura, Sauria, Testudines
Jeude: *see* **Lidth de Jeude**, T[heodorus] W[illem] van
Jewell, Tony R. • ? • New Zealand • Sauria
Jhering: *see* **Ihering**, Hermann von
Ji Qiang • ? • China • *anura
Ji Shuan • ? • China • *anura
Jiang Yaoming • 12 January 1932 • China • Anura, Sauria, Serpentes
Jim, Jorge • ? • Brazil • Anura
Jiménez, Alfonso • ? • Costa Rica • Serpentes
Jiménez de la **Espada**, **Marcos** • 5 March 1831–1898 October 3 • Spain • Anura, Caudata
Jiménez-Fuentes, Emiliano • ? • Spain • *testudines
Jintakune, Piboon • 30 August 1960 • Thailand • (Serpentes)
Jockusch, Elizabeth L[acy] • ? • USA • Caudata
Joger, Ulrich • ? • Germany • Sauria, Serpentes, Testudines
Johnson, Jerry D[ouglas] • 1 September 1947 • USA • Anura, Caudata, (Serpentes)
Johnson, Murray L[eathers] • 16 October 1914–1995 July 24 • USA • Serpentes
Johnson, **R. D. O.** • ?–? • USA / Colombia / USA • (Anura)
Johnston, C[laud] Stuart • 31 December 1900–1939 July 24 • USA • *testudines
Johnston, Gregory R[onald] • 28 September 1962 • Australia • Anura
Johnstone, R[onald] E[ric] • 1949 • Australia • (Sauria)
Joleaud, [Jean] L[éonce François] • 1880–1938 • France • *crocodilia
Jolly, (sir) Gordon Gray (lt.-general) • 1886–1962 March 1 • England / British India • (Serpentes)
Jones, Clyde J[oe] • 3 March 1935 • USA • Caudata
Jones, J. Paul • ?–? • USA • Sauria
Jones, Kirkland Lee ("Kirk") • 1 October 1941 • USA • Anura, Sauria
Jones, Richard Evan • 13 May 1940 • USA • (Sauria)
Jonet, Simon • 13 November 1902–1987 January 29 • Belgium / Portugal • *crocodilia
Jong, Jan Kornelis de • 6 April 1895–1972 August 6 • Netherlands / East Indies / Netherlands • Sauria, Serpentes
Jonnès: *see* **Moreau de Jonnès**, Alexandre
Jopson, Harry Gorgas Michener • 23 June 1911 • USA • Caudata
Jordan, David Starr • 19 January 1851–1931 September 19 • USA • ([Anura], Caudata)
Jörg, Erwin • 10 January 1917–1977 February 11 • Germany • *sauria
Jörgens, Dirk • ? • ? • Anura
Joseph-Ouni, Mehdi • ? • USA • Testudines
Joshy, S. Hareesh • ? • India • Anura
Joss, Jean M. P. • ? • Australia • (Sauria)
Jourdan, (? Claude) • ?–? • France • *rhynchocephalia, Serpentes
Jouve, Stéphane • ? • France • *crocodilia
Joynson, H. W. • ? • (? England) / Siam • (Serpentes)
Judd, Walter S. • ? • USA • Serpentes
Juliá-Zertuche, Jordi • 11 March 1918–1985 June 25 • México • Serpentes
Juncá, Flora Acunã • ? • Brazil • Anura, Serpentes
Jungfer, Karl-Heinz • ? • Germany • Anura

K

Kadivar, Fatemeh • ? • USA • Anura
Kail, Johann Alois • 21 April 1856–? • Austria • *crocodilia
Kaiser, Christine M[ary] (*born* Dwyer, *wife of* Hinrich Kaiser) • 24 January 1964 • USA • Anura
Kaiser, Hinrich (*husband of* Christine M. Kaiser) • 3 February 1965 • Germany / USA • Anura, Serpentes
Kälin, Josef Alois • 1 August 1903–1965 November 13 • Switzerland • *crocodilia
Kallert, Eduard • 1888–? • Germany • (Caudata)
Kalm, Pehr (= Peter) • 6 March 1716–1779 November 16 • Finland / Sweden • (Anura)
Kamei, Tadao • 1925 • Japan • *crocodilia
Kami, Hagi Gholi • ? • Iran • Serpentes
Kamita, Tsuneichi • ? • Japan • Sauria
Kammerer, Paul • 17 August 1880–1926 September 23 • Austria • ([Sauria])
Kampen, P[ieter] N[icolaas] van • 30 June 1878–1937 July 3 • Netherlands • Anura, (Sauria)
Kanberg, Hans • ? • Germany • (Anura), Testudines
Kaneko, Naotomo • 1963 • Japan • *testudines
Kaplan, Moises [Daniel] • ? • USA • Anura
Karaman, Stanko L. • 1889–1959 May 17 • Yugoslavia • Anura, Caudata, Sauria
Karg, Josef Maximilian • ?–(1786–1805)–? • Germany • (*serpentes)
Karges, John Paul • 22 April 1955 • England / USA • Sauria
Karl, Hans-Volker • 30 November 1955 • Germany • *testudines
Karlin, Alvan A. • 3 May 1950 • USA • Caudata
Karsen, Stephen J. • ? • USA • (Caudata)
Kasahara, Sanae • ? • Brazil • Anura
Kashchenko (= Kaščenko), Nikolai Feofanovich • ("25 April") 7 May 1855–1935 March 29 • Russia / USSR • Anura, Sauria
Kaska, Yakup • 2 January 1970 • Turkey • Sauria
Kattan, Gustavo H. • 26 July 1953 • Colombia *and* USA • Anura
Kattinger, Emil • 1902 • Germany • Sauria
Kaudern, Walter [Alexander] • 24 March 1881–1942 July 16 • Sweden • Sauria, Serpentes
Kauffeld, Carl [Frederick] (*husband of* Isabelle Hunt Kauffeld *from* ? *to* ?) • 17 April 1911–1974 July 10 • USA • Serpentes
Kauffeld, Isabelle **Hunt**: *see* **Conant**, Isabelle [de Peyster] Hunt
Kaup (= Caup), Johann Jakob • 20 April 1803–1873 July 4 • Germany • (Gymnophiona), Sauria
Kauri, Hans • 30 May 1906–1999 January 30 • Russia (Estonia) / Sweden / Norway • Caudata

Kawano, Usaburo • ? • Japan • Anura
Kay, Fenton **R**[ay] • 10 October 1942 • USA • Sauria
Keck, Michael **B**[rian] • ? • USA • Anura
Keferstein, Wilhelm [Moritz] • 7 June 1833–1870 January 25 • Germany / Italy / Germany • Anura, Caudata
Keim, Lauren **D.** • ? • Australia • Sauria
Keiserling: *see* **Keyserling**, Alexander [Friedrich Michael Leberecht Nikolaus Arthur] Andreevich (graf von)
Keith, Ronalda • ? • USA • Anura
Kelaart, Edward Frederick (= Edvard Frederik, major) • 21 November 1819–1860 August 31 • Ceylon / England / Gibraltar / Ceylon • Anura, Sauria, Serpentes
Keller, Claudia • ? • Brazil • Testudines
Kellogg, [Arthur] Remington • 5 October 1892–1969 May 8 • USA • Anura, *crocodilia
Kendrick, Peter **G.** • ? • Australia • Sauria
Kennedy, Joseph Patrick • 9 March 1932 • USA • (Anura)
Kennerson, Kim John • ? • Australia • (Sauria)
Kennett, Rod (= Rodney Malcolm Kennett) • 16 April 1963 • Australia • Testudines
Kennicott, Robert • 13 November 1835–1866 May 13 • USA • Serpentes
Kenny, **J**[ulian] **S.** • ? • Trinidad • [Sauria]
Kent, Sally **A.** • ? • Australia • Anura
Kent: *see also* **Saville-Kent**, **W**[illiam] (*born* William Savill Kent)
Kerster, Harold **W**[illiam] • 1932 • USA • Sauria
Kessler, Karl Fedorovich • ("19 November") 1 December 1815–1881 March 15 ("March 3") • Russia • Caudata, Sauria
Ketchersid, Chesley **A**[rthur] ("Ches") • 16 October 1941–2004 March 2 • USA • Sauria
Key, George [Franklin] • 11 April 1942–1999 August 7 • USA • (Sauria)
Keyserling (= Keiserling), Alexander [Friedrich Michael Leberecht Nikolaus Arthur] Andreevich (graf von) • 15 August 1815–1891 May 8 • Latvia / Russia / Estonia • Anura, (Caudata)
Kezer, [Leonard] James ("Jim") • 17 October 1908–2002 January 3 • USA • (Caudata)
Khac: *see* **Le Khac Quyet**
Khan, **A. Q.** • ? • Pakistan • Serpentes
Khan, Muhammad **R. Z.** • ? • Pakistan • Sauria
Khan, Muhammad Sharif • 11 August 1939 • Pakistan • Anura, Sauria, Serpentes
Khare, **M. K.** • ? • India • Anura
Kharin, **V. E.** • ? • USSR • Serpentes
Khonsue, Wichase • ? • Thailand / Japan / Thailand • Anura
Khosatzky (= Khozatskij *or* Chosatzky), Leo (= Lev) **I**[saakovich] • 13 May 1913 • Russia / USSR • *anura, *Testudines
Kiester, **A**[lan] Ross • ? • USA • Sauria
Kiew Bong Heang • ? • Malaysia • Anura
Kimakowicz, Moritz von • ?–(1888)–? • Romania • Sauria
King, Dennis [R.] • ?–? • Australia • Sauria
King, **F**[rederic] Wayne • 30 May 1936 • USA • Sauria
King, [Fred] Willis • 24 May 1908–1998 June 5 • USA • Caudata
King, Max • ? • Australia • Sauria
King, Philip Parker (rear-admiral) • 13 December 1791–1856 February 26 • Norfolk Island / England / Australia • ([Sauria], Serpentes)
Kinghorn, **J**[ames] **R**[oy] • 12 October 1891–1983 March 4 • Australia • Anura, Sauria, Serpentes
Kinkelin, [Georg] Friedrich • 15 July 1836–1913 August 13 • Germany • (Serpentes), *serpentes
Kirițescu (= Kiritzesco), Constantin • 3 September 1876–1965 August 12 • Romania • Caudata
Kirk, (sir) John • 19 December 1832–1922 January 15 • Scotland / East Africa / England • (Gymnophiona, Sauria, Serpentes)
Kirtisinghe, **P**[arakrama] • 12 February 1903–1981 June 15 • Ceylon • Anura
Kirtland, **J**[ared] **P**[otter] • 10 November 1793–1877 December 10 • USA • Serpentes
Kishida, Kyukichi • 25 August 1888–1968 October 4 • Japan • Sauria
Kisteumacher, Geraldo • ? • Brazil • Anura
Kiyasetuo • ? • (? India) • Anura
Kizirian, David **A**[lan] • ? • USA • Anura, Sauria, Serpentes
Klappenbach [-Childe], Miguel **A**[ngel] • 4 May 1920–2000 January 25 • Uruguay • Anura, Sauria
Klauber, Laurence **M**[onroe] ("Larry") • 21 December 1883–1968 May 8 • USA • (Caudata), Sauria, Serpentes, (Testudines)
Klausewitz, Wolfgang • 20 July 1922 • Germany • Sauria
Klaver, Charles **J. J.** • 12 December 1946 • Netherlands • Sauria
Klembara, Josef • ? • Czechoslovakia / Slovakia • *sauria
Klemmer, Konrad • 1930 • Germany • (Anura), Sauria, Serpentes
Klewen, Reiner [Franz] • 1956 • Germany • Caudata
Klingel, Gilbert **C**[larence] • 4 October 1908–1983 May 16 • USA • Sauria
Klinikowski, Ronald **F.** • ? • USA • (Anura), Sauria
Kloss, Cecil Boden • 28 March 1877–1949 August 19 • England / British Malaya / England • (Anura, Sauria, Serpentes)
Kloss, Gertrud Rita • ? • Brazil • [(Anura)]
Kluge, Arnold **G**[irard] • 27 July 1935 • USA • *anura, Sauria
Knaap-van Meeuwen, Maria Sophia • ? • Netherlands • Serpentes
Kneeland, Samuel • 1 August 1821–1888 September 27 • USA • Caudata
Kneller, Mathias • ? • Germany • Anura
Knight, James **L.** • ? • USA • Serpentes
Knight, **R.** Alec • ? • USA • Sauria
Knoepffler, [Louis]-Philippe • 11 March 1926–1984 November 15 • France • Serpentes
Knop, **R.** • ? • Germany • Anura
Knox, Frederick John • 3 April 1794–1873 August 5 • Scotland / New Zealand • Sauria
Kobel, Hans Rudolf • 18 August 1936 • Switzerland • Anura
Koch, Carl • 1 April 1894–1970 June 14 • Germany • (Sauria)
Koch, Carl (= Karl Jakob Wilhelm) • 1 June 1827–1882 April 18 • Germany • Anura
Kochva, Elazar • 16 June 1926 • USSR (Bessarabia) / Israel • Sauria

Koelliker (= Kölliker), [Rudolf] **A**lbert (ritter) von • 6 July 1817–1905 November 2 • Switzerland / Germany • (Sauria)
Kofron, **C**hristopher **P**. • ? • USA / Zimbabwe / Australia • Serpentes
Köhler, **G**unther • ? • Germany • Anura, Caudata, Sauria, Serpentes
Köhler, **J**örn • 23 June 1970 • Germany • Anura, Caudata, Sauria
Köhler, **R**[ichard] • ? • New Zealand • *testudines
Kok, **P**hilippe **J**[acques] **R**[obert] • 20 December 1970 • Congo (DRC) / Belgium • Anura, Sauria, Serpentes
Ko Ko Gyi • ?–? • Burma / USA / Myanmar • (Serpentes)
Kokubum, **M**arcelo **N**[ogueira] de **C**[arvalho] • 2 October 1974 • Brazil • Anura
Kölliker: *see* **Koelliker**, [Rudolf] **A**lbert (ritter) von
Kolombatovič, **G**juro (= Georg) • 1843–1908 • (? Croatia) • (Anura), Caudata, Sauria
Kopp, **K**átia **A**lcione • ? • Brazil • Sauria
Kopstein, **F**elix • 4 June 1893–1939 April 14 • Austria / Dutch East Indies / Netherlands • Sauria, Serpentes
Kormos, Theodor (= Tivadar) • 10 November 1881–1946 August 25 • Hungary • *sauria, *testudines
Koshy, **M**ammen • ? • India • Anura, Sauria
Koslowsky (= Kozlovski), **J**ulio **G**ermán • 15 September 1866–1923 September 23 • Russia / Argentina • Anura, Sauria, Serpentes
Kosswig, **C**urt • 30 October 1903–1982 March 29 • Germany / Turkey / Germany • (Caudata)
Kostin, **A**natoly **A**[ndreevich] • ? • (? USSR) / Manchuria • Anura, Caudata
Kosuch, **J**oachim • 1 January 1966 • Germany • Anura, Sauria
Kotsakis, **T**assos • 1945 • Italy • *sauria
Kottelat, **M**aurice • ? • Switzerland / Singapore • Anura
Kou Zhitong • 2 October 1935 • China • Anura, Caudata, Serpentes
Kraglievich, **J**orge **L**ucas • 3 August 1886–1932 March 13 • Argentina • *sauria
Kramer, **E**ugen • 9 January 1921 • Switzerland • Serpentes
Krasavtsev (= Krassawzeff), **B**oris [Arkadyevich] • 3 December 1909–1943 June • Russia / USSR • Serpentes
Krasovsky (= Krassowsky), **D**. **B**. • ? • USSR • Sauria
Krassawzeff: *see* **Krasavtsev**, **B**oris [Arkadyevich]
Krassowsky: *see* **Krasovsky**, **D**. **B**.
Krauczuk, **E**rnesto [Ruben] • ? • Argentina • (Anura)
Kraus, **F**red (= Edward Frederick) • 12 June 1959 • USA • Anura, Caudata, Sauria, Serpentes
Kraus, **R**egina • ? • Germany • *crocodilia
Krauss, [Christian] **F**erdinand [Friedrich] von • 1812–1890 • Germany • (Testudinata)
Krausse, **A**nton [Hermann] • 1878–1929 • (? Germany) / Sardinia • Sauria
Krebs, **S**alome **L**itwin • ? • USA • Caudata
Krefft, [Johann Ludwig Louis] **G**erard (*uncle of* P. Krefft) • 17 February 1830–1881 February 19 • Germany / Australia • Anura, Crocodilia, (Sauria), Serpentes, Testudines
Krefft, **P**aul (*nephew of* G. Krefft) • 12 June 1872–1945 September 17 • Germany • Caudata
Kreyenberg, **M**artin • ?–? • Germany • (Serpentes)
Krieg, **H**ans • 18 June 1888–1970 October 5 • Germany • (Anura, Sauria)
Krinitsky, **I**van [Andreevich] (= Jean Krynicki) • ("2 July") 13 July 1797–1838 October 12 ("September 30") • Russia • Anura, Sauria
Kripalani, **M**. **B**. • ? • India • Anura
Krishnamurthy, **S**. **V**. • ? • India • Anura
Kroll, **J**ames **C**larence • 5 November 1946 • USA • Sauria
Kroon, **C**harles • 11 September 1941 • USA • Serpentes
Krysko, **K**enneth **L**. • ? • USA • Serpentes
Kubykin, **R**udolf **A**[lexandrovich] • 16 August 1937–2001 August 1 • USSR / Kazakhstan • Sauria
Kuch, **U**lrich • 18 November 1972 • Germany • Serpentes
Kudryavtzev, **S**ergei **V**. • ? • (? Russia) • Serpentes
Kues, **B**arry **S**[tephen] • 15 March 1946 • USA • *crocodilia
Kuhl, **H**einrich • 17 September 1797–1821 September 14 • Germany / Netherlands / Dutch East Indies (Java) • Anura, Sauria, Serpentes, Testudines
Kuhn, **O**skar • 1908–1990 • Germany • *anura, *crocodilia, *sauria, *serpentes
Kükenthal, **W**illy [Georg] • 4 August 1861–1922 August 20 • Germany • (Sauria, Serpentes)
Kulagin, **N**. **M**. • ?–? • Russia • (Sauria)
Kuliga, **P**aul • 1878–1948 • (? Germany) • (Sauria)
Kumlutas, **Y**usuf • 8 October 1963 • Turkey • Sauria
Kupriyanova, **L**arissa **A**[ndreyevna] • 1943 • USSR • Sauria
Kuramoto, **M**itsuru • 24 October 1934 • Japan • Anura
Kushakevich (= Kušakevič), **S**ergius • ?–? • Russia • (Sauria)
Kuss, **S**iegfried [Ernst] • ? • Germany • *testudines
Küster (= Kuester), **H**einrich **C**arl • 1807–1876 • Germany • Anura
Kuznetsov (= Kusnezov), **V**alentin **V**asilevich • 1923 • USSR • *testudines
Kwet, **A**xel • 5 February 1965 • Germany • Anura
Kyi Soe Lwin • ? • Burma / Myanmar • Anura, [Sauria]
Kyi: *see also* **Sai Wunna Kyi**
Kyriakopoulou-Sklavounou, **P**asqualina • ? • Greece • Anura

L

Labra [Lillo], **M**aria **A**ntonieta • 9 December 1963 • Chile • Sauria
Lacepède, **B**ernard-**G**ermain-**É**tienne [de la Ville-sur-Illon] (comte de) • 26 December 1756–1825 October 6 • France • (Caudata), Sauria, Serpentes, (Testudines)
Lacerda, **J**oão **B**aptista de • 12 July 1846–1915 August 6 • Brazil • Serpentes
Lachman, **E**sther • ? • Israel • [Sauria]
Lacordaire, [Jean] **T**héodore • 1 February 1801–1870 July 18 • France / Belgium • (Crocodilia)
Laemmert, **A**ndreas • ? • Germany • *crocodilia
Lafrentz, **K**arl [August E.] • ?–1945 April 26 • Germany • (Anura), Caudata
Lahanas, **P**ete **N**. • ? • USA • Gymnophiona, Serpentes
Lai June-shiang • ? • Taiwan • Anura
Laidlaw, **F**rank **F**ortescue • 1876–1963 • England • Serpentes
Lakim, **M**aklarin • ? • Malaysia • Sauria
Lamagni, **L**uca • 9 November 1962 • Italy • Testudines
Lamar, **W**illiam **W**[ylly, III] • 18 June 1950 • USA • Anura, Sauria, (Serpentes), Testudines

La Marca, **E**nrique • ? • Venezuela • Anura, Sauria
Lamb, **J**. • ?–(1911)–? • Australia • Anura
Lamb, **T**rip (= Alfred Colquitt Lamb, III) • 7 October 1954 • USA • Sauria
Lambe, **L**awrence **M**[orris] • 27 August 1863–1919 March 12 • Canada • *crocodilia, *serpentes
Lambert, **M**ichael **R**[oderick] **K**[irkby] • 3 November 1941–2004 July 18 • England / Ireland / England • Sauria
Lambiris, **A**ngelo **J**ohn **L**eonis • ? • South Africa • Anura
Lamborot, **M**adeleine • ? • Chile • Sauria
Lamotte, **M**axime [Georges] • 26 June 1920 • France • Anura, (Gymnophiona)
Lampe, **E**duard • ?–(1902)–? • Germany • [Serpentes]
Lancini V., **A**bdem **R**amón • ? • Venezuela • Sauria, Serpentes
Landbeck, **C**hristian **L**udwig • 11 December 1807–1890 September 3 • (? Germany) • [Sauria]
Lando, **R**ichard **V**. • ? • USA • Serpentes
Landy, **M**acreay **J**[ohn] • 1937 • USA • Serpentes
Lane, **H**enry **H**iggins • 17 February 1878–1965 June • USA • (Sauria)
Lang, **H**erbert [Otto Henry] • 24 March 1879–1957 May 29 • Germany / USA • (Anura, Sauria)
Langebartel, **D**avid **A**. • 1929 • USA • Anura, *Sauria, Serpentes
Langer, **M**ax **C**ardoso • 7 September 1973 • Brazil / England / Brazil • *testudines
Langhammer, **J**ames **K**[enneth] • 9 March 1936 • USA • Serpentes
Langone, **J**osé **A**. • ? • Uruguay • Anura
Langston, **W**ann, Jr. • 10 July 1921 • USA • *crocodilia
Lankester, (sir) **E**[dwin] **R**ay • 15 May 1847–1929 August 15 • Scotland / England • [Anura, Sauria]
Lannom, **J**oseph **R**[obert], Jr. • ? • USA • (Serpentes)
Lantz, **L**ouis **A**[médée] • 20 March 1886–1953 February 3 • France / Russia / England • Caudata, Sauria
Lanza, **B**enedetto • 24 May 1924 • Italy • Anura, Caudata, Sauria, Serpentes, (Testudines)
Laport, **F**rançois [Louis Nompar de Caumont, comte de Castelnau] • 25 December 1810–1880 February 4 • England / North America / South America / South Africa / Asia / Australia • (Sauria, Serpentes)
Lapparent de Broin, **F**rance de • 5 October 1938 • France • *Testudines
Lara-Góngora, **G**uillermo • 3 August 1956 • México • Caudata, Sauria
Largen, **M**alcolm **J**. • ? • (? England /) Ethiopia / England • Anura
La Rivers, **I**ra [John] • 1 May 1915–1977 October 11 • USA • *anura
Larsell, **O**lof • 13 March 1886–1964 April 8 • Sweden / USA • (Caudata)
Larsen, **K**enneth **R**[ex] • ? • USA • Sauria
Lartet [de Sansan], **É**douard [Armand Isidore Hippolyte] • 15 April 1801–1871 January 28 • France • *anura, *caudata, *sauria, *serpentes, *testudines
Lataste, [Vital-] **F**ernand • 5 March 1847–1934 January 6 • France / Chile / France • Anura, Caudata, Sauria, Serpentes, Testudines
Lathrop, **A**my • ? • Canada • Anura, Sauria, Serpentes
Latifi, **M**ahmoud • 1930 • Iran • (Serpentes)
Latreille, **P**ierre **A**ndré • 29 November 1762–1833 February 6 • France • Anura, Caudata, Crocodilia, Sauria, Serpentes, Testudines
Laube, **G**ustav **C**arl • 9 January 1839–1923 April 12 • Bohemia / Czechoslovakia • *caudata, *testudines
Laufe, **L**eonard **E**[dward] • 22 July 1924–1998 May 30 • USA • Sauria, Serpentes
Laurent, **R**aymond **F**[erdinand Louis-Philippe] • 16 May 1917–2005 February 3 • Belgium / Africa / USA / Argentina • Anura, Sauria, Serpentes, Testudines
Laurenti, **J**osephus **N**icolaus (= Joseph Nicolas) • 4 December 1735–1805 February 17 • Austria • Anura, Caudata, Crocodilia, Sauria, Serpentes
Lavilla, **E**steban **O**. • ? • Argentina • Anura
Lavocat, **R**ené • ? • France • *crocodilia
Lawrie, **B**ruce **C**. • ? • Australia • Sauria
Lawson, **D**wight **P**[aul] • ? • USA • Gymnophiona
Lawson, **R**obin • 24 January 1933 • England / USA • Serpentes
Lazcano [-Villareal], **D**avid • ? • México • Serpentes
Lazell, **B**rian • ? • Australia • (Sauria)
Lazell, **J**ames **D**[raper], Jr. ("Skip") • 5 September 1939 • USA • Caudata, *Sauria, Serpentes
Lazzarini, **A**lfredo • ?–? • Italy • Caudata
Le Khac **Q**uyet • ? • Vietnam • Serpentes
Le Nguyen **N**gat • ? • Vietnam • Anura
Leach, **W**illiam **E**lford • 2 February 1791–1836 August 25 • England • Crocodilia, Sauria, Serpentes, (Testudines)
Leakey, **J**onathan [Harry Erskine] • 4 November 1940 • Kenya • (Serpentes)
LeBreton, **M**atthew • 10 November 1973 • Australia / Cameroon • Sauria
Le Conte, **J**ohn **E**atton (major, *father of* John Lawrence Le Conte) • 22 February 1784–1860 November 21 • USA • Anura, Caudata, Testudines
Le Conte, **J**ohn **L**awrence (*son of* John Eaton Le Conte) • 13 May 1825–1883 November 15 • USA • (Serpentes)
Lee, **A**nthony **K**[ingston] ("Tony") • 1933 • Australia • Anura
Lee, **D**aniel **K**. • ? • USA • Sauria
Lee, **J**ulian **C**[arlee] • 7 September 1943 • USA • (Anura)
Leenhardt, **H**enri • ? • France • *sauria
Legler, **J**ohn **M**[arshall] • 9 September 1930 • USA • (Anura), Serpentes, Testudines
Lehmann-V[alencia], [Federico] **C**arlos • 23 March 1914–1974 August 15 • Colombia • (Anura, Sauria, Serpentes)
Lehr, **E**dgar • 30 June 1969 • Germany / USA • Anura, Sauria, Serpentes, Testudines
Lehrs, **P**hilipp • 9 August 1881–1956 April 20 • Germany • Sauria
Lei Yizhen • 1944 • China • *testudines
Leidy, **J**oseph • 9 September 1823–1891 April 29 • USA • *crocodilia, *sauria, *testudines
Leighton, **G**erald **R**owley • 12 December 1868–1953 September 8 • England / New Zealand / Scotland / England • (Serpentes)
Leitão: *see* **Carvalho**, **A**ntenor **L**[eitão] de
Leith-Adams: *see* **Adams**, **A**ndrew **L**eith
Lema, **T**hales de • ? • Brazil / Canada / Brazil • Serpentes

Lemos-Espinal, Julio A[lberto] • 17 June 1959 • México • (Anura), Sauria, Serpentes
Lenglet, Georges L[ouis] • 16 February 1949 • Belgium • Anura, Sauria
Lenk, Peter • ? • Germany • Testudines
Lenz, Harald Othmar • 27 February 1798–1870 January 13 • Germany • Serpentes
Lenz, Heinrich [Wilhelm Christian] • 1846–1913 • Germany • (Sauria)
Leong Tzi-Ming • 17 October 1972 • Singapore • Anura, Sauria, Serpentes
Lepekhin, Ivan [Ivanovich] (= Iwan Lepechin) • 10 September 1740–1802 April 6 • Russia • Serpentes
Leschen, Richard A. B. • ? • New Zealand • Sauria
Lescure, Jean • 1932 • France • Anura, Gymnophiona
Lesson, R[ené] P[rimevère] • 20 March 1794–1849 April 28 • France • Anura, Caudata, Crocodilia, Sauria, Serpentes, Testudines
Lessona, Michele • 20 September 1823–1894 July 20 • Italy • Anura, (Sauria)
Lesueur, Charles-Alexandre • 1 January 1778–1846 December 12 • France / USA / France • (Anura), Sauria, Serpentes, Testudines
Leuckart, F[riedrich] S[igismund] • 26 August 1794–1843 September 25 • Germany • *Caudata, Sauria
Levins, Richard • 1 June 1930 • USA • (Sauria)
Leviton, Alan E[dward] • 11 January 1930 • USA • Anura, Caudata, Sauria, Serpentes
Lewis, Charles Bernard, Jr. • 9 April 1913 • USA / Jamaica • (Anura, Sauria)
Lewis, Thomas Howard • 28 July 1919 • USA • Sauria
Leybold, Friedrich (= Federico) • 1827–1879 • Germany / Chile • Serpentes
Li Cheng • ? • China • Sauria
Li Dejun • 23 November 1936 • China • [Anura], Sauria, Serpentes
Li Fanglin • ? • China • Anura
Li Jinling • ? • China • *crocodilia, *sauria
Li Shengquan • 20 March 1955 • China • Anura, Sauria
Li Shushen • ? • China • Anura
Li Siming • 20 January 1946 • China • Anura, Sauria
Li (*not* Liu) Youheng • ? • China • *anura, *testudines
Li Zhixun • 24 November 1929 • China • Testudines
Lian: *see* **Tan** Fui-Lian
Liang Yun-sheng • ? • Taiwan • Anura, Sauria
Lichtenstein, [Martin] Hinrich [Carl] • 10 January 1780–1857 September 3 • Germany / South Africa / Germany / *died at sea* • Anura, Sauria, Serpentes
Lidth de Jeude, T[heodorus] W[illem] van • 1 February 1853–1937 May 29 • Netherlands • Anura, Sauria, Serpentes, Testudines
Lieb, Carl S[ears] • 27 May 1949 • USA • Sauria, Serpentes
Liebus, Adalbert • 1876–1946 • Czechoslovakia • *testudines
Liem, David S. (*formerly* Sioe Sing Liem) • 17 September 1937 • Java / Australia / USA • Anura
Liénard [de Lamivoie], Élizé (= Élysée, père) • ?–1876 • Mauritius • *sauria
Lilljeborg (*changed from* Liljeborg *about 1860*), Wilhelm (*or* Vilhelm) • 6 October 1816–1908 July 24 • Sweden / USA / Sweden • Crocodilia, (Serpentes)
Lim Boo-Liat • ? • Malaysia • Anura, (Sauria), Serpentes
Lim Kok-Peng (= Kelvin Lim) • 29 August 1966 • Singapore • Anura, Sauria
Lim Liang-Jim • ? • Singapore • Sauria
Lima, Albertina P. • ? • Brazil • Anura
Lima-Verde, J[osé] S[antiago] • ? • Brazil • Gymnophiona, Sauria, Serpentes
Limeses, Celia E. • 1918–1974 January 3 • Argentina • Anura
Limpus, Colin James ("Col") • ? • Australia • Testudines
Lin Jun-tsong • ? • Taiwan • Serpentes
Lin Jun-yi (= Edgar Lin) • ? • Taiwan • (Anura), Sauria
Lindaker, Johanne Thaddeus • ?–(1791)–? • (? Austria) • Serpentes
Lindholm, Wilhelm [= Wassili *or* Vassily] A[dolfovitch] • 1874–1935 • Russia / USSR • (Anura), Sauria, Serpentes, Testudines, (*testudines)
Liner, Ernest A[nthony] • 11 February 1925 • USA • Sauria, Serpentes
Link, Heinrich Friedrich • 2 February 1767–1851 January 1 • Germany • Sauria, Serpentes, Testudines
Linnaeus, Carl (*which he latinized as* Carolus) (*in 1761 he changed his name to* Carl von Linné *when he was given nobility*) • ("13 May") 23 May 1707–1778 January 10 • Sweden / Netherlands / Sweden • Anura, Caudata, Crocodilia, Gymnophiona, Sauria, Serpentes, Testudines
Lioy, Paolo • 31 July 1834–1911 January 27 • Italy • *crocodilia
Lips, Karen R[enee] • ? • USA • Anura
L'Isle: *see* **Isle** [du Dreneuf], Arthur de la
List, James C[arl] • 6 July 1926 • USA • Anura
Littlejohn, Murray J[ohn] • 1932 • Australia • Anura
Litvinov, Nikolay A[ntonovich] • 15 April 1951 • USSR / Russia • [Serpentes]
Liu Baohe • ? • China • Anura
Liu Ch'eng-chao (= Chengzhao, *husband of* **Hu** Shuqin) • 12 September 1900–1976 April 9 • China • Anura, Caudata
Liu Hsien-t'ing • 1921 • China • *testudines
Liu Huining • ? • China (Hong Kong) • Anura
Liu Jisheng • 11 June 1937 • China • [Anura]
Liu Wanzhao • ? • China • Anura
Liu Youheng: *see* **Li** Youheng
Liu Yuezhen • 4 October 1933 • China • Sauria
Liu Zhijun • ? • China • Caudata
Loader, Simon P[aul] • 30 March 1976 • England • Anura, Gymnophiona
Lobo, Fernando [José] • 9 May 1963 • Argentina • Anura, Sauria
Lockington, William N[eale] • 1840–1902 • England / USA / England • Serpentes
Löding, Henry Peder • 11 September 1869–1942 February 26 • Denmark / USA • (Caudata, Serpentes)
Loennberg: *see* **Lönnberg**, [Axel Johan] Einar
Loftus-Hills, Jasper J. • ?–1974 June 11 (*age* 28) • Australia / USA • Anura
Long, Stephen Harriman (colonel) • 30 December 1784–1864 September 4 • USA • [Anura, Caudata, Sauria, Serpentes]
Longley, George • ?–1945 • Australia • (Sauria)
Longley, Glenn, Jr. • 2 June 1942 • USA • Caudata

Longman, **H**eber **A**lbert • 24 June 1880–1954 February 16 • England / Australia • *crocodilia, Sauria, Serpentes, Testudines

Longmore, **R**ichard • ? • Australia • (Serpentes)

Lönnberg (= Loennberg), [Axel Johan] **E**inar • 24 December 1865–1942 November 21 • Sweden • Anura, Sauria, Serpentes

Loomis, **F**rederic **B**rewster • 22 November 1873–1937 July 28 • USA • *crocodilia, *sauria, *testudines

Loomis, **R**ichard **B**iggar • 18 June 1925 • USA • (Anura), Sauria, Serpentes

López-Jurado, **L**uis **F**elipe • ? • Spain (Canary Islands) • Sauria

López-Luna, **M**arco **A**ntonio • ? • México • Serpentes

López Seoane: *see* **Seoane** [y Pardo-Montenegro], **V**ictor **L**ópez

Lortet, **L**ouis [-Charles] • 22 August 1836–1909 December 26 • France • Testudines

Lötters, **S**tefan • 3 November 1969 • Germany • Anura, Sauria

Loumont [-Vigny], **C**atherine • 6 December 1942 • Switzerland • Anura

Loveridge, **A**rthur • 28 May 1891–1980 February 16 • Wales / East Africa / USA / Saint Helena • Anura, Gymnophiona, Sauria, Serpentes, Testudines

Lovich, **J**effrey **E**[dward] • ? • USA • Testudines

Low, **T**imothy • 1956 • Australia • Sauria

Lowe, **C**harles **H**[erbert, Jr.] • 16 April 1920–2002 September 13 • USA • Anura, Caudata, Sauria, (*sauria), Serpentes

Lu Jun-chang • ? • USA • *crocodilia

Lu Qingwu • ? • China • *testudines

Lucas, **A**[rthur] **H**[enry] **S**[hakespeare] • 7 May 1853–1936 June 10 • England / Australia • Anura, Sauria, Serpentes

Lucas, **S**pencer **G**[eorge] • 25 April 1955 • USA • *crocodilia, *serpentes

Lüderwaldt: *see* **Luederwaldt**, **H**ermann

Ludescher, **C**laudia • ? • Australia / Tuvalu (South Pacific) • Sauria

Ludwig, **R**udolph [August Birminhold Sebastian] • 24 October 1812–1880 December 11 • Germany • *crocodilia

Lue Kuang-yang • ? • Taiwan • Anura, Sauria

Luederwaldt, **H**ermann • 23 January 1865–1934 August 5 • Pomerania / Brazil • (Anura), Testudines

Lugeon, **M**aurice • 10 July 1870–1953 October 23 • France / Switzerland • *testudines

Lugli, **L**uciana • ? • Brazil • Anura

Lukina, **G**[alina] **P**[anteleymonovna] • 5 June 1930 • USSR • Sauria

Lull, **R**ichard **S**wann • 6 November 1867–1957 April 22 • USA • *crocodilia

Lunau, **H**innerk (= Carl Hinrich Friedrich) • 15 May 1894–1984 November 9 • Germany • *caudata

Lundberg, **M**ikael [Per Bruno] • 3 May 1963 • Sweden / Germany • Anura, Serpentes

Lungu, **A**[lexandr] **N**[ikolayevich] • 1 September 1936 • USSR • [*sauria], *testudines

Lunney, **D**aniel • ? • Australia • (Sauria)

Lütken, **C**hristian **F**rederik • 4 October 1827–1901 February 6 • Denmark • Anura, Gymnophiona, Sauria, Serpentes

Lutz, **A**dolpho (*father of* B. Lutz) • 18 December 1855–1940 October 6 • Brazil • Anura, (Testudines)

Lutz, **B**ertha [Maria Julia] (*daughter of* A. Lutz) • 2 August 1894–1976 September 16 • Brazil • Anura, (Sauria)

Lutz, **D**aniela • 8 August 1959 • Austria • Sauria

Lutzmann, **N**icolà • ? • Switzerland • Sauria

Lwin: *see* **Kyi Soe Lwin**

Lydekker, **R**ichard • 25 July 1849–1915 April 16 • England • *crocodilia, *testudines

Lynch, **J**ames **F**[rancis] • 19 November 1942–1998 March 26 • USA • Caudata

Lynch, **J**ohn **D**[ouglas] • 30 July 1942 • USA / Colombia • *Anura, Sauria

Lynch, **S**hannon **C**olleen • ? • USA • *testudines

Lynn, **W**illiam **G**ardner • 26 December 1905–1990 October 26 • USA • Anura, ([Sauria]), *testudines

Lyon, **M**arcus **W**ard, Jr. • 5 February 1875–1942 May 19 • USA • (Serpentes)

M

Ma Desan • ? • China • Anura

Ma Jifan • 23 October 1940 • China • Serpentes

Mabee, **W**illiam **B**ruce • 1 January 1897–1985 September 4 • USA • (Caudata)

Macarovici, **N**eculai • ? • Romania • *testudines

Maccagno, **A**ngiola **M**aria • ? • Italy • *crocodilia

MacClelland (= M'Clelland), **J**ohn • 1805–1875 • British India • (Serpentes)

MacCulloch, **R**oss **D**[ouglas] • 29 September 1948 • Canada • Anura, Sauria, Serpentes

Macdonald, **J**[ames] **R**[eid] • 22 August 1918–2004 February 4 • USA • *sauria

Macdonald, **M. A.** • ? • Scotland • Sauria

MacDougall, **T**homas **B**[aillie] ("don Tomás") • 9 December (? 1895 *or* 1896)–1973 January 17 • Scotland / USA / México • (Anura, Caudata, Sauria, Serpentes)

Macey, **J. R**obert • ? • USA • Anura, Sauria

Machado, **O**ttilio • ? • Brazil • (Anura), Serpentes

Machordom [-Barbé], **A**[nnie] • ? • Spain • Anura

Mackey, **J**ames **P**[enrose] • 28 February 1930 • USA • Anura

Macklot, **H**einrich **C**hristian • 20 October 1799–1832 May 12 • Germany / Netherlands / Dutch East Indies • (Anura), Sauria, (Serpentes)

Mackness, **B**rian **S.** • ? • Australia • *crocodilia, *serpentes, *testudines

Macleay, (sir) **W**illiam [John] (*cousin of* William Sharp Macleay) • 13 June 1820–1891 December 7 • Scotland / Australia • Anura, Sauria, Serpentes

Macleay, **W**illiam **S**harp (*cousin of* William John Macleay) • 21 July 1792–1865 January 26 • England / France / Cuba / Australia • Sauria

Macola, **G. S.** • ? • Argentina • Sauria

Maduwage, **K**alana **P.** • ? • Sri Lanka • Sauria

Magalhães, **O**ctávio de • ? • Brazil • Serpentes

Maglio, **V**incent **J**oseph • 2 October 1942 • USA • Serpentes

Maha, **G**eorge **C**[hristopher] • ? • USA • [Caudata]

Mahendra, **B**eni **C**haran • ?–? • India • Serpentes

Mahoney, **M**eredith **J**[ean] • 2 May 1971 • USA • Caudata

Mahony, **M**ichael **J.** • ? • Australia • Anura

Main, **A**lbert **R**[ussell] ("Bert") • 1919 • Australia • Anura, (Sauria)

Major, [Charles Immanuel] **F**orsyth • 1848–1923 March 25 • England • (Anura)
Maki, **M**oichiro • 20 February 1886–1959 April 19 • Japan / Taiwan / Japan • Caudata, Sauria, Serpentes
Maldonado-Koerdell, **M**anuel • 25 January 1908–(? 1972 *or* 1973) • México • Caudata
Malfay, **J**ean-**P**ierre • ? • France • *sauria
Malhotra, **A**nita • 16 February 1965 • India / England / Scotland / Wales • Serpentes
Malkin, **B**orys • ? • Switzerland • Anura
Malkmus, **R**udolf • ? • (? Germany) / Portugal / Germany • Caudata, Sauria
Malnate, **E**dmond **V**[irgil] • 16 July 1916–2003 November 6 • USA • Serpentes
Malonza, **P**atrick **K**[enyatta] • ? • Kenya • Gymnophiona, Sauria
Mamede: *see* **Costa**, **E**lizabeth **M**aria **M**amede da
Manamendra-Arachchi, **K**elum **N**alinda • ? • Sri Lanka • Anura, Sauria
Manera de Bianco, **T**eresa • ? • Argentina • *testudines
Maneyro [-Landó], **R**aúl [Eduardo] • 7 December 1969 • Uruguay • Anura
Manjunatha Reddy, **A**. **H**. • ? • India • Anura
Mann, **W**illiam **M**[ontana] • 1 July 1886–1960 October 10 • USA • (Caudata, Sauria, Serpentes, Testudines)
Manriquez [-Soto], **G**erman [Raul] • 15 June 1962 • Ukraine / Chile • Sauria
Mansueti, **R**omeo [John] • 18 May 1923 • USA • (Anura)
Mansukhani, **M**. **R**. • ? • India • Anura
Mantell, **G**ideon **A**lgernon • 3 February 1790–1852 November 10 • England • (*testudines)
Manthey, **U**lrich • 1946 • Germany • Anura, Sauria, Serpentes
Manzani, **P**aulo **R**oberto • 20 August 1952 • Brazil • Sauria
Mao Jierong • 25 October 1921 • China • Sauria
Mara, **W**il (= William P.) • 30 November 1966 • USA • (Serpentes)
Marais, **J**ohan (= Johannes Jacobus) • 13 June 1957 • South Africa • Sauria
Marca: *see* **La Marca**, **E**nrique
March, **D**ouglas **D**. **H**. • ?–1939 April 3 (*age* 52) • USA • (Serpentes)
Marchand, **L**ewis **J**. • ? • USA • Testudines
Marinho-Lima, **A**[lmira] **C**laudia • ? • Brazil • Sauria
Markezich, **A**llan **L**[ouis] • 15 May 1949 • USA • Sauria, Serpentes
Marlow, **R**onald **W**illiam • 9 February 1949 • USA • Caudata
Marques, **O**távio **A**[ugusto] **V**[uolo] • ? • Brazil • Serpentes
Marquis, **O**livier • 2 February 1978 • France • Anura
Marsh, **O**thniel **C**harles • 29 October 1831–1899 March 18 • USA • *crocodilia, *sauria, *serpentes
Marshall, **J**eremy **L**[ee] • ? • USA • Caudata
Marshall, **J**oe **T**ruesdell, Jr. • 15 February 1918 • USA • Sauria
Marshall, **L**arry **G**. • ? • USA • *caudata
Marsterson, **S**. **P**. • ? • Australia • Sauria
Martens, **E**duard [Carl] von (= Karl Eduard von Martens) • 18 April 1831–1904 August 14 • Germany • Anura, Caudata, Sauria, Serpentes
Martin, **A**ngus **A**[nderson] • 20 April 1940 • South Africa / Australia • Anura
Martín, **A**urelio • ? • Spain • Sauria
Martin, **H**andel **T**ongue • 1862–1931 January 15 • USA • ([*caudata])
Martin, **J**ames **E**[dward] • 8 October 1949 • USA • *sauria
Martin, **K**eith • ? • Australia • (Sauria)
Martin, **P**aul **S**[chultz] • 22 August 1928 • USA • Sauria
Martin, **R**obert **L**awrence • 18 November 1933 • USA • Sauria
Martin, **W**illiam [Charles Linnaeus] • 1798–1864 February 15 • England • Sauria, Serpentes
Martín del Campo [y Sánchez], **R**afael • 3 January 1910–1987 December 25 • México • Sauria
Martínez, **V**íctor • ? • Panamá • Anura
Martínez-Oliver, [María] **I**nés • ? • Argentina • Sauria
Martínez-Solano, **I**ñigo • ? • Spain • Anura
Martín-Frias, **E**liezer • 1 July 1939 • México • Caudata
Martín-Hidalgo, **A**urelio • ? • Canary Islands • Sauria
Martins, **M**árcio [Roberto Costa] • ? • Brazil / USA / Brazil • Anura, Serpentes
Martius, **C**arl **F**riedrich **P**hilipp von • 17 April 1794–1868 December 13 • Germany / Brazil / Germany • Serpentes
Martof, **B**ernard [Stephen] • 21 August 1920–1978 November 28 • USA • Caudata
Martori, **R**icardo **A**[rmando] • ? • Argentina • Anura, Sauria
Maruska, **E**dward **J**[oseph] • 19 February 1934 • USA • Caudata
Marx, **H**ymen ("Hy") • 27 June 1925–2007 January 25 • USA • Sauria, Serpentes
Marx, **K**evin **W**[all] • 2 October 1923 • USA • Anura
Maslin, **T**[homas] **P**aul, Jr. • 27 October 1909–1984 February 26 • USA • Anura, Caudata, Sauria, Serpentes
Mason, **G**eorge **E**. • ?–(1888)–? • (? British India) • Serpentes
Massalongo, **A**bramo **B**artolomeo • 13 May 1824–1860 May 25 • Italy • Anura, Caudata, Sauria, Serpentes
Massary, **J**ean-**C**hristophe de • ? • France • Anura
Masterson: *see* **Marsterson**, **S**. **P**.
Matalas, **B**essie **L**ouise (*wife of* Max K. Hecht *from* 1947) • 4 December 1922 • USA • Serpentes
Mateo [-Miras], **J**osé **A**ntonio • ? • Spain (Canary Islands) • Sauria
Mather, **W**[illiam] **W**[illiams] • 24 May 1804–1859 February 25 • USA • [Serpentes]
Matheron, [Pierre] **P**hilippe [Émile] • 18 October 1807–1900 December 31 • France • *crocodilia
Mathers, **S**andra • 18 October 1957 • USA • Sauria
Matschie, [Georg Friedrich] **P**aul • 11 August 1861–1926 March 7 • Germany • Anura, Sauria, Serpentes
Matsui, **M**asafumi • 21 February 1950 • Japan • Anura, Caudata, (Sauria)
Matsui, **T**akaji • 24 March 1925 • Japan • Anura, Sauria
Matsumoto, **E**iji • ? • Japan • *crocodilia
Matsumoto, **H**ikoshichiro • ? • Japan • *testudines
Matthes, **B**enno • 15 September 1825–1911 April 30 • Silesia / Germany / USA • Caudata
Matthew, **W**illiam **D**iller • 19 February 1871–1930 September 24 • USA • (*caudata, *sauria), *testudines
Mattioli, **F**abio • ? • Italy • Anura, Sauria
Mauduyt [de la Grève], **L**ubin • 3 January 1782–1870 June 18 • France • Serpentes
Mausfeld, **P**atrick • ? • Germany • Sauria

Mawson, **P**atricia **M**. • ?–? • Australia • (Sauria)
Maximilian: *see* **Wied** [-Neuwied], [Alexander Philipp] **M**aximilian (prinz von)
Maxson, **L**inda [Ellen] **R**[esnick] • 24 April 1943 • USA • Anura, [Caudata]
May: *see* **von May**, **R**udolf
Mayer, **A**[ugust] **F**[ranz] **J**[oseph] **C**[arl] • 1787–1865 • Germany • Anura
Mayer, **W**erner • 27 May 1943 • Austria • Sauria
Mayorga, **H**oracio • ? • Puerto Rico • Anura
McCann, [Yule Mervyn] **C**harles • 4 December 1899–1980 November 29 • India / New Zealand • Sauria
McCarthy, **C**olin **J**[ohn] • 18 February 1951 • England • Anura, Serpentes
McCauley, **R**obert **H**enry, Jr. • 13 June 1913–1979 October • USA • Sauria
McConkey, **E**dwin **H**[enry] • 1930 • USA • Sauria
McCord, **W**illiam **P**[atrick] • 3 December 1950 • USA • Testudines
McCoy, **C**larence **J**[ohn], Jr. ("Jack") • 25 July 1935–1993 July 7 • USA • Caudata, Sauria, Testudines
McCoy (= M'Coy), (sir) **F**rederick • (? 1817 *to* 1823–1899 May 13 • Ireland / Australia • Anura, Sauria, Serpentes
McCoy, **M**ichael • ? • Solomon Islands / Australia • Sauria
McCracken, **S**hawn **F**. • ? • USA • Anura
McCrady, **E**dward, Jr. • 19 September 1906–1981 July • USA • Caudata
McCranie, **J**ames **R**[andall] • 10 October 1944 • USA • Anura, Caudata, Sauria, Serpentes
McCrystal, **H**ugh **K**[reyer] • ? • USA • Sauria
McCulloch, **A**llan **R**[iverstone] • 20 June 1885–1925 September 1 • Australia • Anura, Testudines
McDiarmid, **R**oy **W**[allace] • 18 February 1940 • USA • Anura, Sauria, Serpentes
McDonald, **K**eith **R**. • ? • Australia • Anura, Sauria
McDowell, **S**amuel **B**[oker, Jr.] • 13 September 1928 • USA • Serpentes
McGregor, **R**ichard **C**[rittenden] • 24 February 1871–1936 December 30 • Australia / USA / Philippines • (Sauria)
McGrew, **P**aul **O**rman • 27 August 1909–1983 January • USA • (*crocodilia), [*sauria, *serpentes]
McGuire, **J**immy **A**[dair] • ? • USA • Anura, Sauria, Serpentes
McKenzie, **N**orman **L**eslie • ? • Australia • (Sauria)
McKinney, **C**harles **O**ran • 10 August 1942 • USA • Sauria
McKown, **R**onald **R**[ay] • 1 September 1940–2005 February 18 • USA • Testudines
McLachlan, **G**eoff **R**. • ? • South Africa • Sauria
McLain, **R**obert **B**aird • 16 August 1877–1946 November 15 • USA • Serpentes
M'Clelland: *see* **MacClelland**, **J**ohn
McMahon, (sir) [Arthur] **H**enry (colonel) • 28 November 1862–1949 December 29 • England / British India / England • (Serpentes)
McMillan, **R**obert **P**eter • 10 August 1921 • Australia • (Sauria)
McMorris, **J**ohn **R**obert • 17 December 1946 • USA • Testudines
McPhee, **D**avid **R**. • ? • Australia • (Sauria)
Mead, **J**im **I**. (= James Irving) • 4 August 1952 • USA • *crocodilia
Mead, **L**ouise **S**[outher] • ? • USA • Caudata
Meade-Waldo, [Edmund] **G**ustavus [Bloomfield] • 1855–1934 February 24 • England • Anura
Means, **D**[onald] **B**ruce • 9 March 1941 • USA • Caudata
Mearns, **E**dgar **A**lexander • 11 September 1856–1916 November 1 • USA • (Anura, Sauria)
Measey, **G**. **J**ohn • ? • England / France • Gymnophiona
Mebs, **D**ietrich • ? • Germany • Serpentes
Mecham, **J**ohn **S**[tephen] • 29 February 1928 • USA • *Anura
Mechler, **B**. • ? • Switzerland / ? • Sauria
Medem [-Medem], **F**ederico (= Friedrich Johann, graf von) • 29 August 1912–1984 May 1 • Latvia / Germany / Switzerland / Colombia • (Anura, Caudata), Crocodilia, (Sauria, Serpentes), Testudines
Medina, **C**armen **J**ulia • ? • Venezuela • *crocodilia
Medway (lord): *see* **Cranbrook**, **G**athorne **G**athorne-**H**ardy (5th earl of, *formerly* lord Medway)
Meegaskumbura, **M**adhava [Priyantha] • ? • Sri Lanka / USA • Anura
Meek, **S**eth **E**ugene • 1 April 1859–1914 July 6 • USA • Anura, Sauria, Serpentes
Meerwarth, **H**ermann • 9 October 1870–1943 April 9 • Germany • Anura, Sauria
Megirian, **D**irk • 6 July 1958 • Switzerland / Australia • *crocodilia, *testudines
Méhelÿ, **L**ajos (= Ludwig von) • 24 August 1862–(? 1946 *or* 1952 *or* 1953) • Hungary • Anura, (*anura), Caudata, Sauria, Serpentes
Mehl, **M**aurice **G**[oldsmith] • 25 December 1887–1966 March 30 • USA • *crocodilia
Mehnert, **E**rnst • 21 February 1864–1902 November 17 • Russia / Estonia / Germany • Testudines
Meier, **H**arald • 14 November 1922 • Germany • (Anura), Sauria
Meik, **J**esse **M**[alcolm] • 30 December 1974 • USA • Anura, Sauria
Meinhardt, **D**aniel **J**[oseph] • ? • USA • Anura
Meise, **W**ilhelm • 12 September 1901–2002 August 24 • Germany • Serpentes
Meisner, [Carl] **F**riedrich [August] • 6 January 1765–1825 February 12 • Switzerland • Sauria, Serpentes
Melin, **D**ouglas **E**dvard • 4 October 1895–1946 March 29 • Sweden • Anura
Mell, **R**udolf [Emil] • 16 February 1878–1970 January 12 • Germany / China / Germany • (Anura, Sauria), Serpentes
Mellado: *see* **Pérez** **M**ellado, **V**alentín
Mello-Leitão, **C**andido [Firmino] de • 17 July 1886–1948 December 14 • Brazil • Anura, Serpentes
Mena-[González], **C**ésar **E**nrique • 7 December 1947 • USA • Serpentes
Mendelson, **J**oseph **R**., III • 7 April 1964 • USA • Anura, Caudata, Sauria, Serpentes
Mendelssohn, **H**einrich • 31 October 1910–2002 November 19 • Germany / Israel • Anura, Sauria
Mendoza-Quijano, **F**ernando • 3 April 1957 • México • Serpentes
Menegon, **M**ichele • ? • Italy • Anura, Sauria
Ménétriés, **E**douard • ("2 October") 14 October 1802–1861 April 22 ("April 10") • France / Brazil / Russia • Sauria, Serpentes

Mengden, **G**regory **A**. • ? • USA / Australia / USA • (Serpentes)
Menzies, **J**[ames] **I**. • ? • England / Sierra Leone / Papua New Guinea • Anura
Mercadal [de Barrio], **I**sabel **T**eresa • 3 October 1954 • Argentina • Anura
Mercurio, **V**incenzo • 6 July 1978 • Italy / Germany • Serpentes
Meredith, **R**obert **W**. • ? • USA • *sauria
Merian, **M**aria **S**ibylla • 2 April 1647–(? *about* 1717) • Germany / Friesland / Netherlands / Surinam / Netherlands • (Anura, Sauria)
Merrem, **B**lasius • 4 February 1761–1824 February 23 • Germany • Anura, Caudata, Crocodilia, Sauria, Serpentes, Testudines
Merriam, **C**[linton] **H**art • 5 December 1855–1942 March 19 • USA • (Sauria)
Mertens, **R**obert [Friedrich Wilhelm] • 1 December 1894–1975 August 23 • Russia / Germany • Anura, (*Caudata), Gymnophiona, *Sauria, Serpentes, Testudines
Mesquita, **D**aniel **O**liveira • ? • Brazil • Sauria
Messel, **H**arry • 3 March 1922 • Canada / Australia • (Sauria)
Meszoely, **C**harles **A**[ladar] **M**[aria] • 24 April 1933 • Hungary / USA • *sauria, (Serpentes)
Metaxà, **L**uigi • 1778–1842 November 24 • Italy • Sauria, Serpentes
Metcalf, **Z**eno **P**ayne • 1 May 1885–1956 January 5 • USA • (Caudata)
Metcalfe, **D**ean **C**arlyle • ? • Australia • (Serpentes)
Methuen, **P**aul **A**yshford (fourth baron) • 29 September 1886–1974 January 7 • England • Anura, Sauria, Serpentes
Metter, **D**ean **E**[dward] • 1 August 1932–2001 June 23 • USA • (Testudines)
Meuschen, **F**riedrich **C**hristian • 1719–1790 • Germany • Serpentes
Meyen, **F**[ranz] **J**[ulius] **F**[erdinand] • 28 June 1804–1840 September 2 • Germany • [Anura, Sauria]
Meyer, **A**dolf **B**ernhard • 11 October 1840–1911 February 5 • Germany • Anura, Sauria, Serpentes, Testudines
Meyer, [Christian Erich] **H**ermann von • 3 September 1801–1869 April 2 • Germany • *caudata, *crocodilia, *rhynchocephalia, *serpentes, *testudines
Meyer, **F**riedrich **A**lbrecht **A**nton • 29 June 1768–1795 November 29 • Germany • Crocodilia, Sauria
Meyer, **J**ohn **R**[aymond] • ? • USA • Anura, (Sauria), Serpentes
Meyer de Schauensee, **R**odolphe • 4 January 1901–1984 April 24 • Italy / USA • (Serpentes)
Meylan, **P**eter **A**[ndré] • 28 May 1953 • USA • *anura, *sauria, *serpentes, *Testudines
Michaelsen, [Johann] **W**ilhelm • 1860–1937 • Germany • [Anura, Sauria, Serpentes]
Michahelles, [Georg Christoph] **C**arl [Wilhelm] • 5 May 1807–1834 August 15 • Germany / Greece • Anura, Caudata, (Sauria), Serpentes, Testudines
Michaud, **E**dward **J**[oseph, III] • ? • USA • Serpentes
Mijares-Urrutia, **A**braham • ? • Venezuela • Anura, Sauria
Mikan (= Mickan), **J**ohann **C**hristian • 5 December 1769–1844 December 28 • Czechoslovakia • Gymnophiona, Sauria, Serpentes, Testudines
Mikšić, **S**onja • ? • Yugoslavia • Caudata
Millard, **W**alter **S**. • ?–1952 March 24 (*age* 85) • England / British India / England • (Serpentes)
Miller, **B**rian • ? • Australia • (Sauria)
Miller, **G**errit **S**[mith], Jr. • 6 December 1869–1956 February 24 • USA • Anura, *sauria
Miller, **J**ohn **S**ebastian • (*about* 1715)–(*about* 1790) • Germany / England • Testudines
Miller, **L**oye [Holmes] • 13 October 1874–1970 April 6 • USA • (*testudines)
Miller, **T**racy **J**. • ? • USA • Serpentes
Millet [de la Turtaudière], **P**ierre **A**imé • 1 May 1783–1873 • France • Anura, Serpentes
Millot, **J**acques • 9 July 1897–1980 January 23 • France • Anura, (Sauria)
Milne-Edwards, **H**enri (*originally* Henri-Milne Edwards) • 23 October 1800–1885 July 29 • Belgium / France • Sauria, [Serpentes]
Milner, **A**ndrew **R**[obert] • 12 June 1947 • England • *anura, *caudata
Milstead, **W**illiam **W**[right] • 12 June 1928–1974 November 29 • USA / Brazil / USA • Anura, *testudines
Min Mi Sook • ? • South Korea • Caudata
Minter, **L**eslie **R**. ("Les") • ? • South Africa • Anura
Minton, **M**adge **R**utherford (*born* Madge Alice Shortridge Rutherford, *wife of* Sherman A. Minton, Jr.) • 22 March 1920–2004 November 7 • USA / Pakistan / USA • (Serpentes)
Minton, **S**herman **A**[nthony], Jr. (*husband of* Madge Rutherford Minton) • 24 February 1919–1999 June 15 • USA / Pakistan / USA • Sauria, Serpentes
Minuth, **W**alter • ? • Germany • Sauria
Miralles, **A**urélien • 7 April 1978 • France • Sauria
Miranda, **M**arta **E**. • ? • Argentina • Serpentes
Miranda Ribeiro, **A**lípio do • 21 February 1874–1939 January 8 • Brazil • Anura, Caudata
Mishra, **V**ijay **P**rakash • 1942 • India • *crocodilia
Misuri, **A**lfredo • 1886–? • Italy • *testudines
Mitchell, **A**. **J**. **L**. • ? • South Africa • Sauria
Mitchell, **B**. **L**. • ? • Nyasaland • (Anura, Sauria)
Mitchell, **F**rancis **J**ohn • 8 August 1929–1970 February 23 • Australia • Sauria, Serpentes
Mitchell, **R**obert **W**etsel • 25 April 1933 • USA • Caudata
Mitchell, **S**[ilas] **W**eir • 15 February 1829–1914 January 4 • USA • (Serpentes)
Mitchill, **S**amuel **L**[atham] • 20 August 1764–1831 September 7 • USA / Scotland / USA • Caudata
Mitsukuri, **K**akichi • 1 December 1858–1909 September 17 • Japan • (Sauria)
Mittermeier, **R**ussell **A**[lan] • 8 November 1949 • USA • (Anura), Testudines
Mittleman, **M**[yron] **B**[udd] • 19 December 1918 • USA • Anura, Caudata, Sauria, Serpentes
Mivart, **S**[ain]t **G**eorge [Jackson] (*pseudonym* = D'Arcy Drew) • 30 November 1827–1900 April 1 • England • Anura, Caudata, (Sauria)
Miyata, **K**enneth [Ichiro] • 1951–1983 October 14 • USA • Anura, Sauria, (Serpentes)
Miyazaki, **K**oji • 4 June 1929 • Japan • Caudata
Młynarski, **M**arian • 29 January 1926 • Poland • [*caudata], (*crocodilia), *serpentes, *Testudines

Mocquard, **F**rançois • 27 October 1834–1917 March 19 • France • Anura, Sauria, Serpentes

Modigliani, **E**lio • 13 August 1860–1932 August 6 • Italy • Sauria

Moellendorff: *see* **Möllendorff**, **O**tto **F**ranz von

Mojsisovics, [Felix Georg Herman] **A**ugust (edler von Mojsvár) • 18 November 1848–1897 August 27 • Austria • [Serpentes], Testudines

Mole, **R**ichard **R**ichardson • ?–? • (? England) / Trinidad • [(Sauria)]

Moler, **P**aul **E**[dmunds] • 25 March 1945 • USA • Anura

Molina [-Rodríguez], **C**ésar **R**amón • ? • Venezuela • Anura

Molina [y González], **J**uan **I**gnacio (abbot) • 20 July 1740–1829 September 12 • Chile / Italy • Sauria

Moll, **E**dward **O**[wen] • 30 November 1939 • USA • Serpentes

Moll, **K**arl **E**hrenbert (freiherr) von • 21 December 1760–1838 February 1 • Austria / Germany • [Anura]

Möllendorff, **O**tto **F**ranz von • 1848–1903 • Germany • (Anura), Caudata, (Sauria, Serpentes)

Moller, **A**dolpho **F**rederico • ?–? • Portugal • (Anura, Caudata)

Molnar, **R**alph **E**. • 28 June 1943 • Australia • *crocodilia, *testudines

Monard, **A**lbert • 2 September 1886–1952 September 27 • Switzerland • Anura, Sauria, Serpentes

Monguillot, **J**ulio **C**. • ? • Argentina • Sauria

Monks, **S**arah **P**. • ?–? • USA • (Anura)

Montague, **P**aul **D**. • ?–1918 • England / New Caledonia / England • Serpentes

Montalenti, **G**iuseppe • 13 December 1904–1990 July 2 • Italy • (Anura)

Montanucci, **R**ichard **R**[oman] • 13 May 1944 • Philippines / USA • Sauria

Montero, **R**icardo • ? • Argentina • Sauria

Montgomery, **J**ohn • ? • Australia • (Serpentes)

Montrouzier, (père) • ?–(1860)–? • France • Serpentes

Moodie, **R**oy **L**ee • 30 July 1880–1934 February 16 • USA • (Anura)

Moody, **R**ichard **T**[homas] **J**[ones] • 3 October 1939 • England • *testudines

Moody, **S**cott **M**[ichael] • 13 March 1949 • USA • *sauria

Mook, **C**harles **C**raig • 7 May 1887–1966 October 10 • USA • *crocodilia

Moore, **G**eorge **A**[zro] • 12 September 1899–1998 November 17 • USA • Caudata

Moore, **J**ohn **A**[lexander] • 27 June 1915–2002 May 26 • USA / Australia / USA • Anura

Moore, **J**[ohn] **E**[dward] **S**[alvin] • 1873–? • England • (Anura)

Moore, **J**ohn **P**ercy • 17 May 1869–1965 March 1 • USA • Caudata

Mooser, **O**swald (= Oswaldo Mooser-Barendun) • 23 December 1903–1983 July 14 • Switzerland / México / Switzerland • *testudines

Moraes, **A**. • ?–? • Portugal • *crocodilia

Morafka, **D**avid **J**[oseph] • ?–2004 January 13 • USA • (Anura)

Morales [-Mondoñedo], **V**íctor **R**[aul] • 25 January 1957 • Perú / Puerto Rico / Perú • Anura

Morando [-Avila], **M**ariana • 21 December 1970 • Argentina / USA / Argentina • Sauria

Morath: *see* **Stemmler-Morath**, **C**arl

Moravec, **J**iří • 8 August 1958 • Czechoslovakia / Czech Republic • Anura, Sauria

Moreau de Jonnès, **A**lexandre • 19 March 1778–1870 March 28 • France • Sauria

Morelet, [Pierre Marie] **A**rthur (chevalier) • 26 August 1809–1892 October 9 • France • (Anura, Caudata, Crocodilia, Sauria, Serpentes)

Moreno, **A**na **G**arcía • ? • Spain • Serpentes

Moreno, **L**uis **V**. • ? • Cuba • Anura, Sauria, (Serpentes)

Moret, **L**éon [Marie Louis] • 4 July 1890–1972 November 22 • France • *testudines

Mori, **T**amezo • 1 June 1884–1962 July 18 • Japan / Korea *and* China / Japan • Caudata

Morice, **A**lbert • 1848–1877 • France / Indochina • Serpentes

Moritz, **C**raig [Charles] • ? • Australia / USA / Australia / USA • Sauria

Morris, **I**an [James] • 17 February 1951 • Australia • (Serpentes)

Morton, **S**amuel **G**eorge • 26 January 1799–1851 May 15 • USA • *crocodilia, (Testudines)

Mosauer, **W**alter • 5 May 1905–1937 August 10 • Austria / USA / México • (Caudata), Sauria, (Serpentes)

Mosimann, **J**ames **E**[mile] • 26 October 1930 • USA • Testudines

Mostert, **D**avid **P**. • ? • South Africa • Sauria

Moszkowski, **M**ax • ?–? • Germany • (Anura, Sauria)

Motte, **M**artha • ? • Paraguay • Serpentes

Mou, **Y**ung-**p**ing • ? • Taiwan / France • Anura

Moulton, **J**ohn **C**oney (captain) • 1886–1926 • England / Sarawak / British India • (Sauria)

Mount, **R**obert **H**[ughes] • 25 December 1931 • USA • Sauria, Testudines

Mouton, **P**[ieter] le Fras **N**[ortier] • 4 April 1953 • South Africa • Sauria

Moyer, **D**avid **C**. • ? • Tanzania • Anura, (Sauria)

Msuya, **C**harles **A**. • ? • Tanzania • Anura

Mueses-Cisneros, **J**onh **J**airo • 8 August 1980 • Colombia • Anura

Mukherjee, **D**ebanik • ? • India • Sauria

Müller, **F**riedrich (= Fritz) • 8 May 1834–1895 March 10 • Switzerland • Anura, Sauria, Serpentes

Müller, **H**endrik • ? • England • Gymnophiona

Müller, **J**ohann **W**ilhelm (baron) • 4 March 1824–1866 October 24 • Germany • [Serpentes]

Müller, **J**ohannes [Peter] • 14 July 1801–1858 April 28 • Germany • Anura, Serpentes

Müller, **L**orenz • 18 February 1868–1953 February 1 • Germany • Anura, Caudata, *crocodilia, Sauria, Serpentes, Testudines

Müller, **O**tto **F**rederik • 11 March 1730–1784 December 26 • Denmark • Anura

Müller, **P**aul • 11 October 1940 • Germany • Sauria

Müller, **S**alomon • 7 April 1804–1864 December 29 • Germany / Dutch East Indies / Netherlands • Crocodilia, Serpentes, Testudines

Mumpuni [Sancoyo] • 25 April 1960 • Indonesia • Anura

Muñoz, **A**nna **M**. • ? • USA • Anura

Muñoz [-Saravia], **A**rturo • ? • Bolivia • Serpentes

Munsterman, **H**arold **E**[rnest] • ? • USA • Anura, Serpentes

Murphy, **J**ames **B**[ernard] • 2 November 1939 • USA • Serpentes
Murphy, **J**ohn **C**[harles] • 20 February 1947 • USA • Serpentes
Murphy, **R**obert **C**ushman • 29 April 1887–1973 March 19 • USA • (Sauria)
Murphy, **R**obert **W**[ard] • 22 March 1948 • USA / Canada • Anura, Sauria, Serpentes
Murray, **A**ndrew • 19 February 1812–1878 January 10 • England • Crocodilia
Murray, **J**ames **A.** • 1865–1914 • British India • Anura, Sauria, Serpentes, Testudines
Murray, **K**eith **F.** • 20 August 1916–2002 June 13 • USA • Caudata
Murray, **L**eo **T**ildon • 4 May 1902–1958 March 2 • USA • (Testudines)
Murray, **P**eter **F.** • ? • Australia • *crocodilia, *testudines
Murthy, **T. S. N.** (= Thakkallapati Suryanarayana) • ? • India • (Anura), Sauria
Muskhelishvili (= Mushkelischwili *or* Musxelišvili), **T**eymuras **A**lexandrovich • ? • USSR • Sauria
Mussett, **F**rances • ? • England • *anura
Musters, **C**[ornelis] **J**[ohannes] **M**[aria] • ? • Netherlands • Sauria
Myers, **C**harles **W**[illiam] • 4 March 1936 • USA • Anura, Caudata, Sauria, Serpentes
Myers, **G**eorge **S**[prague] • 2 February 1905–1985 November 4 • USA • Anura, Caudata, Sauria

N

Nabhitabhata, **J**arujin • ? • Thailand • Anura, Sauria
Náder, **R**umio • ? • Argentina • Serpentes
Nagai, **K**amehiko • ? • Japan • Serpentes
Nakamura, **K**enji • 23 November 1903–1984 July 30 • Japan • Sauria
Nambu, **H**isao • ? • Japan • Caudata
Namiye, **M**otoyoshi • ("15 February") 13 March 1854–1918 May 24 • Japan • (Anura), Sauria
Nankinov, **D.** (? Dimitar Nikolov) • ? • (? Bulgaria) • *testudines
Napoli, **M**arcelo **F**elgueiras • ? • Brazil • Anura
Narayan Rao, **C. R.** • 15 August 1882–1960 January 2 • India • Anura, (Gymnophiona), Serpentes
Nardo, **G**iovanni **D**omenico • 4 March 1802–1877 April 7 • Italy • Testudines
Narmandakh, **P.** • ? • USSR • *testudines
Narvaes, **P**atricia • 13 May 1969 • Brazil • Anura
Nascetti, **G**iuseppe • 1 May 1950 • Italy • Anura, Caudata
Nascimento, **F**rancisco **P**aiva do • ? • Brazil • Sauria, Serpentes
Nascimento, **L**uciana **B**[arreto] • ? • Brazil • Anura
Natsuno, **T**etsuya • ? • Japan • Serpentes
Natterer, **J**ohann • 9 November 1787–1843 June 17 • Austria • (Anura), Crocodilia, (Sauria, Serpentes)
Nauman, **R**ichard **S**[teven] • 26 April 1966 • USA • Caudata
Navarro-Barón, **J**osé [**A**lberto] • 5 May 1946 • Chile • Anura
Navarro [-Pacheco], **N**ils • ? • Cuba • Sauria
Navás, **L**onginos • 1858–1938 • Spain • *caudata
Naylor, **B**ruce **G**[ordon] • 19 August 1950 • Canada • *caudata
Nečas, **P**etr • ? • Czech Republic • Sauria
Necker, **W**alter **L**[udwig] • 27 December 1913–1979 December 17 • Germany / USA • (Anura, Sauria)
Neill, **W**ilfred **T**[rammell, Jr.] • 12 January 1922–2001 February 19 • USA • Anura, Caudata, Serpentes, Testudines
Nelson, **C**raig **E**[ugene] • 21 May 1940 • USA • Anura
Nelson, **E**dward **W**illiam • 8 May 1855–1934 May 19 • USA • (Anura, Sauria, Serpentes, Testudines)
Němec, **J.** • ? • Czechoslovakia • *anura
Nesov (= Nessov), **L**[ev] **A**[lexandrovich] • 1947–1995 • USSR / Russia • *anura, *crocodilia, *sauria, *testudines
Nesterov, **P. V.** • ?–? • Russia • Caudata
Netting, **M**[orris] **G**raham • 3 October 1904–1996 August 26 • USA • Anura, Caudata, Serpentes
Neumann, **O**scar [**R**udolph] • 3 September 1867–1946 May 17 • Germany / USA • Sauria
Neuwied: *see* **Wied** [-Neuwied], [Alexander Philipp] **M**aximilian (prinz von)
Nevill, **H**ugh • 1848–1897 • Ceylon • Sauria
Nevo, **E**viatar • 2 February 1929 • Israel • *anura
Newman, **D**onald **G.** • ? • New Zealand • (Sauria)
Newman, **W**alter **B.** • ? • USA • Caudata
Ngat: *see* **Le** Nguyen **N**gat
Ngo, **A**ndre [**D**a] • 22 November 1971 • USA / Canada • Sauria
Nguyen Quang **T**ruong • ? • Vietnam • Anura, Caudata, Sauria, Serpentes
Nguyen Van **S**ang • ? • Vietnam • Sauria
Nicéforo María, (hermano, *formerly* Antoine Rouaire Siauzade) • 29 February 1888–1980 February 24 • France / Colombia • (Anura, Caudata, Gymnophiona, Sauria), Serpentes
Nicholls, **G**eorge **E**dward • 1878–? • England / Australia • (Sauria)
Nicholls, **J**esse **C.**, Jr. • 11 August 1907–2000 February 11 • USA • Caudata
Nicholls, **L**ucius • 22 January 1885–1969 September 3 • England / Ceylon / Singapore • Sauria, Serpentes
Nichols, **J**ohn **T**[readwell] • 11 June 1883–1958 November 10 • USA • (Sauria, Serpentes)
Nicholson, **K**irsten **E.** • ? • USA • Sauria
Nieden (*not* Neiden), **F**ritz (= Friedrich Erich Gottlieb) • 3 March 1883–1952 August 16 • Germany • Anura, Gymnophiona, Sauria
Nielsen, **E**igil [Hans Aage] • 1910 • Denmark • *testudines
Nieto-Montes de Oca, **A**drián • ? • México • Sauria, Serpentes
Nikoloff, **I**van • ? • Bulgaria • *crocodilia
Nikolsky, **A**[leksandr] **M**[ikhailovich] • ("18 February") 3 March 1858–1942 December 8 • Russia / USSR • Anura, Caudata, Sauria, Serpentes, Testudines
Nilson, **G**öran (= Goeran) • 15 February 1948 • Sweden • Anura, Sauria, Serpentes
Nilsson, **S**ven • 8 March 1787–1883 November 30 • Sweden • Anura, (Testudines)
Ninni, **A**lessandro **P**ericle (conte) • 4 April 1837–1892 December 7 • Italy • Anura, Serpentes
Nitsche, **H**einrich • 14 February 1845–1902 November 8 • Germany • (Serpentes)
Nixon, **A. M. A.** • ? • India • Sauria
Nixon, **C**harles **W**illiam • 28 September 1925 • USA • Serpentes

Noble, **G**[ladwyn] **K**ingsley (*husband of* Ruth Crosby Noble) • 20 September 1894–1940 December 9 • USA • *Anura, *Caudata, (Gymnophiona), Sauria

Noble, **R**uth **C**rosby (*wife of* G. Kingsley Noble) • 15 March 1897–1988 March 15 • USA • (Anura)

Nodder, **F**rederick **P**olydore • (*about* 1773)–1800 • England • Anura, Caudata, Serpentes

Nogales, **M**anuel • ? • Spain • Sauria

Nogales-Sornosa, **F**ernando • ? • Ecuador • Anura

Noll, **F**riedrich **C**arl (= Karl) • 22 September 1832–1893 January 14 • Germany • (Sauria)

Noonan, **B**rice **P**[atrick] • ? • USA • Anura

Nootpand: *see* **Nutphand**, **W**irot

Nopsca, **F**erencz (= Franz, báró) • 3 May 1877–1933 April 25 • Romania / Austria • *crocodilia, *sauria, *testudines

Nordmann, **A**lexander [Davidovich] von • ("12 May") 24 May 1803–1866 June 25 • Finland / Russia • Sauria, Serpentes

Norell, **M**ark **A**[llen] • 26 July 1957 • USA • *crocodilia

Norris, **K**enneth **S**[tafford] • 11 August 1924–1998 August 16 • USA • Sauria, Serpentes

Norsham [Suhaina] **Y**[aakob] • 10 April 1972 • Malaysia • Anura, Sauria, Serpentes

Northland [-Leppe], **I**rma [Margot] • 15 January 1946 • Chile • Anura

Noulet, **J**ean-**B**aptiste • 1 May 1802–1890 May 24 • France • *crocodilia, *testudines

Novo-Rodríguez, **J**ulio • ? • Cuba • Anura

Núñez-Cepeda, **H**erman [Antonio] • 6 September 1953 • Chile • Sauria

Núñez [-Navarro], **J**osé **J**[oel] • 8 September 1967 • Chile • Anura

Nussbaum, **R**onald **A**[rchie] • 9 February 1942 • USA • Anura, Caudata, Gymnophiona, Sauria, Serpentes

Nutphand (= Nootpand *or* Nutaphand), **W**irot • 5 November 1932–2005 June 5 • Thailand • Anura, Sauria, Serpentes, Testudines

Nydam, **R**andall **L**[awrence] • 5 February 1969 • USA • *sauria

O

Oakes, **N**icholas • ? • Australia • (Sauria)

Oates, **F**rank (= Francis) • 6 April 1840–1875 February 5 • England / southern Africa • (Anura, [Serpentes])

Obst, **F**ritz **J**ürgen • 3 April 1939 • Germany • Caudata, Sauria, (Testudines)

Ochoa [-Cámara], **J**osé **A**ntonio • ? • Perú • (Anura)

Ochoa-M., **O**scar • ? • Perú • Anura, (Sauria)

Ochoterena [y Mendieta], **I**saac • 28 November 1885–1950 April 11 • México • (Sauria)

Odierna, **G**aetano • ? • Italy • Sauria

Oelrich, **T**homas **M**[ann] • 27 May 1924 • USA • *testudines

Oeser, **R**ichard • 6 July 1891–1974 July 4 • Germany • (Anura)

Oftedal, **O**lav **T**[onnes] • ? • USA • Sauria

Ogilby, **J**[ames] **D**ouglas • 16 February 1853–1925 August 11 • Ireland / England / USA / Australia • Anura, Sauria, Serpentes, Testudines

Ogren, **L**arry **H.** • ? • USA • Sauria, Serpentes

Ohler, **A**nnemarie (*wife of* Alain Dubois) • 11 February 1960 • Austria / France • Anura, Sauria

Okada, **Y**aichiro • 24 June 1892–1976 April 28 • Japan • *Anura, Sauria, Serpentes

Okafuji, **G**oro • ? • Japan • *testudines

Okazaki, **H**iroko • ? • Japan • *testudines

Oken (= Ockenfuss), **L**orenz (*pen name* = Oken) • 1 August 1779–1851 August 11 • Germany / Switzerland • Anura, Sauria, Serpentes

Oldham, **J**onathan **C**[lark] • 6 November 1939–1991 November 21 • USA • Serpentes

Oldham, **R**obert **S.** • ? • (? England) / Nigeria / England • Anura

Olgun, **K**urtuluş • ? • Turkey • Serpentes

Oliver, **J**ames **A**[rthur] • 1 January 1914–1981 December 2 • USA • Sauria, Serpentes

Oliver, **P**aul **M.** • ? • Australia • Anura

Oliver, **W**alter **R**eginald **B**rook • 7 September 1883–1957 May 16 • Tasmania / New Zealand • (Sauria)

Olivier, **G**uillaume-**A**ntoine • 19 January 1756–1814 October 1 • France • Sauria, Serpentes, Testudines

Olsen, **S**tanley **J**[ohn] • 24 June 1919–2003 December 23 • USA • (*serpentes)

Olson, **D**eanna **H**[elen] ("Dede") • ? • USA • Caudata

Olson, **E**verett **C**[laire] • 6 November 1910–1993 November 27 • USA • *sauria

Olson, **R**[upert] **E**arl • 30 June 1934 • USA / Haiti • Sauria

Onderstall, **D**avid • ? • South Africa • Sauria

O'Neill, **E**ric **M**[ichael] • 9 December 1971 • USA • Anura

O'Neill, **F. M**ichael • 25 August 1940 • USA • *crocodilia

O'Neill, **R**obert **C.** • ? • USA • *sauria

Ong, **P**erry **S.** • 2 October 1960 • Philippines / Australia / Philippines • Anura

Oo: *see* **San Lwin Oo**

Oort, **P.** van • ?–? • Java • [Anura]

Oppel, [Nikolaus] **M**ichael • 7 December 1782–1820 February 16 • Germany / France / Germany • Crocodilia, Sauria, Serpentes

Orbigny, [Alcide] **C**harles [Victor Dessalines] de (*brother of* A. d'Orbigny) • 1806–1876 • France • [Serpentes]

Orbigny, **A**lcide [Dessalines] de (*brother of* C. d'Orbigny) • (? 6 *or* 7 *or* 9) September 1802–1857 June 30 • France • (Anura), Sauria, (Serpentes, Testudines)

Orcés (-Villagomez), **G**ustavo • 31 July 1903–1999 March 22 • Ecuador • (Anura), Sauria, (Serpentes)

Orcutt, **C**harles **R**ussell • 27 April 1864–1929 August 24 • USA • (Anura, Sauria, Serpentes)

Orejas-Miranda, **B**raulio **R**[ubí] • 19 February 1933–1985 May 8 • Uruguay • (Anura), Serpentes

Orellana, **A**ndrés • ? • Venezuela • Anura

Orlov, **N**ikolai **L**[utseranovich] (*husband of* Natalia B. Ananyeva) • 12 June 1952 • USSR / Russia • Anura, Sauria, Serpentes

Ormsby, **A**nthony **I.** ("Tony") • ? • Australia • (Sauria)

Ortale, **S**tefano • ? • Italy • Testudines

Ortega [-Coloma], **F**rancisco • 8 April 1965 • Spain • *crocodilia

Ortenburger, **A**[rthur] **I**[rving] • 8 June 1898–1961 November 18 • USA • Serpentes

Ortiz-Z[apata], **J**uan **C**arlos • 16 February 1945 • Chile / France / Chile • Anura, Sauria

Orton, **G**race **L**[ouise] • 13 March 1916–2003 September 10 • USA • (Serpentes)
Orton, **J**ames • 21 April 1830–1877 September 25 • USA • (Serpentes)
Osbeck, **P**ehr (= Peter) • 9 May 1723–1805 December 23 • Sweden • Anura
Osborn, **H**enry **F**airfield • 8 August 1857–1935 November 6 • USA • (Crocodilia), *crocodilia, (Serpentes)
Osborne, **W**illiam S. ("Will") • ? • Australia • Sauria
Osenegg, **K**irsten • ? • Germany • Sauria
O'Shaughnessy, **A**rthur **W**illiam **E**dgar • 14 March 1844–1881 January 30 • England • Sauria
Oshima, **M**asamitsu • 21 June 1884–1965 June 26 • Japan • Sauria, Serpentes
Osman Hill, **W**. **C**. • ? • England • Anura, (Serpentes)
Ostrovskikh, **S**ergei **V**. • ? • (? Russia) • Serpentes
Ota, **H**idetoshi • 25 September 1959 • Japan • Sauria, Serpentes
Otsu, **T**akashi • ? • Japan • Anura
Otsuka, **H**iroyuki • ? • Japan • *testudines
Otth, [Carl] **A**dolf (= Adolphe) • 2 April 1803–1839 May 16 • Switzerland • Anura
Ottley, **J**ohn **R**[andall] • 10 May 1949 • USA • Sauria, Serpentes, Testudines
Ouboter, **P**aul **E**. • ? • Netherlands • Sauria
Oudart (= Oudard), **P**aul [-Louis] • 1796–? • France • Sauria
Oudemans, **J**ohannes **T**heodorus • 22 November 1862–1934 February 20 • Netherlands • Sauria, Testudines
Ouwens, **P**eter **A**. (major) • ?–? • Netherlands / East Indies • (Anura), Sauria, Testudines
Owen, **J**. **S**. • ? • Sudan • (Serpentes)
Owen, (sir) **R**ichard • 20 July 1804–1892 December 18 • England • *crocodilia, Rhynchocephalia, (Sauria), *sauria, (Serpentes), *serpentes, *Testudines
Owens, **D**avid **W**. • ? • USA • Testudines
Özeti, **N**eclâ • 22 December 1932 • Turkey • Anura

P

Pacheco, **J**oviano **A**. **A**. • ? • Brazil • *testudines
Pack, **L**loyd **E**., Jr. • ? • USA • Sauria
Packard, **A**[lpheus] **S**[pring, Jr.] • 19 February 1839–1905 February 14 • USA • Caudata
Packard, **E**arl **L**[eroy] • 10 April 1885–1983 January 20 • USA • *testudines
Padial [-Fregenal], **J**osé **M**[anuel] • 20 February 1977 • Spain / Bolivia / Spain • Anura, Sauria
Padua (*not* Pauda): *see* **Almeida**, **A**ntonio de **P**adua
Pakenham, **R**. **H**. **W**. • ? • (? England / Zanzibar) • (Anura, Sauria)
Palacios-**A**rribas, **F**ernando • ? • Spain • Sauria
Palisot, **A**mbroise-**M**arie-**F**rançois-**J**oseph (baron de Beauvois) • 27 July 1752–1820 January 21 • France • Serpentes
Pallas, **P**eter **S**imon • ("22 September") 3 October 1741–1811 September 20 ("September 8") • Germany / Russia / Germany • Anura, Sauria, Serpentes, Testudines
Palmer, **W**illiam • 1 August 1856–1921 April 8 • England / USA • *testudines
Palmer, **W**illiam **M**[cKee] • 6 December 1935 • USA • Sauria
Pan Junhua • 11 November 1919 • China • (Testudines)
Panha, **S**omsak • ? • Thailand • Anura
Paolillo-O., **A**lfredo • ? • Venezuela • Anura, Testudines
Papenfuss, **T**heodore **J**[ohnstone] • 22 July 1941 • USA • Anura, Caudata, Sauria, Serpentes
Paredes-Recalde, **A**rturo • ? • Ecuador • Anura
Parenzan, **P**ietro • 10 January 1902–1992 November 26 • Croatia / Italy • Testudines
Parham, **J**ames **F**ord ("Jim") • 11 June 1974 • USA • *Testudines
Parker, **F**red (= Frederick Stanley) • 18 December 1941 • Australia • Anura, Sauria, (Serpentes, Testudines)
Parker, **H**[ampton] **W**[ildman] • 5 July 1897–1968 September 2 • England • *Anura, Gymnophiona, Sauria, Serpentes
Parmelee, **J**effrey **R**[obert] • ? • USA • Anura
Parmley, **D**ennis [Craig] • ? • USA • *anura, *serpentes
Parra-Olea, **G**abriela • 4 October 1967 • México • Caudata
Parry, **C**. **R**. • ? • South Africa • Anura
Parsons, **J**ames • March 1705–1770 April 4 • England • (Sauria)
Pasquier: *see* **Du Pasquier**, **L**ouis
Passmore, **N**eville **I**an • 11 September 1947 • South Africa • Anura
Passos, **P**aulo • ? • Brazil • Serpentes
Pasteur, **G**eorges • 26 February 1930 • France • Anura, Sauria
Patrizi, **S**averio **F**. (marchese di Montoro) • 11 January 1902–1957 February 21 • Italy • (Sauria)
Pattabiraman, **R**. • ? • India • Anura
Patterson, **B**ryan • 10 March 1909–1979 December 1 • England / USA • *crocodilia
Patterson, **G**eoffrey **B**. • ? • New Zealand • Sauria
Pauler, **I**ngo • ? • Germany • Testudines
Paulian, **R**enaud • ? • Madagascar • Sauria
Paulson, **D**ennis **R**oy • 29 November 1937 • USA • (Anura), Sauria, Serpentes
Pauwels, **O**livier **S**[ylvain] **G**[érard] • 1 September 1971 • Belgium • Anura, Sauria, Serpentes
Pavan, **D**ante • 3 February 1971 • Brazil • Anura, Sauria
Pavlov, **A**leksey **V**[ladilenovich] • 25 July 1968 • USSR / Russia • Serpentes
Pavlov (= Pavloff), **P**. **A**. • ? • USSR / China • Caudata, Testudines
Pavlovsky, **E**[hvgeny] **N**[ikanorovich] • ("22 February") 5 March 1884–1965 May 27 • Russia / USSR • (Sauria)
Pawar, **S**amraat **S**[hashikant] • 1 September 1975 • India / USA • Serpentes
Peabody, **F**rank **E**[lmer] • 28 August 1914–1958 June 27 • USA • *caudata
Peabody, **R**obert **B**[raynard] • ? • USA • Caudata
Peale, **T**itian **R**[amsay] • (? 2 *or* 17) November 1799–1885 March 13 • USA • Sauria
Pearse, **A**rthur **S**perry • 15 March 1877–1956 December 11 • USA • (Anura)
Peccinini-**S**eale, **D**enise • ? • Brazil • Sauria
Peers, **B**. • ? • South Africa • (Sauria, Testudines)
Péfaur, **J**aime **E**. • 1944 • Venezuela • Anura, Sauria
Pehani, **H**ubert • 1900 • Yugoslavia • Caudata
Peixoto, **O**swaldo **L**uiz • ? • Brazil • Anura
Pellegrin, **J**acques • 12 June 1873–1944 August 12 • France • Sauria, Serpentes

Pellegrino, **K**atia **C**ristina **M**achado • ? • Brazil • Sauria
Pemberton, **A**aron **J**. • ? • USA • Anura
Peña-**P**rieto, **J**ohann **A**. • ? • Colombia • Anura
Pengilley, **R**oss **K**. • ? • Australia • (Anura, Sauria)
Penna-**V**arela, **M**ario • 1 December 1951 • Chile • Anura
Pennisi, **M**aria **G**razia • ? • Italy • Testudines
Peracca, **M**ario **G**iacinto (conte) • 21 November 1861–1923 May 23 • Italy • Anura, Caudata, Sauria, Serpentes
Perälä, **J**armo [Seppo Johannes] • 7 June 1961 • Finland / England / Finland • Testudines
Péres, **A**yrton **K**., Jr. • ? • Brazil • Sauria
Pereyra [-Ruiz], **E**nrique **A**drian • ? • Chile / Argentina • Sauria
Pérez, **C**ristian **H**[ernán] **F**[ulvio] • ? • Argentina • Sauria
Pérez: *see* **Ferrari-P**érez, **F**ernando
Pérez-Higareda, **G**onzalo • 3 November 1932 • México • Sauria, Serpentes
Pérez-Mellado, **V**alentín • ? • Spain • Sauria
Pérez-Ramos, **E**dmundo • ? • México • Caudata, Sauria
Pérez-Santos, **C**arlos • ? • Spain • Serpentes
Peris, **S**. **V**. • ? • Spain • Anura
Perkins, **C**larence **B**asil • 11 November 1888–1955 April 28 • USA • (Serpentes)
Pernetta, **J**ohn **C**. • ? • Fiji / Papua New Guinea • Sauria
Péron, **F**rançois • 22 August 1775–1810 December 14 • France • Anura, Sauria, (Serpentes)
Perret, **J**ean-**L**uc • ? • Switzerland • Anura, Sauria, Serpentes
Perry, **G**eorge • ?–(1810-1811)–? • England • Testudines
Persson, **P**er **O**ve • ? • Sweden • *crocodilia
Pesantes-**S**egura, **O**mar **C**. • ?–2004 June • Brazil / Perú • (Anura)
Peters, **G**ünther • 10 July 1932 • Germany • (Anura), Sauria, Serpentes
Peters, **J**ames **A**[rthur] • 13 July 1922–1972 December 18 • USA • Anura, Sauria, *Serpentes, (Testudines)
Peters, **K**arl **F**erdinand • 13 August 1825–1881 November 7 • Germany • *testudines
Peters, **W**ilhelm [Karl Hartwig] (= Guilielmus Carl Hartvig) • 22 April 1815–1883 April 20 • Germany • Anura, Caudata, Gymnophiona, Sauria, Serpentes, Testudines
Peterson, **H**arold **W**. • ? • USA • Caudata
Peterson, **M**agnus • ? • Australia • (Sauria)
Pethiyagoda, **R**ohan • ? • Sri Lanka • Anura, Sauria
Petit, **G**abriel • 21 February 1870–? • France • (Sauria)
Petranka, **J**ames **W**[alker] • 31 December 1951 • USA • Caudata
Pfeffer, **G**eorg **J**ohann • 1854–1931 • Germany • Anura, Sauria, Serpentes
Pfrender, **M**ichael **E**[ugene] • 13 November 1960 • USA • Caudata
Phelan, **R**obert **L**ee • ? • USA • Sauria
Philipp, **K**ai **M**. • ? • Germany • Sauria
Philippen, **H**ans-**D**ieter • 11 June 1957 • Germany • Testudines
Philippi, **R**udolph **A**mandus • 14 September 1808–1904 July 23 • Germany / Chile • Anura, Sauria, Serpentes, Testudines
Phillips, **S**tephen • ? • Australia • (Sauria)
Phipson, **H**. **M**. • ?–? • British India • (Serpentes)
Pianka, **E**ric **R**[odger] (*husband of* Helen Dunlap Pianka *from* 1965 *to* 1980) • 23 January 1939 • USA / Australia / USA • (Sauria)
Pianka, **H**elen [Louise] **D**unlap: *see* **Dunlap**, **H**elen **L**ouise
Piatt, **J**ean [Barnett] • 22 September 1909–1994 July • USA • Sauria
Picado-Twight, **C**lodomiro • 17 January 1887–1944 May 16 • Costa Rica • (Anura, Caudata, Serpentes)
Picard, **J**acques-**J**. • ? • Belgium • Anura
Pickersgill, **M**artin • 25 April 1956 • England / South Africa • Anura
Pickford, **M**artin • ? • France • *crocodilia
Pictet [de la Rive], **F**rançois **J**ules • 27 September 1809–1872 (? March *or* May) 15 • Switzerland • (*sauria, Serpentes), *testudines
Pidoplichko, **I**[van] **G**[rigorevich] • 1905–1975 • Russia / USSR • *testudines
Pienaar, **U**ys de **V**illiers • ? • South Africa • Sauria
Pieper, **H**arald • 21 February 1942 • Germany • Caudata
Pifano [Capdeville], **F**élix • 1912 • Venezuela • (Serpentes)
Pilgrim, **G**uy **E**llcock • 1875–1943 September 15 • Barbados / England / British India • *crocodilia, (*testudines)
Pillai, **R**. **S**. • ? • India • Anura
Pimenta, **B**runo **V**[ergueiro] **S**[ilva] • 10 April 1978 • Brazil • Anura
Pincheira-**D**onoso, **D**aniel • 2 March 1979 • Chile • Sauria
Ping Chih (= Bing Zhi, *originally* Chai Ping-chih) • 9 April 1886–1965 February 21 • China • (Anura), *testudines
Pintak, **T**homas • ? • Germany • Anura
Pinto, **R**oberta **R**. • Brazil • ? • Serpentes
Piombo, **R**oberta • 16 January 1974 • Italy • Testudines
Pisanetz, **E**hvgeny **M**[aksimovich] • 22 August 1949 • USSR / Ukraine • Anura
Pitman, **C**harles **R**[obert] **S**[enhouse] (captain) • 19 March 1890–1975 September 18 • India / Kenya / Uganda / England • (Anura), Serpentes
Piton, **L**ouis **E**. • ? • France • *crocodilia
Piveteau, [Jacques Honoré] **J**ean [Marie] • 23 September 1899–1991 March 7 • France • (*sauria)
Pizarro, **J**oão **J**oaquim • ?–(1876)–? • Brazil • Anura
Plancy: *see* **Collin de Plancy**, **V**ictor [Émile Marie Joseph]
Plane, **M**ichael • 1933 • Australia • *serpentes
Plath, **M**irco • ? • Germany • Anura
Platz, **J**ames **E**[rnest] • 4 September 1943 • USA • Anura
Pledge, **N**eville **S**[tewart] • 7 November 1939 • Australia • *sauria
Poche, **F**ranz • ?–1945 • Austria • Serpentes
Poelman, **E**rik **H**[erman] • 21 April 1980 • Netherlands • Anura
Poey [y Aloy], **F**elipe • 26 May 1799–1891 January 28 • Cuba • [Testudines]
Poggesi, **M**arta • 26 November 1945 • Italy • Sauria
Poglayen [-Neuwall], **I**vo • 16 January 1921–2003 March 19 • USA • Sauria
Poiret, **J**ean-**L**ouis-**M**arie • 1755–1834 April 7 • France • (Caudata)
Polisar, **J**ohn [R.] • ? • USA • Caudata
Pombal, **J**osé **P**[erez], Jr. • 31 March 1966 • Brazil • Anura
Pomel, [Nicolas] **A**uguste • 21 September 1821–1898 August 2 • France / Algeria • *caudata, *crocodilia, *sauria, *serpentes, *testudines
Ponce-Campos, **P**aulino • 10 April 1964 • México • Serpentes
Ponce [-Mejía], **E**lías • ? • Perú • Anura

Pope, Clifford **H**[illhouse] (*husband of* Sarah Haydock Pope *from* 1929 *to* 1974) • 11 April 1899–1974 June 3 • USA / China / USA • Anura, Caudata, Sauria, Serpentes, Testudines
Pope, Sarah **H**aydock [Davis] (*wife of* Clifford H. Pope *from* 1929 *to* 1974) • ? • USA • Caudata, Serpentes
Porras, **L**ouis [William] • 21 January 1948 • Costa Rica / USA • Anura, (Sauria)
Porter, **A**nn **P**. • ? • USA • Sauria, Serpentes
Porter, **K**enneth **R**[aymond] • 20 October 1931 • USA • Anura
Portis, **A**lessandro • 17 January 1853–1931 December 21 • Italy • *anura, *serpentes, *testudines
Porto, **M**arcovan • ? • Brazil • Serpentes
Posada-**A**rango, **A**ndrés • 11 February 1839–1923 March 13 • Colombia • Caudata, Serpentes
Positano-Spada, **D**omenico • ?–(1892)–? • Italy • Sauria
Potter, **F**loyd **E**., Jr. • 6 December 1925–1987 July 9 • USA • Caudata, Serpentes
Power, **J**[ohn] **H**[yacinth] • 2 November 1884–1964 December 21 • South Africa • Anura, Sauria, Testudines
Powers, **A**rnold **L**. • ? • USA • Serpentes
Poynton, **J**[ohn] **C**[harles] • ? • South Africa • Anura
Prado, **A**lcides • ? • Brazil • Serpentes
Prado, **G**ustavo **M**[achado] • ? • Brazil • Anura
Pramuk, **J**ennifer **B**. • ? • USA • Anura
Prangner, **P**ater **E**ngelbert • 12 September 1812–? • Austria • *crocodilia
Prasad, **K**. **N**. • ? • India • *testudines
Prashanth, **P**. • ? • India • Gymnophiona
Prater, **S**tanley **H**enry • 12 March 1890–1960 October 12 • British India / England • Serpentes
Prato: *see* **Del Prato**, **A**lberto
Pražák, **J**. **P**rokop • 22 June 1870–1904 July 15 • Czechoslovakia • Caudata
Pregill, **G**regory **K**[ent] • 27 November 1946 • USA • *sauria
Preston, **R**obert **E**. • ? • USA • *testudines
Preud'homme de Borre: *see* **Borre**, **A**lfred **P**reud'homme de
Price, **A**ndrew **H**[oyt] • 12 May 1951 • USA • Caudata
Price, **L**lewellyn **I**vor • 9 October 1905–1980 March 14 • Brazil / USA / Brazil • *crocodilia, (*gymnophiona), *sauria, *testudines
Price, **R**obert **M**[ark] • 14 June 1951 • USA • Serpentes
Prigioni, **C**arlos **M**a. • ? • Uruguay • Anura
Pringle, **J**ohn **A**dams • 31 October 1910 • South Africa • Serpentes
Pritchard, **P**eter **C**[harles] **H**[oward] • 26 June 1943 • England / USA • Testudines
Prjevalsky: *see* **Przhevalsky**, **N**ikolai [Mikhailovich]
Procter, **J**oan **B**[eauchamp] • 5 August 1897–1931 September 20 • England • Anura, Sauria, Serpentes, (Testudines)
Provancher, **L**éon [Abel] (l'abbé) • 10 March 1820–1892 March 23 • Canada • Caudata
Proy, **C**hristian • ? • Austria • Anura
Prudente, **A**na **L**úcia da **C**osta • ? • Brazil • Serpentes
Przhevalsky (= Prżewalski *or* Prjevalsky), **N**ikolai [Mikhailovich] (general) • ("31 March") 12 April 1839–1888 November 1 ("October 20") • Russia • [(Anura, Sauria), Serpentes]
Púgener, **L**[ourdes] **A**nalía • ? • USA • *anura
Pugliese, **A**driana • ? • Brazil • Anura
Puorto, **G**iuseppe • ? • Brazil • Serpentes
Purschke, **C**arl **A**rthur • ?–(1885)–? • Austria • *testudines
Putnam, **F**[rederick] **W**[ard] • 16 April 1839–1915 August 14 • USA • (Serpentes, *testudines)
Pyburn, **W**illiam **F**rank ("Billy") • 22 March 1927 • USA • Anura, (Sauria, Serpentes)
Pyles, **R**ebecca **A**[nn] • 22 July 1954 • USA • Anura

Q

Qu Wenyuan • 4 September 1936 • China • Caudata
Queiroz: *see* **de Queiroz**, **K**evin
Quelch, **J**[ohn] **J**[oseph] • 1854–? • (? England) / British Guiana • (Anura)
Quesnel, **V**ictor **C**. • ? • Trinidad • [Sauria]
Quevedo, **A**lonso • ? • Colombia • Anura
Quijano: *see* **Mendoza-Q**uijano, **F**ernando
Quinn, **H**ugh **R**. • 1 October 1946 • USA • Serpentes
Quintana-**C**ardona, **J**osep • 22 February 1969 • Spain (Menorca Island) • *serpentes
Quinteros, **S**ebastián • ? • Argentina • Sauria
Quiroga, **H**oracio • 31 December 1878–1937 February 19 • Uruguay / Argentina • (Serpentes)
Quoy, **J**ean-**R**ené-**C**onstant • 10 November 1790–1869 July 4 • France • Sauria, Testudines
Quyet: *see* **Le K**hac **Q**uyet

R

Rabb, **G**eorge **B**[ernard] • 2 January 1930 • USA • Anura, Caudata, Sauria, Serpentes, Testudines
Rabor, **D**ioscoro **S**. • 18 May 1911 • Philippines • Sauria
Rada [-García], **M**arco [Antonio] • 23 April 1981 • Colombia • Anura
Radcliffe, **C**harles **W**ilson • 20 January 1941 • USA • Serpentes
Raddi, **G**iuseppe • 9 July 1770–1829 September 6 • Italy / Rhodes • Anura, Sauria, Serpentes
Radovanović, **M**ilutin • 15 March 1900–1968 April 20 • Yugoslavia / Namibia • Caudata, Sauria
Radtkey, **R**ay **R**[ichard] • ? • USA • Sauria
Raffles, (sir) [Thomas] **S**tamford • 5 July 1781–1826 July 5 • England / Malaya / Java / Sumatra / Singapore / England • Serpentes
Rafinesque [Schmaltz], **C**onstantine **S**amuel • 22 October 1783–1840 September 18 • Turkey / France / Italy / USA • Anura, Caudata, Sauria, Serpentes, Testudines
Rage, **J**ean-**C**laude • 1 March 1943 • France • (*anura), *sauria, *serpentes
Ramanamanjato, **J**ean **B**. • ? • Madagascar • Sauria
Ramanantsoa, **G**. **A**. • ? • Madagascar • Sauria
Ramanna, **B**. **S**. • ?–? • India • Anura
Ramírez-Bautista, **A**urelio • ? • México • Serpentes
Ramírez-Leyton, **G**ilmar **M**. • ? • Chile • Sauria
Ramírez-Pinilla, **M**artha **P**atricia • ? • Colombia / Argentina / México / Colombia • (Sauria)
Ramón de la Sagra: *see* **Sagra**, **R**amón de la
Ramos Costa, **A**na **M**aria **M**. • ? • Brazil • Sauria
Ramsay, **E**dward **P**ierson • 3 December 1842–1916 December 16 • Australia • Anura, Sauria, Serpentes, Testudines

Ramsden, Charles **T**[heodore] • 11 February 1876–1951 August 24 • Cuba • Sauria
Ramsey, Louis **W**[eete] • 9 September 1904–1991 March • USA • Serpentes, Testudines
Ranawana, **K**ithsiri **D**. • ? • Sri Lanka • (Serpentes)
Rancés-Caicedo, **J**osé • ? • Colombia • Sauria
Rand, **A**[ustin] **S**tanley • 29 September 1932–2005 November 14 • USA / Brazil / Panama (Canal Zone) / USA • (Anura), Caudata, Sauria, Serpentes
Randrianirina, **J**asmin **E**. • ? • Madagascar • Anura, Sauria
Rankin, **P**eter **R**obert • ?–1979 January 2 • Australia / New Caledonia • Sauria, (Testudines)
Rao Dingqi • ? • China • Anura, Serpentes
Rao: *see* **Narayan Rao**, **C**. **R**.
Rapp, **W**ilhelm [Ludwig] von • 3 June 1794–1868 November 11 • Germany • Anura, (Serpentes)
Raselimanana, Achille **P**. • ? • Madagascar • Sauria
Rasmussen, **A**rne **R**edsted • ? • Denmark • Serpentes
Rasmussen, **J**ens **B**[ødtker] • 16 April 1947–2005 May 3 • Denmark • Serpentes
Rasmussen, **T**homas **E**[arl] • ? • USA • *rhynchocephalia
Rastegar-Pouyani, **E**skandar • ? • Iran • Sauria
Rastegar-Pouyani, **N**asrullah • ? • Iran • Sauria
Rathke, [Martin] **H**einrich • 25 August 1793–1860 September 15 ("September 3") • Poland / Estonia / Kaliningrad • Caudata, (Sauria)
Ratnikov, **V**iatcheslav **Y**[urievich] • 13 November 1957 • USSR / Russia • *anura, [Serpentes]
Rau, Charles **S**[hirtcliff] • ? • USA • Sauria
Rauhe, **M**ichael • ? • Germany • *crocodilia
Raun, Gerald **G**[eorge] • 14 July 1932 • USA • (Sauria)
Ravichandran, **M**[asagoundanur] **S**[engodan] • 23 September 1958 • India • Anura, Gymnophiona
Raw, Lynn **R**. **G**. • ? • South Africa • Anura, Sauria, Serpentes
Rawlinson, **P**[eter] **A**[lan] • 10 December 1942–1991 April 11 • Australia / Krakatoa (Anak Krakatau) • (Anura), Sauria
Raxworthy, Christopher **J**[ohn] • 5 February 1964 • England / USA • Anura, Caudata, Sauria, Serpentes
Ray, Clayton **E**[dward] • 6 February 1933 • USA • [*testudines]
Ray, **H**arish **C**handra • 1906 • India • Serpentes
Ray, **P**ranjalendu • ? • India • Anura
Razumovsky, **G**rigorii (= Grégoire de Razoumowsky, graf) • 10 November 1759–1837 June 3 • Russia / Switzerland • Anura, Caudata, Sauria, Serpentes
Rebouças Spieker, **R**egina • ? • Brazil • Sauria
Reddell, **J**ames **R**. • ? • USA • Caudata
Reddy: *see* **Manjunatha Reddy**, **A**. **H**.
Redi, **F**rancesco • 19 February 1626–1667 March 1 • Italy • (Serpentes)
Redkozubov, **O**leg **I**gorevich • 29 July 1952 • USSR • *testudines
Reeder, **T**od **W**[est] • ? • USA • Sauria
Reese, **R**obert **W**[illiam] • 3 August 1917–1990 January 19 • USA • Caudata
Regal, **P**hilip **J**[oe] • 2 December 1939 • USA • Caudata
Regenass, **U**rs • ? • USA • Serpentes
Rehák, **I**van • ? • Czechoslovakia • Serpentes
Rehman, **H**afizur • ? • Pakistan • Sauria
Reichenbach, **H**[einrich] **G**[ottlieb] **L**udwig • 8 January 1793–1879 March 17 • Germany • Caudata
Reichenow, **A**nton • 1 August 1847–1941 July 6 • Germany • Anura, Sauria, Serpentes
Reichle, **S**teffen • ? • Germany • Anura
Reid, **J**ack **R**obert • 13 July 1933 • USA • (Sauria)
Reig, **O**svaldo **A**. • 1929 • Argentina • *Anura
Reimann, **M**ichael **J**. • ? • Germany • (Testudines)
Reinach, **A**lbert (baron) von • 7 November 1832–1905 January 12 • Germany • *testudines
Reinhardt, **J**ohan [Christopher Hagemann] (*father of* Johannes Theodor Reinhardt) • 19 December 1776–1845 October 31 • Norway / Denmark • Serpentes
Reinhardt, **J**ohannes **T**heodor (*son of* Johan Reinhardt) • 3 December 1816–1882 October 23 • Denmark • Anura, Gymnophiona, Sauria, Serpentes
Reinwardt, **C**aspar **G**eorg **C**arl • (? 5 June *or* 3 July) 1773–1854 March 6 • Netherlands • (Anura), Serpentes
Reis, **S**érgio **F**[urtado] dos • 26 June 1952 • Brazil • Anura
Reischek, **A**ndreas • 15 September 1845–1902 April 3 • Austria / New Zealand / Austria • (Rhynchocephalia)
Rendahl, [Carl] **H**ialmar • 26 December 1891–1969 May 2 • Sweden • Gymnophiona, Sauria, Serpentes
Renjifo [-Rey], **J**uan **M**anuel • 6 September 1948 • Colombia • Anura, Serpentes
Renous, **S**abine • 2 July 1942 • France • Gymnophiona
Resetar, **A**lan [R.] • 7 June 1954 • USA • Serpentes
Resnikova: *see* **Reznikova**, **M**.
Restrepo, **A**na **E**ugenia • ? • Colombia • Caudata
Restrepo [-Toro], **J**orge [Humberto] • 22 December 1958 • Colombia • Anura, Serpentes
Retzius, **A**nders **J**ohan • 3 October 1742–1821 October 6 • Sweden • Caudata, (Testudines)
Reuss, **A**dolph • 1804–1878 May 7 • Germany / USA • Anura, Sauria, Serpentes
Reuss, **T**. • ? • Germany • Serpentes
Révoil, **G**eorges • 1852–1894 • France • [Sauria]
Rey, **L**aura **R**aquel • ? • Argentina • Serpentes
Reynolds, **R**obert **P**[aul] ("Bob") • 16 March 1950 • USA • Anura, Serpentes
Reznikova (= Resnikova), **M**. • ? • USSR • Sauria
Rhodin, **A**nders **G**. **J**. • ? • USA • [Anura], Testudines
Riabinin, **A**natoly Nikolaevich • ? • USSR • *testudines
Ribeiro, **L**uiz **F**. • ? • Brazil • Anura
Ribeiro: *see also* **Miranda R**ibeiro, Alípio do
Rice, **F**rank **L**eon • ?–1925 November 4 • USA • Anura
Richards, Christina **M**[acgregor] • 5 August 1929 • USA • Anura
Richards, Stephen **J**. • ? • Australia • Anura
Richardson, Charles **H**[oward], Jr. • 7 July 1887–1977 January 9 • USA • Sauria
Richardson, (sir) **J**ohn • 5 November 1787–1865 June 5 • Scotland / England • Sauria, (Serpentes)
Richmond, **N**eil **D**[wight] • 17 November 1912–1992 February 14 • USA • (Caudata), Serpentes
Richmond, **R**ollin **C**[harles] • 31 May 1944 • USA • Anura
Rickard, **E**dwina • ? • Australia • Sauria
Rickart, **E**ric **A**[llan] • 7 February 1950 • USA • *sauria
Ridley, **H**enry **N**icholas • 10 December 1855–1956 October 24 • England • [(Sauria)]

Riemer, **W**illiam **J**[ohn] • 25 January 1924–2001 February 4 • USA • Caudata, Serpentes
Rieppel, **O**livier [Cedric] • 21 September 1951 • Switzerland / USA • (Sauria), *sauria, *testudines
Riess, **J**ürgen • ? • Germany • *crocodilia
Rios [-Rios], **J**[ehan] **N**inon • ? • Bolivia • Anura
Risch, **J**ean-**P**aul • ? • France / Luxemburg • Anura, Caudata
Risso, **A**ntoine • 8 April 1777–1845 August 25 • France • Anura, Sauria, Serpentes, Testudines
Ristori, **G**iuseppe • 1856–1905 December 29 • Italy • *crocodilia
Rivas-Fuenmayor, **G**ilson • ? • Venezuela • Anura, Sauria, Serpentes
Rivero, **R**amón **A**. • ? • Venezuela • Anura
Rivero-Blanco, **C**arlos **V**. • ? • Venezuela • Anura, Sauria
Rivero [-Quintero], **J**uan **A**[rturo] • 5 March 1923 • Venezuela / Puerto Rico • Anura, Sauria
Rivers, **J**ames **J**ohn • 1824–1913 • England / USA • (Sauria), Testudines
Rivière, **É**mile • 1835–1922 • France • Anura
Roa-Trujillo, **S**onia **H**. • ? • Colombia • Anura
Robb, **J**oan • 11 December 1921 • New Zealand • Sauria, Serpentes
Roberts, **J**[ohn] **D**ale • ? • Australia • Anura
Robertson, **J**esse **S**. • ? • USA • *testudines
Robertson, **P**eter • ? • Australia • (Sauria)
Robertson, **W**illiam **B**[eckwith] • 22 August 1924–2000 January 28 • USA • Serpentes
Robinson, **D**ouglas **C**[lark] • ? • USA / Costa Rica • Caudata
Robinson, **H**erbert **C**[hristopher] • 4 November 1874–1929 May 30 • England / British Malaya / England • (Anura, Sauria), [Testudines]
Robinson, **P**[amela] **L**[amplugh] • ? • England • (*rhynchocephalia)
Robison, **W**ilbur **G**erald, Jr. • 27 December 1933–2004 March 18 • USA • Sauria
Roček, **Z**byněk • 16 August 1945 • Czechoslovakia / Czech Republic • *sauria, *serpentes
Rocha, **C**arlos **F**rederico **D**[uarte] da • 22 October 1958 • Brazil • Sauria
Rochebrune, **A**lphonse [Amédée] **T**rémeau de • 18 September 1836–1912 April 23 • France / French West Africa / France • Anura, *Sauria, *Serpentes, Testudines
Rodda, **G**ordon **H**. • ? • USA • Sauria
Rodeck, **H**ugo **G**[eorge] • 19 September 1902–2004 September 28 • USA • (Sauria)
Rödel, **M**ark-**O**liver • 22 December 1965 • Germany • Anura, Sauria, Serpentes
Rodgers, **T**homas **L**athan • 4 July 1909 • USA • Sauria
Rodhain, [Alphonse-Hubert-] **J**érôme • 25 January 1876–1956 September 26 • Belgium / Belgian Congo / Belgium • (Serpentes)
Rodirigo, **R**oshan **K**. • ? • Sri Lanka • Anura
Rodrigues, **J**oão **B**arbosa • 1842–1909 • (? Brazil) • *crocodilia
Rodrigues, **M**iguel **T**refaut [Urbano] • 4 October 1953 • Portugal / France / Brazil • Anura, Sauria, Serpentes
Rodrigues da Cunha: *see* **Cunha**, **O**svaldo **R**odrigues da
Rodríguez, **D**aniel • ? • Perú • Anura
Rodríguez [-Bayona], **L**ily **O**. • ? • Perú / France / Perú • Anura
Rodríguez [-Gómez], **A**riel [Ricardo] • ? • Cuba • Anura
Rodríguez: *see also* **Novo Rodríguez**, **J**ulio
Roedel: *see* **Rödel**, **M**ark-**O**liver
Roemer, [Karl] **F**erdinand • 5 January 1818–1891 December 14 • Germany • *serpentes
Roesel von Rosenhof: *see* **Rösel von Rosenhof**, **A**ugust **J**ohann
Roesler: *see* **Rösler**, **H**erbert
Roger, **O**tto • ?–? • Germany • *sauria, *testudines
Rogers, **K**arel **L**ouise [Kokx] • 28 June 1947 • USA • *caudata, *serpentes, *testudines
Roig, **V**irgilio **G**erman • 27 September 1930 • Argentina • Anura, Sauria
Roitberg, **E**vgeny **S**. • ? • Germany • Sauria
Román, **B**enigno • ? • Upper Volta • Serpentes
Romano-Hoge, **S**ilvia **A**lma [R. W. de Lemos] (*wife of* **A**lphonse R. Hoge) • ? • Brazil • Serpentes
Romer, **A**lfred **S**herwood • 28 December 1894–1973 November 5 • USA • (*anura)
Romer, **J**[ohn] **D**[udley] • 9 September 1920–1982 March 15 • England / Hong Kong / England • Anura
Römer, **M**iguel • ? • Venezuela • Serpentes
Romiti, **M**assimo • 14 August 1948 • Italy • Sauria
Romoli Sassi, **A**nna • 26 July 1940 • Italy • Sauria
Ron, **S**antiago **R**. • October 1970 • Ecuador / USA • Anura
Rooij, "Nelly" de (= Petronella Johanna, mrs. Breijer-de Rooij) • 30 July 1883–1964 June 10 • Netherlands • Sauria, Serpentes
Rooijen: *see* **Van Rooijen**, **J**ohan
Room, **P**eter **M**[ichael] • 3 July 1947 • Australia • (Sauria)
Roonwal, **M**ithan **L**al • 18 September 1908 • India • Anura
Roopsind, **I**ndranee • ? • Guyana • Anura
Roosevelt, **C**ornelius [Van Schaack] • 23 October 1915 • USA • Sauria
Rose, **A**nthony **B**arclay ("Tony") • ? • Australia • (Sauria)
Rose, **F**rancis **L**[ewis] • 20 December 1935 • USA • Caudata, *testudines
Rose, **W**alter • 21 August 1884–1964 September 13 • England / Ireland / South Africa • Anura, Sauria
Rösel von Rosenhof, **A**ugust **J**ohann • 30 March 1705–1759 March 27 • Germany • (Anura)
Rosén, **N**ils [Valfrid] • 11 June 1882–1970 May 1 • Sweden • Sauria, Serpentes
Rosenberg, **H**ans • ? • Germany • (Sauria), Serpentes
Roshchin, **V**[ladimir] **V**[ictorovich] • ? • USSR • [Sauria]
Rösler, **H**erbert • ? • Germany • Sauria
Ross, **C**harles **A**. • ? • USA • Serpentes
Rossman, **D**ouglas **A**[thon] (*husband of* Nita Jane Rossman) • 4 July 1936 • USA • Caudata, Serpentes
Rossman, **N**ita **J**ane [Kuster] (*wife of* Douglas A. Rossman) • 18 March 1936 • USA • (Serpentes)
Rossmann, **T**orsten • ? • Germany • *crocodilia
Rostombekov, **V**. • ? • USSR • (Sauria), Serpentes
Rothschild, [Lionel] **W**alter (second baron) • 8 February 1868–1937 August 27 • England • (Anura), Sauria, (Serpentes), *Testudines
Roubieu, **G**uillaume **J**oseph • 1757–1834 • France • Sauria
Roule, **L**ouis • 20 December 1861–1942 August 3 • France • (Sauria, Serpentes)

Roux, **J**ean • 5 March 1876–1939 December 1 • Switzerland • Anura, Sauria, Serpentes
Roux-Estève, **R**olande • ? • France • (Anura), Sauria, Serpentes
Rovereto, **G**aetano (marchese) • 1870–1952 • Italy / Argentina / Italy • *crocodilia, *sauria, *testudines
Roxo, **M**athias **G**onsalves de **O**liveira • ? • Brazil • *crocodilia
Roy, **P**ranjalendu • ? • India • Anura
Roze, **J**anis **A**[rnolds] • 31 October 1926 • Latvia / Germany / Venezuela / USA • (Anura), Gymnophiona, Sauria, Serpentes
Rudolf, **V**olker **H**[einz] **W**[ilhelm] • 3 July 1976 • Germany / USA • Anura
Rueda-Almonacid, **J**osé **V**icente • 30 August 1954 • Colombia • Anura, Sauria
Ruedas [-Martin], **L**uis **A**[lberto] • ? • USA • Sauria
Ruibal, **R**odolfo • 27 October 1927 • Cuba / USA • Anura, Sauria
Ruiz, **I**. **R**. **G**. • ? • Portugal • Anura
Ruíz-C[arranza], **P**edro **M**[iguel] • 3 February 1932–1998 September 12 • Colombia • Anura, (Sauria)
Ruiz-García, **F**. • ? • Cuba • Anura
Rummler, **H**ans-Jörg • ? • Germany • Testudines
Rumph, **W**illiam **T**[homas] • 24 March 1947 • USA • Caudata
Rüppell, **E**duard [Wilhelm Peter Simon] • 20 November 1794–1884 December 10 • Germany • Anura, Crocodilia, Sauria, Serpentes, Testudines
Rusconi, **C**arlos • 2 November 1898–1969 • Argentina • *crocodilia, *sauria
Rusconi, **M**auro • 18 November 1776–1849 March 27 • Italy • Caudata
Russell, **A**nthony **P**[atrick] • 10 September 1947 • England / Canada • *crocodilia, Sauria
Russell, **P**atrick • 6 February 1727–1805 July 2 • England / Syria / British India / England • (Sauria, Serpentes)
Russell, **R**ichard **W**estphal • 10 February 1929 • USA / Canada • Caudata
Russo, **P**aul [V.] • 22 September 1965 • USA • Serpentes
Ruthe, **J**ohann **F**riedrich • 16 April 1788–1859 August 24 • Germany • Caudata
Ruthven, **A**lexander **G**[rant] • 1 April 1882–1971 January 18 • USA • Anura, Caudata, Sauria, Serpentes
Ryabov, **S**ergei **A**. • ? • (? Russia) • Serpentes
Rzepakovsky, **V**. **T**. • ? • USSR • Sauria

S

Sá: *see* **de Sá**, **R**afael **O**[mar]
Sabaneyev (= Sabanejew), **L**. **L**. (? A. A.) • ? • Russia • Sauria, Serpentes
Sacco, **F**ederico • 5 February 1864–1948 October 4 • Italy • *crocodilia
Sacherer, **J**anice **M**arie • ? • USA • Serpentes
Sackett, **J**. **T**ownsend • ? • USA • Sauria, Serpentes
Sadlier, **R**oss **A**[llen] • 24 January 1955 • Australia • Sauria
Sáfadez: *see* **Fernández-Sáfadez**, **I**ngrid (*not* Fernádez)
Sage, **R**ichard **D**[avid] • ? • USA • (Sauria)
Sager, **A**bram • 22 December 1810–1877 March 6 • USA • Caudata
Sagra, **R**amón de la • 7 December 1801–1871 May 25 • Spain / Cuba / Switzerland • Anura, (Sauria), Serpentes, [Testudines]
Sahi, **D**eep **N**arain • ? • India • Sauria
Sahlberg, **J**ohn **R**einhold • 1845–1920 • Finland • Sauria
Sahni, **A**shok • 1941 • India • *crocodilia
Saint Girons, **H**ubert [Jean-Marie Jacques] • 13 May 1926–2000 April (? 17 *or* 18) • France • Serpentes
Saint-Hilaire: *see* **Geoffroy Saint-Hilaire, É**tienne [-François]
Sai Wanna Kyi • ? • Burma / Myanmar • [Anura], Serpentes
Sakurai, **A**tsushi • ? • Japan • (Anura)
Saldaña-de la **R**iva, **L**ucia • ? • México • Caudata, Sauria
Salisbury, **S**teven **W**. ("Steve") • 16 February 1972 • Australia • *crocodilia
Sallaberry-A[yerza], **M**ichel [Pedro] • 23 May 1954 • Chile • Anura, Sauria
Salvador, **A**lfredo • ? • Spain • Anura, Gymnophiona, Sauria
Salvado [y Rotea], [Lucas José] **R**osendo (bishop of Victoria) • 1 March 1814–1900 December 29 • Spain / Australia / Italy • Serpentes
Salvidio, **S**ebastiano • ? • Italy • Anura, Sauria, Testudines
Salvin, **O**sbert • 25 February 1835–1898 June 1 • England • (Anura, Caudata), Serpentes, (Testudines)
Samarawickrama, **V**. **A**. **M**. **P**radeep **K**. (*brother of* Prasanna Samarawickrama) • ? • Sri Lanka • Serpentes
Samarawickrama, **V**. **A**. **P**rasanna (*brother of* Pradeep Samarawickrama)• ? • Sri Lanka • Serpentes
Sambhu, **H**emchandranauth • ? • Guyana • Anura
Samec, **S**issi • ? • Austria • Sauria
Sampelayo, **P**rimitivo **H**ernández • 1880–1959 • Spain • *testudines
Sanchez, **D**iego **E**[manuel] **A**[rruda] • 20 September 1980 • Brazil • Anura
Sánchez-Herráiz, **M**. **J**. • ? • Spain • Anura
Sánchez-Herrera, **O**scar • 6 August 1954 • México • Sauria
Sánchez-Pacheco, **S**antiago **J**osé • 6 April 1981 • Colombia • Anura
Sanchíz, **F**rancisco de **B**orja • 1949 • Spain • *Anura, [*caudata, *serpentes]
Sancoyo, **M**umpuni • ? • Indonesia • Anura
Sanders, **K**ate **L**. • ? • Wales • Serpentes
Sanders, **O**ttys [E.] • 25 April 1903–1993 March 13 • USA • Anura
Sanderson, **I**van **T**erence • 30 January 1911–1973 February • Scotland / England / USA • (Anura)
Sandner-Montilla, **F**ernando • ? • Venezuela • Serpentes
Sang: *see* **Nguyen V**an **S**ang
San Lwin Oo • ? • Burma / Myanmar • Serpentes
Santos-Costa, **M**aria **C**ristina dos • ? • Brazil • Serpentes
Sanyal, **D**. **P**. • ? • India • Sauria, Serpentes
Sarasin, **F**ritz (= Carl Friedrich, *cousin of* P. Sarasin) • 3 December 1859–1942 March 23 • Switzerland / Ceylon / southeast Asia / Switzerland • Sauria, (Serpentes)
Sarasin, **P**aul [Benedict] (*cousin of* F. Sarasin) • 11 December 1856–1929 April 7 • Switzerland / Ceylon / southeast Asia / Switzerland • (Sauria, Serpentes)
Sarkar, **A**nurup **K**umar • ? • India • Anura
Sasa [-Marin], **M**ahmood (*or* Mahmoud) • ? • Costa Rica / USA / Costa Rica • Sauria, Serpentes

Sato, **I**kio • 22 November 1903–1945 August 11 • Japan • Caudata

Sattorov, **T**. • ? • USSR • Sauria

Satunin, **K**[onstantin] **A**[lekseevich] • ("20 May") 1 June 1863–1915 November 23 ("November 10") • Russia • (Serpentes)

Sauer de Ávila-Pires: *see* **Ávila-Pires**, **T**eresa **C**[ristina] **S**[auer] de

Sauter, **H**ans • 1871–1948 • (? Austria) / Formosa • (Anura, Sauria, Serpentes)

Sauvage, [Louis César] **H**enri **É**mile • 22 September 1842–1917 January 3 • France • Anura, Caudata, Sauria, Serpentes

Sauzier, **T**héodore • ?–(1892)–? • Mauritius / France • *testudines

Savage, **J**ay **M**[athers] • 26 August 1928 • USA • Anura, Caudata, Gymnophiona, Sauria, (*sauria), Serpentes

Savage, **R**on **F**. • ? • USA • Sauria

Savi, **P**aolo • 11 July 1798–1871 April 5 • Italy • Caudata

Savigny, [Marie] **J**ules **C**ésar [le Lorgne de] • 5 April 1777–1851 October 5 • France • (Anura, Sauria, [Serpentes])

Saville-Kent, **W**[illiam] (*born* William Savill Kent) • 10 July 1845–1908 October 11 • England / Australia / England • (Serpentes)

Savitzky, **A**lan **H**[oward] • 23 June 1950 • USA • Sauria, Serpentes

Sawaya, **P**aulo • 11 September 1903–1995 November 6 • Brazil • Gymnophiona

Sawaya, **R**icardo **J**[annini] • ? • Brazil • Anura, Serpentes

Sawyer, **G**len **T**[homas] ("Tom") • 8 September 1933–2004 January 10 • USA • *crocodilia

Saxe, **L**eroy **H**allowell, Jr. • 8 October 1917 • USA • Sauria

Say, **T**homas • 27 June 1787–1834 October 10 • USA • Anura, Caudata, Sauria, Serpentes, Testudines

Sazima, **I**van • ? • Brazil • Anura, Serpentes

Scalabrini, **P**edro • 21 December 1848–1916 April 24 • Italy / Argentina • *crocodilia, *sauria

Scanlon, **J**ohn **D**. • ? • Australia • *crocodilia, *serpentes

Ščerbak: *see* **Shcherbak**, **N**ikolai **N**ikolaevich

Schaefer, **N**. • ? • South Africa • Serpentes

Schaeffer, **B**obb • 27 September 1913–2004 June 2 • USA • *anura

Schaffer, **H**erbert • 14 February 1936 • Austria • *testudines

Schargel, **W**alter **E**[rnesto] • ? • USA • Sauria, Serpentes

Schätti, **B**eat [August] • 2 October 1953 • Switzerland • Serpentes

Schatzinger, **R**ichard [Allen] • ? • USA • *sauria

Schauensee: *see* **Meyer de Schauensee**, **R**odolphe

Scheben, **L**eonard • ? • Germany / German Southwest Africa • (Anura, Sauria)

Schenkel, **E**hrenfried • 1869–1953 • Switzerland • Sauria, Serpentes

Scheuchzer, **J**ohann **J**acob • 2 August 1672–1733 June 23 • Switzerland • (*caudata, Serpentes)

Schick, **S**usanne • 2 August 1977 • Germany • Anura

Schilthuis, **L**. • ?–(1889)–? • Netherlands • Anura

Schimmenti, **G**iovanni • ? • Italy • Anura, Sauria

Schinz, **H**einrich **R**udolf • 30 March 1777–1861 March 8 • Switzerland • Anura, Sauria, Serpentes, Testudines

Schiøtz (= Schioetz), **A**rne • ? • Denmark • Anura

Schlegel, **H**ermann • 10 June 1804–1884 January 17 • Germany / Netherlands • Anura, Caudata, Crocodilia, Sauria, Serpentes, Testudines

Schleich, **H**[ans] **H**ermann • ? • Germany • *Sauria, *serpentes, *Testudines

Schlosser, **J**ohann **A**lbert • ?–1769 • Netherlands • Sauria

Schlosser, **M**ax • 5 February 1854–1932 October • Germany • *caudata, *serpentes

Schlüter, **A**ndreas • 8 February 1949 • Germany • Anura

Schmacker, **B**ernhard • (*about* 1852)–1896 March 26 • Germany • (Anura, Serpentes, Testudines)

Schmelcher, **D**oris • ? • Germany • Sauria

Schmid, **M**ichael • 21 August 1948 • Germany / Venezuela / Germany • Anura

Schmidt, **A**lfred **A**. • 23 December (?) • Germany • (Caudata, Sauria, Serpentes)

Schmidt, [Eduard] **O**[scar] • 21 February 1823–1886 January 17 • Germany • Anura

Schmidt, **F**[ranklin] **J**[ames] **W**[hite] (*brother of* Karl P. Schmidt) • 25 July 1901–1935 August 8 • USA • (Anura, Caudata), Serpentes

Schmidt, **H**ermann • 3 November 1892–? • Germany • *testudines

Schmidt, **K**. • ? • ? • Sauria

Schmidt, **K**arl **P**[atterson] (*brother of* F. J. W. Schmidt) • 19 June 1890–1957 September 26 • USA • Anura, Caudata, *Crocodilia, *Sauria, Serpentes, *Testudines

Schmidt, **P**hilipp (= Moses Paul Philipp Friedrich) • 18 December 1800–(? 1869 August 6 *or* 1873) • Germany • Anura, Serpentes

Schmidtler, **J**osef **F**riedrich (*son of* Josef Johann Schmidtler) • 6 May 1942 • Germany • Anura, (Caudata), Sauria, Serpentes

Schmidtler, **J**osef **J**ohann (*father of* Josef Friedrich Schmidtler) • 15 August 1910–1983 October 24 • Germany • (Caudata), Sauria, Serpentes

Schmitz, **A**ndreas • ? • Switzerland • Anura, Sauria

Schnee, [Karl Gustav] **P**aul • 1865–? • Germany • Testudines

Schneider, **B**ert • 1944 • Germany • Anura, Serpentes

Schneider, **C**hristopher **J**. ("Chris") • 27 June 1959 • USA • Anura, Sauria

Schneider, **G**ustav, Jr. • ?–? • (? East Indies) • (Sauria)

Schneider, **H**ans • ? • Germany • Anura

Schneider, **J**ohann **G**ottlob [Theaenus] • 18 January 1750–1822 January (? 12 *or* 13) • Germany • Anura, Caudata, Crocodilia, Sauria, Serpentes, Testudines

Schneider, [Karl] **C**amillo • 1867–? • (? Austria) • (Serpentes)

Schnitzer, **E**duard (*alias* Emin Pasha) • 28 March 1840–1892 October (? 20 *or* 23) • Prussia / Ottoman Empire / equatorial Africa • (Sauria, Serpentes)

Schnurrenberger, **H**ans • ?–1964 October 6 • Switzerland / Libya / Nepal • (Serpentes)

Schoepff, Ioannis Davidis: *see* **Schöpf**, **J**ohann **D**avid

Schomburgk, [Moritz] **R**ichard • 5 October 1811–1891 March 24 • Germany / Australia • [(Anura, Sauria), Serpentes, Testudines]

Schönecker, **P**atrick • ? • Germany • Sauria

Schöpf, **J**ohann **D**avid (= Ioannis Davidis Schoepff) • 8 March 1752–1800 September 10 • Germany / USA / Germany • Testudines

Schöttler, **T**homas • ? • Germany • Caudata

Schouteden, **H**enri • 3 May 1881–1972 November 15 • Belgium • (Anura, Sauria, Serpentes)

Schrank, **F**ranz von **P**aula von • 21 August 1747–1835 December 22 • Germany • Anura, Caudata, Sauria

Schreber, **J**ohann **C**hristian **D**aniel von • (? 16 *or* 17) January 1739–1810 December 10 • Germany • Anura

Schreiber, **E**gid • 1836–1913 • Austria • Anura, Caudata, Sauria, Serpentes, Testudines

Schreibers, **C**arl (= Karl Franz Anton, ritter von) • 15 August 1775–1852 May 21 • Austria • (Caudata, Sauria)

Schreitmüller, **W**ilhelm • 1870–? • Germany • (Sauria)

Schubotz, **H**ermann • 1881–? • Germany • (Anura), Sauria, (Serpentes)

Schulte, **J**ames **A**[llen], II • 22 July 1973 • USA • Sauria

Schulte, **R**ainer • ? • Germany / Perú • Anura

Schulz, **K**laus-**D**ieter • ? • Germany • (Serpentes)

Schulze, **E**rwin • 1861–? • Germany • Anura

Schüttler, **B.** • ? • (? Germany) • Sauria

Schüz, **E**rnst • 24 October 1901 • Germany • Sauria

Schwaner, **T**erry **D**[ouglas] • 14 March 1944 • USA / American Samoa / Australia / USA • Sauria, (Serpentes)

Schwartz, **A**lbert • 13 September 1923–1992 October 18 • USA • Anura, Caudata, Sauria, Serpentes, Testudines

Schweigger, **A**ugust **F**riedrich • 8 September 1783–1821 June 28 • Germany / France / Sicily • Testudines

Schweizer, **H**ans • 11 September 1891–1975 July 20 • Switzerland • (Sauria, Serpentes)

Schweizerbarth [-Roth], **E**lise **M**elitta von • ?–? • Germany • Caudata

Schwenk, **K**urt • ? • USA • Sauria

Sclater, **W**illiam **L**utley • 23 September 1863–1944 July • South Africa / England • Anura, Sauria, Serpentes

Scolaro, **J**osé **A**lejandro • 9 January 1946 • Argentina • Sauria

Scopoli, **G**iovanni **A**ntonio • (? 3 *or* 13) June 1723–1788 May 8 • Italy • Serpentes

Scortecci, **G**iuseppe • 2 November 1898–1973 October 18 • Italy • Anura, Sauria, Serpentes

Scott, **E. O. G.** • ? • Australia • Anura

Scott, **E**lizabeth • ? • South Africa / USA • Anura

Scott, **W**illiam **B**erryman • 12 February 1858–1947 March 29 • USA • *crocodilia

Scrocchi [Manfrini], **G**ustavo **J**[osé] • 25 April 1956 • Argentina • Serpentes

Scudday, **J**ames **F**[ranklin] • 16 September 1929 • USA • Sauria

Seabra, **A**nthero **F**rederico de • 1876–? • Portugal • (Anura)

Seba, **A**lbert • ("2 May") 12 May 1665–1736 May 13 ("May 2") • Netherlands • (Sauria, Serpentes, Testudines)

Secoy, **D**iane **M**[arie] • 31 October 1938 • USA / Canada • Sauria

Seeley, **H**arry **G**ovier • 18 February 1839–1909 January 8 • England • *crocodilia, *testudines

Seeliger, **L**illian **M.** • ? • USA • Testudines

Seetzen (*not* Sentzen), **U**lrich **J**asper • 30 January 1767–1811 (? September *or* November) • Germany / Arabic lands • Anura, Serpentes

Seidlitz, **W**ilfried von • 1880–? • Germany • *crocodilia

Seiffert, **J**ürgen • ? • Germany • *anura

Seigel, **R**ichard **A**[llyn] • 2 August 1954 • USA • Sauria

Seipp, **R**obert • ? • Germany • Anura, Sauria

Seligmann, **H**ervé • ? • Israel • Sauria

Seliškar, **A**lbin • 1896–? • Yugoslavia • Caudata

Sellards, **E**lias **H**oward • 2 May 1875–1961 February 4 • USA • *crocodilia, *testudines

Semenov, **D**[mitry] **V**[aleryianovich] • November 1952 • USSR / Russia • Sauria

Semon, **R**ichard [Wolfgang] • 1859–1918 • Germany • [(Anura, Sauria), Testudines]

Semper, **C**arl [= Karl Gottfried] • 6 July 1832–1893 May 29 • Germany / Philippine *and* Palau islands / Germany • (Sauria, Serpentes)

Senanayake, [Falika] **R**anil • ? • Ceylon • Anura

Señaris, **J**[osefa] **C**elsa • ? • Venezuela • Anura, Sauria

Sengoku, **S**howichi • ? • Japan • Sauria

Sentzen: *see* **Seetzen**, **U**lrich **J**asper

Seoane [y Pardo-Montenegro], **V**ictor **L**ópez • 29 September 1834–1900 July 14 • Spain • Anura, Caudata, Sauria, Serpentes

Sequeira, **E**duardo [Enrique Vieira Coelho de] • 1861–? • Portugal • (Caudata)

Serié, **P**edro • ? • Argentina • Serpentes, (Testudines)

Serna-D[íaz], **M**arco **A**[ntonio] (hermano) • 11 July 1936–1991 December 31 • Colombia • Anura, Sauria

Serra-Cobo, **J**ordi • 12 July 1958 • Spain / France / Spain • Anura

Serventy, **D**ominic **L**ouis ("Dom") • 28 March 1904–1988 August 8 • Australia • (Sauria, Serpentes)

Seshachar, **B. R.** • ?–? • India • Gymnophiona

Sever, **D**avid **M**[ichael] • 21 February 1948 • USA • Caudata

Severtsov (= Sewertzow), **N**[ikolai] **A**[lekseevich] • ("24 October") 5 November 1827–1885 February 7 ("January 26") • Russia • Sauria

Shaffer, **H**[oward] **B**radley • 26 October 1953 • USA • Anura, Sauria, Testudines

Shah, **K**aran **B**[ahadur] • ? • Nepal • Sauria, (Serpentes)

Shang Gaus • ? • Taiwan • Sauria

Shannon, **F**rederick **A**[lbert] (*husband of* Frances L. Humphrey *from* 1964 *to* 1965) • 4 May 1921–1965 August 31 • USA • Anura, Caudata, Sauria, Serpentes

Sharma, **R**amesh **C**handra • ? • India • Sauria

Shaw, **C**harles **E**[dward] ("Chuck") • 31 July 1918–1971 May 30 • USA • Sauria

Shaw, **G**eorge • 10 December 1751–1813 July 22 • England • Anura, Caudata, Gymnophiona, Sauria, Serpentes, *Testudines

Shcherbak (= Ščerbak *or* Szczerbak), **N**ikolai **N**ikolaevich (= Mykola Mykolaiovych) • 31 October 1927–1998 January 27 • USSR / Ukraine • Anura, Sauria

Shea, **G**lenn **M**[ichael] • 5 February 1961 • Australia • Anura, Sauria

Sheehy, **C**oleman **M.**, III • ? • USA • Anura

Sheil, **C**hristopher **A**[ndrew] • ? • USA • Anura, Serpentes

Shein: *see* **Awan Khwi Shein**

Shelford, **R**obert **W**alter **C**ampbell • 1872–1912 • British Malaya • Serpentes

Shen Youhui • 16 January 1932 • China • Anura
Shenbrot, **G**eorgyi **I**sidorovich • 19 October 1950 • USSR • Sauria
Sheplan, **B**ruce **R**. • ? • USA • (Sauria), Serpentes
Sheppard, **R**evett • ?–(1804)–? • England • Sauria, Serpentes
Sherbrooke, **W**ade **C**[utting] • 4 June 1941 • USA • Serpentes
Sherman, **F**ranklin, Jr. • 2 November 1877–1947 June 23 • USA • (Caudata)
Shi Haitao • ? • China (Hainan) • Sauria
Shikama, **T**okio • 1912–1978 • Japan • *crocodilia, *testudines
Shine, **R**ichard ("Rick") • 7 June 1950 • Australia • (Sauria)
Shipachev, **V**. **G**. • ?–? • Russia • Anura
Shpinar: *see* **Špinar**, **Z**deněk **V**.
Shreve, **B**enjamin • 7 March 1908–1985 July • USA • Anura, Sauria, Serpentes
Shubin, **N**eil **H**. • ? • USA • *caudata
Shufeldt, **R**obert **W**ilson (major) • 1 December 1850–1934 January 21 • USA • Testudines
Shumakov, **O**leg **V**. • ? • Russia • Serpentes
Shuvalov, **V**. **F**. • ? • USSR • *testudines
Siebenrock, **F**riedrich • 20 January 1853–1925 January 28 • Austria • (Sauria, Serpentes), Testudines
Siebold, **C**arl **T**heodor **E**rnst von (*cousin of* P. F. von Siebold) • 16 February 1804–1885 April 7 • Germany • (Serpentes)
Siebold, **P**hilipp **F**ranz [Jonkheer] von (*cousin of* C. T. E. von Siebold) • (? 7 *or* 17) February 1796–1866 October 18 • Germany / Netherlands / Japan / Germany • [Anura, (Caudata)], (Serpentes), [Testudines]
Sierra de **S**oriano, **B**lanca • ? • Uruguay • Sauria
Sill, **W**illiam **D**udley • ? • Argentina • *crocodilia
Silva, **A**njana • ? • Sri Lanka • Sauria
Silva, **H**élio **R**[icardo] da • ? • Brazil • Anura, Gymnophiona
Silva, **N**elson **J**orge da • 21 December 1960 • Brazil / USA • Serpentes
Silva, **T**hais **F**igueiredo **S**antos (*wife of* Marco Antônio de Freitas) • ? • Brazil • Sauria
Silva, **V**erônica [de Novaes e] • ? • Brazil • Sauria
Silva Coutinho, **J**oão **M**artino (major) da • ?–1889 • Brazil • Testudines
Silva-Haad, **J**uan **J**. • ? • Colombia • Serpentes
Silvano, **D**ébora **L**[eite] • ? • Brazil • Anura
Silverstone, **P**hilip **A**[rthur] • ? • USA • Anura
Simmons, **J**ohn **E**[dward] • 15 April 1951 • USA • Anura
Simmons, **R**obert **S**[tanley] • 30 May 1924–1985 September 28 • USA • (Sauria), Serpentes
Simon, **M**artin [Patrick] • 15 October 1951 • USA • Sauria
Simpson, **G**eorge **G**aylord • 16 June 1902–1984 October 6 • USA • *crocodilia, *rhynchocephalia, *serpentes, *testudines
Singh, **G**. • ? • India • *crocodilia
Sinitsin, **D**emetrius **T**heodorovich (= Demetrius Theodore Tidy *or* Dmitriy Fedorovich Sinitsin) • 23 February 1871–1937 • Russia / USA • Sauria
Sison, **R**ogelio **V**. • ? • Philippines • (Sauria), Serpentes
Sites, **J**ack **W**[alter], Jr. • 6 August 1951 • USA • Sauria
Sivan, **N**aomi • ? • Israel • Sauria, Serpentes
Sjöstedt, [Bror] **Y**ngve • 3 August 1866–1948 • Sweden • (Anura), Sauria, Serpentes
Skilton, **A**very **J**udd • 1802–1858 • USA • Caudata, Sauria
Skinner, **M**orris **F**[redrick] • 14 September 1906–1989 December 15 • USA • (*sauria, *serpentes), [*testudines]
Skuk, **G**abriel • ? • Uruguay • Sauria
Slack, **J**ohn **H**amilton • ?–1874 August 27 • USA • [Sauria]
Slater, **J**ames **A**[lexander] • 10 January 1920 • USA • Sauria
Slater, **J**ames **R**[odenburg] • 11 June 1890–1989 February 28 • USA • Anura, Caudata
Slater, **K**enneth **R**. • ? • Australia • (Anura, Sauria), Serpentes
Slavens, **F**rank **L**[eo] • 19 August 1947 • USA • (Serpentes)
Slevin, **J**oseph **R**[ichard] • 13 September 1881–1957 February 15 • USA • (Anura), Sauria, Serpentes
Slipp, **J**ohn **W**[esley] • 22 July 1917–2005 March 11 • USA • Caudata
Sloane, (sir) **H**ans • 16 April 1660–1753 January 11 • England • (Sauria, *testudines)
Slowinski, **J**oseph **B**[runo] ("Joe") • 15 November 1962–2001 September 12 • USA / Myanmar • (Sauria), Serpentes
Smedley, **N**orman • ? • Singapore • Anura, Serpentes
Smets, **G**érard (l'abbé) • 1857–? • Belgium • *testudines
Smith, (sir) **A**ndrew • 3 December 1797–1872 August 11 • Scotland / South Africa / England • Anura, Crocodilia, Sauria, Serpentes, Testudines
Smith, **C**harles **C**linton • 30 April 1910–1966 • USA • (Anura)
Smith, **E**ric **N**[elson] • 18 September 1969 • USA / Guatemala / USA • Anura, Caudata, Serpentes
Smith, **H**obart **M**[uir] (*husband of* Rozella B. Smith) • 26 September 1912 • USA • Anura, Caudata, Sauria, Serpentes, Testudines
Smith, **H**ugh **M**c**C**ormick • 21 November 1865–1941 September 28 • USA • (Anura, Sauria, Serpentes)
Smith, (sir) **J**ames **E**dward • 2 December 1759–1828 March 17 • England • [Caudata]
Smith, **J**ohn **A**lexander • 1818–1883 • Scotland • Sauria
Smith, **L**awrence **A**lec • 1944 • Australia • (Sauria), Serpentes
Smith, **M**alcolm **A**[rthur] • 1875–1958 July 22 • England / Siam / England • Anura, Caudata, Sauria, Serpentes, Testudines
Smith, **M**eredith **J**. • ? • Australia • *serpentes
Smith, **P**hilip **W**[ayne] • 2 December 1921–1986 October 11 • USA • Anura, Sauria, Serpentes, Testudines
Smith, **R**ozella [Pearl Beverly] **B**[lood] (*wife of* Hobart M. Smith) • 18 May 1911–1987 December 15 • USA • (Anura), Sauria, Serpentes
Smith, **S**arah **A**. • ? • Australia / USA • Sauria
Smith, **T**arleton **F**[riend] • ? • USA • Anura, Serpentes
Snyder, **D**avid **H**[ilton] • 24 June 1938–2004 May 8 • USA • Anura
Soares, **A**na **H**[erminia Simões de] **B**[ello] • ? • Brazil • Sauria
Sobolevsky, **N**. **I**. • ? • USSR • Sauria
Sochurek, **E**rich • 26 April 1923–1987 November 21 • Austria • Caudata, Sauria, Serpentes
Sofianidou, **T**heodora • ? • Greece • Anura
Soini, **P**ekka • ? • Perú • (Sauria)
Solano [de Chacín), **H**aydeé • ? • Venezuela • Anura, Gymnophiona
Solórzano [López], **A**lejandro • 29 July 1958 • Costa Rica • Serpentes
Soman, **P**[rabhakar] **W**[aman] • ? • India • Anura

Song Mingtao • 3 July 1937 • China • Caudata, Sauria, Serpentes, Testudines

Sonneman, **N**eil • ? • Australia • (Serpentes)

Sonnini [de Manoncourt], **C**harles [Nicolas] **S**igisbert • 1 February 1751–1812 May 29 • France • Anura, Caudata, Sauria, Serpentes, Testudines

Sordelli, **F**erdinando • 12 December 1837–1916 January 17 • Italy • Serpentes

Souleyet, **L**ouis **F**rançois **A**uguste • 8 January 1811–1852 October 7 • France • Serpentes

Sound, **P**eter • 1968 • India / Germany • Sauria

Sousa, **F**rancisca [G.] de • ? • Panamá • Sauria

Souza, **J**esus **R**. **D**. • ? • Brazil • Anura

Souza-**F**ilho, **J**onas **P**ereira de • ? • Brazil • *crocodilia

Sowerby, **A**rthur de **C**arle • 8 July 1885–1954 August 16 • China • (Anura), Caudata, (Sauria)

Spada: *see* **Positano-Spada**, **D**omenico

Spanò, **S**. • ? • Italy • Sauria

Sparrman, **A**nders • 27 February 1748–1820 August 9 • Sweden • Sauria, Serpentes

Spawls, **S**tephen [Marcus Philip] • 20 February 1953 • England / Ethiopia • Sauria, Serpentes

Speir, **F**rancis, Jr. • 1856–? • USA • *crocodilia

Spence, **J**. **M**. • ? • South Africa • Sauria

Spencer, (sir) [Walter] **B**aldwin • 23 June 1860–1929 July 14 • England / Australia / Chile • Anura, (Sauria)

Spengel, **J**ohann **W**ilhelm • 19 February 1852–1923 • Germany • (Anura)

Spieler, **R**ichard **A**rno • 8 April 1932 • USA • Sauria

Špinar (= Shpinar), **Z**deněk **V**[lastimil] • 4 April 1916–1995 August 14 • Czechoslovakia / Czech Republic • *anura, (*sauria)

Spinks, **P**hillip **Q**. • ? • USA • Testudines

Spirandelli-**C**ruz, **E**lieth **F**loret • ? • Brazil • Anura

Spiteri, **D**avid **E**mmanuel • 2 February 1946 • Malta / USA • Serpentes

Spix, **J**ohann **B**aptist von • 9 February 1781–1826 May 15 • Germany • Anura, Crocodilia, Sauria, ([Serpentes]), Testudines

Spoecker, **P**eter **D**. • 15 April 1941–2005 June 27 • USA • (Sauria)

Spotorno [-Oyarzun], **A**ngel **E**[nrique] • 9 December 1943 • Chile / USA / Chile • Sauria

Spriggs, **M**atthew [James Thomas] • 14 October 1954 • England / Australia • *crocodilia

Springer, **S**tewart • 5 June 1906–1991 August 23 • USA • Sauria

Spurrell, **H**[erbert] **G**[eorge] **F**[laxman] • 1877–1918 • England • (Anura, Sauria, Testudines)

Srinivasa, **R**. • ? • India • Gymnophiona

Stadelman, **R**aymond **E**[dward] • 27 August 1907–1991 August 13 • USA • (Anura, Serpentes)

Stadie, **C**hristian • ? • Germany • Anura

Staesche, **K**arl • 24 August 1902–1981 March 4 • Germany / Brazil • *testudines

Stafford, **P**eter **J**[eremy] • 6 April 1961 • England • Serpentes

Stahl, **A**gostin (= Augustín) • 1842–1917 • Puerto Rico • (Serpentes)

Stanley, **A**rthur • ? • (? England) / China • Serpentes

Stanley, **W**[illiam] **T**. ("Bill") • ? • USA • Anura

Stansbury, **H**oward • 8 February 1806–1863 April 17 • USA • [(Sauria), Serpentes]

Starrett, **P**riscilla **H**[ollister Latham] ("Holly") • 13 November 1929–1997 October 8 • USA • Anura

Steadman, **D**avid **W**[illiam] • 18 August 1951 • USA • *crocodilia, *sauria

Stebbins, **R**obert **C**[yril] ("Bob") • 31 March 1915 • USA • Caudata, Sauria

Steel, **T**homas • 8 September 1858–1925 August 17 • Scotland / Australia • [Serpentes]

Steenstrup, **J**[ohannes] **J**apetus **S**m[ith] • 8 March 1813–1897 June 20 • Norway / Denmark • Anura

Stefano: *see* **De Stefano**, **G**iuseppe

Steindachner, **F**ranz • 11 November 1834–1919 December 10 • Austria • Anura, Caudata, Sauria, Serpentes, (Testudines)

Steiner, **H**ans **M**. • ? • Austria • Caudata

Steinfartz, **S**ebastian • ? • Germany • Caudata

Steinitz, **H**einz • 26 April 1909–1971 April 28 • Germany / Israel • Anura

Steinlein, **C**laus • ? • Germany • Anura

Steiof, **C**hristian • ? • Germany • Anura

Stejneger (*changed from* Steineger), **L**eonhard [Hess] • 30 October 1851–1943 February 28 • Norway / USA • Anura, Caudata, Gymnophiona, Sauria, Serpentes, Testudines

Stemmler [-Gyger], **O**thmar • 30 May 1934 • Switzerland • Serpentes

Stemmler-**M**orath, **C**arl • 4 April 1904 • Switzerland • (Serpentes)

Štěpánek, **O**takar • 24 August 1903 • Czechoslovakia • Sauria, Serpentes

Stephens, **F**rank • 2 April 1849–1937 October 5 • USA • (Serpentes)

Sternberg, **C**harles **H**[azelius] (*father of* C. M. Sternberg *and* G. F. Sternberg) • 15 June 1850–1943 July 20 • USA / Canada • (*crocodilia)

Sternberg, **C**[harles] **M**[ortram] (*son of* C. H. Sternberg *and brother of* G. F. Sternberg) • 18 September 1885–1981 September 8 • USA / Canada • *crocodilia

Sternberg, **G**eorge **F**[ryer] (*son of* C. H. Sternberg *and brother of* C. M. Sternberg) • 26 August 1883–1969 • USA • (*sauria)

Sternfeld, **R**ichard • 8 February 1884–1943 • Germany • Anura, Sauria, Serpentes

Stevens, **M**argaret **S**keels • ? • USA • *sauria

Stevens, **N**. **C**. • ? • (? Australia) • [Sauria]

Stevens, **R**. **A**ndrew • ? • Malawi / England • Anura, Serpentes

Stevini • ?–(1835)–? • (? Russia) • Serpentes

Stewart, **M**argaret **M**[cBride] ("Meg") • 6 February 1927–2006 August 2 • USA • (Anura)

Steyn, **W**illem **J**[ohannes] • 12 October 1926–1968 June 26 • South Africa • Sauria

St. Girons: *see* **Saint Girons**, **H**ubert [Jean-Marie Jacques]

Stickel, **W**illiam **H**[enson] • 8 November 1912–1996 February 11 • USA • (Anura, Sauria), Serpentes

Stimson, **A**ndrew **F**[rancis] • 7 December 1940 • England • Sauria

Stipanicic, **P**edro **N**[icolas] • 6 June 1921 • Argentina • *anura

Stirling, (sir) **E**dward **C**harles • 8 September 1848–1919 March 20 • Australia • Sauria, Serpentes

Stöck, **M**atthias • ? • Germany • Anura
Stocks, **R**upert • ? • Papua New Guinea • Anura
Stoddart, ? (colonel) • ?–? • British India • (Sauria)
Stokes, **J**ohn **L**ort (admiral) • 1812–1885 June 11 • England • (Sauria, [Serpentes])
Stoliczka, **F**erdinand • ("28 May") 9 June 1838–1874 June 19 • Moravia / Austria / British India • Anura, Sauria, Serpentes
Stone, **W**itmer • 22 September 1866–1939 May 24 • USA • (Sauria)
Storer, **D**avid **H**umphreys • 26 March 1804–1891 September 10 • USA • Anura, Caudata, Serpentes
Storer, **T**racy **I**[rwin] • 17 August 1889–1973 June 25 • USA • Caudata
Storm, **R**obert **M**[acLeod] • 9 July 1918 • USA • (Caudata)
Storr, **G**[len] **M**[ilton] • 22 December 1921–1990 June 26 • Australia • Sauria, Serpentes
Storrs, **G**lenn **W**[illiam] • 12 January 1957 • USA • *crocodilia
Strand, **E**mbrik • 1876–1947 • Germany • Anura, Sauria
Strauch, **A**lexander [Johann August] (= Aleksandr Aleksandrovich) • ("1 March") 13 March 1832–1893 August 14 • Russia • Anura, Caudata, Sauria, Serpentes, Testudines
Straughan, **I**an **R**[othwell] • 23 March 1938 • Australia / USA • Anura
Straus, **A**dolf • 1904–1987 • Germany • (*testudines)
Strecker, **J**ohn **K**[ern, Jr.] • 10 July 1875–1933 January 9 • USA • Anura, (Serpentes)
Stritzke, **R**üdiger • ? • Germany • *sauria
Stromer [von Reichenbach], [Karl Heinrich] **E**rnst [Wolfgang] (freiherr) • 12 June 1871–? • Germany • *testudines
Stuart, **B**ryan **L**[ynn] • ? • USA • Anura, Caudata, Sauria
Stuart, **L**[aurence] **C**[ooper] • 9 July 1907–1983 May 27 • USA / Guatemala / USA • Anura, Caudata, Sauria, Serpentes
Stucki-Stirn, **M**artin **C**[hristian] • 17 August 1935 • Switzerland • Serpentes
Studer, **T**heophil • 27 November 1845–1922 February 12 • Switzerland • (Anura)
Stuebing, **R**obert **B**[utler] • 15 November 1946 • USA / Malaysia / USA • Anura, Sauria, Serpentes
Stuhlmann, **F**ranz [L.] • 29 October 1863–1928 November 19 • Germany • (Anura, Serpentes)
Stull, **O**live **G**riffith (= mrs. Loy Erman Davis) • 10 February 1905–1968 June 15 • USA • Serpentes
Sturm, **J**acob • 21 March 1771–1848 November 28 • Germany • Sauria
Sturn, **A**nn (= Anna Marie Sturn Skalka) • 2 July 1938 • USA • *testudines
Stutchbury, **S**amuel • 15 January 1798–1859 February 12 • England / Australia / England • Gymnophiona, Sauria
Su Chengye • 2 March 1943 • China • Anura, Sauria
Suarez, **J**osé **M**artin • ? • Brazil • *testudines
Suárez-Mayorga, **Á**ngela **M**arcela • ? • Colombia • Anura
Suchow: *see* **Sukhov**, **G**eorgi **F**edorovich
Suckow, **G**eorg **A**dolph • 28 January 1751–1813 May 13 • Germany • Sauria, Serpentes, Testudines
Sukhanov (= Sukanov), **V**ladimir **B**orisovich • ? • USSR • *testudines
Sukhov, **G**eorgi **F**edorovich • 1898–1942 • Russia / USSR • Sauria
Sukumaran, **J**eet • 2 April 1972 • Malaysia / USA • Anura
Sulimski, **A**ndrzej • 1926 • Poland • *sauria
Sullivan, **C**harles **D**[onald] • 19 September 1938 • USA • Caudata
Sullivan, **R**obert **M**[ichael] • 4 August 1951 • USA • *sauria, *serpentes
Sumichrast, [Adrien-Louis-Jean-] **F**rançois (= Francisco) • 15 October 1828–1882 September 26 • Switzerland / México • (Anura, Sauria, Serpentes)
Summers, **K**yle • ? • USA • Anura
Sumontha, **M**ontri • ? • Thailand • Sauria
Sun Ailin • 25 November 1930 • China • *serpentes
Suneja, **I**. **J**. • ? • India • *crocodilia
Supriatna, **J**atna • ? • Indonesia • Sauria
Sutherland, **S**truan **K**[eith] • 17 June 1936–2002 January 11 • Australia • (Sauria)
Swain, **T**om **A**[lfred] • 10 October 1941 • USA • Sauria
Swainson, **W**illiam • 8 October 1789–1855 December 7 • England / New Zealand • Sauria, Serpentes
Swan, **L**awrence **W**[esley] ("Larry") • 9 March 1922–1999 May 5 • India / USA • [(Anura)]
Swan, **M**ichael • 1956 • Australia • (Sauria)
Swan, **S**teven **R**. • ? • Vietnam • Anura
Swanson, **S**tephen • ? • Australia • (Sauria)
Swinhoe, **R**obert (consul) • 1 September 1836–1877 October 28 • India / England / China / England • (Anura, Sauria, Serpentes, Testudines)
Swinton, **W**illiam **E**[lgin] • 30 September 1900–1994 June 12 • Scotland / England / Canada • *crocodilia, *rhynchocephalia, *serpentes, *testudines
Switak, **K**arl **H**[einz] • 30 June 1938 • Germany / USA • (Sauria)
Sworder, **G**. **H**umphrey • ?–? • British India • (Sauria)
Swynnerton, **C**[harles] **F**[rancis] **M**[assey] • 3 December 1877–1938 June 8 • British East Africa • (Sauria)
Symeonidis, **N**ikolaus • 1930 • Greece • *testudines
Szalai, **T**ibor • ? • Hungary • *testudines
Szczerbak: *see* **Shcherbak**, **N**ikolai **N**ikolaevich
Szeliga-Mierzeyevski, **W**. • ? • (? Lithuania) • Caudata
Szyndlar, **Z**bigniew [Andrzej] • 3 February 1952 • Poland • *serpentes, [*caudata]

T

Tada: *see* **Di Tada**, **I**smael **E**[duardo]
Taddei, **A**ntonio • ? • Italy • Sauria
Tagliaferro, **N**apoleone • 1843–? • Malta • *testudines
Tago, **K**atsuya • 26 May 1877–1943 December 19 • Japan • (Anura), Caudata
Takahashi, **A**kio • 14 September 1973 • Japan • *testudines
Takahashi, **S**eiichi • 7 March 1874–1932 March 29 • Japan / Taiwan • Serpentes
Takara, **T**etsuo • 25 June 1913 • Japan • Serpentes
Takeda, **N**obuyuki • ? • Japan • Sauria
Tamsitt, **J**[ames] **R**[ay] • 22 November 1928 • USA / Canada *and* Colombia / USA • (Anura)
Tan Fui-**L**ian (*wife of* Robert F. Inger) • 1951 • Malaysia / USA • Anura
Tandy, **J**ocelyn • ? • USA • Anura

Tandy, [Robert] **M**ills • ? • USA • Anura
Tang Zhenjie • 28 August 1934 • China • Anura
Tang Ziming • 6 August 1932 • China • Caudata
Tang Ziyin • 12 February 1927 • China • Caudata
Tanner, **V**asco **M**[yron] (*brother of* Wilmer W. Tanner) • 29 October 1892–1989 (? April 25 *or* May 14) • USA • Sauria, Serpentes
Tanner, **W**ilmer **W**[ebster] (*brother of* Vasco M. Tanner)• 17 December 1909 • USA / Netherlands / USA • Anura, Caudata, Sauria, Serpentes
Tanzer, **E**rnest **C**laude • ?–1971 January 23 • USA • (Serpentes)
Tao Hsi-jen • ? • Taiwan • *testudines
Taphorn, **D**onald **C**. • ? • Venezuela • Sauria, (Serpentes)
Tarashchuk (= Taraščuk), **V**ladimir **I**vanovich • 1922–1987 • USSR • *testudines
Tarlo: *see* **Halstead** [Tarlo], **L**[ambert] **B**[everly]
Tasman, **K**enneth [Robert] (father) • 13 June 1890–1968 May • England / Rhodesia • (Sauria)
Tasnim, **R**ashida • ? • Pakistan • Anura, Sauria
Tatarinov, **L**eonid **P**etrovich • 12 November 1926 • USSR • *anura
Taufik, **A**gustinus • ? • Indonesia • Anura
Taylor, **E**dward **H**[arrison] • 23 April 1889–1978 June 16 • USA / Philippines / USA • *Anura, *Caudata, Gymnophiona, *Sauria, Serpentes, Testudines
Taylor, **H**arry **L**[eonard] • ? • USA • Sauria
Taylor, **W**. **E**dgar • ?–? • USA • (Serpentes), Testudines
Taylor, **W**alter **P**enn • 31 October 1888–1972 March • USA • (Sauria)
Tchernov, **E**itan • ? • Israel • *crocodilia
Tchernov: *see also* **Chernov**, **S**ergius **A**leksandrovich
Tchibozo, **S**everin • ? • Bénin • Sauria
Tchikvadze: *see* **Chkhikvadze**, **V**iacheslav **M**ikhailovich
Telfair, **C**harles • (*about* 1777)–1833 July 14 • Ireland / Mauritius • (Sauria)
Telford, **S**am **R**ountree, Jr. • 25 August 1932 • USA • Serpentes
Temminck, **C**[oenraad] **J**[acob] • 31 March 1778–1858 January 30 • Netherlands • Anura, Caudata, [Crocodilia], (Sauria, [Serpentes]), Testudines
Tennent, (sir) **J**ames **E**merson • 7 April 1804–1869 March 6 • Ireland / Ceylon / England • [(Sauria)]
Tepper, **J**[ohann] **G**[ottlieb] **O**tto • 19 April 1841–1923 February 16 • Prussia / Australia • Sauria
Terentjev, **P**aul (= Pavel Victorovich) • 23 December 1903–1970 December 30 • Russia / USSR • Anura, Sauria
Test, **F**rederick **C**leveland (*uncle of* Frederick H. Test) • 14 June 1869–1956 July 8 • USA • Anura
Test, **F**rederick **H**[arold] (*nephew of* Frederick Cleveland Test) • 16 April 1912 • USA • Anura
Tewari, **B**. **S**. • ? • India • *testudines
Teynié, **A**lexandre • ? • France • Anura, Sauria
Thales de Lema: *see* **Lema**, **T**hales de
Thanh: *see* **Bui N**gok **T**hanh *and* **Vu N**goc **T**hanh
Theobald, **W**illiam • 1829–1908 March 31 • British India / England • Anura, Crocodilia, Sauria, Serpentes, Testudines
Thienemann, **F**riedrich **A**ugust **L**udwig • (? 22 *or* 25) December 1793–1858 June 24 • Germany • Sauria
Thin Thin • ? • Burma / Myanmar • Anura, [Sauria], Serpentes
Thireau, **M**ichel • 13 February 1943 • France • Sauria
Thomas, **A**lcide • ?–(1855)–? • France • Anura
Thomas, **B**ruce **W**illiam • 6 July 1947 • New Zealand • Sauria
Thomas, [John Paul] **R**ichard • 2 May 1938 • USA / Puerto Rico • Anura, Caudata, Sauria, Serpentes
Thomas, **R**obert **A**[llen] • 10 April 1946 • USA • Serpentes
Thomé, **J**oão **C**arlos **A**lciati • ? • Brazil • (Anura)
Thominot, **A**lexandre • ?–(1878–1889)–? • France • Anura, Sauria
Thompson, **F**red **G**[ilbert] • 13 November 1934 • USA • (Anura), Sauria, Serpentes
Thompson, **H**elen **B**[eulah]: *see* **Gaige**, **H**elen [Beulah] **T**[hompson]
Thompson, **J**oseph **C**heesman • 1874–? • USA • Caudata, Sauria, Serpentes
Thomson, **D**onald [Finlay Ferguson] • 26 June 1901–1970 May 12 • Australia • Serpentes
Thomson, **S**cott **A**[ndrew] • ? • Australia • *Testudines
Thon, **T**heodor • 14 May 1792–1838 November 17 • Germany • Serpentes
Thorn, **R**obert • 24 August 1925 • Luxemburg • Caudata
Thornton, **W**[ilmot] **A**[rnold] • 27 December 1921 • USA • Anura
Thorpe, **R**oger **S**[tephen] • 18 August 1947 • England / Wales • Serpentes
Thuna, **M**indy • ? • ? • [Serpentes]
Thunberg, **K**arl **P**eter • 11 November 1743–1828 August 8 • Sweden • Serpentes, Testudines
Thurow, **G**ordon **R**[ay] • 13 February 1929–2000 February 1 • USA • Caudata
Tian Wanshu • 15 January 1930 • China • Anura, Caudata
Tidy: *see* **Sinitsin**, **D**emetrius **T**heodorovich (= Demetrius Theodore Tidy *or* Dmitriy Fedorovich Sinitsin)
Tiedemann, **F**ranz • 14 January 1949 • Austria • Sauria
Tiedemann, **F**riedrich • 23 August 1781–1861 January 22 • Germany • (Sauria)
Tien: *see* **Dao V**an **T**ien
Tihen, **J**oseph **A**[nton] • 20 November 1918–2002 November 12 • USA • *Anura, *caudata, Sauria, (*serpentes)
Tilak, **R**aj • 1935 • India • Anura
Tilbury, **C**olin **R**. • ? • South Africa • Sauria
Tillack, **F**rank • (? 1963 *or* 1964) • Germany • Serpentes
Tilley, **S**tephen **G**[eorge] • 21 July 1943 • USA • Caudata
Ting, **H**an-po: *see* **Ding**, **H**anbo
Tinkle, **D**onald **W**[ard] • 3 December 1930–1980 February 21 • USA • Sauria, Serpentes, Testudines
Tinsley, **R**ichard **C**. • ? • England • Anura
Tirant, **G**ilbert • ?–(1885)–? • French Indochina • Serpentes, Testudines
Titus, **T**om **A**[lan] • ? • USA • Sauria
Tiwari, **K**. **K**. • ? • India • Sauria, Serpentes
Tjaturadi, **B**urhan • ? • Indonesia (West Papua, *formerly* Irian Jaya) • Anura
Toal, **K**evin **R**., III • ? • USA • Anura
Toft, **C**atherine **A**[nn] • 11 October 1950 • USA • Anura
Toit, **C**ornelius **A**lbertus du • 1910 • South Africa • (Anura)
Tokar, **A**. **A**. • ? • USSR • Serpentes
Tolley, **K**rystal **A**[nne] • 15 August 1963 • USA / South Africa • Sauria

Tomasini, **O**tto (ritter) von • ?–(1889)–? • Bosnia • (Caudata), Sauria
Tong, **H**aiyan • ? • France • *testudines
Toriba, **M**ichihisa • 26 March 1950 • Japan • Serpentes
Tornier, **G**ustav • 9 May 1858–1938 April 25 • Germany • Anura, Gymnophiona, Sauria, Serpentes, (Testudines)
Torres, **F**rank • ? • Puerto Rico • Sauria
Torres-Carvajal, [Lenin] **O**mar • 12 April 1975 • Ecuador / USA • Sauria
Torres-Perez, Fernando [Leon] • 24 April 1971 • Chile / USA • Anura
Toula, **F**ranz • 1845–1920 • Austria • *crocodilia, *testudines
Tourneville, **A**lbert • ?–? • France • Caudata, (Sauria)
Townsend, **C**harles **H**[askins] • 29 September 1859–1944 January 28 • USA • Sauria
Toyama, **M**asanao • ? • Japan • (Sauria), Serpentes
Tracy, **C**. **R**ichard ("Dick") • 24 May 1943 • USA • (Sauria)
Traill, **T**homas **S**tewart • 29 October 1781–1862 July 30 • England • Serpentes
Trapido, **H**arold • 10 December 1916–1991 July 25 • USA • Anura, Caudata, Serpentes
Trauth, **S**tanley **E**[lwood] • 5 September 1948 • USA • Sauria
Trebbau [M.], **P**edro • 20 May 1929 • Germany / Venezuela • (Serpentes), Testudines
Trefaut Rodrigues: *see* **Rodrigues**, **M**iguel **T**refaut [Urbano]
Trémeau de Rochebrune: *see* **Rochebrune**, **A**lphonse [Amédée] **T**rémeau de
Treviño Saldaña, **C**arlos **H**umberto • ? • México • Serpentes
Trevisan, **P**ierluigi • 11 April 1943 • Italy • Caudata
Tripepi, **S**andro • 19 February 1951 • Italy • Testudines
Tristram, **H**enry **B**aker (reverend) • 11 May 1822–1906 March 8 • England • (Testudines)
Troedsson, **G**ustaf [Timoteus] • 9 May 1891–1954 December 12 • Sweden • *crocodilia
Troost, **G**erard • 15 March 1776–1850 August 14 • Netherlands / USA • Serpentes, Testudines
Troschel, **F**ranz **H**ermann (*or* Herrmann *but not* Hugo) • 10 October 1810–1882 November 4 • Germany • Anura, Sauria, Serpentes, Testudines
Trouessart, **É**[douard-] **L**[ouis] • 25 August 1842–1927 June 30 • France • (Caudata)
Troxell, **E**dward **L**[effingwell] • 15 April 1884–1972 September 21 • USA • *crocodilia
True, **F**rederick **W**[illiam] • 8 July 1858–1914 June 25 • USA • (Anura)
Trueb, **L**inda (*wife of* William E. Duellman) • 13 January 1942 • USA • Anura
Truong: *see* **Nguyen Q**uang **T**ruong
Ts'ai Ming-**c**hang • 25 October 1937 • China • Anura
Tsarevsky (= Čarevskij *or* Zarevsky), **S**ergey **F**edorovich • 1887–? • Russia / USSR • Anura, Sauria, Serpentes
Tsaruk, **O**leg **I**[vanovich] • 30 May 1957 • USSR • Sauria
Tschernov: *see* **Chernov**, **S**ergius **A**leksandrovich
Tschudi, **J**ohann **J**akob von • 25 July 1818–1889 (? August *or* October) 8 • Switzerland / Austria • Anura, Caudata, Sauria, Serpentes, Testudines
Tschumakov (= Chumakov), **I**van **S**ergeevich • ? • USSR • *sauria
Tuckey, **J**ames (? **H**ingston *or* **K**ingston) • August 1776–1816 October (? 4 *or* 14) • Ireland / England / Congo • [Serpentes]
Tun: *see* **Hla Tun**
Tuniyev (= Tunijew *or* Tuniév), **B**oris **S**[akovich] • 20 December 1956 • USSR / Russia • Sauria, Serpentes, Testudines
Tunner, **H**einz **G**. • 24 October 1940 • Austria • Anura
Turbott, [Evan] **G**raham • 1914 • New Zealand • Anura, (Sauria)
Turner, **A**ndrew **A**[lexander] • 17 March 1969 • South Africa • Anura
Tweedie, **M**ichael **W**[illmer] **F**[orbes] • 2 September 1907–1993 March 25 • England / Venezuela / Malaya / England • (Anura), Sauria, Serpentes
Twente, **J**ohn **W**[esley], Jr. • 18 December 1926 • USA • *sauria, (*testudines)
Twitty, **V**ictor **C**[handler] • 5 November 1901–1967 March 22 • USA / Germany / USA • Caudata
Tyler, **M**ichael **J**[ames] • 27 March 1937 • England / Australia • *Anura
Tymowska, **J**anina • ? • Switzerland • Anura
Tytler, **R**obert [Francis Christopher Alexander] (major general) • ?–1916 October 19 • England / British India • (Anura), Sauria, (Serpentes)

U

Uéno, **S**hun-ichi • 8 December 1930 • Japan • Sauria
Ugarte, **C**ristina **A**[licia] • ? • USA • Sauria
Ulasievicz, **W**. • ? • (? Lithuania) • Caudata
Ulber, **T**homas • ? • Germany • Sauria
Underwood, **G**arth [Leon] • 16 July 1919–2002 October 15 • England / Jamaica / Trinidad / England • Sauria, (Serpentes)
Unterstein, **W**alter • ? • Germany • (Anura), Caudata
Urbano, **R**. **F**. • ? • USA • Sauria
Urbina-B., **M**edardo • ? • Chile • Sauria
Uribe-Peña, **Z**eferino • ? • México • (Sauria, Serpentes)
Urich, **F**rederick **W**illiam • 1870–1937 July 22 • Trinidad • (Anura), [Sauria]
Ustach, **P**aul **C**[onstantine] • 23 January 1968 • USA • Anura
Uthmöller, **W**olfgang • ? • Germany • Sauria, Serpentes
Utiger, **U**rs • ? • Switzerland • Serpentes
Üzüm, **N**azan • ? • Turkey • Serpentes
Uzzell, **T**homas **M**[arshall], Jr. • 6 April 1932 • USA • Anura, Sauria

V

Vaillant, **L**éon [-Louis] • 11 November 1834–1914 November 24 • France • (Anura), Caudata, *crocodilia, Sauria, (Serpentes), *Testudines
Valdés de la Osa, **A**. • ? • Cuba • Anura
Valdujo, **P**aula **H**. • ? • Brazil • Sauria
Valencia, **J**osé • ? • Chile • Anura
Valenciennes, **A**chille • 9 August 1794–1865 April 13 • France • Anura, Caudata, *crocodilia, Sauria, *Testudines
Valentine, **B**arry **D**[ean] ("B. O.") • 6 June 1924 • USA • Caudata

Valladares [-Faundes], **J**ohn **P**ablo • 23 August 1972 • Chile • Sauria
Vallan, **D**enis • 10 October 1965 • Switzerland • Anura
Valverde [-Gómez], **J**osé **A.** • ? • Spain • Sauria
van Beneden: *see* **Beneden**, **P**ierre **J**oseph van
van Beurden, **E**ric **K.** • ? • Australia • Anura
van Breukelen, **F**rank • ? • USA • Serpentes
Vance, **T**homas **L.** • ? • USA • Sauria
Vancea, **S**tefan • ? • Romania • Sauria, *testudines
Van Couvering, **J**udith **A**nne **H**arris • ? • USA / Canada • *crocodile
van Dam: *see* **Dam**, **G. P. F.** van
Vandelli, **D**omenico (= Domingos) • 8 July 1735–1816 June 27• Italy / Portugal • Sauria, Testudines
Van Denburgh, **J**ohn • 23 August 1872–1924 October 24 • USA • Anura, Caudata, Sauria, Serpentes, *Testudines
Vanderduys, **E**ric • ? • Australia • Sauria
Vanderhaege, **M**aurice • ? • France • (Testudines)
Van der Hoeven: *see* **Hoeven**, **J**an van der
Van der Straaten, **M**ichael ("Mike") • ? • Australia • (Serpentes)
Van Dijk, [David] **E**duard ("Eddie") • 14 September 1925 • South Africa • (Anura)
Van Dijk, **P**eter **P**aul • ? • Netherlands / Ireland / Thailand / Malaysia • (Testudines)
van Frank, **R**ichard • ? • USA • *caudata
Van Gelder, **R**ichard **G**eorge • 17 December 1928 • USA • Sauria
Van Hasselt: *see* **Hasselt**, **A**lexander **W**illem **M**ichiel van *and* **J**ohan **C**oenraad van
Van Heurn, **W**illem **C**ornelis • 20 February 1887–1972 June 6 • Netherlands / Dutch East Indies / Netherlands • (Sauria, Serpentes)
van Kampen: *see* **Kampen**, **P**[ieter] **N**[icolaas] van
Van Lidth de Jeude: *see* **Lidth de Jeude**, **T**heodorus **W**illem van
Vanni, **S**tefano • 22 April 1951 • Italy • Caudata, Serpentes
Vannini, **J**ay **P.** • ? • Guatemala • Sauria
Van Rooijen, **J**ohan • ? • Netherlands • Serpentes
van Wyk, **J. H.** • ? • South Africa • Sauria
Vanzolini, **P**aulo **E**mílio • 25 April 1924 • Brazil • (Anura), *Sauria, *Serpentes
Van Zyl, **J. H. M.** • ? • South Africa • (Sauria)
Varela J., **M**anuel • ? • México • Serpentes
Varona, **L**uis **S.** • ? • Cuba • *crocodilia, Sauria
Vashetko, **E**milia **V**[iktorovna] • 1942 • USSR • Sauria
Vasse, **D**enis [Pierre] • 28 April 1965 • France • *crocodilia
Vasudevan, **K**arthikeyan • 23 June 1972 • India • Anura
Vasvári, **M**iklós • ? • Hungary • Sauria
Vaughan, **R**[amah] **K**athryn • ? • USA • Serpentes
Vaz-Ferreira, **R**aúl • 1918 • Uruguay • Sauria
Vedmederya, **V**alery **I**osiphovich • 1946 • USSR • Sauria, Serpentes
Veith, **G**eorg (hauptmann) • 1875–1925 • Austria • (Sauria, Serpentes)
Veith, **M**ichael [Karl-Heinz] • 18 November 1957 • Germany / Netherlands • Anura
Velasco, **J**osé **M**aria • 1840–1912 • México • Caudata
Velásquez-Álvarez, **Á**lvaro **A**ndrés • 2 June 1980 • Colombia • Anura
Velázquez-Solis, **V**ictor **M.** • ? • México • (Sauria), Testudines
Vélez-R[odríguez], **C**laudia **M**[aría] • 5 December 1965 • Colombia / Chile • Anura
Vellard, **J**ehan **A**lbert • 18 July 1900–1996 July 28 • Tunisia / France / Brazil / Perú / Argentina • Anura, Serpentes
Veloso-M[artínez], **A**lberto [Rafael] • 29 October 1940 • Chile • Anura, (Sauria)
Vences, **M**iguel • 24 April 1969 • Germany / Netherlands / Germany • Anura, Sauria, Serpentes
Venczel, **M**árton • ? • Romania • *anura
Venegas [-Ibáñez], **P**ablo **J.** • ? • Perú • Anura
Venning, **F**rancis **E**smond **W**ingate (brigadier) • 26 January 1882–1970 August 28 • Ceylon / England • (Serpentes)
Venzmer, **G**erhard • 1 June 1893 • Germany • Serpentes, Testudines
Verdade, **V**anessa **K**ruth • 16 December 1974 • Brazil • Anura, Sauria
Veronese [-Schein], **L**auren • ? • Brazil • Sauria
Verrastro, **L**aura • ? • Brazil • Sauria
Verrill, **A**ddison **E**[mory] • 9 February 1839–1926 December 10 • USA • (Sauria)
Versilin, **N. N.** • ? • USSR • *testudines
Vesey FitzGerald, [L.] **D**esmond [E.] **F.** • 1910–1974 May 3 • Arabia / East Africa • (Sauria)
Vestergren, **G**reta • ? • Sweden • Gymnophiona, Serpentes
Vial, **J**ames **L**[eslie] • 19 December 1924 • USA • Caudata
Vianna, **A.** • ? • Portugal • *crocodilia
Vicente, **L**uis **A.** • ? • Portugal • Sauria
Vidal, **N**icolas • ? • France • Serpentes
Videla, **F**ernando • ? • Argentina • Sauria
Vieira, **G**ustavo **H**[enrique] **C**[alazans] • ? • Brazil • Sauria
Vieites, **D**avid **R.** • 15 May 1975 • USA • Caudata
Viets, **B**rian **E.** • ? • USA • Sauria
Vigle, **G**regory **O**[wen] • 9 September 1958 • USA • Anura
Vigny: *see* **Loumont** [-Vigny], **C**atherine
Vijaya, **J**aganathan ("Viji") • ?–1987 April (28 years old) • India • (Testudines)
Vijayakumar, **S. P.** • ? • India (Andaman Islands) • Serpentes
Vilaró-Diaz, **J**uan • ?–? • Cuba • Testudines
Villada, **M**anuel **M**aría • 1841–1924 • México • Serpentes
Villanueva: *see* **Bocquentin-V**illanueva, **J**[uan]
Villa-Rivas, **J**aime [Dolan] • 28 March 1944 • Nicaragua / USA • (Sauria), Serpentes
Villavicencio, **J**osé • ? • Argentina • Sauria
Villiers, **A**ndré • 9 April 1915–1983 June 8 • France / French West Africa / France • (Anura), Serpentes
Villiers, **A**therton **L**[ilford] de: *see* **De Villiers**, **A**therton **L**[ilford]
Villiers, C[ornelius] G[erhardus] S[tephanus] de: ***see*** **De Villiers, C[ornelius] G[erhardus] S[tephanus]**
Viña: *see* **Bertolotto**, **C**arolina **E**lena **V**iña
Vinciguerra, **D**ecio • 23 May 1856–1934 October 5 • Italy • Sauria, Serpentes
Vindum, **J**ens **V**[erner] • 28 August 1954 • Denmark / USA • Sauria
Vinson, **J**ean-**M**ichel (*son of* Jean Vinson) • ? • Mauritius • Sauria
Vinson, [Joseph Lucien] **J**ean (*father of* Jean-Michael Vinson) • 1906–1966 • Mauritius • Sauria

Viosca, [Paul] **P**ercy, Jr. • 24 June 1892–1961 August 27 • USA • Anura, Caudata, (Testudines)

Vis: *see* **De Vis**, **C**[harles] **W**[alter]

Visser, **J**ohn • ? • South Africa • Sauria

Vitt, **L**aurie **J**[oseph] • 20 August 1945 • USA • Sauria, (Serpentes)

Voeltzkow, **A**lfred • 1860–1947 • Germany • [(Anura, Sauria, Serpentes), Testudines]

Vogel, **G**ernot • 1963 • Germany • Sauria, Serpentes

Vogel, **Z**deněk • 10 November 1913–1986 December 9 • Czechoslovakia • (Testudines)

Vogl, **C**ornelius (padre) • 1884–1953 • Venezuela • (Anura, Sauria, Testudines)

Vogt, **C**arl [Christopher] • 5 July 1817–1895 May 5 • Germany / Switzerland • Crocodilia

Vogt, **R**ichard **C**[arl] • ? • USA / México • Sauria, Serpentes

Vogt, **T**heodor • 22 April 1881–? • Germany • Anura, Sauria, Serpentes, Testudines

Voigt, **F**riedrich **S**iegmund • 1 October 1781–1850 December 10 • Germany • Sauria

Volz, **W**alter • 17 December 1875–1907 • Switzerland • (Anura)

von May, **R**udolf • 1975 • Perú / USA • Anura

Voris, **H**arold **K**[night] • 5 October 1940 • USA • Serpentes

Vosmaer, **A**ernout • 23 October 1720–1799 January 15 • Netherlands • (Sauria, *testudines)

Vosseler, **J**ulius • 16 December 1861–1933 September • Germany / German East Africa / Germany • (Sauria, Serpentes)

Vrcibradic, **D**avor • ? • Brazil • Sauria

Vu Ngoc **T**hanh • ? • Vietnam • Sauria

Vuillemin, **S**imone • ? • Madagascar • Testudines

W

Waga, **A**ntoni [**S**tanisław] • 8 May 1799–1890 November 23 • Poland • Caudata

Wagenaar Hummelinck, **P**[ieter] • 13 January 1907–2003 March 24 * Netherlands * Sauria

Wager, **V**incent **A**. • ? • South Africa • Anura

Wagler, **J**ohann [Georg] • 28 March 1800–1832 August 23 • Germany • Anura, Caudata, Crocodilia, Sauria, Serpentes, Testudines

Wagner, **P**hilipp [Rüdiger] • 29 November 1973 • Germany • Sauria

Waite, **E**dgar **R**[avenswood] • 5 May 1866–1928 January 19 • England / Australia / New Zealand / Australia • Serpentes

Wake, **D**avid **B**[urton] (*husband of* Marvalee H. Wake) • 8 June 1936 • USA • Caudata, (Sauria)

Wake, **M**arvalee **H**[endricks] (*wife of* David B. Wake) • 31 July 1939 • USA • *Gymnophiona, (Sauria)

Walbaum, **J**ohann **J**ulius • 30 June 1724–1799 August (? 21 *or* 26) • Germany • Anura, Testudines

Walker, **C**harles **F**[rederic] • 27 December 1904–1979 February 16 • USA • Anura, Caudata, Sauria, Serpentes

Walker, **C**yril **A**lexander • 8 February 1939 • England • *testudines

Walker, **D**on • ? • USA • *serpentes

Walker, **J**ames **M**artin • 21 October 1938 • USA • Sauria

Walker, **M**yrl **V**[incent] • 20 March 1903–1985 May • USA • *sauria

Walker, **W**arren **F**[ranklin], Jr. • 27 September 1918 • USA • Serpentes

Wall, **F**rank (colonel) • 21 April 1868–1950 May 19 • British India / England • (Sauria), Serpentes

Wallace, **H**oward **K**[eefer] • 7 May 1907–1994 February 15 • USA • (Caudata)

Wallach, **V**an [Stanley Bartholomew] • 6 April 1947 • USA • *Serpentes

Walley, **H**arlan **D**. • 7 November 1932 • USA • Sauria

Wallis, **K**urt • ? • Austria • Serpentes

Waltl, **J**oseph • ?–(1832–1849)–? • Germany • (Caudata)

Wandolleck, **B**enno [Edward Max Julius] • 18 April 1864–? • Germany • Anura, Sauria

Wang, **B**ao-zhong • ? • China • *crocodilia

Wang Chin-**s**hiang • ? • Taiwan • Anura, Sauria

Wang Yuan • ? • China • *anura, *caudata

Wang Yuezhao • ? • China • Sauria

Ward, **J**oseph **P**[atric] • 1944 • USA • Testudines

Ward, **T**. **M**elbourne • ? • Australia • (Sauria)

Wardell-Johnson, **G**rant [Wesley] • 15 January 1957 • Australia • Anura

Warren, **E**rnest • 1871–1946 January 29 • England / South Africa / England • (Sauria, Serpentes)

Warren, **J**ames **W**[olfe] • ? • USA / Australia • Serpentes

Watkins-Colwell, **G**regory **J**[on] • ? • USA • Sauria

Watling, **D**ick • ? • England • Sauria

Watson, **G**illian **W**. • 3 July 1938 • England / Rhodesia / England • Serpentes

Watson, **G**raeme **F**[ield] • 17 December 1942 • Australia • Anura

Weaver, **W**[illiam] **G**[lenn], Jr. • 1936 • USA • *testudines

Webb, **G**rahame **J**[ohn] **W**[arren] • 20 September 1947 • Australia • (Crocodilia)

Webb, **P**hilip **B**arker • 10 July 1793–1854 August 31 • England • [Sauria]

Webb, **R**obert **G**[ravem] • 18 February 1927 • USA • Anura, Caudata, Sauria, Serpentes, Testudines

Webber, **P**aul [Nagle] • 25 August 1949 • England / Australia • Sauria

Weber, **J**ay **A**. • ? • USA • (Anura)

Weber, **M**ax **W**ilhelm **C**arl (*or* Max Carl Wilhelm Weber)• 5 December 1852–1937 February 7 • Germany / Netherlands • Sauria, Serpentes

Webster, **T**homas **P**[reston, III] • 14 July 1947–1975 November 9 • USA • Anura, (Caudata), Sauria

Weed, **A**lfred **C**[leveland] • 28 July 1881–? • USA • Anura

Weekes, [Hazel] **C**laire • 1904–1990 • Australia • (Sauria)

Weems, **R**obert **E**[dwin] • 22 January 1947 • USA • *testudines

Wegner, **T**heodor [Hubert] • 9 September 1880–1934 November 15 • Germany • *testudines

Wei Feng • ? • China • *testudines

Wei Jinlai • 22 November 1940 • China • Anura

Weig, **B**lair **S**. • ? • USA • *testudines

Weigel, **C**hristian **E**hrenfried von • 24 May 1748–1831 August 8 • Germany • Serpentes

Weigel, **J**ohn • ? • Australia • (Anura), Sauria

Weigelt, **J**ohannes • 1890–1948 • Germany • (*caudata, *crocodilia), *sauria

Weijenbergh: *see* **Weyenbergh**, **H**endrik

Weinland, **D**avid **F**riedrich • 30 August 1829–1915 September 16 • Germany • Anura, Sauria

Weismann, [Friedrich Leopold] **A**ugust • 17 January 1834–1914 November 5 • Germany • (Caudata)

Weiss, **C**hristian **S**amuel • 26 February 1780–1856 October 1 • Germany • *testudines

Weithofer, [K.] **A**nton • ?–(1888)–? • Austria • *sauria

Weitzel, **K**arl • ? • Germany • *anura, *crocodilia

Welch, **K**enneth **R**[eginald] **G**[eorge] • ? • England • Sauria

Wellborn, **V**era • 1909 • Germany • Caudata

Weller, **W**orth **H**amilton • 28 May 1913–1931 June 22 (*age* 18!) • USA • Caudata, Serpentes

Welles, **S**amuel **P**aul • 9 November 1907–1997 August 6 • USA • *rhynchocephalia

Wellington, **C**[lifford] **R**oss • ? • Australia • Anura, Crocodilia, Sauria, Serpentes, Testudines

Wellman, **J**ohn • 6 December 1937 • USA • Sauria

Wells, **R**ichard **W**[alter] • ? • Australia • Anura, Crocodilia, Sauria, Serpentes, Testudines

Wellstead, **C**arl **F**[rederick] • 1951 • USA • *sauria

Welter, **W**ilfred **A**[ugust] • 29 March 1906–1939 December 20 • USA • (Caudata)

Wen Yetang • 27 August 1924 • China • Caudata, Serpentes

Werler, **J**ohn **E**[rnest] • 11 June 1922–2004 March 21 • Germany / USA • Anura, Caudata, Sauria, Serpentes

Wermuth, **H**einz • 5 January 1918–2002 December 28 • Germany • Sauria, (Serpentes), Testudines, (*testudines)

Werneck, **F**ernanda [de Pinho] • ? • Brazil • Sauria

Werner, **C**hrista • ? • Germany • *caudata

Werner, **F**ranz [Josef Maria] • 15 August 1867–1939 February 28 • Austria • Anura, Caudata, Crocodilia, Gymnophiona, Sauria, Serpentes, Testudines

Werner, **Y**ehudah **L**[eopold] • 16 June 1931 • Germany / Israel • Sauria, Serpentes, (Testudines)

Werning, **H**eiko • ? • ? • [Sauria]

Westersheimb: *see* **Wettstein** [-Westersheimb], **O**tto

Westphal, **F**rank • 27 June 1930 • Germany • *caudata

Wetmore, [Frank] **A**lexander • 18 June 1886–1978 December 7 • USA • (Anura, Sauria, Serpentes)

Wettstein [-Westersheimb], **O**tto • 7 August 1892–1967 July 10 • Austria • *Anura, Gymnophiona, Sauria, Serpentes, Rhynchocephalia, (Testudines)

Weyenbergh (= Weijenbergh), **H**endrik • 6 December 1842–1885 July 25 • Netherlands / Argentina / Netherlands • Anura

Weyer, **D**ora • ? • USA / Costa Rica / Liberia • (Anura), Serpentes

Weygoldt, **P**eter • 1933 • Germany • Anura

Wheeler, **G**eorge **M**[ontague] (major) • 9 October 1842–1905 May 3 • USA • [Anura], (*crocodilia), [Sauria, Serpentes]

Wheeler, **W**ard **C**[hristopher] • 12 December 1963 • USA • Anura

Wheeler, **W**illiam **M**orton • 19 March 1865–1937 April 19 • USA • (Sauria)

Whipple: *see* **Wilder**, **I**nez **L**[uanne] (*maiden name* Whipple)

Whistler, **D**avid **P**[aul] • 15 June 1940 • USA • *serpentes

Whitaker, **A**nthony **H**[ume] ("Tony") • 5 September 1944 • England / New Zealand • Sauria

Whitaker, **N**ikhil (*son of* R. Whitaker) • 9 September 1978 • India • (Serpentes)

Whitaker, **R**omulus [Earl, II] ("Rom" *father of* N. Whitaker) • 23 May 1943 • USA / India • Serpentes

White, **A**rthur **W**.• ? • Australia • Testudines

White, **G**ilbert (reverend) • 18 July 1720–1793 June 26 • England • [(Testudines)]

White, **J**ohn • (*about* 1750 *or* 1756 *or* 1757)–1832 February 20 • England / Australia • Anura, Sauria

White, **T**heodore **E**lmer • 8 December 1905–1977 September 2 • USA • *crocodilia, *sauria

Whiteside, **D**. **I**. • ? • England • *rhynchocephalia

Whiting, **M**artin **J**. • ? • South Africa • Sauria

Whymper, **E**dward • 27 April 1840–1911 September 16 • England • (Anura, Serpentes)

Wickramasinghe, **L**. **J**. **M**endis • ? • Sri Lanka • Anura, Sauria

Wiederhecker, **H**elga **C**[orrêa] • ? • Brazil • Sauria

Wiedersheim, **R**obert [Ernst Eduard] • 21 April 1848–1923 July 12 • Germany • Caudata, (Sauria)

Wied [-Neuwied], [Alexander Philipp] **M**aximilian (prinz von) • 23 September 1782–1867 February 3 • Germany • Anura, Sauria, Serpentes, Testudines

Wiegmann, **A**rend **F**riedrich **A**ugust [Heinrich] • 2 June 1802–1841 January 15 • Germany • Anura, Sauria, Serpentes, Testudines

Wieland, **G**[eorge] **R**[eber] • 24 January 1865–1953 January 18 • USA • *testudines

Wiens, **J**ohn **J**[oseph] • ? • USA • Anura, Caudata

Wiest, **J**ohn **A**[lton], Jr. • ? • USA • Serpentes

Wiggins, **I**ra **L**[oren] • 1 January 1899–1987 November 28 • USA • (Sauria)

Wijesena, **N**alin **M**. • ? • Sri Lanka • Serpentes

Wilcox, **T**homas **P**. • ? • USA • Caudata

Wild, **E**rik **R**ussell • ? • USA • Anura

Wilder, **H**arris **H**[awthorne] (*husband of* Inez L. Wilder) • 7 April 1864–1928 February 27 • USA / Germany / USA • (Anura, Serpentes)

Wilder, **I**nez **L**[uanne] (*born* Inez Luanne Whipple, *wife of* Harris H. Wilder) • 19 May 1871–1929 April 29 • USA • (Anura, Caudata)

Wiley, **G**race **O**live (*born* Grace Olive Koontz) • 1883–1948 July 20 • USA • (Sauria)

Wilhoft, **D**aniel **C**[harles] • 16 November 1930 • USA • Sauria

Wilkinson, **J**effery **A**[lan] • 20 January 1958 • USA • Anura

Wilkinson, **M**ark • 26 July 1963 • USA / England • Anura, Gymnophiona

Willey, **A**rthur • 9 October 1867–1942 December 26 • England / USA / Ceylon / Canada • ([Serpentes])

Willi, **J**ohanna • ? • ? • Anura

Williams, **E**rnest **E**[dward] • 7 January 1914–1998 September 1 • USA • (Anura), Sauria, (Serpentes), *Testudines

Williams, **J**orge **D**[aniel] • 10 October 1953 • Argentina • Sauria

Williams, **K**enneth **L**[ee] • 4 September 1934 • USA • Anura, Sauria, Serpentes

Williamson, **T**homas **E**[dward] • ? • USA • *crocodilia

Willis, **P**aul **M**[ichael] **A**[rthur] • 2 December 1963 • England / Australia • *crocodilia

Williston, **S**amuel **W**endell • 10 July 1851–1918 August 30 • USA • *testudines

Wilms, Thomas **M**.• ? • Germany • Sauria
Wilson, Larry **D**avid • 10 April 1940 • USA • Anura, Caudata, Sauria, Serpentes
Wilson, **R**ichard **L**[eland] • ? • USA • *anura, *sauria, *serpentes
Wilson, Stephen **K**arl • ? • Australia • (Sauria), Serpentes
Wilson, Vivian **J**. • ? • Rhodesia • (Serpentes)
Wiman, Carl [Johan Josef Ernst] • 10 March 1867–1944 June 15 • USA / Sweden • *testudines
Win: *see* **Htun Win**
Wink, **M**ichael • 10 April 1951 • Germany • Testudines
Winterl, **J**akob József • 15 April 1739–1809 • Hungary • (Caudata)
Wistuba, Eloisa **M**[aria] • ? • Brazil • Anura
Witschi, **E**mil • 18 February 1890–1971 June • Switzerland / USA • Anura
Witte, **G**aston-**F**rançois de • 12 June 1897–1980 June 1 • Belgium / Belgian Congo / Belgium • Anura, Sauria, Serpentes
Witten, Geoffrey **J**. • ? • Australia • Sauria
Wogan, Guinevere **O**. **U**. • ? • USA • Anura
Wolf, Alan **J**[effrey] • ? • USA • Anura
Wolf, Johann • 26 May 1765–1824 February (? 12 *or* 16) • Germany • Sauria
Wolf, **S**iegfried • ? • Germany • Anura
Wolterstorff, Willy (= Wilhelm Georg) • 16 June 1864–1943 January 21 • Germany • Anura, Caudata, (Sauria)
Wombey, **J**ohn **C**. • ? • Australia • Sauria
Wong, **A**nna • ? • Malaysia • Anura
Wood, **G**[ilbert] Congdon • 14 November 1915–2000 October 9 • USA • Serpentes
Wood, **R**oger Conant • 20 May 1941 • USA • *testudines
Wood, **S**earles **V**alentine • 14 February 1798–1880 October 26 • England • *crocodilia
Wood, **W**allace **F**. • 1910 • USA • Caudata
Wood, **W**illiam **W**. • ?–(1825)–? • USA • Caudata
Woodall, **H**arold **T**. • ?–World War II • USA • Caudata
Woodbury, **A**ngus **M**[unn] (*father of* D. M. Woodbury) • 11 July 1886–1964 August 1 • USA • Sauria, Serpentes
Woodbury, **D**ixon **M**[iles] (*son of* A. M. Woodbury) • 6 August 1921–1991 July 29 • USA • Serpentes
Woodin, **W**illiam **H**[artman, III] • 16 December 1925 • USA • Serpentes
Woodley, **J**eremy **D**. • ? • Jamaica • Anura
Worrell, **E**ric [Frederic Arthur] • 27 October 1924–1987 July 12 • Australia • Sauria, Serpentes, (Testudines)
Worthington, **E**[dgar] **B**arton • 13 January 1905 • England • (Serpentes)
Worthington, **R**ichard **D**[ane] • 20 September 1941 • USA • Caudata
Worthy, **T**revor **H**[enry] • 3 January 1957 • New Zealand • *anura, *crocodilia, *sauria
Wouters, **G**eorges [Florent] • 1935 • Belgium • (*crocodilia)
Wranik, **W**olfgang • ? • Germany • Sauria
Wright, **A**lbert **H**azen (*husband of* Anna A. Wright) • 15 August 1879–1970 July 4 • USA • Anura, (Caudata, Sauria, Serpentes)
Wright, **A**nna **A**llen (*wife of* Albert H. Wright) • 4 March 1882–1964 December 5 • USA • Anura, (Serpentes)
Wright, **J**ohn **W**[illiam] • 9 July 1936 • USA • Anura, Caudata, Sauria, Serpentes
Wright, **M**argaret **R**[uth] • 24 March 1913 • USA • Caudata
Wu Guanfu • 12 May 1935 • China • Anura, Caudata
Wu Hsien-wen • March 1900–1985 April 3 • China • Anura
Wu Lu • 15 October 1915–1983 December 2 • China • [Anura]
Wu Shuhui • 26 July 1936 • China • Caudata
Wu Weitang • ? • China • *testudines
Wu Xiao-Chun • ? • China / Canada • *crocodilia
Wucherer, **O**tho (= Otto Eduard Heinrich) • 7 July 1820–1874 May 7 • Portugal / England / Brazil • (Anura, Sauria), Serpentes
Wurfbain (= Wurffbain), **J**ohann **P**aul • 13 December 1655–1711 January 17 • Germany • (Caudata)
Wussow, **W**. • ? • (? Prussia) • Testudines
Wüster, **W**olfgang • 17 February 1964 • Germany / Switzerland / England / Scotland / Wales • Serpentes
Wynn, **A**ddison **H**[artwell] • 9 November 1955 • USA • Caudata

X

Xantus, **J**ohn (= János Xántus, *alias* Louis de Vésey) • 5 October 1825–1894 December 13 • Hungary / USA / Hungary • (Sauria)
Xavier, **F**rançoise • 13 July 1939 • France • Anura
Xing Yiling • 8 April 1961 • China • Caudata
Xu Runhua • 26 August 1931 • China • Anura
Xu Yubin • ? • China • *testudines

Y

Yaakob: *see* **Norsham** [Suhaina] **Y**[aakob]
Yambun, **P**aul [Imbun] • ? • Malaysia • Anura, Sauria
Yanev, **K**ay **P**[hilbrick] • ? • USA • Caudata
Yáñez [-Valenzuela], **J**osé [Lautaro] • 18 April 1951 • Chile • Sauria
Yang Datong • 12 March 1937 • China • Anura, Caudata, Gymnophiona, Sauria, Serpentes
Yang Fuhua • 24 August 1930 • China • Anura, Caudata
Yang Jungjien • ? • China • *anura
Yang Rongsheng • ? • China • Caudata, Sauria
Yang, **S**uh **Y**ung • 1934 • South Korea / USA / South Korea • Caudata
Yang Zhongjian: *see* **Young C**hung**c**hien (= Yang Zhongjian)
Yarkov, **A**lexander **A**. • ? • USSR / Russia • *crocodilia
Yarrow, **H**enry **C**rècy ("Harry") • 19 November 1840–1929 July 2 • USA • Anura, Sauria, Serpentes
Yatkola, **D**aniel **A**. • 1947–1976 March 13 • USA • *sauria, (*serpentes)
Yazdani, **G**. **M**. • ? • India • Anura
Ye Changyuan • 14 November 1938 • China • Anura, Caudata
Ye Xiang**k**ui (= Yeh Hsiang-k'uei) • 10 October 1926 • China • *crocodilia, *testudines
Yefimov: *see* **Efimov**, **M**ichael **B**orisovich
Yeh Hsiang-**k**'uei: *see* **Ye X**iang**k**ui
Yeriomtschenko, **V**alery **K**onstantinovich • 1949 • USSR • Sauria
Yilmaz, **C**an • ? • Turkey • Serpentes

Yilmaz, **İ**rfan • 24 September 1953 • Turkey • Anura
Yoshpa, **M**ichael • ? • USA • Anura
Youmans, **T**imothy **M.** • ? • USA • Sauria
Young **C**hung**c**hien (= Yang Zhongjian) • 1 June 1897–1979 January 15 • China / Germany / China • *caudata, *crocodilia, *testudines
Young, **J**eanne **E**[lise] • 17 September 1969 • USA / Australia • Anura
Youngson, **W**illiam **K**enneth • ? • Australia • (Sauria)
Yu **Z**hiwei • 7 January 1937 • China • Caudata
Yuan **H**ong • 16 January 1955 • China • Serpentes
Yuztis, **E**nrique • ? • Venezuela • Anura

Z

Zaher, **H**ussam [El Dine] • ? • Brazil • Sauria, Testudines
Zaldívar-Riverón, **A**lejandro • ? • México • Sauria
Zammit-Maempel, **G**eorge • 1925 • Malta • *sauria
Zanella, **N**oeli • ? • Brazil • Serpentes
Zangerl, **R**ainer • 19 November 1912–2004 December 27 • USA • *crocodilia, *Testudines
Zao: *see* **Zhao Ermi**
Zarevsky: *see* **Tsarevsky**, **S**ergey **F**edorovich
Zarudny (= Zaroudnoi), **N**ikolai **A**lekseevich • ("13 September") 25 September 1859–1919 March 17 • Russia • (Sauria, Testudines)
Zatz, **M**ariana **G.** • ? • Brazil • Sauria
Zavattari, **E**doardo • 21 October 1883–1972 February 17 • Italy • (Sauria)
Zbyszewski, **G**eorges • 22 October 1909–1999 March 1 • Russia / France / Portugal • *crocodilia
Zeng **X**iaomao • ? • China • Caudata
Zenker, **G.** • ?–? • (? Germany) • (Anura, Serpentes)
Zenneck, **J**onathan Adolf Wilhelm • 1871–1959 • Germany • Serpentes
Zerova, **G**alina **A**natolievna • 14 January 1958 • USSR • *sauria
Zhang **F**u**j**i • 3 January 1946 • China • Caudata, Serpentes
Zhang **L** • ? • China • *crocodilia
Zhang **M**engwen: *see* **Chang** **M**ang-**v**en
Zhang **M**ing**h**ua • 19 January 1947 • China • *Testudines
Zhang **M**ingwan • ? • China • Anura
Zhang **Y**uxia • 27 September 1931 • China • Anura
Zhao (= Djao *or* Zao) **E**r**m**i (*pen name* = Chuan Dong) • 30 January 1930 • China • Anura, Caudata, Sauria, Serpentes
Zhao (= Chao) **K**en**t**ang • 1 September 1932 • China • Sauria
Zheng (= Cheng) **J**i • 3 February 1931 • China • Serpentes
Zhou **K**aiya • 8 October 1932 • China • Sauria
Zhou **M**ingzhen: *see* **Chow** **M**ing-**c**hen
Ziegler, **T**homas • 1970 • Germany • Anura, Sauria, Serpentes
Zietz, **A**mandus **H**einrich **C**hristian • (? 1839 *or* 1874–1921 *or* 1922) • Germany / Australia • Sauria, Serpentes
Zietz, **F**rederick **R**obert • ? • Australia • Sauria
Zigno: *see* **De Zigno**, **A**chille (barone)
Zimmermann, **E**lke (*daughter of* Helmut Zimmermann) • 1958 • Germany • Anura
Zimmermann, **H**elmut (*father of* Elke Zimmermann) • 1926 • Germany • Anura
Zinniker, **W**alter • 29 May 1930 • Switzerland • Sauria, (Serpentes)
Zittel, **K**arl **A**lfred (ritter) von • 25 September 1839–1904 January 5 • Germany • *caudata, *rhynchocephalia, *serpentes
Zong **Y**u • 29 December 1936 • China • Serpentes
Zonkov, **J**ordan • ? • Bulgaria • Caudata, Serpentes
Zug, **G**eorge **R**[obert] • 16 November 1938 • USA • (Anura), Sauria, Serpentes, *testudines, (Testudines)
Zugmayer, **E**rich [Johann Georg] • 16 May 1879–1938 • Austria / Germany • (Anura), Sauria
Zuñiga **U.**, **O**scar • ? • Chile • Sauria
Zweifel, **R**ichard **G**[eorge] • 5 November 1926 • USA • *Anura, (Gymnophiona), Sauria, Serpentes

ACADEMIC LINEAGES OF DOCTORAL DEGREES IN HERPETOLOGY SECOND EDITION

by Ronald Altig

THE PROGRESSION OF THE MAJOR PROFESSOR–STUDENT association through time results in academic lineages, genealogies, or schools. A tabulation of such information was first presented at the First World Congress of Herpetology at Canterbury, England along with two other compilations. These chapters were published in a book by the Society for the Study of Amphibians and Reptiles (e.g., Altig, 1989). With additional searching and the passage of some 18 years, the listing is now about two and a half times larger (≈ 1450 vs. 3810). Series 1 in the listing for the United States increased from 106 to 224 steps, and if the supreme effort were made to actually complete the entire list, it would surely at least triple.

This seemingly simple suite of information is often quite difficult to obtain. In general, people did not respond well to queries about the project, especially outside the U.S. where the concept of "major professor" (=adviser, mentor, promoter) is less common, and there are a lot of erroneous memories. Many names remain to be entered, particularly from generations before about 1990, and coverage of universities outside of the U.S. is grossly incomplete. As one goes further back in time, the linkages obviously become less well defined. For example, if one uses close affiliations and mentoring influences to track backwards from the early parts of U.S. Series 1, one could make this lineage from the 17th Century: M. Malpighi → A. M. Valsalva → G. B. Morgagni → A. Scarpa → I. Döllinger → L. Agassiz and onward as shown below (Anderson et al., 1988).

As a criterion, I looked for herpetological projects done for the doctoral requirements, extensive herpetological publications, or graduation of students in herpetology after the doctorate. I used the doctoral degree only as a historical marker with no intent to limit or define someone's expertise or exclude the many contributions made by those not holding a doctoral degree. Also, the scope of major professors is becoming more diffuse with time because many students are now being directed by persons in adjacent fields, like conservation biology and community ecology.

The scope of acceptance was considerably wider in this edition than in the first; with genetics, development, molecules of all sorts, and physiology being used commonly in many studies, it is increasingly inappropriate to draw artificial boundaries. Once a lineage appeared, it was carried as far as possible as a historical record even if all persons were not herpetologists. The lineage of a person who went to another country for the doctorate appears under that country and not under his or her original or eventual location. Notes at the start of the listing for each country alert the reader to these cases.

Information for this revision was obtained largely from Dissertation Abstracts (UMI ProQuest; http://wwwlib.umi.com/dissertations/search). This on-line database is now limited to the two most recent years, but full access can be had through universities that subscribe to ProQuest. Thus, additions primarily involve years since about 1990 when DI started consistent postings of the names of major professors from U.S. universities. Dissertations from outside the U.S. seldom contain the name of the major professor.

Countries are listed alphabetically, and lineages for each country are presented as a series of keys. Lineages within countries are first listed by decreasing length, and single-step lineages are listed alphabetically in the last series of each country. Find the desired name in the index and locate that entry by following the codes: country (2 upper-case letters), series (number before the decimal) and couplets (number after the decimal; for example: Marvalee H. Wake = US 1.24). Names of U.S. states in university names are abbreviated by U.S. postal abbreviations. Person's names enclosed in brackets signify that an earned doctorate was not obtained. Names in the index are alphabetized by absolute letter order in the surname and then initials without regard for capitalization, diacritical marks, punctuation, or spaces.

Additional information and corrections are eagerly solicited; the positive responses to the first published edition can continue productively only with the help of the entire community. Send to: Ronald Altig, Department of Biological Sciences, P.O. Drawer GY, Mississippi State University, Mississippi State, MS 39762 USA - raltig@biology.msstate.edu. I will pass this information along to that young, exuberant volunteer who wishes to continue the historical search.

A specific listing of those who aided this project would be excessively long, so I extend a general note of thanks and appreciation to all persons who tolerated my numerous requests. Errors surely exist, but the information is true to the extent of the information I was able to obtain.

Searches for information stopped on 1 November 2006. A short bibliography, including the literature cited, of similar sources of information follows.

BIBLIOGRAPHY

ALTIG, R. 1989. Academic Lineages of Doctoral Degrees in Herpetology, p. 179–202. *In* K. Adler (ed.), Contributions to the History of Herpetology. Soc. Study Amphib. Reptiles, Contr. Herpetol., vol. 5.

ANDERSON, J. M., D. W. Corson, and P. W. Gilbert. 1990. Howard Bemhardt Adelmann, p. 3–8. *In* Memorial Statements, Cornell University Faculty 1988–1989. Ithaca, New York.

CARD, W., and J. B. Murphy. 2000. Lineages and Histories of Zoo Herpetologists in the United States. Soc. Study Amphib. Reptiles, Herpetol. Circ., 27: 1–49.

HALL, P. 1995. On the Origin of Theses. New Sci., 14 October 1995: 49–50.

JONES, J. K., Jr. 1991. Genealogy of Twentieth-Century Systematic Mammalogists in North America: The Descendants of

Joseph Grinnell, p. 48–56. *In* M. A. Mares and D. J. Schmidly (eds.), Latin American Mammalogy: History, Biodiversity, and Conservation. Univ. Oklahoma Press, Norman.
ROCKE, A. J., and A. J. Ihde. 1979. A Badger Chemist Genealogy. Jour. Chem. Educ., 56: 93–95.
RODRÍGUEZ-ROBLES, J. A., D. A. Good, and D. B. Wake. 2003. Brief History of Herpetology in the Museum of Vertebrate Zoology, University of California, Berkeley, with a List of Type Specimens of Recent Amphibians and Reptiles. Univ. California Publ. Zool., 131: xv, 1–96.
SPRUGEL, D. G. 1980. A "Pedagogical Genealogy" of American Plant Ecologists, Bull. Ecol. Soc. Amer., 61: 197–200.

ACADEMIC LINEAGES

ARMENIA (AM)

See: **Russia** 1.6, 11.55

ARGENTINA (AR)

See: **Belgium** 1.2–1.4 and **Italy** 1.2–1.3

SERIES 1

001. FERNANDEZ DE RECONDO, M.
Maggese, M. C. ....1986.... B Aires 2
002. MAGGESE, M. C.
Echeverría, D. D. ....1987.... B Aires
Úbeda, C. A. ....1999.... B Aires

SERIES 2

001. SCHNACK, J. A.
Basso, N. ....1988.... La Plata 2
Kehr, A. ....1987.... La Plata
002. BASSO, N.
Lajmanovich, R. ....1998.... La Plata

SERIES 3

001. BONPARTE, J. F.
Powell, J. E. ....1986.... Tucuman
002. PISANÓ, A.
Ieandro, M. ....1995.... B Aires
Miranda, A. ....19??.... B Aires
Montero, R. ....19??.... B Aires
Paz, D. ....1987.... B Aires
003. REIG, O.
Baez, A. M. ....19??.... B Aires
Marcicano, C. ....1994.... B Aires
Peri, S. ....1994.... La Plata

AUSTRALIA (AU)

See: United Kingdom 9.2 and **United States** 1.105, 1.161, 1.185, 1.206, 1.213, 3.9, 33.3, 80.2

SERIES 1

001. WARING, H. W.
Main, A. R. ....1956.... W Aust 2
002. MAIN, A. R.
Blackwell, J. M. ....1974.... W Aust
Bradshaw. S. D. ....1965.... W Aust 3
Bull, C. M. ....1973.... W Aust 4
Burbidge, A. A. ....1968.... W Aust
Lee, A. K. ....1965.... W Aust 5
Littlejohn, M. J. ....1958.... W Aust 6
Storr, G. M. ....1960.... W Aust
003. BRADSHAW, S. D.
Baverstock, P. R. ....19??.... W Aust
004. BULL, C. M.
Andrews, R. H. ....1981.... Flinders
Belan, I. ....1989.... Flinders
Chilton, N. B. ....1989.... Flinders
Dubas, G. ....1987.... Flinders
Odendaal, F. ....1981.... Flinders
Petney, T. N. ....1981.... Flinders
Williamson, I. ....1988.... Flinders
Yeatman, E. ....1988.... Flinders
005. LEE, A. K.
Straughn, I. R. ....1966.... Queensland
Wilson, K. J. ....1971.... Monash
006. LITTLEJOHN, M. J.
Gartside, D. F. ....1973.... Melbourne
Humphries, R, B. ....1979.... Aust Nat
Loftus-Hills, J. J. ....1971.... Melbourne
MacNally, R. C. ....1982.... Melbourne
Martin, A. A. ....1968.... Melbourne
Robertson, J. G. M. ....1982.... Aust Nat
Sherwin, W, B. ....1984.... Melbourne
Watson, G. F. ....1974.... Melbourne
Woodruff, D. S. ....1972.... Melbourne

SERIES 2

001. CAUGHLEY, G. J.
Magnusson, W. E. ....1979.... Sydney 2
002. MAGNUSSON, W. E.
Azevedo-Ramos, C. ....1995.... E Campinas (BR)
Bayliss, P. ....1996.... Queensland
Campos, Z. M. ....2002.... Minas Gerais (BR)
Galatti, U. ....1996.... Pesquisas (BR)
Hero, M.-J. ....1991.... Griffith
Lima, A. P. ....1996.... Pesquisas (BR)...... 3
Silveira, R. ....2001.... Pesquisas (BR)
003. LIMA, A. P.
Menin, M. ....2005.... Pesquisas (BR)

SERIES 3

001. DAVIDSON, J.
Andrewartha, H. G. ....19??.... Adelaide............... 2

002. ANDREWARTHA, H. G.
Green, B. ....1969.... Adelaide
King. R. D. ....1973.... Adelaide

SERIES 4

001. ?
Edmonds, S. J. ....1958.... Adelaide................ 2
002. EDMONDS, S. J.
Bennett, L. J. ....1971.... Adelaide

SERIES 5

001. THULBORN, T.
Georges, A. ....1982.... Queensland 2
002. GEORGES, A.
Kennett, R. ....1994.... Queesland

SERIES 6

001. BARWICK, R. E.
Bustard, H. R. ....1966.... Aust Nat
Carter, D. ....1992.... Aust Nat
002. BLAIR, D.
Barton, D. P. ....1995.... James Cook
003. BURNSTOCK, G.
Bell, C. ....1965.... Melbourne
004. DAWBIN, W. H.
Clarke, C. J. ....19??.... Sydney
Hill, L. G. ....19??.... Sydney
005. HAYMAN, D. L.
King, M. ....19??.... Adelaide
006. KIKKAWA, J. T.
Ingram, G. J. ....1986.... Queensland
007. MORITZ, C.
FitzSimmons, N. N. ....1997.... Queensland
Rosenblum, E. B. ....2005.... CA Berkeley (US)
008. SHAW, D.
Sarre, S. ....1994.... Queensland
009. SMYTH, M. E.
Henzell, R. P. ....1971.... Adelaide
Sharrad, R. D. ....1979.... Adelaide
010. TYLER, M. J.
Burton, T. C. ....1983.... Adelaide
Davies, M. ....1987.... Adelaide
Horton, P. ....1986.... Adelaide

AUSTRIA (AS)

SERIES 1

001. REISINGER, E.
Fachbach, G. ....1965.... Graz 2
002. FACHBACH, G.
Faber, H. ....1991.... Graz

SERIES 2

001. VERSLUYS, J.
Haas, G. ....1928.... Vienna.................. 2
002. HAAS, G.
Kochva, E. ....19??.... Hebrew (IS)
Nevo, E. ....19??.... Hebrew (IS)
Werner, Y. L. ....19??.... Hebrew (IS)

AZERBAIJAN (AZ)

See: Russia 1.5, 1.6, 7.2

BELARUS (BL)

See: **Russia** 1.15

BELGIUM (BE)

SERIES 1

001. BRIEN, P.
Laurent, R. F. ....1940.... Libre 2
002. LAURENT, R. F.
Abdala, V. ....19??.... Tucuman (AR)
Fabrezi, M. ....1993.... Tucuman (AR)
Lavilla, E. O. ....1983.... Tucuman (AR)...... 3
Scrocchi Manfrini, G. J. ...1987.... Tucuman (AR)...... 4
003. LAVILLA, E. O.
Lobo, F. J. ....1994.... Tucuman (AR)
Manzano, A. S. ....199?.... Tucuman (AR)
Vera Candioti, M. F. ....2006.... Tucuman (AR)
004. SCROCCHI MANFRINI, G. J.
Avila, L. J. ....1996.... Rio Cuarto (AR)
Giraudo, A. ....199?.... Cordoba (AR)
Moro, S. ....199?.... Tucuman (AR)

BRAZIL (BR)

See: **Australia** 2.2–2.3 and **United States** 1.218, 2.12, 2.20-2.22, 25.3

SERIES 1

001. FROEHLICH, C. G.
Jim, J. ....1981.... São Paulo 2
Sazima, I. ....1980.... São Paulo 3
002. JIM, J.
Cais, A. ....1992.... E Paulista
Del-Grande, M. L. ....2002.... E Paulista
Martins, I. A. ....2001.... E Paulista
Rossa-Feres, D. C. ....1997.... E Paulista
Stevaux, M, N. ....2002.... E Paulista
003. SAZIMA, I.
Cardoso, A. J. ....1986.... E Campinas........... 4
Di Bernardo, M. ....1999.... E Paulista 7
Marques, O. A. V. ....1998.... São Paulo 5
Martins, M. ....1994.... E Campinas........... 6
Martins, M. B. ....1998.... Católica RG Sul
004. CARDOSO, A. J
Haddad, C. F. B. ....1991.... E Campinas........... 8
Arzabe, C. ....1998.... Fed Paraná
005. MARQUES, O. A. V.
Pizzatto, L. ....2006.... E Campinas
006. MARTINS, M.
Brasileiro, C. A. ....2004.... São Paulo
Cechin, S. Z. ....1999.... Católica RG Sul
Hartmann, P. A. ....2005.... E Paulista
Jerozoljnski, A. ....2005.... São Paulo
Nogueira, C. C. ....2006.... São Paulo

Oliveira, M. E. ....2004.... São Paulo
Sawaya, R. J. ....2004.... São Paulo
007. DI BERNARDO, M.
Costa, M. C. S. ....2003.... Católica RG Sul
Oliveira, R. B. ....2005.... Católica RG Sul
008. HADDAD, C. F. B.
Baldissera, F. A., Jr. ....2001.... E Paulista
Bastos, R. P. ....1996.... E Paulista 10
Castanho, L. M. ....2000.... E Paulista
D'Heursel, A. ....2004.... E Paulista
Garcia, P. C. A. ....2003.... E Paulista
Hartmann, M. T. ....2004.... E Paulista
Pombal, J. P., Jr. ....1995.... E Paulista 9
Prado, C. P. A. ....2004.... E Paulista
Souza, M. B. ....2003.... E Paulista
009. POMBAL, J. P., JR.
Abrunhosa, P. A. ....2005.... Fed Rio de Janeiro
Wogel, H. ....2005.... Fed Rio de Janeiro
010, BASTOS, R. P.
Guimaraes, L. D. ....2005.... Fed Goiás...............
Telles, M. P. C. ....2005.... Fed Goiás

SERIES 2

001. MENDES, E. G.
Abe, A. S. ....1977.... São Paulo 2
002. ABE, A. S.
Andrade, D. V. ....1998.... E Paulista
Araujo, A. B. ....1992.... E Campinas........... 3
Bernarde, P. S. ....2004.... E Paulista
Cruz Neto, A. P. ....1998.... E Paulista
Monteiro, L. R. ....1999.... E Paulista
Rabello, G. H. ....2002.... E Campinas
Rocha, C. F. D. ....1992.... E Campinas........... 4
Souza, F. L. ....1999.... E Paulista
Van Sluys, M. ....1995.... E Campinas........... 5
Vianna, V. O. ....1999.... E Paulista
003. ARAUJO, A. B.
Brandão, R. A. ....2002.... Brasília
Faria, R. G. ....2005.... Brasília
004. ROCHA, C. F. D.
Barros, M. C. ....2002.... E Rio de Janeiro
Dias, E. J. R. ....2006.... E Rio de Janeiro
Hatano, F. H. ....2004.... E Rio de Janeiro
Kieffer, M. C. ....2003.... E Campinas
Vrcibradic, D. ....2001.... E Campinas
005. VAN SLUYS, M.
Galdino, C. A. B. ....2004.... E Rio de Janeiro

SERIES 3

001. FERNANDES, G. W.
Eterovick, P. C. ....2002.... Fed Minas Gerais
002. GOMES, N. M. B.
Molina, F. B. ....1995.... São Paulo
Pexioto, O. L. ....1987.... São Paulo
003. LEMA, T.
Cappellari, L. H. ....2005.... Católica RG Sul
D'Agostini, F. M. ....2005.... Católica RG Sul
Ferreira, V. L. ....2000.... Católica RG Sul
Outeiral, A. B. ....2006.... Católica RG Sul
Prudente, A. L. C. ....1998.... Fed Paraná
Strüssmann, C. ....2003.... Católica RG Sul
Yuki, R. N. ....1998.... Fed Paraná
Zanella, N. ....2004.... Católica RG Sul
004. REIS, S. F.
Alves, A. C. R. ....2004.... E Paulista
005. VASCONCELOS NETO, J.
Giaretta, A. A. ....1999.... E Campinas

CANADA (CA)

See: United States 1.87,1.154, 2.9, 2.24, 2.26, 69.2, 76.2

SERIES 1

001. LINDSEY, C. C.
Stewart, K. W. ....1966.... Brit Col 2
002. STEWART, K. W.
Cook, F. R. ....1978.... Manitoba
Gregory, P. T. ....1974.... Manitoba 3
003. GREGORY, P. T.
Charland, M. B. ....1991.... Victoria
Davis, T. M. ....1996.... Victoria
Lawson, P. A. ....1991.... Victoria
Ovaska, K. ....1987.... Victoria
Rutherford, P. L. ....2002.... Victoria

SERIES 2

001. ANDERSON, R. C.
Baker, M. R. ....1978.... Guelph
002. ATKINSON, B. G.
Dhanarajan, Z. C. ....1979.... W Ontario
Merrifield, P. A. ....1979.... W Ontario
Saleem, M. ....1988.... W Ontario
003. BOAG, P. T.
Campbell, B. N. ....1997.... Queens
Galbraith, D. A. ....1991.... Queens
004. BOUTIN, S.
Larsen, K. ....1993.... Alberta
005. BROOKS, R.
Obbard, M. ....1979.... Guelph
006. EFFORD, I. E.
Calef, G. W. ....1971.... Brit Col
007. FALLS, J. B.
Gibson, A. R. ....197?.... Toronto
008. GIBBS, H. L.
Prosser, M. R. ....1999.... McMaster
009. HOAR, W. S.
Neish, I. C. ....1970.... Brit Col
010. JONES, D. R.
Langille, B. L. ....1975.... Brit Col
Milsom, W. K. ....1979.... Brit Col
West, N. H. ....1974.... Brit Col
011. KASINSKY, H. E.
Byrd, E. D., Jr. ....1973.... Brit Col
012. LICHT, L. E.
Grant, K. P. ....1997.... York
013. LIVERSAGE, R. A.
Horbay, G. L. ....1986.... Toronto

014. LOUGHEED, S.
Koscinski, D. ....2003.... Queens
015. MCKEOWN, B. A.
Eagleson, G. W. ....1978.... Simon Fraser
016. MCMILLEN, D. B.
Peek, W. D. ....1976.... W Ontario
017. MROSOVSKY, N.
Godfrey, M. H. ....1997.... Toronto
018. RANDALL, D. J.
Toews, D. P. ....1969.... Brit Col
019. REISZ, R. R.
deBarga, M. ....2001.... Toronto
Laurin, M. ....1994.... Toronto
Modesto, S. P. ....1996.... Toronto
020. SECOY, D. M.
Vincent, T. K. ....1976.... Regina
021. STEVENS, E. D.
Renaud, J. M. ....1982.... Guelph
022. STOREY, K. B.
Churchill, T. A. ....1993.... Carleton
de Cross, J. N. A. ....2003.... Carleton
Grundy, J. E. ....1996.... Carleton
McNally, J. D. ....2003.... Carleton
Willmore, W. G. ....1997.... Carleton
Wu, S.-B. ....1999.... Carleton
023. TITMAN, R.
Walde, A. D. ....1998.... McGill
024. WEATHERHEAD, P. J.
Brown, G. P. ....1997.... Carleton
Prior, K. A. ....1997.... Carleton

DENMARK (DE)

SERIES 1

001. BOHR, C.
Krogh, A. ....19??.... ? 2
002. KROGH, A.
Jörgensen, C. B. ....19??.... A Krogh 3
003. JÖRGENSEN, C. B. ....
Budtz, P. ....19??.... A Krogh
Larsen, E. H. ....19??.... A Krogh
Larsen, L. O. ....19??.... A Krogh
Rosenkilde, P. ....19??.... A Krogh
Spies, I. ....19??.... A Krogh

SERIES 2

001. STEENBERG, C. M.
Bræstrup, F. W. ....19??.... Copenhagen 2
Volsøe, H. ....1944.... Copenhagen 3
002. BRÆSTRUP, F. W.
Rasmussen, J. B. ....1977.... Copenhagen
003. VOLSØE, H.
Schiøtz, A. ....1967.... Copenhagen

FINLAND (FI)

SERIES 1

001. EKMAN, G.
Toivonen, S. ....1939.... Helsinki

FRANCE (FR)

See: **United States** 1.161

SERIES 1

001. CUVIER, G.
Milne-Edwards, H. ....18??.... ? 2
002. MILNE-EDWARDS, H.
Vaillant, L. ....1865.... Paris

SERIES 2

001. LAMOTTE, M.
Amiet, J.-L. ....19??.... Paris
Barbault, R. ....19??.... Paris
Busse, K. ....19??.... Paris
Dubois, A. ....1976.... Paris 2
Dzieduszycka, S. ....19??.... Paris
Gipouloux, J.-D. ....19??.... Paris
Xavier, F. ....19??.... Paris
Züber-Vogeli, M. ....19??.... Paris
002. DUBOIS, A.
Delorme, M. ....2004.... MNHN Paris
Grosjean, S. ....2004.... MNHN Paris

SERIES 3

001. DE RICQLES, A.
Castanet, J. ....1982.... Paris 2
002. CASTANET, J.
Caetano, M. E. ....1988.... Lisboa (PT)

SERIES 4

001. BLANC, C. P.
Ineich, I. ....1987.... Montpellier
Squalli, H. ....1991.... Montpellier

GEORGIA (GA)

See: **Russia** 9.1, 11.12, 11.14, 11.21, 11.40

GERMANY (GE)

See: **United States** 1.1

SERIES 1

001. SPEMANN, H.
Barth, L. G. ....19??.... Freiburg 2
002. BARTH, L. G.
Corner, M. ....1963.... Columbia (US)
Moore, J. A. ....1940.... Columbia (US)...... 3
Sze, L. C. ....1952.... Columbia (US)
Zwilling, E. ....1941.... Columbia (US)
003. MOORE, J. A.
Ruibal, R. ....1956.... Columbia (US)...... 4
Salthe, S. N. ....1963.... Columbia (US)...... 5
Volpe, E. P. ....1953.... Columbia (US)...... 6
004. RUIBAL, R.
Brown, H. A. ....1966.... CA Riverside (US)
Dimmitt, M. A. ....1975.... CA Riverside (US)
Evans, K. J. ....1964.... CA Riverside (US)

Hillyard, S. D. ....1974 ....CA Riverside (US)
Philibosian, F. R. ....1972 ....CA Riverside (US)
Pomeroy, L. V. ....1981 ....CA Riverside (US)
Rose, B. R. ....1981 ....CA Riverside (US)
Simon, C. A. ....1973 ....CA Riverside (US) .. 7
Weintraub, J. D. ....1968 ....CA Riverside (US)

005. SALTHE, S. N.
Kaplan, R. H. ....1978 ....CUNY (US)
Levy, P. L. ....1972 ....CUNY (US)

006. VOLPE, E. P.
Curtis, S. K. ....1968 ....Tulane (US)
Dasgupta, S. ....1962 ....Tulane (US)
DeGruy, C. V. ....1974 ....Tulane (US)
Gebhardt, B. M. ....1967 ....Tulane (US)
Rosenbaum, P. A. ....1980 ....Tulane (US)
Roux, K. H. ....1974 ....Tulane (US)
Skinner, J. M. ....1966 ....Tulane (US)
Turpen, J. B. ....1973 ....Tulane (US)............ 8
Zettergren, L. D. ....1972 ....Tulane (US)

007. SIMON, C. A.
Bissinger, B. E. ....1983 ....CUNY (US)

008. TURPEN, J. B.
Smith, P. B. ....1984 ....PA St (US)

SERIES 2

001. ?
Herre, W. ....1930 ....Halle........................ 2

002. HERRE, W.
Böhme, W. ....1971 ....Kiel......................... 3

003. BÖHME, W.
Auliya, M. ....2003 ....Bonn
Becker, H.-O. ....1993 ....Bonn
Blosat, B. ....1998 ....Bonn
Brandstätter, F. ....1995 ....Saarbruecken
Dirksen, L. ....2000 ....Bonn
Etti, S. ....2004 ....Bonn
Glaw, F. ....1999 ....Bonn
Handwerk, J. ....1998 ....Bonn
Heimes, P. ....1994 ....Bonn
Hinkel, H. ....1993 ....Mainz
Jogler, U. ....19??.....Bonn
Kneitz, S. ....1998 ....Bonn
Koehler, J. ....2000 ....Bonn
Kölpin, T. ....2005 ....Bonn
Lang, M. ....1989 ....Bonn
Lenz, S. ....1995 ....Bonn
Lötters, S. ....2000 ....Bonn
Mausfeld, P. ....2003 ....Bonn
Osenegg-Leconte, K. ......1995 ....Bonn
Schmidt-Loske, K. ....2004 ....Bonn
Schmitz, A. ....2003 ....Bonn
Schunke, A. ....2005 ....Bonn
Vences, M. ....2000 ....Bonn........................ 4
Ziegler, T. ....2000 ....Bonn

004. VENCES, M.
Vieites, R. D. ....2004 ....Vigo (SP)

SERIES 3

001. BAUR, E.
Kosswig, C. ....1927 ....Hamburg 2

002. KOSSWIG, C.
BAŞOĞLU, M. ....1942....Istanbul (TU) 3

003. BAŞOĞLU, M.
Atatür, M. K.......................19??.... Aegean (TU)
Baran, I. ....1969.... Aegean (TU)
Özeti, N. ....1963.... Aegean (TU)

SERIES 4

001. HAECKER, V.
Eisentraut, M. ....1925.... Halle 2

002. EISENTRAUT, M.
Wermuth, H. ....1950.... Humboldt............. 3

003. WERMUTH, H.
Jelden, D. ....1984.... Heidelberg
Kilias, R. ....1958.... Humboldt
Peters, G. ....1960.... Humboldt

SERIES 5

001. KRONE, W.
Schmid, M. ....1978.... Ulm 2

002. SCHMID, M.
de Almeida, C. G. ....19??.... ?
Feichtinger, W. ....19??.... ?

SERIES 6

001. MÜLLER, J.
Haeckel, E. ....18??.... Berlin 2
Peters, W. ....1838.... Berlin

002. HAECKEL, E.
Goeldi, E. A. ....1883.... Jena
von Bedriaga, J. ....1875.... Jena

SERIES 7

001. BALTZER, F.
Stauffer, E. ....1945.... Hamberg

002. BORN, ?
Gebhardt, W. ....1894.... Breslau

003. CHUN, C.
Mertens, R. ....1915.... Leipzig

004. ENGELS, W.
Kwet, A. ....2000.... Erhard Karls

005. KOEPCKE, H.-W.
Schlüter, A. ....1984.... Hamburg

006. KRAMER, G.
Medem, F. ....1942.... Tübingen

007. LINSENMAIR, K. E.
Rödel, M.-O. ....1998.... Würzburg

008. MAIER, W.
Haas, A. ....1994.... Tübingen

009. MÜLLER, P.
Monzel, M. ....2005.... Trier

010. NEUMANN, D.
Sinsch, U. ....1983.... Cologne

011. REINWARDT, C.
Schlegel, H. ....1832.... Jena

012. RENSCH, R.
Fritz, U. ....1996.... Hohenheim
Rahmann, H. ....19??.... Wilhelms-Münster

013. SCHNEIDER, H.
Akef, M. S. A. ....1984.... Bonn
Alfs, B. ....1977.... Bonn
Brzoska, J. ....1979.... Bonn
Heinzmann, U. ....1969.... Bonn
Hertwig, I. ....1985.... Bonn
Knorr, A. ....1975.... Bonn
Lörcher, K. ....1968.... Bonn
Manz, R. ....1974.... Bonn
Mohneke, R. ....1981.... Bonn
Müller, B. ....1983.... Bonn
Obert, H.-J. ....1972.... Bonn
Radwan, N. M. M. ....1987.... Bonn
Rustama, A. ....1977.... Bonn
Rüth, E. ....1973.... Bonn
Schmid, E. ....1976.... Bonn
Schmidtke-Führer, U. ....1986.... Bonn
Schneider-Lowitz, B. ....1982.... Bonn
Semmler, M. ....1977.... Bonn
Wahl, M. ....1968.... Bonn
Walkowiak, W. ....1979.... Bonn
Weber, E. ....1973.... Bonn
014. SCHÖFFL, F.
Kienie, M. S. ....2005.... Eberhard Karl
015. SEMPER, C.
Sarasin, F. ....1883.... Würzburg
016. STREIT, B.
Köhler, G. ....1995.... JW Goethe
Lehr, E. ....2001.... JW Goethe
017. TAUTZ, D.
Steinfartz, S. ....2003.... Cologne
Thomas, M. ....2006.... Cologne

HUNGARY (HU)

SERIES 1

001. MÉHELY, L.
Bolkay, S. J. ....1909.... Budapest
von Fejérváry, G. J. ....1917.... Budapest

INDIA (IN)

See: United States 1.173

SERIES 1

001. NADAKARNI, V. B.
Goudar, B. Y. M ....1978.... Karnatak 3
Saidapur, S. K. ....1974.... Karnatak 2
002. SAIDAPUR, S. K.
Dheeraj, K. V. ....2007.... Karnatak
Gramapurohit, N. P. ....2001.... Karnatak
Hiragond, N. C. ....2002.... Karnatak
Hoque, B. ....1995.... Karnatak
Hussain, M. ....1983.... Karnatak
Kanamadi, R. D. ....1982.... Karnatak 4
Kupwade, V. A. ....1989.... Karnatak
Kurian, T. ....1984.... Karnatak
Pancharatna, M. ....1984.... Karnatak 5
Pramoda, S. ....1984.... Karnatak
Prasadmurthy, Y. S. ....1988.... Karnatak
003. GOUDAR, B. Y. M.
Dodamani, L. S. ....1990.... Karnatak
Gaitonde, S. G. ....1982.... Karnatak
004. KANAMADI, R. D.
Hiramath, C. R. ....1992.... Karnatak
Jirankali, C. S. ....1992.... Karnatak
Kadadevaru, G. C. ....2002.... Karnatak
005. PANCHARATNA, K.
Chandran, P. S. ....2000.... Karnatak
Kulkarni, J. T. ....1997.... Karnatak
Kumbar, S. M. ....2002.... Karnatak

SERIES 2

001. SHANBHAG, A.
Prasad, B. S. K. ....1992.... Karnatak
Radder, R. S. ....2001.... Karnatak

ISRAEL (IS)

See: Austria 2.2

ITALY (IT)

SERIES 1

001. BECCARI, N.
Cei, J. M. ....1930.... Pisa 2
002. CEI, J. M.
Barbieri, F. D. ....1954.... Tucuman (AR) 3
Bertini, F. ....1961.... Tucuman (AR)
Cohen, R. ....1963.... Cuyo (AR)
Crespo, E. ....1980.... Lisboa (PT) 4
Rengel, D. ....1953.... Tucuman (AR)
003. BARBIERI, F. D.
Cabada, M. O. ....1973.... LaPlata (AR)
Fernandez, M. R. ....1967.... Tucuman (AR)
Legname, A. H. ....1964.... Tucuman (AR)
Manes, M. E. ....1976.... Cordoba (AR)
Miceli, D. C. ....1979.... Tucuman (AR)
Raisman, J. S. ....1975.... Tucuman (AR)
Salomon, H. L. ....1968.... Tucuman (AR)
Sanchez, S. O. ....1983.... Tucuman (AR)
004. CRESPO, E. G.
Vincente, L. A. ....1988.... Lisboa (PT)

SERIES 2

001. LESSONA, M.
Peracca, M. G. ....1887.... Turin

JAPAN (JA)

SERIES 1

001. HIRAIWA, Y.
Kawamura, T. ....1941.... Hiroshima 2
Sato, I. ....1941.... Hiroshima
002. KAWAMURA, T.
Nishioka, M. ....1963.... Hiroshima 3
Kuramoto, M. ....1965.... Hiroshima
003. NISHIOKA, M.
Kashiwagi, A. ....1989.... Hiroshima
Miura, I. ....1995.... Hiroshima

Ohta, S. ....1988.... Hiroshima
Okumoto, H. ....1983.... Hiroshima
Ryuzaki, M. ....1980.... Hiroshima
Sumida, M. ....1981.... Hiroshima
Ueda, H. ....1980.... Hiroshima

SERIES 2

001. TASUMI, M.
Hikida, T. ....1993.... Kyoto 2
Matsui, M. ....1984.... Kyoto 3
002. HIKIDA, T.
Yasukawa, Y. ....1997.... Kyoto
003. MATSUI, M.
Kamezaki, N. ....1997.... Kyoto
Tanaka-Ueno, T. ....1998.... Kyoto

SERIES 3

001. KOMAI, T.
Maki, M. ....1932.... Kyoto
Mori, T. ....1936.... Kyoto
Nakamura, K. ....1935.... Kyoto 2
002. NAKAMURA, K.
Ueno, S. ....1960.... Kyoto
Fukada, H. ....1961.... Kyoto

SERIES 4

001. MOTOMURA, I.
Iwasawa, H. ....1961.... Tohoku 2
002. IWASAWA, H.
Kobayashi, T. ....1989.... Niigata
Sato, T. ....1993.... Niigata

SERIES 5

001. SASAKI, M.
Kohno, S. ....1977.... Hokkaido 2
002. KOHNO, S.
Kuro-o, M. ....1990.... Tohoku

SERIES 6

001. AOKI, J.
Ihara, S. ....2001.... Yokohama Nat
002. AOTO, T.
Uchiyama, M. ....1982.... Hokkaido
003. HIDAKA, T.
Mori, A. ....1993.... Kyoto
004. MAKINO, S.
Seto, T. ....1965.... Hokkaido
005. MIYASHITA, K.
Hasegawa, M. ....1991.... Tokyo Metro
Kusano, T. ....1981.... Tokyo Metro
Fukuyama, K. ....1992.... Tokyo Metro
Yabe, T. ....1993.... Tokyo Metro

KAZAKHSTAN (KA)

See: **Russia** 3.1, 11.11, 11.31

KYRGYZSTAN (KY)

See: **Russia** 8.2

LATVIA (LA)

See: **Russia** 1.6, 11.39

MALAYSIA (MA)

See: United States 1.134

MEXICO (ME)

See: **United States** 1.158, 4.8, 4.11, 8.4

SERIES 1

001. FERRUSQUIA, I.
Casas-Andreu, G. ....1982.... UNAM 2
002. CASAS-ANDREU, G.
Gadseden-Esparza, H. ....1987.... UNAM
Mendez-de la Cruz, F. R. ....1989.... UNAM 3
003. MENDEZ-DE LA CRUZ, F. R.
Hernández-Gallegos, O. .....2004.... UNAM
Manriquez-Moran, N. L. ....2002.... UNAM
Rodriguez Romero, F. D. J. ..2004.... UNAM

SERIES 2

001. EZCURRA, E.
Flores Villela, O. ....1991.... UNAM 2
002. FLORES VILLELA, O.
Goyenechea-Mayer, I. ....2000.... UNAM

MOLDAVIA (MO)

See: **Russia** 1.8

NETHERLANDS (NE)

SERIES 1

001. ?
Tinbergen, N. ....1932.... Leiden 2
002. TINBERGEN, N.
Cullen, J. M. ....1956.... Oxford (UK) 3
003. CULLEN, J. M.
Arnold, E. N. ....19??.... Oxford (UK)
Halliday, T. R. ....1972.... Oxford (UK) 4
004. HALLIDAY, T. R.
Hosie, C. ....1992.... Open (UK)
Raxworthy, C. T. ....1989.... Open (UK) 5
Verrell, P. A. ....1987.... Open (UK) 6
005. RAXWORTHY, C. T.
Le, M. D. ....2006.... Columbia
006. VERRELL, P. A.
Fronzuto, J. A. ....2000.... WA St (US)
Mabry, M. L. ....2004.... WA St (US)

SERIES 2

001. OOMEN, H. C. J.
Baars, J. J. H. ....19??.... Cath Nijmegen
Backx, R. J. C. M. M. ....19??.... Cath Nijmegen
Berkes, P. A. ....19??.... Cath Nijmegen
Cools, P. J. M. ....19??.... Cath Nijmegen

Engelen, E. W. I. .... 19?? .... Cath Nijmegen
Fournier, M. .... 19?? .... Cath Nijmegen
Gardeniere, J. M. .... 19?? .... Cath Nijmegen
Hendrikx, W. M. L. .... 19?? .... Cath Nijmegen
Hoedemaekers, H. C. M. .. 19?? .... Cath Nijmegen
Hommels, T. J. M. .... 19?? .... Cath Nijmegen
Kalkhoven, J. T. R. .... 19?? .... Cath Nijmegen
Leeman, J. F. .... 19?? .... Cath Nijmegen
Plooy, F. X. .... 19?? .... Cath Nijmegen
Sax, H. .... 19?? .... Cath Nijmegen
Strijbosch, H. .... 19?? .... Cath Nijmegen 2
van Biejeren, J. .... 19?? .... Cath Nijmegen
van Gelder, J. J. .... 19?? .... Cath Nijmegen 3
van Venrooy, T. M. .... 19?? .... Cath Nijmegen
Wijffels, C. C. B. M. .... 19?? .... Cath Nijmegen
Wijnands, H. E. J. .... 1979 .... Cath Nijmegen 4
Willems, J. F. W. .... 19?? .... Cath Nijmegen

002. STRIJBOSCH, H.
Allard, J. L. M. .... 19?? .... Cath Nijmegen
Berends, J. E. .... 19?? .... Cath Nijmegen
Beuningen, H. .... 19?? .... Cath Nijmegen
Beyloos, B. .... 19?? .... Cath Nijmegen
Bogaerts, S. .... 19?? .... Cath Nijmegen
Bonnemayer, J. J. A. M. ... 19?? .... Cath Nijmegen
Broen, A. J. J. .... 19?? .... Cath Nijmegen
Broers, J. L. V. .... 19?? .... Cath Nijmegen
Christiaans, M. M. C. .... 19?? .... Cath Nijmegen
Clerx, P. M. J. .... 19?? .... Cath Nijmegen
Couwenberg, J. G. M. .... 19?? .... Cath Nijmegen
Creemers, R. .... 19?? .... Cath Nijmegen
Dietvorst, P. J. M. .... 19?? .... Cath Nijmegen
Frigge, P. A. J. .... 19?? .... Cath Nijmegen
Gremmen, W. H. E. .... 19?? .... Cath Nijmegen
Hagelen, E. S. M. .... 19?? .... Cath Nijmegen
Hanssen, T. J. G. M. .... 19?? .... Cath Nijmegen
Heereveld, P. J. T. L. .... 19?? .... Cath Nijmegen
Helmer, W. .... 19?? .... Cath Nijmegen
Huisman, M. J. H. .... 19?? .... Cath Nijmegen
Klompen, J. S. H. .... 19?? .... Cath Nijmegen
Kobusen, V. W. M. .... 19?? .... Cath Nijmegen
Lamberts, J. .... 19?? .... Cath Nijmegen
Lenders, A. J. W. .... 19?? .... Cath Nijmegen
Lenders, H. J. R. .... 19?? .... Cath Nijmegen
Marijnissen, J. W. W. .... 19?? .... Cath Nijmegen
Martens, J. G. W. .... 19?? .... Cath Nijmegen
Meeuwsissen, P. C. .... 19?? .... Cath Nijmegen
Middelburg, J. J. M. .... 19?? .... Cath Nijmegen
Mulder, J. .... 19?? .... Cath Nijmegen
Nuyten, K. .... 19?? .... Cath Nijmegen
Osstrik, P. J. B. .... 19?? .... Cath Nijmegen
Peeters, H. T. M. .... 19?? .... Cath Nijmegen
Peters, A. .... 19?? .... Cath Nijmegen
Schellekens, A. G. A. .... 19?? .... Cath Nijmegen
Scholte, P. .... 19?? .... Cath Nijmegen
Smeets, D. F. C. M. .... 19?? .... Cath Nijmegen
Spaargaren, J. J. .... 19?? .... Cath Nijmegen
Stempkens, T. P. M. .... 19?? .... Cath Nijmegen
Stommen, P. L. A. .... 19?? .... Cath Nijmegen
Stortelder, L. J. M. .... 19?? .... Cath Nijmegen
van Buggenum, H. J. M. .... 19?? .... Cath Nijmegen
van Cauwenberghe, A. P. M. . 19?? .... Cath Nijmegen
van Cauwenberghe, P. W. M. . 19?? .... Cath Nijmegen
van der Kraan, P. .... 19?? .... Cath Nijmegen
van der Reijnders, R. A. G. M. 19?? .. Cath Nijmegen
van der Rijst, H. .... 19?? .... Cath Nijmegen
van der Ven, M. .... 19?? .... Cath Nijmegen
van der Winden, J. .... 19?? .... Cath Nijmegen
van Dijk, W. P. .... 19?? .... Cath Nijmegen
van Hout, C. G. J. M. .... 19?? .... Cath Nijmegen
van Levels, P. J. .... 19?? .... Cath Nijmegen
van Nies, J. H. M. .... 19?? .... Cath Nijmegen
van Piggelen, D. C. G. .... 19?? .... Cath Nijmegen
van Rooy, P. T. C. J. .... 19?? .... Cath Nijmegen
van Schraven, G. A. M. .... 19?? .... Cath Nijmegen
van Wezel, H. A. T. M. .... 19?? .... Cath Nijmegen
Vergeer, L. H. T. .... 19?? .... Cath Nijmegen
Verhoeven, K. J. F. .... 19?? .... Cath Nijmegen
Voesenek, L. A. C. J. .... 19?? .... Cath Nijmegen

003. VAN GELDER, J. J.
Hemelaar, A. S. M. .... 1989 .... Cath Nijmegen .. 5

004. WIJNANDS, H. E. J.
Giele, F. .... 19?? .... Cath Nijmegen
Govers, K. T. A. M. .... 19?? .... Cath Nijmegen
Husslage, H. J. .... 19?? .... Cath Nijmegen
Kamminga, H. .... 19?? .... Cath Nijmegen
Kleuters, J. P. J. .... 19?? .... Cath Nijmegen
Las, J. A. C. F. .... 19?? .... Cath Nijmegen
Reuvers, M. G. H. .... 19?? .... Cath Nijmegen
Schellinx, J. A. A. .... 19?? .... Cath Nijmegen
van den Oord, J. F. .... 19?? .... Cath Nijmegen
van Gorp, P. .... 19?? .... Cath Nijmegen

005. HEMELAAR, A. S. M.
Driessen, B. .... 19?? .... Cath Nijmegen
Eeken, R. .... 19?? .... Cath Nijmegen
Hermans, J. C. H. M. .... 19?? .... Cath Nijmegen
Kniets, F. .... 19?? .... Cath Nijmegen
Krasowski, M. .... 19?? .... Cath Nijmegen
Moonen, M. P. H. M. .... 19?? .... Cath Nijmegen
Munninghoff, M. .... 19?? .... Cath Nijmegen
Peeters, M. H. T. .... 19?? .... Cath Nijmegen
Scheres, H. M. E. J. .... 19?? .... Cath Nijmegen
Sluypers, P. J. M. .... 19?? .... Cath Nijmegen
van der Staak, A. L. H. C. .... 19?? .... Cath Nijmegen
van Schijndel, F. J. J. .... 19?? .... Cath Nijmegen

SERIES 3

001. ?
Boschma, H. .... 1920 .... Amsterdam 2

002. BOSCHMA, H.
Parker, H. W. .... 1949 .... Leiden

SERIES 4

001. DE JONG, J. K.
Brongersma, L. D. .... 1934 .... Amsterdam 2

002. BRONGERSMA, L. D.
Hoogmoed, M. S. .... 1973 .... Leiden 3

003. HOODMOED, M. S.
Avila-Pires, T. C. S. .... 1996 .... Rijksuniversiteit

SERIES 5

001. ?
Dullemeijer, P. ....1956....Leiden..................2
002. DULLEMEIJER, P.
de Jongh, H. J. ....1968....Leiden

SERIES 6

001. ENGEL, H.
Hillenius, D. ....1959....Amsterdam
002. STOCK, Y. H.
Arntzen, J. W. ....1991....Amsterdam
Blommers-Schlösser, R. M. A. 1980....Amsterdam

NEW ZEALAND (NZ)

See: **United States** 1.184

PAKISTAN (PA)

SERIES 1

001. MUFTI, S. A.
Khan, M. S. ....1991....Punjab

POLAND (PO)

SERIES 1

001. HOYER, H.
Grodziński, Z. ....1922....Jagellonian...........2
002. GRODZIŃSKI, Z.
Juszczyk, W. ....1945....Jagellonian...........3
Młynarski, M. ....1956....Jagellonian...........4
Szarski, H. ....1937....Jagellonian...........5
003. JUSZCZYK, W.
Guzik, M. ....1983....Pedagog Acad
Schimscheiner, L. ....1983....Pedagog Acad
Swierad, J. ....1977....Pedagog Acad
Zakrzewski, M. ....1974....Pedagog Acad
Zamachowski, W. ....1967....Pedagog Acad
Zysk, A. ....1973....Pedagog Acad
004. MŁYNARSKI, M.
Kowalewski, L. ....1969....Jagellonian
Szyndlar, Z. ....1982....Acad Sci Krakow
005. SZARSKI, H.
Czopek, G. ....1963....Copernicus
Czopek, J. ....1954....Copernicus...........6
Płytycz, B. ....1975....Jagellonian
Poczopko, P. ....1959....Copernicus
Rafinski, J. N. ....1973....Jagellonian
Strawinski, S. ....1956....Copernicus
Szymura, J. M ....1975....Jagellonian
Witalinska, L. ....1972....Jagellonian
006. CZOPEK, J.
Andrzejewski, H. ....1969....Copernicus
Francio-Krassowska, A. ....1975....Copernicus
Przystalska, A. ....1978....Copernicus
Wilczynskja, B. ....1977....Copernicus

SERIES 2

001. NUSBAUM-HILAROWICZ, J.
Hirschler, J. ....19??....Jan Casimir..........2
002. HIRSCHLER, J.
Sembrat, K. ....1925....Jan Casimir...........3
003. SEMBRAT, K.
Kielbowna, L. ....1969....Krakow
Orska, J. ....1948....Wroclaw................4
004. ORSKA, J.
Lecyk, M. ....1961....Krakow
Pawlowska-Indyk, A. ....1975....Krakow

SERIES 3

001. SIEDLECKI, M.
Skowdon, S. ....1922....Jagellonian............2
002. SKOWDON, S.
Srebro, Z. ....1961....Medic Acad...........3
003. SREBRO, Z.
Sliwa, L. ....1982....Medic Acad
Sura, P. ....1985....Medic Acad

SERIES 4

001. KILARSKI, W.
Kordylewski, L. ....1975....Jagellonian
002. KOSCIELSKI, B.
Ogielska, M. ....1982....Krakow
003. RAABE, Z.
Dobrowolska, H. ....1969....Warszawa
004. SMRECZYNSKI, S.
Rzechak, K. ....1964....Jagellonian
005. URBANSKI, J.
Berger, L. ....1961....Poznan

PORTUGAL (PT)

See: **France** 3.2 and **Italy** 1.2, 1.4

RUSSIA AND PREVIOUSLY ASSOCIATED REPUBLICS (RR)

IPEE = Institute of Problems in Ecology and Evolution

SERIES 1

001. BOGDANOV, M.
Nikolsky, A. M. ....1887....St. Petersburg........2
002. NIKOLSKY, A. M.
Chernov, S. A. ....1926....Kharkov (UR).......3
Laister, A. F. ....19??....Kharkov (UR)
Terentjev, P. V. ....1930....Leningrad.............4
003. CHERNOV, S. A.
Alekperov, A. M. ....19??....Leningrad.............5
Darevsky, I. S. ....1957....Leningrad.............6
Emelianov, A. A. ....1944....Leningrad
004. TERENTJEV, P. V.
Ischchenko, V. G. ....1967....Leningrad.............7
Khozatsky, L. I. ....19??....St. Petersburg........8
Schwarts, S. S. ....1946....Leningrad.............9
005. ALEKPEROV, A. M.
Aliev, T. R. O. ....1974....Baku (AZ)
Velieva, Z. D. ....1975....Baku (AZ)
006. DAREVSKY, I. S.
Agasyan, A. ....1989....Leningrad

Aliyev, T. A. ....19??....Baku (AZ)
Bakradze, M. A. ....1977....Leningrad
Berdibaeva, Z. S. ....1970....Leningrad
Caune, I. A. ....1988....Latvian Mus (LA)
Chan, K. ....1984....Leningrad
Danielyan, F. D. ....1968....Erevan (AM)
Flyaks, N. L. ....1989....Leningrad
Khonyakina, Z. P. ....1964....Daghestan
Kireev, V. A. ....19??....Leningrad
Kupriyanova, L. A. ....1970....Leningrad
Lada, G. A. ....1993....Leningrad
Misjura, A. N. ....1989....Leningrad
Negmedzyanov, V. ....19??....Leningrad
Orlov, N. B. ....19??....Leningrad
Pizchelawri, V. P. ....1990....Moscow
Roitberg, E. S. ....1989....Daghestan
Saparova, I. S. ....1993....Ashgabat (TM)
Shimansky, A. M. ....1970....Leningrad
Tran, K. ....1967....Leningrad
Tsellarius, A. Y. ....1985....?
Tuniev, B. ....1988....Leningrad
Yakovlev, V. A. ....1985....Leningrad
Zaune, L. ....19??....Leningrad
Zellarius, A. ....1986....Leningrad

007. ISCHCHENKO, V. G.
Kuranova, V. N. ....1998....Sverdlovsk

008. KHOZATSKY, L. I.
Ananjeva, N. B. ....1972....Leningrad............ 10
Borkhvardt, V. G. ....1968....St. Petersburg...... 11
Borkin, L. J. ....1986....Leningrad............ 12
Juriev, K. B. ....197?....St. Petersburg
Kotlyarevskaja, V. A. ....1969....Novosibirsk
Nesov, L. A. ....1977....Leningrad............ 13
Tofan, V. E. ....1967....Moldavia (MO)

009. SCHWARTS, S. S.
Dobrinsky, L. N. ....1961....Sverdlovsk.......... 14
Gatiyatullina, E. Z. ....1975....Sverdlovsk
Kovaltchook, L. A. ....1977....Sverdlovsk
Pikulik, M. M. ....1977....Sverdlovsk.......... 15
Pyastolova, O. A. ....1970....Sverdlovsk.......... 16
Shchupak, E. L. ....1970....Sverdlovsk.............
Suzumova, L. M. ....1959....Sverdlovsk.......... 17
Toporkova, L. Y. ....1965....Sverdlovsk

010. ANANJEVA, N. B.
Danilov, I. ....1999....Leningrad

011. BORKHVARDT, V. G.
Cherepanov, G. O. ....1986....St. Petersburg
Kovalenko, E. E. ....1985....Leningrad
Malashichev, Y. B. ....1999....St. Petersburg

012. BORKIN, L. J.
Litvinchuk, S. ....1998....Leningrad

013. NESOV, L. A.
Averjanov, A. O. ....1991....St. Petersburg

014. DOBRINSKY, L. N.
Sokolova, T. M. ....1978....Sverdlovsk

015. PIKULIK, M. M.
Kosova, L. V. ....1996....Minsk (BL)

016. PYASTOLOVA, O. A.
Ivanova, N. L. ....1975....Sverdlovsk
Tarkhnishvili, D. N. ......1986.... Sverdlovsk

017. SUZUMOVA, L. M.
Ledentsov, A. V. ....1990.... Sverdlovsk.................
Vershinin, V. L. ....1983.... Sverdlovsk.................

SERIES 2

001. SEVERTZOV, A. N.
Schmalhausen, I. I. ....1916.... Moscow.................... 2

002. SCHMALHAUSEN, I. I.
Balinsky, B. I. ....19?? .... Kiev (UR)................. 3
Regel, E. D. ....1969.... Leningrad

003. BALINSKY, B. I.
Fabian, B. C. ....1969.... Witwatersrand (SA)
Passmore, N. I. ....1978.... Witwatersrand (SA).. 4

004. PASSMORE, N. I.
Backwell, P. R. Y. ....1991.... Witwatersrand (SA)
Bishop, P. ....1994.... Witwatersrand (SA)
Dyson, M. L. ....1989.... Witwatersrand (SA)
Picker, M. ....1982.... Witwatersrand (SA)
Telford, S. R. ....1983.... Witwatersrand (SA)

005. LEBEDKINA, N. S.
Severtzov, A. S. ....1968.... Moscow.................... 6

006. SEVERTZOV, A. S.
Lyapkov, S. M. ....1992.... Moscow
Salomatina, N. I. ....1977.... Moscow
Surova, G. S. ....1985....Moscow

SERIES 3

001. DOMBROVSKY, B. A.
Dilmukhamedov, M. E. 1975 ....Kazakh (KA) 2

002. DILMUKHAMEDOV, M. E.
Dujsebayeva, T. N. ......1986.... Kazan (TR)
Khromov, V. A. ....1979.... Kazan (TR)
Malkhov, D. V. ....1985.... Kazan (TR)

SERIES 4

001. MATVEYEV, B. S.
Iordansky, N. N. ....1968.... Moscow.................... 2

002. IORDANSKY, N. N.
Morozov, N. N. ....1979.... Moscow

SERIES 5

001. NAUMOV, S. P.
Kondrashyov, S. L. ......1976.... Moscow
Markooze, V. K. ....1965.... Moscow
Vassiljev, B. D. ....1969.... Moscow.................... 2

002. VASSILJEV, B. D.
Smirnov, S. V. ....1977.... Moscow

SERIES 6

001. POPOV, V. A.
Astradomov, V. I. ....1973.... Kazan (TR)
Garanin, V. I. ....1965.... Kazan (TR)............... 2
Shaldybin, S. L. ....1974.... Kazan (TR)

002. GARANIN, V. I.
Ganeev, I. G. ....1981.... IPEE
Pantchenko, I. M. ....1984.... Moscow
Pavlov, A. V. ....1998.... Kazan (TR)

SERIES 7

001. RAGOZINA, M. N.
Korneva, L. G. .... 1968.... Tashkent (UZ)
Zusman, I. N. .... 1965.... Moscow 2
002. ZUSMAN, I. N.
Iskenderov, T. M. .... 1978.... Baku (AZ)
Khmelevskaya, N. V. .... 1972.... Moscow

SERIES 8

001. ?
Szczerbak, N. N. .... 1962.... Kiev (UR)............ 2
002. SZCZERBAK, N. N.
Badmajeva, V. I. .... 1983.... Kalmyk
Dotsenko, I. L. .... 1986.... Kiev (UR)
Erjomtschenko, V. .... 1984.... Kiev (UR)
Golubev, M. L. .... 1982.... Kiev (UR)
Gontscharenko, A. .... 1980.... Uman (UR)
Jeriomtschenko, V. K. .... 1984.... Frunze (KY)
Kotenko, T. I. .... 1983.... Kiev (UR)
Pisanets, E. .... 1978.... Melitopol UR)
Sattorov, T. .... 1981.... Kiev (UR)
Shcherban, M. .... 1976.... Uzhgorod (UR)
Tarashchuk, S. .... 1987.... Kiev (UR)
Tertishinkov, M. .... 1972.... Stavropol
Trofimov, A. G. .... 1981.... Kiev (UR)
Zabroda, S. N. .... 1983.... Kiev (UR)

SERIES 9

001. TATARINOV, L. P.
Chkhikvadze, V. M. .... 1972.... Tbilissi (GA)......... 2
Ivakhnenko, M. F. .... 1973.... Moscow
Kalandadze, N. N. .... 1974.... Moscow
Shishkin, M. A. .... 1969.... Moscow
002. CHKHIKVADZE, V. M.
Zerova, G. A. .... 1993.... Kiev (UR)

SERIES 10

001. YABLOKOV, A. V.
Baranov, A. S. .... 1984.... Kiev (UR)
Smirina, E. M. .... 1976.... Moscow
Streltsov, A. B. .... 1981.... Moscow
Turutina, L. V. .... 1979.... Moscow
Veletsky, A. V. .... 1987.... Sverdlovsk
Zakharov, V. M. .... 1979.... Moscow 2
002. ZAKHAROV, V. M.
Chubinishvili, A. T. .... 1997.... Moscow

SERIES 11

001. ABDUSSALYAMOV, I. A.
Said-Aliyev, S. A. .... 1975.... Moscow
002. AGARKOV, G. B.
Solukha, B. V. .... 1975.... Moscow
003. AIRAPETYANTS, E. S.
Safarov, K. M. .... 1971.... Tadjik (TD)
004. BANNIKOV, A. G.
Belova, Z. V. .... 1965.... Moscow
Melkumyan, L. S. .... 1973.... Sverdlovsk
005. BARKAGAN, Z. S.
Mitelman, S. S. .... 1969.... Altai
006. BAZHANOV, V. S.
Kusnetsov, V. V. .... 1963 .... Kazan (TR)
007. BROMLEY, G. F.
Korotkov, Y. M. .... 1974 .... Vladivostok
008. BOGDANOV, O. P.
Lim, V. P. .... 1969 .... Tashkent (UZ)
Makeev, V. M. .... 1969 .... Tashkent (UZ)
Persoyanova, L. A. .... 1972 .... Tashkent (UZ)
Sharifov, F. G. .... 1974 .... Kuban
Utemisov, O. E. .... 1974 .... Sverdlovsk
Vashetko, E. V. .... 1974 .... Tashkent (UZ)
Zinyakova, I. P. .... 1967 .... Taskkent (UZ)
009. BULAKHOV, V. L.
Bobylev, Y. P. .... 1980 .... Dnepropetrovsk (UR)
Konstantinova, N. F. .... 1980 .... Dnepropetrovsk (UR)
010. DERYUGIN, K. M.
Tsarevskiy, S. F. .... 19??..... St Petersburg
011. DOLGUSHIN, I. A.
Iskakova, K. I. .... 1962 .... Alma-Ata (KA)
012. ERISTAVI, K. D.
Krassilnikov, E. N. .... 1964 .... Tbilissi (GA)
013. GEPTNER, V. G.
Orlova, V. F. .... 1975 .... Moscow
014. GOVARDOWSKY, V. I.
Chkheidze, N. I. .... 1988 .... Tbilisi (GA)
015. FLINDT, V. E.
Bondarenko, D. A. .... 1994 .... Moscow
Bozhansky, A. T. .... 1986 .... Moscow
016. ILJICHYOV, V. D.
Levenko, B. A. .... 1973 .... Moscow
017. IVANTER, E. V.
Cherlin, V. A. .... 1989 .... Leningrad
Khilkov, T. N. .... 1998 .... Petrozavodsk
018. KARAMYAN, A. I.
Alexandrova, T. A. .... 1976 .... Leningrad
019. KAZNELSON, Z. S.
Ziyade, A. M. .... 1974 .... Leningrad
020. KISTYAKOVSKY, A. B.
Lyalyushko, D. M. .... 1973 .... Kiev (UR)
Pashchenko, Y. .... 1968 .... Kiev (UR)
021. KOBAKHIDZE, D. N.
Muskhelishvili, T. A. .... 1967 .... Tbilisi (GA)
022. KOROCHKIN, L. I.
Birsten, V. J. .... 1987 .... Apatity
023. KOROSOV, A. V.
Chilkov, T. N. .... 1998 .... Petrozavodsk
024. KOSTENKO, V. A.
Belova, V. T. .... 1973 .... Vladivostok
025. KRIVOLUTSKY, D. A.
Leontjeva, O. A. .... 1990 .... IPEE
026. LOGVINENKO, V. M.
Pryalkina, T. I. .... 1989 .... Leningrad
027. LOZAN, M. N.
Kozar, F. V. .... 1987 .... Moscow
028. MAKAROV, P. V.
Zakharov, A. M. .... 1968 .... Leningrad
029. MAKSIMOV, A. A.
Vaetapetov, L. G. .... 1980 .... Novosibirsk
030. MAKSUDOV, I. K.
Zemlyanova, N. A. .... 1964 .... Tashkent (UZ)

031. MARKOV, G. S.
Radchenko, N. M. ....1976.... Kiev (UR)
Vakker, V. G. ....1970.... Alma-ata (KA)
032. MAZHUGA, P. M.
Pegeta, V. P. ....1969.... Kiev (UR)
033. NURGELDYEV, O. N.
Ataev, C. ....1969.... Ashkhabad (TM)
034. NYUBERG, N. D.
Orlov, O. Y. ....1966.... Moscow
035. OKULOVAN, M.
Guseva, A. Y. ....1998.... Moscow
036. OLEJNIKOV, N. S.
Lukina, G. P. ....1966.... Rostov
037. PETROV, V. S.
Lebedinsky, A. A. ....1984.... Moscow
Pisarenko, S. S. ....1987.... Moscow
Sharygin, S. A. ....1980.... Sverdlovsk
038. POPOV, V. V.
Oganessyan, R. O. ....1969.... Moscow
039. PRUSAJTE, Y.
Gajzauskene, I. I. ....1971.... Vilnius (LA)
040. ROSTOMBEKOV, V. N.
Balavadze, E. S. ....1966.... Tbilissi (GA)
041. RUSTAMOV, A. K.
Atajev, C. ....1993.... Turkmen (TU)
Ateva, A. A. ....1981.... Kiev (UR)
Khabibulov, M. R. ....1990.... Leningrad
Shammakov, S. ....1988.... Ashkhabad (TM)
042. SAKHIBOV, D. N.
Malikov, M. ....1971.... Tashkent (UZ)
043. SHAKHMARDANOV, Z. A.
Kurbanov, Z. M. ....1981.... Daghestan
044. SHAPOSHNIKOV, L. V.
Zharkova, V. K. ....1973.... Ryazan
045. SOLOMONOV, N. G.
Naumov, Y. S. ....1981.... Yakutian
046. STRAUTMAN, F. I.
Kushniruk, V. A. ....1968.... Chernovtsy (UR)
047. TEMBOTOV, A. K.
Shebzukhova, E. A. ....1969.... Naltchik
048. TOKIN, B. P.
Aziz, F. K. ....1978.... Leningrad
049. TURAKULOV, Y. K.
Davlyatov, Y. ....1967.... Tashkent (UZ)
050. VASLJEV, B. D.
Polynova, G. V. ....1987.... Moscow
051. VOROBJEVA, E. I.
Chugunova, T. Y. ....1988.... IPEE
Kuzmin, S. L. ....1987.... IPEE
Semenov, D. V. ....1987.... IPEE
052. VORONOV, A. G.
Yushkov, R. A. ....1997.... Moscow
053. VORONTSOV, N. N.
Sokolowsky, V. V. ....1975.... Vladivostok
054. ZAKHIDOV, T. Z.
Munkhbayar, K. ....1973.... Tashkent (UZ)
055. ZASSUKHIN, D. N.
Mkrtchyan, Z. A. ....1967.... Erevan (AM)

SOUTH AFRICA (SA)

See: **Russia** 2.3–2.4

SERIES 1

001. LANG, A.
de Villiers, C. G. ....1922.... Zurich (SL) 2
002. DE VILLIERS, C. G.
Badenhorst, N. C. ....19??.... Stellenbosch
Barry, T. H. ....1955.... Stellenbosch
Boonstra, L. D. ....1933.... Stellenbosch
Brink, H. E. ....1935.... Stellenbosch
Crompton, A. W. ....1953.... Stellenbosch
de Jager, E. F. J. ....1937.... Stellenbosch
de Vos, C. M. ....1937.... Stellenbosch
du Toit, C. A. ....1934.... Stellenbosch.......... 3
Hoffman, A. C. ....1939.... Stellenbosch
Jurgens, J. D. ....1970.... Stellenbosch.......... 4
Latsky, L. ....1930.... Stellenbosch
Malan, M. E. ....1946.... Stellenbosch.......... 5
Nel, P. S. ....1930.... Stellenbosch.......... 6
Skinner, J. A. ....1959.... Stellenbosch
Smit, A. L. ....1952.... Stellenbosch
Swanepoel, J. H. ....1969.... Stellenbosch
van der Merwe, N. J. ....1943.... Stellenbosch
van der Westhuizen, C. M. 1960.... Stellenbosch
van Dijk, D. E. ....1959.... Stellenbosch.......... 7
van Eeden, J. A. ....1950.... Stellenbosch
003. DU TOIT, C. A.
Ryke, P. A. J. ....1952.... Potchefstromm
004. JURGENS, J. D.
van Pletzen, A. ....1952.... Stellenbosch.......... 8
005. MALAN, M. E.
Oelofsen, B. ....19??.... ? 9
van den Heever, J. A. ....1988.... ?
Visser, J. G. ....19??.... ?
006. NEL, P. S.
Heideman, N. ....19??.... Stellenbosch
007. VAN DIJK, D. E.
Channing, A. ....1977.... Pietermaritzburg .. 10
008. VAN PLETZEN, A.
Kok, D. J. ....192?.... Orange Free St.... 11
009. OELOFSEN, B.
Baard, E. H. W. ....1990.... Stellenbosch
Rubidge, B. ....19??.... Stellenbosch
010. CHANNING, A.
Wilson, B. A. ....1994.... ?
011. KOK, D. J.
du Preez, L. ....1994.... Orange Free St
van der Linde, T. C. ....1983.... Orange Free St

SERIES 2

001. BURSELL, E.
Loveridge, J. P. ....1989.... Zimbabwe (ZI)...... 2
002. LOVERIDGE, J. P.
Damstra, K. S. J. ....1988.... Zimbabwe (ZI)
Games, I. P. ....1991.... Zimbabwe (ZI)
Hutton, J. M. ....1984.... Zimbabwe (ZI)

SERIES 3

001. HAUGHTON, S. H.
Kitching, J. A. ....19??....?

002 LOUW, G.
van Wyk, J. H. ....1992....Cape Town

003. VAN DER HORST, C. J.
Patterson, N. F. ....1939....?

004. ZWARENSTEIN, H.
Gitlin, G. ....1940....Cape Town
Gunn, J. W. C. ....192?....Cape Town
Rimer, G. E. G. ....1937....Cape Town
Schrire, V. ....1940....Cape Town
Shapiro, B. G. ....1935....Cape Town
Shapiro, H. A. ....1933....Cape Town
Zoond, A. ....19??....Cape Town

SERIES 4

001. ERASMUS, T.
Els, S. ....19??....Port Elizabeth

002. JARVIS, J. U. M.
Leonard, C. J. ....198?....Cape Town

003. POYNTON, J
Broadley, D........................1967....Natal

SPAIN (SP)

See: Germany 2.4

SERIES 1

001. SANTOS, A.
Ponz, F. ....1942....Barcelona.............. 2

002. PONZ, F.
Nadal, J. ....1965....Barcelona.............. 3

003. NADAL, J.
Llorente, G. A. ....1984....Barcelona.............. 4
Montori, A. ....1988....Barcelona

004. LLORENTE, G. A.
Carretero, M. A. ....1993....Barcelona

SERIES 2

001. BERNIS, F.
Salvador, A. ....1979....Complutense......... 2

002. SALVADOR, A.
Valentin, P. M. ....1981....Salamanca

SERIES 3

001. GARCIA DEL CID, F.
Balcells, E. ....1956....Barcelona.............. 2

002. BALCELLS, E.
Martínez Rica, J. P. ....1973....Barcelona.............. 3

003. MARTÍNEZ RICA, J. P.
Bea, A. ....1987....Barcelona
Cirer, A. M. ....1987....Barcelona
Martinez Castilla, A. ....1989....Barcelona
Sanuy, D. ....1990....Barcelona

SERIES 4

001. CASTROVIEJO, J.
de la Riva, I. ....1993....Complutense

002. GADEA, E.
Vives Balmaña, M. V. ....1982....Barcelona

SWEDEN (SD)

SERIES 1

001. LINNAEUS, C.
Forsskål, P. ....1760....Uppsala
Hasselquist, F. ....1750....Uppsala
Kalm, P. ....1745....Uppsala
Lofling, P. ....1749....Uppsala
Osbeck, P. ....1754....Uppsala
Schreber, J. C. D. ....1760....Uppsala
Thunberg, C. P. ....1770....Uppsala

SWITZERLAND (SL)

See: South Africa 1.1

SERIES 1

001. ?
Hadorn, E. ....1949....Zurich 2

002. HADORN, E.
Walder, P. ....1950....Zurich
Zeller, C. ....1956....Zurich

SERIES 2

001. FISCHBERG, M.
Loumont, C. ....19??....Geneva

002. HOTZ, H.
Beerli, P. ....1993....Zurich

003. KRAMER, E.
Regenass, U. ....19??....Basel

004. YUNG, E.
Roux, J. ....1899....Geneva

005. ZISWILER, V.
Dusej, G. ....19??....Zurich
Schätti, B. ....19??....Zurich
Zeller, C. ....1956....Zurich

TADJIKISTAN (TD)

See: **Russia** 11.3

TAIWAN (TA)

See: **United States** 1.157

TATARSTAN (TR)

See: **Russia** 3.2, 6.1–6.2, 11.6

TURKEY (TU)

See: **Germany** 3.1–3.2 and **Russia** 11.41

TURKMENISTAN (TM)

See: **Russia** 1.6, 11.33, 11.41

UNITED KINGDOM (UK)

See: **Netherlands** 1.2–1.4

SERIES 1

001. ?
    Bellairs, A. A. ....1942.... Univ College......... 2
002. BELLAIRS, A. A.
    Bryant, S. V. ....19??.... London.................. 3
    Evans, S. E. ....19??.... London.................. 4
    Maderson, P. F. ....1962.... London.................. 5
003. BRYANT, S. V.
    Carlson, M. R. J. ....2000.... CA Irvine (US)
    Hayamizu, T. F. ....1993.... CA Irvine (US)
    Iten, L. E. ....1975.... CA Irvine (US)
    Krasner, G. N. ....1981.... CA Irvine (US)
    Metscher, B. D. ....1996.... CA Irgine (US)
    Mullen, L. M. ....1995.... CA Irvine (US)
    Muneoka, K. ....1983.... CA Irvine (US)
    Torok, M. A. ....1997.... CA Irvine (US)
004. EVANS, S. E.
    Sprackland, R. G. ....1995.... London
005. MADERSON, P. F.
    Contard, P. C. ....1987.... CUNY (US)

SERIES 2

001. ELLINGER, P.
    Coulson, R. A. ....19??.... London.................. 2
002. COULSON, R. A.
    Dessauer, H. C. ....1952.... LA St (US)............ 3
    Herbert, J. ....19??.... LA St (US)
003. DESSAUER, H. C.
    Densmore, L. D., III ....1981.... LA St (US)............ 4
    DiMaggio, A. ....1961.... LA St (US)
    Jiminez, J. M. ....1963.... LA St (US)
    Rosenquist, J. W. ....1969.... Tulane (US)
    Shochat, D. ....1975.... LA St (US)
    Sutton, D. E. ....1969.... LA St (US)
004. DENSMORE, L. D., III
    Dever, J. A. ....2000.... TX Tech (US)
    Knight, R. A., Jr. ....1991.... TX Tech (US)
    Ray, D. A. ....2002.... TX Tech (US)
    Wilkinson, J. A. ....1997.... TX Tech (US)

SERIES 3

001. LACK, D.
    Ashmore, N. P. ....19??.... ? 2
002. ASHMORE, N. P.
    Rothstein, S. I. ....1970.... Yale (US) 3
003. ROTHSTEIN, S. I.
    Blaustein, A. R. .... 1976 .. CA Santa Barbara (US).. 4
004. BLAUSTEIN, A. R.
    Belden, L. K. ....2001.... OR St (US)
    Hatch, A. C. ....2003.... OR St (US)
    Hokit, D. G. ....1994.... OR St (US)
    Kiesecker, J. M. ....1997.... OR St (US)
    Lefcort, H. G. ....1993.... OR St (US)
    Olson, D. H........................1988.... OR St (US)

SERIES 4

001. DE BEER, G. R.
    Ford, P. ....1955.... London
    Fox, H. ....1951.... London.................. 2
002. FOX, H.
    Turner, S. C. ....1970.... London.................. 3
003. TURNER, S. C.
    Touey, J. ....1987.... Portsmouth

SERIES 5

001. ELTON, C.
    Murdoch, E. ....19??.... Oxford................... 2
002. MURDOCH, W.
    Sih, A. .... 1980 .. CA Santa Barbara (US) .. 3
003. SIH, A.
    Kats, L. B. ....1989.... KY (US)
    Storfer, A. T. ....1997.... KY (US)

SERIES 6

001. HUGHES, G.
    Shelton, G. ....1958.... Cambridge 2
002. SHELTON, G.
    Boutilier, R. ....1981.... East Anglia
    Burggren, W. ....1976.... East Anglia............ 3
    Jones, D. ....1965.... East Anglia
003. BURGGREN, W. A.
    Crossley, D. A., II ....1999.... N TX
    Hou, P.-C. L. ....1991.... MA
    Infantino, R. L., Jr. ....1992.... MA
    Jia, X.-X. ....1991.... MA
    Kimmel, P. B. ....1990.... MA
    Lou, P. C. L. ....1992.... MA
    Pinder, A. ....1985.... MA
    Quinn, D. ....1982.... MA
    Territo, P. ....1996.... NV-Las Vegas

SERIES 7

001. MEDEWAR, P.
    Brent, L. ....19??.... ? 2
002. BRENT, L.
    Branch, W. R. ....1971.... Southampton

SERIES 8

001. ?
    Beebee, T. J. C. ....1972.... Sussex 2
002. BEEBEE, T. J. C.
    Denton, J. S. ....1991.... Sussex

SERIES 9

001. CHITTY, D.
    Smyth, M. B. ....1964.... Oxford................... 2
002. SMYTH, M. B.
    Braysher, M. L. ....1972.... Adelaide (AU)
    Henzell, R. P. ....1972.... Adelaide (AU)
    Roberts, J. D. ....1976.... Adelaide (AU)

SERIES 10

001. ?
    Clarke, B. T. ....1988.... Nat Acad Awards .. 2

002. CLARKE, B. T.
Smith, D. T. ....1989.... ?

SERIES 11

001. UNDERWOOD, G.
Thorpe, R. S. ....19??.... London.................. 2

002. THORPE, R. S.
Brown, R. P. ....1991.... Aberdeen
Daltry, J. C. ....1995.... Aberdeen
Gardner, A. ....1985.... Aberdeen
Malhotra, A. ....1992.... Aberdeen
Wüster, W. ....1992.... Aberdeen

SERIES 12

001. WATSON, D. M.
Newth, D. ....19??.... London.................. 2
Whitear, M. ....1951.... London

002. NEWTH, D.
Blackler, A. ....1965.... London
Hamilton, L. ....1965.... London

SERIES 13

001. COE, M.
Cherry, M. I. ....1989.... Oxford
Frazier, J. G. ....19??.... Oxford

002. DAVIES, N. B.
Arak, A. ....1982.... Cambridge

003. GRUNEBERG, H.
Truslove, G. ....1951.... London

004. OWEN, R. W.
Riley, J. ....19??.... Leeds
Tinsley, R. ....19??.... Leeds......................

005. SIMNETT, J.
Oates, C. ....197?.... Newcastle

UNITED STATES OF AMERICA (US)

See: Australia 6.7; **Germany** 1.2-1.8; **Netherlands** 1.6; and **United Kingdom** 1.3, 1.5, 2.2-2.4, 3.2–3.4, 5.2–5.3

SERIES 1

001. OKEN, L./VON HUMBOLDT, A.
Agassiz, L. ....1829.... Munich (GE).......... 2
von Tschudi, J. J. ....18??.... Munich (GE)

002. AGASSIZ, L.
[Jordan, D. S]....................1872.... Cornell.................. 3
Whitman, C. O. ....1???.... Harvard.................. 4
Wilder, B. G. ....1868.... Harvard.................. 5

003. JORDAN, D. S.
Eigenmann, C. H. ....1889.... IN
Forbes, S. ....1884.... IN.................. 6
Grinnell, J. ....1913.... Stanford.................. 7
Myers, G, S. ....1933.... Stanford.................. 8

004. WHITMAN, C. O.
Adams, C. C. ....1908.... Chicago.................. 9

005. WILDER, B. G.
[Gage, S. H.]....................1877.... Cornell.................. 10
Reed, H. G. ....1899.... Cornell.................. 11
Wright, A. H. ....1908.... Cornell.................. 12

006. FORBES, S.
Eddy, S. ....1929.... IL.................. 13
Welch, P. S. ....1913.... IL.................. 14

007. GRINNELL, J.
Burt, W. H. ....1930.... CA Berkeley.......... 15
Davis, W. B. ....1938.... CA Berkeley.......... 16
Dice, L. R. ....1915.... CA Berkeley.......... 17
Fitch, H. S. ....1937.... CA Berkeley.......... 18
Hall, E. R. ....1928.... CA Berkeley.......... 19
Leopold, A. S. ....1944.... CA Berkeley.......... 20
Miller, A. H. ....1930.... CA Berkeley.......... 21
Storer, T. I. ....1925.... CA Berkeley

008. MYERS, G. S.
Alcala, A. C. ....1966.... Stanford
Anderson, S. C. ....1966.... Stanford
Banta, B. H. ....1961.... Stanford
Brown, W. C. ....1950.... Stanford
Cliff, F. S. ....1954.... Stanford
Funkhauser, A. ....1957.... Stanford
Funkhauser, J. W. ....1951.... Stanford
Leviton, A. E. ....1960.... Stanford
Maslin, T. P. ....1945.... Stanford.................. 22
Mead, G. W., Jr. ....1953.... Stanford.................. 23
Savage, J. M. ....1955.... Stanford.................. 24
Schuierer, F. W. ....1966.... Stanford
Weitzman, S. H. ....1960.... Stanford

009. ADAMS, C. C.
Ruthven, A. G. ....1906.... MI.................. 25

010. GAGE, S. H.
Kingsbury, B. F. ....1895.... Cornell.................. 26

011. REED, H. G.
McMullen, E. C. ....1936.... Cornell

012. WRIGHT, A. H.
Axtell, H. ....1947.... Cornell
Bishop, S. C. ....1925.... Cornell.................. 27
Brown, E. E. ....1940.... Cornell
Cowles, R. B. ....1929.... Cornell.................. 28
Funkhouser, W. D. ....1916.... Cornell
Gilbert, P. W. ....1940.... Cornell.................. 29
Gordon, K. L. ....1936.... Cornell.................. 30
Hamilton, W. J., Jr. ....1930.... Cornell.................. 31
Harper, F. ....1925.... Cornell
Liu, C.-C. ....1934.... Cornell
Livezey, R. L. ....1946.... Cornell
McCauley, R. H., Jr. ....1940.... Cornell
Raney, E. C. ....1939.... Cornell.................. 32
[Schmidt, K. P.] ....1917.... Cornell.................. 33
Trapido, H. ....1943.... Cornell

013. EDDY, S.
Breckenridge, W. J. ....1941.... MN.................. 34
Nelson, W. F. ....1963.... MN
Underhill, J. C. ....1957.... MN.................. 35

014. WELCH, P. S.
Chandler, D. C. ....1934.... MI.................. 36
Rogers, J. S. ....1931.... MI.................. 37

015. BURT, W. H.
Schwartz, A. ....1952.... MI

016. DAVIS, W. B.
Baldauf, R. J. ....1956.... OK St.................. 38
Degenhardt, W. G. ....1960.... TX A&M.................. 39

Advisor / Student	Year	Institution	No.
017. DICE, L. R.			
Blair, W. F.	1938	MI	40
018. FITCH, H. S.			
Burkett, R. D.	1969	KS	
Clark, D. R., Jr.	1968	KS	41
Hall, R. J.	1969	KS	
Legler, J. M.	1959	KS	42
Loomis, R. B.	1956	KS	
Malaret, L.	1985	KS	
Platt, D. R.	1966	KS	
Plummer, M. V.	1975	KS	
Regan, G. T.	1972	KS	
Seigel, R. A.	1984	KS	
Smith, A. K.	1973	KS	
Waltner, R. C.	1979	KS	
Webb, R. G.	1960	KS	
019. HALL, E. R.			
Cockrum, E. L.	1951	KS	43
Findley, J. S.	1955	KS	44
020. LEOPOLD, A. S.			
Jones, R. E.	19??	?	45
021. MILLER, A. H.			
Eaton, T. H., Jr.	1933	CA Berkeley	46
Johnston, R. F.	1955	CA Berkeley	47
Mayhew, W. W.	1953	CA Berkeley	48
Salt, G. W.	1951	CA Berkeley	49
Test, F. H.	1940	CA Berkeley	50
022. MASLIN, T. P.			
Beargie, K. M.	1971	CO	
Campbell, J. B.	1970	CO	
Cuellar, O.	1969	CO	
Ferner, J. W.	1972	CO	
Knopf, G. N.	1966	CO	
McCoy, C. J.	1965	CO	
Pennock, L. A.	1966	CO	
Reese, R. W.	1969	CO	
Walker, J. M.	1966	CO	51
Werth, R. J.	1969	CO	
023. MEAD, G. W., JR.			
Musick, J. A.	1969	Harvard	52
024. SAVAGE, J. M.			
Bickford, D. P.	2001	U Miami	
Chen, S.-H.	2001	U Miami	
Crother, B. I.	1989	U Miami	
Deweese, J. E.	1976	S CA	
Donnelly, M. A.	1987	U Miami	53
Emerson, S. B.	1971	S CA	
Gaudin, A. J.	1969	S CA	
Guyer, C.	1986	U Miami	54
Harris, R. T.	1973	S CA	
Hayes, M. P.	1991	U Miami	
Heyer, W. R.	1968	S CA	
Johnston, G. R.	1996	U Miami	
Kluge, A. G.	1964	S CA	55
Krempels, D. M.	1989	U Miami	
Lahanas, P. N.	1992	U Miami	
Lieberman, S. S.	1982	S CA	
Lips, K. R.	1995	U Miami	
McDiarmid, R. W.	1968	S CA	
Meyer, J. R.	1969	S CA	
Miyamoto, M. M.	1982	S CA	56
Morafka, D. J.	1973	S CA	
Nicholson, K. E.	2001	U Miami	
Presch, W. F., Jr.	1970	S CA	
Scott, N. J., Jr.	1969	S CA	57
Silverstone, P. A.	1971	S CA	
Slowinski, J. B.	1991	U Miami	
Talbot, J. J.	1976	S CA	
Vial, J. L.	1965	S CA	
Wake, D. B.	1964	S CA	58
Wake, M. H.	1967	S CA	59
Werman, S. D.	1986	U Miami	
Young-Downey, A. R.	1990	U Miami	
025. RUTHVEN, A. G.			
Bailey, J. R.	1940	MI	60
Blanchard, F. N.	1919	MI	61
Hartweg, N. E.	1934	MI	62
Hubbs, C. L.	1927	MI	63
Oliver, J. A.	1941	MI	
Ortenburger, A. I.	1925	MI	64
Stuart, L. C.	1933	MI	
Stull, O. G.	1929	MI	
van Tyne, J.	1928	MI	65
Walker, C. F.	1935	MI	66
026. KINGSBURY, B. F.			
Adelmann, H. B.	1924	Cornell	
027. BISHOP, S. C.			
Grobman, A. B.	1943	Rochester	67
Tihen, J. A.	1945	Rochester	68
028. COWLES, R. B.			
Brattstrom, B. H.	1959	CA Los Angeles	69
Lowe, C. H., Jr.	1951	CA Los Angeles	70
Norris, K. S.	1959	CA Los Angeles	71
Stebbins, R. C.	1945	CA Los Angeles	72
029. GILBERT, P. W.			
Evans, H. Edward	1950	Cornell	73
Snyder, R. C.	1948	Cornell	74
030. GORDON, K. L.			
Storm, R. M.	1948	OR St	75
031. HAMILTON, W. J., JR.			
Barbour, R. W.	1949	Cornell	76
Harris, L. E., Jr.	1954	Loma Linda	
Jopson, H. G.	1938	Cornell	
Oplinger, C. S.	1963	Cornell	
Stewart, M. M.	1956	Cornell	77
Tilger, G. M.	1963	Cornell	
Whitaker, J. O., Jr.	1962	Cornell	
032. RANEY, E. C.			
Hecht, M. K.	1953	Cornell	78
Kezer, L. J.	1949	Cornell	79
033. SCHMIDT, K. P.			
Inger, R. F.	1954	Chicago	80
034. BRECKENRIDGE, W. J.			
Bellis, E. D.	1957	MN	81
Chelberg, D. A.	1964	MN	
Dawley, C. W.	1944	MN	
035. UNDERHILL, J. C.			
Fishbeck, D. W.	1968	MN	

Fleming, P. L. ....1976....MN
Hedeen, S. E. ....1970....MN
Schmid, W. D. ....1962....MN.......... 82

036. CHANDLER, D. C.
Trama, F. B. ....1957....MI.......... 83

037. ROGERS, J. S.
Carr, A. F., Jr. ....1938....FL.......... 84
Goin, C. J. ....1946....FL.......... 85

038. BALDAUF, R. J.
Robinson, D. C. ....1968....TX A&M

039. DEGENHARDT, W. G.
Aldridge. R. D. ....1973....NM
Applegarth, J. S. ....1979....NM
Christiansen, J. L. ....1969....NM
Hardy, L. M. ....1969....NM
Jacob, J. S. ....1977....NM
Jones, K. L. ....1974....NM
Lucchino, R. ....1973....NM
Seidel, M. ....1973....NM
Sugarman, R. ....1975....NM
Williams, S. R. ....1976....NM

040. BLAIR, W. F.
Awbrey, F. T. ....1965....TX.......... 86
Axtell, R. W. ....1958....TX
Bogart, J. P. ....1969....TX.......... 87
Brown, L. E. ....1967....TX.......... 88
Brownell, J. A. ....1971....TX
Follansbee, L. A. ....1954....TX
Fouquette, M. J., Jr. ....1959....TX.......... 89
Gerhardt, H. C., Jr. ....1970....TX.......... 90
Greding, E. J. ....1968....TX
Grubb, J. C. ....1971....TX
Guttman, S. I. ....1967....TX.......... 91
Heredia-Cano, J. F. ....1973....TX
Hunsaker, D., II ....1960....TX
Hunt, H. W. ....1971....TX
Jameson, D. L. ....1952....TX.......... 92
Kennedy, J. P. ....1958....TX
Laughlin, H. E. ....1966....TX
Lindsay, H. L., Jr. ....1958....TX
Low, B. S. ....1967....TX.......... 93
Mares, M. A. ....1973....TX
Martin, R. F. ....1969....TX
Martin, W. F. ....1970....TX
McAllister, W. ....1961....TX
McKown, R. R. ....1972....TX
Meacham, W. R. ....1958....TX
Mecham, J. S. ....1955....TX.......... 94
Michaud, T. C. ....1959....TX
Milstead, W. W. ....1953....TX
Nelson, C. E. ....1966....TX.......... 95
Pettus, D. ....1956....TX.......... 96
Pierce, J. R. ....1968....TX
Porter, K. R. ....1962....TX
Pyburn, W. F. ....1956....TX
Ralin, D. B. ....1970....TX
Rogers, J. S ....1971....TX
Sage, R. D. ....1974....TX
Tandy, R. M. ....1972....TX
Thornton, W. A. ....1954....TX
Wasserman, A. O. ....1956....TX

041. CLARK, D. R., JR.
Ballinger, R. E. ....1971....TX A&M.......... 97
Fleet, R. R., II ....1972....TX A&M

042. LEGLER, J. M.
Berry, J. F. ....1978....UT
Bull, J. J. ....1977....UT.......... 98
Feuer, R. C. ....1966....UT
Moll, E. O. ....1969....UT
Winokur, R. M. ....1973....UT

043. COCKRUM, E. L.
Patton, J. L. ....1969....AZ.......... 99

044. FINDLEY, J. S.
Fleharty, E. D. ....1963....NM

045. JONES, R. E.
Austin, H. B. ....1988....CO
Cuellar, H. S. ....1972....CO
Duvall, D. J. ....1980....CO.......... 100
Fawcett, J. D. ....1975....CO
Fitzgerald, K. T. ....1982....CO
Gerrard, A. M. ....1974....CO
Guillette, L. J., Jr. ....1981....CO.......... 101
La Greek, F. T. ....1975....CO
Lopez, K. H. ....1992....CO
Rand, M. S. ....1991....CO
Roth, J. J. ....1974....CO
Summers, C. H. ....1987....CO.......... 102
Tokarz, R. R. ....1977....CO

046. EATON, T. H., JR.
Callison, G. L. ....1969....KS

047. JOHNSTON, R. F.
Packard, G. C. ....1966....KS.......... 103

048. MAYHEW, W. W.
Bartlett, P. N. ....1970....CA Riverside
Wright, S. J. ....1968....CA Riverside

049. SALT, G. W.
Willard, D. E. ....1966....CA Davis

050. TEST, F. H.
Bachman, M. D. ....1964....MI
Beiswenger, R. E. ....1972....MI
Carpenter, C. C. ....1951....MI.......... 104
Delzell, D. E. ....1958....MI
Dole, J. W. ....1963....MI
Heatwole, H. F. ....1960....MI.......... 105
Martof, B. S. ....1951....MI.......... 106
Sexton, O. J. ....1956....MI.......... 107

051. WALKER, J. M.
Abuhteba, R. M. ....1990....AR
Arnett, J. T. ....1974....AR
Britton, J. M. ....1986....AR
Cordes, J. E. ....1991....AR
Crowell, R. L. ....1981....AR
Cundall, D. L. ....1974....AR.......... 108
Durham, M. S. ....1980....AR
Ireland, P. H. ....1971....AR
Killibrew, F. C. ....1976....AR
Kuss, B. D. ....1986....AR
Reagan, D. P. ....1972....AR

052. MUSICK, J. A.
Bartol, S. M. ....1999....William & Mary
Keinath, J. A. ....1993....William & Mary
Mansfield, K. L. ....2006....William & Mary
053. DONNELLY, M. A.
Ugarte, C. A. ....2004....FL Int
Watling, J. I. ....2005....FL Int
054. GUYER, C.
Boback, S. M. ....2005....Auburn
Butterfield, B. P. ....1996....Auburn
Reed, R. N. ....2001....Auburn
055. KLUGE, A. G.
Arnold, S. J. ....1972....MI 109
Ast, J. C. ....2002....MI
Bookstein, F. L. ....1977....MI
Fox, G. M. ....2006....MI
Heying, H. E. ....2001....MI
Jaslow, A. P. ....1982....MI
Kerfoot, W. C. ....1972....MI 110
Kraus, E. F. ....1987....MI
Moody, S. M. ....1980....MI 111
Summers, K. ....1990....MI
Tolson, P. J. ....1982....MI
Wolf, A. J. ....1996....MI
056. MIYAMOTO, M. M.
Liu, F.-G. R. ....2005....FL
057. SCOTT, N. J., JR.
Jennings, R. D. ....1991....NM
Lotter, C. F. ....1975....CN
Mitchell, S. L. ....1988....NM
Reynolds, R. P. ....1978....NM
Woodward, B. D. ....1981....NM
058. WAKE, D. B.
Alberch, P. ....1980....CA Berkeley 112
Busack, S. D. ....1985....CA Berkeley
Cadle, J. E. ....1982....CA Berkeley
Collazo, A. ....1990....CA Berkeley
Darda, D. M. ....1988....CA Berkeley
Deban, S. M. ....1997....CA Berkeley
de Querioz, K. ....1989....CA Berkeley
Edwards, J. L. ....1976....CA Berkeley 113
Good, D. A. ....1985....CA Berkeley
Graybeal, A. ....1995....CA Berkeley
Hanken, J. ....1980....CA Berkeley 114
Houck, L. D. ....1977....CA Berkeley 115
Jackman, T. R. ....1993....CA Berkeley
Jockusch, E. L. ....1996....CA Berkeley 116
Kuchta, S. R. ....2002....CA Berkeley
Larson, L. A. ....1982....CA Berkeley
Lombard, R. E. ....1971....Chicago............ 117
Lynch, J. F. ....1974....CA Berkeley
Mahoney, M. J. ....1999....CA Berkeley
Maiorana, V. C. ....1974....CA Berkeley
Marks, S. B. ....1995....CA Berkeley
Mueller, R. L. ....2005....CA Berkeley
Parks, D. S. ....2000....CA Berkeley
Parra-Olea, G. ....1999....CA Berkeley
Peterson, J. ....1973....Chicago
Schneider, C. J. ....1993....CA Berkeley
Sessions, S. K. ....1985....CA Berkeley
Shafland, J. L. ....1968....Chicago
Staub, N. L. ....1989....CA Berkeley
Stewart, J. R. ....1976....CA Berkeley
Sweet, S. S. ....1978....CA Berkeley 118
Tan, A.-M. ....1993....CA Berkeley
Vredenburg, V. T. ....2002....CA Berkeley
Yanev, K. P. ....1978....CA Berkeley
059. WAKE, M. H.
Bauer, A. M. ....1986....CA Berkeley
Bemis, W. E. ....1983....CA Berkeley 119
Dickie, R. ....1999....CA Berkeley
Erickson, G. M. ....1997....CA Berkeley
Hetherington, T. E. ....1979....CA Berkeley 120
Klosterman, L. L. ....1983....CA Berkeley
Lappin, A. K. ....1999....CA Berkeley
Meyer, A. ....1988....CA Berkeley
Olsen, W. M. ....1998....CA Berkeley
Schwenk, K. ....1984....CA Berkeley
Sheen, J. P.-Y. ....2001....CA Berkeley
060. BAILEY, J. R.
Bruce, R. C. ....1968....Duke
Davis, D. M. ....1969....Duke
Hutchison, V. H. ....1959....Duke 121
Murphy, T. D. ....1963....Duke
Osgood, D. W. ....1968....Duke
061. BLANCHARD, F. N.
Burt, C. E. ....1930....MI
Clark, H. D. ....1941....MI
Clay, W. M. ....1937....MI
George, I. D. ....1940....MI
Gloyd, H. K. ....1937....MI
Stickel, E. L. F. ....1949....MI
062. HARTWEG, N. E.
Brown, B. C. ....1955....MI
Cagle, F. R. ....1943....MI 122
Dammann, A. E. ....1961....MI
Dowling, H. G. ....1951....MI 123
Downs, F. L. ....1965....MI
Duellman, W. E. ....1956....MI 124
Etheridge, R. E. ....1960....MI
Mosimann, J. E. ....1956....MI
Orton, G. L. ....1944....MI
Peters, J. A. ....1952....MI
063. HUBBS, C. L.
Walker, B. W. ....1949....CA San Diego... 125
064. ORTENBURGER, A. I.
Holzapfel, R. A. ....1935....OK
065. VAN TYNE, J.
Brodkorb, P. ....1936....MI 126
066. WALKER, C. F.
Adler, K. K. ....1968....MI 127
Green, N. B. ....1952....OH St
Healy, W. R. ....1966....MI
Martin, P. S. ....1956....MI
Rabb, G. B. ....1957....MI
Starrett, P. H. ....1968....MI
Uzzell, T. M., Jr. ....1962....MI
Wilbur, H. M. ....1971....MI 128
Zug, G. R. ....1969....MI

067. GROBMAN, A. B.
- Auffenberg, W. .... 1956....FL.......... 129
- Highton, R. T. .... 1956....FL.......... 130

068. TIHEN, J. A.
- Chantell, C. J. .... 1965....Notre Dame

069. BRATTSTROM, B. H.
- Gilbert, M. .... 1998....CA Fullerton

070. LOWE, C. H., JR.
- Beck, D. D. .... 1991....AZ
- Bezy, R. L. .... 1970....AZ
- Braun, E. J. .... 1969....AZ
- Cole, C. J. .... 1969....AZ
- Cross, J. K. .... 1979....AZ
- Felger, R. S. .... 1966....AZ
- Ferguson, J. H. .... 1965....AZ
- Gates, G. O. .... 1963....AZ
- Goldberg,S. R. .... 1970....AZ
- Halpern, E. A. .... 1970....AZ
- Heath, W. G. .... 1962....AZ
- Hinds, D. S. .... 1970....AZ
- Krizman, R. D. .... 1972....AZ
- Lardner, P. J. .... 1969....AZ
- Larson, E. .... 1986....AZ
- McCord, R. D. .... 1997....AZ
- Robinson, M. D. .... 1972....AZ
- Rosen, P. C. .... 2000....AZ
- Stropes, S. L. .... 1975....AZ
- Tremor, J. W. .... 1962....AZ
- Wright, J. W. .... 1965....AZ

071. NORRIS, K. S.
- Asplund, K. K. .... 1968....CA Los Angeles
- Barrett, R. .... 1969....CA Los Angeles
- Brown, T. W. .... 1970....CA Los Angeles
- Campbell, H. W. .... 1967....CA Los Angeles
- Evans, W. E. .... 1975....CA Los Angeles
- Heath, J. E. .... 1962....CA Los Angeles
- Porter, W. P. .... 1966....CA Los Angeles.. 131
- Pough, F. H., Jr. .... 1968....CA Los Angeles.. 132
- Shallenberger, E. W. .... 1970....CA Los Angeles
- Warren, P. P. .... 1966....CA Los Angeles

072. STEBBINS, R. C.
- Anderson, J. D. .... 1960....CA Berkeley.......... 133
- Anderson, P. K. .... 1958....CA Berkeley
- Berry, K. H. .... 1972....CA Berkeley
- Brown, A. G. .... 1971....CA Berkeley
- Brown, C. W. .... 1970....CA Berkeley
- Bury, R. B. .... 1972....CA Berkeley
- Fox, R. W., Jr. .... 1950....CA Berkeley
- Glaser, H. S. R. .... 1960....CA Berkeley
- Gorman, J, B., Jr. .... 1954....CA Berkeley
- Hendrickson, J. R. .... 1952....CA Berkeley.......... 134
- Karlstrom, E. L. .... 1956....CA Berkeley
- LaPointe, J. L. .... 1966....CA Berkeley.......... 135
- Marlow, R. W. .... 1979....CA Berkeley
- McGinnis, S. M., Jr. .... 1965....CA Berkeley
- Papenfuss, T. J. .... 1979....CA Berkeley
- Reimer, W. J. .... 1956....CA Berkeley
- Rosenthal, G. M. .... 1954....CA Berkeley
- Ruth, S. B. .... 1977....CA Berkeley
- Tollestrup, K. .... 1979....CA Berkeley
- Turner, F. B. .... 1957....CA Berkeley
- Wilhoft, D. C. .... 1962....CA Berkeley..... 136
- Zweifel, R. G. .... 1954....CA Berkeley

073. EVANS, H. EDWARD
- Blackburn, D. G. .... 1985....Cornell

074. SNYDER, R. C.
- Frazzetta, T. H. .... 1963....WA.......... 137
- Larsen, J. H., Jr. .... 1963....WA.......... 138
- Pianka, E. R. .... 1965....WA.......... 139
- Waters, J. F. .... 1969....WA

075. STORM, R. M.
- Altig, R. G. .... 1969....OR St.......... 140
- Beatty, J. J. .... 1979....OR St
- Briggs, J. L. .... 1971....OR St
- Brodie, E. D., Jr. .... 1969....OR St.......... 141
- Clothier, G. W. .... 1971....OR St
- Clover, R. C. .... 1975....OR St
- Cohen, N. W. .... 1956....OR St
- Darrow, T. D. .... 1967....OR St
- Dumas, P. C. .... 1953....OR St.......... 142
- Ferguson, D. E. .... 1957....OR St.......... 143
- Freiberg, R. E. .... 1954....OR St
- Haertel, J. D. .... 1970....OR St
- Hanlin, H. G. .... 1980....OR St
- Johnson, O. W. .... 1965....OR St
- MacKenzie, D. S. .... 1970....OR St
- Nussbaum, R. A. .... 1972....OR St.......... 144
- O'Hara, R. K. .... 1981....OR St
- Pimentel, R. A. .... 1952....OR St
- Roberts, J. O. .... 1970....OR St
- Smith, J. M. .... 1967....OR St
- Stewart, G. R. .... 1964....OR St
- Sype, W. E. .... 1975....OR St
- Taylor, J. T. .... 1977....OR St

076. BARBOUR, R. W.
- Batson, J. D. .... 1965....KY
- Ernst, C. H. .... 1969....KY.......... 145
- Petranka, J. W. .... 1982....KY

077. STEWART, M. M.
- Townsend, D. S. .... 1984....SUNY
- Woolbright, L. L. .... 1985....SUNY

078. HECHT, M. K.
- Cook, C. P. .... 1996....CUNY
- Hinderstein, B. .... 1969....CUNY
- Kropach, C. N. .... 1973....CUNY
- LaDuke, T. C. .... 1991....CUNY
- Madden, R. C. .... 1975....CUNY
- Malone, B. S. .... 1979....CUNY

079. KEZER, J.
- Leon, P. E. .... 1974....OR

080. INGER, R. F.
- Airame, S. .... 1999....Chicago
- Bacon, J. P., Jr. .... 1970....Chicago
- Frogner, K. .... 1975....Chicago
- Liem, D. S. .... 1969....IL
- Schwalm, P. A. .... 1981....Chicago
- Voris, H. K. .... 1969....Chicago
- Wassersug, R. J. .... 1973....Chicago.......... 146

081. BELLIS, E. D.
- Barthalmus, G. T. ....197O... PA St
- Ehlert, G. W. ....1973.... PA St
- Pluto, T. G. ....1983.... PA St
- Sherbine, K. B. ....1969.... PA St

082. SCHMID, W. D.
- Boernke, W. E. ....1972.... MN
- Coiner, S. J. ....1978.... MN

083. TRAMA, F. B.
- Dorando, S. L. S. ....1978.... Rutgers

084. CARR, A. F., JR.
- Belkin, D. A. ....1961.... FL
- Bjorndal, K. A. ....1979.... FL..................... 147
- Caldwell, D. K. ....1957.... FL
- Christman, S. P. ....1975.... FL
- Corn, M. J. ....1981.... FL
- Crenshaw, J. W., Jr. ....1955.... FL..................... 148
- Dietz, D. C. ....1979.... FL
- Ehrenfeld, D. W. ....1966.... FL..................... 149
- Goodman, D. E. ....1971.... FL
- Gourley, E. V. ....1969.... FL
- Hirth, H. F. ....1962.... FL..................... 150
- Jackson, C. G. ....1964.... FL
- Jackson, J. F. ....1972.... FL
- Meylan, A. B. ....1984.... FL
- Mortimer, J. A. ....1981.... FL
- Mount, R. H. ....1961.... FL..................... 151
- Pritchard, P. C. ....1969.... FL
- Rossman, D. A. ....1962.... FL..................... 152

085. GOIN, C. J.
- Jones, D. A. ....1967.... FL
- Wilder, C. D., Jr. ....1962.... FL

086. AWBREY, F. T.
- Maxson, L. R. ....1973.... San Diego St..... 153

087. BOGART, J. P.
- Chek, A. A. ....2000.... Guelph (CA)
- Green, D. M. ....1981.... Guelph (CA)..... 154
- Rye, L. A. ....2000.... Guelph (CA)

088. BROWN, L. E.
- Gray, R. H. ....1971.... IL St
- Moll, D. L. ....1977.... IL St
- Thrall, J. H. ....1972.... IL St

089. FOUQUETTE, M. J., JR.
- Bowker, R. W. ....1988.... AZ St
- Hulse, A. C. ....1974.... AZ St
- Platz, J. E. ....1974.... AZ St
- Renaud, M. L. ....1977.... AZ St
- Sullivan, B. K. ....1983.... AZ St 155
- Vitt, L. J. ....1976.... AZ St 156
- Wylie. S. R. ....1981.... AZ St

090. GERHARDT, H. C.
- Bee, M. A. ....2001.... MO
- Davis, M. S. ....1988.... MO
- Diekamp, B. ....1990.... MO
- Doherty, J. A. ....1982.... MO
- Humfeld, S. C. ....2003.... MO
- Keller, M. J. ....2000.... MO
- Marshall, V. T. ....2003.... MO
- Ptacek, M. B. ....1991.... MO
- Tanner, S. D. ....1995.... MO
- Welch, A. M. ....2000.... MO

091. GUTTMAN. S. I.
- Karlin, A. A. ....1978.... Miami U
- Merkle, D. A. ....1975.... Miami U
- Romano, M. A. ....1984.... Miami U

092. JAMESON, D. L.
- Mackey, J. P. ....1958.... OR
- Quinn, H. R. ....1987.... Houston

093. LOW, B. S.
- Heinen, J. T. ....1992.... MI

094. MECHAM, J. S.
- Flury, A. G. ....1972.... TX Tech

095. NELSON, C. E.
- Barron, J. M. ....1995.... IN
- Bowden, R. M. ....2001.... IN
- Clark, P. J. ....2000.... IN
- Cortright, S. A. ....1987.... IN
- Etchberger, C. R. ....1991.... IN
- Freedburg, S. ....2003.... IN
- Klukowski, M. ....1998.... IN
- Lin, E. ....1975.... IN...................... 157
- Morris, M. R. ....1987.... IN
- Murphy, N. L. ....2006.... IN
- Viets, B. E. ....1993.... IN
- Williams, P. K. ....1973.... IN

096. PETTUS, D.
- Corn, P. S. ....1982.... CO St
- Gillis, J. E. ....1975.... CO St
- Hess, J. B. ....1970.... CO St
- Matthews, T. ....1968.... CO St
- Post, D. D. ....1972.... CO St
- Spencer, A. W. ....1964.... CO St
- Tordoff, W., III..................1971.... CO St

097. BALLINGER, R. E.
- Jones, S. M. ....1983.... NB
- Lemos-Espinal, J. A. ....1993.... NB..................... 158
- Rowe, J. W. ....1993.... NB
- Sandridge, J. W. ....2001.... NB
- Schwartz, S. ....1981.... NB
- Smith, G. R. ....1995.... NB

098, BULL, J. J.
- Pfennig, D. W. ....1989.... TX..................... 159
- Smith, C. R. ....1989.... TX
- White, R. B. ....1991.... TX

099. PATTON, J. L.
- Case, S. M. ....1976.... CA Berkeley
- Luke, C. A. ....1989.... CA Berkeley
- Rainey, W. ....1983.... CA Berkeley

100. DUVALL, D. J.
- Graves, B. M. ....1988.... WY
- Hayes, W. K. ....1991.... WY

101. GUILLETTE, L. J., JR.
- Baldwin, K. S. ....1998.... FL
- Barbeau, T. R. ....2004.... FL
- Cox, M. C. ....1994.... FL
- Crain, D. A. ....1997.... FL
- Demarco, V. G. ....1991.... FL
- Masson, G. R. ....1995.... FL

Matter, J. M. ....1996....FL
Palmer, B. D. ....1990....FL
Rooney, A. A. ....1998....FL
102. SUMMERS, C. H.
Korzan, W. J. ....2004....SD
103. PACKARD, G. C.
Cagle, K. D. ....1994....CO St
104. CARPENTER, C. C.
Black, J. H. ....1973....OK
Brackin, M. F. ....1975....OK
Bussjaeger, L. J. ....1971....OK
Clarke. R. F. ....1963....OK
Gillingham, J. C. ....1976....OK
Griffith, S. J. ....1966....OK
Jenssen, T. J. ....1969....OK 160
Krupa, S. S. ....1987....OK
Leuck, B. E. ....1980....OK
Lynn, R. T. ....1963....OK
Mahmoud, I. Y. ....1960....OK
Marcellini, D. L. ....1970....OK
McDaniel, S. J. ....1966....OK
Ogilvie, P. W. ....1966....OK
Paulissen, M. A. ....1986....OK
Preston, W. B. ....1970....OK
Purdue, J. R. ....1974....OK
Sassaman, J. F. ....1974....OK
Snow, J. E. ....1984....OK
Tyler, J. D. ....1969....OK
105. HEATWOLE, H. F.
Burns, G. ....1984....New Eng (AU)
Daniels, C. ....1989....New Eng (AU)
Firth, B. ....1974....New Eng (AU)
Fraser, S. ....1985....New Eng (AU)
Frear, L. A. ....2000....NC St
Horton, D. ....1973....New Eng (AU)
Miller, J. ....1983....New Eng (AU)
Parmenter, C. J. ....1976....New Eng (AU)
Shine, R. ....1976....New Eng (AU).. 161
Simbotwe, M. ....1985....New Eng (AU)
Taylor, J. ....1984....New Eng (AU)
Webb, G. ....1972....New Eng (AU)
Zimmerman, K. ....1989....New Eng (AU)
106. MARTOF, B. S.
Bush, F. M. ....1962....GA
Forester, D. C. ....1974....NC St
Perrill, S. A. ....1973....NC St
Thompson, E. F. ....1959....GA
Ward, J. P. ....1980....NC St
107. SEXTON, O. J.
Axtell, C. B. ....1973....WA U
Bizer, J. R. ....1977....WA U
Hathaway, L. M. ....1969....WA U
Marion, K. R. ....1970....WA U 162
Ochs, K. F. ....1970....WA U
Phillips, C. A. ....1989....WA U
Seale, D. B. ....1973....WA U
108. CUNDALL, D. L.
Deufel, A. ....2003....Lehigh
Reinert, H. K. ....1983....Lehigh
109. ARNOLD. S. J.
Abell, A. J. ....1996....Chicago
Brodie, E. D., III ....1991....Chicago............ 163
Bronikowski, A. M. ....1997....Chicago
Geist, N. R. ....1999....OR St
Kephart, D. G. ....1981....Chicago
Manier, M. K. ....2005....OR St
O'Steen, S. ....1995....Chicago
Shaffer, H. B. ....1982....Chicago............ 164
Wildy, E. L. ....2000....OR St
110. KERFOOT, W. C.
McNaught, A. S. ....1993....MI
111. MOODY, S. M.
Paez, V. P. ....1995....OH
112. ALBERCH, P.
Burke, A. C. ....1989....Harvard
113. EDWARDS, J. L.
Jayne, B. C. ....1985....MI St
114. HANKEN, J.
Carl, T. F. ....2000....CO
Gross, J. B. ....2005....Harvard
Jackson, K. ....2002....Harvard
Jennings, D. H. ....1997....CO
Seufert, D. W. ....1994....CO
115. HOUCK, L. D.
Adams, E. M. ....2004....OR St
Dyal, L. A. ....2005....OR St
Palmer, C. A. ....2004....OR St
116. JOCKUSCH, E. L.
Richmond, J. Q. ....2005....CN
117. LOMBARD. R. E.
Sperry, D. G. ....1980....Chicago
118. SWEET. S. S.
Christopher, S. V. ....2004....CA Santa Barbara
Fontenot, C. L., Jr. ....2002....CA Santa Barbara
119. BEMIS, W. E.
Richmond, A. M. ....1999....MA-Amherst
120. HETHERINGTON, T. E.
Anderson, N. L. ....2002....OH St
Caprette, C. L. ....2005....OH St
Coupe, B. H. ....2005....OH St
Kim, J. N. ....1993....OH St
Lindquist, E. D. ....1997....OH St
Porej, D. ....2004....OH St
121. HUTCHISON, V. H.
Braun, B. A. ....1984....OK
Deen, C. M. ....2006....OK
deLuque, O. ....1969....RI
do Amaral, J. P. S. ....2001....OK
Engbretson, G. A. ....1975....OK
Erskine, D. J. ....1981....OK
Graham, T. E......................1972....RI
Gratz, R. K. ....1976....OK
Guimond, R. W. ....1972....RI
Hazard, E. S. ....1981....OK
Kosh, R. J. ....1970....RI
Lutterschmidt, W. I. ....1997....OK
Maness, J. D. ....1979....OK
Marvin, G. ....1996....OK

McIlvaine, P. M. .... 1971.... RI
Miller, K. .... 1978.... OK
Paulson, B. K. .... 1986.... OK
Rawding, R. S. .... 1992.... OK
Rebach, J. O. .... 1970.... RI
Sievert, L. M. .... 1988.... OK
Songdahl, J. H. .... 1971.... RI
St. Clair, R. C. .... 1995.... OK
Tilden, A. R. .... 1994.... OK
Tu, M.-C. .... 1994.... OK
Vinegar, A. .... 1968.... RI
Whitford, W. G. .... 1964.... RI 165

122. CAGLE, F. R.
Boyer, D. R. .... 1958.... Tulane
Chaney, A. H. .... 1958.... Tulane
Dobie, J. L. .... 1966.... Tulane 166
Gordon, R. E. .... 1956.... Tulane 167
Shoop, C. R. .... 1963.... Tulane 168
Tinkle, D. W. .... 1956.... Tulane 169

123. DOWLING, H. G.
Al-Murani, A. S. .... 1990.... NY
Anderson, K. .... 1986.... NY
Brummer, B. J. .... 1980.... NY
DeFina, A. V. .... 1981.... NY
Estol, C. O. .... 1981.... NY
Jenner, J. V. .... 1981.... NY
Kelly, P. .... 1993.... NY
Lazell, J. D., Jr. .... 1970.... RI
Muehlbauer, E. I. .... 1987.... NY
Pinou, T. .... 1993.... NY
Price, R. M. .... 1981.... NY
Pytel, B. A. .... 1985.... NY
Watanabe, M. E. .... 1980.... NY

124. DUELLMAN, W. E.
Berger, T. J. .... 1985.... KS
Burger, W. L. .... 1971.... KS
Burrowes, P. A. .... 1997.... KS
Caldwell, J. P. .... 1974.... KS 170
Campbell, J. A. .... 1983.... KS 171
Coloma Roman, L. A. .... 1997.... KS
Crump, M. L. .... 1974.... KS 172
Dutta, S. K. .... 1985.... KS 173
Echternacht, A. C. .... 1970.... KS 174
Edwards, S. R. .... 1974.... KS
Fritts, T. H. .... 1972.... KS
Frost, D. R. .... 1988.... KS 175
Gyi, K. K. .... 1965.... KS
Joglar, R. L. .... 1986.... KS
Kizirian, D. A. .... 1994.... KS ....
Lee, J. C. .... 1977.... KS 176
Lynch, J. D. .... 1969.... KS 177
Mendelson, J. R., III .... 1997.... KS
Montanucci, R. R. .... 1972.... KS 178
Morrison, M. E. .... 1994.... KS
Myers, C. W. .... 1970.... KS
Nieto-Montes de Oca, A. .... 1994.... KS
Parmalee, J. R. .... 1998.... KS
Pefaur, J. E. .... 1978.... KS
Roble. S. M. .... 1985.... KS
Santiago Ron, M. A. .... 1998 .... KS
Savitzky, A. H. .... 1979 .... KS 179
Schwaner, T. D. .... 1979 .... KS
Titus, T. A. .... 1992 .... KS
Trueb, L. .... 1968 .... KS 180

125. WALKER, B. W.
Barlow, G. W. .... 1958 .... CA Los Angeles .. 181

126. BRODKORB, P.
Holman, J. A. .... 1961 .... FL 182

127. ADLER, K. K.
Able, D. J. .... 1996 .... Cornell
Clark. R. W. .... 2004 .... Cornell
Grafe, T. U. .... 1995 .... Cornell
Phillips, J. B. .... 1985 .... Cornell 183
Rodda, G. H. .... 1982 .... Cornell
Ryan, M. J. .... 1982 .... Cornell 184
Villa, J. D. .... 1978 .... Cornell
Waldman, B. .... 1983 .... Cornell 185

128. WILBUR, H. M.
Alford, R. A. .... 1986 .... Duke 186
Bernardo, J. .... 1991 .... Duke
Chambers, R. C. .... 1984 .... Duke
Church, D. R. .... 2004 .... VA
Crespi, E. J. .... 2001 .... VA
Dakin, S. F. .... 1978 .... Duke
Freeman, S. L. .... 2003 .... VA
Grover, M. C. .... 1996 .... VA
Gustafson, M. P. .... 1994 .... Duke
Harris, P. M. .... 1992 .... Duke
Harris, R. N. .... 1986 .... Duke
Leimberger, J. D. .... 1989 .... Duke
McCollum, S. A. .... 1993 .... Duke
Morin, P. J. .... 1982 .... Duke 187
Pearman, P. B. .... 1991 .... Duke
Pechmann, J. H. K. .... 1994 .... Duke
Resetarits, W. J. .... 1988 .... Duke 188
Rissler, L. J. .... 2000 .... VA
Ruthig, G. R. .... 2006 .... VA
Steinwascher, K. F. .... 1978 .... Duke
Travis, J. A. .... 1980 .... Duke 189
van Buskirk, J. .... 1990 .... Duke

129. AUFFENBERG, W.
Auth, D. L. .... 1980 .... FL
Iverson, J. B., III .... 1978 .... FL
Jackson, D. R. .... 1977 .... FL
Meylan, P. A. .... 1985 .... FL
Taylor, R. W., Jr. .... 1982 .... FL
Weaver, W. G, Jr. .... 1967 .... FL

130. HIGHTON, R. T.
Danstedt, R. T., Jr. .... 1972 .... MD
Fellers, G. M. .... 1976 .... MD
Fraser, D. F. .... 1974 .... MD
Hass, C. A. .... 1990 .... MD
Hedges, S. B. .... 1988 .... MD 190
Hurlbutt, H. .... 1963 .... MD
Jaeger, R. G. .... 1969 .... MD 191
Peabody, R. B. .... 1978 .... MD
Schlitter, D. .... 1976 .... MD
Worthington, R. D. .... 1968 .... MD

131. PORTER, W. P.
Curtin, C. G. ..... 1995.... WI
Muth, F. A. ..... 1980.... WI
Tracy, C. R. ..... 1972.... WI 192
Waldschmidt, S. R. ..... 1984.... WI

132. POUGH, F. H.
Beuchat, C. A. ..... 1982.... Cornell
Mautz, W. J. ..... 1979.... Cornell
Preest, M. R. ..... 1993.... Cornell 193
Taigen, T. L. ..... 1981.... Cornell 194
Wells, K. D. ..... 1976.... Cornell 195
Wilson, R. E. ..... 1976.... Cornell

133. ANDERSON, J. D.
Graham, R. E. F. ..... 1971.... Rutgers
Hassinger, D. D. ..... 1972.... Rutgers

134. HENDRICKSON, J. R.
Berry, P. Y. ..... 19??.... Malaya (MA)
Hooi, C. W. ..... 19??.... Malaya (MA)
Owens, D. W. ..... 1976.... AZ 196
Wood, J. R., Jr. ..... 1974.... AZ

135. LAPOINTE, J. L.
Price, A. H. ..... 1986.... NM

136. WILHOFT, D. C.
DePari, J. A. ..... 1988.... Rutgers
Garrick, L. D. ..... 1973.... Rutgers
Greenberg, N. B. ..... 1973.... Rutgers 197
Hotaling, E. C. ..... 1990.... Rutgers
Kinneary, J. J. ..... 1992.... Rutgers
Marchisin, A. ..... 1980.... Rutgers
Mele, J. A. ..... 1980.... Rutgers
Walker, R. F. ..... 1971.... Rutgers

137. FRAZZETTA, T. H.
Buth, D. G. ..... 1978.... IL 198
Kardong, K. V. ..... 1972.... IL 199
Lawson, G. R. ..... 1997.... IL
Wyneken, J. ..... 1988.... IL

138. LARSEN, J. H., JR.
Beneski, J. T. ..... 1986.... WA St
Davitt, C. M. ..... 1987.... WA St
Guthrie, D. J. ..... 1975.... WA St
Hall, J. A. ..... 1993.... WA St
Herrington, R. E. ..... 1985.... WA St
Laughran, L. J. ..... 1980.... WA St
Miller, B. T. ..... 1989.... WA St
Wickham, M. G. ..... 1972.... WA St
Williams, T. A. ..... 1984.... WA St
Wilson, A. G., Jr. ..... 1993.... WA St
Zalisko, E. J. ..... 1987.... WA St

139. PIANKA, E. R.
Haydon, D. T. ..... 1992.... TX
Heger, N. A. ..... 2000.... TX
Hodges, W. L. ..... 2002.... TX
Jennings, W. B. ..... 2002.... TX
Perry, G. ..... 1995.... TX
Radtkey, R. R. ..... 1993.... TX
Ramakrishnan, L. ..... 2000.... TX
Schall, J. J. ..... 1976.... TX

140. ALTIG, R. G.
Bonner, L. A. ..... 1988.... MS St
Fortman, J. R. ..... 1973.... MS St
Smith, H. C. ..... 1975.... MS St
Taylor, C. T. ..... 1992.... MS St
Thomas, R. B. ..... 1999.... MS St
Tomson, O. H. ..... 1974.... MS St

141. BRODIE, E. D., JR.
Coss, H. T. ..... 1974.... Clemson
Cupp, P. V., Jr. ..... 1974.... Clemson
Dodd, C. K., Jr. ..... 1974.... Clemson 200
Foley, D. H., III ..... 2002.... UT ST
Hanifin, C. T. ..... 2005.... UT St
Hileman, K. S. ..... 1995.... TX-Arlington
Mushinsky, H. R. ..... 1973.... Clemson 201
Steele, C. A. ..... 2001.... UT St
Ustach, P. C. ..... 2002.... UT ST

142. DUMAS, P. C.
Metter, D. E. ..... 1963.... ID 202

143. FERGUSON, D. E.
Culley, D. D., Jr. ..... 1968.... MS St 203
Landreth, H. F. ..... 1966.... MS St
McKeown, J. P. ..... 1968.... MS St
Murphy, G. G. ..... 1970.... MS St
Taylor, D. H. ..... 1970.... MS St 204

144. NUSSBAUM, R. A.
Burke, R. L. ..... 1994.... MI
Ducey, P. K. ..... 1988.... MI
Feaver, P. E. ..... 1977.... MI
Hover, E. L. ..... 1982.... MI
Lehtinen, R. M. ..... 2003.... MI
Wu, S.-H. ..... 1994.... MI
York, D. S. ..... 1995.... MI

145. ERNST, C. H.
Akre, T. S. B. ..... 2003.... George Mason
Boucher, T. P. ..... 1999.... George Mason
Brophy, T. R. ..... 2003.... George Mason
Creque, T. R. ..... 2001.... George Mason
Hartsell, T. D. ..... 2001.... George Mason
Leuteritz, T. E. J. ..... 2002.... George Mason
Pollio, C. A. ..... 2002.... George Mason
Wilgenbusch, J. C. ..... 2000.... George Mason
Wilson, G. L. ..... 2003.... George Mason
Wilson, T. P. ..... 2002.... George Mason

146. WASSERSUG. R. J.
Breden, F. J. ..... 1982.... Chicago
Rot-Nikcevic, I. ..... 2004.... Dalhousie

147. BJORNDAL, K. A.
Bouchard, S. S. ..... 2004.... FL
Carthy, R. R. ..... 1996.... FL 205
Horikoshi, K. ..... 1992.... FL
Moran, K. L. ..... 2003.... FL
Pryor, G. S. ..... 2003.... FL
Tiwari, M. ..... 2004.... FL
Witherington, B. E. ..... 1992.... FL

148. CRENSHAW, J. W., JR.
Newcomer, R. J. ..... 1968.... MD

149. EHRENFELD, D. W.
Barzilay, S. S. ..... 1980.... Rutgers
Camhi, M. D. ..... 1993.... Rutgers

150. HIRTH, H. F.
Parker, W. S. ..... 1974.... UT

151. MOUNT, R. H.
Shealy, R. M. ....1973.... Auburn
Trauth, S. E. ....1980.... Auburn
Weatherby, C. A. ....1982.... Auburn
152. ROSSMAN, D. A.
Blaney, R. M. ....1971.... LA St
Boundy, J. J. ....1999.... LA St
Keiser, E. D., Jr. ....1967.... LA St................ 206
Lawson, R. ....1985.... LA St
Thomas, J. P. R. ....1976.... LA St
Vaeth, R. H. ....1985.... LA St
Varkey, A. ....1973.... LA St
Williams, K. L. ....1970.... LA St
Wilson, L. D. ....1968.... LA St
153. MAXSON, L. R.
Hay, J. M. ....1998.... PA St
Heise, P. J. ....1998.... TN
Lopez, Y. J. ....1994.... PA St
Maha, G. C. ....1982.... IL
Moberg, K. D. ....1989.... IL
154. GREEN, D. M.
Gray, H. M. ....2001.... McGill (CA)
Kaiser, H. ....1994.... McGill (CA)
Smith, M. A. ....2004.... McGill (CA)
155. SULLIVAN, B. K.
Gergus, E. W. A. ....1999.... AZ St
Hamilton, P. S. ....2005.... AZ St
Kwiatkowski, M. A. ....2001.... AZ St
156. VITT, L. J.
Anderson, R. A. ....1986.... CA Los Angeles
Durtsche, R. D. ....1999.... OK
Gier, P. J. ....1997.... OK
Roth, E. D. ....2005.... OK
Sartorius, S. S. ....2002.... OK
Secor, S. M. ....1992.... CA Los Angeles
157. LIN, E.
Chou, W.-H. ....1997.... Tunghai (TA)
158. LEMOS-ESPINAL, J. A.
Zuñiga-Vega, J. J. ....2005.... UNAM (ME)
159. PFENNIG, D. W.
Loeb, M. L. G. ....2001.... NC Chapel Hill
160. JENSSEN, T. A.
Lovern, M. B. ....2000.... VA Tech
Orrell, K. S. ....2002.... VA Tech
Passek, K. M. ....2002.... VA Tech
161. SHINE, R.
Armstrong, D. ....1990.... Sydney (AU)
Blomberg, S. P. ....1995.... Sydney (AU)
Borges-Landaez, P. ....1999.... Sydney (AU)
Broomhall, S. ....2003.... Sydney (AU)
Caley, M. J. ....1991.... Sydney (AU)
Doughty, P. ....1996.... Sydney (AU)
Downes, S. J. ....1999.... Sydney (AU)
Fitzgerald, M. ....2004.... Sydney (AU)
Houston, D. L. ....1992.... Sydney (AU)
James, C. D. ....1989.... Sydney (AU)
Kearney, M. ....2003.... Sydney (AU)
Keogh, J. S. ....1996.... Sydney (AU)..... 207
Langkilde, T. ....2005.... Sydney (AU)
Lourdais, O. ....2003.... CEBC (FR)
O'Connor, D. ....2003.... Sydney (AU)
Pearson, D. J. ....2002.... W Aust Mus
Phillips, B. ....2004.... Sydney (AU)
Qualls, C. P. ....1995.... Sydney (AU)
Qualls, F. J. ....1996.... Sydney (AU)
Schwarzkopf, L. ....1991.... Sydney (AU)
Wall, M. ....2005.... Sydney (AU)
Webb, J. K. ....1996.... Sydney (AU)
Whitaker, P. B. ....1999.... Sydney (AU)
162. MARION, K. R.
Blackwell, E. A. ....2003.... AL-Birmingham
Moore, D. S. ....1997.... AL-Birmingham
163. BRODIE, E. D., III
Agrawal, A. F. ....2003.... IN
Allen, C. E. ....2001.... IN
Brandt, J. M. ....2003.... IN
Ridenhour, B. J. ....2004.... IN
Welter, S. M. ....2004.... IN
164. SHAFFER, H. B.
Benard, M. F. ....2005.... CA Davis
Davidson, C. ....2000.... CA Davis
Engstrom, T. N. ....2003.... CA Davis
Fisher, R. N. ....1995.... CA Davis
Fitzpatrick, B. M. ....2004.... CA Davis
Lind, A. J. ....2005.... CA Davis
165. WHITFORD, W. G.
Delson, J. H. ....1974.... NM St
166. DOBIE, J. L.
Fahey, K. M. ....1987.... Auburn
Scott, A. F. ....1976.... Auburn
167. GORDON, R. E.
Harrison, J. R., III ....1964.... Notre Dame
MacMahon, J. A. ....1964.... Notre Dame
Snyder, D. H. ....1972.... Notre Dame
Werner, J. K. ....1968.... Notre Dame
168. SHOOP, C. R.
Agardy, M. T. ....1989.... RI
Doty, T. L. ....1978.... RI
Hudson, D. ....1984.... RI
Nyman, S. ....1985.... RI
Schwartz, M. L. ....2001.... RI
Thompson, N. B. ....1981.... RI
169. TINKLE, D. W.
Collins, J. P. ....1975.... MI 208
Devine, M. C. ....1977.... MI
Dunham, A. E. ....1978.... MI 209
Ferguson, G. W. ....1969.... MI 210
Flaspohler, R. J. ....1970.... MI
Hirshfield, M. ....1976.... MI
Howe, H. F. ....1977.... MI
McKinney, C. O. ....1969.... MI
Ruby, D. E. ....1976.... MI
Smith, D. C. ....1977.... MI
Tilley, S. G. ....1970.... MI
Vinegar, M. E. ....1973.... MI
170. CALDWELL, J. P.
Leary, C. J. ....2005.... OK
Richter, S. C. ....2004.... OK

171. CAMPBELL, J. A.
Doan, T. M. .... 2002.... TX-Arlington
Fernandes, R. .... 1995.... TX-Arlington
Gutberlet, R. L., Jr. .... 1998.... TX-Arlington
Harvey, M. B. .... 1997.... TX-Arlington
Idris, O. N. .... 2004.... TX-Arlington
Keck, M. B. .... 1998.... TX-Arlington
Lawson, D. P. .... 1999.... TX-Arlington
Sasa Marin, M. .... 2000.... TX-Arlington
Smith, B. E. .... 1996.... TX-Arlington
Smith, E. N. .... 2001.... TX-Arlington
Spencer, C. L. .... 2003.... TX-Arlington
172. CRUMP, M. L.
Etheridge, K. .... 1986.... FL
Freed, A. N. .... 1982.... FL
Hensley, F. R. .... 1994.... FL
Pounds, J. A. .... 1987.... FL
173. DUTTA, S. K.
Pandav, B. .... ???? .... Utkal (IN)
Vasudevan, K. .... ???? .... Utkal (IN)
174. ECHTERNACHT, A. C.
Campbell, T. S. .... 2000.... TN
Gerber, G. P. .... 2000.... TN
Jones, R. L. .... 1982.... TN
Michaud, E. J., III .... 1990.... TN
Minesky, J. J. .... 1999.... TN
Mitchell, J. C. .... 1982.... TN
Redmond, W. H., Jr. .... 1985.... TN
Smith, K. G. .... 2005.... TN
Thompson, J. F. .... 1984.... TN
Vincent, T. C. .... 2002.... TN
Wade, J. K. .... 1981.... TN
Wilson, M. A. .... 1987.... TN
175. FROST, D. R.
Faivovich, J. .... 2005.... Columbia
Grant, T. .... 2006.... Columbia
176. LEE, J. C.
Barr, B. R. .... 1997.... U Miami
177. LYNCH, J. D.
Hoger, C. E. .... 1976.... NB
Kaplan, M. D. .... 1993.... NB
Korky, J. K. .... 1977.... NB
Kruse, K. C. .... 1978.... NB
178. MONTANUCCI, R. R.
Platt, S. G. .... 1996.... Clemson
179. SAVITZKY, A. H.
Allsteadt, J. R. .... 2003.... Old Dominion
Cross, C. L. .... 1998.... Old Dominion
Mebert, K. .... 2003.... Old Dominion
Velhagen, W. R. A., Jr. ...... 1995.... Duke
180. TRUEB, L.
Andersen, M. L. .... 1978.... KS
Cannatella, D. C. .... 1985.... KS 211
da Silva, H. R. .... 1998.... KS
Ford, L. S. .... 1989.... KS
Greenbaum, E. .... 2006.... KS
Maglia, A. M. .... 2000.... KS
Meinhardt, D. J. .... 2002.... KS
Pramuk, J. B. .... 2004.... KS
Pregill, G. K. .... 1979.... KS
Pugener, L. A. .... 2002.... KS
Pyles, R. A. .... 1988.... KS
Sheil, C. A. .... 2003.... KS
Wild, E. R. .... 1997.... KS
181. BARLOW, G. W.
Baylis, J. R. .... 1969.... CA Berkeley 212
Stamps, J. A. .... 1971.... CA Berkeley 213
182. HOLMAN, J. A.
Andrews, K. D. .... 2000.... MI St
Fay, L. P. .... 1984.... MI St
Rogers, K. L. .... 1975.... MI St
Sullivan, R. M. .... 1980.... MI St
Wilson, V. V. .... 1975.... MI St
183. PHILLIPS, J. B.
Deutschlander, M. .... 1998.... IN
Dommer, D. .... 2007.... VA Tech
Fischer, J. .... 2000.... IN
Wente, W. .... 2001.... IN
184. RYAN, M. J.
Bernal, X. .... 2007.... TX
Billy, A. J. .... 1986.... TX
Deinart, E. .... 1997.... TX
Ellingson, J. M. .... 1994.... TX
Engeszer, R. .... 2006.... TX
Gridi-Papp, M. .... 2003.... TX
Hews, D. K. .... 1990.... TX
Johnson, D. .... 2001.... TX
Kime, N. M. .... 2001.... TX
Paige, R. .... 2007.... TX
Phelps, S. M. .... 1999.... TX
Rosenthal, G. .... 2000.... TX
Thompson, R. .... 1998.... TX
Wagner, W. E., Jr. ...... 1991.... TX
Warkentin, K. M. .... 1998.... TX
Warren, P. .... 2000.... TX
185. WALDMAN, B.
Barribeau, S. .... 2006.... Canterbury (NZ)
Bos, D. .... 2005.... Canterbury (NZ)
McKinnon, J. S. .... 1994.... Harvard
Sadic, E. .... 2007.... Canterbury (NZ)
Villinger, J. .... 2006.... Canterbury (NZ)
186. ALFORD, R. A.
Cohen, M. P. .... 1995.... James Cook (AU)
Crossland, M. R. .... 1997.... James Cook (AU)
Gyuris, E. .... 1993.... James Cook (AU) .. 214
Hearnden, M. N. .... 1992.... James Cook (AU)
Kurrini, M. D. .... 2005.... James Cook (AU)
Lynch, T. .... 1999.... James Cook (AU)
Retallick, R. W. R. ...... 2001.... James Cook (AU)
Rowley, J. J. R. .... 2006.... James Cook (AU)
Woodhams, D. C. .... 2003.... James Cook (AU)
187. MORIN, P. J.
Kurzava, L. M. .... 1993.... Rutgers
Lawler, S. P. .... 1991.... Rutgers 215
Pehek, E. L. .... 1994.... Rutgers
188. RESETARITS, W. J.
Chalcraft, D. R. .... 2002.... IL

189. TRAVIS, J.
Blouin, M. S. .... 1989 .... FL St 216
Gascon, C. .... 1990 .... FL St
Gunzburger, M. S. .... 2004 .... FL St

190. HEDGES, S. B.
Blair, J. E. .... 2005 .... PA St

191. JAEGER, R. G.
Akin, J. A. .... 1998 .... LA-Lafayette
Anthony, C. D. .... 1995 .... LA-Lafayette
Beachy, C. K. .... 1992 .... LA-Lafayette
Buchanan, B. W. .... 1993 .... LA-Lafayette
Devlin, D. J. .... 2004 .... LA-Lafayette
Gabor, C. R. .... 1997 .... LA-Lafayette
Gillette, J. R. .... 2003 .... LA-Lafayette
Guffey, C. A. .... 1998 .... LA-Lafayette
Holland, D. .... 1992 .... LA-Lafayette
Lancaster, D. L. .... 1994 .... LA-Lafayette
Marshall, J. L. .... 2000 .... LA-Lafayette
Mathis, S. A. .... 1989 .... LA-Lafayette
Moore, M. K. .... 1997 .... LA-Lafayette
Peterson, M. G. .... 2001 .... LA-Lafayette
Prosen, E. D. .... 2004 .... LA-Lafayette
Simons, R. R. .... 1992 .... LA-Lafayette
Swart, C. C. .... 2003 .... LA-Lafayette
Taylor, R. C. .... 2004 .... LA-Lafayette
Walls, S. C. .... 1991 .... LA-Lafayette
Wicknick, J. A. .... 1995 .... LA-Lafayette
Wise, S. E. .... 1995 .... LA-Lafayette

192. TRACY, C. R.
Christian, K. A. .... 1981 .... CO
Espinosa. R. E. .... 2002 .... NV Reno
Nussear, K. E. .... 2004 .... NV Reno
O'Connor, M. P. .... 1989 .... CO St
Peters, E. L. .... 1994 .... CO St
Simandle, E. T. .... 2002 .... NV Reno
Snell, H. L. .... 1984 .... CO St 217
Turner, J. S. .... 1982 .... CO St
Zimmerman, L. C. .... 1989 .... CO St

193. PREEST, M. R.
Paterson, A. V. .... 1999 .... Miami

194. TAIGEN, T. L.
Navas, C. A. .... 1995 .... CN
Ressel, S. J. .... 1993 .... CN

195. WELLS, K. D.
Bevier, C. R. .... 1995 .... CN
Dawley, E. .... 1984 .... CN
Freidenburg, L. K. .... 2003 .... CN
Given, M. F. .... 1987 .... CN
Ibáñez, R. .... 1991 .... CN
O'Brien, J. A. .... 2002 .... CN
Owen, P. C. .... 2003 .... CN
Robakiewicz, P. E. .... 1992 .... CN

196. OWENS, D. W.
Comuzzie, D. K. .... 1987 .... TX A&M
Craven, K. S. .... 2001 .... TX A&M
Dutton, P. H. .... 1995 .... TX A&M
Grassman, M. A .... 1984 .... TX A&M
Kalb, H. J. .... 1999 .... TX A&M
Moon, D.-Y. .... 1992 .... TX A&M
Plotkin, P. T. .... 1994 .... TX A&M
Rostal, D. C. .... 1991 .... TX A&M
Scott, T. P. .... 1996 .... TX A&M
Shaver, D. J. .... 2000 .... TX A&M
Valverde, R. A. .... 1996 .... TX A&M
Wibbells, T. R. .... 1988 .... TX A&M 218

197. GREENBERG, N. B.
Font, E. .... 1988 .... TN
Kramer, M. .... 1989 .... TN

198. BUTH, D. G.
Colli, G. R. .... 1996 .... CA Los Angeles .. 219

199. KARDONG, K. V.
Mackessy, S. P. .... 1989 .... WA St
Smith, T. L. .... 2002 .... WA St

200. DODD, C. K., JR.
Johnson, S. A. .... 2001 .... FL
Knapp, C. R. .... 2005 .... FL
Smith, L. L. .... 1999 .... FL

201. MUSHINSKY, H. R.
Delis, P. R. .... 2001 .... S FL
Jansen, K. P. .... 2001 .... S FL
Keller, C. A. .... 2005 .... S Fl
Michot, T. C. .... 1981 .... LA St
Wilson, D. S. .... 1996 .... S FL
Witz, B. W. .... 1994 .... S FL

202. METTER, D. E.
Bradford, J. .... 1973 .... MO
Huser, C. F. .... 1986 .... MO
Peterson, C. L. .... 1985 .... MO
Powell, R. .... 1984 .... MO

203. CULLEY, D. D., JR.
Mayeaux, M. H. .... 1994 .... LA St

204. TAYLOR, D. H.
DeRosa, C. .... 1977 .... Miami U
Ford, N. B. .... 1979 .... Miami U
Nixdorf, W. L. .... 1998 .... Miami U
Nowak, R. T. .... 1983 .... Miami U
Rice, T. M. .... 1998 .... Miami U

205. CARTHY. R. R.
Lamont, M. M. .... 2002 .... FL

206. KEISER, E. D., JR.
Austin, R. M., Jr. .... 1997 .... MS
Morgan, E. C. .... 1973 .... LA-Lafayette

207. KEOGH, J. S.
Chapple, D. G. .... 2004 .... Aust Nat (AU)
Osborne, L. .... 2004 .... Aust Nat (AU)
Stapley, J. .... 2004 .... Aust Nat (AU)
Wong, B. B. M. .... 2003 .... Aust Nat (AU)

208. COLLINS, J. P.
Brunner, J. .... 2004 .... AZ St
Harvey, E. .... 2002 .... AZ St
Holomuzski, J. P. .... 1986 .... AZ St
Jones, T. R. .... 1989 .... AZ St
Zerba, K. E. .... 1989 .... AZ St
Ziemba, R. E. .... 1998 .... AZ St

209. DUNHAM, A. E.
Angilletta, M. J., Jr. .... 1998 .... PA
Beaupre, S. J. .... 1993 .... PA 220
Grant, B. W. .... 1988 .... PA

Haenel, G. J. ....1995.... PA
Kemp, S. J. ....2004.... PA
Newman, R. A. ....1987.... PA
Niewiarowski, P. H. ....1992.... PA
Robson, M. A. ....2000.... PA
Roosenburg, W. M. ....1992.... PA.................... 221
Sears, M. W. ....2001.... PA
Watkins, G. G. ....1995.... PA
Winters, R. S. ....2001.... PA
Zaradic, P. M. A. ....2003.... PA

210. FERGUSON, G. W.
Bohlen, C. H. ....1976.... KS St
Cooper, W. E., Jr. ....1972.... KS St
Derickson, W. K. ....1974.... KS St

211. CANNATELLA, D. C.
Brown, R. M. ....2004.... TX
Glusenkamp, A. G. ....2001.... TX
LaDuc, T. J. ....2003.... TX
McGuire, J. A. ....1998.... TX.................... 222
Poe, S. J. ....2000.... TX
Reeder, T. W. ....1993.... TX
Wiens, J. J. ....1995.... TX.................... 223
Yeh, J. J. ....2001.... TX

212. BAYLIS, J. R.
Martins, E. P. ....1992.... WI

213. STAMPS, J. A.
Greene, C. M. ....2001.... CA Davis
Ono, K. A. ....1981.... CA Davis
Troyer, K. E. ....1982.... CA Davis

214. GYURIS, E.
Andre, J. ....???? .... James Cook (AU)
Fuentes, M. ....???? .... James Cook (AU)
Hazel, J. ....???? .... James Cook (AU)

215. LAWLER, S. P.
Leyse, K. E. ....2005.... CA Davis

216. BLOUIN, M. S.
Hoffman, E. A. ....2004.... OR St
Monsen, K. J. ....2003.... OR St

217. SNELL, H. L.
Altamirano B., M. A. ....2002.... NM
Fitzgerald, L. A. ....1993.... NM.................... 224
Jordan, M. A. ....1990.... NM

218. WIBBELLS, T. R.
Geis, A. A. ....2004.... AL-Birmingham
Murdock, C. A. ....2004.... AL-Birmingham

219. COLLI, G. R.
Mesquita, D. O. ....2005.... Brasília (BR)
Pérez, A. K., Jr. ....2003.... Brasília (BR)
Teixeira, R. D. ....2003.... E Campinas (BR)
Vieira, G. H. C. ....2005.... Brasília (BR)
Wiederhecker, H. C. ....2004.... Brasília (BR)

220. BEAUPRE, S. J.
Hill, J. G., III ....2004.... AR
Montgomery, C. E. ....2005.... AR
Pilgrim, M. A,...................2005.... AR
Zaidan, F., III ....2001.... AR

221. ROOSENBERG, W. M.
Ford, D. K. ....2005.... OH

222. MCGUIRE, J. A.
Burbrink, F. T. ....2000.... LA St

223. WIENS, J. J.
Stephens, P. R. ....2005.... SUNY-Stony Brook

224. FITZGERALD, L. A.
Dayton, G. H. ....2005.... TX A&M
Saenz, D. ....2004.... TX A&M

SERIES 2

001. OSBORN, H. F.
Gaffney, E. S. ....19??.... Columbia
Gregory, W. K. ....1910.... Columbia 2

002. GREGORY, W. K.
Camp, C. L. ....1923.... Columbia 3
Noble, G. K. ....1922.... Columbia 4
Romer, A. S. ....1921.... Columbia 5
Williams, E. E. ....1950.... Columbia 6

003. CAMP, C. L.
Estes, R. D. ....1960.... CA Berkeley 7
Gregory, J. T. ....1938.... CA Berkeley 8

004. NOBLE, G. K.
Evans, L. T. ....1936.... Harvard

005. ROMER, A. S.
Baird, D. D. ....1955.... Harvard
Carroll, R. L. ....1963.... Harvard 9
Dunkle, D. H. ....1939.... Harvard
Olson, E. C. ....1934.... Chicago............... 10
Parsons, T. S. ....1957.... Harvard
Rivero, J. A ....1953.... Harvard
Thomson, K. S. ....1963.... Harvard 11
Vanzolini, P. E. ....1951.... Harvard 12

006. WILLIAMS, E. E.
Blake, J. A. ....1981.... Harvard
Gans, C. ....1957.... Harvard 13
Gorman, G. C. ....1968.... Harvard 14
Greer, A. E., Jr. ....1969.... Harvard
Hall, W. P., III ....1973.... Harvard
Hertz, P. E. ....1977.... Harvard
Holt, R. D. ....1979.... Harvard
Huey, R. B. ....1976.... Harvard 15
Kiester, A. R. ....1975.... Harvard
Mayer, G. C. ....1989.... Harvard
Miyata, K. I. ....1980.... Harvard
Moermond, T. C. ....1974.... Harvard 16
Rand, A. S. ....1961.... Harvard
Roughgarden, J. D. ....1971.... Harvard 17
Schoener, T. W. ....1970.... Harvard
Seger, J. A. ....1980.... Harvard
Wood, R. C. ....1971.... Harvard

007. ESTES, R. D.
Meszoely, C. A. ....1967.... Boston

008. GREGORY, J. T.
Bramble, D. M. ....1972.... CA Berkeley

009. CARROLL, R. L.
Anderson, J. S. ....2001.... McGill (CA)
Lin, K. ....1994.... McGill (CA)
Meckett, D. ....1996.... McGill (CA)
Reynoso-Rosales, V. H. ....1997.... McGill (CA)

010. OLSON, E. C.
Bader, R. S. ....1954.... Chicago.............. 18
Bolt, J. R. ....1968.... Chicago

Maclean, W. P., III ....1969.... Chicago
Robinson, J. A. ....1976.... CA Los Angeles

011. THOMSON, K. S.
Padian, K. ....1980.... Yale 19

012. VANZOLINI, P. E.
Caramaschi, U. ....1989.... São Paulo (BR) ... 20
Carvalho, C. M. .... 1992 ... Pesquisas Amazônia (BR)
Carvalho-e-Silva, A. M. P. T. .. 1997 .. São Paulo (BR)
Cruz, C. A. G. ....1987.... São Paulo (BR) ... 21
Rebouças-Spicker, R. L. ...1972.... São Paulo (BR)
Rodrigues, M. T. ....1984.... São Paulo (BR) ... 22

013. GANS, C.
Carrier, D. R. ....1988.... MI 23
Moon, B. D. ....1998.... MI
Rosenberg, H. I. ....1988.... SUNY-Buffalo 24
Scanlon, T. ....1982.... MI
Taub, A. M. ....1965.... SUNY-Buffalo

014. GORMAN, G. C.
Adest, G. A. ....1978.... CA Los Angeles
Drewes, R. C. ....1981.... CA Los Angeles
Harwood, R. H., Jr. ....1978.... CA Los Angeles
Lieb, C. S. ....1981.... CA Los Angeles .. 25
Murphy, R. W. ....1982.... CA Los Angeles .. 26
Wyles, J. S. ....1980.... CA Los Angeles

015. HUEY, R. B.
Adolph, S. C. ....1987.... WA
Sinervo, B. R. ....1988.... WA 27
Tsuji, J. S ....1986.... WA
van Berkum, F. H. ....1985.... WA
Wilson, B. S. ....1990.... WA
Zamudio, K. R. ....1996.... WA

016. MOERMOND, T. C.
Crouse, D. T. ....1985.... WI

017. ROUGHGARDEN, J. D
Dial, R. ....1992.... Stanford

018. BADER. R. S.
Underhill, D. K. ....1967.... IL 28

019. PADIAN, K.
Angielczyk, K. D. ....2003.... CA Berkeley
Gauthier, J. ....1984.... CA Berkeley 29
Hutchinson, J. R. ....2001.... CA Berkeley
Parham, J. F. ....2003.... CA Berkeley
Skulan, J. L. ....1999.... CA Berkeley

020. CARAMASCHI, U.
Arata, P. C. R. ....1996.... E Campinas (BR)
Borges-Nojosa, D. M. .. 2002 .. Fed Rio de Janeiro (BR)
Feio. R. N. .... 2002 .. Fed Rio de Janeiro (BR)
Napoli, M. F. .... 2000 .. Fed Rio de Janeiro (BR)
Nascimento, L. B. .... 2003 .. Fed Rio de Janeiro (BR)
Souza, E. M. F. .... 2001 .. Fed Rio de Janeiro (BR)
Weber, L. N. .... 2004 .. Fed Rio de Janeiro (BR)

021. CRUZ, C. A. G.
Sousa, B. M. ....2000 .. Rural Rio de Janeiro (BR)

022. RODRIGUES, M. T.
Bertoluci, J. A. ....1996.... São Paulo (BR)
Carvalho-e-Silva, S. P. ......1994.... São Paulo (BR)
Jünca, F. A. ....1999.... São Paulo (BR)
Martins, J. M. ....1997.... São Paulo (BR)
Narvaes, P. ....2003.... São Paulo (BR)
Rocha, P. L. B. ....1998.... São Paulo (BR)
Sugliano, G. O. S. ....1999.... São Paulo (BR)
Silva, V. X. ....2000.... São Paulo (BR)
Verdade, V. K. ....2006.... São Paulo (BR)

023. CARRIER, D. R.
Lee, D. V. ....2005.... Utah
Ritter, D. A. ....1998.... Brown

024. ROSENBERG, H. I.
Young, B. A. ....1989.... Calgary (CA)

025. LIEB, C. S.
Powell, R. L. ....2003.... TX-El Paso

026. MURPHY, R. W.
Fu, J. ....1999.... Toronto (CA)
Griffith, H. ....1991.... Toronto (CA)
Lowcock, L. A. ....1992.... Toronto (CA)

027. SINERVO, B. R.
Calsbeek, R. G. ....2001.... CA Santa Cruz
Comendant, T. L. ....2002.... CA Santa Cruz

028. UNDERHILL, D. K.
Tabachnick, W. J. ....1974 ... Rutgers

029. GAUTHIER, J.
Clarke, J. A. ....2002.... Yale
Gishlick, A. D. ....2002.... Yale
Joyce, W. G. ....2004.... Yale
Maisano, J. A. ....2000.... Yale
Marshall, C. L. ....2000.... Yale
Smith, K. ....2006.... Yale
Tsuihiji, T. ....2004.... Yale

SERIES 3

001. ?
Redfield, A. C. ....1917.... Harvard 2

002. REDFIELD, A. C.
Clarke, G. C. ....19?? .... ? 3

003. CLARKE, G. C.
Bartholomew, G. A. ....1947.... Harvard 4

004. BARTHOLOMEW, G. A.
Dawson, W. R. ....1953.... CA Los Angeles 5
Lillywhite, H. B. ....1970.... CA Los Angeles 6
Lustick, S. ....1960.... CA Los Angeles 7
MacMillen, R. E. ....1960.... CA Los Angeles 8
Seymour, R. S. ....1972.... CA Los Angeles 9

005. DAWSON, W. R.
Bennett, A. F. ....1971.... MI 10
Buttemer, W. A. ....1981.... MI
Carey, C. ....1976.... MI ....
Crawford, K. M. ....1988.... MI
DeWitt, C. B. ....1963.... MI
Dunson, W. A. ....1965.... MI 11
Gatten, R. E. ....1973.... MI
Lasiewski, R. C. ....1962.... MI 12
Licht, P. ....1964.... MI 13
Minnich, J. E. ....1968.... MI
Moberly, W. R., III ....1966.... MI
Shoemaker, V. H. ....1964.... MI 14
Templeton, A. R. ....1972.... MI 15

006. LILLYWHITE, H. B.
Gregory, L. F. ....2002.... FL
Kam, Y.-C. ....1992.... FL

McCoy, J. J. .... 1987.... FL
Roberts, J. .... 1980.... FL
Smits, A. W. .... 1984.... FL
Zippel, K. C. .... 2000.... FL

007. LUSTICK, S.
Butler, J. A. .... 1978.... OH St
Hinko, A. .... 1980.... OH St

008. MACMILLEN, R. E.
Case, T. J. .... 1974.... CA Irvine
Kaufmann, J. S. .... 1992.... CA Irvine

009. SEYMOUR, R. S.
Booth, D. .... 1981.... Adelaide (AU)
Thompson, M. B. .... 1983.... Adelaide (AU)

010. BENNETT, A. F.
Adams, B. A. .... 1987.... CA Irvine
Cullum, A. J. .... 1997.... CA Irvine
Garland, T., Jr. .... 1985.... CA Irvine 16
Gleeson, T. T. .... 1979.... CA Irvine 17
John-Alder, H. B. .... 1983.... CA Irvine
McLister, J. D. .... 1997.... CA Irvine
Watkins, T. B. .... 1997.... CA Irvine

011. DUNSON, W. A.
Freda, J. .... 1985.... PA St
Horne, M. T. .... 1994.... PA St
Laposata, M. M. .... 1998.... PA St
Mazzotti, F. J. .... 1983.... PA St
Rowe, C. L. .... 1994.... PA St
Sadinski, W. J., Jr. .... 1991.... PA St
Stokes, G. D. .... 1981.... PA St
Warner, S. C. .... 1994.... PA St

012. LASIEWSKI, R. C.
Snyder, G. K. .... 1970.... CA Los Angeles
Weathers, W. W. .... 1969.... CA Los Angeles

013. LICHT, P.
DeNardo, D. F. .... 1994.... CA Berkeley 18
Denver, R. J. .... 1989.... CA Berkeley 19
Feder, M. E. .... 1977.... CA Berkeley
Hayes, T. B. .... 1994.... CA Berkeley 20
Hubbard, G. M. .... 1985.... CA Berkeley
Hugenberger, J. L. .... 1997.... CA Berkeley
Mendonça, M. T. .... 1985.... CA Berkeley 21
Porter, D. A. .... 1986.... CA Berkeley
Ruben, J. A. .... 1975.... CA Berkeley
Stamper, D. L. .... 1992.... CA Berkeley

014. SHOEMAKER, V. H.
Jones, R. M. .... 1978.... CA Riverside
Kingsbury, B. A. .... 1991.... CA Riverside
McClanahan, L. L., Jr. .... 1966.... CA Riverside
Nagy, K. A. .... 1971.... CA Riverside 22
Stinner, J. N. .... 1980.... CA Riverside

015. TEMPLETON, A. R.
Routman, E. J. .... 1990.... WA U
Strasburg, J. L. .... 2004.... WA U

016. GARLAND, T., JR.
Bonine, K. E. .... 2001.... WI
Bulova, S. J. .... 1994.... WI
Dohm, M. R. .... 1994.... WI

017. GLEESON, T. T.
Donovan, E. R. .... 2005.... CO
Hancock, T. V. .... 2001.... CO
Wagner, E. L. .... 1996.... CO

018. DENARDO, D. F.
Taylor, E. N. .... 2005.... AZ St

019. DENVER, R. J.
Boorse, G. C. .... 2004.... MI
Glennemeier, K. A. .... 2000.... MI

020. HAYES, T. B.
Barber, P. H. .... 1998.... CA Berkeley
Buchholz, D. R. .... 1999.... CA Berkeley
Liu, R. .... 2002.... CA Berkeley
Noriega, N. C. .... 2001.... CA Berkeley

021. MENDONÇA, M. T.
Ward, C. K. .... 2005.... Auburn

022. NAGY, K. A.
Brown, T. K. .... 1999.... CA Los Angeles
Fusari, M. H. .... 1982.... CA Los Angeles
Henen, B. T. .... 1995.... CA Los Angeles
Karasov, W. H. .... 1981.... CA Los Angeles .. 23
Peterson, C. C. .... 1993.... CA Los Angeles

023. KARASOV, W. H.
Jung, R. E. .... 1996.... WI

SERIES 4

001. ?
Allen, A. A. .... 1910.... Cornell 2

002. ALLEN, A. A.
Dilger, W. C. .... 1955.... Cornell 3
Hensley, M. M. .... 1951.... Cornell 4
Tanner, J. T. .... 1940.... Cornell 5
Welter, W. A. .... 1932.... Cornell

003. DILGER, W. C.
Wiewandt, T. .... 1977.... Cornell

004. HENSLEY, M. M.
Gibbons, J. W. .... 1967.... MI St 6
Nelson, D. H. .... 1974.... MI St
Premo, D. B. .... 1985.... MI St
Wacasey, J. W. .... 1961.... MI St

005. TANNER, J. T.
Bickert, J. H. .... 1970.... TN
Orr, L. P. .... 1962.... TN 7

006. GIBBONS, J. W.
Benabib, M. .... 1991.... GA 8
Buhlmann, K. A. .... 1998.... GA
Burke, V. J. .... 1995.... GA
Camp, C. D. .... 1985.... GA
Lamb, A. C. .... 1986.... GA
Lovich, J. E. .... 1990.... GA
Mayack, D. T. .... 1988.... GA
Semlitsch, R. D. .... 1984.... GA 9

007. ORR, L. P.
Bobes, M. .... 1973.... Kent St
Brodman, R. D. .... 1992.... Kent St
Collier, A. .... 2001.... Kent St
Davic, R. D. .... 1983.... Kent St
Fitzpatrick, L. C. .... 1970.... Kent St 10
Keen, W. H. .... 1975.... Kent St
Kleeberger, S. R. .... 1982.... Kent St
Maple, W. T. .... 1974.... Kent St

Matson, T. O. ....1988.... Kent St
Shaefer, G. C. ....1972.... Kent St
White, E. M. ....2003.... Kent St

008. BENABIB, M.
Ramírez-Bautista, A. ....1995.... UNAM (ME) 11

009. SEMLITSCH, R. D.
Birchfield, G. L. ....2002.... MO
Boone, M. D. ....2000.... MO
Bridges, C. M. ....1999.... MO
Crawford, J. A. ....2007.... MO
Harper, E. B. ....2007.... MO
James, S. M. ....2005.... MO
Johnson, J. R. ....2005.... MO
Mills, N. E. ....2002.... MO
Parris, M. J. ....1999.... MO
Rothermel, B. B. ....2003.... MO
Ryan, T. J. ....2000.... MO

010. FITZPATRICK, L. C.
McAllister, C. T. ....1989.... N TX St

011. RAMÍREZ-BAUTISTA, A.
Balderas-Valdivia, C. J. ....2004.... UNAM (ME)

SERIES 5

001. MCCLUNG, C.
Taylor, E. H. ....1926.... KS 2

002. TAYLOR, E. H.
Reeve, W. ....1950.... KS
Smith, H. M. ....1936.... KS 3
Tanner, W. W. ....1950.... KS 4

003. SMITH, H. M.
Bell, E. L., II ....1954.... IL
Brandon, R. A. ....1962.... IL 5
Chrapliwy, P. S. ....1964.... IL
Hammerson, G. ....1981.... CO
Landy, M. J. ....1970.... IL
Langebartel, D. A. ....1958.... IL
Liem, K. F. ....1969.... IL 6
List, J. C. ....1956.... IL 7
Northcutt, G. ....1968.... IL
Oldham, J. ....1975.... CO
Olsen, R. E. ....1978.... CO
Scudder, K. M. ....1982.... CO
Smith, P. W. ....1953.... IL 8
Swain, T. A. ....1977.... CO
Taylor, H. L. ....1983.... CO
White, F. N. ....1955.... IL

004. TANNER, W. W.
Avery, D. F. ....1970.... Brig Young
Burkholder, G. L. ....1973.... Brig Young
Cox, D. C. ....1976.... Brig Young
Larsen, K. R. ....1973.... Brig Young
Smith, N. M. ....1972.... Brig Young

005. BRANDON, R. A.
Besharse, J. C. ....1973.... S IL
Buckley, L. J. ....1997.... S IL
Carr, J. L. ....1991.... S IL
Johnston, G. F. ....1990.... S IL
Labanick, G. M. ....1978.... S IL
Morris, M. A. ....1987.... S IL
Reilly, S. M. ....1986.... S IL 9
Williams, A. A. ....1976.... S IL
Wortham, J. W., Jr. ....1975.... S IL

006. LIEM, K. F.
Brainerd, E. L. ....1991.... Harvard 10
Irish, F. J. ....1987.... Harvard
Rose, C. ....1992.... Harvard

007. LIST, J. C.
Atkinson, M. J. ....1985.... Ball St
Branham, A. E. ....1977.... Ball St
Kramer, D. C. ....1971.... Ball St
Miller, D. E. ....1985.... Ball St
Mosher, D. D. ....1982.... Ball St
Shepherd, J. L. ....1973.... Ball St
Smith, D. L. ....1971.... Ball St

008. SMITH, P. W.
Liem, S. S. ....1969.... IL

009. REILLY, S. M.
Larson, P. M. ....2003.... OH
McBrayer, L. D. ....2002.... OH

010. BRAINERD, E. L.
Azizi, E. ....2005.... MA-Amherst
Kley, N. J. ....2001.... MA-Amherst

SERIES 6

001. PARK, O.
Hairston, N. G., Sr. ....1948.... Northwestern 2

002. HAIRSTON, N. G., SR.
Brockelman, W. Y. ....1968.... MI
Dundee, H. A. ....1958.... MI 3
Gill, D. E. ....1971.... MI 4
Nishikawa, K. C. ....1985.... NC-Chapel Hill..... 5
Organ, J. A. ....1960.... MI
Smith, C. K. ....1988.... NC-Chapel Hill
Southerland, M. T. ....1985.... NC-Chapel Hill
Stenhouse, S. L. ....1984.... NC-Chapel Hill
VanDevender, R. W. ....1975.... MI

003. DUNDEE, H. A.
Bayless, L. E. ....1966.... Tulane
Hart, D. R. ....1979.... Tulane
Rose, F. L. ....1965.... Tulane 6
Sever, D. M. ....1974.... Tulane

004. GILL, D. E.
Bervan, K. A. ....1981.... MD
Mock, B. A. ....1983.... MD

005. NISHIKAWA, K. C.
Anderson, C. W. ....1996.... N AZ
Gray, L. A. ....1997.... N AZ
Monroy, J. A. ....2004.... N AZ
O'Reilly, J. C. ....1998.... N AZ
Wiltenmuth, E. B. ....1997.... N AZ

006. ROSE, F. L.
Armentrout, D. ....1973.... TX Tech
Judd, F. W. ....1973.... TX Tech
Long, D. R. ....1986.... TX Tech
Selcer, K. W. ....1986.... TX Tech

SERIES 7

001. ?
Bagnara, J. T. ....1956.... IA 2

002. BAGNARA, J. T.
Briggs, N. J. ....1977.... AZ
Frost, J. S. ....1976.... AZ.................... 3
Frost-Mason, S. K. ....1978.... AZ.................... 4
Morrisett, F. W. ....1986.... AZ
Potter, S. J. W. ....1970.... AZ
Sherbrooke, W. C. ....1988.... AZ
Taylor, J. D. ....1967.... AZ
003. FROST, J. S.
Hillis, D. M. ....1985.... KS.................... 5
004. FROST-MASON, S. K
Dean, A. D. ....1992.... KS
Morrison, R. L. ....1993.... KS
Thibaudeau, D. G. ....1991.... KS
005. HILLIS, D. M.
Brown, R. M. ....2004.... TX
Chippendale, P. T. ....1995.... TX.................... 6
de Sá, R. O. ....1991.... TX
Holder, M. T. ....2001.... TX
Holloway, A. K. ....2004.... TX
Huelsenbeck, J. P. ....1995.... TX
Krejca, J. K. ....2005.... TX
Mable, B. K. ....1996.... TX
Smith, C. R. ....1989.... TX
006. CHIPPENDALE, P. T.
Baldwin, A. S. ....2002.... TX-Arlington
Noonan, B. P. ....2004.... TX-Arlington

SERIES 8

001. ?
Hess, E. H. ....1948.... John Hopkins........ 2
002. HESS, E. H.
Burghardt, G. M. ....1966.... Chicago................ 3
003. BURGHARDT, G. M.
Andreadis, P. T. ....1998.... TN
Bock, B. C. ....1984.... TN
Bowers, B. B. ....1992.... TN
Davis, M. ....1981.... TN
Drummond, H. M. ....1980.... TN.................... 4
Dugan, B. A. ....1980.... TN
Froese, A. D. ....1974.... TN
Gove, D. ....1978.... TN
Greene, H. W. ....1977.... TN.................... 5
Halloy de Grosee, M. ....1995.... TN
Krause, M. A. ....2000.... TN
Lyman Henley, L. P. ....1991.... TN
Mehta, R. S. ....2006.... TN
Rivas, J. A. ....1999.... TN
Savitsky, B. A. ....1989.... TN
Schwartz, J. M. ....1989.... TN
Scudder, L. M. ....1983.... TN
Waters, R. M. ....2000.... TN
Weldon, P. J. ....1983.... TN
004. DRUMMOND, H. M.
Manjarrez-Silva, J. ....2003.... UNAM (ME)
005. GREENE, H. W.
Autumn, K. ....1995.... CA Berkeley
Carothers, J. H. ....1987.... CA Berkeley
Huang, W.-S. ....2004.... Cornell
Jaksic, F. M. ....1982.... CA Berkeley
Losos, J. B. ....1989.... CA Berkeley 6
Reiserer, R. S. ....2001.... CA Berkeley
Roberts, W. E. ....1994.... CA Berkeley
Rodríguez Robles, J. A. ....1998.... CA Berkeley
Seib, R. L. ....1985.... CA Berkeley
006. LOSOS, J. B.
Butler, M. A. ....1998.... WA U
Creer, D. A. ....2001.... WA U
Glor, R. E. ....2004.... WA U
Irschick, D. J. ....1996.... WA U
Leal, M. ....2000.... WA U
Parks, J. W., III ....2001.... WA U

SERIES 9

001. ?
Borror, D. J. ....1935.... OH St.................... 2
002. BORROR, D. J.
Alexander, R. D. ....1956.... OH St.................... 3
Kolson, R. A. ....1973.... OH St
003. ALEXANDER, R. D.
Howard, R. D. ....1977.... MI.................... 4
Pace, A. E. ....1972.... MI
Sherman, C. K. ....1980.... MI
004. HOWARD, R. D.
Harris, W. E. ....2002.... Purdue
King, R. B. ....1985.... Purdue.................. 5
Whiteman, H. H. ....1994.... Purdue
005. KING. R. B.
Bittner, T. D. ....2000.... N IL

SERIES 10

001. ?
Dillon, L. S. ....1954.... TX A&M 2
002. DILLON, L. S.
Dixon, J. R. ....1962.... TX A&M 3
Putnam, J. L. ....1970.... TX A&M
003. DIXON, J. R.
Camper, J. D. ....1990.... TX A&M
Dean, R. H. ....1984.... TX A&M
Forstner, M. R. J. ....1995.... TX A&M
Gallagher, D. S., Jr. ....1979.... TX A&M
Greene, B. D. ....1993.... TX A&M
Hayes, L. ....1992.... TX A&M
Hendricks, F. S. ....1975.... TX A&M
Johnson, J. D. ....1984.... TX A&M
Ketchersid, C. A. ....1974.... TX A&M
Kofron, C. P. ....1980.... TX A&M
Kroll, J. C. ....1973.... TX A&M
Lavín-Murcio, P. A. ....1998.... TX A&M
Mather, C. M. ....1976.... TX A&M
Mora, J. M. ....1991.... TX A&M
Rivero-Blanco, C. V. ....1979.... TX A&M
Saxon, J. G. ....1970.... TX A&M
Scuddy, J. F. ....1971.... TX A&M
Sites, J. W., Jr. ....1980.... TX A&M 4
Thomas, R. A. ....1976.... TX A&M
Vaugh, R. K. ....1991.... TX A&M
Wiest, J. A., Jr. ....1978.... TX A&M

004. SITES, J. W., JR.
Arévalo, E. .... 1992.... Brig Young
de Silva, N. J., Jr. .... 1995.... Brig Young
Marshall, J. C. .... 2004.... Brig Young
Reed, J. M. .... 1991.... Brig Young............ 5
Whiting, A. S. .... 2004.... Brig Young
005. REED, J. M.
Homan, R. N. .... 2003.... Tufts
Regosin, J. V. .... 2003.... Tufts

SERIES 11

001. [DOBZHANSKY, T.] .... 1921.... Kiev (UR)
Ayala, F. J. .... 1964.... Columbia 2
002. AYALA, F. J.
Avise, J. C. .... 1975.... CA Davis 3
003. AVISE, J. C.
Bermingham, E. L. .... 1986.... GA
Bowen, B. W. .... 1992.... GA.......................... 4
Douglas, M. E. .... 1978.... GA.......................... 5
Karl, S. A. .... 1992.... GA
Pearse, D. E. .... 2001.... GA
Walker, L. D. .... 1998.... GA
004. BOWEN, B. W.
Kinlaw, A. E. .... 2007.... FL
005. DOUGLAS, M. E.
Holycross, A. T. .... 2002.... AZ St

SERIES 12

001. ?
Gorbman, A. .... 1940.... CA Berkeley 2
002. GORBMAN, A.
Clark, N. B. .... 1965.... Columbia
Ergezen, S. S. .... 1976.... WA
Kaye, N. W. .... 1960.... Columbia
Norris, D. O. .... 1966.... WA........................ 3
Solursh, M. .... 1969.... WA
003. NORRIS, D. O.
Carr, J. A. .... 1988.... CO
Gern, W. A. .... 1976.... CO
Moore, F. L. .... 1974.... CO.......................... 4
Norman, M. F. .... 1985.... CO
Pancak, M. K. .... 1987.... CO
Platt, J. E. .... 1974.... CO.......................... 5
Snow, G. E. .... 1977.... CO
004. MOORE, F. L.
Boyd, S. K. .... 1987.... OR St
Lowry, C. .... 1995.... OR St
Orchinik, M. .... 1992.... OR St
Petersen, S. .... 1984.... OR St
Propper, C. R. .... 1989.... OR St
Zoeller, T. .... 1984.... OR St
005. PLATT, J. E.
Ibrahim, R. H. .... 1989.... Denver
Rao, A. K. .... 1986.... Denver

SERIES 13

001. ?
Lehrman, D. S. .... 1954.... NY 2
002. LEHRMAN, D. S.
Crews, D. .... 1973.... Rutgers.................. 3
003. CREWS, D.
Bergeron, J. M. .... 1997.... TX
Coomber, P. .... 1995.... TX
Day, E. B. .... 1999.... TX
Fleming, A. A. .... 2002.... TX
Garstka, W. R. .... 1982.... Harvard
Lindzey, J. K. .... 1990.... TX
Mason, R. T. .... 1987.... TX......................... 4
Rhen, T. E. .... 2000.... TX
Sakata, J. T. .... 2002.... TX
Tousignant, A. J. .... 1993.... TX
Wade, J. S. .... 1992.... TX
Wennstrom, K. L. .... 1997.... TX
Whittier, J. M. .... 1986.... TX
Woolley, S. C. .... 2002.... TX
Young, L. J. .... 1994.... TX
004. MASON, R. T.
Greene, M. J. .... 1998.... OR St
LeMaster, M. P. .... 2002.... OR St
Moore, I. T. .... 2000.... OR St

SERIES 14

001. ?
McVaughn, R. .... 1935.... PA.......................... 2
002. MCVAUGH, R.
Denton, M. F. .... 1971.... MI 3
003. DENTON, M. F.
Olmstead, R. G. .... 1988.... WA........................ 4
004. OLMSTEAD, R. G.
Goebel, A. M. .... 1996.... CO

SERIES 15

001. ?
Weiss, P. .... 1942.... NY 2
002. WEISS, P.
Kollros, J. J. .... 1948.... Chicago................ 3
003. KOLLROS, J. J.
Baird, J. J. .... 1957.... IA
Bovbjerg, A. M. .... 1961.... IA
Decker, R. S. .... 1970.... IA
Eichler, V. B. .... 1969.... IA
Flanigan, N. S. .... 1958.... IA
Harrison, J. L. .... 1958.... IA
Just, J. J. .... 1968.... IA.......................... 4
Kaung, H. C. .... 1971.... IA
Martin, E. W. .... 1962.... IA
Naughten, J. C. .... 1971.... IA
Payne, W. F. .... 1958.... IA
Petsetsky, I. .... 1959.... IA
Pollack, E. D. .... 1970.... IA
Race, J., Jr. .... 1960.... IA
Reynolds, W. A. .... 1962.... IA
Shell, L. C. .... 1959.... IA
Stephens, L. B., Jr. .... 1957.... IA
004. JUST, J. J.
Forman, L. J. .... 1979.... KY

SERIES 16

001. ?
Wilson, E. O. .... 1955.... Harvard................. 2
002. WILSON, E. O.
Simberloff, D. S. .... 1969.... Harvard................. 3
003. SIMBERLOFF, D. S.
Means, D. B. .... 1975.... FL St...................... 4
004. MEANS, D. B.
Brock, O. G. .... 1980.... FL St

SERIES 17

001. ?
Carlisle, H. J. .... 1964.... WA......................... 2
002. CARLISLE, H. J.
Crawshaw, L. I. .... 1970.... CA Santa Barbara .. 3
003. CRAWSHAW, L. I.
O'Connor, C. S. .... 1993.... Portland St

SERIES 18

001. CODY, M.
Regal, P. J. .... 1968.... CA Los Angeles..... 2
002. REGAL, P. J.
Karns, D. R. .... 1984.... MN
Lang, J. W. .... 1977.... MN 3
003. LANG, J. W.
Baumgardt, L. L. .... 1997.... ND

SERIES 19

001. DANIEL, J. F.
Eakin, R. M. .... 1935.... CA Berkeley 2
002. EAKIN, R. M.
Kemp, N. E. .... 1941.... CA Berkeley 3
003. KEMP, N. E.
Allen, A. C., Jr. .... 1970.... MI
Dugan, T. S. .... 1978.... MI
Lee, P. A. .... 1965.... MI

SERIES 20

001. FORBES, W. T. M.
Chermock, R. L. .... 1947.... Cornell.................. 2
002. CHERMOCK, R. L.
Boyles, J. M. .... 1966.... AL
Cliburn, J. W. .... 1960.... AL......................... 3
003. CLIBURN, J. W.
Brode, W. E. .... 1969.... S MS
Crain, J. L. .... 1966.... S MS
Donovan, L. A. .... 1980.... S MS
Meade, L. E. .... 1993.... S MS
Moody, J. G. .... 1993.... S MS
Nalepa, B. R. .... 1993.... S MS
Schwab, D. S. .... 1972.... S MS
Woods, J. E. .... 1968.... S MS

SERIES 21

001. ?
Hudson, G. E. .... 1936.... NB 2
002. HUDSON, G. E.
Dunlap, D. G. .... 1955.... WA St.................... 3
003. DUNLAP, D. G.
Anderson, D. W. ....1967.... SD
Timken, R. L. ....1968.... SD

SERIES 22

001. ?
Keeton, W. T. ....1958.... Cornell................. 2
002. KEETON, W. T.
Howland, H. C. ....1968.... Cornell................. 3
003. HOWLAND, H. C.
Ballarino, J. ....1982.... Cornell
Fleischman, L. J. ....1986.... Cornell

SERIES 23

001. LETTVIN, J.
Capranica, R. R. ....1965.... MA I Tech............ 2
002. CAPRANICA, R. R.
Constantine-Paton, M. ......1976.... Cornell
Narins, P. M. ....1976.... Cornell................. 3
Rose, G. ....1983.... Cornell
Zakon, H. ....1981.... Cornell
003. NARINS, P. M.
Zelick, R. D. ....1984.... CA Los Angeles

SERIES 24

001. ?
Meslow, E. C. ....1970.... WI......................... 2
002. MESLOW, C.
Morrison, M. L. ....198?.... OR ST.................. 3
003. MORRISON, M. L.
Welsh, H. H., Jr. ....1993.... CA Berkeley

SERIES 25

001. REEDER, W. G.
Ream, C. H. ....1967.... WI
Riechert, S. E. ....1973.... WI......................... 2
Vogt, R. C. ....1978.... WI......................... 3
002. RIECHERT, S. E.
Middendorf, G. A. ....1979.... TN
003. VOGT, R. C.
Fachin-Teran, A. R. C. ......2000.... Pesquisas (BR)

SERIES 26

001. ?
Sealander, J. A., Jr. ....1949.... IL 2
002. SEALANDER, J. A., JR.
Elick, G. E. ....1970.... AR
Spotila, J. R. ....1970.... AR........................ 3
003. SPOTILA, J. R.
Bell, B. A. ....2005.... Drexel
Curtin, A. J. ....2006.... Drexel
Elwood, J. R. L. ....2003.... Drexel
Hertler, H. ....2002.... Drexel
Leslie, A. J. ....1997.... Drexel
Penick, D. N. ....1996.... Drexel
Steyermark, A. C. ....1999.... Drexel
Wallace, B.P. ....2005.... Drexel

SERIES 27

001. ?
Dusi, J. L. ....1950.... OH St........ 2
002. DUSI, J. L.
Folkerts, G. W. ....1968.... Auburn........ 3
Fugler, C. M. ....1967.... Auburn
003. FOLKERTS, G. W.
Caldwell, R. S. ....1977.... Auburn

SERIES 28

001. PARK, T.
Cole, L. C. ....1944.... Chicago........ 2
002. COLE, L. C.
Hurlbert, S. H. ....1968.... Cornell
Sheldon, A. L. ....1966.... Cornell........ 3
003. SHELDON, A. L.
Daugherty, C. H. ....1979.... MT

SERIES 29

001. ?
Spratt, N. T., Jr. ....1940.... Rochester........ 2
002. SPRATT, N. T., JR.
Browder, L. W. ....1967.... MN
McKinnell, R. G. ....1959.... MN........ 3
003. MCKINNELL, R. G.
Carlson, R. G. ....1977.... MN
Ellinger, M. S. ....1976.... MN
King, D. R. ....1979.... MN
Rollins, L. A. ....1977.... MN
Sujarit, V. K. ....1982.... MN

SERIES 30

001. ?
Rich, E. R. ....1954.... Chicago........ 2
002. RICH, E. R.
King, F. W. ....1966.... MI........ 3
003. KING, F. W.
Hall, P. M. ....1990.... FL
Janes, D. E. ....2004.... FL
Krysko, K. L. ....2001.... FL
Seijas, A. E. ....1998.... FL
Thorbjarnarson, J. B. ....1991.... FL
Verdade, L. M. ....1997.... FL
Vliet, K. A. ....1987.... FL

SERIES 31

001. ?
Ricklefs, R. E. ....1967.... PA........ 2
002. RICKLEFS, R. E.
Reznick, D. N. ....1980.... PA........ 3
003. REZNICK, D. N.
Halama, K. J. ....1998.... CA Riverside
Morey, S. R. ....1994.... CA Riverside

SERIES 32

001. ?
Terborgh, J. W. ....1963.... Harvard........ 2
002. TERBORGH, J. W.
Toft, C. A. ....1977.... Princeton........ 3
003. TOFT, C. A.
Nabors, L. K. ....2005.... CA Davis

SERIES 33

001. ?
Twitty, V. C. ....1929.... Yale........ 2
002. TWITTY, V. C.
Coates, M. L. ....1967.... Stanford........ 3
DeLanney, L. E. ....1940.... Stanford
Heath, H. D. ....1950.... Stanford
Ray, J. C. ....1945.... Stanford
Triplett, E. L. ....1956.... Stanford
003. COATES, M.
Roberts, J. D. ....1976.... Adelaide (AU)

SERIES 34

001. DE ROOS, R. M.
Gist, D. H. ....1968.... MO........ 2
Jacobson, E. R. ....1975.... MO........ 3
Piper, G. D. ....1970.... MO
Varma, M. M. ....1971.... MO
002. GIST, D. H.
Gribbins, K. M. ....2003.... Cincinnati
003. JACOBSON, E. R.
Maxwell, L. K. ....2001.... FL
Origgi, F. C. ....2001.... FL

SERIES 35

001. CHILD, C. M.
Shelford, V. E. ....1907.... Chicago........ 2
002. SHELFORD, V. E.
Kendeigh, S. C. ....1930.... IL........ 3
003. KENDEIGH, S. C.
Bresler, J. B. ....1957.... IL
Heckrotte, C. ....1960.... IL

SERIES 36

001. GIESE, A. C.
Lawrence, J. M. ....1966.... Stanford........ 2
002. LAWRENCE, J. M.
Diehl, W. J. ....1983.... S FL........ 3
003. DIEHL, W. J.
Hranitz, J. M. ....1993.... MS St

SERIES 37

001. ?
Klopfer, P. H. ....1958.... Yale........ 2
002. KLOPFER, P. H.
Fairchild, L. ....1979.... Duke
Hailman, J. P. ....1964.... Duke........ 3
Petras, K. M. ....1995.... Duke
003. HAILMAN, J. P.
Northern, P. T. ....1970.... WI
Sustare, B. D. ....1976.... WI

SERIES 38

001. ?
McNab, B. K. ....1962.... WI........ 2
002. MCNAB, B. K.
Krakauer, T. H. ....1970.... FL

Ultsch, G. R. ....1972....FL........................3
Wygoda, M. L. ....1982....FL
003. ULTSCH, G. R.
Crowder, W. C. ....1997....AL
Reese, S. A. ....2002....AL

SERIES 39

001. ?
Odum, E. P. ....1939....IL 2
002. ODUM, E. P.
de la Cruz, A. A. ....1965....GA.......................3
003. DE LA CRUZ, A. A.
Lue, K.-Y. ....1976....MS St

SERIES 40

001. ?
Templeton, J. R. ....1959....OR........................2
002. TEMPLETON, J. R.
Claussen, D. L. ....1971....MT 3
003. CLAUSSEN, D. L.
Costanzo, J. P. ....1988....Miami U
Finkler, M. S. ....1998....Miami U
Jiang, S. ....1991....Miami U
Manis, M. L. ....1985....Miami U
Sugalski, M. T. ....2000....Miami U

SERIES 41

001. ?
Allendorf, F. W. ....1975....WA.......................2
002. ALLENDORF, F. W.
Funk, W. C. ....2004....MT

SERIES 42

001. ?
Aronson, L. R. ....1945....NY 2
002. ARONSON, L. R.
Kunstadt, L. P. ....1977....NY

SERIES 43

001. AGASSIZ, A.
Barbour, T. ....1911....Harvard................2
002. BARBOUR, T.
Bailey, J. W. ....1927....Harvard
Dunn, E. R. ....1921....Harvard

SERIES 44

001. ?...
Baird, I. L. ....1960....Harvard................2
002. BAIRD, I. L.
White, J. S. ....1979....PA St

SERIES 45

001. BLACK, I.
Brick, I. ....1960....NY 2
002. BRICK, I.
Leblanc, J. M. ....1977....NY
Rivera-Colon, M. ....1984....NY
Twerksy, L. H. ....1981....NY
Vlasto, G. S. ....1985....NY

SERIES 46

001. ?
Bothner, R. C. ....1959....Fordham................2
002. BOTHNER, R. C.
Cerrone, T. H. ....1983....St Bonaventure

SERIES 47

001. CALDER, W. A., III
Schwalbe, C. R. ....1981....AZ.........................2
002. SCHWALBE, C. R.
Bounds, D. L. ....1996....AZ
Nichols, W. J. ....2003....AZ
Seminoff, J. A. ....2000....AZ

SERIES 48

001. ?
Conaway, C. H. ....1955....WI 2
002. CONAWAY, C. H
Philibert, R. H. ....1964....MO
Wilkinson, R. F., Jr. ....1965....MO

SERIES 49

001. ?
Dent, J. N. ....1941....John Hopkins........2
002. DENT, J. N.
Benson, D. G., Jr. ....1965....VA

SERIES 50

001. ?
DeVillez, E. J. M. ....1964....IL 2
002. DEVILLEZ, E. J. M.
Justis, C. S. ....1985....Miami U

SERIES 51

001. ?
Dickinson, J. C., Jr. ....1951....FL.........................2
002. DICKINSON, J. C., JR.
Wharton, C. ....1958....FL

SERIES 52

001. DODD, J. L.
Tanner, G. W. ....1979....TX A&M 2
002. TANNER, G. W.
Babbitt, K. J. ....1996....FL.........................3
003. BABBITT, K. J.
La Fiandra, E. M. ....2006....NH

SERIES 53

001. ?
DuShane, G. P. ....1934....Yale 2
002. DUSHANE, G. P.
Steaner, S. P. ....1946....Chicago

SERIES 54

001. ?
Eckhardt, R. A. ....1969....Catholic.................2
002. ECKHARDT, R. A.
Risley, M. S. ....1977....CUNY

SERIES 55

001. ?
Edds, M. V. ....1943.... Yale 2
002. EDDS, M. V.
Luckenbill, L. M. ....1958.... Brown

SERIES 56

001. ?
Elinson, R. P. ....1970.... Yale 2
002. ELINSON, R. P.
Beckham, Y. M. ....2003.... Toronto (CA)
Callery, E. M. ....1999.... Toronto (CA)
Holowacz, T. A. ....1995.... Toronto (CA)

SERIES 57

001. ?
Esch, G. W. ....1963.... OK 2
002. ESCH, G. W.
Bourque, J. E. ....1974.... Wake Forest

SERIES 58

001. EVANS, H. ENSIGN
Valentine, B. D. ....1960.... Cornell 2
002. VALENTINE, B. D.
Handler, C. C. ....1971.... OH St
Juterbock, J. E. ....1979.... OH St
Rubenstein, N. M. ....1969.... OH St

SERIES 59

001. ?
Evans, W. L. ....1956.... TX 2
002. EVANS, W. L.
Woods, W. H. ....1973.... AR

SERIES 60

001. ?
Fischer, M. S. ....1969.... CA Berkeley 2
002. FISCHER, M. S.
Maxwell, E. S. ....1977.... MA

SERIES 61

001. ?
Fisher, S. K. ....1969.... Purdue 2
002. FISHER. S. K.
Bernstein, S. A. ....1985.... CA Santa Barbara

SERIES 62

001. ?
Formanowicz, D. R., Jr. ....1982.... SUNY 2
002. FORMANOWICZ, D. R., JR.
DeVito, J. ....2003.... TX-Arlington
Gerson, M. M. ....2004.... TX-Arlington
O'Connell, D. J. ....1998.... TX-Arlington

SERIES 63

001. ?
Frye, B. E. ....1956.... Emory 2
002. FRYE, B. E.
Bartell, M. H. ....1969.... MI
Brown, P. R. S. ....1968.... MI
Farrar, E. S. ....1972.... MI
Noonen, R. A. ....1977.... MI
Snyder, B. W. ....1970.... MI

SERIES 64

001. ?
Golley, F. B. ....1958.... MI St 2
002. GOLLEY, F. B.
Smith, G. C. ....1971.... GA

SERIES 65

001. ?
Gordon, M. S. ....1958.... Yale 2
002. GORDON, M. S.
Bishop, L. G. ....1965.... CA Los Angeles
Hillman, S. ....1975.... CA Los Angeles
McNabb, R. A. ....1968.... CA Los Angeles
Putnam, R. W. ....1978.... CA Los Angeles

SERIES 66

001. ?
Grant, P. ....1953.... Columbia 2
002. GRANT, P.
Longley, A. J. W. ....1974.... OR

SERIES 67

001. ?
Greenwald, L. I. ....1970.... Duke 2
002. GREENWALD, L. I.
Munford, J. W. ....1978.... OH St

SERIES 68

001. ?
Haas, C. A. ....1990.... Cornell 2
002. HAAS, C. A.
Bishop, D. C. ....2005.... VA Tech

SERIES 69

001. HASLER, A. D.
Doepke, P. A. ....1970.... WI
Nursall, J. R. ....1953.... WI 2
002. NURSALL, J. R.
Jande, S. S. ....1974.... Alberta (CA)

SERIES 70

001. ?
Humphries, A. A., Jr. ....1953.... Princeton 2
002. HUMPHRIES, A. A., JR.
del Pino, E. M. ....1972.... Emory
Freeman, S. B. ....1967.... Emory
Hughes, W. N. ....1958.... Emory

SERIES 71

001. ?
Jacobs, L. L. ....1977.... AZ 2
002. JACOBS, L. L.
Head, J. J. ....2002.... S Methodist
Kobayashi, Y. ....2004.... S Methodist
Lu, J. ....2004.... S Methodist

SERIES 72

001. ?
    Kilham, S. S. ....1971.... Duke 2
002. KILHAM, S. S.
    Hunte-Brown, M. E. ....2006.... Drexel

SERIES 73

001. KING, J. R.
    Peterson, C. R. ....1982.... WA St................... 2
002. PETERSON, C. R.
    Bartelt, P. E. ....2000.... ID St
    Cobb, V. A. ....1994.... ID St
    Cossel, J. O., Jr. ....2003.... ID St
    Dorcas, M. E. ....1995.... ID St
    Merriam, J. L. ....2002.... ID St
    Patla, D. A. ....1997.... ID St
    Pilliod, D. S. ....2001.... ID St

SERIES 74

001. MARSHALL, W.
    Tester, J. R. ....1960.... MN....................... 2
002. TESTER, J. R.
    Ewert, M. A. ....1969.... MN

SERIES 75

001. ?
    McCauley, W. J. ....1955.... S CA..................... 2
002. MCCAULEY, W. J.
    Campbell, C. E ....1974.... AZ

SERIES 76

001. ?
    M'Closkey, R. T. ....1970.... CA Irvine
002. M'CLOSKEY, R. T.
    Baia, K. A. ....198?.... Windsor (CA)
    Deslippe, R. ....1989.... Windsor (CA)
    Hecnar, S. J. ....1997.... Windros (CA)
    Seburn, C. N. L. ....1990.... Windsor (CA)

SERIES 77

001. ?
    Malvestuto, S. P. ....1978.... Auburn................. 2
002. MALVESTUTO, S. P.
    Henword, T. A. ....1987.... Auburn

SERIES 78

126. ?
    Massare, J. A. ....1984.... John Hopkins........ 2
002. MASSARE, J. A.
    Callaway, J. M. ....1989.... Rochester
    Sheldon, M. A. ....1996.... Rochester

SERIES 79

001. MILLER, R. S.
    Fox, S. ....1973.... Yale...................... 2
002. FOX, S.
    Bontrager, S. K. ....1980.... OK St
    Bruner, M. A. ....2002.... OK St
    Cline, G. R. ....1990.... OK St
    McCoy, J. K. ....1995.... OK St
    Rostker, M. A. ....1983.... OK St
    Shipman, P. A. ....2002.... OK St
    Stovall, R. H. ....1982.... OK St

SERIES 80

001. MORRIS, R. W.
    Grigg, G. C. ....1968.... OR......................... 2
002. GRIGG, G. C.
    Courtice, G. P. ....1978.... Sydney (AU)
    Kirshner, D. ....1985.... Sydney (AU)
    Seebacher, F. ....1994.... Queensland (AU)
    Stokes, E. E. ....1986.... Sydney (AU)
    Taplin, L. ....1982.... Sydney (AU)
    Thompson, J. ....1977.... Sydney (AU)
    Van Burden, E. ....1977.... Sydney (AU)
    Wright, J. C. ....1985.... Sydney (AU)

SERIES 81

001. ?
    Percival, H. F. ....1972.... Clemson................ 2
002. PERCIVAL, H. F.
    Maskey, T. M. ....1989.... FL
    Rice, K. G. ....1996.... FL
    Schmid, J. R. ....2000.... FL

SERIES 82

001. PAINE, R. T.
    Power, M. E. ....1981.... WA....................... 2
002. POWER, M. E.
    Kupferberg, S. J. ....1996.... CA Berkeley
    Sabo, J. L. ....2000.... CA Berkeley

SERIES 83

001. PRANGE, H. D.
    Ackerman, R. A. ....1975.... FL.......................... 2
002. ACKERMAN, R. A.
    Davis, T. A. ....1987.... IA St
    Lott, D. B. ....1998.... IA St
    Rimkus, T. A. ....1996.... IA St

SERIES 84

001. ROBERTS, J. F.
    Wiley, J. E. ....1981.... NC St.................... 2
002. WILEY, J. E.
    Bonner, D. S. ....2005.... E Carolina

SERIES 85

001. ROSE, S. M.
    Akin, G. C. ....1965.... Tulane
    Clarke, B. J. ....1974.... Tulane
    Colville, J. E. ....1968.... Tulane
    Dial, N. A. ....1960.... IL
    Donaldson, D. J. ....1968.... Tulane.................. 2
    Richards, C. M. ....1957.... IL
    Westly, S. K. ....1979.... Tulane
002. DONALDSON, D. J.
    Dunlap, M. K. ....1979.... TN
    Wertz, R. L. ....1978.... TN

SERIES 86

001. SCHLEIDT, W. M.
    Madison, D. M. ....1971....MD.......................2
002. MADISON, D. M.
    Lamoureux, V. S. ....2000....SUNY-Binghamton
    Maerz, J. C., III.................2000....SUNY-Binghamton
    McGowan, E. M. ....2003....SUNY-Binghamton
    Petokas, P. J. ....1987....SUNY-Binghamton
    Rohr, J. R. ....2002....SUNY-Binghamton
    Sullivan, A. M. ....2004....SUNY-Binghamton

SERIES 87

001. SCHECHTMAN, A. M.
    Nace, G. W. ....1950....CA Los Angeles....2
002. NACE, G. W.
    Gill, S. J. ....1972....MI
    Glass, L. E. ....1958....Duke
    Hegennauer, J. C. ....1970....MI
    Levinson, J. W. ....1979....MI
    Mohanty-Hejmadi, P. ....1970....MI
    Ostrovsky, D. S. ....1970....MI
    Rittschof, D. ....1975....MI

SERIES 88

001. ?
    Smith, L. D. ....1964....IN.........................2
002. SMITH, L. D.
    Anderson, D. M. ....1977....Purdue
    Hollinger, T. G. ....1974....Purdue
    Kawata, E. E. ....1986....Purdue
    Shih, R. J. T. ....1975....Purdue
    Taylor, M. A. ....1986....Purdue
    Williams, M. A. ....1978....Purdue

SERIES 89

187. ?
    Smalley, A. E. ....1959....GA........................2
002. SMALLEY, A. E.
    Cheng, H. Y. ....1984....Tulane

SERIES 90

190. ?
    Solbrig, O. T. ....1959....CA Berkeley 2
002. SOLBRIG, O. T.
    Jasienski, M. ....1992....Harvard

SERIES 91

001. ?
    Stahnke, H. L. ....1940....IA St......................2
002. STAHNKE, H. L.
    Johnson, B. D. ....1967....AZ St

SERIES 92

001. STEPHAN, W.
    Glenn, T. C. ....1997....MD........................2
002. GLENN, T. C.
    Davis, L. M. ....2002....SC
    Hauswaldt, S. ....2004....SC

SERIES 93

001. ?
    Stockard, A. H. ....1932....MI.........................2
002. STOCKARD, A. H.
    Fitch, K. L. ....1957....MI

SERIES 94

001. ?
    Stocum, D. L. ....1968....PA..........................2
002. STOCUM, D. L.
    Kim, W. S. ....1985....IL

SERIES 95

001. ?
    Tassava, R. A. ....1968....MI St.....................2
002. TASSAVA, R. A.
    Bantle, J. ....1974....OH St
    Christensen, R. N. ....2001....OH St
    Garling, D. J. ....1981....OH St
    Goldhamer, D. ....1987....OH St
    Loyd, R. M. ....1978....OH St
    Manson, J. M. ....1974....OH St
    Mescher, A. L. ....1985....OH St
    Olsen, C. L. ....1982....OH St
    Onda, H. ....1991....OH St
    Salley, J. D. ....1981....OH St
    Yang, E. ....1992....OH St

SERIES 96

001. ?
    Thorson, T. B. ....1953....WA........................2
002. THORSON, T. B.
    Suhr, J. D. ....1976....NB

SERIES 97

001. ?
    Tipton, S. R. ....1933....Duke 2
002. TIPTON, S. R.
    Turner, J. E. ....1970....TN

SERIES 98

001. ?
    Tompkins, R. ....1968....IN..........................2
002. TOMPKINS, R.
    Miller, M. W. ....1986....Tulane

SERIES 99

001. ?
    Tracey, M. L., Jr. ....1971....Brown 2
002. TRACEY, M. L., JR.
    Meshaka, W. E. ....1994....FL Int

SERIES 100

001. ?
    Tucker, V. A. ....1963....CA Los Angeles....2
002. TUCKER, V. A.
    Greenwald, O. E. ....1970....Duke

SERIES 101

001. UTZ, W. R.
Hallam, T. G. ....1965.... MO........................ 2
002. HALLAM, T. G.
Hom, C. L. ....1986.... TN

SERIES 102

001. ?
Wilczynski, W. ....1978.... MI 2
002. WILCZYNSKI, W. ....
Allison, J. D. ....1992.... TX-Austin
Burmeister, S. S. ....2001.... TX-Austin
Chu, J. C. ....1998.... TX-Austin
Fox, J. H. ....1993.... TX-Austin
Lynch, K. S. ....2005.... TX-Austin
McClelland, B. E. ....1994.... TX-Austin
Venator, K. R. ....1999.... TX-Austin
Yang, E.-J. ....2002.... TX-Austin

SERIES 103

001. ?
Vanable, J. W., Jr. ....1962.... Rockefeller 2
002. VANABLE, J. W., JR.
Fangboner, R. F. ....1962.... Purdue
McGarry, M. P. ....1969.... Purdue
McGinnis, M. E. ....1985.... Purdue
Mortensen, R. D. ....1967.... Purdue

SERIES 104

001. ?
Weigert, R. C. ....1962.... MN........................ 2
002. WIEGERT, R. G.
Frazer, N. B. ....1983.... GA
Richardson, J. I. ....1982.... GA

SERIES 105

001. WERNER, E. E.
Peacor, S. D. ....2000.... MI
Relyea, R. A. ....1998.... MI
Schiesari, L. C. ....2004.... MI
Skelly, D. K. ....1992.... MI 2
Yurewicz, K. L. ....2002.... MI
002. SKELLY, D. K.
Halverson, M. A. ....2005.... Yale

SERIES 106

001. ?
Wilde, C. E., Jr. ....1949.... Princeton............... 2
002. WILDE, C. E., JR.
Horton, K. M. ....1969.... PA

SERIES 107

001. ?
Wyss, A. R. ....1989.... Columbia 2
002. WYSS, A. R.
Whatley, R. L. ....2005.... CA Santa Barbara

SERIES 108

001. WILLOWS, A. O. D.
Lohmann, K. J. ....1998.... WA........................ 2
002. LOHMANN, K. .J
Avens, L. I. ....2003.... NC-Chapel Hill
Irwin, W. P. ....2004.... NC-Chapel Hill

SERIES 109

001. AIELLO, E.
Blando-Hoegler, C. F. ....1974.... Fordham
002. ALBRIGHT, J. T.
Michaels, J. E. ....1970.... Boston
003. ALLEN, J. M.
Visentin, L. P. ....1969.... MI
004. BAGBY. R. M.
Fisher, B. A. ....1974.... TN
005. BAJER, A. S.
Rider, C. L. ....1977.... OR
006. BATES, J. M.
Carnaval, A. C. O. Q. ....2004.... Chicago
007. BALL, G. H.
Telford, S. R., Jr. ....1964.... CA Los Angeles
008. BANKS, E. M.
Moehn, L. D. ....1971.... IL
009. BART, J. B.
Campbell, E. W. ....1996.... OH St
010. BASKETT, T. S.
Schroeder, E. E. ....1966.... MO
011. BEAM, H. W.
Carpenter, S. J. ....1964.... IA St
012. BECKETT, D. C.
Himes, J. G. ....2002.... S MS
013. BENNETT, H. J.
Lauter, F. H. ....1959.... LA St
Rabalais, F. C. ....1967.... LA St
014. BERNSTEIN, R.
Serena, M. L. ....1980.... CO
015. BERTKE, E. M.
Lemanski, L. F. ....1971.... AZ St
Mays, C. E. ....1968.... AZ St
Nickerson, M. A. ....1969.... AZ St
016. BIESELE, J. J.
Barker, K. R. ....1966.... TX
Man-Li, S. Y. ....1966.... TX
Yew, M. S. ....1966.... TX
017. BINFORD, M. W.
Conway, K. M. ....2004.... FL
018. BLACK, C. P.
Teubner, V. A. ....1986.... Toledo
019. BRIGGS, C. J.
Rachowicz, L. J. ....2005.... CA Berkeley
020. BOYDEN, C. A.
Frair, W. F. ....1962.... Rutgers
021. BOWERS, M.
Bowne, D. R. ....2002.... VA
022. BRADBURY, J. W.
Alberts, A. C. ....1989.... CA San Diego
023. BRADLEY, G. L.
Hunsicker, G. R. ....1987.... Loma Linda
024. BREITENBACH, R.
Magliola, L. ....1983.... MO

025. BRIGGS, R.
Etkin, L. D. ....1977.... IN
VanDeusen, E. B. ....1973.... IN
026. BRISBIN, I. L.
Standora, E. A. ....1982.... GA
027. BROTHERS, A. J.
Pierce, K. E. ....1984.... CA Berkeley
028. BROWN, M. T.
Surdick, J. A. ....2005.... FL
029. BROWNING, H. C.
Watson, J. T. ....1974.... TX-Houston
030. BULLOCK, T. H.
Hartline, P. H. ....1969.... CA San Diego
031. BURGER, J.
Garber, S. D. ....1994.... Rutgers
032. BURDICK, C. J.
Fruchter, J. ....1982.... CUNY
033. CALHOUN, A. J. K.
Baldwin, R. F. ....2005.... ME
Kolozsvary, M. B. ....2003.... ME
034. CALLARD, I. P.
Custodia, L. N. ....2001.... Boston
Dubois, W. ....1982.... Boston
Duggan, A. E. ....2000.... Boston
Giannoukos, G. ....1995.... Boston
Kitana, N. ....2005.... Boston
Rie, M. T. ....2000.... Boston
035. CARLSON, J. G.
Cole, M. B., Jr. ....1969.... TN
036. CARROLL, D.
Garrett, J. E. ....1984.... UT
Lam, B. S. ....1981.... UT
Young, D. B. ....1982.... UT
037. CASE, E. C.
White, T. ....1936.... MI
038. CASEY, T. M.
Gemmell, D. J. ....1994.... Rutgers
039. CHIASSON, R. B.
Brown, R. H. ....1965.... AZ
Kleese, W. C. ....1981.... AZ
040. COLE, R.
Weigmann, D. L. ....1982.... MI St
041. COOK, J. B.
Fletemeyer, J. R. ....1991.... Fl Int
042. COOPER, E. L.
Klempau, A. E. ....1983.... CA Los Angeles
043. CROWDER, L. B.
Hart, K. M. ....2005.... Duke
Snover, M. L. ....2002.... Duke
044. CROWLEY, P. H.
Niedzwiecki, J. H. ....2005.... KY
045. DALTON, H. C.
Zimmerman, S. B. ....1958.... NY
046. DANE, B.
Auger, P. J. ....1989.... Tufts
047. DEANGELIS, D. L.
Richards, P. M. ....2003.... U Miami
048. DELLMAN, H. D.
Chang, A. C. ....1981.... IA St
049. DUNN, P. O.
Casper, G. S. ....2003.... WI-Milwaukee
050. EASON, P.
Attum, O. A. ....2004.... Louisville
051. ENDLER, J. A.
Greenberg, D. B. ....2002.... CA Santa Barbara
052. ENGEL, W. L.
Alexander, D. G. ....1966.... NC
053. EVANS, F. C.
deBenedictus, P. A. ....1970.... MI
054. EVANS, H. J.
Kerns, R. T. ....1997.... VA C'wealth
Stefansson, S. ....1990.... VA C'wealth
055. FANKHOUSER, G.
Bachmann, K. ....1964.... Princeton
056. FENTRESS, J. C.
Mates, J. W. ....1973.... OR
057. GALL, J.
Battin, W. J. ....1956.... MN
058. GEBHARDT, L. P.
de St. Jeor, S. C. ....1969.... UT
059. GERHARDT, J.
Gimlich, R. L. ....1985.... CA Berkeley
060. GILBERT, C. H.
VanDenburgh, J. ....1897.... Stanford
061. GRANDA, A. M.
Sisson, D. F. ....1986.... DE
062. GRAY, P.
Wurst, G. Z. ....1974.... Pittsburgh
063. GRAZIADEI, P. P. C.
Dehan, R. S. ....1972.... FL St
Stout, R. P. ....1979.... FL St
064. GRIFFIN, C. R.
Whitlock, A. L. ....2002.... MA-Amherst
065. GUTSKE, W. H. N.
Britson, C. A. ....1996.... Memphis
Mullin, S. J. ....1998.... Memphis
066. GUYNN, D. C., JR.
Baughman, W. M. ....2000.... Clemson
Russell, K. R. ....2000.... Clemson
067. HAAS, H. J.
Gerdy, J. R. ....1975.... S IL
068. HADLEY, N.
Burns, T. ....1969.... AZ St
Congdon, J. D. ....1977.... AZ St
069. HAGADORN, I. R.
Sedirec, M. J. ....1977.... NC
070. HALL, F. G.
McCutheon, F. H. ....1936.... Duke
071. HAMBERGER, V.
Muchmore, W. B. ....1950.... WA U
072. HAMBURGH, M.
Greenhouse, G. A. ....1969.... CUNY
Kim, Y. S. ....1979.... NY
Saland, L. C. ....1968.... CUNY
073. HARGITT, G. T.
McManus, M. I. ....1935.... Duke
074. HERMAN, C. A.
Robleto, D. O. ....1987.... NM St

075. HERRERA, A. A.
Regnier, M. ....1991.... S CA
076. HERRNKIND, W. F.
Brock, O. G. ....1980.... FL St
077. HIBBARD, E.
Ulshafer, R. J. ....1978.... PA St
078. HILLEMAN, H. H.
Thornburg, K. L. ....1972.... OR St
079. HOBERTON, R. L.
Cash, W. B., III ....2000.... MS
080. HOLMES, S. J.
Atsatt, S. R. ....1931.... CA Berkeley
081. HOPSON, J. A.
Beck, A. L. ....2004.... Chicago
Blob, R. W. ....1998.... Chicago
Clark, J. M. ....1986.... Chicago
Fisk, R. M. ....1980.... Chicago
O'Keefe, F. R. ....2000.... Chicago
Panko, L. J. ....2001.... Chicago
Sidor, C. A. ....2000.... Chicago
082. HOROWITZ, S. B.
Century, T. J. ....1972.... PA
083. HOWARD, W. E.
Moore, R. G. ....1976.... CA Davis
084. JACOBS, A. G.
Berlinrood, M. ....1969.... TX
Burnside, M. B. ....1968.... TX
Etheridge, A. L. ....1968.... TX
Fullilove, S. L. ....1968.... TX
085. JACOBSON, M.
Beach, D. H. ....1977.... U Miami
086. JAMESON, E. W., JR.
Allison, A. ....1979.... CA Davis
Benes, E. S. ....1966.... CA Davis
Simon, M. P. ....1982.... CA Davis
087. JANSON, C. H.
Valenzuela Castro, M. N. ..1999.... SUNY
088. JANZEN, F. J.
Morjan, C. L. ....2000.... IA St
089. JANVOY, J., JR.
Dollahon, N. R. ....1971.... NB
090. JEFFERY, W. R.
Buchanan, E. S. A. ....1988.... TX
Capco, D. G. ....1980.... TX
091. JOHNSON, D. M.
Diller, L. V. ....1981.... ID
092. JONES, A. W.
Powders, V. N. ....1967.... TN
093. JONES, G. S.
Wallach, V. ....1998.... Northeastern
094. KALTENBACH, J. C.
Fry, A. E. ....1969.... MA
095. KAUFMAN, J. H.
Dalton, S. A. ....1979.... FL
Fowler, L.E ....1983.... FL
096. KENT, G. C., JR.
Ridgway, P. M., Jr. ....1971.... LA St
097. KIMMEL, C. B.
Jacoby, J. ....1979.... OR
098. KING, T. J.
DiBernadino, M. A. ....1962.... PA
099. KINSEY, A. C.
Blair, A. P. ....1940.... IN
100. KINGSLEY,? ....
Higgins, G. M. ....1919.... IL
101. KIRBY, G. S.
Dmytrenko, G. M. ....1979.... Tulane
102. KLEIN, P. A.
Coberley, S. S. ....2002.... FL
Herbst, L. H. ....1995.... FL
103. KOPAC, M. J.
Kiremidjian, L. ....1970.... NY
104. LABARBERA, M.
Magwene, P. M. ....1999.... Chicago
Socha, J. J. ....2002.... Chicago
105. LAGUSHA, J.
Grismer, L. L. ....1994.... Loma Linda
106. LANDRY, A. M., JR.
Coyne, M. S. ....2000.... TX A&M
Loop, K. A. ....1996.... TX A&M
Metz, T. L. ....2004.... TX A&M
Wang, H.-C. ....2005.... TX A&M
107. LANNERS, H. N.
Lowichik, M. ....1988.... Tulane
108. LARIMER, J. L.
Parker, R. B. ....1965.... TX
109. LARSON, A.
Kozak, K. H. ....2005.... WA U
Macey, J. R. ....1999.... WA U
Schulte, J. A., II ....2001.... WA U
Townsend, T. M. ....2002.... WA U
Weisrock, D. W. ....2003.... WA U
110. LAUDER, G. V.
Gillis, G. B. ....1997.... CA Irvine
111. LEHMAN, H. E.
Anderson, L. G. ....1959.... NC
Desaix, P. ....1977.... NC
Harris, T. M. ....1962.... NC
112. LEHMAN, T. M.
Tomlinson, S. L. ....1997.... TX Tech
113. LEONE, C. A.
Pearson, D. D. ....1965.... KS
114. LOISELLE, B.
Escalona, T. ....2003.... MO-St Louis
115. LUTZ, P. L.
Lutcavage, M. E. ....1987.... U Miami
Milton, S. L. ....1994.... U Miami
Perez-Pinzon, M. A. ....1991.... U Miami
116. LYSER, K. M.
Vargas-Bodas, P. ....1975.... CUNY
117. MAHER, M. H.
Waddick, J. W. ....1971.... KS
118. MCDOWELL, S. B.
Crumly, C. R. ....1984.... Rutgers
Freeman, H. L. ....1977.... Rutgers
119. MEAD, R. A.
Highfill, D. R. ....1975.... ID
120. MEIER, A. H.
Ferrell, B. R. ....1979.... LA St

121. MELLGREN, R. L.
Keathley, V. L. ....2004.... TX-Arlington
122. MENAKER, M.
Underwood, H. A., Jr. ....1972.... TX-Austin
123. MERRIAM, H. G.
Clark, W. T., Jr. ....1969.... TX
124. MICHENER, C. M.
Andrews, R. M. ....1971.... KS
125. MIENTS, R. H.
Carver, F. J. ....1974.... NB
126. MILLER, G. C.
Stuart, M. D. ....1985.... NC St
127. MILLS, K. S.
Funkhouser, A. B. ....1966.... OK
128. MINNIC, D. E.
Wilde, W. S. ....1937.... MN
129. MITCHELL, R. W.
Rudolph, D. C. ....1980.... TX A&M
130. MOCK, D.
Haugen, L. F. ....2002.... OK
131. MOORE, C.
Reynolds, A. E. ....1941.... Chicago
132. MOORE, M. C.
Knapp, R. ....1996.... AZ St
Marler, C. A. ....1988.... AZ St
Thompson, C. W. ....1992.... AZ St
133. MORGAN, T. H.
Boring, A. M. ....1910.... Bryn Mawr
King, H. D. ....1899.... Bryn Mawr
134. MORRIS, J. G.
Cassler, C. L. ....2000.... FL I Tech
Giovanetto, L. A. ....1992.... FL I Tech
135. MOUNT, J.
Yarnell, S. M. ....2005.... CA Davis
136. MOULTON, M. P.
Campbell, C. L. ....2003.... FL
137. MOUSSEAU, T. A.
Litzgus, J. D. ....2003.... SC
138. NACHTRIEB, H. F.
Johnson, C. E. ....1912.... MN
139. NEESS, J. C.
Meyer, D. E. ....1959.... WI
Urban, E. K. ....1964.... WI
140. NICOLL, C. S.
Tarpey, J. F. ....1986.... CA Berkeley
141. NUTTALL, R. P.
Gordon, B. K. ....1981.... Emory
142. O'LEARY, M. A.
Hill, R. V. ....2004.... SUNY
143. OPPENHEIM, R. W.
Haverkamp, L. J. ....1983.... NC
144. OTIS, D. L.
Kinkead, K. E. ....2004.... Clemson
145. PEACHEY, L. D.
Claude, P. ....1968.... PA
146. PEADON, A. M.
Piatka, B. A. ....1967.... Catholic
147. PORTER, J. W.
Eckert, K. L. ....1988.... GA
148. PRATT, I.
Anderson, G. A. ....1964.... OR St
149. RASMUSSEN, H.
Lipton, S. A. ....1977.... PA
150. RICHARDS, A.
Bragg, A. N. ....1937.... OK
Johnson, V. O. ....1940.... OK
Trowbridge, M. S. ....1939.... OK
151. RICHMOND, M.
Morreale, S. J. ....1999.... Cornell
152. RIEDESEL, M. L.
Sturbaum, B. A. ....1972.... NM
153. RIGGS, A. F.
Watt, K. W. K. ....1976.... TX
154. ROBERTS, J. L.
Lillo, R. S. ....1979.... MA
155. ROOT, T.
Swimmer, J. Y. B. ....1987.... MI
156. ROZE, J.
Bernal Carlo, A. ....1991.... CUNY
157. SASSAMAN, C. A.
Simovich, M. A. ....1995.... CA Riverside
158. SCHULTZ, V.
Fish, J. L. ....1972.... WA St
159. SCHULTZE, H. P.
Cloutier, R. ....1989.... KS
Cunningham, C. R. ....1993.... KS
160. SELF, J. T.
Deakins, D. E. ....1973.... OK
McKnight, T. J. ....1959.... OK
161. SHAVER, J. R.
Throneberry, J. B. ....1962.... MI St
162. SHIVERS, C. A.
James, W. S. ....1980.... TN
163. SHUMWAY, W.
Sanders, J. M. ....1931.... IL
164. SHURE, D. J.
Orser, P. N. ....1974.... Emory
165. SIMONS, T. R.
Bailey, L. L. ....2002.... NC St
166. SIMPSON, S. D., JR.
Bayne, E. B. K. ....1976.... Northwestern
167. SLADE, D. S.
Lewis, J. J. ....1985.... Northwestern
168. SMITH, M. H.
McAlpine, S. K. ....1992.... GA
169. SONNEBORN, D. R.
Phillips, C. R. ....1979.... WI
170. SORENSON, M. W.
Prieto, A. A. ....1971.... MO
171. SPILLETT, J. J.
Hammer, D. A. ....1973.... UT St
172. SPITSE, K.
Osentoski, M. F. ....2001.... U Miami
173. SPITZER, N. C.
Baccaglini, P. I. ....1978.... CA San Diego
O'Dowd, D. K. ....1985.... CA San Diego
174. STEIN, R. A.
Thornhill, G. M. ....1985.... OH St

175. STEINBOCK, H. B.
Davison, J. A. ....1955.... MN
Stephenson, W. K. ....1955.... MN
176. STETSON, D.
Walton, H. C. ....2005.... OH St
177. STIVEN, A. E.
Miller, J. R., II ....2002.... NC-Chapel Hill
178. SULLIVAN, J. M.
Carstens, B. C. ....2004.... ID
179. TAYLOR, J. H.
Chambers, J. C. ....1982.... FL St
180. THEIL, E. C.
Shull, G. E. ....1983.... NC St
181. TORRES, J. J.
Foley, A. M. ....1998.... S FL
182. TORREY, H. B.
Johnson, M. E. ....1912.... CA Berkeley
183. VANATTA, J. C.
Fitzgerrel, W. W. ....1974.... TX-Dallas
184. VOGEL, T.
Walker, R. J. ....1999.... MI St
185. WALLACE, R. A.
Gottleib, T. A. ....1982.... TN
Wiley, H. S. ....1979.... TN
186. WALLACE, R. L.
Howard, J. H. ....1979.... ID
187. WARNER, R. E.
Rubin, C. S. ....2000.... IL
188. WEINBERGER, N. M.
Cruikshank, S. J. ....1997.... CA Irvine
Galvan, V. V. ....2001.... CA Irvine
Goodman, D. A. ....1969.... CA Irvine
McLin, D. E., III ....2001.... CA Irvine
189. WHITE, B. A.
Lyman, D. F. ....1989.... CN
190. WINGFIELD, J. C.
Dunlap, K. D. ....1993.... WA
191. WILEY, R. H.
Hall, S. P. ....1987.... NC
192. WILLIS, D. L.
Lappenbusch, W. L. ....1969.... OR St
Valett, B. B. ....1970.... OR St
193. WITHERS, P. C.
Abts, M. L. ....1985.... Portland St
194. WITSCHI, E.
Sehe, C. T. ....1957.... IA St
195. WYMAN, R.
Frank, J. A. ....2004.... SUNY
196. YOUNG, F. N.
Thurow, G. R. ....1955.... IN
197. ZUCKER, N. D.
Carpenter, G. C. ....1994.... NM St
Hager, S. B. ....1998.... NM St

UKRAINE (UR)

See: **Russia** 1.2, 2.2, 8.1–8.2, 9.2, 10.1, 11.9, 11.20, 11.31–11.32, 11.41–11.42, 11.46, 11.49, 11.54 and **United States** 11.1

UZBEKISTAN (UZ)

See: **Russia** 7.1, 11.8, 11.30

ZIMBABWE (ZI)

See: South Africa 2.1–2.2

INDEX TO LINEAGES

Country Codes:

AM = Armenia
AR = Argentina
AS = Austria
AU = Australia
AZ = Azerbaijan
BE = Belgium
BL = Belarus
BR = Brazil
CA = Canada
GA = Georgia
DE = Denmark
FI = Finland
FR = France
GA = Georgia
GE = Germany
HU = Hungary
IN = India
IS = Israel
IT = Italy
JA = Japan
KA = Kazakhstan
KY = Kyrgyzstan
LA = Latvia
MA = Malaysia
ME = Mexico
NE = Netherlands
NZ = New Zealand
PA = Pakistan
PO = Poland
PT = Portugal
RR = Russia
SA = South Africa
SD - Sweden
SL = Switzerland
SP = Spain
TA = Taiwan
TM = Turkmenistan
TR = Tatarstan
TU = Turkey
UK = United Kingdom
UR = Ukraine
US = United States
UZ = Uzbekistan
ZI = Zimbabwe

A

B

C

D

E

F

G

H

I

J

K

N

O

P

Q

R

S

T

U

V

W

X

Y

Z

PUBLICATIONS OF THE
SOCIETY FOR THE STUDY OF AMPHIBIANS AND REPTILES

SOCIETY PUBLICATIONS may be purchased from:
Breck Bartholomew, SSAR Publications Secretary
P. O. Box 58517
Salt Lake City, Utah 84158–0517, USA

Telephone and Fax: area code (801) 562–2660
E-mail: ssar@herplit.com
Web: www.ssarherps.org

SSAR members receive a 10% discount on all prices listed below. Prices are effective through December 2007. Make checks payable to "SSAR." Overseas customers must make payment in USA funds using a draft drawn on American banks or by International Money Order. All persons may charge to MasterCard, American Express, Discover Card, or VISA (please provide account number and expiration date).

Shipping and Handling Costs

• *Shipments inside the USA*: Shipping costs are in addition to the price of publications. Add an amount for shipping of the first item ($4.00 for a volume costing $15.00 or more or $3.00 if it costs less than $15.00) plus an amount for any additional items ($3.00 each for a volume costing over $15.00 and $2.00 for each item costing less than $15.00).

• *Shipments outside the USA*: Determine the cost for shipments inside USA (above) and then add 6% of the total cost of the order.

• *Large prints* (marked *): Shipments only inside the USA, add $5.00 for any quantity. Regrettably, prints cannot be shipped outside the USA due to their large size.

CONTRIBUTIONS TO HERPETOLOGY

Book-length monographs, comprising taxonomic revisions, results of symposia, and other major works. Prepublication discount to Society members. Missing volumes are out-of-print and *no longer available from the Society*.

Vol. 2. *The Turtles of Venezuela,* by Peter C. H. Pritchard and Pedro Trebbau. 1984. Covers half of the turtle species of South America. 414 p., 48 color plates (25 watercolors by Giorgio Voltolina and 165 photographs of turtles and habitats) measuring 8½ × 11 inches, keys, 16 maps. Regular edition, clothbound $50.00; patron's edition, two leatherbound volumes in cloth-covered box, signed and numbered by authors and artist $300.00.

Vol. 6. *Snakes of the* Agkistrodon *Complex: A Monographic Review*, by Howard K. Gloyd and Roger Conant. 1990. Comprehensive treatment of four genera: *Agkistrodon*, *Calloselasma*, *Deinagkistrodon*, and *Hypnale*. Includes nine supplementary chapters by leading specialists. 620 p., 33 color plates (247 photographs of snakes and habitats), 20 uncolored plates, 60 text figures, keys, 6 charts, 28 maps. Clothbound $75.00. (*Also*: separate set of the 33 plates, in protective wrapper. $30.00)

Vol. 7. *The Snakes of Iran*, by Mahmoud Latifi. 1991. Review of the 60 species of Iranian snakes, covering general biology, venoms, and snake bite. Appendix and supplemental bibliography by Alan E. Leviton and George R. Zug. 167 p., 22 color plates of snakes (66 figures), 2 color relief maps, 44 species range maps. Clothbound $25.00.

Vol. 9. *Herpetology: Current Research on the Biology of Amphibians and Reptiles*. 1992. Proceedings of the First World Congress of Herpetology (Canterbury, 1989), edited by Kraig Adler and with a foreword by H.R.H. Prince Philip, Duke of Edinburgh. Includes the plenary lectures, a summary of the congress, and a list of delegates with their current addresses. 225 p., 28 photographs. Clothbound $28.00.

Vol. 11. *Captive Management and Conservation of Amphibians and Reptiles*, by James B. Murphy, Kraig Adler, and Joseph T. Collins (eds.). 1994. Includes chapters by 70 leading specialists. Foreword by Gerald Durrell. 408 p., 35 photographs, 1 color plate. Clothbound $58.00.

Vol. 12. *Contributions to West Indian Herpetology*, by Robert Powell and Robert W. Henderson (eds.). 1996. Results of a Society-sponsored symposium, including research chapters by 59 authors and a checklist of species with complete citations. Foreword by Thomas W. Schoener. 457 p., 98 photographs (70 in color), index. Clothbound $60.00.

Vol. 13. *Gecko Fauna of the USSR and Contiguous Regions*, by Nikolai N. Szczerbak and Michael L. Golubev. 1996. Covers the systematics, natural history, and conservation of the gecko fauna of the former Soviet Union and related species in surrounding regions from Mongolia through Pakistan, the Middle East, and northern Africa. 245 p., 24 colored and 60 black-and-white photographs, spot distribution maps, bibliography, index. Clothbound $48.00.

Vol. 14. *Biology of the Reptilia, vol. 19 (Morphology G)*, by Carl Gans and Abbot S. Gaunt (eds.). 1998. Chapters by 11 authors cover the major organs situated in the coelom: lungs, heart, liver, and spleen. 660 p., 145 figures, indices. Clothbound $58.00. (A complete list of the earlier volumes in this series [vols. 1–18], with names of publishers, is given in the book.)

Vol. 15. *The Lizards of Iran*, by Steven C. Anderson. 1999. Comprehensive summary, including systematics, natural history, and distribution. 450 p., 103 maps, 190 color photographs of lizards and habitats. Clothbound $65.00.

Vol. 16. *Slithy Toves: Illustrated Classic Herpetological Books at the University of Kansas in Pictures and Conversations*, by Sally Haines. 2000. A treasure trove of some of the finest illustrations of amphibians and reptiles ever produced, dating from the 16th to early 20th centuries. 190 p., 84 color photographs. Stiff paper cover $60.00.

Vol. 17. *The Herpetofauna of New Caledonia*, by Aaron M. Bauer and Ross A. Sadlier. French translations by Ivan Ineich. 2000. 322 p., 47 maps, 63 figures, 189 color photographs of animals and habitats. Clothbound $60.00.

Vol. 18. *The Hylid Frogs of Middle America*, expanded edition, by William E. Duellman. 2001. Review of the 165 hylid species from México through Panamá, with paintings by David M. Dennis. Foreword by David B. Wake. 1180 p., 443 figures and maps, 94 plates (46 in color). Clothbound in 2 volumes $125.00. (*Also*: separate set of the 46 color plates, in protective wrapper $45.00. Some of the original watercolor drawings by Mr. Dennis are available for sale. A pricelist may be obtained on request to the Publications Secretary.)

Vol. 19. *The Amphibians of Honduras*, by James R. McCranie and Larry David Wilson. 2002. Comprehensive summary of 116 species, including systematics, natural history, and distribution. Foreword by Jay M. Savage. 635 p., 126 figures, 33 tables, 154 color photographs of animals and habitats. Clothbound $60.00.

Vol. 20. *Islands and the Sea. Essays on Herpetological Exploration in the West Indies*, by Robert W. Henderson and Robert Powell (eds.). 2003. Personal recollections of field experiences by 31 scientists who are or were active in the West Indies. 312 p., 316 photographs, 14 maps, indices. Clothbound $48.00.

Vol. 21. *Contributions to the History of Herpetology, volume 2*, by Kraig Adler, John S. Applegarth, and Ronald Altig. 2007. Biographies of 285 leading herpetologists (with portraits and signatures), index to 3604 authors in taxonomic herpetology, and academic lineages of 3810 herpetologists. Worldwide coverage. 396 p., 269 portraits, color frontispiece. Clothbound $65.00.

FACSIMILE REPRINTS IN HERPETOLOGY

Exact reprints of classic and important books and papers. Most titles have extensive new introductions by leading authorities. Prepublication discount to Society members. Missing titles are out-of-print and *no longer available from the Society*.

ANDERSON, J. 1896. *Contribution to the Herpetology of Arabia*. Introduction and new checklist of Arabian amphibians and reptiles by Alan E. Leviton and Michele L. Aldrich. 160 p., illus. (one plate in color), map. Clothbound $25.00.

BARBOUR, T. and C. T. RAMSDEN. 1919. *The Herpetology of Cuba*. The first comprehensive review of Cuban amphibians and reptiles with a complete descriptions of all species. Foreword by Rodolfo Ruibal. 200 p., 15 plates, measuring 8½ × 11 inches. Clothbound $55.00.

BOURRET, R. 1941. *Les Tortues de l'Indochine*. The definitive review of the turtles and tortoises of Indochina, covering an area from northeast India and the Malay Peninsula to Vietnam. Introduction by Indraneil Das. 264 p., 48 uncolored and 6 colored plates. Clothbound $65.00.

COPE, E. D. 1864. *Papers on the Higher Classification of Frogs*. Reprinted from Proceedings of the Academy of Natural Sciences of Philadelphia and Natural History Review. 32 p. $3.00.

—. 1871. *Catalogue of Batrachia and Reptilia Obtained by McNiel in Nicaragua; Catalogue of Reptilia and Batrachia Obtained by Maynard in Florida*. 8 p. $1.00.

COWLES, R. B., and C. M. BOGERT. 1944. *A Preliminary Study of the Thermal Requirements of Desert Reptiles*. With extensive review of recent studies by F. Harvey Pough. Reprinted from Bulletin of American Museum of Natural History. 52 p., 11 plates. $5.00.

ESPADA, M. JIMÉNEZ DE LA. 1875. *Vertebrados del Viaje al Pacifico: Batracios*. A major taxonomic work on South American frogs. Introduction by Jay M. Savage. 208 p., 6 plates, maps. Clothbound $20.00.

FAUVEL, A.-A. 1879. *Alligators in China*. Original description of *Alligator sinensis*, including classical and natural history. 42 p., 3 plates. $5.00.

FERGUSON, W. 1877. *Reptile Fauna of Ceylon*. First comprehensive summary of the herpetofauna of Sri Lanka. Introduction by Kraig Adler. 48 p. $8.00.

FITZINGER, L. 1826 & 1835. *Neue Classification der Reptilien* and *Systematische Anordnung der Schildkröten*. Important nomenclatural landmarks for herpetology, including Amphibia as well as reptiles; world-wide in scope. Introduction by Robert Mertens. 110 p., folding chart. Clothbound $30.00.

FRANCIS, E. T. B. 1934. *The Anatomy of the Salamander*. Standard text of salamander anatomy, including osteology, myology, nervous system, and internal organs. Introductions by F. J. Cole and James Hanken. 465 p., 25 highly detailed plates, color frontispiece. Clothbound $60.00.

GRAY, J. E. 1825. *A Synopsis of the Genera of Reptiles and Amphibia*. Reprinted from Annals of Philosophy. 32 p. $3.00.

GRAY, J. E., and A. GÜNTHER. 1845–1875. *Lizards of Australia and New Zealand*. The reptile section from "Voyage of H.M.S. Erebus and Terror," together with Gray's 1867 related book on Australian lizards. Introduction by Glenn M. Shea. 82 p., 20 plates (measuring 8½ × 11 inches). Clothbound $20.00. (*Also*: set of the 20 plates, in protective wrapper $12.00.)

GÜNTHER, A. 1885–1902. *Biologia Centrali-Americana. Reptilia and Batrachia*. The standard work on Middle American herpetology with 76 full-page plates measuring 8½ × 11 inches (12 in color). Introductions by Hobart M. Smith, A. E. Gunther, and Kraig Adler. 575 p., photographs, maps. Clothbound $50.00. (*Also*: separate set of the 12 color plates, in protective wrapper $18.00.)

HOLBROOK, J. E. 1842. *North American Herpetology*. Five volumes bound in one. The classic work by the father of North American herpetology. Exact facsimile of the definitive second edition, including all 147 plates, measuring 8½ × 11 inches (20 reproduced in full color). Introduction and checklists by Richard and Patricia Worthington and by Kraig Adler. 1032 p. Clothbound $60.00.

KIRTLAND, J. P. 1838. *Zoology of Ohio* (herpetological portion). 8 p. $1.00.

LECONTE, J. E. 1824–1828. *Three Papers on Amphibians*, from the Annals of the Lyceum of Natural History, New York. 16 p. $2.00.

MCLAIN, R. B. 1899. *Contributions to North American Herpetology* (three parts). 28 p., index. $2.00.

ORBIGNY, A. D' [and G. BIBRON]. 1847. *Voyage dans l'Amérique Méridionale*. This extract comprises the complete section on reptiles and amphibians from this voyage to South America. 14 p., 9 plates measuring 8½ × 11 inches. $3.00.

PERACCA, M. G. 1882–1917. *The Life and Herpetological Contributions of Mario Giacinto Peracca (1861–1923)*. A collection of 64 herpetological titles, with descriptions of 22 new species of amphibians and 52 of reptiles. Introduction, annotated bibliography, and synopsis of taxa by Franco Andreone and Elena Gavetti. 550 p. Clothbound $55.00.

PETERS, W. 1838–1883. *The Herpetological Contributions of Wilhelm C. H. Peters (1815–1883)*. A collection of 174 titles, world-wide in scope, and including the herpetological volume in Peters' series, "Reise nach Mossambique." Biography, annotated bibliography, and synopsis of species by Aaron M. Bauer, Rainer Günther, and Meghan Klipfel. 714 p., 114 plates, 9 photographs, maps, index. Clothbound $75.00.

SCHMIDT, K. P., and G. K. NOBLE. 1919–1923. *Contributions to the Herpetology of the Belgian Congo*. Essential reference for the Congo rain forest and Sudanese savanna. Introductions by Donald G. Broadley and John C. Poynton. 780 pages, 141 photographs, maps, indices. Clothbound $65.00.

SHAW, G. 1802. *General Zoology, vol. 3: Amphibia*. Herpetological section from the first world summary of amphibians and reptiles in English. Introduction by Hobart M. Smith and Patrick David. 1014 p., 140 plates. Clothbound $75.00.

SMITH, A. 1826–1838. *The Herpetological Contributions of Sir Andrew Smith*. A collection of 10 shorter papers, including many descriptions of new South African amphibians and reptiles. Biography, concordance of names, and annotated bibliography by William R. Branch and Aaron M. Bauer. 83 p. $10.00.

SOWERBY, J. DeC., E. LEAR, and J. E. GRAY. 1872. *Tortoises, Terrapins, and Turtles Drawn From Life*. The finest atlas of turtle illustrations ever produced. Introduction by Ernest E. Williams. 26 p., 61 full-page plates (measuring 8½ × 11 inches). Clothbound $40.00.

STEJNEGER, L. 1907. *Herpetology of Japan and Adjacent Territory*. Introduction by Masafumi Matsui. Also covers Taiwan, Korea, and adjacent China and Siberia. 684 pages, 35 plates, 409 text figures, keys, index. Clothbound $58.00.

TSCHUDI, J. J. von. 1838. *Classification der Batrachier*. A major work in systematic herpetology, with introduction by Robert Mertens. 118 p., 6 plates. $18.00.

—. 1845. *Reptilium Conspectus*. New reptiles and amphibians from Peru. 24 p. $2.00.

VanDENBURGH, J. 1914. *The Gigantic Land Tortoises of the Galapagos Archipelago*. The most extensive review of Galapagos tortoises. Foreword by Peter C. H. Pritchard. 290 pages, 205 photographs, maps, index. Clothbound $55.00.

WAITE, E. R. 1929. *The Reptiles and Amphibians of South Australia*. Introduction by Michael J. Tyler and Mark Hutchinson. 282 p., color plate, portrait, 192 text figures including numerous photographs. Clothbound $35.00.

WRIGHT, A. H., and A. A. WRIGHT. 1962. *Handbook of Snakes of the United States and Canada, Volume 3, Bibliography*. Cross-indexed bibliography to Volumes 1 and 2. 187 p. Clothbound $18.00.

CATALOGUE OF AMERICAN AMPHIBIANS AND REPTILES

Loose-leaf accounts of taxa (measuring 8½ × 11 inches) prepared by specialists, including synonymy, definition, description, distribution map, and comprehensive list of literature for each taxon. Covers amphibians and reptiles of the entire Western Hemisphere. Issued by subscription. Individual accounts are not sold separately.

CATALOGUE ACCOUNTS:
- Complete set: Numbers 1–840, $450.00.
- Partial sets: Numbers 1–190, $65.00.
  - Numbers 191–410, $75.00.
  - Numbers 411–840, $310.00.

INDEXES TO ACCOUNTS:
- Numbers 1–400: Cross-referenced, 64 pages, $6.00.
- Numbers 401–600: Cross-referenced, 32 pages, $6.00.
- Numbers 601–800: Cross-referenced, 36 pages, $6.00.

IMPRINTED POST BINDER: $35.00. (*Note*: one binder holds about 200 accounts.)

SYSTEMATIC TABS: Ten printed tabs for binder, such as "Class Amphibia," "Order Caudata," etc., $6.00 per set.

HERPETOLOGICAL CONSERVATION

A series of book-length monographs, including symposia, devoted to all aspects of the conservation of amphibian and reptiles. Prepublication discount to Society members.

Vol. 1. *Amphibians in Decline: Canadian Studies of a Global Problem*, by David M. Green (ed.). 1997. Chapters by 52 authors dealing with population dispersal and fluctuations, genetic diversity, monitoring of natural populations, as well as effects of temperature, acidity, pesticides, UV light, forestry practices, and disease. 351 p., numerous photographs, figures, and tables; index. Paperbound $40.00.

Vol. 2. *Ecology, Conservation, and Status of Reptiles in Canada*, by Carolyn N. L. Seburn and Christine A. Bishop (eds.). 2007. Chapters by 42 authors dealing with ecology and conservation. Appendices on the conservation strategy and the current status of reptiles in Canada. 256 p., numerous photographs, figures, and tables; index. Clothbound $40.00.

HERPETOLOGICAL REVIEW AND H.I.S.S. PUBLICATIONS

The Society's official newsletter, international in coverage. In addition to news notes and feature articles, regular departments include regional societies, techniques, husbandry, life history, geographic distribution, and book reviews. Issued quarterly as part of Society membership or separately by subscription. All numbers are paperbound as issued and measure 8½ × 11 inches. In 1973, publications of the Herpetological Information Search Systems (News-Journal and Titles and Reviews) were substituted for Herpetological Review; content and format are the same.

Volumes 31–37 (2000–2006), four numbers in each volume, $6.00 per number. All previous volumes and numbers are out-of-print and *no longer available from the Society*.

Cumulative Index for Volumes 1–7 (1967–1976), 60 pages, $5.00.

Cumulative Index for Volumes 1–17 (1967–1986), 90 pages, $8.00.

H.I.S.S. Publications: News-Journal, volume 1, numbers 1–6, and Titles and Reviews, volume 1, numbers 1–2 (all of 1973–1974), complete set, $10.00.

Index to Geographic Distribution Records for Volumes 1–17 (1967–1986), including H.I.S.S. publications, 44 pages, $6.00.

JOURNAL OF HERPETOLOGY

The Society's official scientific journal, international in scope. Issued quarterly as part of Society membership. All numbers are paperbound as issued, measuring 7 × 10 inches.

Volumes 34–40 (2000–2006), four numbers in each volume, $9.00 per number. All previous volumes and numbers are out-of-print and are *no longer available from the Society*.

Cumulative Index for Volumes 1–10 (1968–1976), 72 pages, $8.00.

HERPETOLOGICAL CIRCULARS

Miscellaneous publications of general interest to the herpetological community. All numbers are paperbound, as issued. Prepublication discount to Society members. Missing numbers are out-of-print and *no longer available from the Society*.

No. 8. *A Brief History of Herpetology in North America Before 1900* by Kraig Adler. 1979. 40 p., 24 photographs, 1 map. $3.00.

No. 10. *Vernacular Names of South American Turtles* by Russell A. Mittermeier, Federico Medem, and Anders G. J. Rhodin. 1980. 44 p. $3.00.

No. 11. *Recent Instances of Albinism in North American Amphibians and Reptiles* by Stanley Dyrkacz. 1981. 36 p. $3.00.

No. 13. *Silver Anniversary Membership Directory*, including addresses of all SSAR members, addresses and publications of the herpetological societies of the world, and a brief history of the Society. 1983. 56 p., 4 photographs. $3.00.

No. 14. *Checklist of the Turtles of the World with English Common Names* by John Iverson. 1985. 14 p. $3.00.

No. 15. *Cannibalism in Reptiles: A World-Wide Review* by Joseph C. Mitchell. 1986. 37 p. $4.00.

No. 17. *An Annotated List and Guide to the Amphibians and Reptiles of Monteverde, Costa Rica* by Marc P. Hayes, J. Alan Pounds, and Walter W. Timmerman. 1989. 70 p., 32 figures. $5.00.

No. 18. *Type Catalogues of Herpetological Collections: An An-notated List of Lists* by Charles R. Crumly. 1990. 50 p. $5.00.

No. 21. *Longevity of Reptiles and Amphibians in North American Collections* (2nd ed.) by Andrew T. Snider and J. Kevin Bowler. 1992. 44 p. $5.00.

No. 22. *Biology, Status, and Management of the Timber Rattlesnake* (Crotalus horridus): *A Guide for Conservation* by William S. Brown. 1993. 84 p., 16 color photographs. $12.00.

No. 23. *Scientific and Common Names for the Amphibians and Reptiles of Mexico in English and Spanish / Nombres Científico y Comunes en Ingles y Español de los Anfibios y los Reptiles de México*, by Ernest A. Liner. Spanish translation by José L. Camarillo R. 1994. 118 p. $12.00.

No. 24. *Citations for the Original Descriptions of North American Amphibians and Reptiles*, by Ellin Beltz. 1995. 48 p. $7.00.

No. 26. *Venomous Snakes: a Safety Guide for Reptile Keepers*, by William Altimari. 1998. 28 p. $6.00.

No. 27. *Lineages and Histories of Zoo Herpetologists in the United States*, by Winston Card and James B. Murphy. 2000. 49 p., 53 photographs. $8.00.

No. 28. *State and Provincial Amphibian and Reptile Publications for the United States and Canada*, by John J. Moriarty and Aaron M. Bauer. 2000. 56 p. $9.00.

No. 29. *Scientific and Standard English Names of Amphibians and Reptiles of North America North of Mexico, with Comments Regarding Confidence in Our Understanding*, by the Committee on Standard English and Scientific Names (Brian I. Crother, chair). 2000 [2001]. 86 p. $11.00.

No. 30. *Amphibian Monitoring in Latin America: a Protocol Manual / Monitoreo de Anfibios en America Latina: Manual de Protocolos*, by Karen Lips, Jamie K. Reaser, Bruce E. Young, and Roberto Ibañez. 2001. 121 p. $13.00.

No. 31. *Herpetological Collecting and Collections Management.* (revised ed.), by John E. Simmons. 2002. 159 p. $16.00.

No. 32. *Conservation Guide to the Eastern Diamondback Rattlesnake*, Crotalus adamanteus, by Walter W. Timmerman and William H. Martin. 2003. 74 p., color photographs. $13.00.

No. 33. *Chameleons: Johann von Fischer and other Perspectives*, by James B. Murphy. 2005. 123 p., many color plates. $13.00.

No. 34. *Synopsis of Helminths Endoparasitic in Snakes of the United States and Canada*, by Carl H. Ernst and Evelyn M. Ernst. 2006. 86 p. $9.00.

No. 35. *A Review of Marking and Individual Recognition Techniques for Amphibians and Reptiles*, by John W. Ferner. 2007. 72 p. $11.00.

No. 36. *Society for the Study of Amphibians and Reptiles: A Fifty Year History 1958 to 2007*, by John J. Moriarty and Breck Bartholomew. 2007. 64 p. $10.00.

PUBLICATIONS OF THE OHIO HERPETOLOGICAL SOCIETY

OHS (1958–1966) was the predecessor organization to the Society for the Study of Amphibians and Reptiles. All of its publications are now out-of-print and *no longer available from the Society.*

OTHER MATERIALS AVAILABLE FROM THE SOCIETY

The following color prints and brochures may be purchased from the Society. (*Extra postage required; see "Shipping and Handling Costs" at the beginning of this list.)

*SILVER ANNIVERSARY COMMEMORATIVE PRINT. Full-color print (11½ × 15 inches) of a Gila Monster (*Heloderma suspectum*) on natural background, from a watercolor by David M. Dennis. Issued as part of Society's 25th Anniversary in 1982. Edition limited to 1000. $4.00 each or $3.00 in quantities of 10 or more.

*GOLDEN ANNIVERSARY COMMEMORATIVE PRINT. Full-color poster (12¾ × 17¼ inches) of a Gray Treefrog (*Hyla versicolor*) from a photograph by David M. Dennis. Issued as part of Society's 50th Anniversary in 2007. Edition limited to 2000. $4.00 each or $3.00 in quantities of 10 or more.

*WORLD CONGRESS COMMEMORATIVE PRINT. Full-color print (11½ × 15 inches) of an Eastern Box Turtle (*Terrapene carolina*) in a natural setting, from a watercolor by David M. Dennis. Issued as part of SSAR's salute to the First World Congress of Herpetology, held at Canterbury, United Kingdom, in 1989. Edition limited to 1500. $4.00 each or $3.00 in quantities of 10 or more.

GRANTS AND AWARDS FOR HERPETOLOGISTS, by Joan C. Milam. 1997. A detailed listing, with descriptions and addresses, for about 100 research award programs. 106 p. $8.00.

•

PRODUCTION SPECIFICATIONS

Formatting of Text: Cornell University, Ithaca, New York, USA (Timothy D. Perry, Dawn A. Potter). The type used for the text was set in Times using Adobe InDesign 3.0 and an Apple Macintosh G3 computer.

Printing of Text and Binding: Thomson-Shore, Inc., Dexter, Michigan, USA (Angela M. Fugate, Jeff Rhoades, John E. Sinclair). The text is printed on 70-pound Huron Matte paper, a recycled stock. The hardbound copies are covered in Roxite C cloth (vellum finish) over 98-point binder's boards with 80-pound Rainbow Textured endpapers and reinforced cloth hinges. The softbound copies are covered in 90-pound Carnival Cordwain stock. All copies are Smyth sewn.

Paper: All paper in this book is acid- and groundwood-free. It meets the guidelines for permanence and durability of the Committee on Publication Guidelines for Book Longevity of the Council on Library Resources.

Date and Place of Publication: 11 July 2007 in Saint Louis, Missouri, USA.

Total Number of Copies Printed: 2600.